Advances in Intelligent Systems and Computing

Volume 236

T0091458

Series editor

Janusz Kacprzyk, Warsaw, Poland

For further volumes:
http://www.springer.com/series/11156

About this Series

The series "Advances in Intelligent Systems and Computing" contains publications on theory, applications, and design methods of Intelligent Systems and Intelligent Computing. Virtually all disciplines such as engineering, natural sciences, computer and information science, ICT, economics, business, e-commerce, environment, healthcare, life science are covered. The list of topics spans all the areas of modern intelligent systems and computing.

The publications within "Advances in Intelligent Systems and Computing" are primarily textbooks and proceedings of important conferences, symposia and congresses. They cover significant recent developments in the field, both of a foundational and applicable character. An important characteristic feature of the series is the short publication time and world-wide distribution. This permits a rapid and broad dissemination of research results.

Advisory Board

Chairman

Nikhil R. Pal, Indian Statistical Institute, Kolkata, India
e-mail: nikhil@isical.ac.in

Members

Emilio S. Corchado, University of Salamanca, Salamanca, Spain
e-mail: escorchado@usal.es

Hani Hagras, University of Essex, Colchester, UK
e-mail: hani@essex.ac.uk

László T. Kóczy, Széchenyi István University, Győr, Hungary
e-mail: koczy@sze.hu

Vladik Kreinovich, University of Texas at El Paso, El Paso, USA
e-mail: vladik@utep.edu

Chin-Teng Lin, National Chiao Tung University, Hsinchu, Taiwan
e-mail: ctlin@mail.nctu.edu.tw

Jie Lu, University of Technology, Sydney, Australia
e-mail: Jie.Lu@uts.edu.au

Patricia Melin, Tijuana Institute of Technology, Tijuana, Mexico
e-mail: epmelin@hafsamx.org

Nadia Nedjah, State University of Rio de Janeiro, Rio de Janeiro, Brazil
e-mail: nadia@eng.uerj.br

Ngoc Thanh Nguyen, Wroclaw University of Technology, Wroclaw, Poland
e-mail: Ngoc-Thanh.Nguyen@pwr.edu.pl

Jun Wang, The Chinese University of Hong Kong, Shatin, Hong Kong
e-mail: jwang@mae.cuhk.edu.hk

B. V. Babu · Atulya Nagar
Kusum Deep · Millie Pant
Jagdish Chand Bansal
Kanad Ray · Umesh Gupta
Editors

Proceedings of the Second International Conference on Soft Computing for Problem Solving (SocProS 2012), December 28–30, 2012

Volume II

 Springer

Editors

B. V. Babu
Institute of Engineering and Technology
JK Lakshmipat University
Jaipur
Rajasthan
India

Atulya Nagar
Department of Computer Science
Liverpool Hope University
Liverpool
UK

Kusum Deep
Department of Mathematics
Indian Institute of Technology Roorkee
Roorkee
Uttaranchal
India

Millie Pant
Department of Paper Technology
Indian Institute of Technology Roorkee
Roorkee
India

Jagdish Chand Bansal
Department of Applied Mathematics
South Asian University
New Delhi
India

Kanad Ray
Umesh Gupta
Institute of Engineering and Technology
JK Lakshmipat University
Jaipur
Rajasthan
India

ISSN 2194-5357   ISSN 2194-5365   (electronic)
ISBN 978-81-322-1601-8   ISBN 978-81-322-1602-5   (eBook)
DOI 10.1007/978-81-322-1602-5
Springer New Delhi Heidelberg New York Dordrecht London

Library of Congress Control Number: 2013955048

Printed on acid-free paper

Springer is part of Springer Science+Business Media (www.springer.com)

Foreword

I am delighted that the Second International Conference on "Soft Computing for Problem Solving (SocProS 2012)" was organized by the Institute of Engineering and Technology of our University from December 28–30, 2012.

l. Zadeh, founder of soft computing had stated that soft computing differs from conventional (hard) computing in that, unlike hard computing, it is tolerant of imprecision, uncertainty, partial truth and approximation. In effect, the role model for soft computing is the human mind. The guiding principle of soft computing being: Exploit the tolerance for imprecision, uncertainty, partial truth and approximation to achieve tractability, robustness and low solution cost. Every single one of these methods can be called soft computing method.

The soft computing models are based on human reasoning and are closer to human thinking. Soft computing is not a melange; it can be viewed as a founding component for the emerging field of conceptual intelligence. It is the fusion of methodologies assigned to model and enable solutions to real-world problems which are not modelled or too difficult to model mathematically. The core of soft computing comprises neural networks, fuzzy logic and genetic computing. In the coming years, soft computing will play a more important role in many areas, including software engineering. Soft computing is still growing. It is still somewhat in the formative stage and the definite components that comprise soft computing have not yet been designed. More new sciences are still merging into soft computing. SocProS 2012 was aimed to reach out to such development.

I am happy to know that Springer is publishing the Proceedings of this conference as AISC book series. I am confident that the readers will be benefited by the research inputs provided by delegates during the conference. This book will certainly provide an enriching and rewarding experience to all for academic networking.

I congratulate the editorial team and Springer for meticulously working on all the details of the book. I also congratulate all the authors of research papers for their contribution to SorProS 2012.

Jaipur, September 27, 2013   Upinder Dhar

About JK Lakshmipat University

Set up in the pink city of Jaipur, India and spread over a sprawling 30 acre campus, JK Lakshmipat University (JKLU) offers state-of-the-art academic infrastructure and world class faculty for conducting Graduate, Post Graduate and Ph.D. programmes in the mainstream disciplines of Management and Engineering. Promoted by JK Organisation, one of India's leading industrial conglomerates with a rich heritage of over 100 years, the University offers students an open, green and high-tech learning environment, combining the serene settings of the Gurukuls of yesteryears with the technological advancements of the new age.

The Institute of Management offers MBA (Full-Time Residential), MBA (Family Business and Entrepreneurship), MEA (Master of Educational Administration), a 5-year integrated dual degree (BBA + MBA) and Ph.D. programme in Management, besides PG Diploma in Tourism Administration and PG Diploma in Family Business & Entrepreneurship. Specializations include Finance and Accounting, Marketing, HRM, International Business and Information Technology. In order to balance conceptual framework with the industry practices, the delivery of each course is done in close consultation with the corporate world.

The University has also set up a 'Management Development Centre' for practicing Managers aiming to build capabilities through short-term programmes. It also organizes faculty development programmes for the benefit of academia.

The Institute of Engineering and Technology offers 4 year B.Tech, 2 year M.Tech, 5 year integrated dual degree (B.Tech + MBA) and (B.Tech + M.Tech), and Ph.D. programme in Chemical Engineering, Civil Engineering, Computer Science Engineering, Electrical Engineering, Electronics and Communications Engineering, Information Technology and Mechanical Engineering.

The curriculum is designed to enrich the students with the knowledge and relevant skills to prepare them not only to face the contemporary world but also to make them future ready to effectively perform in leadership roles assigned to them. The curriculum is updated to integrate changes that are taking place in the business environment.

JKLU is visualized to emerge as a premier institution of higher learning with global standards; it has tied up with a few renowned universities, such as University of Houston (USA), Hanyang University (South Korea), St. Cloud State University (USA), University of Wales (UK) and Szechenyi Istvan University (Hungary) for cooperation in the field of Faculty Development, Students Exchange and Research. The University has also signed an MoU with IBM India Ltd. for establishing a 'Centre of Technology Excellence' for undertaking technology development projects involving faculty members and students.

JKLU has been set up under Rajasthan Private Universities Act by 'Lakshmipat Singhania Foundation for Higher Learning'.

Preface

Unlike hard computing, the guiding principle of soft computing is exploiting the tolerance for imprecision, uncertainty, partial truth, and approximation to achieve tractability, robustness, and low solution cost, the role model being the human mind. The areas that come under the purview of Soft Computing include Fuzzy Logic, Neural Computing, Evolutionary Computation, Machine Learning and Probabilistic Reasoning, with the latter subsuming belief networks, chaos theory and parts of learning theory. The successful applications of soft computing and its rapid growth suggest that the impact of soft computing will be felt increasingly in the coming years. Soft computing is likely to play a very important role in science and engineering, but eventually its influence may extend much farther.

After the successful completion of the First International Conference on Soft Computing for Problem Solving in 2011 (SocProS 2011) at IIT Roorkee, the 2nd of the series, Second International Conference on Soft Computing for Problem Solving (SocProS 2012), was held in JK Lakshmipat University (JKLU), Jaipur, during December 28–30, 2012. It had been a matter of great privilege and pleasure to organize SocProS 2012 at JKLU, Jaipur. The response had been overwhelming and heartwarming. We had received 312 paper submissions of which 200, i.e., 64 % had been accepted for presentation. The review process included double review of submitted papers by an International team of reviewers using Easy Chair as online conference management software. There were 20 Technical Sessions including two Special Sessions, besides five Keynote Addresses by eminent Academicians and Scientists from different parts of the world that made SocProS 2012 an enriching experience to all the participants. In addition, there was one Workshop on Game Programming and one Special Tutorial on Brain like computing as a part of this conference. There were 144 Programme Committee members and 20 International Advisory Committee members from across the world associated with SocProS 2012.

We would like to express our sincere gratitude to Dr. Madhukar Gupta, Divisional Commissioner, Government of Rajasthan for gracing the occasion as the Chief Guest for the Inaugural Session.

We would also like to extend our heartfelt gratitude to all Programme Committee and International Advisory Committee members. We sincerely thank the Keynote Speakers Prof. Dipankar Dasgupta from University of Memphis—USA,

Prof. Anirban Bandyopadhyay from National Institute of Material Sciences (NIMS)—Japan, Prof. Ravindra Gudi from IIT Bombay—Mumbai, Prof. D. Nagesh Kumar from IISc—Bangalore and Prof. Ajoy Ray from Bengal Engineering and Science University (BESU)—Howrah.

We are grateful to Prof. Anirban Bandyopadhyay for conducting a Special Tutorial also in addition to delivering Keynote Address. We thank Mr. Arijit Bhattacharyya from Virtual Infocom—Kolkata for conducting the Workshop.

We thank the invited guests for accepting our invitation and also for chairing the technical sessions. We thank Dr. Kannan Govindan and Dr. P. C. Jha for delivering the Invited Talks. We express sincere thanks to the entire national and local organizing committee members for their continuous support and relentless cooperation right from the conception to execution in making SocProS 2012 a memorable event. We thank all the participants who had presented their research papers and attended the conference. A special mention of thanks is due to our student volunteers for the spirit and enthusiasm they had shown throughout the duration of the event, without which it would had been difficult for us to organize such a successful event.

Thanks are due to Springer for Publishing the Conference Proceedings. We hope that the Proceedings will prove helpful towards understanding about Soft Computing in teaching as well as in their research and will inspire more and more researchers to work in this interesting, challenging and ever growing field of Soft Computing.

We are thankful to the JKLU family for the support given in making this mega event successful. We sincerely hope that the delegates had certainly taken home several pleasant memories of SocProS 2012 with them.

We are honoured and it has been a proud privilege to be associated with this Second International Conference on SocProS 2012 as its General Chairs.

Dr. B. V. Babu
Dr. Atulya Nagar
Dr. Kusum Deep
Dr. Millie Pant
Dr. Jagdish Chand Bansal
Dr. Kanad Ray
Dr. Umesh Gupta

Committees

Organizing Committee

Patron	Dr. Upinder Dhar, JK Lakshmipat University, Jaipur, India
General Chairs	Dr. B. V. Babu, JK Lakshmipat University, Jaipur, India Dr. Atulya Nagar, Liverpool Hope University (LHU), Liverpool, UK
	Dr. Kusum Deep, IIT Roorkee, Roorkee, India
Programme Committee Chairs	Dr. Millie Pant, IIT Roorkee, Roorkee, India
	Dr. Jagdish C. Bansal, South Asian University, New Delhi, India
Local Organizing Chairs	Dr. Kanad Ray, JK Lakshmipat University, Jaipur, India
	Dr. Umesh Gupta, JK Lakshmipat University, Jaipur, India
Special Session Chairs	Dr. Millie Pant, IIT Roorkee, Roorkee, India
	Dr. Jagdish C. Bansal, ABV-IIITM, Gwalior, India
	Mr. Alok Agarwal, JK Lakshmipat University, Jaipur, India
Publicity Chairs	Dr. Sandeep Kumar Tomar, JK Lakshmipat University, Jaipur, India
	Dr. Sonal Jain, JK Lakshmipat University, Jaipur, India
Best Paper Chairs	Dr. Ravidra Gudi, IIT Bombay, Mumbai, India
	Dr. D. Nagesh Kumar, IISc, Banglore, India
	Dr. P. C. Jha, Delhi University, New Delhi, India
Steering Committee	Dr. Atulya Nagar
	Dr. B. V. Babu
	Dr. Kusum Deep
	Dr. Millie Pant
	Dr. Jagdish C. Bansal

International Advisory Committee

Prof. Zbigniew Michalewicz	University of Adelaide, Adelaide, Australia
Dr. Gurvinder S. Baicher	University of Wales, UK
Prof. Lipo Wang	Nanyang Technological University, Singapore
Prof. Patrick Siarry,	Université de Paris, France
Prof. Michael N. Vrahatis	University of Patras, Greece
Prof. Helio J. C. Barbosa	Federal University of Juiz de Fora, Brazil
Prof. S. K. Singh	CSIR—Central Building Research Institute, Roorkee, India
Prof. Wei-Chiang Samuelson Hong	Oriental Institute of Technology, Taiwan
Prof. P. K. Kapur	Amity University, Noida, India
Prof. Samrat Sabat	University of Hyderabad, India
Prof. S. S. Rao	University of Miami, Florida, USA
Prof. Pramod Kumar Singh	ABV-IIITM, Gwalior, India
Prof. Montaz Ali	Witwatersrand University, Johannesburg, South Africa
Prof. Andries P. Engelbrecht	University of Pretoria, South Africa
Prof. Suresh Chandra	Indian Institute of Technology New Delhi, New Delhi, India
Prof. Kalyanmoy Deb	Indian Institute of Technology Kanpur, Kanpur, India
Prof. D. Nagesh Kumar	Indian Institute of Science, Bangalore, India
Prof. Nirupam Chakraborti	Indian Institute of Technology Kharagpur, Kharagpur, India
Prof. K. C. Tan	National University of Singapore, Singapore
Prof. Roman R. Poznanski	Universiti Tunku Abdul Rahman (UTAR), Malaysia

Programme Committee

Dr. Abhijit Sanyal	Dr. K. K. Shukla
Dr. Abhay Jha	Dr. K. P. Singh
Dr. Abhijit Sarkar	Dr. Kusum Deep
Dr. Abhinay Pandya	Dr. Lalit Awasthi
Dr. Adel Aljumaily	Dr. Laxman Tawade

Dr. Aitorrodriguez Alsina
Dr. A. J. Umbarkar
Dr. Akila Muthuramalingam
Dr. Amit Pandit
Dr. Andres Muñoz
Dr. Anil Parihar
Dr. Antonio Jara
Dr. Anupam Singh
Dr. Anuradha Fukane
Dr. Aram Soroushian
Dr. Arnab Nandi
Dr. Arshin Rezazadeh
Dr. Ashish Gujarathi
Dr. Ashraf Darwish
Dr. Ashwani Kush
Dr. AsokeNath
Dr. Atulya Nagar
Dr. Ayoub Khan
Dr. B. V. Babu
Dr. Balaji Venkatraman
Dr. Balakrishna Maddali
Dr. Banani Basu
Dr. Bharanidharan Shanmugam
Dr. Bratin Ghosh
Dr. Carlos Fernandezllatas
Dr. Chu-Hsing Lin
Dr. Ciprian Dobre
Dr. D. Nagesh Kumar
Dr. Dante Tapia
Dr. Dipika Joshi
Dr. Dipti Gupta
Dr. Eduard Babulak
Dr. Farhad Nematy
Dr. Francesco Marcelloni
Dr. G. Shivaprasad
Dr. G. R. S. Murthy
Dr. Gauri S. Mittal
Dr. Gendelman Oleg
Dr. Gurvinder Singh-Baicher

Dr. Leonard Barolli
Dr. Manjaree Pandit
Dr. Manoj Saxena
Dr. Manoj Thakur
Dr. Mansaf Alam
Dr. Manu Augustine
Dr. Mario Koeppen
Dr. Mehul Raval
Dr. Millie Pant
Dr. Mohammad Ahoque
Dr. Mohammad Reza Nouri Rad
Dr. Mohammed Abdulqadeer
Dr. Mohammed Rokibul Alam Kotwal
Dr. Mohdabdul Hameed
Dr. Mourad Abbas
Dr. Mrutyunjaya Panda
Dr. Munawar A. Shaik
Dr. Musrrat Ali
Dr. Ninansajeeth Philip
Dr. Nitin Merh
Dr. O. P. Vyas
Dr. Philip Moore
Dr. Poonam Sharma
Dr. Pramod Kumar Singh
Dr. Punam Bedi
Dr. Pushpinder Patheja
Dr. Radha Thangaraj
Dr. Rajesh Sanghvi
Dr. Ram Ratan
Dr. Ramesh Babu
Dr. K. C. Raveendranathan
Dr. Ravi Sankar Vadali
Dr. Razibhayat Khan
Dr. Ritu Agarwal
Dr. Rodger Carroll
Dr. Sami Habib
Dr. Sandeep Kumar Tomar
Dr. Sanjeev Singh
Dr. Shaojing Fu

Contents

Contents

Part VIII
Soft Computing and Web Technologies (SCWT)

Heirarchy of Communities in Dynamic Social Network

S. Mishra and G. C. Nandi

Abstract Discovering the hierarchy of organizational structure can unveil significant patterns that can help in network analysis. In this paper, we used Enron email data which is well-known benchmarked data set for this sort of research domain. We derive a hierarchical structure of organization by calculating the individual score of each person based on their frequency of communication via email using page rank algorithm. After that, a communication graph is plotted that shows power of each individual among themselves. Experimental results showed that this approach was very helpful in identifying primal persons and their persistent links with others over the period of months.

Keywords Dynamic social network analysis · Social network analysis · Hicrarchal structure.

1 Introduction

A network structure is the perfect epitome that provides a formal way of representing data that emphasizes the association between entities. This representation has a substantial importance that gives the insight of knowledge into the data. Since for the work to be done many entities these days are interconnected and behaviors of individual entity reflect the function of whole system to a large extent. The entity could be people, organization [1], computer nodes [2]. Networks are primarily studied in mathematical framework, i.e., graph [3].

S. Mishra (✉) · G. C. Nandi
Robotics and AI Lab, Indian Institute of Information Technology,
Allahabad, India
e-mail: seema.mishra6@gmail.com

G. C. Nandi
e-mail: gcnandi@iiita.ac.in

B. V. Babu et al. (eds.), *Proceedings of the Second International Conference on Soft Computing for Problem Solving (SocProS 2012), December 28–30, 2012,* Advances in Intelligent Systems and Computing 236, DOI: 10.1007/978-81-322-1602-5_84, © Springer India 2014

In modern era, social network analysis is proliferated area of research, has been in existence for quite some time and experiencing a surge in popularity to understand the behavior of the users at individual and group level [1, 2]. Understanding the behavior of individual social networking methods assuage the analysts to revealing hidden patterns from social communication. In order to model the social network mathematically, most popular data structure typically known as graphs are used where the nodes depict the individual or group of person, or event or organization, etc, and each link/edge represents connection/relationship between two individual. Social network analysis attempts to understand the network and its components like nodes (social entities commonly known as actor or event) and connections (interconnection, ties, and links). It has main focus of analyzing individuals and their interdependent relationships among them rather than individuals and their attributes as we deal in conventional data structure.

2 Dynamic Network Analysis

Versatile power of social network is being applied to mining pattern of social interaction in wide ranging applications including: disease modeling [4] information transmission, behavior analysis [5, 6], and business management and behavior analysis. Network analysis also came in to picture as its practical applications in intelligence and surveillance [7] and has become popularized paradigm to uncover antisocial netwo:k such like criminal, terrorist, and fraud network majorly after the tragically event of September 11, 2001 which has shattered the whole world.

Social interaction could be in any form that depends on the type of data available [8]. It might be verbal or written communication (cell phones, emails, and blogs chatting), scientific collaboration (co-authorship network, citation network), browsed websites, and group of animals.

This mathematical network model is very successful in analysis of social network but major drawback is that it may miss the temporal aspect of interaction because social interaction is inherently dynamic in nature. The static model of interaction can give the information that could be inaccurate and decision made based merely on this contributed information might lead analyst to false position of analysis.

Several shortcomings can be highlighted when dealing with static model of social network that could forbid acknowledging the casual relationship of pattern of social interaction [8]:

- What is the rate of spreading diseases while modeling diseases and who is the central person whom should be vaccinated to control spread of it among group of person.
- What are the causes and consequences of social structure evolution?

Dynamic social network analysis is emerging research area that play a crucial role to fill gap between traditional social network analysis and time domain. Dynamic study of network includes classical network analysis, link analysis, and multiagent systems.

Dynamic network analysis facilitates the analysis of multiple types of nodes (multinode) and multiple types of links (multiplex) simultaneously. On the contrary, static network analysis can only focus at most two mode data and analyze one type of link at a time. There are several characteristics of dynamic network:

- Nature of nodes are dynamic, there properties changes with respect to time.
- Deals with meta-network.
- Network evolution is consequent of agent-based modeling.

2.1 Community Detection

A community is subpart of whole network between which intercommunity interaction is relatively frequent and strong than intracommunity interaction. It can be in any form for example group, subgroup, and cluster. It may; (a) citation network represents related papers on single topics, (b) Web pages on related topics.

Community detection is a classical problem in social network analysis. Commonly, it can be the problem of identifying subgraphs of original graph and called vertex sparsifier [9]. These small networks uphold the relevant information of original group. Four levels of analyses are being conducted in community identification as shown in Fig. 1:

2.2 Analysis of Previous Work

Hierarchical methods for community detection falls into two categories: agglomerative and divisive. In former case each node is assumed to be a community and repetitively group together. Similarly, in later case initially whole network is considered as a community and divided subsequently into smaller one. Most methods

Fig. 1 Level of analysis of community detection

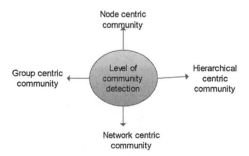

are graph clustering and partition. Distance-based structural equivalence [10] uses distance metrics to identify similar entities. In graph partition methods several algorithms has been proposed [11, 12]. Newman-Girvan method and spectral clustering methods [13, 14] uses a notion of modularity and utilizes edge betweenness metric to divide into groups.

In analyzing dynamic pattern, many methods use the temporal snapshots of interaction over the times [15, 16].

2.3 Discovering Hierarchy of Group

Before analyzing community hierarchy we define several basic terminologies. Hierarchy of community provides the power of each individual in a group. If somehow we know this chain we can find the leader of group. Regarding this we used well-known algorithm Page Rank which calculate the individual score I_score of each person to represent importance of person. Higher the score of person more powerful the person is.

Definition 1 If $P = \{p_1, p_2, \ldots, p_k\}$ be the collection of person involved in a commuinication. For any member p_i and p_j, if I_score$(p_i) \geq$ I_score(p_j) then p_i is more powerful than p_j

A. Data Set

We performed the experiments on Enron data set. Since email communication data has become a practical source for research in network analysis like social network. Mostly the experiments are carried out on the artificial data due to the non-availability of real life communication data. The Enron email data set [17] has become a benchmark for this sort of research domain in network analysis. This data set was made public and posted on Web by the Federal Energy Regulatory Commission during its investigation for fraud happened in company, in order to make it test bed for validating and testing the efficacy of methodologies developed for counter-terrorism, fraud detection, and link analysis.

Data is about 150 users communication mostly senior managers organized into folder where nodes are people and edges are email communication between them. But this data set has still lots of issues regarding integrity issue and duplicate messages issue. We preprocessed the data sets in to socio matrix of 12 months from January to December and finally draw a graph of interaction over the months.

B. Experiments and Analysis

In this section, we evaluate the capability of the proposed approach on discovering organizational structure and to exploring evolution of organizational structure in a dynamic social network. The implementation was done in DEV C++. The experiments were conducted on a 2.1GHz PC with Core(TM)2 Dual-Core Pentium 4 processor with 2 GB RAM.

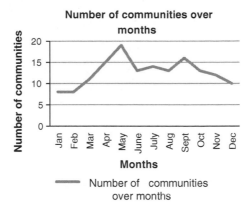

Fig. 2 Number of communities over months

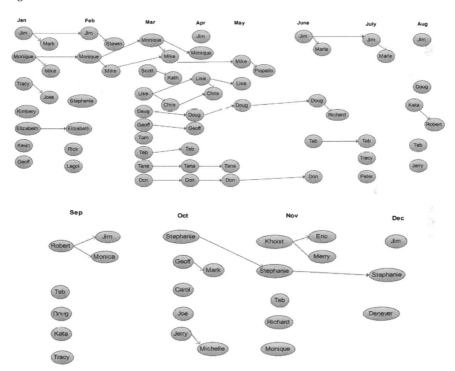

Fig. 3 Evolution of community from months Jan to Dec

On examining the results in Fig. 2, we analyzed the number of grouping in the month of May was maximum. Figure 3 shows that Jim was the person who headed the group from Jan to Feb. Monique was leader throughout the months form Jan to April.

3 Conclusion and Future Work

In this paper we introduced a concept of hierarchy of positions in group by taking temporal interaction data of 12 months in organization that shows how the position of group members changes when people joining and leaving the group. In future, we can also model the changes over time and detect event occurrences as consequence of this change. We also will improve the integrity issues of preprocessed Enron data results of experiments.

References

1. Wasserman, S., Faust, K.: Social Network Analysis: Methods Application. Cambridge University Press, New York (1994)
2. Wellman, B.: Computer networks as social networks. Sci. Mag. **293**, 2031–2034 (2001)
3. Robert, A.H., Riddle, M.: Introduction to Social Network Methods. University of California, Riverside (published in digital form http://faculty.ucr.edu/~hanneman/) (2005)
4. Kretzschmar, M., Morris, M.: Measures of concurrency innetworks and the spread of infectious disease. Math. Biosci. **133**, 165–195 (1996)
5. Baumes, J., Goldberg, M., Magdon-Ismail, M., Wallace. W., Discovering hidden groups in communication networks. In: Proceedings of the 2nd NSF/NIJ Symposium on Intelligence and Security Informatics (2004)
6. Tyler, J., Wilkinson, D., Huberman, B.: Email asspectroscopy: Automated discovery of community structure within organizations. In: Proceedings of the 1st International Conference on Communities and Technologies (2003)
7. Baumes, J., Goldberg, M., Magdon-Ismail, M., Wallace, W.: Discovering hidden groups in communication networks. In: Proceedings 2nd NSF/NIJ Symposium on Intelligence and Security Informatics (2004)
8. Berger-Wolf, T.Y., Saia, J.: A framework for analysis of dynamic social networks. In: Proceedings of the KDD'06, pp. 523–528 (2006)
9. Moitra, A.: Approximation algorithms for multicommoditytype problems with guarantees independent of the graph size. In: Proceedings of the FOCS, pp. 3–12 (2009)
10. Santo, F., Castellano, C.: Community structure in graphs, chapter of Springer's Encyclopedia of Complexity and System Science (2008)
11. Chekuri, C., Goldberg, A., Karger, D., Levin, M., Stein, C.: Experimental study of minimum cut algorithms. In: Proceedings of the 8th SAIM Symposium on Discreet Algorithm, pp. 324–333 (1997)
12. Andrew, Y., Wu., et al.: Mining scale-free networks using geodesic clustering. In: Proceedings of the KDD'04, pp. 719–724 (2004)
13. Newman, M.E.J., Girvan, M.: Finding and evaluating community structure in networks. Phys. Rev. E **69**, 026113 (2004)
14. Newman, M.E.J.: Modularity and community structure in networks. PNAS **103**(23), 8577–8582 (2006)
15. Zhou, D., Councill, I., Zha, H., Lee Giles, C.: Discovering temporal communities from social network documents. In: Proceedings of the ICDM'07, pp. 745–750 (2007)
16. Tantipathananandh, C., Berger-Wolf, T., David Kempe, A.: Framework for community identification in dynamic social networks. In: Proceedings of the KDD'07, pp. 717–726 (2007)
17. The original dataset can be downloaded from William Cohen's web page http://www-2.cs.cmu.edu/~enron/

SLAOCMS: A Layered Architecture of SLA Oriented Cloud Management System for Achieving Agreement During Resource Failure

Rajkumar Rajavel and Mala T

Abstract One major issue of cloud computing is developing Service Level Agreement (SLA)-oriented cloud management system, because situations like resource failures may lead to the violation of SLA by the provider. Several research works has been carried out regarding cloud management system were the impact of SLA is not properly addressed in the perspective of resource failure. In order to achieve SLA in such circumstance, a novel-layered architecture of SLA-oriented Cloud Management System (SLAOCMS) is proposed for service provisioning and management which highlight the importance of various components and its impacts on the performance of SLA- based jobs. There are two components such as Task Scheduler and Load Balancer which are introduced in SLA Management Framework to achieve SLA during resource failure. So, an SLA Aware Task Scheduling Algorithm (SATSA) and SLA Aware Task Load Balancing Algorithm (SATLB) are proposed in the above components to improve the performance of SLAOCMS by successfully achieving the SLA's of all the user jobs. The results of traditional and proposed algorithms are compared in the scenario of resource failure with respect to violations of SLA-based jobs. Moreover, SLA negotiation framework is introduced in application layer for supporting personalized service access through the negotiation between the service consumer and service provider.

Keywords Cloud management · SLA management · Task scheduling · Task load balancing · Achieving SLA

R. Rajavel (✉)
Anna Centenary Research Fellow, Department of IST, Anna University, Chennai, India
e-mail: rajkumarprt@gmail.com

Mala T
Assistant Professor, Department of IST, Anna University, Chennai, India
e-mail: malanehru@annauniv.edu

B. V. Babu et al. (eds.), *Proceedings of the Second International Conference on Soft Computing for Problem Solving (SocProS 2012), December 28–30, 2012*, Advances in Intelligent Systems and Computing 236, DOI: 10.1007/978-81-322-1602-5_85, © Springer India 2014

1 Introduction

Cloud Computing provides differentiated services to the users on demand with respect to the expectation of service quality on a pay for usage basis. One major issue in this area is cloud management which requires the management of services such as Infrastructure as a Service (IaaS), Platform as a Service (PaaS), Software as a Service (SaaS), and Storage as a Service in the specified layers. In cloud management several challenges such as scalability, multiple levels of abstraction, federation, sustainability, and dynamism are addressed to adapt in future [6]. A taxonomical spectrum of cloud computing framework describes the various components of cloud where SLA Management is addressed as one of the major component in the management service [11].

Existing cloud management systems does not support the SLA-based service provisioning with respect to all the available parameters of service. Because any malfunction such as resource failure and VM failure may lead to serious trouble for the providers to meet the SLAs of all the user jobs. The major challenges of SLA-oriented resource management such as architecture framework, SLA-based scheduling policies, and SLA resource allocators were addressed [3]. In the future perspective, cloud management seems to have most daunting and challenging issues, because improper management without any precautions will swipe out the customer data and leads to the provider's SLA violation. Supporting of SLAs with multiple objectives is addressed as one of the important requirement of future cloud management [4]. Above challenges, issues and requirements motivates our research work to develop a novel SLAOCMS for supporting SLA-based service provisioning and management in future trends of cloud.

Objective of this SLAOCMS is to provide SLA-based cloud service provisioning and management during the resource failure by using an efficient Task Scheduler and Task Load Balancer components. In order to test the efficiency of the proposed SLAOCMS, SLA-based jobs are submitted to unified resource layer by insisting resource failure (limited resource availability). As an impact of introducing the SATS and SATLB algorithms in the proposed system, the SLA is achieved for all jobs by rearranging its sequence within the resource in case of compatible deadline and migrating to other resources in case of incompatible deadline. In case of limited resource availability, only SATS algorithm can be effectively used to achieve the SLA of user jobs. This situation may impact the SATLB algorithm, to migrate only the moderate amount of workload to the free resource available in the environment. Hence, the achievement of jobs SLA during resource failure is probably in the hands of resource availability present at that moment. Moreover, the proposed research work of SLAOCMS system addresses various open research issues and challenges in specific to SLA Management activity.

2 Related Work

In this section, traditional model of cloud management systems, and its drawbacks are addressed such as Iaas, PaaS, SaaS, and Storage as a Service management. The SLA-based service virtualization architecture is proposed to avoid SLA violation under changing workload, system malfunction, hardware and software failure [8]. A service-oriented policy-driven IaaS Management system is proposed to improve the performance of services running on IaaS [14]. A market-oriented hierarchical scheduling strategy is proposed for SwinDeW-C cloud workflow system using the service-level scheduling algorithm and meta-heuristic-based scheduling algorithm [13]. One phase and two phase algorithms are proposed to overcome the problems of workflow scheduling in IaaS cloud [1].

An Analysis is made regarding the security mechanism and SLA for the development of geoprocessing services in cloud PaaS which helps to manipulate the geographic information [9]. A problem of limited resource capacity is considered to serve the SLA-based SaaS request for the cloud resource management. The knapsack problem model employs the virtual infrastructure to obtain the heuristic solution [2]. In order to effectively use the cloud resource an SLA-based admission control and scheduling algorithm is proposed for the benefit of SaaS provider [12]. Paper [4], discuss the importance of managing SLA activity in cloud were SLA Management is noted as one of the research challenge in future cloud management. In order to reduce the quality of service violation, an energy efficient resource management framework is proposed for cloud service provisioning [7]. In the previous research work hierarchical scheduling is used for prioritizing the deadline-based job in both cloud and cluster controller level [10]. Hence SLA management framework is included in application, platform and unified resource layers of proposed SLAOCMS where SLA aware scheduler and load balancer are highlighted as important component to manage the SLA of all the user jobs.

3 SLA-Oriented Cloud Management System

In this section, an SLAOCMS is introduced as shown in Fig. 1 to support the personalized service provisioning and achieving SLA in the future cloud. This cloud management system consists of four layers such as fabric, unified resource, platform, and application. A service layer is introduced as a sub layer of application layer which supports the consumer and provider to negotiate with respect to SLA parameters for providing personalized cloud service access. The service layer contains the components like SLA template, SLA negotiation framework, negotiation protocol, request handler, service classifier, and information manager. An SLA template is a pattern for creating agreement which contains the information like name of the service and parties, context, terms and negotiation constraints. A negotiation strategy is followed by the SLA negotiation framework for making communication with the consumer

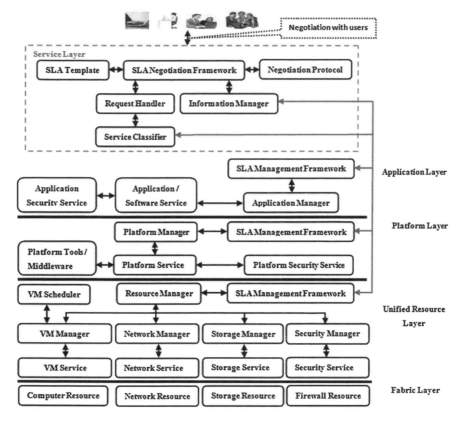

Fig. 1 Architecture of SLAOCMS

or broker by exploiting the negotiation protocol and SLA template. Moreover, the information manager is invoked during the negotiation process for collecting the service availability and its reservation information. The request handler will obtains the negotiated request and placed in the request handler queue for further processing. Then the request is periodically pulled by the service classifier, to identifying its service type and forwards the request to SLA management framework of concern layer. The structure of SLA management framework is shown in Fig. 2 which includes the components such as service integrator and manager, task scheduler, task load balancer, SLA manager, agreement generator, service provisioner, dispatcher, SLA monitor, notification generator, SLA terminator, and utility manager.

This management framework contains the SLA Manager who is responsible for managing all the SLA-related activities of cloud services by using the above components in the lifecycle. The reservation queue is maintained in the reservation manager which contains all the committed jobs to execute in the reserved time. In the existing system, this reservation queue follows the FCFS scheduling by executing the task one

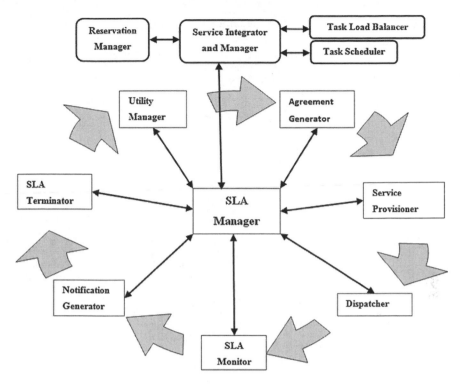

Fig. 2 Structure of SLA management framework in unified resource layer

by one in the same sequence as present in the queue. This type of system may leads to SLA violation of tasks during the VM failure in the Data Center (DC). Because of this failure, the scheduled task may be delayed due to shortage of VM available in the DC. In this scenario proposed SLAOCMS invokes the task scheduler and load balancer component to avoid the SLA violations of the submitted job which is present in the reservation queue. These two components exploit the SLA Aware Task Scheduling Algorithm (SATSA) and SLA Aware Task Load Balancing Algorithm (SATLBA) as shown in Algorithm 1 and 2. Here, the Estimated Completion Time (ECT) of any deadline-based jobs present in the Deadline Job List (DJL) size 'n' can be computed using the Eq. (1). This equation will compute the ECT of Deadline Job (DJ) at the position 'i' DJ_i by summing of all the jobs Completion Time (CT) present in the $DJL_{1 to i}$.

$$ECT(DJL_i) = \sum CT(DJ_1) + CT(DJ_2) + \cdots CT(DJ_i) \qquad (1)$$

Algorithm 1 SLA Aware Task Scheduling Algorithm

Begin
    Get the set of Virtual Machine (VM) L from the Data Center DC_i
    Assign the reserved deadline RD to each job
    Get the set of Jobs J from the Reservation Queue (RQ)
    for all VM l_i • **L do**
      Ping the VM
      Add to Active Virtual Machine (AVM) set AL
    end for
    for all AVM al_i • **AL do**
      if AL.length() == L.length() **then**
        Submit the job in FCFS fashion
      else if AL.length() < L.length()
        Get the set of Deadline based Job List (DJL)
        Compute the Estimated Completion Time (ECT) of DJL
        Align the DJL in the order in which the ECT is less as first
        for all djl_i • **DJL do**
          if $ECT(djl_i) > RD(djl_i)$ **then**
            Increment the position of djl_i from the RQ
            Compute the Estimated Completion Time (ECT) of DJL_i
            if $ECT(DJL_i) > RD(DJL_i)$ **then**
              Invoke **Algorithm 2**
            else
              Refresh the RQ position
            end if
          else if $ECT(djl_i) <= RD(djl_i)$
            Keep the job in same position
          end if
        end for
      end if
    end for
end

Algorithm 2 SLA Aware Task Load Balancing Algorithm

Begin
    Get the set of Available Data Center (ADC) in service provider side
    Get the DJL_i from the Reservation Queue (RQ)
    for all adc_j • **ADC do**
      Compute the $ECT(DJL_i)$ in the ADC
      if $ECT(djl_i) < RD(djl_i)$ **then**
        Migrate to RQ of ADC
      end if
    end for
end

The main objective of the SATSA is to prioritizing the task present in the reservation queue based on the deadline specified in the agreement. At the maximum level, this algorithm will avoid the violation of task by prioritization. In case of more VM failure in the DC may be difficult to avoid the violation of the entire task. Because of the availability of VM in the DC, it is possible to prioritize only limited number of task. So to over such situation, SATLBA is invoked by this algorithm to migrate the extra task (which is not able to prioritize) to other DC which is capable of completing the task within the stipulated time.

4 Performance Evaluation

The simulation of cloud environment with the resource provisioning and management is implementation using the cloudsim toolkit [5]. For the sake of simplicity, experimental setup is simulated with 3 DC with 5 VM. Assume the tasks present in the DC1 with Reserved Deadline (RD) and Execution Time (ET) of each task is in the sequence as shown in Table 1.

The performance evaluation of proposed SATSA and SATLBA is evaluated in this experimental setup. Assume there is no task executed in the reservation queue of DC1 at the initial stage (all the VM are free). First the experiment is simulated for traditional FCFS scheduling algorithm, SATSA without SATLBA and SATSA with SATLBA in the unified resource layer without any resource failure. Similarly, the experiment is simulated with 1, 2, and 3 VM failures. The number of task violation is marked for all the simulated cases in the graph as shown in Fig. 3.

Table 1 Task Information present in DC1 reservation queue

Task ID	ET (min)	RD (min)
1 – 5	30	30
6 – 10	30	60
11 – 15	30	NIL
16 – 20	30	NIL
21 – 25	30	150
26 – 30	30	180
31 – 35	30	NIL
36 – 40	30	NIL
41 – 45	30	270
46 – 50	30	300
51 – 55	30	NIL
56 – 60	30	NIL

Fig. 3 Performance evaluation of FCFS, SATSA and SATSA

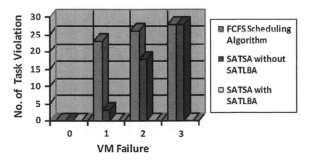

5 Conclusion and Future Work

Thus from the performance graph, it is clear that the proposed SATSA with SATLBS gives no SLA violation of task with respect to number of VM failures. The task violation gradually increases with respect to number of VM failures for the case of SATSA without SATLBA. But in the case of FCFS scheduling algorithm, number of task violation drastically increases with respect to number of VM failures. Hence the proposed SATSA with SATLBA in the SLAOCMS helps to achieve SLA during resource failure by migrating task from one DC to another.

In future, Automated Dynamic SLA Negotiation Framework will be proposed to support the personalized service access in the cloud environment. This framework constitutes the negotiation strategy of service provider, trusted third party broker and service consumer for describing the communication process involved between the negotiating parties. In addition, the framework derives the mathematical model using the game theory approach to providing optimal solution for bargaining between negotiation parties.

Acknowledgments We like to sincerely thank the Anna University, India for their financial support to this research work under the scheme of Anna Centenary Research Fellowship.

References

1. Abrishami, S., Naghibzadeh, M., Epema, D.: Deadline-constrained workflow scheduling algorithms for IaaS Clouds. Future Genera. Comput. Syst. **29**, 158–169 (2012)
2. Aisopos, F., Tserpes, K., Varvarigou, T.: Resource management in software as a service using the knapsack problem model. Int. J. Prod. Econ. **141**, 465–477 (2011)
3. Buyya, R., Garg, S.K., Calheiros, R.N.: SLA-Oriented resource provisioning for cloud computing: challenges, architecture, and solutions. In: IEEE International Conference on Cloud and Service, Computing, pp. 1–10 (2011)
4. Cook, N., Milojicic, D., Talwar, V.: Cloud management. J. Internet Serv. Appl. **3**, 67–75 (2012)
5. Calheiros, R.N., Ranjan, R., Beloglazov, A., De Rose, C.A.F., Buyya, R.: CloudSim: a toolkit for modeling and simulation of cloud computing environments and evaluation of resource provisioning algorithms. Softw. Pract. Experience **41**(1), 23–50 (2011)
6. Forell, T., Milojicic, D., Talwar, V.: Cloud management: challenges and opportunities. In: IEEE International Parallel and Distributed Processing, Symposium, pp. 881–889 (2011)
7. Guazzone, M., Anglano, C., Canonico, M.: Energy-efficient re-source management for cloud computing infrastructures. In: Third IEEE International Conference on Coud Computing Technology and Science, pp. 424–431 (2011)
8. Kertesz, A., Kecskemeti, G., Brandic, I.: An interoperable and self-adaptive approach for SLA-based service virtualization in heterogeneous cloud environments. In: Rana O., Corradi A. (eds.) Special Issue of Future Generation Computer Systems on Management of Cloud Systems, Elsevier (2012)
9. Ludwig, B., Coetzee, S.: Implications of security mechanisms and Service Level Agreements (SLAs) of Platform as a Service (PaaS) clouds for geoprocessing services. Appl. Geomatics **5**(1), 25–32 (2012)
10. Rajavel, R., Mala, T.: Achieving service level agreement in the cloud environment using job prioritization in hierarchical scheduling. In: International Conference on Information Systems

Design and Intelligent Applications, Advances in Intelligent and Soft Computing, vol. 132, pp. 547–554 (2012)
11. Rimal, B.P., Choi, E.: A service-oriented taxonomical spectrum, cloudy challenge sand opportunities of cloud computing. Int. J. Commun. Syst. **25**, 796–819 (2012)
12. Wu, L., Garg, S.K., Buyya, R.: SLA-based admission control for a software-as-a-service provider in cloud computing environments. J. Comput. Syst. Sci. **78**, 1280–1299 (2012)
13. Wu, Z.Liu, X. Ni, Z., Yuan, D., Yang, D.: A market-oriented hierarchical scheduling strategy in cloud workflow systems. J. Supercomput. **63**(1), 256–293 (2013)
14. Xiao-xiang, L., Mei-na, S., Jun-de, S.: Research on service-oriented policy-driven IAAS management. J. Chin. Univ. Posts Telecommun. **18**(1), 64–70 (2011)

Semantic Search in E-Tourism Services: Making Data Compilation Easier

Juhi Agarwal, Nishkarsh Sharma, Pratik Kumar, Vishesh Parshav, Anubhav Srivastava, Rohit Rathore and R. H. Goudar

Abstract After the advancement of the internet technology, user can get any information on tourism. Tourism is the world's largest and fastest growing industry. It contains so many things like accommodation, food, events, transportation package, etc. So information must be reliable because tourism product is intangible in nature. Customer cannot physically evaluate the service until he/she physically experienced but there are some areas where a greater measure of intelligence is required. The Semantic Web did a lot of work to enhance the Web by enriching its content with semantic data. E-Tourism is a good candidate for such enrichment, since it is an information-based business. In this paper, we are constructing E-Tourism ontology to provide intelligent tourism service. The algorithm is designed to integrate data from different reliable sources and structure properly in tourism knowledge base for efficiently searching the data.

Keywords Semantic web · Ontology · SPARQL

J. Agarwal (✉) · N. Sharma · P. Kumar · V. Parshav · A. Srivastava · R. Rathore · R. H. Goudar
Graphic Era University, Dehradun, India
e-mail: juhiiagrawal@gmail.com

N. Sharma
e-mail: nishkarsh4@gmail.com

P. Kumar
e-mail: pratikkumar938@gmail.com

V. Parshav
e-mail: vishparshav1@gmail.com

A. Srivastava
e-mail: anubhav.v.sri@gmail.com

R. H. Goudar
e-mail: rhgoudar@gmail.com

B. V. Babu et al. (eds.), *Proceedings of the Second International Conference on Soft Computing for Problem Solving (SocProS 2012), December 28–30, 2012*, Advances in Intelligent Systems and Computing 236, DOI: 10.1007/978-81-322-1602-5_86, © Springer India 2014

1 Introduction

After the advancement of the internet technology, user can get any information on tourism very easily. Tourism is a global information-intensive industry that has a long chain of stakeholders including service providers, marketers, managers, and consumers. So E-Tourism sites are making our journey comfortable but the tourism product is intangible in nature. Customer cannot physically evaluate the service until he/she physically experienced. So the information should be accurate and believable. The Semantic Web did a lot of work to enhance the Web by enriching its content with semantic data. E-Tourism is a good candidate for such enrichment, since it is an information-based business [1]. Semantic search uses ontology [2]. Ontology is a good approach. It is a formal conceptualization of a particular domain that is shared by a group of people [3]. Ontology is used in knowledge-based systems as conceptual frameworks for providing, accessing, and structuring information in a comprehensive manner [4]. Retrieval of keyword and text matching in the retrieval engine can be successful only through Semantic Web. In this paper, we are constructing E-Tourism ontology for intelligent tour service where we can search the data semantically using Semantic Search Engine. It can improve the process of searching for the perfect tourism package by analyzing the user interest with the help of ontology.

2 Ontology Creation for E-Tourism Service

2.1 Domain Ontology for Tourist Information

It could be observed that the tourism sector is an area in which ontology is playing very important role. Ontology essentially consists of a vocabulary of terms in a domain of interest and their meanings. This includes definition of concepts, the properties of the concepts, and the interrelationship between concepts [5, 6]. Ontology creation for E-Tourism service is given in Fig. 1. Figure 3 is an ER-diagram showing relationship and properties of the classes. The class hierarchy is as follows:

Thing: All the other classes are the sub class of this class. In other words, this is the Super Class containing all the other classes.

Club: The tourists may be interested in places like Clubs and Disco, etc., for night life, leisure activities. This class includes sub classes Casino, PubBar, and Disco.

Shopping Place: This class includes information about the shopping places available at a destination. It is further sub classed into Mall, Market, and Shop.

Object properties are used to link objects of different classes and tell about their relation. The object properties are as follows:

Belongs to State: This property links the objects of City class with the objects of State class to tell a particular city belongs to which State.

Fig. 1 E-tourism ontology

Has Climate: This property links the objects of Destination class with the Climate class. It tells about the climate that a particular place has Fig. 2.

2.2 Creation of Intelligent Tourist Information Service Using Semantic Search Engine

There are four phases for development and implementation of Intelligent Tourist Information Service using Semantic Search Engine, which are explained below:

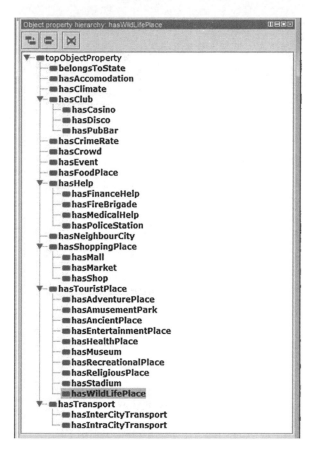

Fig. 2 Object properties

2.2.1 Prepare Backbone Structure for Capturing and Representing Knowledge in Tourism Domain. India Tourism Ontology Creation

- Identifying the concepts of the domain.
- Identify the attributes or properties of these concepts.
- Identify the relationship between these concepts and defining the properties for these relationships.
- Defining the rules or axioms for the domain on the basis of which inferences could be drawn.

2.2.2 Create the India Tourism Knowledge Base Using Ontology

This crucial phase will follow the given sequence of steps:

- Identify all the authentic sources of information available on the Web about the India Tourism.

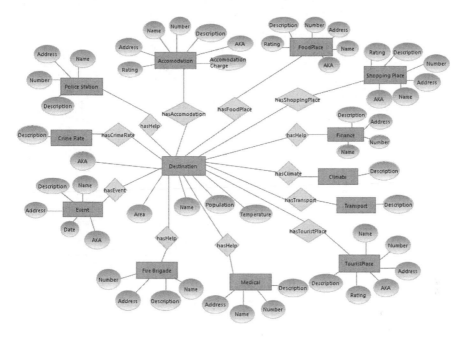

Fig. 3 ER-diagram for E-tourism. The destination is connecting different classes means it is showing relationship and properties from destination to different classes

- Identify additional sources of information to make the tourism information as rich as possible by the tourism department in various forms.
- Validating and integrating the information to populate the knowledge base Fig. 4.

2.2.3 Create the Semantic Search Mechanism

- User enters the search query in the normal English language, i.e., similar to the search query of keyword based search engines like Google.
- This keyword query needs to be expanded on the basis of the created Tourism ontology to add a number of semantically related terms to it.
- The expanded query will be converted to a semantic query.
- Semantic query will be fired and necessary information will return to user Fig. 5.

2.2.4 Creating the User-Interface and the E-Tourism Portal

- Out of the huge information possible for the tourism domain, the most important information should be presented to the user directly i.e., through the use of menus and tabs the broad classes of information will be available.

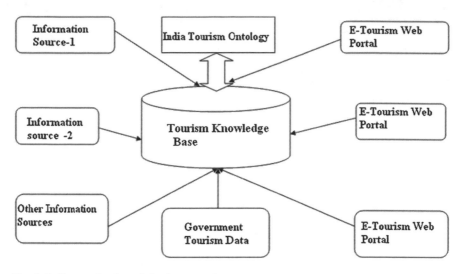

Fig. 4 Indian tourism knowledge base creation

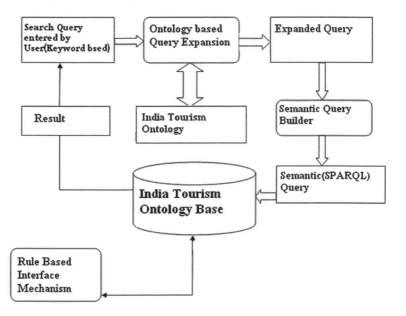

Fig. 5 Various semantic search steps to return the result to the user

- A semantic search interface will be presented to the user which will act as the gateway to the comprehensive India Tourism related information. The various links containing the information within the portal or referring to the other sites will be returned through this search.
- The facilities of booking will be provided through the web portal.

- The system will keep track of the type of users so that the information specific to the user interest could be provided. There will be a separate login for the tourism and the other players in the tourism domain like tour-guide.

3 Searching Algorithm

Example of searching:

- User enters the keyword " Dehradoon ".
- After parsing (removing spaces from the search query), the keyword becomes "Dehradoon". The keyword is converted to lowercase.
- The keyword is then matched to the data element that is linked to the objects using has Name property in the ontology base.
- If the match is found (which is not true in this case), the SPARQL query is built upon the selection of the user using the keyword.
- If the match is not found (which is true in this case), the keyword is matched with the synonyms of all the objects. If again, the match is not found, user is displayed a message "No Records Found".
- If the synonym match is found, then the SPARQL query is built upon the selection of the user using the keyword.
- The query is then executed and the result is displayed to the user (Figs. 6, 7, 8, 9).

4 Semantic Matching Algorithm Based on Mathematical Model

Question: Semantic Search
Answer: Let query=q
Parsed query=p
Let S be the set of Synonyms stored in O such that $S = \{s_1, s_2, s_3 \ldots \ldots s_n\}$
Let the ontology dataset be O,
Let the result set be R.
$R \neq \phi$, If $p \in O$, result found.
$R = \phi$, if $p \notin O$
If $R = \phi$, then
$R = \phi$, if $p \notin S$
$R \neq \phi$, If $p \in S$, result found.

It searches data according to the keyword. If keyword matches with the ontology then it returns the result, but if the result set is null, then the search is performed according to the synonyms of the keyword, i.e., the keyword p is searched in the synonym set S. If p matches any keyword in S, it means keyword exists in the ontology base and the result is returned according to that matched synonym.

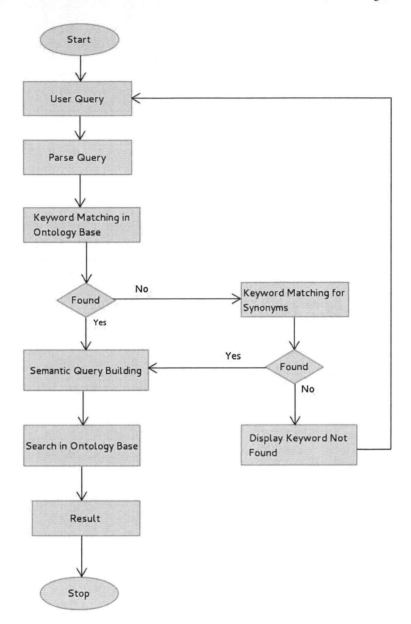

Fig. 6 Workflow of searching the data semantically

```
1.    search()
2.    {  q ← read query;
3.    q ← parse(q);
4.    q ← tolower(q);
5.    match ← keyword_match_name(q);
6.    if(match==true)
7.    then
8.        build SPARQL query
9.        result ← execute_SPARQL_query(s);
10.       display result;
11.   else
12.   then
13.       match ← keyword_match_synonyms(q);
14.       if(match==true)
15.       then
16.           build SPARQL query
17.           result ← execute_SPARQL_query(s);
18.           display result;
19.       else
20.       then
21.           display "No records found"       }
```

Fig. 7 Algorithm for searching data

```
1.    boolean keyword_match_for_name(query q)
2.    {   triple ←   "select ?subject
3.    where
4.    ?subject hasName q";
5.    res ← execute_SPARQL_query(triple);
6.    if(result_num_rows(res)>0)
7.      return true;
8.    else
9.      return false;     }
```

Fig. 8 Algorithm for searching data according to keyword

```
1.    boolean keyword_match_for_synonyms(query q)
2.    {   triple ← "select ?subject
3.    where
4.    ?subject AKA q";
5.    res ← execute_SPARQL_query(triple);
6.    if(result_num_rows(res)>0)
7.      return true;
8.    else
9.    return false;     }
```

Fig. 9 Algorithm for searching data according to synonyms (semantically)

5 Conclusion

It could be observed that the tourism sector is an area in which ontology can be applied anywhere [7]. The tourism domain is a decent area for new information and communication technologies by assisting users and agencies with quick information searching, integrating, recommending, and various intelligent E-Tourism services [8]. We introduced a creation of intelligent tourism information service using semantic search engine based on E-Tourism and Domain Ontology for Tourist Information for intelligent tourist information services. In addition, the ontology will play an important role as they promise a shared and common understanding of tourism and travel concepts that reaches across people and application systems. So semantic search is making searching very easier than keyword search [9]. Ontology based search can be made on tourism sites by constructing tourism ontology and information can be generated accurately according to the tourist interest using ontology.

References

1. Yan, Z.: Ontology and semantic management system: state-of-the-arts analysis. In: Proccedings of the IADIS International Conference, ISBN: 978-972-8924-44-7, pp. 111–115 (2007)
2. Werther, H.: Intelligent systems in travel and tourism. IEEE Intell. Syst. **17**(6) (2003)
3. Park, H., Yoon, A., Kwon H.-C.: Task model and task ontology for intelligent tourist information service. Int. J. u- e- Serv. Sci. Technol. **5**(2), 47–58 (2012)
4. Cardoso, J.: Developing dynamic packaging systems using semantic web technologies. Trans. Inf. Sci. Appl. **3**(4), 729–736 (2006)
5. Mohsin, A.: Tourist attitudes and destination marketing—the case of Australia's Northern Territory and Malaysia. Tourism Manag. **26**, 723–732 (2005)
6. Jun, S.H., Vogt, C.A., MacKay, K.J.: Relationships between travel information search an travel product purchase in pretrip contexts. J. Travel Res. **45**(3), 266–274 (2007)
7. Daramola, O., Adigun, M., Ayo, C.: Building an ontology-based framework for tourism recommendation services. Information and Communication Technologies in Tourism, Amsterdam, pp. 135–147 (2009)
8. Zhou, L., Zhang, D.: An ontology-supported misinformation model: toward a digital misinformaton library. IEEE Trans. Syst. Man Cybern. A. Syst. Humans **37**(5), 804–813 (2007)
9. Damljanovic, D., Devedzic, V.: Applying Semantic Web to E-tourism. Chapter X, Springer, Berlin, pp. 243–263 (1993)

Deep Questions in the "Deep or Hidden" Web

Sonali Gupta and Komal Kumar Bhatia

Abstract The Hidden Web is a part of the Web that consists mainly of the information inside databases, i.e., anything behind an interactive electronic form (search interfaces), which cannot be accessed by the conventional Web crawlers [1, 2, 8]. However, there have been well-defined, effective, and efficient methods for accessing Deep Web contents. One of these methods for accessing the Hidden Web employs an approach similar to 'traditional' crawling but aims at extracting the data behind the search interfaces or forms residing in databases. The paper brings insight into the various steps, a crawler must perform to access the contents in the Hidden Web. We structure the problem area and analyze what aspects have already been covered by previous research and what needs to be done.

Keywords WWW · Hidden web · Surface web · Hidden web crawler

1 Introduction

The growth of the World Wide Web (WWW) has been phenomenal over the years [8, 10, 11]. Surface Web refers to the abundant web pages that are static, typically having, outgoing links to other web pages, and incoming links which allow them to be reached from other pages, creating a spider-web like system of interconnected data; whereas the Hidden Web (HW) consists of unlinked data and refers to the Web pages created dynamically as the result of a specific search. The Hidden or Deep Web consists mainly of information inside databases, i.e., anything behind an

S. Gupta (✉) · K. K. Bhatia
Department of Computer Engineering, YMCA University of Science and Technology,
Faridabad, India
e-mail: Sonali.goyal@yahoo.com

K. K. Bhatia
e-mail: Komal_bhatia1@rediffmail.com

B. V. Babu et al. (eds.), *Proceedings of the Second International Conference on Soft Computing for Problem Solving (SocProS 2012), December 28–30, 2012*, Advances in Intelligent Systems and Computing 236, DOI: 10.1007/978-81-322-1602-5_87, © Springer India 2014

interactive electronic search form interface with most of it elicited by the HTTP form submission. Examples of Hidden web content include directories and collection of patents, scientific and research articles, holiday booking interfaces, etc. Estimates of the size of the Hidden Web differ, but some place it at up to 500 times the size of the traditional surface Web [1, 3, 5, 7].

The Hidden Web though hidden and not accessible through traditional document-based search engines, is a huge and distributed repository of data lying in databases which has to be accessed by some means. Methods must exist to prove the expediency of the source [8]. There are two basic approaches to access the contents in the HW:

1. Crawling/Trawling or Surfacing: It refers to the crawler's activity of collecting in the background as much relevant, interesting fraction of the data as possible and updating the search engine's index. This approach has the main advantage of best fit with the conventional search engine technology.
2. Virtual Data Integration: It refers to the creation of vertical search engines for specific domains where APIs will be used to access Hidden Web sources at time and construct the result pages based on their responses. Since external API calls need to be made by the search engine, this approach is traditionally slower than crawling.

The major goal of the paper is to describe the research problems associated with Hidden Web crawlers and analyze the existing research in the context of the research problems so as to identify and bring outstanding issues to the forefront.

2 Background (Search Engines/Crawlers)

Finding or Searching information on the Web has become an important part of our daily lives and about 550 million Web searches are performed every day [10, 11]. The tools that have been used to find information on the Web are typically known as Web Search Engines [2, 10, 11]. Figure 1 illustrates the activities and the corresponding components or elements of a basic search engine.

The various activities performed by a search engine can be divided into: *Crawling* by which a search engine gathers pages from the WWW; *Indexing* which is building a data structure that will allow quick searching of the text [11]; or "the act of assigning index terms to documents" where an index term is a (document) word whose semantics helps in remembering the document's main themes [11]; *Query Processing which* includes receiving a query from the user, searching the index or database for relevant entries, and presenting the results to the user. The component responsible for the process of crawling is known as a Crawl Engine or more typically a Web crawler [2, 11], whereas the element responsible for building the search engine's index is termed as the Index Engine or typically as an Indexer.

The increasing prevalence of online databases has influenced the structure of the web crawlers and their capabilities for information access through search form

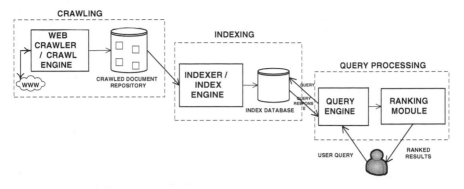

Fig. 1 Elements of a basic search engine

interfaces [2]. So, the paper discusses crawling, distinguishing on the basis of Surface and Hidden web crawl. The common belief is that over 1 million search engines currently operate on the WWW [7, 11] most of which cover only a small portion of the Web in their indices or databases. This coverage can be increased by either of the following means:

1. Employing multiple search engines: A search system that uses other search engines to perform the search and combines their search results is generally called a meta-search engine.
2. Enhancing the crawler's Capability: Since the HW comprises a major part of the Web (almost $\approx 80\%$) developing hidden Web crawlers has clearly become the next frontier for information access on the Web.

3 Hidden Web Crawler

Figure 2 illustrates the sequence of steps that take place when a user wants to access the contents in any Hidden Web resource. The user has to fill out a query form for retrieving documents that have been dynamically generated from the underlying database [3, 4].

Figure 3 illustrates the difference in the sequence of steps undertaken by any crawler to access the Hidden Web's informational content.

A Hidden Web crawler starts the same as the Surface web crawler by downloading the required web page, but then later it requires a lot of analysis and intelligence to extract information from the hidden web. The Surface Web Crawlers can record the address of a search front page but can tell nothing about the contents of the database [1, 5, 8].

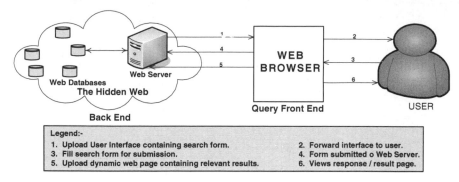

Fig. 2 User interaction with a search form interface

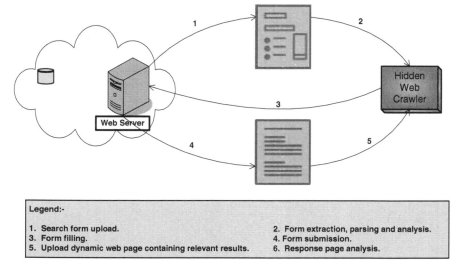

Fig. 3 A crawler interacting with the search form interface

4 Research Problems

Making a comprehensive crawl of the HW does not seem practical due to the two fundamental reasons [1, 3, 4, 8]: *Scale* The Hidden Web is unprecedented in many ways; unprecedented in size and content quality; unprecedented in the lack of coordination in its creation (distributed nature), and unprecedented in the diversity of backgrounds and motives of its participants. *Restricted Search form interfaces*: Access to the Hidden Web databases is provided only through restricted search interfaces, perceived to be used by humans [3, 4]. This raises the non-trivial problem of "training" the crawler for an appropriate use of the restricted interface to extract relevant content. Below are presented the two sub-problems or steps in the present scenario that suggest likely directions:

1. Resource discovery: In order to overcome the problem of Scale, the crawler must be trained to carry a crawl of only the relevant sources (effective crawling) rather than carrying a comprehensive crawl of the Hidden Web (exhaustive crawling) [3, 4, 8, 16, 17]. This requires the crawler to first locate the sites containing search form interfaces and then select the relevant subset from it. And as the Hidden Web data sources are growing continuously at a high rate, selecting the subset of relevant sources will prove not only cost-effective but also effective in time and make the crawler less prone to errors.

2. Content extraction: The task of harvesting information lying behind the form interfaces of the selected Hidden Web sources depends largely on the way, the crawler is able to understand and model the search form interface so as to come up with meaningful queries to issue to the search interface for probing the database behind it. The crawler must then be able to extract the data instances from the retrieved result pages. This problem of Content Extraction poses significant challenges and the solution lies in the three steps process comprising: Understanding the Search Form interfaces, automatically filling them and Information extraction.

The three steps together form the following basic modules of the system: *Form Analyzer* that will analyze each and every downloaded page to see if it can be used as a search page to retrieve information or not. It basically checks whether a web page is query able, has some form fields or not; *Form Parser* that extracts the fields from the search form and passes them on to the form processor for filling; *Form Processor* that fills in the various form fields by assigning appropriate values and finally submits the form for retrieving result pages; *Result Analyzer* that will analyze all the result pages obtained by the crawler after form execution, in order to get the required information.

5 The State of the Art in Hidden Web

In this section we discuss the previous work in the area grouped by the problem domains. Research on Hidden Web search can be dated back to the 1980s. Since then, substantial progress has been made in different sub-problems of crawling and accessing the Hidden Web.

5.1 Resource Discovery

The goal of any focused crawler is to select links that lead to documents that have been identified as relevant to the topic of interest and hence addresses the resource discovery problem, The work on focused crawling [14, 16, 17] describes the design of topic-specific crawlers for the Surface Web which complements our problem, as same resource discovery techniques can be used to identify target sites for a Hidden Web crawler. The work in [14] discusses a best-first focused crawler which uses a page classifier to guide the search. Unlike an exhaustive crawler which follows each link in a page in a breadth first manner, this crawler gives priority to links that belong to pages classified as relevant. In domains that are not very narrow the number of irrelevant links can still be very high, the strategy can lead to suboptimal harvest rates and an improvement to which was proposed in [17].

An issue that remains with these focused crawlers is that they may miss relevant pages by only crawling pages that are expected to give immediate benefit. In order to address this limitation, certain strategies have been proposed that train a learner by collecting features from paths leading to a page, as opposed to just considering the contents of a page [17–19]. Reference [19] extends the idea in [14] and presents a focused crawling algorithm that builds a model for the context within which topically relevant pages occur on the Web. Another extension of the focused crawler idea is presented in [18] using a reinforcement learning algorithm to develop an efficient crawler for building domain-specific search engines.

Finally, it is worth pointing out that there are directories specialized on hidden-Web sources, e.g., [1, 20] that organize pointers to online databases in a searchable topic hierarchy and hence can be used as seed points for the crawl.

5.2 Content Extraction

Extracting content from the Hidden Web has received a lot of attention to date [3–6, 12, 15]. Most approaches to information retrieval in the Hidden Web are focused on understanding the various semantics associated with the form elements and automatically filling them as they are the only entry points to the Hidden Web.

Reference [3] presents an architectural model for a task-specific semi-automatic Hidden-Web crawler. The main focus of their work is to learn Hidden-Web query interfaces, not to generate queries automatically. A significant contribution of this work is the label matching approach that identifies elements of a form based on layout position, not proximity within the underlying HTML code. When analyzing forms, HiWE only associates one text to each form field according to a set of heuristics that take into account the relative position of the candidate texts with respect to the field (texts at the left and at the top are privileged), and their font sizes and styles. To learn how to fill in a form, HiWE matches the text associated with each form field and the labels associated to the attributes defined in its LVS. In this process, HiWE has the following restriction: it requires the LVS table to contain an attribute definition matching with each unbounded form field.

Many approaches exist that rely on filling forms [4, 6, 15] automatically. The main focus of the work in [4] is to generate queries automatically without any human intervention in order to crawl all the content behind a form. New techniques are proposed to automatically generate new search keywords from previous results, and to prioritize them in order to retrieve the content behind the form, using the minimum number of queries. The problem of extracting the full content behind a form has been also addressed in [6]. They have proposed a domain-independent approach for automatically retrieving the data behind a given Web form. The approach to gather data is based on two phases: first the responses from the web site of interest are sampled and then if necessary methodically try all possible queries until either a fix point of retrieved data has arrived or all possible queries have exhausted. They have developed a prototype tool that brings the user into the process when an automatic decision becomes hard to make. These techniques focus on coverage, i.e., retrieve as big a portion of the site's content as possible.

The hidden web can also be accessed using the meta-search paradigm instead of the crawling paradigm. This body of work is often referred to as meta-searching or database selection problem over the Hidden Web. In meta-search systems, a query from the user is automatically redirected to a set of underlying relevant sources, and the obtained results are integrated to return a unified response. Data integration is the problem of combining data from various web databases sources to provide the users with a unified view of data [8]. One of the main tasks to formalize the design of a data integration system is to establish the mapping between the Web database sources and a global schema. The meta-search approach is more lightweight than the crawling approach, since it does not require indexing the content from the sources. Nevertheless, the users will get higher response times since the sources are queried in real-time.

6 Open Issues in Hidden Web Crawling

A critical look at the available literature indicates that the following issues need to be addressed while designing the framework for any fully automatic crawler for the Hidden Web. Most of the research to date has focused on the last issue. Little attention has been made to the first two questions of scalability and synchronization:

1. There exists a variety of Hidden Web sources that provide information about the multitude of topics/domains [1, 7, 20]. The continuous growth of information about the WWW [8, 10] and hence the domain-specific information with ever-increasing number of domain areas pose a challenge to crawler's performance. The crawl of the portion of the web for a particular domain must be completed within the expected time. This download rate of the crawler is limited by the underlying resources. An open challenge is the design a crawler that scales its performance according to the increase in the information on the WWW and number of domains. These scalability limitations stem from search engines' attempt to crawl the whole Web, and to answer any query from any user.

2. Decentralizing the crawling process is clearly a more scalable approach and bears the additional benefit that crawlers can be driven by a rich context (topics, queries, user profiles) within which to interpret pages and select the links to be visited. However, a rigorous focus only on scalability can be costly; of course,

the system must also coordinate information coming from multiple sources, not all of which are under the control of the same organization. The pattern of communication is many-to-many, with each server talking to multiple clients and each client invoking program on multiple servers.

3. As the number of data sources is growing continuously at a very high rate, it is very tedious, time-consuming, and error-prone to process the search interfaces in web-based applications. An important objective of any Hidden Web crawler is to build an internal representation of these search forms [4–6] that supports efficient form processing and interface matching techniques so as to fully automate the process.

7 Conclusion and Future Work

A move in the Web structure from hyperlinked graph in the past to electronic form-based search interfaces of present day, represent the biggest challenge a Web crawler needs to tackle with. Despite the Web's great success as a technology and the significant amount of computing infrastructure on which it is built, it remains as an entity, surprisingly unstudied. Users need and want better access to the information on the Web. We believe that Hidden Web crawling is an increasingly important and fertile area to explore as such a crawler will enable indexing, analysis, and mining of Hidden Web content, akin to what is currently being achieved with the Surface Web. The paper provides a look at some of the technical challenges that must be overcome to model the Web as a whole, keep it growing and understand its continuing social impact. The topic of concerns as mentioned in the paper are further exacerbated by the rapid growth of Hidden Web content, fueled by the success of social networking online, the proliferation of Web 2.0 content and the profitability of the companies that steward in this new era. We look forward to continuing this promising line of research. One of the main objectives of our work will remain as the design of a crawler whose performance can be scaled up by adding additional low-cost processes and using them to run multiple crawls in parallel.

References

1. Bergman, M.K.: The deep web: Surfacing hidden value. J. Electron. Publ. 7(1), 1174–1175 (2001)
2. Sherman, C., Price, G.: The Invisible Web: Uncovering Information Sources Search Engines Can't See. CyberAge Books, Medford (2001)
3. Raghavan, S., Garcia-Molina, H.: Crawling the hidden web. In: 27th International Conference on Very large databases (Rome, Italy, September 11–14: VLDB'01), pp. 129–138. Morgan Kaufmann Publishers Inc., San Francisco (2001)
4. Ntoulas, A., Zerfos, P., Cho, J.: Downloading textual hidden web content through keyword queries. In: 5th ACM/IEEE Joint Conference on Digital Libraries (Denver, USA, Jun 2005)

JCDL05, pp. 100–109 (2005)
5. Barbosa, L., Freire, J.: Siphoning hidden-web data through keyword-based interfaces. In: SBBD, 2004, Brasilia, Brazil, pp. 309–321 (2004)
6. Liddle, S.W., Embley, D.W., Scott, D.T., Yau, S.H.: Extracting data behind web forms. In: 28th VLDB Conference 2002, HongKong, China, pp. 38–49 (2002)
7. Chang, K.C.-C., He. B., Li. C., Patel, M., Zhang, Z.: Structured databases on the web: Observations and implications. SIGMOD Rec. *33*(3), 61–70 (2004)
8. Gupta, S., Bhatia, K.: Exploring 'hidden' parts of the web: The hidden web. In: Lecture notes in Electrical Engineering, Proceedings of the International Conference ArtCom 2012, pp. 508–515, Springer, Heidelberg (2012)
9. Gupta, S., Bhatia, K.: A system's approach towards domain identification of web pages. In: Proceedings of the Second IEEE International Conference on Parallel, Distributed and Grid Computing (India, December 6–8, 2012) PDGC'12, IEEE Xplore
10. Lawrence, S., Giles, C.L.: Accessibility of information on the web. Nature **400**, 107–109 (1999)
11. Baeza-Yates, R., Ribeiro-Neto, B.: Modern Information Retrieval, 2nd edn. Addison-Wesley-Longman, Boston (1999)
12. Ipeirotis, P.G., Gravano, L., Sahami, M.: Probe, count, and classify: Categorizing hidden-web databases. In: Proceedings of the ACM SIGMOD International Conference on Management of Data, pp. 67–78, Santa Barbara, CA, USA, May (2001)
13. Wang, W., Meng, W., Yu, C.: Concept hierarchy based text database categorization. In: Proceedings of International WISE Conference, pp. 283–290, China, June (2000)
14. Chakrabarti, S., van den Berg, M., Dom, B.: Focused crawling: A new approach to topic-specific web resource discovery. In: Proceedings of the 8th International WWW Conference (1999)
15. Zhang, Z., He, B., Chang, K.C.-C.: Light-weight domain-based form assistant: Querying web databases on the fly. In: Proceedings of the 31st Very Large Data Bases Conference (2005)
16. McCallum, A., Nigam, K., Rennie, J., Seymore.K.: Building domain-specific search engines with machine learning techniques. In: Proceedings of the AAAI Spring Symposium on Intelligent Agents in Cyberspace (1999)
17. Chakrabarti, S., Punera, K., Subramanyam, M.: Accelerated focused crawling through online relevance feedback. In Proceedings of WWW, pp. 148–159 (2002)
18. Rennie, J., McCallum, A.: Using reinforcement learning to spider the web efficiently. In Proceedings of ICML, pp. 335–343 (1999)
19. Diligenti, M., Coetzee, F., Lawrence, S., Giles, C.L., Gori.M.: Focused crawling using context graphs. In: Proceedings of the 26th International Conference on Very Large Databases, pp. 527–534 (2000)
20. Profusion's search engine directory. http://www.profusion.com/nav

Combined and Improved Framework of Infrastructure as a Service and Platform as a Service in Cloud Computing

Poonam Rana, P. K. Gupta and Rajesh Siddavatam

Abstract Cloud computing is based on five attributes: multiplexing, massive scalability, elasticity, pay as you go, and self provisioning of resources. In this paper, we describe various cloud computing platforms, models, and propose a new combined and improved framework for Infrastructure as a Service (IAAS) and Platform as a Service (PAAS). As we know, that PAAS Framework has certain desirable characteristics that are important in developing robust, scalable, and hopefully portable applications like separation of data management from the user interface, reliance on Cloud Computing standards, an Integrated Development Environment, Life cycle management tools but it also has some drawbacks like the PAAS platform such as in Google Application engine, a large number of web servers catering to the platform are always running. This paper proposes an architecture which combines IAAS and PAAS framework and remove the drawbacks of IAAS and PAAS and describes how to simulate the cloud computing key techniques such as data storage technology (Google file system), data management technology, Big Table as well as programming model, and task scheduling framework using CLOUDSIM simulation tool.

Keywords PAAS · IAAS · Virtualization · Google application engine

P. Rana
KIET School of Engineering and Technology, Ghaziabad , UP 201206, India
e-mail: doli238rana@gmail.com

P. K. Gupta (✉) · R. Siddavatam
Department of CSE and IT, Jaypee University of Information Technology,
Waknaghat, Solan , HP 173234, India
e-mail: pradeep1976@yahoo.com

R. Siddavatam
e-mail: srajesh@juit.ac.in

B. V. Babu et al. (eds.), *Proceedings of the Second International Conference on Soft Computing for Problem Solving (SocProS 2012), December 28–30, 2012*, Advances in Intelligent Systems and Computing 236, DOI: 10.1007/978-81-322-1602-5_88, © Springer India 2014

1 Introduction

A cloud computing platform dynamically configures, reconfigures the servers as needed. Virtualization is the key that enables cloud computing with increased utilization, compared with deploying, installing, and maintaining traditional forms of servers for on a task per server basis [1, 2]. Linux, Apache, and other programming languages like C++, Python, and Java as well as PHP have been widely adopted for Cloud Computing supported by many vendors. Cloud Computing delivers infrastructure, platform, and software as service in a pay as you go model to consumers. These services in industries are referred to as IAAS, PAAS, SAAS (Software as a service). IAAS [3, 4] is the capability provided to the consumer for processing storage networks and other fundamental computing resources where the consumer is able to deploy and run arbitrary software, which can include operating system and applications and limited control of select networking components. PAAS is the capability to deploy onto the cloud infrastructure consumer created or acquired using programming languages and tools supported by the provider [3, 4]. The consumer does not manage or control the underlying cloud infrastructure including network servers, operating system, or storage but has control over the deployed applications and possible application hosting environment, which can include operating systems and applications. Unlike server based application development, cloud based applications development is focused on splitting two things which is also required by every application too. First, every application needs computing power to manipulate the data. Second, data storage that can also be done at different levels: shared and nonshared.

2 Related Work

There are various service providers and various platforms or framework available for cloud computing. Anandasivam and Weinhardt [5] inspected the problem of how Cloud service providers would decide to accept or reject requests for services when the resources offering these services become scarce and proposed a decision support policy called Customized Bid-Price Policy (CBPP). Armbrust et al. [6] have redefined the cloud computing and reduces the confusion by clarifying the various terms. They have also quantified comparison between cloud computing and conventional computing and identified the top technical and non-technical obstacles and opportunities of cloud computing. Böhm et al. [7] have considered different point of view to look at the cloud computing from an IT provisioning perspective and further examined the evolution from outsourcing cloud computing as a new IT deployment paradigm. Buyya et al. [8] suggested that the cloud computing powers the next generation data centers and enables application service providers to lease data center capabilities for deploying applications. They have proposed an extensible simulation toolkit that enables modeling and simulation of cloud computing environment which is known as CLOUDSIM. In [9] Buyya et al. presented the vision of computing for twenty-first century and delivered the vision of computing utilities. They have also stated the importance of cloud computing and provides the architecture for creating market-oriented clouds by leveraging technologies such as virtual machines. In [10] Yadav has presented the latest vision of cloud computing and identifies various commercially available cloud services promising to deliver the infrastructure on demand and provides architecture and description about various types of clouds.

There are number of companies that offer cloud computing services like Amazon [11, 12] offers something called Amazon Elastic Cloud (EC2). Amazon Web Services (AWS) provide infrastructure as a service offerings in the cloud for organizations requiring computing, power, storage, and other services. Kumar et al. [13] discusses about the various aspects of cloud computing and focuses on the security and trust to share the data for developing cloud computing applications in a distributed environment. They have also provided the concept of utility cloud to be used by the persons.

3 Cloud Models and Implementation Issues

The Google cloud also known as Google Application Engine (GAE) [14, 15], is a (PAAS) offering. Figure 1 describes PAAS offering which hides the actual execution environment from users. Instead a software platform is provided along with an SDK for users to develop applications and deploy them on the cloud [16, 17]. The PAAS platform is responsible for executing the applications including servicing external service requests, as well as running scheduled jobs in the applications. By making the actual execution servers transparent to the user, a PAAS platform is able to share application servers across users who need lower capacities, as well as automatically scale resources allotted to applications that experience heavy loads. Normally users upload their code in any programming language along with all required files and are stored with the GAE [18].

Resource usage for an application is measured in terms of web requests served and CPU hour actually spent on executing requests or batch jobs. The desirable characteristics of PAAS are [1, 18]:

- PAAS application can be made globally available 24×7 but can change only when accessed.
- Deploying applications in GAE is free within usage limits, thus applications can be developed and tried out free and incurs cost only when actually accessed by a large volume of requests.
- The PAAS model enables GAE to provide such a free service because applications do not run in dedicated virtual machines. Deployed application which is not accessed merely, consumes storage for its code, data, and expands no CPU cycles.

GAE allows a user to run web applications written using the Python Programming language, other than supporting the Python standard library [19]. We have

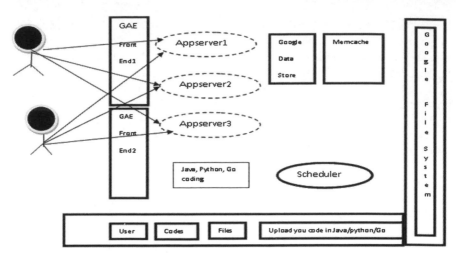

Fig. 1 Google application engine PAAS framework

developed our application using Java development tools and standard APIs. Our application interacts with the environment using the Java Servlet standard and common web application technologies such as Java Server Pages (JSP) [9]. Ease of use and supporting tools are advantages of GAE for Java and other cloud computing solutions. Google data store is a distributed object store where objects of all GAE applications are maintained and also provides a NOSQL schema-less object data store with a GQL query engine and atomic transactions. You have choice between two data storage options differentiated by their availability and consistency guarantees. Memcache [15, 20], provides your application with a high performance in memory key value cache that is accessible by multiple instances of your application. Memcache is useful for data that does not need the persistence transactional features of the data, stored as temporary data or data copied from the data, stored to the cache for high speed access.

4 Proposed Combined Framework of IAAS and PAAS

First, we have compared the existing IAAS and PAAS cloud models from an economic perspective. The PAAS model described by the Google Application Engine and Microsoft Azure cloud offerings can exhibit economic advantages as compared to an IAAS model for certain classes of applications. In Fig. 2, we have described the combined framework of IAAS and PAAS. We know that as far as the cloud computing is concerned, the cost factors also come into the account and there are various system matrices which measures the total cost. These measurements are an aggregate of one or more web servers in infrastructure capacity and measure the system level statistics.

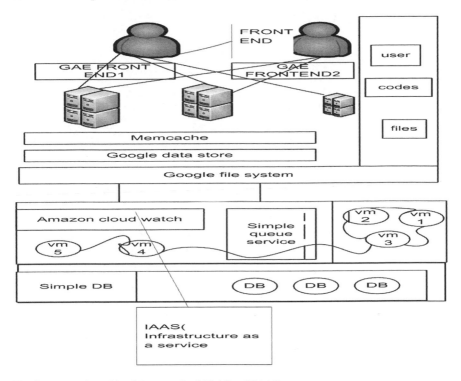

Fig. 2 Proposed combined framework of IAAS and PAAS

The PAAS model exemplified by the Google Application Engine and Microsoft Azure Cloud offerings can exhibit economic advantages as compared to an IAAS model for certain classes of application. Consider a web application that needs to be available 24 × 7, but where the transaction volume is highly unpredictable and can vary rapidly. With IAAS model a minimum number of servers would be provisioned at all times to ensure availability of the web service [21]. While in PAAS model such as Google Application Engine, deploying the application costs nothing. As usage increases beyond the free limits, charges begin to increase and a well-engineered application scale to meet the demand. So, both the IAAS and PAAS suffer from disadvantages. Further, taking additional severs for an application takes a finite amount of time, such as 15 or 20 min; the minimum capacity that needs in IAAS needs to account for this delay by organizing and paying for excess bandwidth capacity even in the cloud environment. But in the PAAS platform such as GAE, supports a large number of serving platforms, thus every users application code is available to all severs via the distributed file system. Sudden rises in demand are automatically routed to free web servers by the load balancer and ensures minimal performance degradation. The additional overhead induced is loading the code and data from the file system when such a web server handles a request to a new application for the first time.

5 Simulation of the Proposed Architecture

The simulation of the proposed framework in IAAS is being done by CLOUDSIM toolkit 3.0 released in Jan 2011 [8, 22]. CLOUDSIM is an extensible simulation toolkit that enables modeling and simulation of cloud computing environment for difficult applications and service models [21]. CLOUDSIM toolkit also supports modeling and creation of one or more virtual machines on a simulated node of a data center with jobs and their mappings to the virtual machines. For quantifying the scheduling performance and allocation policies in a real cloud environment, different applications and service models are available under different conditions [8]. An alternative is the utilization of the simulation tools that open the possibility of evaluating the hypothesis prior to software development in an environment where one can reproduce tests. Tests can be of various types. A machine instance (physical or virtual) is primarily defined by four essential resources (1) CPU (2) Memory (RAM) (3) Disk (4) Network connectivity. Each of these resources can be measured by tools that are operating system specific but for which tools that are their counterparts exists for all operating systems [8, 23]. Examining the four parameters, the server system metrics do not give you enough information to do meaningful capacity planning. Load testing gives an answer to many questions: (1) Maximum load that the current system can support. (2) Which resources represent the bottleneck in the current system that limits the system performances? This parameter is referred to as resource cutting depending upon a server's configuration. Now, what performance measurement tool should be used is the main goal to create a set of resource utilization curves.

6 Results

In simulation configuration, we tried to reconstruct virtual machines and cloudlet on a E-series Intel core 350 m, processor 2.26 GHz, Window 7 Home Basic (64-bit), using Eclipse workbench and JDK 1.6 (standard edition). We tried to illustrate the proposed architecture using data centers with one host each and run the cloudlets on them. We have to describe various parameters with the help of CLOUDSIM toolkit for simulating our proposed architecture. The snapshots of the simulation are illustrated below in Table 1.

The Table 1 shows that our proposed architecture is successfully simulated and below are the results given for one of our executed scenario.

```
Starting CloudSimExample2...
Initializing...
Starting CloudSim version 3.0
Datacenter_0 is starting...
Broker is starting...
Entities started.
```

Table 1 Results of proposed framework

S. No.	Scenario	Status	Result
1	Data center with one host and two cloudlets on it	Successful	71.2 ms
2	Data center with two hosts and two cloudlets on it	Successful	224.8 ms
3	Two Data centers with two hosts and run cloudlets of two users on them	Successful	35.6 ms of each user
4	Scalable simulation	Successful	6156.3 ms of each user
5	How to pause and resume the simulation	Successful	The simulation is paused for 5 s
6	How to create a simulation entity using global broker entity	Successfully created	–

0.0: Broker: Cloud Resource List received with 1 resource(s)
0.0: Broker: Trying to Create VM #0 in Datacenter_0
0.0: Broker: Trying to Create VM #1 in Datacenter_0
0.1: Broker: VM #0 has been created in Datacenter #2, Host #0
0.1: Broker: VM #1 has been created in Datacenter #2, Host #0
0.1: Broker: Sending cloudlet 0 to VM #0
0.1: Broker: Sending cloudlet 1 to VM #1
1000.1: Broker: Cloudlet 0 received
1000.1: Broker: Cloudlet 1 received
1000.1: Broker: All Cloudlets executed. Finishing...
1000.1: Broker: Destroying VM #0
1000.1: Broker: Destroying VM #1
Broker is shutting down...
Simulation: No more future events
CloudInformationService: Notify all CloudSim entities for shutting down.
Datacenter_0 is shutting down...
Broker is shutting down...
Simulation completed..........

7 Conclusion

In this paper, we have proposed a novel combined framework using the PAAS and IAAS to produce a hybrid architecture using Google Application Engine as front end and IAAS elastic compute service as back end and then simulate our architecture using CLOUDSIM toolkit. As we know, that heavier application IAAS is better suited, so it is better if we combine them and get all the advantages of PAAS framework as well as IAAS framework. Therefore, we proposed an architecture that is combination of IAAS and PAAS architecture. Here PAAS provides the web front end and IAAS provides the power when needed. The performance statistics of the load balanced requests can also be monitored by cloud watch and used by auto scale to add or remove servers from the load balanced users. Using these tools the users can configure a scalable architecture that also adjusts resource consumption. The future work can be resource allocation and resource provisioning can be done, to develop such architecture which is economically lower as compared to the proposed hybrid architecture of IAAS and PAAS.

References

1. Sosinky, B.: Cloud Computing Bible, pp. 1–528. Wiley, India (2011)
2. Roy, G.M., Saurabh, S.K., Upadhyay, N.M., Gupta, P.K.: Creation of virtual node, virtual link and managing them in network virtualization. In: IEEE 2011 World Congress on Information and Communication Technologies (WICT), pp. 738–742. Mumbai, India (2011)
3. Peter M., Timothy G.: The NIST Definition of Cloud Computing, Tech Report National Institute of Standards and Technology. pp. 1–7. (2011)
4. Info. Apps. Gov. : What are the Services. http://info.apps.gov/content/what-are-services
5. Anandasivam, A., Weinhardt, C.: Towards an Efficient Decision Policy for Cloud Service Providers. In: Proceedings of the ICIS, Paper 40 (2010)
6. Armbrust, M., Fox, A., Griffith, R., Anthony, D., Katz, R., Konwinski, A., Lee, G., Patterson, D., Rabkin, A., Stoica, I., Zaharia, M.: A view of cloud computing. Commun. ACM **53**(4), 50–58 (2010)
7. Böhm, M., Leimeister, S., Riedl, C., Krcmar, H.: Cloud Computing - Outsourcing 2.0 or a new Business Model for IT Provisioning. Springer Application, Management. pp. 31–56. Gabler, GmbH (2011)
8. Buyya, R., Ranjan, R., Rodrigo, N.: Modelling and Simulation of scalable cloud computing environment and the cloudsim toolkit: challenges and opptuties. In: IEEE International Conference on High Performance computing & simulation, pp. 1–11. (2009)
9. Buyya, R., Yeo, C.S., Venugopal, S.: Market oriented cloud computing, vision hype and reality for delivering it services as computing utilities. In: The 10th IEEE conference on high performance computing and communications, pp. 5–13. (2008)
10. Yadav, S.S.: Cloud a computing Infrastructure on demand. In: International conference on computer engineering and technology (ICCET), pp. 423–426. (2010)
11. Mather, T., Kumaraswami, S., Latif, S.: Cloud Security and Privacy an Enterprise respective on Risks and Compliance. O' Reilly Media, Inc, Canada (2009)
12. Amazon. : Amazon web services. http://aws.amazon.com
13. Pardeep, K., Sehgal, V.K., Chauhan, D.S., Gupta, P.K., Diwakar, M.: Effective ways of secure, private and trusted cloud computing. IJCSI Int. J. Comput. Sci. Issues **8**(3), 412–421 (2011)
14. Bedra, A.: Getting started with Google App Engine and Closure. IEEE Internet Comput. **14**(4), 85–88 (2010)
15. Malawski, M., Kuzniar, M., Wojcik, P., Bubak, M.: How to use Google App Engine for Free Computing. IEEE Internet Comput. **17**(1), 50–59 (2011)
16. Baiardi, F., Sgandurra, D.: Securing a community cloud. In: IEEE 30th International Conference on Distributed Computing Systems Workshops (ICDCSW), pp. 32–41. (2010)

17. Alexander, L., Klems, M., Nimis, J., Tai, S., Sandholm, T.: What's inside the Cloud? An architectural map of the cloud landscape. In: Proceedings of the, ICSE Workshop on Software Engineering Challenges of Cloud, Computing, pp. 23–31. (2009)
18. Gautam, S.: Enterprise Cloud Computing: Technology, Architecture. Applications. Cambridge University Press, Cambridge (2010)
19. Armando, F., Michael, A., Rean, G.: The Clouds: A Berkely View of Cloud, Computing, UC Berkeley, USA (2009)
20. Google Developers. : What is Google App Engine? https://developers.google.com/appengine/docs/whatisgoogleappengine
21. www.cloudbus.org/.
22. https://developers.google.com/appengine/docs/
23. Ekstrom, J., Bailey, M.: Teaching web deployment with OS-virtualization. In: Proceedings of 2009 ASEE Annual Conference and Exposition, pp. 1–8. (2009)

Web Site Reorganization Based on Topology and Usage Patterns

R. B. Geeta, Shashikumar G. Totad and P. V. G. D. Prasad Reddy

Abstract The behavioral web users' access patterns help website administrator/web site owners to take major decisions in categorizing web pages of the web site as highly demanding pages and medium demanding pages. Human beings act as a spider surfing the web pages of the website in search of required information. Most of the traditional mining algorithms concentrate only on frequency/support of item sets (web pages set denoted as ps in a given web site), which may not bring considerably more amount of profit. The utility mining model focuses on only high utilities item sets (ps). General utility mining model was proposed to overcome weakness of the frequency and utility mining models. General utility mining does not encompass website topology. This limitation is overcome by a novel model called human behavioral patterns' web pages categorizer (HBP-WPC) which considers structural statistics of the web page in addition to support and utility. The topology of the web site along with log file statistics plays a vital role in categorizing web pages of the web site. The web pages of the website along with log file statistics forms a population. Suitable auto optimization metric is defined which provides guidelines for website designers/owners to restructure the website based on behavioral patterns of web users.

Keywords Web mining · Reorganization · Log file · Web site topology

R. B. Geeta (✉)
Department of Information Technology, GMRIT, Rajam, AP, India
e-mail: geetatotad@yahoo.co.in

S. G. Totad
Department of Computer Science, GMRIT, Rajam, AP, India
e-mail: skumartotad@yahoo.com

P. V. G. D. Prasad Reddy
Department of CS and SE, Andhra University, Vizag, AP, India
e-mail: prsadreddy.vizag@gmail.com

B. V. Babu et al. (eds.), *Proceedings of the Second International Conference on Soft Computing for Problem Solving (SocProS 2012), December 28–30, 2012,* Advances in Intelligent Systems and Computing 236, DOI: 10.1007/978-81-322-1602-5_89, © Springer India 2014

1 Introduction

Web is huge repository of information. There has been gigantic development of the world wide web. The information is available throughout the globe in the form of websites. Web mining is the application of data mining. Web mining is a process of extracting useful information from the Web. Yang and others [1] suggested that web mining forms universal set of web structure mining, web usage mining and web content mining. Web structure mining deals with how well the web pages in the web site can be organized, so that most demanding pages can be kept very near to home page.

2 Related Work

To perform any website evaluation, web visitor's information plays an important role, in order to assist this, many tools are available. Li et al. [2] expressed that web mining is a popular technique for analyzing website visitor's behavioral patterns in e-service systems. Jian et al. [3] found that web log mining helps in extracting interesting and useful patterns from the log file of the sever. Shen et al. [4] suggested that HTML documents contain more number of images on the WWW. Such documents' containing meaningful images ensures a rich source of images cluster for which query can be generated. The documents which are highly needed by users can be placed near to the home page of the website. Manoj and Deepak [5] suggested that the development of web mining techniques such as web metrics and measurements, web service optimization, process mining etc... will enable the power of WWW to be realized. Wang et al. [6] found that weakness of both frequency and utility can be overcome by general utility mining model. Miller and Remington [7] revealed that the structure of linked pages has decisive impact factor on the usability. Geeta et al. [8] suggested that the number of pages at a particular level, the number of forward links and the number of backward links to a particular web page reflect the behavior of visitors to a specific page in the website. However Garofalakis [9] pointed out that the number of hit counts calculated from log file is an unreliable indicator of page popularity. Geeta et al. [10] suggested that the topology of the website plays an important role in addition to log file statistics to help users to have quick response. Jia-Ching et al. [11] found that web usage mining helps in discovering web navigational patterns mainly to predict navigation and improve website management. Lee et al. [12, 13] proved that the web behavioral patterns can be used to improve the design of the website. These patterns also could help in improving the business intelligence.

3 Web Structure Mining

Web structure mining concentrates on link structure of the web site. The different web pages are linked in some fashion. The potential correlation among web pages makes the web site design efficient. This process assists in discovering and modeling the link structure of the web site. Generally topology of the web site is used for this purpose. The linking of web pages in the Web site is challenge for web structure mining. The page with in degree high indicates that the page is with valuable content.

4 Web Usage Mining

Web Usage mining helps in understanding the users' behavior while interacting with the website. The main objective of web usage mining is to identify useful patterns and assist site adaption to better suit the users. Log file of server provides the various statistics such as the number of hits to a particular web page tells us about how well a particular web page is popular.

5 Proposed Work

Whenever a client/user requests for a server, communication takes place between client and server [14]. If user requesting page is at leaf level, the user has to go through all intermediary nodes to reach leaf node, meanwhile those many times communications take place between server and client using various resources like bandwidth, server's processor time, client processor time and power spent at both client and server. If a particular web page is located at nth level, all $(n-1)$ entries will be entered into log file of the server. All these parameters can be mathematically modeled as follows. Time spent by server and client is denoted by TSC.

> t_s is the time spent by the server and
> t_c is time spent by the client.

Let CCSC represents how many times communication is carried out between client and server and SL denotes space required to store all log entries. The parameter n represents the level of the web page. BW is used to represent the bandwidth utilized. nbw stands for network bandwidth.

> $\text{TSC} = (n+1)^* \ t_s + (n+1)^* \ t_c \text{ units}$
> $\text{CCSC} = (n+1)^* \ (\text{msg\_generated\_by\_client} + \text{msg\_generated\_by\_server})$
> $\text{SL} = (n+1)^* \ b \text{ bytes}$
> $\text{BW} = (n+1)^* \ \text{nbw}$

The main motto of this work is to minimize all above mentioned parameters, so that users will have quick response, bandwidth can be utilized efficiently, server time and client time can be used effectively and reduction of log file entries thereby saving memory space, reducing power supply and users will have quick response.

5.1 System Design

Whenever web user requests for a particular web page, he leaves the information in the log file. The web pages of the web site along with server's log file details forms the population which helps in decision making. The fitness function is to maximize the number of visitors to the web page/(s) and to minimize log file entries. This approach helps in satisfying users to have quick response.

We start with the formal definition of HBP-WPC as follows.

Let us consider a website of n pages $\{w_p, g_p, t_p\}_{p=1,...,n}$ for binary categorization problem, where $g_p \in \{Excellent, Medium\}$ and $t_p \in (0, 1]$ is the degree of web page w_p belonging to g_p.

The HBP-WPC equation is defined with the following terms.

- $P = \{p_1, p_2, \ldots, p_n\}$ is the web pages lookup table which contains mapping of acronyms for actual web page names $\{w_1, w_2, \ldots w_n\}$. The index n>0 and n is the total number of web pages for the given website.
- $L = \{T_1, T_2, \ldots, T_n\}$ is a transactions database where each transaction $T_i \in$ Log File of the server.
- $S(p_i)$ denotes the ratio of frequency of web page p_i to total frequency of all web pages in all transactions for a particular session threshold.
- $O(p_i)$ is the ratio of out degree of web page p_i to total number of links in the website.
- $I(p_i)$ denotes the ration of in degree of web page p_i to total number of links in the website.
- $D(p_i)$ denotes the ratio of p_i to total number of pages in the web site.
- $L(p_i)$ =level of p_i/total number of pages in that level.
- $Sup(w_i)$, the support count of an web page w_i, is the frequency of occurrence of all transactions containing w_i, for a particular session threshold.
- S, the total number of frequency of occurrences of web pages in all transactions.
- U, the total time spent on different web pages by various transactions in a given threshold.
- $U(p_i)$ represents the ratio of utility of a web page p_i in all transactions over the given threshold to the total utility of all web pages for a given threshold.
- The HBP-WPC of a web page p_i denoted as $WC(p_i)$, is the linear combination of log file statistics and web site topology statistics as shown in Eq. 1

$$WC (p_i) = W (p_iL) + W (p_iS) \qquad (1)$$

W(p_iS) denotes details of the web page such as in degree, out degree, level and number of pages in that level. α and β are grading factors taking all possible values between 0 and 1 which help to assign different combination of weightages for Support and Utility.

- $W(p_iL) = \alpha S(p_i) + (1 - \alpha) U(p_i)$
- $W(p_iS) = \beta O(p_i) + (1 - \beta) I(p_i) + D(p_i) + L(p_i)$
- $G1(p_i, \alpha, \beta) = W(p_iL) + W(p_iS)$ for $i = 1..n$, $\alpha = 0..1$ and $\beta = 0..1$
- $Temp = Avg (G1 (p_i, \alpha, \beta))$
- $G2(p_i, \alpha) = min(diff(G1(p_i, \alpha, \beta), Temp)$
- Let $C = Avg(G2)$. The web page whose (G2) $>= C$, such a page can be considered as highly demanding web page else it can be considered as low demanding web page.

5.2 Experimental Results

Figure 1 shows structure of the website. The website contains seven web pages A, B, C, D, E, F and G with support and utility for each web page. The page B has 2 forward links 1 backward link, the level of page is 2 and number of pages at level 2 is 3. The Page B has utility 5 and frequency 8 etc... total utility of the web site is 213 and total Support is 101.

As per Garafalokis relative access (RA) is calculated using absolute access (AA) of each web page obtained from log file of server as shown in Eq. 2.

$$RA [i]\ \alpha\ AA [i] \qquad (2)$$

i stands for index of web page. The Relative Access of web page is directly proportional to AA. The constant of proportionality is replaced with K. K is defined as follows

$$RA [i] = K [i] \,{}^* AA [i]$$

Fig. 1 Structure of the website with utility and support

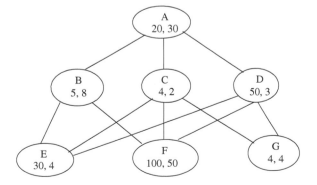

K [i] = d [i] + n [i] / [i] where d-depth of page, n-number of pages as level i and r-number of references to page i from other pages. The value of RA for each web page is shown in Table 3. The heap tree is generated based RA. Six nodes were swapped to obtain heap tree in 3 iterations. The web pages which should not be moved can be made stable. With the heap tree generation all demanding pages will be brought very near to home page. Geeta and others suggested constant of proportionality K in Eq. 2 as follows

$$GRA [i] = K [i] {}^{*}AA [i] \qquad (3)$$

K [i] = L [i] + FL [i] /BL [i] where i-index of a web page, L-level of page i, FL[i]-number of forward links from web page i and BL[i]-number of backward links to page i. The value of GRA calculated for each web page is shown in Table 3. The heap tree is generated based on RA. The WC calculated for each web page using Eq. 1 is also shown in Table 3. Using this approach, the highest grade pages at 2nd level are compared with its successors and swapped. After this next highest grade web page is considered and compared with its children, the page with highest WC is swapped with parent and so on.

α is the grading factor for support β is the grading factor for in degree and out degree. When α is 0 utility is given the importance, when $\alpha = 1$ support is given the importance. When β is 0 in degree is given the importance and when β is 1 out degree is given the importance. For each page G is calculated for all possible values of α and β ranging from 0 to 1 at the step of 0.1. Calculate G1 and G2 using the above formulas. The G1 for each web page is as shown in Table 1.

Table 2 shows values of G2 for each web page. Consider average of G2 of all web pages. The average G2 for above table is 0.14869. If G2 (p_i) > average of G2 consider the page as excellent/highly demanding web page and such a page can be moved near to the home page. To decide which pages are highly demanding and which pages are low demanding, fuzzy approach can be applied. This algorithm returns the values between 0 and 1. The web page whose G2 is greater than average (G2),

Table 1 G1 for all web pages

A	B	C	D	E	F	G
1.3731	1.4455	1.4863	1.7022	1.7057	2.0344	1.5382
1.3934	1.4511	1.4864	1.6817	1.6956	2.0369	1.5403
1.4137	1.4566	1.4865	1.6612	1.6855	2.0395	1.5424
1.4340	1.4622	1.4866	1.6407	1.6754	2.0420	1.5445
1.4543	1.4678	1.4867	1.6202	1.6652	2.0446	1.5465
1.4746	1.4734	1.4868	1.5997	1.6551	2.0472	1.5486
1.4949	1.4789	1.4869	1.5792	1.6450	2.0497	1.5507
1.5153	1.4845	1.4870	1.5587	1.6349	2.0523	1.5528
1.5356	1.4901	1.4871	1.5382	1.6247	2.0548	1.5549
1.5559	1.4957	1.4872	1.5177	1.6146	2.0574	1.5570
1.5762	1.5012	1.4873	1.4972	1.6045	2.0599	1.5590

Table 2 G2 for each web page

Web page	G2
A	0.1432
B	0.1364
C	0.1352
D	0.1361
E	0.1557
F	0.1978
G	0.1483

Table 3 Values obtained using Eqs. 1, 2 and 3

Web Page	G2(HBP-WPC)	RA	GRA
A	Fixed	Stable	Fixed
B	0.1364	40–II	32–II
C	0.1352	10	8
D	0.1361	15	15–III
E	0.1557–II	16	12
F	0.1978–I	200–I	150–I
G	0.1483–III	18–III	12

such a web page can be treated as highly demanding web page, otherwise it will be medium demanding web page. With such an approach all highly demanding pages can be moved near to home page. The web pages E and F are moved to upper level. The web page F is interchanged with B and E is interchanged with C. This information can be provided to the web site administrator for reorganization of the website. The G2 values can be compared with the values obtained from Eqs. 2 and 3.

Table 3 shows ranking of web pages for all three approaches. HBP-WPC tries to scale all the values between 0 and 1 and accurate results can be obtained, whereas the values calculated using RA and GRA take all possible integer values. HBC-WPC gives importance for web pages whose in-degree is high compared to out degree. So HBC-WPC is efficient and provides accurate result compared to other two approaches. Based on this ranking web site owners/designers can go for reorganization of web pages of the web site.

6 Conclusion

The contributions of this paper are as follows. First, this model helps in best utilization of web sources such as client, server and communication media. Second, since the highly demanding pages are brought near to the home page, users will have quick response. The control messages generated between server and client to have communications are minimized along with power consumption. Third, whenever client requests for HTTP transaction the entry will be made in the log file. Since most demanding pages are very near to home page, entries in log file are minimized. Client's and server's process time is minimized. Fourth, highly demanding pages can be kept in high speed servers', medium pages can be clustered and stored in low speed servers. This work aims to provide sophisticated metric, robust, useful techniques and fundamental basis for high conformity website reorganization routine and applications. In future work, we will investigate techniques for improvement in the metric and the procedure adopted.

References

1. Yang, Q., Zhang, H.: Web-log mining for predictive caching. IEEE Trans. Knowl. Data Eng. **15**(4), 1050–1053 (2003)
2. Li, Y., Zhang, C., Zhang, H.: Cooperative strategy web-based data cleaning. Appl. Artifi. Intell. **17**(5–6), 443–460 (2003)
3. Pei, J., Han, J., Mortazavi-Asl, B., Zhu, H.: Mining access patterns effciently from web logs. In: Pacific-Asia Conference on Knowledge Discovery and Data Mining (PAKDD'00), Kyoto, Japan, pp. 396–407, April 2000
4. Shen, H.T., Ooi, B.C., Tan, K.: Giving meanings to WWW, ACM SIGM Multimedia, L.A., pp. 39–47 (2000)
5. Mano, M., Deepak, G.: Semantic web mining of un-structured data: challenges and opportunities. Int. J. Eng. **5**(3), 268–276 (2011)
6. Wang, J., Liu, Y., Zhou, L., Shi, Y., Zhu X.: Pushing frequency constraint to utility mining model. In: ICCS, LNCS, vol. 4489, pp. 685–692. Springer, Heidelberg (2007)
7. Miller, C.S., Remington, R.W.: Implications for information architecture. Human Comput. Interact. J. IEEE Web Int. **19**(3), 225–271 (2004)
8. Geeta, R.B., Shashikumar G.T., PrasadReddy, PVGD.: Optimizing user's access to web Pages, RJooiJA. Trans. World Wide Web-Spring **8**(1), 61–66 (2008)
9. Garofalakis, Web Site optimization using page popularity. IEEE Int. Comput. **3**940, 22–29 (1999)
10. Geeta, R.B, Shashikumar G.T., PrasadReddy PVGD.: In: Conference, Topological Frequency Utility Mining Model Springer International, SocPros 11, pp. 505–508 (2011)
11. Ying, J.-C., Tseng, V.S. Yu, P.S.: In: IEEE International Conference on Data Mining Workshops. IEEE Computer Society (2009)
12. Lee, Y.S., Yen, S.J., Hsiegh, M.C.: A lattice-based framework for interactively and incrementally mining web traversal patterns. Int. J. Web Inf. Syst. 197–207 (2005)
13. Lee, Y.S., Yen, S.J., Tu, G.H., Hsieh, M.C.: Mining traveling and purchasing behaviors of customers in electronic commerce environment. In: Proceedings of the EEE'04, pp. 227–230 (2004)
14. Geeta, R.B., Shashikumar G.T., PrasasdReddy PVGD.: Manager-members dis tributed software development reference model. In: IEEE International Advanced Computing Conference IACC 2009 Patiala, 6–7 March 2009

Web Search Personalization Using Ontological User Profiles

Kretika Gupta and Anuja Arora

Abstract In web, users with different interest and goal enter queries to the search engine. Search engines provide all these users with the same search results irrespective of their context and interest. Therefore, the user has to browse through many results most of which are irrelevant to his goal. Personalization of search results involves understanding the user's preferences based on his interaction and then re-ranking the search results to provide more relevant searches. We present a method for search engine to personalize search results leading to better search experience. In this method, a user profile is generated using reference ontology. The user profile is updated dynamically with interest scores whenever, he clicks on a webpage. With the help of these interest scores in the user profile, the search results are re-ranked to give personalized results. Our experimental results show that personalized search results are effective and efficient.

Keywords Personalization · Ontological user profile · Re-ranking

1 Introduction

The amount of information in world wide web has seen a phenomenal increase in the past years. In 1994, one of the first web search engines had to index 110,000 web pages approximately. Today, search engines need to deal with more than 25 billion documents. Search results retrieved by internet search engines display the same result irrespective of who has queried. A user looking for "apple" maybe interested in apple

K. Gupta (✉) · A. Arora
Department of Computer Science, Jaypee Institute of Information Technology,
Noida, India
e-mail: kretika@hotmail.com

A. Arora
e-mail: anuja.arora@jiit.ac.in

B. V. Babu et al. (eds.), *Proceedings of the Second International Conference on Soft Computing for Problem Solving (SocProS 2012), December 28–30, 2012,* Advances in Intelligent Systems and Computing 236, DOI: 10.1007/978-81-322-1602-5_90, © Springer India 2014

as a fruit instead of apple the company. A user has to go through irrelevant search results before he finds his required results. This irrelevant information is due to the one size fits all policy of the search engines [1]. Identical queries from different users with different interest generate same search results. Another main reason of irrelevant search results is ambiguity in query. Ambiguity can be attributed to polysemy, existence of many meanings for a single word, and synonymy, existence of many words with the same meaning. Ontology is defined as an explicit specification of conceptual categories and relationships between them [2]. Therefore, to personalize the search results, a user profile is required to map the user interest. Re-ranking of webpages is done using user profile. Many approaches have been developed to personalize web search. User preference based on the analysis of past click history was discussed in detail by Pretschner and Gauch [3] and Sugiyama et al. [4]. Short-term personalization based on a current user session was discussed by Sriram et al. [5].

2 Methodology

Reference ontology is built by using Open Directory project. A user profile is generated by annotating interest scores in the concepts provided by the reference ontology. The interest scores in the user profile created is updated dynamically whenever he clicks on a webpage. With the help of the user interest the search results are re-ranked.

2.1 User Profile Generation

The User profile is an instance of reference domain ontology. The reference domain ontology is created with the help of a web directory, Open Directory Project (ODP) [6]. A portion of ODP has been shown in Fig. 1. In this, the concepts are

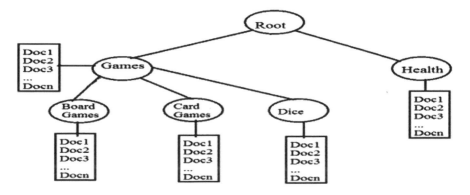

Fig. 1 Portion of an ontological profile. Each node has documents associated with it

annotated with an interest score which is updated dynamically each time the user clicks on a webpage. Open directory project is considered as the "largest human-edited directory of the web". The data structure is organized in Directed Acyclic Graph. Each category has a set of documents associated with it which were used as a training set for classification. Text classification is required to find out under which category the content of the webpage lies in. For text classification, all the documents classified under one category in the ODP structure is merged under one super document. Whenever a user clicks on a webpage, a page vector is computed and then compared with each category's vector in the DAG to calculate the similarities. Trajkova and Gauch [7] have calculated the similarity between Web pages visited by the user and the concepts in an ontology. The page vector is computed with the help of the title of the web page, Metadata Description Unigrams, and Metadata Keywords Unigrams associated with the webpage [8].

2.2 Updating User Profile

The User Profile for a given user saves his interests in the particular categories determined by the ODP structure. The user does not have to choose his interest areas explicitly [9]. This is automatically generated using various features which will be further discussed. The user profile is dynamic and keeps updating over time. As, whenever a user clicks on given link, the interest score is determined and updated. Since the profile is dynamically updated it takes into consideration the changing interests of a user.

Interest score is calculated with the help of the time spent, length, and subject similarity of the webpage. Time denotes the user's duration of viewing the webpage, length denotes the number of characters in a webpage. Subject similarity denotes the similarity between the webpage's content and the category defined by the ODP structure. As shown in Fig. 2.

Sim (d, c_i) refers to the similarity of match between the content of document (d) and category (c_i) defined by ODP. Adjustment of the interest of a user in category (c_i) is $\delta(i, c_i)$. The interest score is updated with the help of the following equation, according to [3].

$$\delta(i, c_i) = \log(\text{time}/(\log \text{length})) * \text{Sim}(d, c_i) \tag{1}$$

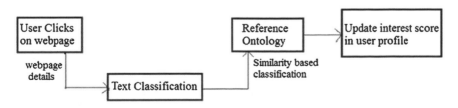

Fig. 2 Updating user profile

It can be noted that the above equation takes length into less consideration as the users can tell from a glance that the webpage is not relevant and move on to the next webpage swiftly irrespective of the length.

2.3 Re-ranking Search Results

Web search API: many commercial search engines have provided their API's so third party tools can access their search results (index). Google custom search API is used to retrieve search results for a query given by the user. These search results are retrieved with their index and are then used to re-rank web pages according to the interest scores in the generated user profile of that user.

The pages are re-ranked by a similarity matching function that computes the similarity of the retrieved result's document with each concept in the user profile's ontology to find the best matching concept.

$$\text{CSim}(\text{UserProfile}_i, \text{Result}_j) = \sum_{k=1}^{N} \text{wp}_{i,k} * \text{wd}_{j,k} \tag{2}$$

where,

Wp$_{i,k}$ represents the weight of concept k in the user profile,
Wd$_{j,k}$ represents the weight of concept k in the result j.

As Google applies its own PageRank algorithm, to rank websites based on their importance, we have incorporated Google's original ranking score as well. This will keep a check that we do not miss important webpages.

FinalRank(UserProfile$_i$, Result$_j$)
$$= \gamma * \text{CSim}(\text{UserProfile}_i, \text{Result}_j) + (1 - \gamma)\,\text{GRank}(\text{Result}_j) \tag{3}$$

where GRank is the original rank. γ is used to combine the two ranking measures. We consider γ as 0.5 to give equal weightage to both the ranking mechanisms. If γ is 0, ranking will be done based on Google search results and if γ is 1 the ranking is done purely according to context. Each time, a user clicks on the links of the search results; the interest score is updated dynamically to determine the user's preferences. This has been represented in Fig. 3.

3 Experiments

To evaluate the effectiveness of personalized search results we need to find:

Research Question 1: (RQ1): Do the interest scores for individual concepts in onto logical profile converge?

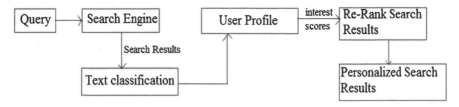

Fig. 3 Re-ranking results

Research Question 2: (RQ2): Can the interest scores maintained by the onto logical profile be used to re-rank Web search to give personalized search results?

3.1 Experiment 1

With this experiment we want to evaluate RQ1, if the rate of increase in the user's interest scores for all categories stabilizes over incremental updates [10]. The categories are defined by the user's ontology. Each time the user clicks on a webpage the user interest are updated in the ontological user profile. Initially, the interest scores for the categories in the user profile will continue to change rapidly. However, once enough information has been collected and processed, the rate of change interest scores should decrease. Hence, we wanted to find out if over time the concepts with the highest interest scores would become relatively stable or not. For conducting the experiment, 15 users were asked to use the personalized search engines over a period of 20 days. Their user profile was monitored during these days. The number of categories the profiles converged to, changed according to the user, mainly it was in the range of 48 and 180. The Fig. 4 shows the convergence for a sample of 4 users. We can see that over time the user profile converges and becomes stable.

3.2 Experiment 2

In this experiment, we determined if the users found the personalized search results more relevant than standard web search results for RQ2. Experiment has been performed manually.To conduct this comparative experiment, whenever the user clicked on a given webpage for a query, we asked the user to mark the page as relevant or irrelevant. 15 users entered several queries over a period of 20 days. On a single search query, 12 webpages from each of the standard search engine and personalized search engine was randomly presented to the user. Few pages were marked as "both", if they were common to both the search engines. By looking at the log of the user, it was determined how many relevant webpages the users clicked on from each.

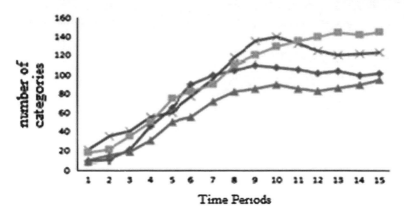

Fig. 4 Convergence of profiles

The proposed personalized search results were 55 % more relevant than the normal search results for the user searches.

4 Conclusion

This paper proposed a method for a search engine to personalize search results based on a user's preferences. The user preferences were mapped to a user profile. It was shown with the help of experiments that over time, the interest got converged. With the help of the user profile, web search results can be re-ranked leading to more relevant results for the users. In future, we plan to optimize our search engine for more relevant results. We would also look into the location based information of user to provide better search results.

References

1. Allan, J., et al.: Challenges in information retrieval and language modeling. ACM SIGIR Forum **37**(1), 31–47 (2003)
2. Sieg, A., Mobasher, B., R. Burke.: Web search personalization with ontological user profiles. In: Proceedings of CIKM (2007)
3. Pretschner, A., Gauch, S.: Ontology based personalized search. In: Proceedings of the 11th IEEE International Conference on Tools with Artificial Intelligence. Chicago, IL, pp. 391–298. IEEE Computer Society (1999)
4. Sugiyama, K., Hatano, K., Yoshikawa, M.: Adaptive web search based on user profile constructed without any effort from user. In: Proceedings of the 13th International Conference on World Wide Web. New York, pp. 675–684. (2004)
5. Sriram, S., Shen, X., Zhai, C.: A session-based search engine. In: Proceedings of SIGIR(2004)
6. Open Directory Project - http://dmoz.org

7. Trajkova, J., Gauch, S.: Improving ontology-based user profiles. In: Proceedings of the Recherched'Information Assiste par Ordinateur, RIAO 2004, pp. 380–389. University of Avignon (Vaucluse), France, April (2004)
8. Chirita, P., Firan, C., Nejdl, W.: Summarizing local context to personalize global web search. In: Proceedings of the 15th ACM International Conference on Information and Knowledge Management, CIKM 2006, pp. 287–296. Arlington, VA, November (2006)
9. Teevan, J., Dumais, S., Horvitz, E.: Personalizing search via automated analysis of interests and activities. In: Proceedings of the 28th Annual International ACM SIGIR Conference on Research and Development in Information Retrieval, SIGIR 2005, pp. 449–456. Salvador, Brazil, August (2005)
10. Liu, F., Yu, C., Meng, W.: Personalized web search for improving retrieval effectiveness. IEEE Trans. Knowl. Data Eng. **16**(1), 28–40 (2004)

Autonomous Computation Offloading Application for Android Phones Using Cloud

Mayank Arora and Mala Kalra

Abstract The usage of smartphones has increased hastily over the past few years. The number of smartphones being sold is much more than the number of PC's due to the smartphone's mobile nature and good connectivity. However, they are still constrained by limited processing power, memory, and Battery. In this paper, we propose a framework for making the applications of these smartphones autonomous enough, to offload their compute intensive parts automatically from the smartphone to the virtual image of the smartphones on the cloud, thus using the unlimited resources of the cloud and enhancing the performance of the smartphones. By using this framework the application developers will be able to increase the capabilities of the smartphones making them even more feature rich.

Keywords Smartphone · Offloading · Cloud · Android

1 Introduction

Cloud can be viewed as a great service which hosts everything, and it may be the data, applications, or any other running programs. A Cloud could be seen as an amorphous collection of computers and servers that could be accessed through the Internet[1–3]. Cloud computing has emerged as the great technology in term of scalability and portability. It has changed our view of carrying data and communication. Cloud services are also very much indulging into mobile networks as most of the smartphones have the capability to support cloud computing environment [4].

M. Arora (✉) · M. Kalra
Computer Science and Engineering Department, NITTTR, Chandigarh, India
e-mail: mayank.nitttr@gmail.com

M. Kalra
e-mail: malakalra2004@yahoo.co.in

B. V. Babu et al. (eds.), *Proceedings of the Second International Conference on Soft Computing for Problem Solving (SocProS 2012), December 28–30, 2012,* Advances in Intelligent Systems and Computing 236, DOI: 10.1007/978-81-322-1602-5_91, © Springer India 2014

Smartphones are mobile phones with advanced computing capability, connectivity, and rich set of functionality. In a nutshell, a smartphone combines the functionalities of a phone, personal digital assistant (PDA), and a small computer. With the increasing popularity and a large number of developers developing applications for smartphones, the users of these phones have started using them for high end 3D gaming, to handle their finances, i.e., Internet banking and as their health and wellness managers (e.g., Eat This, Not That application for Android [5]). These new applications could be very resource exhaustive and the phones have a limited memory, computational power, and battery life. That's why it makes good sense to offload the heavy applications to the Virtual Smartphone running on the cloud, thus saving the actual phone's precious resources [6, 7].

A number of techniques have been proposed to offload the applications of smartphones to the cloud [8–14], including complete offloading of the applications as well as partial offloading of the applications. In these techniques used for offloading, the application is partitioned at the binary level and thus making this partitioning transparent for the application developer. But this has its drawbacks, i.e., first, this process is compute intensive. Second, to make changes at the binary level of an application needs changes in the application loader, which is difficult as well as leads to security vulnerabilities. Furthermore, in the proposed techniques an application called the application partitioner or off loader needs to be installed on the smartphone which makes the partitions and offloads the appropriate partition of other applications to the cloud. The application offloader makes the offloading decision for all the applications in the phone weather small or big in terms of computation required and thus become an overhead on the phone's resources. In this paper, we propose a framework for offloading an application partially, i.e., only the compute intensive, non-interactive part of an application is offloaded. The partitioning is done by the application developer at the time of development of the application and the offloading decision is taken by the application itself thus eradicating the need of making changes at the binary level and the need of application partitioner or offloader. These applications will offload their compute intensive part to the cloud autonomously.

The remainder of this paper is organized as follows. In Sect. 2, we describe the motivation behind the proposed architecture. Sections 3 and 4 outline the review of the proposed architectures in the related research and the challenges faced by the offloading techniques proposed. In Sect. 5 we describe the new offloading architecture to cater the challenges discussed in the above sections. Toward the end, we give the details about the working of our framework. In Sect. 6 the benefits of the proposed architecture are discussed and the paper is concluded in Sect. 7.

2 Motivation

The applications and features of smartphones are increasing day by day because the usage of these feature-rich phones is increasing. People are replacing their laptops and personal computers with these smartphones, thus the demand for processing and memory is increasing. These phones use a battery as their power source which has a limited capacity as compared to plug in devices like personal computers.

Some of the major problems faced by the smartphone users are nowadays listed as follows. First, the applications using heavy graphics, memory, or CPU result in a lot of battery drainage. Second, due to the small size of the phones the processing power, memory, and battery are limited and these phones are not able to perform compute intensive tasks which our laptops or desktops could perform. The solution to these problems is either to increase the size of battery, processing power of the CPU and the size of memory which in turn results in increased size, and cost of the phone or to use the resources of the cloud to execute the heavy applications thus saving the phone's scarce resources.

Cloud computing on the other hand provides computing resources (hardware and software) as a service through Internet. We can use the resources such as memory, processing power in a pay per use environment.

The major motivation for this paper is to use the computing resources provided as a service by the cloud to run the resource exhaustive applications of the smartphones connected to the Cloud through Internet. In the past couple of years some techniques have been proposed to partially or completely offload the applications on to the cloud. We will be discussing those techniques in the next section and the challenges faced by the offloading techniques.

3 Review of Proposed Architectures in the Related Research

Quite a few approaches have been proposed for offloading applications from a smartphone to the cloud, which includes offloading the complete application, offloading an application partially. Related work in the field of offloading applications from android phones to the cloud have been discussed below [8–14].

In 1998, Alexey Rudenko et al. [8] proposed a scheme to enhance the battery life of a laptop through wireless remote processing of power costly tasks. They proposed that the battery life of a laptop could be increased by shifting the power costly tasks on to a server through wireless connectivity or the Internet. This powerful server will perform the tasks as required and send back the results to the laptop saving the laptop from processing the tasks itself and in the meanwhile the laptop will keep on performing the less power costly tasks. This research gave birth to a new idea of remote processing of power costly tasks of the smartphones using the resources of the cloud.

Year 2009 witnessed the proposal of Augmented Smartphone Applications Through Clone Cloud Execution by Byung-Gon Chun and Petros Maniatis [9]. This research proposed to augment the smartphone's capabilities by offloading an application partially or completely to a clone smartphone. A clone is a virtual system on the cloud running the same operating system as that of the phone using hardware from the cloud's pool of hardware. The application is offloaded partially because only the part of the application which is compute intensive is to be offloaded and thus reducing the load on the smartphone. While the compute intensive part is being executed by the clone the actual phone executes the remaining application. After the clone is finished with the execution of the compute intensive part of the application it returns the results to the actual phone. The phone processes the results as required and provides the user with the results. The proposed architecture includes a Controller and a Replicator installed in the actual phone and an Augmenter and Replicator installed on the clone. The Replicator synchronizes the changes in the phone software and state to the clone. The controller offloads the application from the smartphone to the clone and merges back the results from the clone to the phone. The Augmenter running in the clone manages the local execution, and returns a result to the actual phone.

Following the vision provided by the above research, Byung-Gon Chun et al. [10] in the year 2011 implemented an architecture named Clone Cloud for offloading an application partially to its clone in the cloud. This scheme uses a partition analyzer which partitions the application to be offloaded for remote execution. The partition analyzer has a static analyzer which discovers the possible migration points and the constraints for migration and a dynamic profiler to build a cost model for execution and migration. The partition analyzer helps the migration unit to migrate and re-integrate the application at the chosen points. The migration unit comprises of a migrator, node manager, and a partition database. The migrator provides the part of application to be migrated to the node manager which migrates the part to the clone and an entry is made to the partition database which helps in re-integrating the partitioned application.

In the year 2011, a new approach of offloading the applications from android smartphone to the cloud was introduced by Eric Y. Chen and Mistutaka Itoh named Virtual Smartphone over IP [11]. In this approach, the complete application was offloaded from the android smartphone to the cloud. In this approach it was proposed to provide cloud computing environment specifically tailored for smartphone users. This architecture allows users of smartphones to create virtual smartphone images in the cloud and install and run their applications in these images remotely. The user can create a number of smartphone images using a dedicated server for each user.

In 2012, Eric Y. Chen et al. [12] introduced a framework for offloading heavy back-end tasks of a standalone android application to an android virtual machine in the cloud, which is an extension of Virtual Smartphone over IP. This architecture uses android interface definition language in order to offload without modifying the source code. This framework incorporates a dedicated server for each client to offload their application [11]. This architecture divides an application into two parts, i.e., GUI and a compute intensive component and it offloads only the compute intensive component to the android virtual machine. This framework comprises of basically three components a helper tool, a service offloader, and a virtual machine. The helper tool has to be integrated to the application code at the time of development by the developer. The service offloader has to be installed to the android phone by downloading it from the application market and each user needs to have at least one dedicated instance of the android virtual machine which is the virtual image of the phone hosted on the cloud. The helper tool generates two copies of the application, i.e., one for local execution and one for remote execution and calls the service offloader which then analyzes the cost of remote execution and local execution. If the cost of remote execution is less than the cost of local execution then the service is offloaded otherwise it is not offloaded to the virtual android phone.

In 2010, Georgios Portokalidis et al. [13] proposed a new scheme named Paranoid Android to provide security to android phones by applying security checks on remote security servers that host exact replicas of the smartphones in virtual environments. As the remote servers are not constrained with battery or processing power, multiple detection schemes could be applied simultaneously. On the phone, a tracer records all information needed to accurately replay its execution. The recorded execution trace is transmitted to the cloud over an encrypted channel, where a replica of the phone is running on an emulator. On the cloud, a replayer receives the trace and faithfully replays the execution within the emulator.

Another system named MAUI was introduced in 2010 by Eduardo Cuervo et al. [14]. MAUI enables fine-grained energy aware offload of mobile's code to a cloud infrastructure. MAUI uses code portability to create two versions of a smartphone application, one of which runs locally on the smartphone and the other runs remotely in the infrastructure. Managed code enables MAUI to ignore the differences in the instruction set architecture between today's mobile devices (which typically have ARM-based CPUs) and servers (which typically have x86 CPUs). It uses programming reflection combined with type safety to automatically identify the remote able methods and extract only the program state needed by those methods. MAUI profiles each method of an application and uses serialization to determine its network shipping costs (i.e., the size of its state). MAUI combines the network and CPU costs with measurements of the wireless connectivity, such as its bandwidth and latency to construct a linear programming formulation of the code offload problem.

4 Challenges

The main problem which is identified in the previous architectures is that offloading a compute intensive application partially can improve the battery life of a smartphone, but the offloading system will incur some overhead on the phone especially if the offloading decision is taken for a large number of applications when only a few number of applications are actually required to be offloaded.

The architectures discussed in the above section need to make changes in the binary of the application at the time of execution, to make it offloadable. To make changes at the binary level, the program loader has to be changed which may result in security vulnerabilities and to analyze the binary of an application could be compute intensive thus imposing overhead on the phone.

5 Proposed Architecture

In this paper, we present a framework for automated offloading of compute intensive applications of android smartphones to the virtual image of the smartphone on the cloud. An offloading framework is proposed which if used by the developers of the application, will empower the application to offload its compute intensive, non interactive parts based on static analysis to the smartphone image on the cloud. The static analysis is done to make the decision-making more fast and light than the previous techniques.

This framework proposes to make an application autonomous for offloading itself from the smartphone to the smartphone image on the cloud. An application will be divided into two major parts, i.e., a user interactive part which takes the inputs from the user and provides the output to the user and a compute intensive non interactive part which does the computations as shown in the Fig. 1. This framework will empower the application to offload its compute intensive part to the cloud via Internet after analyzing the cost of offloading over the cost of running the application on the phone itself. The analysis will be done using parameters like input size and internet connectivity. By using this framework the developers will empower the applications to offload themselves without the need of some other application to analyze and offload parts of the application.

The Fig. 2 shows the working of the proposed architecture where the offloading decision is taken by the application itself after analyzing the parameters discussed above. According to the offloading decision the application is partially offloaded to the virtual phone on the cloud. The cloud performs the compute intensive tasks and returns the results to the phone. Thus saving the resources of the phone and increasing the capabilities of the phone.

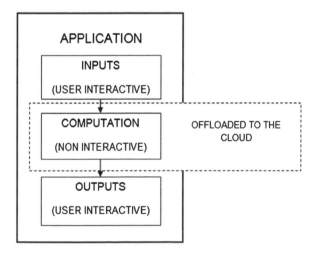

Fig. 1 Partitioning of applications into interactive and non-interactive parts

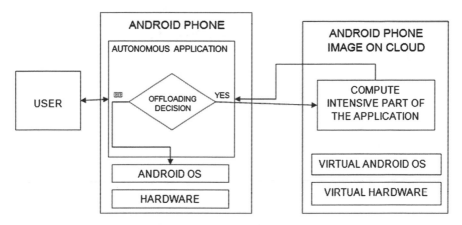

Fig. 2 Framework of autonomous computation offloading application for android phones

6 Advantages of the Proposed Architecture

The major advantage of using this framework is that the offloading decision will be taken by the application itself. This decision making will have to be inserted in the source code while writing the applications which is advantageous over the previous approaches in the following ways. First, the approaches discussed earlier modified the applications at binary level using modifications in the loader and thus increasing the security vulnerabilities. Second, they analyzed each application's binary to offload it, but only a few compute intensive applications need to be offloaded thus causing an overhead on the phone's resources. By using this technique the offloading decision will be taken only for the compute intensive applications which need to be offloaded and the application developers will not use this offloading framework for small applications. This technique will need to modify the applications at the development stage and will not modify the applications binary thus eradicating the need of changing the application loader.

7 Conclusion and Future Work

In this paper, we explored the design of a framework which makes an application autonomous to offload its compute intensive part to the cloud thus saving the resources of the android phone. This framework makes changes in the application at the development time thus eradicating the need to make changes in the application's binary. The application will make the offloading decision using static analysis.

Our future work includes implementation and evaluation of this design, adding dynamic analysis to support the offloading decision and comparing the performance of both frameworks using static analysis and the one using dynamic analysis. We are also planning to incorporate security, privacy, and trust related models [15–18], in the proposed framework.

References

1. Peng, J., Zhang, X., Lei, Z., Zhang, B., Zhang, W., Li, Q.: Comparison of several cloud computing platforms. In: Proceedings of IEEE international conference on Information Science and Engineering, 23–27 (2009)
2. National Institute of Science and Technology.: The NIST definition of cloud computing. http://csrc.nist.gov/publications/nistpubs/800-145/SP800-145.pdf (2010)
3. Buyya, R., Broberg, J., Goscinski, A.: Introduction to cloud computing. In: Cloud Computing: Principles and Paradigms. Wiley Press [Online]. 1–44. http://media.johnwiley.com.au/product_data/excerpt/90/04708879/0470887990-180.pdf (2009)
4. SONG, W., SU, X..: Review of mobile cloud computing. In IEEE 3rd International Conference on Communication Software and Networks, 1–4 (2011)
5. Health and Wellness - Phone Apps. University of California. http://wellness.ucr.edu/Wellness%20Apps%20Resources.pdf
6. Saarinen, A., Siekkinen, M., Xiao, Y., Nurminen, J.K., Kemppainen, M., Hui, P. Smartdiet: offloading popular apps to save energy. ACM SIGCOMM Conference, 297–298 (2012)
7. Android Open Source Project. Philosophy and Goals. Google. http://source.android.com/about/philosophy.html (2012)
8. Rudenko, A., Reiher, P., Popek, G.J., Kuenning, G.H.: Saving portable computer battery power through remote process execution. In: MCCR'98–ACM SIGMOBILE Mobile Computing and Communications Review Newsletter. Vol. 2, no. 1, 19–26 (1998)
9. Chun, B.G., Maniatis, P.: Augmented smartphone applications through Clone Cloud execution. In: Proceedings of 12th conference on Hot topics in operating systems, 8–8 (2009)
10. Chun, B.G., Ihm, S., Manitis, P., Naik M., Patti, A.: Clonecloud: elastic execution between mobile device and cloud. In: Proceedings of 6th Conference on Computer Systems, 301–314 (2011)
11. Chen, E.Y., Itoh, M.: Virtual smartphone over IP. In: Proceedings of IEEE international conference on World of Wireless Mobile and Multimedia Networks, 1–6 (2010)
12. Chen, E., Ogata, S., Horikava, K.: Offloading android applications to the cloud. In: Proceedings of IEEE International Conference on Pervasive Computing and Communications Workshops (PERCOM Workshops), 788–793 (2012)
13. Portokalidis, G., Homburg, P., Anagnostakis K., Bos, H.: Paranoid android: versatile protection for smartphones. In: Proceedings of the 26th Annual Computer Security Applications Conference, 347–356 (2010)
14. Cuervo, E., Balasubramanian, A., Cho, D.K., Wolman, A., Saroiu, S., Chandra, R., Bahl,P.: MAUI: Making Smartphones Last Longer with Code Offload. In Proceedings of the 8th international conference on Mobile systems, applications, and services, 49–62 (2010)
15. Singh S., Bawa, S.: A privacy, trust and policy based authorization framework for services in distributed environments. Inter. J. Comput. Sci. 2(1): 85–92 (2007)
16. Singh, S.,Bawa, S.: A framework for handling security issues in grid environment using web services security specifications. In: Second International Conference on Semantics, Knowledge and Grid, : SKG'06, Guilin,China 68(2008), (2006)
17. Singh, G., Singh, S.: A comparative study of privacy mechanisms and a novel privacy mechanism [Short Paper]. In: ICICS'09 Proceedings of the 11th International Conference on Information and Communication Security, Beijing, China. 346–358 (2009)
18. Singh, S.: Trust based authorization framework for grid services. J. Emerg. Trends Comput. Inf. Sci. 2(3): 136–144 (2011)

Optimizing Battery Utilization and Reducing Time Consumption in Smartphones Exploiting the Power of Cloud Computing

Variza Negi and Mala Kalra

Abstract Over the past few years, the usage and boost of handheld devices such as Personal Digital Assistants (PDAs) and smartphones have increased rapidly and estimates show that they will even exceed the number of Personal Computers (PCs) by 2013. Smartphones enable a rich, new, and ubiquitous user experience, but have limited hardware resources on computation and battery. In this paper, the focus has been made on enhancing the capabilities of smartphones by using cloud computing and virtualization techniques to shift the workload from merely a smartphone to a resource-rich computational cloud environment.

Keywords Cloud computing · Smartphone · Offloading

1 Introduction

According to Morgan Stanley's Internet Trends Report, as shown in Fig. 1, smartphone is forecasted to be a dominant computing platform [2].

Smartphones have emerged as a type of mobile device providing "all-in-one" convenience by integrating traditional mobile phone functionality and the functionality of handheld computers. Now mobile users look up songs by audio samples; analyze, index, and aggregate their mobile photo collections; play games; capture, edit, and upload video; manage their personal health and wellness and also analyze their finances. Smartphones have every capability of a computer as smartphones are capable of producing great computing and processing output.

V. Negi (✉) · M. Kalra
Computer Science and Engineering Department, NITTTR, Chandigarh, India
e-mail: variza9@gmail.com

M. Kalra
e-mail: malakalra2004@yahoo.co.in

B. V. Babu et al. (eds.), *Proceedings of the Second International Conference on Soft Computing for Problem Solving (SocProS 2012), December 28–30, 2012*, Advances in Intelligent Systems and Computing 236, DOI: 10.1007/978-81-322-1602-5_92, © Springer India 2014

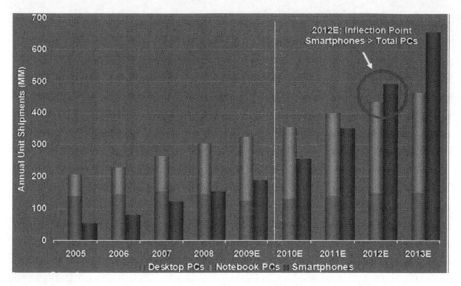

Fig. 1 Global unit shipments of desktop PCs, notebooks, and smartphones, 2005–2013E (estimation) [2]

However, power consumption by these devices is a prime factor when usage of these features comes into play. Power consumption increases dramatically while we are using different features like surfing Internet, watching videos, antivirus scanning, etc. Also the hardware capabilities of smartphones can be compared with or are similar to those of desktop PC's of the mid-1990, many generations of hardware and software behind (see Table 1 vs. Table 2)

Furthermore, there are many APIs installed in smartphones that tend to consume limited battery resources, which in turn results in short battery life time for smartphones. Cloud computing is a rapidly accelerating revolution within IT and will become the default method of IT delivery moving into the future—organizations

Table 1 Specifications of a few high-end smartphones

Phone	CPU (GHz)	RAM (MB)	Battery life (web-browsing)
Nokia lumia	1.4	512	4hr,10min
HTC one X	1.5	1024	4hr,18min
Samsung galaxy III	1.4	1024	5hr,17min

Table 2 Specification of a commodity laptop and a desktop

Computer	CPU (GHz) (GHz)	RAM (GB) (GB)
MAC book pro laptop	2.5, Quad-core	8/16
Origin genesis desktop	4.6, Quad-core	16

would be advised to consider their approach toward beginning a move to the clouds sooner, rather than later [8–10].

Cloud services could provide great means to save this type of power consumption. Much research has been done while considering this vision like Clone including [3–6]. All of the aforesaid papers lay emphasis on the basic idea of maintaining a replica or a mirror image of the smartphone on a Cloud computing infrastructure. With the help of a mirror, we can greatly reduce the workload and virtually expand the resources of the smartphones.

This paper has been organized as follows: The above section has introduced the general background of smartphones and Cloud computing. Also, limitations of smartphones have been pointed out. Next section provides an overlay of the various offloading techniques proposed in the past research studies. In Sect. "Paranoid Android Architecture", the major concerns and issues related to the offloading in smartphones have been laid down. In Sect. "Augmented Smartphone", the methodology to be used by in this particular research work has been discussed and finally a framework for migrating a compute-intensive Smartphone Application to the Cloud Computing Environment has been presented as the Proposed Architecture. Section "Clone Cloud Architecture" concludes as well as points out the future scope. Section "Mirroring Smartphone" presents the acknowledgements.

2 Related Work

2.1 Paranoid Android Architecture

Portokalidis et al. proposed Paranoid Android in [3]. This architecture aims at providing security as just another service to a smartphone by offloading security services on to the Cloud. The basic idea is to run a synchronized replica of the smartphone on a security server in a Cloud. As the server does not have the tight resource constraints of a phone, security checks can be performed on the Cloud that otherwise are too expensive to run on the phone itself. To achieve this, a minimal trace of the phone's execution (enough to permit replaying only) is recorded, which is then transmit to the Cloud-server, where a replica of the phone is running on an emulator. Thus on the Cloud, a replayer receives the trace and faithfully replays its execution within the emulator.

This architecture focuses more on attack detection. The prototype implementation works on the Android platform and is specific to the Android architecture. This architecture can be used to provide offloading by implementing suitable security functions in the Cloud; but always requires an extra overhead of synchronization. This architecture assumes loose synchronization, as it is assumed that the smartphones cannot always be connected to the Internet. Therefore availability is affected, as the security functions in the Cloud are not always available to the smartphones.

2.2 Augmented Smartphone

Distribution of computation between the smartphones and the Cloud resources in the form of clone Cloud architecture has been suggested by Chun et al. [4]. The concept behind this architecture is to seamlessly offload execution from the smartphone to a cloud infrastructure. Resource intensive processes or portions of processes are performed by the smartphone clone in the Cloud and the results are then merged with the state of the smartphone which resumes execution. This process of splitting the computation between the smartphone and its clone is referred to as "augmentation". This augmentation of computation is being done in four steps:

1. Creation of a clone of the smartphone in the Cloud;
2. Periodic or on-demand synchronisation of the smartphone and the clone;
3. Applications can be augmented (whole applications or augmented pieces of an application) automatically or upon request; and
4. Results obtained from the augmented segment of computation from the clone are synchronized with the smartphone.

Clone execution architecture for the smartphone aims at transforming a single machine's execution into a distributed execution. The smartphone and the clone have a replicator which is responsible for synchronizing the clone state with the smartphone (on-demand or periodic) and a controller component on the smartphone that invokes augmented execution in the clone and integrates the results of the computation by the clone with the smartphone.

2.3 Clone Cloud Architecture

The Clone Cloud architecture presents an effective method for computation offloading from the mobile devices. A prototype implementation of the Clone Cloud is demonstrated in [5]. In the Clone Cloud prototype, an application is partitioned by the use of a static analyzer, dynamic profiler, and an optimization solver. The execution (migration and re-integration) points are defined where the application migrates part of its execution to the Cloud. This takes place at a thread level and when the compute intensive thread completes execution in the Cloud; its results are merged into the state of the smartphone. The other threads can still continue to run on the smartphone. Migration decisions are made based on criteria such as current network characteristics, CPU speed, and energy consumption at the smartphone.

The prototype implementation evaluation proves up to 20x speedup and 20x energy reduction. Again in this technique an extra overhead of synchronization comes into play and the phone has to be always kept synchronized with the clone. Furthermore, the details of how a smartphone and its clone are kept synchronized have not been discussed. Therefore, analysis of how a clone would be created and managed needs to be considered if this architecture is to be adopted.

2.4 Mirroring Smartphone

Zhao et al. [6] propose a framework to keep the mirrors of smartphones on a computing infrastructure in a telecommunications network thereby offloading heavy computations to the mirror. The mirror server in the telecommunications network is capable of hosting a large number of virtual machines. Synchronization between the smartphone and the mirror is achieved by replaying all the inputs to the smartphone in the same order at the mirror. Their framework is different from the others in the following two major aspects:

- The Cloud is located in the telecommunication service provider's infrastructure
- A smartphone is connected to its mirror through a 3G network

Modifications to the existing telecommunications network are necessary to enable the forwarding of all the incoming traffic of the smartphone to the mirror server. In this system design, it is considered that the changes in the state of the smartphone are triggered by user or network inputs which produce deterministic outputs. Thus they suggest that synchronisation can be achieved by replaying these inputs in the mirror in the order of their occurrence. Data caching applications and anti-virus scanning are the two services for smartphones that have been identified to benefit from this framework. However, this framework has not been implemented yet. Furthermore, this approach does not consider Wi-Fi and Bluetooth connections and caters only to the Internet connections through 3G.

3 Issues

The major problem identified in the above discussed techniques is that, to make a clone of the phone on the Cloud, the entire phone's data is required to be uploaded on to the Cloud which consumes a lot of bandwidth and time.

Second, whenever the phone's data is updated the same changes are required to be made in the clone and thus an extra overhead due to continuous monitoring and synchronization between the smartphone and the Smartphone's image/clone on the Cloud is leveraged on the phone. Also, for this constant synchronization, a smart phone always has to be connected to the Internet service.

In some of the previous researches, the detail of how a Smartphone and its clone are kept synchronized has not been discussed.

4 Proposed Architecture

In this research, all together a new approach has been proposed which focuses on optimizing the battery utilization and reducing the time consumption by the heavy applications in the smartphones, exploiting the power of Cloud computing. This can be done by completely offloading the compute intensive applications on to the Cloud and thus unleashing the smartphone from doing heavy computational work. Also, there is no need to synchronize the smartphone and the Cloud continuously, which is a major drawback in all the other previously proposed offloading architectures.

CPU and battery consumption measurements will be performed by gadgets available for android in android market. Due to availability of resources, android will be chosen [7]. A compute intensive application in android will be developed. Further, analysis of the time taken and battery utilized by the application on the actual device will be made. These results will be further compared with the results when the application will be uploaded on the Cloud. This off-loading mechanism will benefit in saving battery and demonstrate the concept of API-as-a-service.

Figure 2 illustrates a high-level overview of the proposed system. On the smartphone's side, a small instance module of the actual compute intensive application is deployed within the smartphone's operating system (OS). This small instance collects the smartphone input data, including user keyboard inputs, and transmits them to the mirror server located on the Cloud. The compute intensive application and its workings are designed according to the specification provided by the Service Provider and the manufacturer.

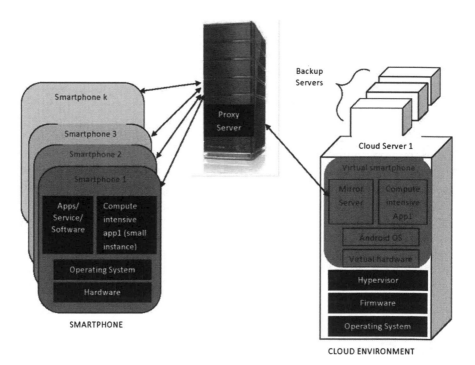

Fig. 2 Framework for migrating compute intensive smartphone application to the cloud computing environment

On the Cloud environment's side, the mirror server is a powerful application server maintaining one or more VMs. Each VM is a mirror to one of the smartphone. The Mirror Server on the Cloud receives the data from the Smartphone's application instance. The communication can be done using any of the communication techniques; here the communication is shown via Internet. The detailed designs and implementations of the other modules are out of the scope of this paper.

5 Conclusion and Future Scope

A detailed analysis of battery and time consumption in a smartphone on running a compute intensive application will be performed, based on measurements of a software tool. It will be shown how the battery drainage varies with increasing computational work. Focus will be on delivering all together a new method in which the whole application has been kept on the Cloud rather than on the phone.

Also the detailed measurements of the same application has been done for the VM running on Cloud, which has depicted lesser execution time and no issues of battery, as the Cloud is always connected to the mains directly. The ultimate aim of this work is to enable a systematic approach for improving power management of mobile devices. Presenting these analysis, future research can be done regarding how the offloading techniques can be made more intelligent and adaptive.

References

1. Gartner: gartner research report. http://www.gartner.com/technology/home.jsp (2011)
2. Stanley: internet trends report (2010)
3. Portokalidis, P.H., Anagnostakis, H.B.: Paranoid android: versatile protection for smartphones, 26th annual computer security applications conference, pp. 347–356 (2010)
4. Chun, P.M: Augmented smartphone applications through clone cloud execution, 12th conference on hot topics in operating systems, pp. 8–8 (2009)
5. Chun, S.I., Manitis, M., Patti: Clonecloud: elastic execution between mobile device and cloud, 6th Conference on computer systems, pp. 301–314 (2011)
6. Zhao, Z.X., Chi, S.Z., Cao.: mirroring smartphones for good: a feasibility study, ZTE, communications (2011)
7. Wikipedia : android (Operating System) http://en.wikipedia.org/wiki/Android_ (operatingsystem) (2012)
8. Kepes: understanding the cloud computing stack SaaS, Paas, IaaS. diversity limited (2011)
9. Sengupta, V.K., Sharma: cloud computing security- trends and research directions, IEEE World Congr Serv, pp. 524–531 (2011)
10. Dillon, C.W., Chang: Cloud computing: issues and challenges, 24th IEEE international conference on advanced information networking and applications, pp. 27–33 (2010)
11. Singh, S.B.: A privacy, trust and policy based authorization framework for services in distributed environments. Int. J. Comput. Sci. 2(1), 85–92 (2007)
12. Singh, S.B.: Design of a framework for handling security issues in grids, international conference on information technology, ICIT'06, pp. 178–179 (2006)

13. Singh, S. B.: A framework for handling security issues in grid environment using web services security specifications, second international conference on semantics, knowledge and grid, 2006, SKG'06, Guilin, China, pp. 68 (2006)
14. Singh, S.B.: A privacy policy framework for grid and web services. Inform. Technol. J. (6), pp. 809–817 (2007)

Part IX
Algorithms and Applications (AA)

On Clustering of DNA Sequence of Olfactory Receptors Using Scaled Fuzzy Graph Model

Satya Ranjan Dash, Satchidananda Dehuri, Uma Kant Sahoo
and Gi Nam Wang

Abstract Olfactory perception is the sense of smell that allows an organism to detect chemical in its environment. The first step in odor transduction is mediated by binding odorants to olfactory receptors (ORs) which belong to the heptahelical G-protein-coupled receptor (GPCR) super-family. Mammalian ORs are disposed in clusters on virtually all chromosomes. They are encoded by the largest multigene family ($\sim$1000 members) in the genome of mammals and *Caenorhabditis elegans*, whereas *Drosophila* contains only 60 genes. Each OR specifically recognizes a set of odorous molecules that share common molecular features. However, local mutations affect the DNA sequences of these receptors. Hence, to study the changes among affected and non-affected, we use unsupervised learning (clustering). In this paper, a scaled fuzzy graph model for clustering has been used to study the changes before and after the local mutation on DNA sequences of ORs. At the fractional dimensional level, our experimental study confirms its accuracy.

Keywords Olfactory receptors · Clustering · Fuzzy graph model · Fractional dimension

S. R. Dash (✉) · U. K. Sahoo
School of Computer Application, KIIT University, Bhubaneswar 751024, India
e-mail: satyaranjan.dash@gmail.com; sdashfca@kiit.ac.in

U. K. Sahoo
e-mail: umakant.iitkgp@gmail.com

S. Dehuri
Department of Systems Engineering, Ajou University, San 5, Woncheon-dong,
Yeongtong-gu , Suwon 443-749, Republic of Korea
e-mail: satch@ajou.ac.kr

G. N. Wang
Department of Industrial Engineering, Ajou University, San 5,Woncheon-dong,
Yeongtong-gu , Suwon 443-749, Republic of Korea
e-mail: gnwang@ajou.ac.kr

B. V. Babu et al. (eds.), *Proceedings of the Second International Conference on Soft Computing for Problem Solving (SocProS 2012), December 28–30, 2012*, Advances in Intelligent Systems and Computing 236, DOI: 10.1007/978-81-322-1602-5_93, © Springer India 2014

1 Introduction

The olfactory system [13] has the notable capability to discriminate a wide range of odor molecules. In humans, smell is rather considered to be an esthetic sense in contrast to most other species, which rely on olfaction to detect food, predators, and mates. Terrestrial animals, including humans, smell air-borne molecules, whereas aquatic animals smell water-soluble molecules with low volatility, such as amino acids. Humans are thought to have a poor olfactory ability compared with other animals such as dog or rodents, and yet they can perceive a vast number of volatile chemicals. Of the millions of volatile molecular species that have been catalogued by chemists, hundreds of thousands of distinct odors can be detected by the human nose. Odorants, typically small organic molecules of less than 400 Da, can vary in size, shape, functional groups, and charge. They include a set of various alcohols, aliphatic acids, aldehydes, ketones and esters; chemicals with aromatic, alicyclic, polycyclic, or heterocyclic ring structures; and innumerable substituted chemicals of each of these types, as well as combinations of them. However, subtle differences in the structure of an odorant, even between two enantiomers, can lead to pronounced modifications in odor quality. In this paper our special focus is on to study the changes of OR before or after local mutations. Through clustering analysis on DNA sequences of OR this work can detect the changes.

Clustering has been a folklore problem in the area like Bioinformatics, data mining, pattern recognition, image analysis, etc. Clustering techniques used in many applications are either dominated by distance based or connectivity based. A few alike algorithms have been used in [3]. However, the other popular category of clustering based on graph theory approach also igniting many researchers for application in bioinformatics field.

Graph-based clustering is the task of grouping the vertices of the graph into clusters taking into consideration the edge structure of the graph in such a way that there should be many edges within each cluster and relatively few between the clusters.

Recent study [4] shows that the fuzzy graph [1, 3] approach is more powerful in cluster analysis than the usual graph theoretic approach due to its ability to handle the strengths of arcs effectively. Fuzzy graph models can be used to solve various practical issues in clustering analysis, network analysis, information theory, database theory, etc. [7]. Obviously, this relies on the fact that these problems should involve uncertainty to get better results. More about fuzzy graph theory can be obtained in [5, 6].

In this paper we have used the scaled version of fuzzy graph model for clustering of DNA sequences of OR. Based on their fractal dimensions, before or after mutations, we study the difference in the clustering of these receptors. To increase the compatibility of the data, instead of considering a 0–1 scale we considered a 0-max scale.

2 Background of the Research

2.1 Clustering

A cluster is a set of nodes. Clustering can be considered the most important unsupervised learning problem; so, as every other problem of this kind, it deals with finding a structure in a collection of unlabeled data. Clustering is the task of assigning a set of nodes into groups, so that the nodes in the equivalent cluster are more related to each other than to those in other clusters. Typical cluster models include:

Connectivity models: build models based on distance connectivity.

- Centroid models: symbolize every cluster by a distinct mean vector.
- Distribution models: clusters are modeled using statistic distributions.
- Density models: defines clusters as associated broad regions in the data space.
- Subspace models: clusters are modeled with both cluster members and relevant attributes.
- Group models: some algorithms do not offer a superior model for their results and just provide the combination in sequence.
- Graph-based models: a subset of nodes in graph such that every two nodes in the subset are connected by an edge can be considered as a prototypical form of cluster.

Clustering's can be roughly distinguished in hard clustering where each object belongs to a cluster or not and soft clustering where each object belongs to each cluster to a certain degree. When a clustering outcome is evaluated based on the records that was clustered itself, this is called internal evaluation. These methods usually assign the best score to the algorithm that produces clusters with high comparison within a cluster and low similarity between clusters. For more information reader may refer [8].

2.2 Fuzzy Graph Model

Fuzzy graph theory is interplay between graph theory and theory of fuzzy sets introduced by Zadeh [2]. Rosenfeld, in his paper Fuzzy Graphs, presented the basic structural and connectivity concepts, while Yeh and Bang introduced different connectivity parameters of a fuzzy graph and discussed their applications in the paper titled Fuzzy relations, Fuzzy graphs, and their applications to clustering analysis.

A graph is a pair $G = (V, E)$ of sets such that E is a subset of V^2, thus, the elements of E are 2-elements subsets of V. If we associate weights w to each of the edges e_i, we get a weighted graph. A fuzzy graph is a pair of functions $G(\sigma, \mu)$ where σ is a fuzzy subset of V and μ is a symmetric fuzzy relation on σ. This has been studied in [4–6] (Figs. 1 and 2).

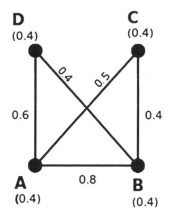

Fig. 1 Fuzzy Graph

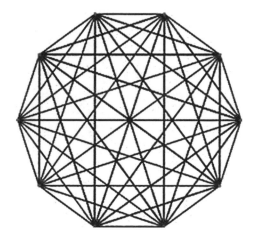

Fig. 2 Fuzzy graph of the Protein Sequence

What σ does is assign a number between 0 and 1 to each node and μ assigns a number between 0 and 1 to each edge present in the fuzzy graph. The following diagram shows a fuzzy graph on 4 vertices and 5 edges with their respective σ and μ values.

Fuzzy graph theory has numerous applications in various fields like clustering analysis, database theory, network analysis, information theory, etc. [7]. Research on this has witnessed an exponential growth in mathematics and its applications, since it fosters the possibility of interdisciplinary research.

2.3 Fractional Dimension

In topological geometry, we have dimension of a point, line, plane, and space as 0, 1, 2, and 3 respectively. This is called the topological dimension. Formally, we call an object as n-dimensional if it can be represented by n independent variables. When we deal with more than one objects grouped together, the dimension of such a collection is the maximum of dimensions of the individual objects. However if this collection is infinite, the dimension grows.

Topological dimensions take care of the issues with dimensionality of regular objects. Nevertheless, this cannot be used to describe the dimensionality of fractals. For example the Koch snowflake has a topological dimension of 1 but is no way just a curve. Also it is not a (part of) plane, hence should not have a dimension of 2. In a way, it seems that it is too big to have a dimension of 1 and small to have a dimension of 2. So, a number between 1 and 2 would be an appropriate indicator of its dimensionality. The formulation of the above idea was the inception of the concept of fractal dimension.

3 Dataset and Preprocessing

We collect the DNA sequences and protein sequences of various Olfactory receptors from the ORDB [9]. Then, we mutated the DNA sequences and converted it into protein sequences using a translator [10]. Now we processed the before and after protein sequences through the protein structure analysis of the Bioinformatics toolbox of MATLAB. Then we processed the obtained graphics files in BENOIT [11], a fractal analysis software. The resultant was a set of fractal dimensions of the various properties of the protein sequence.

4 Our Work

In the initial phase of this work we have prepared the data for cluster analysis. From the prepared dataset we collected the fractional dimensions of 10 OR DNA sequences. We selected the fractal dimensions of the nine properties of each of the proteins. Now each protein is represented by a vector (v_i) with nine parameters.

$$v_i = (\text{fd}_1, \ \text{fd}_2, \ \text{fd}_3, \ \text{fd}_4, \ \text{fd}_5, \ \text{fd}_6, \ \text{fd}_7, \ \text{fd}_8, \ \text{fd}_9)$$

We now use the Euclidean transformation E: $\mathbb{R}^9 \to \mathbb{R}$ to represent each protein vector v_i where,

$$\text{E}(v_i) = \text{length of the vector } v_i = \text{rms} \, [v_i]$$

This is done for each protein sequence before and after mutation. Now each protein sequence has two representative fractal dimensions, one before and one after the mutation. So now we have two representative fractal dimensions of each protein sequence. We then separated the before and after data. We applied the following techniques to both datasets.

First we formed a 10*10 matrix A, such that $A[i, j] = 10000 \left(fd_i - fd_j\right)$ where fd_i is the representative fractal dimension of v_i. The number 10,000 was multiplied to scale the small numbers $(fd_i - fd_j)$ to a perceivable scale. We then found out the cohesive matrix $matB$ from the matrix A. Note that the elements of A are not fuzzy parameters, but we use the same techniques to find $matB$, since scalability has no effect on relative clustering. We then found the α-cut matrix. We store this α-cut matrix as cut $matB(\alpha)$

matrixBefore=

0	0.244396	0.455526	0.788137	1.22191	1.967	1.84422	2.49999	2.7113	4.04374
0.244396	0	0.21113	1.03253	1.46631	1.7226	2.08862	2.25559	2.4669	4.28814
0.455526	0.21113	0	1.24366	1.67744	1.51147	2.29975	2.04446	2.25577	4.49927
0.788137	1.03253	1.24366	0	0.433773	2.75514	1.05609	3.28812	3.49944	3.25561
1.22191	1.46631	1.67744	0.433773	0	3.18891	0.622312	3.7219	3.93321	2.82183
1.967	1.7226	1.51147	2.75514	3.18891	0	3.8112	0.532986	0.744302	6.01074
1.84422	2.08862	2.29975	1.05609	0.622312	3.81122	0	4.34421	4.55552	2.19952
2.49999	2.25559	2.04446	3.28812	3.7219	0.532986	4.34421	0	0.211315	6.54373
2.7113	2.4669	2.25577	3.49944	3.93321	0.744302	4.55552	0.211315	0	6.75504
4.04374	4.28814	4.49927	3.25561	2.82183	6.01074	2.19952	6.54373	6.75504	0

Similar technique is applied to the data obtained from after the mutation.

matrixAfter=

0	2.76695	3.19985	4.04418	3.4556	3.62222	4.49996	1.5668	1.77779	1.95573
2.76695	0	0.432901	1.27724	0.68865	0.855271	1.73301	1.20015	0.989153	0.81122
3.19985	0.432901	0	0.844335	0.255749	0.42237	1.30011	1.63305	1.42205	1.24412
4.04418	1.27724	0.844335	0	0.588586	0.421965	0.455776	2.47739	2.26639	2.08846
3.4556	0.68865	0.255749	0.588586	0	0.166621	1.04436	1.8888	1.6778	1.49987
3.62222	0.855271	0.42237	0.421965	0.166621	0	0.877741	2.05542	1.84442	1.66649
4.49996	1.73301	1.30011	0.455776	1.04436	0.877741	0	2.93316	2.72217	2.54423
1.5668	1.20015	1.63305	2.47739	1.8888	2.05542	2.93316	0	0.210997	0.38893
1.77779	0.989153	1.42205	2.26639	1.6778	1.84442	2.72217	0.210997	0	0.177934
1.95573	0.81122	1.24412	2.08846	1.49987	1.66649	2.54423	0.38893	0.177934	0

5 Simulation Results

In our simulation studies by varying the value of alpha, we found out the alpha cut matrix and then corresponding clusters. For $\alpha = 4$, we found that before mutation the clusters were (1), (2), (3), (4), (5) and (6,7,8,9,10), whereas after clustering it turned out to be (1), (2), (3), (4), (5), (6), (7), (8), (9), and (10). The fuzzy clusters of all levels are summarized in Table 1 (for $\alpha = 1$ to 7).

Table 1 Clustering results for various α cuts

α cut	Clustering before Mutation	Clustering after Mutation
1	(1, 2, 3, 4, 5, 6, 7, 8, 9, 10)	(1, 2, 3, 4, 5, 6, 7, 8, 9, 10)
2	(1, 2, 3, 4, 5, 6, 7, 8, 9, 10)	(1, 2, 3, 4, 5, 6, 7, 8, 9, 10)
3	(1, 2, 3, 4, 5, 6, 7, 8, 9, 10)	(3, 4, 5, 6, 7),(1),(2),(8),(9),(10)
4	(1, 2, 3, 6, 7, 8, 9, 10),(4),(5)	(4, 7),(1),(2),(3),(5),(6),(8),(9),(10)
5	(6, 7, 8, 9, 10),(1),(2),(3),(4),(5)	(1), (2), (3), (4), (5), (6), (7), (8), (9), (10)
6	(6, 7, 8, 9, 10), (1), (2), (3), (4), (5)	(1), (2), (3), (4), (5), (6), (7), (8), (9), (10)
7	(1), (2), (3), (4), (5), (6), (7), (8), (9), (10)	(1), (2), (3), (4), (5), (6), (7), (8), (9), (10)

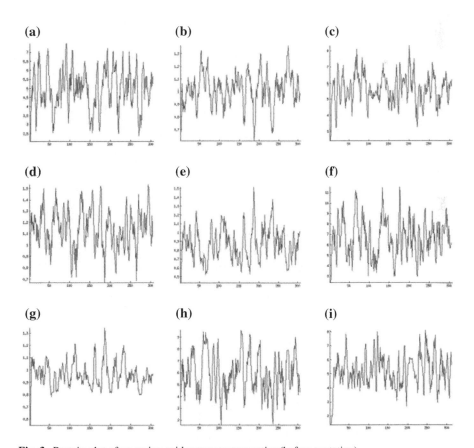

Fig. 3 Protein plot of an amino acid sequence properties (before mutation)

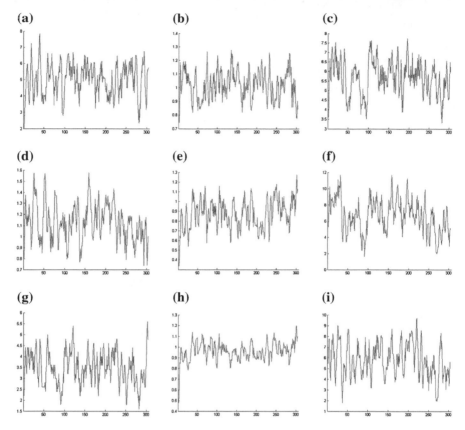

Fig. 4 Protein plot of an amino acid sequence properties (after mutation)

CutMat$B$ (4) =

$$
\begin{pmatrix}
0 & 0 & 0 & 0 & 0 & 0 & 0 & 0 & 0 & 0 \\
0 & 0 & 0 & 0 & 0 & 0 & 0 & 0 & 0 & 0 \\
0 & 0 & 0 & 0 & 0 & 0 & 0 & 0 & 0 & 0 \\
0 & 0 & 0 & 0 & 0 & 0 & 0 & 0 & 0 & 0 \\
0 & 0 & 0 & 0 & 0 & 0 & 0 & 0 & 0 & 0 \\
0 & 0 & 0 & 0 & 0 & 0 & 1 & 1 & 1 & 1 \\
0 & 0 & 0 & 0 & 0 & 1 & 0 & 1 & 1 & 1 \\
0 & 0 & 0 & 0 & 0 & 1 & 1 & 0 & 1 & 1 \\
0 & 0 & 0 & 0 & 0 & 1 & 1 & 1 & 0 & 1 \\
0 & 0 & 0 & 0 & 0 & 1 & 1 & 1 & 1 & 0
\end{pmatrix}
$$

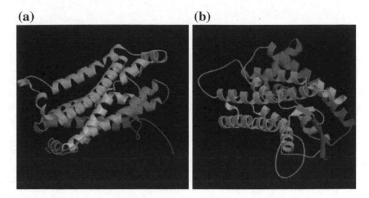

Fig. 5 3D Protein structure of an amino acid before and after mutation

CutMat$A$ (4) =

$$\begin{pmatrix}
0\ 0\ 0\ 0\ 0\ 0\ 0\ 0\ 0\ 0 \\
0\ 0\ 0\ 0\ 0\ 0\ 0\ 0\ 0\ 0 \\
0\ 0\ 0\ 0\ 0\ 0\ 0\ 0\ 0\ 0 \\
0\ 0\ 0\ 0\ 0\ 0\ 0\ 0\ 0\ 0 \\
0\ 0\ 0\ 0\ 0\ 0\ 0\ 0\ 0\ 0 \\
0\ 0\ 0\ 0\ 0\ 0\ 0\ 0\ 0\ 0 \\
0\ 0\ 0\ 0\ 0\ 0\ 0\ 0\ 0\ 0 \\
0\ 0\ 0\ 0\ 0\ 0\ 0\ 0\ 0\ 0 \\
0\ 0\ 0\ 0\ 0\ 0\ 0\ 0\ 0\ 0 \\
0\ 0\ 0\ 0\ 0\ 0\ 0\ 0\ 0\ 0
\end{pmatrix}$$

Figures 3 and 4 shows the difference in the fractal dimensions of the amino acid sequence properties. Figure 5 illustrates the structural changes of before and after local mutation of the Olfactory Receptors.

6 Conclusion

As a result of intensive interdisciplinary research to characterize the mechanisms underlying olfaction, the understanding of sensory systems has impressively grown in recent years. Without loss of generality the study of the changes of ORs before and after local mutation through fuzzy graph model, our simulation result confirms that a local mutation in the DNA sequence of olfactory receptors brings in a huge difference in their clustering. It is also evident from Figure 5 that there has been a change in the structure of the mutated protein.

References

1. Rosenfeld, A.: Fuzzy graphs. In: Zadeh, L.A., Fu, K.S., Shimura, M. (eds.), Fuzzy Sets and Their Applications to Cognitive and Decision Processes, Academic Press, New York, pp. 77–95(1975).
2. Zadeh, L.A.: Fuzzy sets. Inform. Control **8**, 338–353 (1965)
3. Yeh R.T., Bang S.Y., Fuzzy relations, Fuzzy relations, fuzzy graphs and their applications to clustering analysis. In: Zadeh L.A., Fu K.S., Shimura M. (eds.), Fuzzy Sets and Their Applications, Academic Press, pp. 125–149(1975).
4. Mathew, S., Sunitha, M.S.: Node connectivity and arc connectivity of a fuzzy graph. Inform. Sci. **180**, 519–531 (2010)
5. Bhattacharya, P.: Some remarks on fuzzy graphs. Pattern Recogn. Lett. **6**, 297–302 (1987)
6. Bhattacharya, P., Suraweera, F.: An algorithm to compute the maxmin powers and a property of fuzzy graphs. Pattern Recogn. Lett. **12**, 413–420 (1991)
7. Mordeson, J.N.: Nair. Fuzzy graphs and fuzzy hypergraphs. Physica-Verlag, P.S. (2000)
8. Diday, E., Simon, J.C.: Clustering analysis. In: Fu, K.S. (ed.) Digital Pattern Recognition, pp. 47–94. NJ, Springer-Verlag, Secaucus (1976)
9. Olfactory Receptors Database. http://senselab.med.yale.edu/ordb/
10. DNA to protein translation. http://insilico.ehu.es/translate/
11. BENOIT. http://www.trusoft-international.com/
12. Protein Structure Prediction Server. http://ps2.life.nctu.edu.tw/index.php
13. Gaillard, I., Rauquier, S., Giorgi, D.: Olfactory Receptors. ICMLS Cell. Mol. Life Sci. **61**, 456–469 (2004)

Linear Hopfield Based Optimization for Combined Economic Emission Load Dispatch Problem

J. P. Sharma and H. R. Kamath

Abstract In this paper a linear Hopfield model is used to solve the problem of combined economic emission dispatch (CEED). The objective function of CEED problem comprises of power mismatch, total fuel cost and total emission subjected to equality/inequality constraints. In proposed methodology, inclusion of power mismatch in objective function exhibits the ability of attaining power mismatch to any desirable extent and may be employed for large-scale highly constrained nonlinear and complex systems. A systematic procedure for the selection of weighting factor adopted. The proposed method employs a linear input-output model for neurons. The efficacy and viability of the proposed method is tested on three test systems and results are compared with those obtained using other methods. It is observed that the proposed algorithm is accurate, simple, efficient, and fast.

Keywords Economic dispatch · Emission · Power mismatch · Modified price penalty factor · Linear hopfield model

1 Introduction

Emissions produced by fossil-fueled electric power plants release many contaminants like sulfur oxides (SO_2) and oxides of nitrogen (NO_x) and directly effects on human beings, plants and animals etc. In recent years, due to strict environmental regulations, emission control has become one of the important operational objectives of power systems and as result of it thermal power plants cannot only be run at

J. P. Sharma (✉)
The Electrical Engineering, Department of JK, Lakshmipat University, Jaipur, India
e-mail: jpsharma.cseb@gmail.com

H. R. Kamath
The Malwa Institute of Technology, Indore, India
e-mail: rskamath272@gmail.com

B. V. Babu et al. (eds.), *Proceedings of the Second International Conference on Soft Computing for Problem Solving (SocProS 2012), December 28–30, 2012*, Advances in Intelligent Systems and Computing 236, DOI: 10.1007/978-81-322-1602-5_94, © Springer India 2014

absolute minimum fuel cost criterion. So, it become evitable to consider total fuel cost (operating cost) and total pollutant emission as objective function in power system optimization problems. In power system optimization problems, emissions can be considered either in the objective function or treated them as additional constraints, which is called combined economic emission load dispatch (CEED) and reliable and useful planning for power system optimization. The main aim of the combined economic emission load dispatch (CEED) problem is to find an optimal combination of the output power of all the online generating units that simultaneously minimizes both fuel cost and pollutant emissions, while satisfying unit constraints, equality and inequality constraints. The recent literature shows that many mathematical and different heuristic techniques such as Particle Swarm Optimization (PSO), Hopfield Neural Network, Ant Colony Optimization (ACO), Artificial Bee Colony (ABC), Genetic Algorithm (GA) and Differential Evolution (DE) gives solution very easily in terms of machine usage and time of computation etc. In this paper, objective function of CEED problem consists of three terms, power mismatch, total fuel cost and total emission cost. Each term is multiplied by a weighting factor, which represents the relative importance of that term. These weighting factors are named as A, B and C respectively. A systematic procedure for evaluating the value of weighting factors, associated with the various terms of the energy function is developed. This work focuses on emission of nitrogen oxides (NO_x) only, because its control is a significant issue at the global level. The suitability of hopfield neural network [1] for a CEED problem is investigated and a linear function for neuron's input-output characteristics is proposed to reduce computational requirement. Because sigmoidal function for neuron's input-output characteristics utilizes the iterative procedures to solve the optimization problem gives large amount of computational requirement. The proposed method is applied to a three [2, 3] and six [3] generating unit system and the results obtained are compared with those obtained using Hybrid Cultural Algorithm [2], Genetic Algorithm [2], PSO [2], ABC [3], Conventional Method [3], Hybrid Genetic Algorithm [3], Hybrid GTA [3] and Simulated Annealing Algorithm [4]. The computational results reveal that Proposed Hopfield Method is simple, accurate, efficient and straightforward.

2 Multi-Objectives Combined Economic Emission Dispatch

2.1 Economic Objective

$$F_1 = \sum_{i=1}^{n} \left(a_i + b_i P_i + c_i P_i^2 \right) \tag{1}$$

2.2 Environmental Objective

$$F_2 = \sum_{i=1}^{n} \left(\alpha_i + \beta_i P_i + \gamma_i P_i^2 \right) \tag{2}$$

2.3 Problem Formulation

In this paper CEED problem is the combination of three single objective functions like power mismatch, total fuel cost and emission level and it is formulated as to minimize $f(P_m, F_1, F_2)$ subject to the equality and inequality constraints.

$$E = \frac{A}{2} \left[(P_D + P_L) - \sum_{i}^{n} P_i \right]^2 + \frac{B}{2} \left[\sum_{i=1}^{n} \left(a_i + b_i P_i + c_i P_i^2 \right) \right]$$

$$+ \frac{C}{2} \left[\sum_{i=1}^{n} \left(\alpha_i + \beta_i P_i + \gamma_i P_i^2 \right) \right]$$

$$E = \frac{A}{2} P_m^2 + \frac{B}{2} F_1 + \frac{C}{2} F_2 \tag{3}$$

2.3.1 Equality Constraints

$$P_m = P_D + P_L - \sum_{i=1}^{n} P_i \tag{4}$$

where

$$P_L = \sum_{i=1}^{n} \sum_{j=1}^{n} P_i B_{ij} P_j \tag{5}$$

2.3.2 Equality Constraints

$$P_{i\,\text{min}} \leq P_i \leq P_{i\,\text{max}} \tag{6}$$

The value of weighting factor 'C' is equal to modified price penalty factor (h). In this paper a modified price penalty factor [5, 6] is used for the particular load demand and calculated by following equation

$$Z = h_{i1} + \frac{(h_{i2} - h_{i1})}{(p_{\max 2} - p_{\max 1})} * (P_D - P_{\max 1}) \tag{7}$$

In this paper weighting factor 'A' is calculated using Eq. 21 by varying B from 0.4 to 1.25 along with calculated value of weighting factor 'C' by Eq. 7 at a particular load demand.

2.4 Mapping of CEED Problem on to Hopfield Model

In this paper, continuous Hopfield Neural Network is considered and dynamic characteristic of each neuron is described by the following equation [1].

$$\frac{dU_i}{dt} = \sum_{j}^{n} T_{ij} V_i + I_i \tag{8}$$

The energy function of continuous Hopfield model [1] is defined as

$$E = \left(-\frac{1}{2}\right) \sum_{i=1}^{n} \sum_{j=1}^{n} T_{ij} V_i V_j - \sum_{i=1}^{n} I_i V_i \tag{9}$$

By comparing Eq. 3 with Eq. 9, parameters for neurons in terms of weighting factors for aforesaid objective function are defined as follows

$$T_{ii} = -A - Bc_i - C\gamma_i \tag{10}$$

$$T_{ij} = -A \tag{11}$$

$$I_i = A(P_D + P_L) - \frac{B}{2}b_i - \frac{C}{2}\beta_i \tag{12}$$

2.5 Proposed Linear Hopfield Model

The neuron model chosen is a linear input output model as shown in Fig. 1, instead of conventional sigmoid model.

$$P_i = K_{1i}U_i + K_{2i} \ \forall \ U_{\min} \leq U_i \leq U_{\max} \tag{13}$$

Fig. 1 The linear input-output function

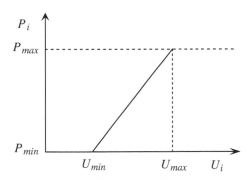

where

$$K_{1i} = \frac{(P_{i\,max} - P_{i\,min})}{(U_{max} - U_{min})} \tag{14}$$

$$K_{2i} = -K_{1i} U_{min} + P_{i\,min}$$

2.6 Computational Expressions for CEED Problem

The dynamic equation [1] of a neuron is

$$\frac{dU_i}{dt} = \sum_{j}^{n} T_{ij} P_j + I_i \tag{15}$$

After substituting different parameter from Eqs. 10, 11 and 12 in Eq. 15, dynamic equation of a neuron is as follows

$$\frac{dU_i}{dt} = A P_m - \frac{B}{2} \left(\frac{dF_1}{dP_i} \right) - \frac{C}{2} \left(\frac{dF_2}{dP_i} \right) \tag{16}$$

putting values from Eqs. 1, 2 and 13 in to Eq. 16, the dynamic Eq. 16 becomes

$$\frac{dU_i}{dt} = A P_m - \frac{B}{2} [b_i + 2c_i (K_{1i} U_i + K_{2i})] - \frac{C}{2} [\beta_i + 2\gamma_i (K_{1i} U_i + K_{2i})] \tag{17}$$

$$= K_{4i} + K_{3i} U_i$$

where

$$K_{3i} = -K_{1i} \left(Bc_i + C\gamma_i \right) = - \left(Bc_i + C\gamma_i \right) \frac{(P_{i \max} - P_{i \min})}{(U_{\max} - U_{\min})}$$

$$K_{4i} = AP_m - \frac{B}{2} b_i - Bc_i K_{2i} - \frac{C}{2} \beta_i - C\gamma_i K_{2i}$$

Neurons input function U_i is obtained by solving Eq. 17 as follows

$$U_i = \left[U_i^0 + K_{4i} \div K_{3i} \right] e^{K_{3i}^t} - \left[K_{4i} \div K_{3i} \right] \tag{18}$$

By putting neurons input function U_i in Eq. 13, the neuron's output function

$$P_i = \left[\left[K_{1i} U_i^0 + K_{2i} \right] - \left(\frac{2K_{AB} P_m - b_i - K_{CB} \beta_i}{2c_i + 2K_{CB} \gamma_i} \right) \right] e^{K_{3i}t}$$
$$+ \left(\frac{2K_{AB} P_m - b_i - K_{CB} \beta_i}{2c_i + 2K_{CB} \gamma_i} \right) \tag{19}$$

where $K_{AB} = \frac{A}{B}$ and $K_{CB} = \frac{C}{B}$

At $t = \infty$ the steady state value of Eq. 19 give optimum power generation

$$P_i^* = \left(\frac{2K_{AB} P_m - b_i - K_{CB} \beta_i}{2c_i + 2K_{CB} \gamma_i} \right) \tag{20}$$

Optimum power generation does not depend on P_i^0 and $(U_{\max} - U_{\min})$, which is reasonable and comprehensible. From Eqs. 13, 20 and 19, a more simple formula for K_{AB} can be written as

$$K_{AB} = \frac{\left[(P_D + P_L) + \frac{1}{2} \sum_{i=1}^n \left\{ \frac{(b_i + K_{CB}\beta_i)}{(c_i + K_{CB}\gamma_i)} \right\} \right]}{\left[P_m \sum_{i=1}^n \left\{ \frac{1}{(c_i + K_{CB}\gamma_i)} \right\} \right]} \tag{21}$$

3 Simulation Results

The applicability and validity of the proposed algorithm has been tested on three different test systems (test system-I [2], test system-II [3] and test system-II [3]) and results are presented in Tables 1, 2, 3, 4, 5, and 6.

Table 1 Best compromise solution by proposed method for combined economic emission dispatch (CEED)-test system-I

Particulars	$P_D = 650\,MW$	$P_D = 700\,MW$	$P_D = 750\,MW$
Power mismatch factor (A)	458488.7621	469375.5820	477456.8377
Fuel cost factor (B)	0.60	0.56	0.52
Price plenty factor (h)	35.1840	33.7478	32.3117
P_1 (MW)	168.0304	181.4098	194.8724
P_2 (MW)	251.3413	271.1990	291.1563
P_3 (MW)	250.6923	270.7759	290.9497
Fuel cost (Rs/hr)	32885.40	35461.14	38094.31
Emission level (Kg/hr)	551.34	651.66	762.36
Total cost (Rs/hr)	52283.84	57453.20	62727.49
Power loss (MW)	20.0640	23.3848	26.9784
Total generation (MW)	670.0640	723.3847	776.9784
Escape time (sec)	0.842	0.606	0.66

Table 2 Comparison of combined economic emission dispatch (CEED) results for test system-I

P_D	Performance	Hybrid CA [2]	Proposed method
650 MW	P_1 (MW)	157.5395	168.0304
	P_2 (MW)	248.7636	251.3413
	P_3 (MW)	264.0454	250.6923
	Fuel cost (Rs/hr)	32876.00	32885.40
	Emission level (Kg/hr)	553.9684	551.34
	Power loss (MW)	20.2448	20.0640
	Total generation (MW)	670.3485	670.0640
700 MW	P_1 (MW)	182.7369	181.4098
	P_2 (MW)	276.0215	271.1990
	P_3 (MW)	264.6389	270.7759
	Fuel cost (Rs/hr)	35466.00	35461.14
	Emission level (Kg/hr)	651.9255	651.66
	Power loss (MW)	23.3498	23.3848
	Total generation (MW)	723.5368	723.3847
750 MW	P_1 (MW)	192.8946	194.8724
	P_2 (MW)	292.1908	291.1563
	P_3 (MW)	291.9240	290.9497
	Fuel cost (Rs/hr)	38089.00	38094.31
	Emission level (Kg/hr)	762.5753	762.36
	Power loss (MW)	27.0066	26.9784
	Total generation (MW)	777.4081	776.9784

Table 3 Comparison of results by different methods for combined economic emission dispatch (CEED) with a demand of 650 MW for test system-I

Performance	Hybrid CA [2]	GA [2]	PSO [2]	Proposed method
Fuel cost (Rs/hr)	32876.00	32888.6	32888.0	32885.40
Emission level (Kg/hr)	553.9684	551.299	551.274	551.34
Power loss (MW)	20.2448	20.0477	20.0466	20.0640

Table 4 Comparison of combined economic emission dispatch (CEED) results for test system-II

P_D	Performance	Conventional [4]	SGA [4]	RGA [4]	SA [4]	ABC [3]	Proposed method
400 MW	Fuel cost (Rs/hr)	20898.83	20801.94	20801.81	20838	20838	20835.53
	Emission level (Kg/hr)	201.50	201.35	201.21	200.22	200.2211	200.27
	Total cost (Rs/hr)	29922	29820	29812	29804.17	29804.2	27024.07
	Power loss (MW)	7.41	7.69	7.39	7.41	7.5681	7.4085
	Escape time (sec)	–	–	–	4.945824	1.895377	1.17
500 MW	Fuel cost (Rs/hr)	25486.64	25474.56	25491.64	25495	25495	25484.50
	Emission level (Kg/hr)	312.00	311.89	311.33	311.16	311.1553	311.55
	Total cost (Rs/hr)	39458	39441	39433	39428.25	39428.3	30014.49
	Power loss (MW)	11.88	11.80	11.70	11.69	11.6937	11.7140
	Escape time (sec)	–	–	–	5.808751	2.340603	0.651
700 MW	Fuel cost (Rs/hr)	35485.05	35478.44	35471.48	35464	35464	35459.54
	Emission level (Kg/hr)	652.55	652.04	651.60	651.58	651.5775	651.59
	Total cost (Rs/hr)	66690	66659	66631	66622.52	66622.52	57450.21
	Power loss (MW)	23.37	23.29	23.28	23.37	23.3664	23.3485
	Escape time (sec)	–	–	–	5.174793	3.770030	0.639

Table 5 Best compromise solution by proposed method for combined economic emission dispatch (CEED)-test system-II

Particulars	$P_D = 400$ MW	$P_D = 500$ MW	$P_D = 700$ MW
Power mismatch factor (A)	318544.5134	265374.9454	466933.7337
Fuel cost factor (B)	0.83	0.77	0.55
Price plenty factor (h)	30.9003	14.54038	33.7494
P_1 (MW)	101.4039	124.4383	181.4495
P_2 (MW)	154.2941	195.3598	271.1577
P_3 (MW)	151.7105	191.9158	270.7413
Total generation (MW)	407.4085	511.7140	723.3485

Table 6 Comparison of combined economic emission dispatch (CEED) results for test system-III

P_D	Performance	Conventional [3]	RGA [3]	HGA [3]	Hybrid GTA [3]	SA [4]	ABC [3]	Proposed method
500 MW	Fuel cost (Rs/hr)	27638.30	27692.1	27695	27613.4	27613	27613	27608.16
	Emission level (Kg/hr)	262.454	263.472	263.37	263.00	263.01	263.012	263.14
	Total cost (Rs/hr)	39159.5	39258.10	39257.5	39158.9	39156.9	39156.9	38620.37
	Power loss (MW)	8.830	10.172	10.135	8.930	8.943	8.9343	8.9550
	Escape time (sec)	–	–	–	–	4.887728	4.11105	0.667
900 MW	Fuel cost (Rs/hr)	48892.90	48567.7	48567.5	48360.9	48351	47045.3	47950.14
	Emission level (Kg/hr)	701.428	694.169	694.172	693.570	693.79	693.791	705.05
	Total cost (Rs/hr)	82436.58	81764.5	81764.4	81529.1	81527.6	81527.6	67609
	Power loss (MW)	35.230	29.725	29.178	28.004	28.01	28.0087	28.5528
	Escape time (sec)	–	–	–	–	5.914351	3.94864	0.629

4 Conclusions

A linear Hopfield Neural model for the solution of CEED problem has been proposed and investigated. Proposed method exhibits the ability of attaining power mismatch to any desirable extent due to consideration of mismatch power in objective function, which is more meaningful for large scale generating system. For applying Hopfield neural network to CEED Problem, proper selection of weighting factors associated with the energy function are required, which is a cumbersome task. This paper employs a systematic procedure for evaluating the value of weighting factors, associated with the various terms of the energy function. The proposed method has been applied to three test systems and it's applicability for solving multi objective generation dispatch problem is showed by obtained results, which is very comparable to the optimization algorithms, such as conventional method, RGA and SGA, Hybrid GA,PSO , ABC and SA algorithm. Thus, proposed algorithm is accurate, simple, efficient, and fast.

References

1. Su, C.T., Chiou, G.J.: A fast-computation hopfield method to economic dispatch. IEEE Trans. Power Syst. **12**(4), 1759 (1997)
2. Bhattacharya, B., Mandal, K.K., Chakraborty, N.: A hybrid cultural approach for combined economic and emission dispatch. Annual IEEE India Conference (INDICON), 16–18 (2011).
3. Gaurav Prasad, D., Hari Mohan, D., Manjaree, P., Panigrahi, B.K.: Artificial bee colony optimization for combined economic load and emission dispatch. Second International Conference on Sustainable Energy and Intelligent System (SEISCON), 340–345 (2011).
4. Kaurav, M.S., Dubey, H.M., Manjaree, P., Panigrahi, B.K.: Simulated annealing algorithm for combined economic and emission dispatch. UACEE Int. J. Adv. Electron. Eng. **1**, 172–178 (2011)
5. Venkatesh, P., Gnanadass, R.: Narayana Prasad, P.: Comparison and application of evolutionary programming techniques to combined economic emission dispatch with line flow constraints. IEEE Trans. Power Syst. **18**(2), 688 (2003)
6. Gupta, A., Swarnkar, K.K., Wadhwani, S., Wadhwani, A.K.: Combined economic emission dispatch problem of thermal generating units using particle swarm optimization. Int. J. Sci. Res. Publ. **2**(7), 1 (2012)

Soft Computing Approach for VLSI Mincut Partitioning: The State of the Arts

Debasree Maity, Indrajit Saha, Ujjwal Maulik and DariuszPlewczynski

Abstract Recent research shows that the partitioning of VLSI-based system plays a very important role in embedded system designing. There are several partitioning problems that can be solved at all levels of VLSI system design. Moreover, rapid growth of VLSI circuit size and its complexity attract the researcher to design various efficient partitioning algorithms using soft computing approaches. In VLSI partitioning, *netlist* is used to optimize the parameters like mincut, power consumption, delay, cost, and area of the partitions. Hence, the Genetic Algorithm is a soft computational meta-heuristic method that has been applied to optimize these parameters over the past two decades. Here in this paper, we have summarized important schemes that have been adopted in Genetic Algorithm for optimizing one particular parameter, called *mincut*, to solve the partitioning problem.

Keywords Genetic algorithm · Mincut · Netlist · Survey · VLSI circuit partitioning

D. Maity
Department of Electronics and Communication Engineering, MCKV Institute
of Engineering, Howrah, Liluah 711204, India
e-mail: debasree.maitysaha@gmail.com

I. Saha (✉) · U. Maulik
Department of Computer Science and Engineering, Jadavpur University,
Kolkata 700032, India
e-mail: indra@icm.edu.pl

D. Plewczynski
Interdisciplinary Centre for Mathematical and Computational Modelling,
University of Warsaw, 02106 Warsaw, Poland
e-mail: darman@icm.edu.pl

B. V. Babu et al. (eds.), *Proceedings of the Second International Conference on Soft Computing for Problem Solving (SocProS 2012), December 28–30, 2012*, Advances in Intelligent Systems and Computing 236, DOI: 10.1007/978-81-322-1602-5_95, © Springer India 2014

1 Introduction

With the rapid increment and development of modern electronic circuit technology and its fabrication make the very large-scale integration (VLSI) circuit so dense that a single silicon chip may contain millions of circuit modules, like transistors, resistors, capacitors, etc. The circuit modules are communicating with each other through physical connections. In order to decrease the complexity of VLSI circuits and improve the circuit delay characteristics, an effective method is the dividing and rule. Moreover, to speed up the design process it is necessary to split the circuit into many subcircuits so that each subcircuit can be designed independently. Minimization of the interconnections, area differences between the subcircuits, cost of the partitioning, delay, power consumption, and maximization of the module compactness in the circuit is the main objectives of circuit partitioning [1, 4].

VLSI is a vast field of electronics system. There are very complicated electronics circuitries made to perform different types of operations that are very effective to simulate before hardware design. The circuit needs partitioning [1, 2] to reduce the complexity of the large circuit, as it is divided into small independent parts. This will make the circuit more compact, reduce the area and fabrication cost [5, 13]. However, in circuit partitioning process, some points should be kept in mind that the individual area of the partitions is minimized to maximize the compactness and the number of interconnections between the partitions are also be minimum to reduce the delay, power consumption. These are the constraints considered as individual functions for partitioning problems. Moreover, the number of partitions is going to be made, depending on problem is more specifically termed as *k-way* or *multi-way* partitioning [6, 7].

To solve the VLSI partitioning problems the circuit description is very important. For this purpose, ISPD'98 [8] benchmark standard already exist for circuit description. In this benchmark, circuits are defined as *Netlist*, indicating the number of components/modules, I/O ports, and nuts, in the circuit. Hence, the circuit partitioning is also termed as *Netlist Partitioning* [9]. Netlist is considered as a hypergraph with vertices corresponding to cells (components/modules/gates) and edges corresponding to nets [2], where *Mincut* [10] indicates the possible minimum number of interconnections between partitions. Here, in this short review, we have considered *mincut* as an object that has been optimized by a well-known soft computing metaheuristic method, called Genetic Algorithm (GA) [14], whereas other parameters, such as delay, power, and area are kept constant. However, the optimization of these parameters has also been reported in [28–31].

2 Mathematical Formulation of VLSI Mincut Problem

The problem of minimizing the interconnection, i.e., mincut, is considered here. Generally, an electronics circuit partitioning problem is just like a graph partitioning problem [11, 12]. A standard electronic circuit can be described by the circuit netlist and the netlist can be represented by a hypergraph $H(V, E)$ with m modules and n nets, where V is a set of nodes and E is a set of nets (connection lines). In a circuit, a net can connect all the modules in the same signal line. However, partitioning is required mainly to minimize the interconnections between the subcircuits. Let there is a circuit that contains m number of modules and n number of nets (connections). This circuit will be partitioned to a specified number of subset k, where $k = 2$ as it is a bipartition problem, to assign the modules that satisfying the specified constraints. Now, if the subsets are denoted by $V_1, V_2, \ldots, V_k$ with S_1 and S_2 are the maximum and areas, then the constraints can be defined as follows:

$$V_1 \cup V_2 \cup \ldots \cup V_k = V$$
$$V_1 \cap V_2 = \varphi; \, S_2 \leq S(V_i) \leq S_1$$

where $S(V_i)$ is the area of subset V_i, $i = 1, 2 \ldots, k$

The graphical representation of a digital circuit is shown in Fig. 1a, b. A possible two-way partitioning of the circuit with nets and modules is also shown in Fig. 1c. In this paper, for single objective partitioning, our aim is to minimize the number of interconnections between two or multipartitions, called *mincut*. Therefore, this partitioning can be modeled mathematically on the basis of *mincut*. The mathematical model for mincut is defined in [3, 10] as:

$$f_{\text{min cut}} = \left(\sum_{j=1}^{n} \sum_{k=1}^{2} y_{jk} \right) \tag{1}$$

where, if net j is in block or subset of k, then $y_{jk} = 1$, otherwise $y_{jk} = 0$.

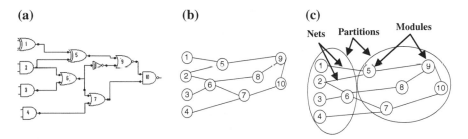

Fig. 1 **a** A digital circuit, **b** Graphical representation of the digital circuit, **c** Possible two-way partitioning indicating nets, modules, and partitions

3 Genetic Algorithm for VLSI Mincut Circuit Partitioning

Genetic Algorithm (GA) [14] is a heuristic search algorithm based on the evolutionary concepts of genetics and natural selection. Figure 2a shows the different steps of GA in flow chart form. In VLSI domain, many different types of research based on GA have been done in past two decades. For example, mincut partitioning [3, 10, 15], multilevel partitioning [16], dynamic embedding partitioning [32], VLSI physical design [2, 4], network partitioning [7, 17], cell placement techniques [33], floor plan designing [34], channel routing [35], VLSI layout optimization [36], network flow-based partitioning [18, 19]. For partitioning the VLSI circuits using GA, different schemes have been observed and those are described in below.

3.1 Initialization of Population

In GA, the population is initialized by different encoding schemes in different papers [3, 13, 20]. In [3], the encoding scheme, called 0-1 encoding, where the bi-partitioning problem allocates the circuit modules into a pair of disjoint blocks or subsets A and B. The 0 values of the genes represent the respective modules are assigned to subset A and when the value is 1, the respective modules are in subset B. Figure 2b shows its pictorial representation. Moreover, it has also been noticed that the integer encoding scheme is used in [3], where each chromosome is encoded with the module number. It is then divided into parts to create disjoin subsets of modules.

In [13], another integer encoding scheme is used, where the modules interconnection information as in the netlist is converted to an adjacency matrix first and then Breath First Search (BFS) search algorithm is applied on the matrix to traverse. Once the BFS order the modules, it is then used to form the chromosome for GA.

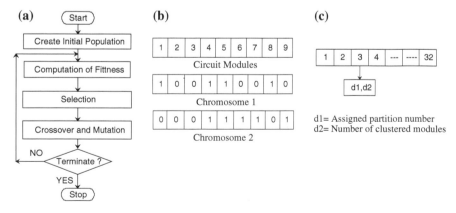

Fig. 2 **a** Flow chart of genetic algorithm, **b** 0-1 Encoding based chromosome, and **c** 32-bit encoded chromosome

It is a 32 bit chromosome containing integers for each gene of the chromosome. Each integer value is encoded to represent the assigned partition number and number of elements clustered to make each chromosome element is shown in Fig. 2c. Using this encoding policy random population is generated with user defined population size. Then for each population cost is calculated.

Another different type of chromosome encoding for bi-partitioning of a circuit is depicted in [20]. Here, the chromosomes are in layered form. Depending on the number of partitions made in the circuit, the layers of the chromosome are formed. Figure 3a shows a 2-layer chromosome structure which represents each individual in the populations. It can be extended for the multi layer chromosomes (for the multi way partitioning).

3.2 Fitness Computation

The fitness functions defined for the mincut partitioning problem that uses above mentioned encoding schemes are stated in Eq. 1–3. In [3], fitness function for binary encoding is defined as follows:

$$f_{mincut} = \left(\sum_{j=1}^{n} \sum_{k=1}^{2} y_{jk} \right) \left(1 - \frac{V_1 - V_2}{V_1 + V_2} \right)^{r_1} \tag{2}$$

where V_1 and V_2 are actual size of the two subsets or blocks, r_1 is a factor that controls the ratio for cut partitioning $(0 \le r_1 \le 2)$, y_{jk} is defined in Eq. (1). However, for integer encoding, fitness function is same as Eq. (1). The fitness function reported in [20] is stated below:

$$f_{mincut} = \left(\frac{NG}{NSG} \right) \tag{3}$$

where, NSG = the number of all connections between the created subgraphs, i.e., external connections and NG = the number of all connections in the graph (constant).

(a)   (b)

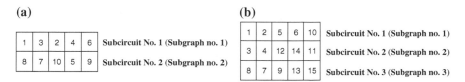

1	2	5	6	10	Subcircuit No. 1 (Subgraph no. 1)

1	3	2	4	6	Subcircuit No. 1 (Subgraph no. 1)
8	7	10	5	9	Subcircuit No. 2 (Subgraph no. 2)

3	4	12	14	11	Subcircuit No. 2 (Subgraph no. 2)
8	7	9	13	15	Subcircuit No. 3 (Subgraph no. 3)

Fig. 3 **a** a 2-layer chromosome representation of a bi-partitioned circuit, **b** 3-layer chromosome representation of a tri-partitioned circuit

3.3 Selection

In the selection process, mostly the Roulette Wheel selection strategy is used [3, 13]. In [20], fan selection (a modified Roulette Wheel selection) operations are used to find the best solution.

3.4 Crossover and Mutation

The crossover and mutation operations performed for the above mentioned encoded chromosomes are different. For 0-1 encoded chromosome, two parents are taken randomly to do the crossover operation with a crossover probability [3]. Here, two-point crossover is used to get a better offspring. In this scheme, according to the mutation probability, some bits are inverted in the binary chromosome randomly. The mutation probability is not constant and varies with the number of circuit modules. The mutation probability is defined in [3] as $M_P = 2/m$.

For integer encoded chromosome, if conventional crossover operation is used, then two problems can occur: (a) possibility of having the same module in both the blocks and (b) some modules may not be in any blocks. Hence, to avoid these problems, partial mapped crossover is used. This is illustrated in Fig. 4. Here, for this encoding policy, the general exchange mutation operation can't be performed, as it gives some invalid results. Therefore, an improved exchange mutation is used here. It first selects a random bit b_1 and then a second bit b_2 by the following formula:

$$b_2 = \begin{cases} \text{rand}(1, m_1) & b_1 > m_1 \\ \text{rand}((m_1 + 1), m) & b_1 \leq m_1 \end{cases} \tag{4}$$

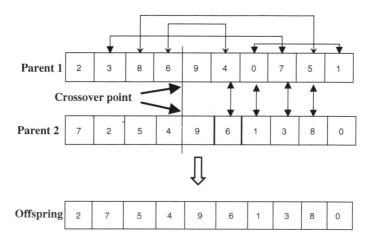

Fig. 4 General partial mapping crossover

where m_1 is the number of modules which are in the first block or subset and m is the total number of modules. As this process is used for all the pool of chromosomes, the mutation probability is kept constant 0.02. In [13] also this type of crossover and mutation is noticed.

In [20], idea of the Partial Mapping Crossover (PMX) [21] operation is used. The crossover mainly has 3 steps. In first step, it selects parent chromosome and the crossover point as shown in Fig. 5a, and after this, the cross point is exchanged, as shown in Fig. 5b. In second step, non-colliding genes, those are not present still in the newly generated off-springs are copied into the respective blank places from the parents, as shown in Fig. 5c, for parent X these are '5' and '3' and for Y it is '3' and '5'. Finally, in third step, the rest blank places in each new off-spring are filled by the module values that are still not present in it, as depicted in Fig. 5d.

Here, mutation operation is done using two different operators. After selecting a gene for mutation, type of mutation operator is randomly selected with equal probability. In case of the first operator, two genes are selected randomly from the parent individual and are exchanged during mutation as shown in Fig. 6a. In second type of operation, a circulation is done on the randomly selected column, shown in Fig. 6b.

Moreover, except mincut optimization for VLSI circuit partitioning using GA, there are other soft computing meta-heuristic techniques have been noticed to solve the same problems using Ant Colony Optimization [22], Particle Swarm Optimization [25], Simulated Annealing [26], and Tabu Search [27].

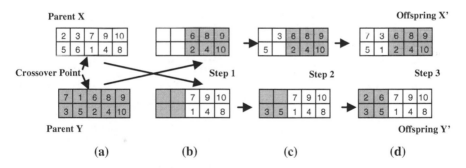

Fig. 5 Crossover operation of two-layered chromosomes

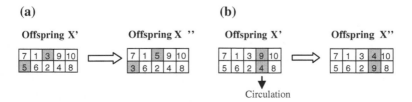

Fig. 6 **a** First mutation operation, **b** second mutation operation

4 Conclusion

In this survey paper, we have discussed the available genetic algorithms for VLSI circuit partitioning using mincut as an objective function. These methods adopt the hypergraph partitioning effectively to reduce the number of interconnections between two partitions in the circuit. It can also be applicable for multi-way partitioning. This short survey shows the different types of encoding, crossover, and mutation process that can be used to solve the mincut problems.

As a scope of further research, apart from mincut, other parameters like area, circuit delay, power consumption, etc., are also needed to study. Moreover, the application of different meta-heuristic methods can also be reviewed in context of single [36] and multiobjective optimization [23, 37] of different parameters of VLSI circuits. Authors are working in this direction.

Acknowledgments Mr. Saha is grateful to the All India Council for Technical Education (AICTE) for providing National Doctoral Fellowship (NDF) to support the work.

References

1. Johannes, F.M.: Partitioning of VLSI circuits and systems. In: Proceedings 33rd ACM/IEEE International Conference on Design Automation, pp. 83–87 (1996).
2. Mazumder, P., Rudnick, E.M.: Genetic Algorithms for VLSI Design, Layout and Test Automation Partitioning. Prentice Hall, New Jercy (1999)
3. Nan, G.F., Li, M.Q., Kou, J.S.: Two novel encoding strategies based genetic algorithms for circuit partitioning. In: Proceedings of 3rd International Conference on Machine Learning and, Cybernetics 4, pp. 2182–2188 (2004).
4. Sherwani, N.: Algorithms for VLSI Physical Design and Automation, 3rd edn. New Delhi, Springer (India) Private Limited (2005)
5. Bui, T.N., Moon, B.R.: A fast and stable hybrid genetic algorithm for the ratio-cut partitioning problem on hypergraphs. In: proceedings of 31st ACM/IEEE International Conference on Design Automation, pp. 664–669 (1994).
6. Tan, X., Tong, J., Tan, P., Park, N., Lombardi, F.: An efficient multi-way algorithm for balance partitioning of VLSI Circuits. In: Proceedings of International IEEE Conference on Computer Design: VLSI in Computers and Processors, pp. 608–613 (1997).
7. Sanchis, L.A.: Multiple-way network partitioning with different cost functions. IEEE Trans. Comput. **42**(12), 1500–1504 (1993)
8. Alpert, C.J.: The ISPD98 circuit benchmark suite. In: Proceedings of International Symposium on Physical Design, pp. 80–85 (1998).
9. Alpert, C.J., Khang, A.B.: Recent directions in netlist partitioning: a survey. Integr. VLSI J. **19**(1–2), 1–81 (1995)
10. Krishnamurthy, B.: An improved min-cut algorithm for partitioning VLSI networks. IEEE Trans. Comput **33**(5), 438–446 (1984)
11. Bui, T.N., Moon, B.R.: Genetic algorithm and graph partitioning. IEEE Trans. Comput. **45**(7), 841–855 (1996)
12. Andreev, K., Racke, H.: Balanced graph partitioning. In: Proceedings of 16th International Annual ACM Symposium on Parallelism in Algorithms and Architectures, pp. 120–124 (2004).
13. Gill, S.S., Chandel, R., Chandel, A.: Genetic algorithm based approach to circuit partitioning. Int. J. Comput. Electr. Eng. **2**(2), 196–201 (2010)

14. Chambers, L.D.: Practical Handbook of Genetic Algorithms. CRC Press, Inc. Boca Raton(1995).
15. Jiang, X., Shen, X., Zhang, T., Liu, H.: An improved circuit-partitioning algorithm based on min-cut equivalence relation. Integr. VLSI J. **36**(1–2), 55–68 (2003)
16. Alpert, C.J., Huang, J.H., Khang, A.B.: Multilevel circuit partitioning. In: proceedings of 34th ACM/IEEE International Conference on Design Automation, pp. 530–533 (1997).
17. Fiduccia, C.M., Mattheyses, R.M.: A linear time heuristic for improving network partitions. In: Proceedings of 19th International Conference on Design Automation, pp. 175–181 (1982).
18. Yang, H., Wong, D.F.: Efficient network flow based min-cut balanced partitioning. Proceedings of IEEE/ACM International Conference on Computer-Aided Design **15**(12), 1533–1540 (1996)
19. Liu, H., Wong, D.F.: Network-flow-based multiway partitioning with area and pin constraints. IEEE Trans. Comput. Aided Des. Integr. Circuits Syst. **17**(1), 50–59 (1998)
20. Slowik, A., Bialko, M.: Partitioning of VLSI circuits on subcircuits with minimal number of connections using evolutionary algorithm. In: Proceedings of International Conference AISC, pp. 470–478(2006).
21. Goldberg, D.E., Lingle, R.,: Alleles, loci and the TSP. In: Proceedings of International Conference on Genetic Algorithms, pp. 154–159 (1985).
22. Gill, S.S., Chandel, R., Chandel, A.: Comparative study of ant colony and genetic algorithms for VLSI circuit partitioning. Eng. Tech. **28**, 890–894 (2009)
23. Sait, S.M., El-Maleh, A.H., Al-Abaji, R.H.: Evolutionary algorithms for VLSI multi-objective netlist partitioning. Eng. Appl. Artif. Intell. **19**(3), 257–268 (2005)
24. Peng, S., Chen, G.L., Guo, W.Z.: A discrete PSO for partitioning in VLSI circuit. In: Proceedings of International Conference on Computational Intelligence and, Software Engineering, pp. 1–4 (2009)
25. Kolar, D., Puksec, J.D., Branica, I.: VLSI circuit partitioning using simulated annealing algorithm. In: Proceedings of IEEE Melecon, pp. 12–15 (2004).
26. Lodha, S.K., Bhatia, D.: Bipartitioning circuits using TABU search. In: Proceedings of 11th IEEE Annual International Conference on ASIC, pp. 223–227 (1998).
27. Rajaraman, R., Wong, D.F.: Optimal clustering for delay minimization. In: Proceedings of 30th ACM/IEEE International Conference on Design Automation, pp. 309–314 (1993).
28. Yang, H., Wong, D.F.: Circuit clustering for delay minimization under area and pin constraints. IEEE Trans. Comput. Aided Des. Integr. Circuits Syst. **16**(9), 976–986 (1997)
29. Vaishnav, H., Pedram, M.: Delay optimal partitioning targeting low power VLSI circuits. IEEE Trans. Comput. Aided Des. Integr. Circuits Syst. **18**(6), 298–301 (1999)
30. Yang, H., Wong, D.F.: Optimal min-area min-cut replication in partitioned circuits. IEEE Trans. Comput. Aided Des. Integr. Circuits Syst. **17**(11), 1175–1183 (1998)
31. Kim, C.K., Moon, B.R.: Dynamic embedding for genetic VLSI circuit partitioning. Eng. Appl. Artif. Intell. **11**(1), 67–76 (1998)
32. Esbensen, H., Mazumder, P.: SAGA: A unification of the genetic algorithm with simulated annealing and its applictaion to macro-cell placement. In: Proceedings of 7th International Conference on, VLSI Design, pp. 211–214 (1994).
33. Cohoon, J.P., Hegde, S.E., Martin, W.N., Richards, D.S.: Distributed genetic algorithms for the floorplan design problem. IEEE Trans. Comput. Aided Des. Integr. Circuits Syst. **10**(4), 483–492 (1991)
34. Lienig, J., Thulasiraman, K.: A genetic algorithm for channel routing in VLSI circuits. Evol. Comput. **1**(4), 293–311 (1993)
35. Schnecke, V., Vornberger, O.: An adaptive parallel genetic algorithm for VLSI layout optimization. In: Proceedings of 4th International Conference on Parallel Problem Solving from Nature III, pp. 859–868 (1996).
36. Maulik, U., Saha, I.: Modified differential evolution based fuzzy clustering for pixel classification in remote sensing imagery. Pattern Recognit. **42**(9), 2135–2149 (2009)
37. Saha, I., Maulik, U., Plewczynski, D.: A new multi-objective technique for differential fuzzy clustering. Appl. Soft Comput. **11**(2), 2765–2776 (2011)

Multimedia Classification Using ANN Approach

Maiya Din, Ram Ratan, Ashok K. Bhateja and Aditi Bhateja

Abstract Digital multimedia data in the form of speech, text and fax is being used extensively. Segregation of such multimedia data is required in various applications. While communication of such multimedia data, the speech, text and fax data are encoded with CVSD coding, Murray code and Huffman code respectively. The analysis and classification of such encoded multimedia from unorganized and unstructured data is an important problem for information management and retrieval. In this paper we proposed an ANN based approach to classify text, speech and fax data. The normalized frequency of binary features of varying length and PCA criterion is considered to select effective features. We use selected features in Back-propagation learning of MLP network for multimedia data classification. The proposed method classifies data efficiently with good accuracy. The classification score achieved for encoded plain data is of the order of 91, 93 and 90 % for speech, text and fax respectively. Also for 30 % distorted data, the classification score obtained is of the order of 78, 80 and 72 % for speech, text and fax respectively.

Keywords Multimedia data · Binary features · PCA criterion · ANN classifier · MLP network · Back-propagation learning

M. Din (✉) · R. Ratan · A. K. Bhateja
Defence Research and Development Organization, Scientific Analysis Group, Delhi, India
e-mail: anuragimd@gmail.com

R. Ratan
e-mail: ramratan_sag@hotmail.com

A. K. Bhateja
e-mail: akbhateja@gmail.com

A. Bhateja
Ambedkar Institute of Advanced Communication Technologies and Research, Delhi, India
e-mail: aditibhateja89@gmail.com

B. V. Babu et al. (eds.), *Proceedings of the Second International Conference on Soft Computing for Problem Solving (SocProS 2012), December 28–30, 2012*, Advances in Intelligent Systems and Computing 236, DOI: 10.1007/978-81-322-1602-5_96, © Springer India 2014

1 Introduction

With emergence of digital communication, most of the signals (speech, text or fax) are being sent over a channel in digital form. Nowadays digital multimedia data is being used extensively even by common people in their communication. In the age of Information Technology, the information flow in binary form between communicating and computing devices. The classification of multimedia data as speech, text and fax is very important before any further analysis in several applications such as steganography and cryptography in information security [1–5] and information indexing and categorization for efficient information management and retrieval systems [6]. For efficient communication or storage of data with adequate signal quality, the data should be encoded with appropriate encoding techniques. The speech is encoded in binary form by Continuously Variable Slope Delta (CVSD) coding is adopted apart from various existing encoding techniques due to its low bit rate and with reasonable speech quality [7]. The English text is converted into binary form using Murray code [8]. The fax messages are represented in binary form using Huffman encoding technique [9].

For classification of data, normally it is represented in pattern space by variety of multidimensional feature vectors. The components of feature vector may be frequency of binary words, correlation values or energy-based measurements etc. We apply N-gram features which have also been applied in pattern recognition problems such as classification of speech and text, and identification of speech encodings [10–12]. The Principal Component Analysis (PCA) helps to find independent features for their use in pattern recognition [13]. The conventional classifiers are less effective when data to be classified is ambiguous or vague and class boundaries are overlapped and not defined clearly. The artificial neural networks are found very suitable to tackle ambiguous situations efficiently [14–17]. In this paper, an ANN based approach is proposed to classify multimedia data. The approach uses normalized frequency of binary features of varying length and PCA criterion to identify and select effective features. We utilize effective features to train Multi Layer Percepton (MLP) Network for classification using Back Propagation Learning [15]. The proposed method classifies multimedia data efficiently with higher classification accuracy.

The paper is organized as follows: We describe encoding for multimedia data in Sect. 2. The detail of data preparation and effective feature selection is discussed in Sect. 3. The Back-propagation learning and MLP network is described briefly in Sect. 4 and classification results obtained are presented in Sect. 5. The paper is concluded in Sect. 6.

2 Multimedia Data Encoding

We represent multimedia data in binary form for its classification. The speech is encoded in binary form by CVSD coding [7] apart from various existing encoding techniques due to its low bit rate and reasonable speech quality. The English Text is converted into binary form using Murray code [8]. Fax data are represented in binary form using Huffman encoding technique [9].

2.1 Speech Coding

The speech is a time varying signal representing audio signals. The speech coding is carried out basically using Waveform coders and Voice coders. Pulse Code Modulation (PCM), Linear Predictive Coding (LPC) and CVSD are commonly used coding techniques in speech encodings [7, 18]. We consider CVSD encoding which represents speech data with adequate speech quality. The CVSD coding is a special type of Adaptive Delta Modulation coding where adaptation is considered at pitch period. It possesses remarkable degree of robustness to bit error rates of order of 1 %. The CVSD encoded speech contains special sequence called idling binary pattern due to pause regions exist in speech utterances.

2.2 Text Encoding

There are various codes used for binary representation of text messages. The American Standard Code for Information Interchange (ASCII), Unicode and Murray Code are commonly used in computer and communication applications. We consider Murray code to represent English text messages in binary form with five-bit length fixed code [8]. Some of the monographs, digraphs and trigraphs of letters occur frequently and some occur rarely in plain English text. The linguistic characteristics are also exists in corresponding binary texts.

2.3 Fax Encoding

Fax is a 2-Dimensional data representing spatial variation of document. Huffman coding is used to encode fax messages in binary form [9]. It is a variable length binary coding scheme based on the probability of symbols. In this coding, the most frequent symbol is represented by least number of bits and least frequent symbol is represented by highest number of bits.

3 Effective Feature Selection

The binary multimedia data is prepared from speech, text and fax messages. We consider 500 speech frames each of 2,500 bits, 500 text messages each of 500 characters and 50 fax messages each of 25,000 bits. These fax messages are divided into frames

Table 1 Effective feature patterns used for classification

Effective features	Feature pattern
Bi-grams	01, 10
Tri-grams	101, 110
Four-grams	1101, 1110
Five-grams	10000, 00001, 00011, 00110
Six-grams	011001
Ten-grams	0000100101, 00110010100
Twelve-grams	000000000000, 111111111111
Thirteen-grams	0101010101010

each of 2,500 bits. This data is distorted randomly at 10, 20 and 30 % distortion levels for its classification study.

We consider 124 features (bi-gram to six-gram), two ten-gram (0000100101, 00110010100) features, two twelve-gram (000000000000, 111111111111) features and one thirteen-gram (0101010101010) binary feature of multimedia data. After careful study of speech, text and fax data and applying PCA criterion, following 15 binary effective feature patterns are identified as shown in Table 1.

4 MLP-ANN Based on Back-Propagation Learning

An artificial neural network is a computational model that is inspired by the structure and/or functional aspects of biological neural networks. It is also known as Neural Network (NN). An ANN consists of an interconnected group of artificial neurons, and it processes information using a connectionist approach. In general an ANN works as an adaptive system that changes its structure according to external or internal information flowing through the network during its learning process. Modern neural networks are applied as non-linear modeling of given data. ANNs are usually used to model complex relationships between inputs and outputs, finding hidden patterns in data and data mining applications.

A Multilayer Perceptron (MLP) Network [10] is a feed-forward artificial neural network that maps sets of input data onto a set of appropriate output. MLP is a modification of the standard linear perceptron, which can distinguish data that is not linearly separable. An MLP consists of multiple layers of nodes, with each layer fully connected to the next one. Except for the input nodes, each node is a neuron with a nonlinear activation function. An ANN having multi layers (Input layer, Hidden layers and Output layers) is called Multilayer ANN as shown in the Fig. 1. MLP utilizes a supervised learning technique for training the network.

There are several learning techniques used to train MLP-ANNs. Back-propagation learning technique is one out of these used to train the network [15]. In this learning, gradient search optimization technique is used to minimize cost function up to least mean square error between target and actual net outputs. An ANN based

Input Layer   Hidden Layers   Output Layer

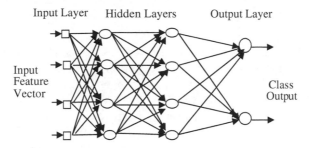

Input
Feature
Vector

Class
Output

Fig. 1 Multilayer artificial neural network

classification approach is developed to classify speech, text and fax data. The normalized frequencies of 15 effective features are computed by dividing frame length, which are used in Back-propagation learning of Multilayer Perceptron (MLP) network having three layers. The number of neurons used in the network layers are 15, 7, 2 (Input layer to output layer respectively) based on Sigmoid Activation Function. The above effective features are used in training and testing of ANN. This network is trained using BP Learning on 900 feature vectors i.e. 300 vectors of each class consisting of clear and distorted data with 10, 20 and 30 % distortion. The variable learning rate used is from 1.8 to 0.5 for achieving required error threshold level. The network is tested on different 150 feature vectors corresponding to clear data and 450 corresponding to distorted data consisting of 150 for each class with 10, 20 and 30 % distortion level.

5 Classification Results

The proposed ANN classification method is applied on both training and test data sets. For test data, the classification results of speech, text and fax in clear and distorted domain is summarized in Table 2.

The classification score achieved through proposed approach for clear data is of the order of 91, 93 and 90 % for speech, text and fax respectively and 78, 80 and 72 % for corresponding 30 % distorted data. The classification score for text data is higher than speech and fax data in clear as well as distorted domain. It is also observed from obtained results that as level of distortion increases the classification accuracy decreases.

Table 2 Classification results

Multimedia test data	Classification (%)		
	Speech	Text	Fax
Clear data	91	93	90
10 % Distorted data	90	92	89
20 % Distorted data	84	86	80
30 % Distorted data	78	80	72

6 Conclusion

An ANN based method has been proposed for classification of plain and distorted multimedia data to segregate speech, text and fax data. The effective binary features of multimedia data have been utilized in the proposed method. The PCA criterion has been applied to select effective features that have been used in Back-propagation Learning of MLP network for efficient and effective classification. The classification score obtained for clear speech, text and fax data is of the order of 91, 93 and 90 % and for 30 % distorted speech, text and fax data is of the order of 78, 80 and 72 % respectively. The proposed ANN method could be used for segregation, analysis and interpretation of multimedia data in various multimedia applications.

References

1. Katzenbeisser, S., Petitcolas, F.A.P.: Information hiding techniques for steganography and digital watermarking. Artech House, London (2000)
2. Ratan, R., Madhavan, V.C.E.: Steganography based information security. IETE Tech. Rev. **19**, 213–219 (2002)
3. Menezes, A., Van Oorschot, P., Vanstone, S.: Handbook of Applied Cryptography. CRC Press, USA (1996)
4. Stallings, W.: Cryptography and Network Security. Prentice Hall, Englewood Cliffs (2011)
5. Becker, H., Piper, F.: Cipher Systems: The Protection of Communication. Northwood Booker, London (1982)
6. Yates, R.B., Nieto, B.R.: Information Retrieval. Addison Wesley, England (1999)
7. Deller, J.R., Hansen, J.H.L , Proakis, J.G.: Discrete-Time Processing of Speech Signals. IEEE Press, New York (2000)
8. Beker, P.W.: Recognition of Patterns: Using Frequency of Binary Words. Springer, New York (1978)
9. Jain, A.K.: Fundamentals of Digital Image Processing. Prentice Hall, USA (1989)
10. Chelba, C., Acero, A.: Discriminative training of N-gram classifiers for speech and text routing. Proceedings of the Eurospeech International Conference ISCA (2003).
11. Maithani, S., Saxena, P.K.: Identification of coding in enciphered speech. Proceedings of 6th International Conference on recent trends in Speech, Music and Allied, Signal Processing (2001).
12. Maithani, S., Maiya, D.: Speech systems classification based on frequency of binary word features. Proceedings of the IEEE International Conference SPCOM-04 (2004).
13. Jolliffe, I.T.: Principal Component Analysis. Springer-Verlag, New York (2002)
14. Katagiri, S. (ed.): Hand Book of Neural Networks for Speech Processing. Artech House, London (2000).
15. Haykin, S.: Neural networks- a comprehensive foundation. Macmillan, New York (2001)
16. Basu, J.K., Bhattacharyya, D., Kim, T.H.: Use of artificial network in pattern recognition. Int. J. Softw. Eng. Appl. **2**(2), 23–34 (2010)
17. Remeikis, N., Skucas, I., Melninkaite, V.: Text categorization using neural networks initialized with decision trees. Informatica **15**(4), 551–564 (2004)
18. Kondoz, A.M.: Digital speech codings for low bit rate communications systems. John Wiley and Sons, New York (1995)

Live Traffic English Text Monitoring Using Fuzzy Approach

Renu, Ravi and Ram Ratan

Abstract Current communication systems are very efficient and being used conveniently for secure exchange of vital information. These communication systems may be misused by adversaries and antisocial elements by capturing our vital information. Mostly, the information is being transmitted in the form of plain English text apart from securing it by encryption. To avoid losses due to leakage of vital information, one should not transmit his vital information in plain form. For monitoring of huge traffic, we require an efficient plain English text identifier. The identification of short messages in which words are written in short by ignoring some letters as in mobile messages is also required to monitor. We propose an efficient plain English text identifier based on Fuzzy measures utilizing percentage frequencies of most frequent letters and least frequent letters as features and triangular Fuzzy membership function. Presented method identifies plain English text correctly even, the given text is decimated/discontinuous and its length is very short, and seems very useful.

Keywords Traffic analysis · Fuzzy approach · Linguistic features · Plain text · Random text · Information security

Renu
Defence Research and Development Organization, Defence Scientific Information
and Documentation Center, Delhi, India
e-mail: renu25686@gmail.com

Ravi
Guru Premsukh Memorial College of Engineering, Guru Gobind Singh
Indraprastha University, Delhi, India
e-mail: talk2ravi@yahoo.in

R. Ratan (✉)
Defence Research and Development Organization, Scientific Analysis Group, Delhi, India
e-mail: ramratan_sag@hotmail.com

B. V. Babu et al. (eds.), *Proceedings of the Second International Conference on Soft Computing*   911
for Problem Solving (SocProS 2012), December 28–30, 2012, Advances in Intelligent Systems
and Computing 236, DOI: 10.1007/978-81-322-1602-5_97, © Springer India 2014

1 Introduction

With the advancement in information technology, efficient communication systems such as mobile and internet are available and being used nowadays even by common people. While communication of sensitive or vital information, it should not be communicated in plain form in any case and should be protected by appropriate safety measures to avoid losses due to capturing of such information by adversaries. The English language is commonly used worldwide in exchange of information. There are following ways to achieve information security—spread-spectrum, steganography and cryptography [1–5]. The spread-spectrum techniques modulate given information over the carriers randomly within the available channel bandwidth. The steganographic techniques conceal the existence of information by hiding it in another ordinary data. The cryptographic techniques conceal the content of information by transforming it into unintelligible form. The cryptographic techniques are being used widely all over the Globe in the area of information security for achieving the confidentiality apart from other security issues. The encryption process transforms plain text into crypt text by distorting the linguistic characteristics. When the English text is encrypted by cryptographic techniques, the obtained crypt text appears as random and there remains no intelligibility of text. The frequency of each letter occurs almost equal in random or encrypted text but in English text some letters occur highly and some letters occur very rarely.

Identification of plain text from such encrypted or random text is required in monitoring of sensitive traffic. The careful monitoring and analysis of traffic for English text are the most important problem to safeguard the communication. Moreover, monitoring and analysis of text have a vital role in combating terrorist activities and cyber crimes. The monitoring of plain text is also required to manage sensitive data securely over computer network. The identification of short messages in which words are written in short by ignoring some letters is also required to monitor mobile messages and analysis of crypts. We should never keep such data in plain form and it should always be kept secure. For monitoring of such huge traffic, the identification method should perform the task with minimum efforts, i.e., minimum computing complexity, time, and memory requirements.

Text mining is a very important field that attempts to extract meaningful information automatically from huge volume of natural language text for particular use [6, 7]. The text mining problems involve assessing the similarity between different texts and grouping them. Automatic text categorization has many practical applications such as indexing for retrieval and organizing and maintaining large catalogs of Web resources. The language identification is a particular application of text categorization. The text categorization or classification is gaining momentum due to its diverse applications such as classification of sensitive text [8], cyber terrorism investigation [9] and E-mail filtering [10]. The n-grams or patterns of n consecutive letters and words are used to identify language of texts [11, 12]. It is seen that the linguistic characteristics of the language have a key role in analysis of English text [13, 14]. The conventional methods of identifying English text become inapplicable specifically when the size of text is less or the text is not continuous and also due to the uncertainty or variability of the linguistic features occurred. A Fuzzy approach based on soft computing is most suitable to tackle such problems better in uncertain and ambiguous situations where lot of difficulties arises in taking right decisions and perform better than neural networks and traditional pattern recognition approaches [15–19]. The Fuzziness or degree of uncertainty pertains due to the inconsistency of linguistic properties appear in text when it is of special type such as discontinuous and its length is short.

In this paper, a Fuzzy method is proposed for identification of English text in which we utilize the percentage occurrence of most frequent and least frequent letters as features and triangular Fuzzy membership function. The triangular Fuzzy membership function has been applied successfully in various pattern recognition problems. The method is also tested for partial text which is discontinuous or decimated text of short length.

The paper is organized as follows: In Sect. 2, we present briefly the linguistic characteristics of English language useful in the analysis of English text. We present the Fuzzy approach for identification of English text in Sect. 3. The performance of proposed Fuzzy method is presented in Sect. 4. Finally, the paper is concluded in Sect. 5 followed by the references.

2 Letter Characteristics of English Text

The English language has 26 symbols/letters (excluding spaces, punctuation marks, etc.). The occurrences of letters, n-grams, digraphs and its reversals, left–right contacts of letters, percentage of vowels and consonants, etc. in the plain-text and in the crypt-text are the only statistical measures which may be helpful. The following most important facts [13] are observed from frequency distributions of letters:

1. Irregular appearance occurred due to the fact that some of the letters are most frequent and others are very rare.
2. Most prominent crests are marked by letters E T O A N I R S H and the most prominent troughs are marked by J K Q X Z.
3. The occurrences of vowels and consonants are as:

Vowels	A E I O U Y	40 %
Consonants		
High frequent	T S R N H	34 %
Medium frequent	D C L M P F W G B V	24 %
Low frequent	J K Q X Z	2 %

4. The relative order of letters according to decreasing order of frequencies is as E T O A N I R S H D L U C M P F Y W G B V K J X Z Q.
5. Not more than two consecutive vowels normally occur (except 'ious').
6. Not more than four consecutive consonants normally occur.

7. Most frequent digraph reversals normally have a vowel.
8. Less frequent letters mostly contact with vowels than with other consonants.
9. The most frequent digraphs according to frequencies are as
   TH ER ON AN RE HE IN ED ND HA AT EN ES OF OR
   NT EA TI TO IT ST IO LE IS OU AR AS DE RT VE.

Above facts have been successfully utilized in the analysis of linguistic properties of texts [11–14, 20, 21]. For online identification of English text, the features to be used should be simple and easy to compute for efficient and accurate results. Some measures such as n-grams [11], words [12], phi measure [13] joint mutual information [20] and index of garbledness [21] have been used in this regard. The frequency of letters, n-grams, words, digraphs, joint mutual information, and index of garbledness are not suited for English text identification when the length of given text is less and the text is discontinuous or decimated. In short and discontinuous English text, the linguistic properties are changed and become useless in categorization and language identification. In random English text, the properties of plain text do not remain and the alphabets appear uniformly.

In this paper, we consider the percentage frequencies of most frequent and least frequent letters for efficient identification of English text in live traffic analysis. Here, the letters $E\ T\ O\ A\ N\ I\ R\ S\ H$ are taken as most frequent letters and the letters $J\ K\ Q$ $X\ Z$ are taken as least frequent letters whose approximately percentage frequencies are 73 and 2 % respectively.

Conventionally, the text is identified as plain-text when $P_{mf} \geq T_{mf}$ and $P_{lf} \leq T_{lf}$. The P_{mf} and P_{lf} are the computed and T_{mf} and T_{lf} are expected percentage values of most frequent and least frequent letters. The selection of proper thresholds is very difficult because of uncertainty in plain English text and we get incorrect results when the value of thresholds is not appropriate.

3 Fuzzy English Text Identification

As Fuzzy approach handles uncertain and redundant data more comfortably to solve ambiguous problems, we do not require the tuning of T_{mf} and T_{lf} exactly for identifying plain English text correctly in suggested method. The value of T_{mf} and T_{lf} is taken 73 and 2 % respectively which are the expected percentage values of most frequent and least frequent letters as mentioned in Sect. 2 for English text. For a given English text, we compute the value of the P_{mf} and P_{lf} features.

For comparing a given pattern with reference pattern, the normal process is to compute the Hamming or the Euclidean distance and to use minimum criteria. Here, we use a triangular fuzzy membership function to obtain membership value (μ) for $P_{mf}(u)$ feature as shown in Fig. 1.

The $T_{mf}(r)$ is the reference value for plain English text and $P_{mf}(u)$ is the computed value in given text to be identified. Similarly, the Fuzzy membership function is also defined for $P_{lf}(u)$. The values $\mu(P_{mf}u))$ and $\mu(P_{lf}(u))$ lie between 0 and 1 and

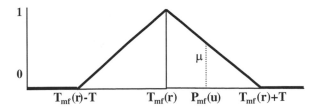

Fig. 1 Fuzzy membership function

is computed using triangular Fuzzy membership function as defined. The value of T used in membership function is chosen suitably. The value of $\mu(P_{\mathrm{mf}}(u))$ and $\mu(P_{\mathrm{lf}}(u))$ for English text is close to 1 and for random text is close to 0.

The similarity score of a given text with reference text is computed by using the concept of fuzzy intersection [16]. The similarity of a given pattern with the reference pattern is given by

$$S = \cap \{\mu(P_{\mathrm{mf}}(u))\}, \{\mu(P_{\mathrm{lf}}(u))\}. \tag{1}$$

According to the definition of fuzzy intersection, the similarity is the minimum value of $\mu(P_{\mathrm{mf}}(u))$ and $\mu(P_{\mathrm{lf}}(u))$.

The overall similarity score can also be obtained by computing the average similarity as given by

$$S = 1/2[\mu(P_{\mathrm{mf}}(u)) + \mu(P_{\mathrm{lf}}(u))]. \tag{2}$$

The decision of identification is taken based on the similarity. The given text is identified as plain English text when the similarity score is near to 1 and it is identified as random or encrypted English text when the similarity score is near to 0.

4 Performance

The Fuzzy approach proposed is applied on various English text of varying length of normal plain text as well as decimated/discontinuous partial text to study the performance. For demonstrating the results obtained, we take normal English text in partial form by decimation of letters for identification. As an illustration, the plain English text, decimated texts, and short messages are given in Table 1.

Here, the decimated texts are unintelligible and look difficult to identify as part of plain English text. Identification of decimated texts is harder compare to normal plain English texts and mobile short messages. The space, punctuation marks, and non-English alphabet letters are excluded and only 26 alphabet letters are considered in identification of given English texts.

Proposed Fuzzy method identify given plain English text correctly with high confidence and almost 100 % accuracy even when the text is discontinuous and its length is very short. The result of text identification for decimated English texts of

Table 1 Plain english text, decimated text and short message

Plain text	The importance of the role played by reprtitions in the analysis of cryptograms is well understood even by the amateur cryptnalyst repetitions in cryptographic text are basically of two sorts casual and accidental casual repetitions…
Decimated text (letter 1, 4, 7, …)	TIOAETREADRETNNEASORTRSWL DSOVBHMECP…
Decimated text (letter 2, 5, 8, …)	HMRNOHOPYBETISTALIFYOAIEU ETDEYEAURT…
Decimated text (letter 1, 8, 15, …)	TRFEBINLCRERVERNEOYPTIFT LICEO…
Short message 1	Wish you a very happy and prosperous new year
Short message 2	Certainly SOCPROS 2012 is to be a grand successful event of JKLU

Table 2 Identification of decimated plain english texts

Decimated text	Length of decimated text	Membership values $\mu(P_{mf}(u))$ and $\mu(P_{lf}(u))$	Similarity scores as (1) and (2)
TRFEBINLCRERVERNEOYPTIFTLICEO	29	0.87, 0.95	0.87, 0.91
HTTPYTTYRALSEACAPNPHACTSADAPN	29	0.87, 0.95	0.87, 0.91
EAHLRIHSYMLTNMRLESTIRAWCNEUES	29	0.92, 0.95	0.92, 0.94
INEAEOEIPSUOBAYYTIOCELOADNSTA	29	0.97, 0.95	0.95, 0.96
MCRYPNASTINOYTPSINGTBLSUATAI	28	0.86, 0.95	0.86, 0.91
PEDEESNOOSDDTETTTCREAYOSCALT	28	1.0, 0.95	0.95, 0.98
OOLDTIAFGWEEHUARIRAXSORACLRI	28	0.90, 0.96	0.90, 0.93
TIOAETREADRETNNEASORTRSWLDSO VBHMECPNYREOIRTRHTTESA	50	0.99, 0.98	0.98, 0.99
HMRNOHOPYBETISTALIFYOA IEUETDEYEAURTASETNNYOAIEABIL	50	1.00, 1.00	1.00, 1.00
EPTCFELLEYPIOIHNYSCPGMSLNRO ENTATRYALTPISCPGPCXRACL	50	0.99, 0.98	0.98, 0.99

short length is shown in Table 2. The performance shows that Fuzzy method presented is efficient for monitoring of traffic for detection and further analysis of plain English text of very short length even when such texts are discontinuous or decimated. The performance of the proposed Fuzzy method for identification of plain English text is not comparable with other methods [11–14] and [20, 21] as these are applicable on larger size of English text.

The presented method can meet the requirement of monitoring and analysis of live traffic sensitive texts and taking preventive measures accordingly to protect adversaries and criminal activities on mobile and internet.

5 Conclusion

The detection of plain English text is one of the most important requirements to monitor and analyze the traffic for preventing transmission of vital information through mobile and internet communication. Moreover, detection and analysis of plain English text are also in demand to intercept the communication of adversaries and antisocial elements for protecting terror and criminal activities by taking appropriate safety measures. The fuzzy method utilizing percentage frequency of most and frequent letters as features and triangular Fuzzy membership function has been proposed in the paper. As per simulation results shown, the proposed method is very efficient and can be applied for online monitoring of huge traffic for identification and further analyze of English text with high accuracy even when it is discontinuous and its length is very short. The Fuzzy method proposed has vast applications in various defence as well as civil applications such as monitoring and analysis of harmful and sensitive short messages being communicated through mobile and internet by adversaries and antisocial elements.

References

1. Stephen, G.W.: Digital Modulation and Coding. Prentice Hall, New Jersey (1996)
2. Katzenbeisser, S., Petitcolas, F.A.P.: Information Hiding Techniques for Steganography and Digital Watermarking. Artech House, London (2000)
3. Ratan, R., Madhavan, V.C.E.: Steganography Based Information Security. IETE Tech. Rev. **19**(4), 213–219 (2002)
4. Menezes, A., Van Oorschot, P., Vanstone, S.: Handbook of Applied Cryptography. CRC Press, USA (1996)
5. Stallings, W.: Cryptography and Network Security. Prentice Hall, Englewood Cliffs (2011)
6. Han, J., Kamber, M.L.: Data Mining: Concepts and Techniques. Elsevier, Amsterdam (2006)
7. Sebastiani, F.: Machine learning in automated text categorization. ACM Comput. Surv. **34**(1), 1–47 (2002)
8. Wong, A.K.S., Lee, J.W.T., Yeung, D.S.: Using complex linguistic features in context - sensitive text classification techniques. Proceedings of the 4th IEEE International Conference on Machine Learning and Cybernetics **5**, 3183–3188 (2005)
9. Simanjuntak, D.A., Ipung, H.P., Lim, C., Nugroho, A.S.: Text classification techniques used to facilitate cyber terrorism investigation. Proceedings of the 2nd IEEE International Conference on Advances in Computing, Control and Telecommunication Technologies (ACT), 198–200 (2010).
10. Upasana, Chakrabarty, S.: A survey of text classification techniques for e-mail filtering. Proceedings of the 2nd IEEE International Conference on Machine Learning and Computing (ICMLC), 32–36 (2010).

11. Cavnar, W.B., Trenkle, J.M.: N-Gram based text categorization, pp. 161–175. Proceedings of the Symposium on Document Analysis and Information Retrieval. Las Vegas, NV pp (1994)
12. Grefenstette, G.: Comparing two language identification schemes. Proceedings of the International Conference on Statistical Analysis of Textual Data JADT-95 (1995).
13. Kullback, S.: Statistical Methods in Cryptanalysis. Aegean Park, CA (1976)
14. Friedman, W.F.: Elements of Cryptanalysis. Aegean Park, CA (1976)
15. Anderson, A., Rosenfield, E.: Neuro Computing. MIT Press, MA (1988)
16. Dubois, D., Prade, H.: Fuzzy Sets and Systems: Theory and Applications. Academic Press, New York (1980)
17. Bezdek, J.C., Pal, S.K.: Fuzzy Models for Pattern Recognition. IEEE Press, New York (1992)
18. Puri, S., Kaushik, S.: A technical study and analysis on fuzzy similarity based models for text classification. Int. J. Data Min. Knowl. Manag. Process 2(2), 1–15 (2012).
19. Ratan, R., Saxena, P.K.: Algorithm for the restoration of distorted text documents. Proceedings of the International Conference on Computational Linguistics, Speech and Document Processing (ICCLSDP) (1998)
20. Yang, H.H., Moody, J.: Feature selection based on joint mutual information. J. Comput. Intell. Met. Appl. Int. Comput. Sci. Conv. **13**, 1–8 (1999)
21. Saxena, P.K., Yadav, P., Mishra, G.: Index of garbledness for automatic recognition of plain english texts. Defence Sci. J **60**(4), 415–419 (2010)

Digital Mammogram and Tumour Detection Using Fractal-Based Texture Analysis: A Box-Counting Algorithm

K.C. Latha, S. Valarmathi, Ayesha Sulthana, Ramya Rathan, R. Sridhar and S. Balasubramanian

Abstract Mammography and X-ray imaging of the breast are considered as the mainstay of breast cancer screening. In the past several years, there has been tremendous interest in image processing and analysis techniques in mammography. The fractal is an irregular geometric object with an infinite nesting of structure of different sizes. Fractals can be used to make models of any objects. The most important properties of fractals are self-similarity, chaos, and non-integer fractal dimension. The fractal dimension analysis has been applied to study the wide range of objects in biology and medicine and has been used to detect small tumors, microcalcification in mammograms, tumors in brain, and to diagnose blood cells and human cerebellum. Fractal theory also provides an appropriate platform to build oncological-related software program because the ducts within human breast tissue have fractal properties. Fractal analysis of mammogram was used for the breast parenchymal density assessment. The fractal dimension of the surface is determined by utilizing the Box-counting method. The Mammograms were collected from HCG Hospital, Bangalore. In this study, a method was developed in the Visual Basic for extracting the suspicious

S. Valarmathi · R. Sridhar · S. Balasubramanian
DRDO-BU-CLS, Bharathiar University, Coimbatore, Tamil Nadu, India
e-mail: valar28aadarsh@gmail.com

R. Sridhar
e-mail: rmsridhar@rediffmail.com

S. Balasubramanian
e-mail: director_research@jssuni.edu.in

A. Sulthana (✉) · K. C. Latha
Water and Health, JSS University, Mysore, Karnataka, India
e-mail: ayeshasulthanaa@gmail.com

K. C. Latha
e-mail: latha_tanvi23@yahoo.com

R. Rathan
JSS University, Anatomy, Mysore, Karnataka, India
e-mail: ramirohith@ymail.com

B. V. Babu et al. (eds.), *Proceedings of the Second International Conference on Soft Computing for Problem Solving (SocProS 2012), December 28–30, 2012*, Advances in Intelligent Systems and Computing 236, DOI: 10.1007/978-81-322-1602-5_98, © Springer India 2014

region from the mammogram based on texture. The fractal value obtained through Box-counting method for benign and malignant breast cancer is combined into a set. An algorithm was used to calculate the fractal value for the extracted image of the mammogram using Box-counting method.

Keywords Fractal dimension Box-counting algorithm · Mammogram · Benign · Malignant · Range and pixel based algorithms

1 Introduction

Mammography and X-ray imaging have largely contributed for the early detection of breast cancer. However, the effectiveness and sensitiveness of digital mammography in detection of breast cancer is currently under investigation. The imaging modality separates image acquisition and image display. Therefore, radiologists have to optimize the efficiency of both processes for treatment or removal of the tumor. A series of heuristic techniques such as filtering and thresholding (texture analysis) automatically detect abnormalities. These methods suffered from lack of robustness when the number of images to be classified is large. Recently, several statistical methods are used to overcome such problems and the primary one is the fractal dimension which is widely applied in biology and medicine; especially this technique is used to detect small tumors and microcalcification in mammograms [1]. Several algorithms were used for fractal dimension in which the Box-counting method is applied to estimate the fractal value of an X-ray (mammogram) image, because the Box-counting dimension provides the information regarding the 3D objects of the tissue composition of the breast mammogram [2]. Hence fractal analysis may be useful as the strategy of the current study allows an algorithm which aims at assisting the radiologist toward fast detection and early diagnosis in the prediction of breast cancer. The fractal is an irregular geometric object with an infinite nesting of structure of different sizes. Fractals can be used to make models of any objects. The most important properties of fractals are self-similarity, chaos and non-integer fractal dimension.

Challenges in the use of fractal methods on images are the limitation of the possible range of power law behavior because the results have been hard to be reproduced by other researchers. Therefore, this approach has been questioned by the medical practioners [3]. Nilsson and Georgsson have reported that the Box-counting method can be applied to estimate the fractal value of an X-ray (mammogram) image, because the Box-counting dimension provides the information regarding the 3D objects of the tissue composition of the breast mammogram [2].

Therefore, fractal analysis may be useful to evaluate mammographically discovered breast masses. The design and analysis strategy of the current study allows an algorithm which aims at assisting the radiologist towards fast detection and early diagnosis in the prediction of breast cancer. With this background, this study aims:

(i) To develop an algorithm for extracting the abnormal region in the breast mammogram
(ii) To calculate the fractal value for the extracted image from the mammogram using Box-counting method
(iii) To clarify the usefulness of using the fractal features (as a texture scale-invariant features) and to classify benign and malignant region in the respective images.

1.1 Fractal Dimension

The concept of fractal was first introduced by Mandelbrot in 1967. Fractal geometry compares the irregular forms, even at different scales since the approximate measure of the dimension D is independent of the unit of measurement [4]. The term "fractal" means break or fragment and is a pattern that reveals greater complexity as it is enlarged. For fractals, the counterparts of the dimensions (0, 1, 2, and 3) are known as fractal dimensions (FD). When applied to a point, a line, a square, or a cube, D simply gives the number of ordinary dimensions needed to describe the object —0, 1, 2, and 3, respectively. The value of fractal dimension is between one to two for a curve and in between two to three for a surface. The larger the D, rougher is the surface. The fractal dimension indicates how the measures of the object change with generalization [5].

Fractal dimension quantifies the metric information in lines and surfaces. The shapes of fractal objects keep invariant under successive magnification or shrinkage of the objects. A variety of procedures, including Box-counting, fractal Brownian motion [6], and fractal interpolation function system have been proposed for estimating the fractal dimension of images.

The fractional Brownian motion model (FBM) with gray-scale variation has shown promise in the medical image texture [6]. The Brownian motion curve concepts was extended to the FBM curve I(x), and $|I(x_2) - I(x_1)|$ having a mean value proportional to $|(x_1 - x_2)|^H$. Thus, in the FBM there is only one parameter of interest, the Hurst coefficient, which can be described as texture features when applied to classify breast tumor images. Considering the topological dimension T_d, for images, the FD can be estimated from the Hurst coefficient $H = T_d - D$. Using this relationship, the FD can be applied for the medical images [7] as:

$$D = \log(n2/n1) / \log(s1/s2) \qquad (1)$$

1.2 Box-Counting Method

The fractal dimension of the surface is determined by utilizing the Box-counting method. This method yields quantitative agreement with the line-segment method. In the Box-counting method, the Hausdorff-dimension was utilized, where the number, $N(\xi)$, of squares of side-length 'ξ' is needed to cover a set of increases like $N(\xi) \ \alpha \ \xi^{-d}$ for $\xi \ \leftarrow \ 0$ for a set of fractal Hausdorff dimension d. From $d = -\log[N(\xi_1)/N(\xi_2)]/\log[\xi_1/\xi_2]$, the fractal dimension can be obtained theoretically [8].

Box-counting method was applied to compute FD for measuring the structure of an object in a digital image and to analyze morphological changes within organs. Fractal analysis based on microcomputer studies of image systems provides many advantages to the understanding of the complex microscopical arrangement of natural objects [9]. The fractal-based computerized image analysis of mammographic parenchymal patterns was used to differentiate women at high risk and low risk for the development of breast cancer and this study was evaluated [10].

2 Data Preparation

The Marathon database (Digital Database for Screening Mammography) associated with South Florida (United States), contains approximately 2,500 studies, each one includes two images of each breast, along with the information of the associated patient (age at time of study, breast density, and abnormalities) and image information (scanner and spatial resolution). Images containing suspicious areas also have associated pixel-level "ground truth" information about the locations and types of suspicious regions. The database is further organized into cases and volumes. A case is a collection of images and information corresponding to one mammography. A volume is simply a collection of cases. Some cases contain more than one cancer in one breast, a cancer in each breast, or a cancer along with other abnormal/suspicious regions. The outlines of all the regions have been transcribed from markings made by an experienced radiologist [11].

3 Methodology

A method was developed in the visual basic for extracting the suspicious region from the mammogram based on texture. For the extracted image, the fractal dimension was calculated using Box-counting method.

3.1 Extraction of Breast Cancerous Surface: Marathon Database and HCG Hospital Bangalore Database

The following steps were used for extracting the abnormal region from the mammogram.

Step 1: Extraction of the boundary from the given mammogram by using an algorithm given in the Appendix I.

Step 2: Find the centroid for a set of boundary points which was extracted through the previous step.

Step 3: Determine the distance of each boundary point from the centroid and also find the highest and lowest distance.

Step 4: With highest distance as a radius, to draw a circle which encompasses all the boundary points and to store all the points of circle.

3.2 Fractal Dimension: Box-Counting Method

The algorithm used to calculate the fractal value for the extracted image of the mammogram using Box-counting method is shown in Appendix II. The red pixels or white pixels in the image cover the border of the object. A well-spaced grid with multiple small boxes of a particular pixel length was superimposed on the cell. Thus the FD of an image can be calculated with the boxes (2, 4, 8, 16, 32, 64, and 128).

4 Results

Computer-aided diagnosis techniques were used to improve the diagnostic accuracy and efficiency for screening the mammogram. For this several algorithms are used to separate the normal and the abnormal region from the mammogram. As mentioned earlier, the identification is a laborious process, while comparing the breast lesions with the normal report, two problems occur in the image processing which are as follows:

(i) To locate the malignant region in the breast mammogram and
(ii) To identify the area of the spread in different regions of the mammogram.

In addition to separate the malignancy from the breast mammogram through image processing, a need for an algorithm arises. Fractal dimension methods are applicable for image processing technique for which Mandelbrot formula was followed:

$$\text{Fractal Dimension} = \log N / \log (1/r) \qquad (2)$$

The algorithm used for FD is based on the above equation where they have used the number of boxes and its length. In this study, images from the oncology institute HCG, Bangalore were obtained to identify a suspicious region in the mammogram, which was illustrated and marked by the radiologist. This image of the breast cancer is used to segment and analyze the fractal dimension. The image scanned from left to right and top to bottom is brought to the system as a JPEG file. The JPEG image was used for further analysis. A method was proposed to extract cancer surface on an image of breast cancer and Box-counting method is used to calculate the fractal dimension for benign and malignant breast cancer.

4.1 Extraction of Breast Cancerous Surface: Marathon Database and HCG, Bangalore Database.

The desired images of the breast cancer mammogram in the JPEG format were opened as a picture file. An algorithm was employed for converting the given image into a black and white image by classifying each pixel with Red Green Blue (RGB) color and the respective positions were stored in an array. The mild abnormality includes, the calcification, well-defined or circumscribed masses, speculated masses, ill-defined masses, asymmetry, and the architectural distortion. The benign and malignant breast mammogram involved in this study was considered as group A and group B for calculating the fractal value. Thus, the image of breast cancer was processed using the above steps and the suspicious area was extracted, which is shown in the Fig. 1

4.2 Fractal Dimension: Box-Counting Method

The fractal value for the extracted breast cancer image from the Marathon Database is presented in Table 1.

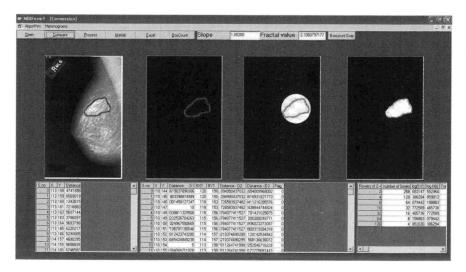

Fig. 1 Extraction of the abnormal region from the mammogram

Table 1 Fractal value of an image—marathon database value for the breast

Powers of 2^r	Number of boxes	log(1/r)	log N(r)	Fractal value
2	264	−0.69315	5.575949	−8.0444
4	59	−1.38629	4.077537	−2.94132
8	12	−2.07944	2.484907	−1.19499

Usually the fractal value of an image obtained will be negative. The calculated fractal value for the eight benign and ten malignant breast mammograms using an algorithm is observed in the Table 2 in which the sample value 2.2297 (Malignant) is shown in the Fig. 2.

Similarly, the fractal values for the benign and malignant in the above table were calculated from their respective slope. With the benign and malignant fractal value, the mean and the standard deviation were calculated to find the significance of the results. The fractal value for the benign mammogram ranges from 2.2685 to 2.5850 and for malignant from 2.0922 to 2.4236. The fractal mean and standard deviation for the benign breast mammogram was found to be 2.4621 ± 0.11384 and for malignant as 2.3028 ± 0.09454, respectively. The result obtained by the above method was used for classifying the unknown breast mammogram into benign or malignant.

The fractal value was calculated for ten benign and ten malignant mammograms based on the number of boxes using the Eq. 1. The fractal value for the benign and malignant mammograms ranges from 2.2759 to 2.6904 and 2.1771 to 2.4663, respectively. The Mean Fractal value for the benign and malignant tumor in the mammograms was found to be 2.44834 ± 0.16039 and 2.35554 ± 0.077322, respectively. Similarly, the output obtained by this method was also used for classifying the mammogram into benign or malignant tumor.

Table 2 Mean fractal value for the breast cancer mammograms—marathon database

S.No	n = 8 Benign	n = 10 Malignant
1	2.5294	2.2297
2	2.3171	2.3068
3	2.5284	2.3767
4	2.4886	2.3342
5	2.4352	2.4236
6	2.5850	2.0922
7	2.5443	2.3317
8	2.2685	2.2458
9		2.3095
10		2.3774
Mean	2.4621	2.3028
SD	0.11384	0.09454

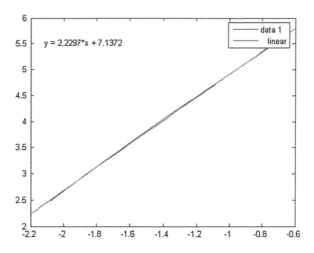

Fig. 2 Least square method of regression analysis — marathon database

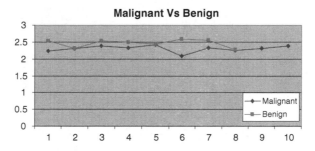

Fig. 3 Fractal value for the benign and malignant mammograms

4.3 Linear Regression and Statistical Method

Log(1/r) and log [N(r)] in X-axis Y-axis co-ordinates was used in MATLAB for the (n) number of boxes, marathon database and presented in Fig. 2. The number of boxes and dimension of boxes in the graph was observed as a straight line which was predicted using Least Square Method of Regression Analysis and the fractal dimension was obtained as the slope of the line as 2.2297. Statistical t-test was calculated between two independent samples of size eight and ten and was found to be as 3.2407. There was a significant difference between benign and malignant at 1 % (2.95) level of significance, respectively. The fractal dimension of the benign and malignant was observed as 2.4621 ± 0.14 and 2.3028 ± 0.09 with 95 % security coefficient in the Table 3.

Using the fractal values obtained from the benign and malignant breast cancer mammogram, a graph was drawn and shown in the Fig. 3

Table 3 Statistical calculations for fractal dimension of benign and malignant

	Parameter	Benign	Malignant
	Mean	2.4621	2.3028
	Stdandard error of mean	0.04025	0.02990
	Median	2.5085	2.3206
	Mode	2.27	2.09
	Stdandar deviation	0.1138	9.454E−02
	Variance	1.296E−02	8.938E−03
	Skewness	−0.941	−1.195
	Standard error of skewness	0.752	0.687
	Kurtosis	−0.490	1.890
	Standard error of kurtosis	1.481	1.334
	Range	0.32	0.33
	Minimum	2.27	2.09
	Maximum	2.59	2.42
	Sum	19.70	23.03
Percentiles	25	2.3466	2.2418
	50	2.5085	2.3206
	75	2.5406	2.3769

Therefore, the methodology adopted above proves that the fractal dimension of benign is greater than malignant breast mammogram.

The FD values of the benign and malignant were subjected to 't' test and found, there is a significant difference between benign and malignant at 1 % probability level, which will reveal that the FD of benign is greater than the malignant breast mammogram.

Determination of the Fractal Box-counting Dimension of a cell surface was applied to derive a quantitative measure for the raggedness of cells or small biological organisms [8]. Analysis of fractal dimension is used to find the roughness value and to locate the suspicious region in the mammogram of the breast cancer [5]. Sign of breast cancer is the appearance of clustered microcalcifications whose individual particles are less than 0.5 mm in diameter with irregular and heterogeneous shape [12]. Fractal properties are used to differentiate the tumor from healthy tissue and for the segmentation within an image, but it is not possible to compare fractal properties between images [13]. Fractal properties alone are not sufficient for effective texture segmentation and suggested the use of fractal features in texture classification [6, 10]. In the study of frequency distribution histogram, a large proportion of the lymphocytes of hairy cell leukemia patients had a fractal dimension exceeding 1.28 [8]. The fractal dimension calculated by applying the Box-counting method for the proximal convoluted tubule of the dog kidney was obtained as 1.33 ± 0.18 with 95 % security coefficient [14]. The metastases of colorectal cancer with fractal dimensions greater than 1.35 were associated with poor survival rate [15].

Fractal dimensions of surface growth patterns in different grades of endometrial adenocarcinoma ranged from 2.318 to 2.383 [16]. These values were greater than the topological dimension of a surface, reflecting a three-dimensional structure. This concludes that the endometrial adenocarcinoma has a fractal structure, and that the Mean FD may differ according to histological grades. Fractal dimension analysis was used to identify the migration in breast cancer cell wound healing assay. In image intensity fluctuation, fractal dimension analysis can be used as a tool to quantify cell migration in terms of cancer severity and treatment responses [17]. Fractal analysis based on Chen's method for the tumor images, the slope of the benign and the malignant line was found to be 0.3644 and 0.3425, respectively [6]. Similar studies were applied for the smear of breast and cervical lesions where they calculated the fractal dimension values as 0.8536 ± 0.1120 for malignant and 0.8403 ± 0.1115 for benign [18]. The Mann-Whitney U test of these two samples shows significant difference at 5 % probability level for cervix, whereas 2 % probability level for breast cancer. Significant difference between the fractal dimension of benign and malignant cells at $p = 0.006$ [19].

5 Conclusion

From the above study it is concluded that the measurement of fractal dimension is helpful in discriminating the malignant and benign breast lesions and to study the screening of classic image morphology based on Euclidean geometry. The algorithms proposed in this study extracted the similar suspicious area. Therefore, any one of the proposed methods can be adopted for easy extraction from breast cancer mammogram in advanced and benign cases.

In this study, fractal dimension has been applied to identify microcalcifications and tumors in the tissues of the body. Through this method, the size, location, and seriousness of the abnormality or suspicious regions are estimated for better diagnosis. However, a more intricate and higher level study, i.e., at cellular level is required for treatment of benign or malignant tumors.

Appendix I

Algorithm for Boundary Extraction from the Mammogram.

```
For i = 0 To Picture1.ScaleWidth
For j = 0 To Picture1.ScaleHeight
tcol1 = GetPixel(Picture1.hdc, i, j)
r1 = tcol1 Mod 256
g1 = (tcol1 Mod 256) \ 256
b1 = tcol1 \ 256 \ 256
If r1 > 200 And g1 < 200 And b1 < 200 Then
tcol1 = vbRed
Else
tcol1 = vbBlack
End If
SetPixel Picture2.hdc, i, j, tcol1
Next j
Next i
```

Appendix II

Algorithm for calculating the fractal value using Box-counting method.

```
Begin
Set N(ε) < −0 for all values of ε
For each pixel, S(x, y), in the image
M = 0
For each cell of size ε
Center a cell of size c on (x, y)
If S(x − ε/2, y − ε/2) to S(x + ε/2, y + ε/2) ≠ 0 then
M = 1
End if
End for
If m = 1, then increment N(ε) by 1
End for
Estimate D as the regression slope of log(ε) versus log(1/N(ε))
```

References

1. Buczko, W., Mikolajczak, D.C.: Shape analysis of MR brain images based on the fractal dimension. Annales UMCS Informatica (AI) **3**, 153–158 (2005)
2. Georgsson, F., Jansson, S., Olsen, C.: Fractal analysis of mammograms. In: Ersboll, B.K., Pedersen, K.S. (eds.) SCIA, LNCS, vol, pp. 92–101. 4522. Springer, Heidelberg (2007)
3. Baish, J.W., Jain, R.K.: Fractals and cancer. Cancer Res. **60**, 3683–3688 (2000)
4. Mandelbrot, B.B.: How long is the coast of britain? statistical self-similarity and fractional dimension. Science **156**(3775), 636–638 (1967)
5. Rejani, Y.A.I., Thamaraiselvi, S.: Digital mammogram segmentation and tumour detection using artificial neural networks. Int. J. Soft Comput. Medwell **3**(2), 112–119 (2008)
6. Chen, C.C., Daponte, J.S., Fox, M.D.: Fractal feature analysis and classification in medical imaging. IEEE Trans. Med. Imaging. **8**(2), 133–142 (1990)
7. Mohamed, A.W., Kadah, M.Y.: Computer aided diagnosis of digital mammograms. IEEE 1, 299–303 (2007). ISSN 4244-1366-4/07.
8. Bauer, W., Mackenzie, D.C.: Cancer detection via determination of fractal cell dimension. Heavy Ion Phys. **14**, 39–46 (2001)
9. Cross, S.S., Cotton, D.W.K.: The fractal dimension may be a useful morphometric discriminant in histopathology. J. Path. **166**, 409–411 (1992)
10. Lee, W.L., Chen, Y.C., Chen, Y.C., Hsieh, K.S.: Unsupervised segmentation of ultrasonic liver images by multiresolution fractal feature vectors. Inf. Sci. **175**(3), 177–199 (2005)
11. Rose, C., Turi, D., Williams, A., Wolstencroft, K., Taylor, C.: University of south florida digital mammography home page. IWDM. http://marathon.csee.usf.edu/Mammography/ Database.html, (2006)
12. Kopans, D.B.: Breast Imaging Medicine, Lippincott Williams and Wilkins, ISBN-10.PS, Philadelphia (1989).
13. Veenland, J., Grashuis, J.L., Van der Meer, F., Beckers, A.L.D., Gelsema, E.S.: Estimation of fractal dimension in radiographs. Med. Phys. **23**, 584–585 (1996)
14. Gil, J., Gimeno, M., Laborda, J., Nuviala, J.: Fractal dimension of dog kidney proximal convoluted tubuli sections by mean Box-counting algorithm. Int. J. Morphol. **24**(4), 549–554 (2006)
15. Metser, U.: 18 F-FDG PET in evaluating patients treated for Metastatic colorectal cancer: can we predict prognosis? J. Nucl. Med. **45**(9), 1428–1430 (2004)
16. Kikuchi, A., Kozuma, S., Yasugi, T., Taketani, Y.: Fractal analysis of surface growth patterns in endometrioid endometrial adenocarcinoma. Gynecol. Obstet. Invest. **58**(2), 61–67 (2004)
17. Sullivan, R., Holden, T., Jr Tremberger, G., Cheung, E., Branch, C., Burrero, J., Surpris, G., Quintana, S., Rameau, A., Gadura, N., Yao, H., Subramaniam, R., Schneider, P., Rotenberg, A., Marchese, P., Flamhlolz, A., Lieberman, D., Cheung, T.: Fractal dimension of breast cancer cell migration in a wound healing assay. Eng. Technol. 34, 25–30 (2008). ISSN 2070-3740.
18. Ohri, S., Dey, P., Nijhawan, R.: Fractal dimension in aspiration cytology smears of breast and cervical lesions. Anal. Quant. Cytol. Histol. **26**(2), 109–112 (2004)
19. Dey, P., Mohanty, K.S.: Fractal dimensions of breast lesions on cytology smears. Diagn. Cytopathol. 29(2), 85–86 (2003).
20. Snedecor, W.G., Cochran, G.W.: Statistical Methods, 7th edn. The IOWA State University Press, Ames (1980)

Approaches of Computing Traffic Load for Automated Traffic Signal Control: A Survey

Pratishtha Gupta, G. N. Purohit and Adhyana Gupta

Abstract Traffic images captured using CCTV camera can be used to compute traffic load. This document presents a survey of the research works related to image processing, traffic load, and the technologies used to re-solve this issue. Results of the implementation of two approaches: morphology-based segmentation and edge detection using sobel operator, which are close to traffic load computation have been shown. Segmentation is the process of partitioning a digital image into its constituent parts or objects or regions. These regions share common characteristics based on color, intensity, texture, etc. The first step in image analysis is to segment an image based on discontinuity detection technique (Edge-based) or similarity detection technique (Region-based). Morphological operators are tools that affect the shape and boundaries of regions in the image. Starting with dilation and erosion, the typical morphological operation involves an image and a structure element. The edge detection consists of creating a binary image from a grayscale image where the pixels in the binary image are turned off or on depending on whether they belong to region boundaries or not. Image processing is considered as an attractive and flexible technique for automatic analysis of road traffic scenes for the measurement and data collection of road traffic parameters. Combined background differencing and edge detection and segmentation techniques are used to detect vehicles and measure various traffic parameters. Real-time measurement and analysis of road traffic flow parameters such as volume, speed and queue are increasingly required for traffic control and management.

P. Gupta (✉) · G. N. Purohit · A. Gupta
Banasthali University, Banasthali, Rajasthan, India
e-mail: pratishtha11@gmail.com

G. N. Purohit
e-mail: gn_purohitjaipur@yahoo.co.in

A. Gupta
e-mail: adhyanagupta@gmail.com

B. V. Babu et al. (eds.), *Proceedings of the Second International Conference on Soft Computing for Problem Solving (SocProS 2012), December 28–30, 2012*, Advances in Intelligent Systems and Computing 236, DOI: 10.1007/978-81-322-1602-5_99, © Springer India 2014

Keywords Image processing · Simulation · Segmentation · Edge detection · Real time · Traffic load computation

1 Introduction

This study describes all the functions and specifications which can be used for Traffic Load Computation using traffic images, which can be further used for real-time image processing for traffic signal control. In this paper, after analyzing existing video object segmentation algorithms, it is found that most of the core operations can be implemented with simple morphological operations. Therefore, with the concepts of morphological image processing element array and stream processing, a reconfigurable morphological image processing accelerator is proposed, whereby the proposed instruction set, the operation of each processing element can be controlled, and the interconnection between processing elements can also be reconfigured.

Field Programmable Gate Array (FPGA) technology has become an alternative for the implementation of software algorithms. The unique structure of the FPGA has allowed the technology to be used in many applications from video surveillance to medical imaging applications. FPGA is a large-scale integrated circuit that can be reprogrammed.

This study comprises of eight sections including the present one, which provides an introduction and objectives of this brief survey same purpose. Section 7 compares the two approaches implemented in Sects. 5 and 6. Finally Sect. 2 presents various image processing approaches. Section 3 presents various research works aimed at computing traffic load. Section 4 presents various technologies used in this field. Section 5 implements one of the morphological approaches for traffic load computation purpose. Section 6 presents the implementation of an approach using sobel operator for the conclusion drawn in Sect. 8.

2 Image Processing Approaches

2.1 Image Segmentation

Shao-Yi Chien and Liang-Gee Chen [1] discussed in this paper, after analyzing existing video object segmentation algorithms, it is found that most of the core operations can be implemented with simple morphology operations. Therefore, with the concepts of morphological image processing element array and stream processing, a reconfigurable morphological image processing accelerator is proposed, whereby the proposed instruction set, the operation of each processing element can be controlled, and the interconnection between processing elements can also be reconfigured.

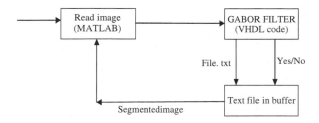

Fig. 1 Use of Gabor filter

Thakur et al. [2] explain tonsillitis, tumor, and many more skin diseases can be detected in its early state and can be cured. Image segmentation is the processes of partitioning a digital image into multiple segments that is sets of pixels Fig. 1.

Ramadevi et al. [3] discussed in this paper image segmentation is to partition an image into meaningful regions with respect to a particular application. Image segmentation is the process of partitioning/subdividing a digital image into multiple meaningful regions or sets of pixels regions with respect to a particular application. The three steps in edge detection process is: (a) filtering, (b) enhancement, and (c) detection. This paper focuses mainly on the Image segmentation using edge operators.

Salem Saleh Al amri et al. [4] have proposed segmentation algorithms based on one of two basic properties of intensity values discontinuity and similarity. First category is to partition an image based on abrupt changes in intensity, such as edges in an image. Second category is based on partitioning an image into regions that are similar according to predefined criteria.

Bo Peng et al. [5] discussed image segmentation as an inference problem, where the final segmentation is established based on the observed image. In addition, a faster algorithm has been developed to accelerate the region merging process, which maintains a nearest neighbor graph (NNG) in each iteration.

2.2 Digital Image Segmentation

Haris Papasaika-Hanusch [6] explains a digital image differs from a photo in that the values are all discrete.

- **Image Enhancement**: Processing an image so that the result is more suitable for a particular application (sharpening or deblurring an out of focus image, highlighting edges, improving image contrast, or brightening an image, removing noise).
- **Image Restoration**: This may be considered as reversing the damage done to an image by a known cause (removing of blur caused by linear motion, removal of optical distortions).
- **Image Segmentation:** This involves subdividing an image into constituent parts, or isolating certain aspects of an image (finding lines, circles, or particular shapes in an image, in an aerial photograph, identifying cars, trees, buildings, or roads).

2.3 Edge Detection

Allin Christe et al. [7] discussed in this paper focuses on processing an image pixel by pixel and in modification of pixel neighborhoods and the transformation that can be applied to the whole image or only a partial region.

Draper et al. [8] explain although computers keep getting faster and faster, there are always new image processing (IP) applications that need more processing than is available. Examples of current high-demand applications include real-time video stream encoding and decoding, real-time biometric (face, retina, and/or fingerprint) recognition, and military aerial and satellite surveillance applications.

2.4 Fuzzy Edge Detection

Sriramakrishnan et al. [9] explain edge is a basic feature of image; edge detection is a process of identifying and locating edges in image which is vital for image segmentation. Distortion, noise, overlaps, and intensity variation are some of the factors, which contributes to edge extraction. Image dentification and segmentation pose a challenge to image retrieval process.

3 Traffic Load Computation

Gupta et al. [10] have also proposed a model capable of managing intelligent traffic system using CCTV cameras and WAN. The proposed model will make the traffic signaling dynamic and automatic as well. Besides this, it will generate the dynamic messages for the users on the message boards to avoid congestion, reduce waiting time, pollution control, accident control, and vehicle tracking. This brief survey presents various approaches for intelligent traffic systems.

Duan et al. [11] discussed due to a huge number of vehicles, modern cities need to establish effectively automatic systems for traffic management and scheduling. This method optimizes speed and accuracy in processing images taken from various positions.

Abhijit Mahalanobis et al. [12] explain the detection and tracking of humans as well as vehicles is of interest. The three main novel aspects of the work presented in this paper are (i) the integration of automatic target detection and recognition techniques with tracking, (ii) the handover and seamless tracking of objects across a network, and (iii) the development of real-time communication and messaging protocols using COTS networking components.

Siyal et al. [13] has presented real-time measurement and analysis of road traffic flow parameters such as volume, speed, and queue are increasingly required for traffic control and management. Many techniques have been proposed to speed up the analysis of road traffic images, and some intelligent approaches have been developed to compensate for the effects of variable ambient lighting, shadows, occlusions, etc., in the road traffic images.

Kastrinaki et al. in [14] present an overview of image processing and analysis tools used in these applications, and we relate these tools with complete systems developed for specific traffic applications. Image processing also finds extensive applications in the related field of autonomous vehicle guidance, mainly for determining the vehicle's relative position in the lane and for obstacle detection.

Koutsia et al. [15] explain traffic control and monitoring using video sensors has recently drawn increasing attention, due to the significant advances in the field of computer vision. The paper presents a real-time vision system for automatic traffic monitoring based on a network of autonomous tracking units that capture and process images from one or more recalibrated cameras.

Ejaz [16] discussed manually configured traditional time-based traffic signals are not categorically efficient in controlling traffic and merely tend optimize a certain traffic condition, which may only occur at a specific time of day. By doing that it may also create a highly unoptimized situation for some other traffic conditions. Sensor-based approaches tend to provide only a limited number of parameters, which may prove to be incomplete for making fully context aware decisions for road signals. Large cities of developing countries face a huge rise in the total number of on-road vehicles over a period of last few years.

Bosman [17] discussed traffic loads are the most important variable actions to be accounted for in the design of road pavements and bridges. Axle mass is probably one of the main factors that determine the effect of traffic loads on pavements. In the case of bridges, it is not only the axle mass, but also the spacing between axles, that determines the effect of traffic loads on the different structural elements.

4 Technologies

Parker et al. [18] explain the goal of this project is to implement histogram equalization algorithm using MATLAB for a real-time processing system on a FPGA. The histogram equalization algorithm was implemented and tested using a known 4×4 array.

Chikalli et al. [19] discussed Histogram is used for automatically determining the threshold for different region in image. (Fig. 2) The histogram is a very important tool in image analysis.

Ali et al. [20] evaluated in Xilinx system generator is a very useful tool for developing computer vision algorithm. Image processing is used to modify the picture, extract information, and change their structure. This paper focuses in the processing

Fig. 2 Use of histogram tool

pixel to pixel of an image and in the modification of pixel neighborhoods and of course the transformation can be applied to the whole image or only a partial region.

Elamaran et al. [21] discussed a real-time image processing algorithm are implemented on FPGA. Implementation of these algorithm on a FPGA is having advantage of using large memory and embedded multipliers.

Chandrashekar et al. [22] discussed enhancing digital image to extract true image is a desired goal in several applications, such transformation is known as image enhancement. Performing the task automatically without human intervention is particularly hard in image processing. FPGA implemented result compared with Matlab experiments and comparisons to histogram equalization are conducted.

Acharya et al. [23] discussed an efficient FPGA-based hardware design for enhancement of color and gray scale image in image and video processing. The approach that is used is adaptive histogram equalization, which works very effectively for image captured under extremely dark environment as well as nonuniform lighting environment where bright regions are kept unaffected and dark object in bright background.

Gribbon et al. [24] explain FPGA as implementation platforms for real-time image processing applications because the structure allows them to exploit spatial and temporal parallelism. High level languages and compilers which automatically extract parallelism from the code do not always produce an efficient mapping to hardware.

Devika et al. [25] explain FPGA technology has become viable target for the implementation of real-time algorithms suited to video image processing applications. The FPGA technologies offer basic digital blocks with flexible interconnections to achieve high speed digital hardware realization Fig. 3.

Anusha et al. [26] discussed the image processing algorithms has been limited to software implementation which is slower due to the limited processor speed.

Fig. 3 Typical steps in image processing algorithms

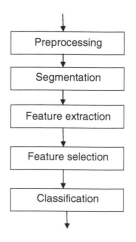

5 Implementation of Morphological Based Segmentation

FPGA technology has become viable target for the implementation of real-time algorithms suited to video image processing applications. The FPGA technologies offer basic digital blocks with flexible interconnections to achieve high speed digital hardware realization.

5.1 Morphological Operations

Morphological operators are defined as combinations of basic numerical operations taking place over an image A and a small object B, called a structuring element. B can be seen as a probe that scans the image and modifies it according to some specified rule. The shape and size of B, which is typically much smaller than image A, in conjunction with the specified rule, define the characteristics of the performed process. Binary mathematical morphology is based on two basic operators: Dilation, and erosion. Both are defined in terms of the interaction of the original image A to be processed, and the structuring element B. The Morphological operations are:

(i) **Dilation Operation**: The basic effect of the operator on a binary image is to gradually enlarge the boundaries of regions of foreground *pixels* (i.e. white pixels, typically). Thus areas of foreground pixels grow in size while holes within those regions become smaller.

(ii) **Erosion Operation**: The basic effect of the operator on a binary image is to erode away the boundaries of regions of foreground pixels (i.e. white pixels, typically). Thus areas of foreground pixels shrink in size, and holes within those areas become larger.

(iii) **Opening Operation**: The basic effect of an opening is somewhat like erosion in that it tends to remove some of the foreground (bright) pixels from the edges of regions of foreground pixels. However it is less destructive than erosion in general. As with other morphological operators, the exact operation is determined by a *structuring element*.

(iv) **Closing Operation**: It tends to enlarge the boundaries of foreground (bright) regions in an image (and shrink background color holes in such regions), but it is less destructive of the original boundary shape. As with other *morphological operators*, the exact operation is determined by a *structuring element*. The effect of the operator is to preserve background regions that have a similar shape to this structuring element, or that can completely contain the structuring element, while eliminating all other regions of background pixels.

(v) **Thinning operation**: It is particularly useful for *skelitonization*. In this mode it is commonly used to tidy up the output of *edge detectors* by reducing all lines to single pixel thickness. Thinning is normally only applied to binary images, and produces another binary image as output. The thinning operation is related

ALGORITHM:

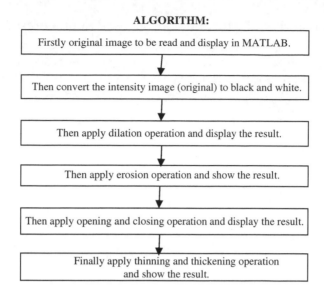

Fig. 4 Steps of morphological approach

to the *hit-and-miss transform*, and so it is helpful to have an understanding of that operator before reading on.

(vi) **Thickening operation**: It has several applications, including determining the approximate *convex hull* of a shape, and determining the *skeleton by zone of influence*. Thickening is normally only applied to binary images, and it produces another binary image as output Fig. 4.

5.2 Experimental Results

The experimental results for Morphological based Segmentation in MATLAB are shown below:

Figure 5. Firstly original image to be read and display in MATLAB.

Figure 6. Then convert the intensity image (original) to black and white.

Figure 7. Then apply dilation operation and display the result.

Figures 8 and 9. Then apply erosion operation and show the result.

Figures 10 and 11. Then apply opening and closing operation and display the result.

Figures 12 and 13. Finally apply thinning and thickening operation and show the result.

Fig. 5 Original image

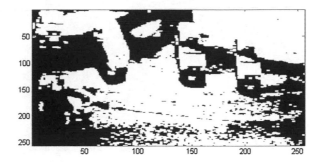

Fig. 6 Original image into black and white

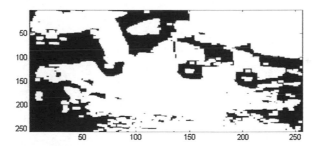

Fig. 7 Dilation operation

6 Implementation of SOBEL Edge Detection on FPGA

The proposed work presents FPGA based architecture for Edge Detection using Sobel operator and uses Histogram method for Segmentation. The data of edge detection is very large so the speed of image processing is a difficult problem. FPGA can overcome it. Sobel operator is commonly used in edge detection. Sobel operator has been researched for parallelism but Sobel operator locating complex edges are not accurate.

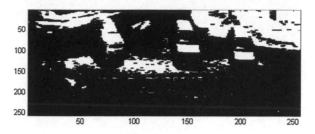

Fig. 8 Erosion operation

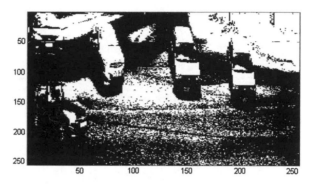

Fig. 9 Erosion operation

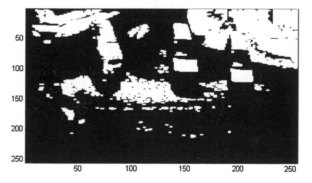

Fig. 10 Opening operation

Edge detection

Edge detection is a method of determining the discontinuities in gray level images. Conventional edge detection mechanisms examine image pixels for abrupt changes by comparing pixels with their neighbors. This is often done by detecting the maximal value of gradient such as Roberts, Prewitt, Sobel, Canny, and so on all of which are classical edge detectors.

Sobel Edge Detection

The Sobel operator is a classic first order edge detection operator computing an approximation of the gradient of the image intensity function. At each point in the

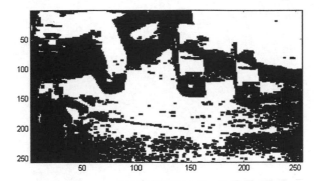

Fig. 11 Closing operation

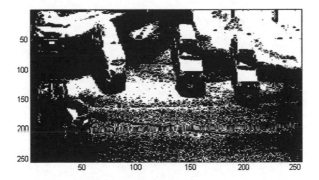

Fig. 12 Thinning operation

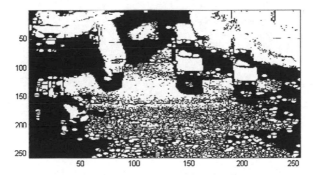

Fig. 13 Thickening operation

image, the result of the Sobel operator is the corresponding norm of this gradient vector. The Sobel operator only considers the two orientations which are 0 and 90 degrees convolution kernels as shown in Fig. 14.

These kernels can then be combined together to find the absolute magnitude of the gradient at each point. The gradient magnitude is given by:

Fig. 14 Convolution kernels
in X and Y direction

-1	-2	-1
0	0	0
1	2	1
Gy		

-1	0	1
-2	0	2
-1	0	1
Gx		

$$|G| = \sqrt{Gx^2 + Gy^2}$$

Typically an approximate magnitude is computed using:

$$|G| = |Gx| + |Gy|$$

This is much faster to compute.

The sobel operator has the advantage of simplicity in calculation.

But the accuracy is relatively low because it only used two convolution kernels to detect the edge of image.

6.1 Experimental Results

The experimental results for image edge detection in MATLAB are shown below:

Figure 15 is the original image for edge detection.

Figure 16 shows gray scale image for edge detection.

Figure 17 shows edge detection result using MATLAB. (Table 1)

Fig. 15 Original image for edge detection

Fig. 16 Gray scale for edge detection

Fig. 17 Result of edge detection

7 Comparison

Table 1 Comparison of morphological and sobel operator approaches

Morphological-based segmentation	SOBEL Edge Detection on FPGA
Segmentation could be also obtained using morphological operations	Sobel operator due to its property of less deterioration in high level of noise
Segmentation subdivides an image into its constituent regions or objects	Edge detection is used to check the Sudden change in images
Segmentation should stop when the objects of interest in an application have been isolated	It analysis and processing the image with the help sobel operator
For example, in the automated inspection of electronic assemblies, interest lies in analyzing images of the products with the objective of determining the presence or absence of specific anomalies, such as missing components or broken connection paths	The execution time for the entire program of edge detection for an image of size 256×256 is few seconds. The execution time for the entire program of edge detection for an image of size 256×256 is few seconds
Image processing algorithms are conventionally implemented in DSP processors and some special purpose processors	To improve the speed and efficiency pipelining can be further done in edge detection
In morphology we have performed erosion, dilation, opening, closing, thinning and thickening	The edge detection is a terminology in image processing particularly in the areas of feature extraction to refer to algorithms which aim at identifying points in a digital image at which the image brightness changes sharply

8 Conclusion

In this paper, we introduced the existing image processing algorithms suitable for real-time traffic load computation and analyzed their outputs. We implemented two important approaches edge detection and morphology based segmentation and provided a qualitative comparison of those two approaches for our requirement. After analyzing existing image object segmentation algorithms, it is found that most of the core operations can be implemented with simple morphology operations. Therefore, with the concepts of morphological image processing element array and stream processing, a reconfigurable morphological image processing accelerator is proposed, whereby the proposed instruction set, the operation of each processing element can be controlled, and the interconnection between processing elements can also be reconfigured.

In summary, the open issue in real-time traffic load computation is to find the most efficient algorithm to compute traffic load in real time. The research direction is that as technologies in hardware are advancing and becoming mature, digital image processing algorithms need to be developed so as to get the real-time benefits of these hardware technologies.

References

1. Chien, S.-Y., Chen, L.-G.: Reconfigurable Morphological Image Processing Accelerator for Video Object Seg-mentation. Signal. Process. Syst. **62**(1), 77–96 (2011)
2. Thakur, R.R., Dixit, S.R., Dr.Deshmukh, A.Y.: VHDL design for image segmentation using gabor filter for disease detection. Int. J. VLSI Design. Commun. Sys. 3(2), 211 (2012).
3. Ramadevi, Y., Sridevi, T., Poornima, B., Kalyani, B.: Segmentation and object recognition using edge detection techniques. Int. J. Comp. Sci. Info. Technol. **2**(6), 153–161 (2010)
4. Al-amri, S.S., Kalyankar, N.V., Khamitkar, S.D.: Image segmentation by using thershod techniques. J. Comput. 2(5), ISSN 2151–9617 (2010).
5. Peng, B., Zhang, L., Zhang, D.: Automatic image segmentation by dynamic region merging. The Hong Kong Polytechnic University, Hong Kong (2010)
6. Papasaika-Hanusch, H.: Digital image processing using matlab. ETH Zurich, Zurich (1967)
7. Mrs. Allin Christe, S., Mr. Vignesh, M., Dr. Kandaswamy, A.: An efficient FPGA implementation of MRI image filtering and tumour characterization using Xilinx system generator. Int. J. VLSI. Des. Comm. Sys. 2(4), (2011).
8. Draper, B.A.: Ross Beveridge, J., Willem Böhm, A.P., Ross, C., Chawathe, M.: Accelerated image processing on FPGAs. IEEE Trans. Image Process. **12**(12), 1543–1551 (2003)
9. Sriramakrishnan, C., Shanmugam, A.: Image Retrieval Optimization Using FPGA Based Fuzzy Segmentation, ISSN 1450–216X 63(1) (2011).
10. Gupta, P., Purohit, G.N., Dadhich, A.: Approaches for intelligent traffic system: a survey. Banastahli University, Jaipur (2012)
11. Duan, T.D., Du Hong, T.L., Phuoc, T.V.: Hoang. Building an automatic vehicle license- plate recognition system. Int. J. Adv. technol, N.V. (2005)
12. Abhijit Mahalanobis, Jamie Cannon, S. Robert, Stanfill, Robert Muise, Lockheed Martin, Network video image pro-cessing for security, Surveillance, and Situational Awareness. Digital. Wireless. Commun. doi:10.1117/12.548981.

13. Siyal, M.Y., Fathi, M., Atiquzzaman, M.: A parallel pipeline based multiprocessor system for real-time measurement of road traffic parameters. Int. J. Imaging. Sys. Technol. **21**(3), 260–270 (2011)
14. Kastrinaki, V., Zervakis, M., Kalaitzakis, K.: A survey of video processing techniques for traffic applications. Image. Vision. Comput. **21**, 359–381 (2003)
15. Koutsia, A., Semertzidi1, T., Dimitropoulos, K., Grammalidis, N.: Intelligent traffic monitoring and surveillance with multiple camras. In: Proceedings of International Workshop on Content-Based Multimedia Indexing (CBMI '08), 125–132 (2008).
16. Ejaz, Z.: Morphological image processing based road traffic signal control system.
17. Bosman, J.: Traffic loading characteristics of south african heavy vehicles.
18. Parker, S.: Ladeji-Osias. Implementing a histogram equalization algorithm in reconfigurable hardware, J.K. (2009)
19. Ms. Chikkali, P S.: FPGA based Image edge detection and segmentation. Int. J. Adv. Eng. Sci. **9**(2), 187–192 (2011).
20. Ali, S.M., Mr. Naveen, Mr. Khayum.: FPGA based design and implementation of image architecture using XILINX system generator, IJCAE, **3**(1), 132–138 (2012).
21. Elamaran, V., Rajkumar, G.: FPGA implementation of point processes using Xilinx system generator **41**(2), (2012).
22. Chandrashekar, M., Naresh Kumar, U., Sudershan Reddy, K., Nagabhushan Raju, K.: FPGA implementation of high speed In: Frared Image Enhancement, ISSN 0975–6450 **1**(3), 279–285 (2009).
23. Acharya, A., Mehra, R., Takher, V.S.: FPGA based non uniform illumination correction in image processing applications Int. J. Comp. Tech. Appl. **2**(2), 349–358 (2009)
24. Gribbon, K. T., Bailey, D. G., Johnston, C.T.: Design patterns for image Processing Algorithm Development on FPGAs.
25. Devika, S.V., Khumuruddeen, S.K., Alekya.: Hardware implementation of Linear and Morphological Image Processing on FPGA. **2**(1), 645–650 (2012).
26. Anusha, G., Dr.JayaChandra Prasad, T., Dr.Satya Narayana, D.: Implementation of SOBEL edge detection on FPGA. **3**(3) (2012).

Comparative Analysis of Neural Model and Statistical Model for Abnormal Retinal Image Segmentation

D. Jude Hemanth and J. Anitha

Abstract Artificial Neural Networks (ANN) are gaining significant importance in the medical field, especially in the area of ophthalmology. Though the performance of ANN is theoretically stated, the practical applications of ANN are not fully explored. In this work, the suitability of Back Propagation Neural Network (BPN) for ophthalmologic applications is highlighted in the context of retinal blood vessel segmentation. The neural technique is tested with Diabetic Retinopathy (DR) images. The performance of the BPN is compared with the k-Nearest Neighbor (k-NN) classifier which is a statistical classifier. Experimental results verify the superior nature of the BPN over the k-NN approach

Keywords Back propagation network · k-Nearest neighbor · Retinal images.

1 Introduction

Eye diseases are mostly gradual in nature which affects the human society to a high extent. The nature of the eye disease can be determined from the affected anatomical structures. Hence, detecting the anatomical structures like blood vessels is mandatory for treatment planning.

The literature survey reveals the variety of techniques available for blood vessel segmentation. Supervised methodologies are used for retinal vessel segmentation in [1]. Morphological approaches-based blood vessel segmentation is implemented in [2]. Blood segmentation using functional and textural features are reported in [3, 4]. The combined approach of line operators and support vector machines

D. J. Hemanth (✉) · J. Anitha
Department of ECE, Karunya University, Coimbatore, India
e-mail: jude_hemanth@rediffmail.com

J. Anitha
e-mail: rajivee1@rediffmail.com

B. V. Babu et al. (eds.), *Proceedings of the Second International Conference on Soft Computing for Problem Solving (SocProS 2012), December 28–30, 2012*, Advances in Intelligent Systems and Computing 236, DOI: 10.1007/978-81-322-1602-5_100, © Springer India 2014

are used for segmentation in [5]. Fuzzy-based techniques are used for segmentation applications in [6]. Filtering approaches are also used for blood vessel extraction in [7]. Statistical approaches are also widely used for vascular detection in retinal images [8]. Ridge-based detection methods are reported in [9]. Wavelet-based segmentation methodologies are also available in the literature [10].

Though many techniques are available, the segmentation efficiency of such techniques is not very high which ultimately limits the practical applications of such systems. This drawback is overcome in this proposed approach in which the automated system is based on Artificial Intelligence (AI). BPN is the segmentation technique proposed in this work and the results are compared with the k-NN technique.

2 Proposed Methodology

The automated system used for this application is shown in Fig. 1.

In this work, 40 retinal images provided by the DRIVE database are used for segmentation. It is then preprocessed by various methods which are followed by feature extraction process. These features are given to BPN for the training process. This trained network will be used to segment the retinal blood vessel from the retinal images. The same procedure is repeated with the k-NN technique except for the difference in the training methodology.

3 Image Pre-Processing

Retinal images usually have pathological noise and various texture backgrounds, light variations, and poor contrast which may cause difficulties in extraction. In order to reduce these imperfections and generate images more suitable for extracting the pixel features demanded in the classification step, a preprocessing comprising the

Fig. 1 Flow diagram of the proposed work

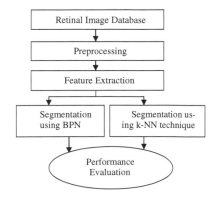

following steps is applied: (1) green channel extraction, (2) gray scale conversion, (3) adaptive histogram equalization, (4) vessel central light reflex removal, and (5) background homogenization.

4 Feature Extraction

Feature extraction is a methodology of dimensionality reduction in which suitable features representing the blood vessel pixels and non-blood vessel pixels are extracted from the pre-processed images. The features used in this work are mean, standard deviation, skewness, kurtosis, energy, and entropy. These features are computed using the following equations prescribed by [11].

5 Segmentation Approaches

In this work, two segmentation techniques are employed. One is the neural technique and the other is the statistical technique.

5.1 Back Propagation Neural Network

BPN is a supervised neural network in which gradient descent method is used to minimize the total squared error of the output computed by the network. The proposed architecture consists of six neurons in the input layer, 15 neurons in the hidden layer, and two neurons in the output layer. Two set of weight matrices are involved in the architecture for the hidden layer and the output layer. In addition to the input vector and output vector, the target vector is given to the output layer neurons. Since BPN operates in the supervised mode, the target vector is mandatory. During the training process, the difference between the output vector and the target vector is calculated and the weight values are updated based on the difference value. After the segmentation process, a two-step post-processing is used: the first step is aimed at filling pixel gaps in detected blood vessels, while the second step is aimed at removing falsely detected isolated vessel pixels. A detailed algorithm is given in [12].

5.2 k-NN Approach

k-nearest neighbor (k-NN) classification makes the classification by getting votes of the k-nearest neighbors. Performance of k-NN classifier depends largely upon the efficient selection of k-nearest neighbors. Classification using an instance-based classifier can be a simple matter of locating the nearest neighbor in instance space and labeling the unknown instance with the same class label as that of the located (known) neighbor. This approach is often referred to as a nearest neighbor classifier. In the k-nearest neighbor (k-NN) algorithm, the classification of a new sample is determined by the class of its k-nearest neighbors. The value of k is selected randomly depending on the availability of the database and the nature of application. After segmentation, a thresholding operation is used as the post-processing step to extract the blood vessel region.

6 Experimental Results and Discussions

The software used for the implementation is MATLAB [13] and the processor specification is 1 GB RAM with 1.66 GHz clock frequency. Initially, the image pre-processing results are shown in Fig. 2.

Figure 2 illustrates the necessity for pre-processing steps. Figure 2h shows the pre-processed output which is much suitable for further processing than Fig. 2a. After pre-processing, the features are extracted from the images. The features are sufficiently different for the blood vessel category and non-blood vessel category. The segmented outputs for the two segmentation techniques are given in Fig. 3.

The qualitative results have verified the superior nature of the BPN over the statistical classifier. The blood vessels are clearly identified in the neural technique but it is hardly visible in the statistical technique. The orientation of k-NN classifier is given in a different direction. A quantitative analysis on the segmented images of

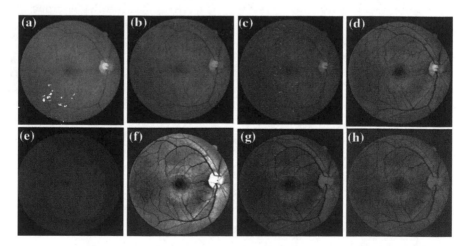

Fig. 2 Illustration of preprocessing process: **a** RGB image, **b** *Green* channel of the RGB image, **c** *Gray* scale image, **d** Adaptive histogram equalized image, **e** Background image, **f** Vessel central light reflex removed image, **g** Shade-corrected image, **h** Homogenized image

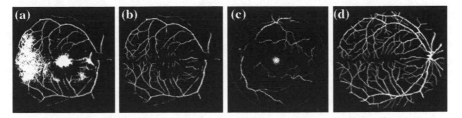

Fig. 3 Sample results: **a** BPN output before postprocessing, **b** BPN output after postprocessing, **c** k-NN output, **d** Target image

Table 1 Performance measures of BPN

Input	Se	Npv	Acc
Image1	0.7474	0.9800	0.9811
Image2	0.5702	0.9591	0.9612
Image3	0.7105	0.9800	0.9809
Image4	0.5779	0.9546	0.9511
Image5	0.6344	0.9626	0.9649
Image6	0.6564	0.9696	0.9713
Image7	0.6662	0.9688	0.9706

the BPN technique is given in Table 1. The BPN algorithm is evaluated in terms of Sensitivity (Se), Negative predictive value (Npv), and Accuracy (Acc). These values are displayed for randomly selected images from DRIVE database.

The ideal values of these performance measures are unity and the values indicated in these tables are closely related to unity which indicates the superior nature of the proposed approach. Thus, the advantages of the neural technique for ophthalmologic applications are verified in this work. The quantitative analysis of the k-NN approach is shown in Table 2.

The quantitative analysis has clearly shown the superior nature of the neural method. Only sample results are shown here but the same procedure has been experimented on all the images. The values of Negative Predictive Value and Accuracy are very low for the statistical technique in comparison to the neural-based approach. The average accuracy of the BPN technique is 97 % whereas the accuracy of the

Table 2 Performance measures of k-NN

Input	Se	Npv	Acc
Image1	0.3271	0.5002	0.5121
Image2	0.3002	0.4772	0.4987
Image3	0.3910	0.4832	0.4952
Image4	0.2512	0.4705	0.4800
Image5	0.3067	0.4812	0.4832
Image6	0.3126	0.4728	0.4325
Image7	0.3775	0.4912	0.4873

conventional technique is only 48 %. The other performance measures are also affected in the same manner. A 50 % increase in efficiency is obtained for the neural-based technique over the conventional statistical technique. Thus, this work has explored the application of neural-based techniques for retinal image analysis.

7 Conclusion

In this work, effort has been taken in exploring the usage of neural network for retinal image segmentation. The proposed BPN method segments the blood vessels in an efficient way so that further processing of this segmented image will help the ophthalmologist to diagnose diseases like diabetic retinopathy and glaucoma. The superior nature of ANN over other conventional techniques is also verified in this work.

Acknowledgments The authors thank Dr. A. Indumathy, Lotus Eye Care Hospital, Coimbatore, India for her help regarding database validation. The authors also thank the Council of Scientific and Industrial Research (CSIR), New Delhi, India for the financial assistance towards this research (Scheme No: 22(0592)/12/EMR-II).

References

1. Martin, D., Aquino, A., Arias, M.E.G., Bravo, J.M.: A new-supervised method for blood vessel segmentation in retinal images by using gray-level and moment invariants-based features. IEEE Trans. Med. Imaging **30**(1), 146–158 (2011)
2. Mendonca, A.M., Campilho, A.: Segmentation of retinal blood vessels by combining the detection of centerlines and morphological reconstruction. IEEE Trans. Med. Imaging **25**(9), 1200–1213 (2006)
3. Narasimha-Iyer, H., Beach, J.M., Khoobehi, B., Roysam, B.: Automatic identification of retinal arteries and veins from dual-wavelength images using structural and functional features. IEEE Trans. Biomed. Eng. **54**(8), 1427–1435 (2007)
4. Li, H., Chutatape, O.: Fundus image features extraction. In: Proceedings of the 22nd Annual EMBS International Conference, Chicago IL, July 2000, pp. 3071–3073.
5. Ricci, E., Perfetti, R.: Retinal blood vessel segmentation using line operators and support vector classification. IEEE Trans. Med. Imaging **26**(10), 1357–1365 (2007)
6. Kande, G.B., Savithri, T.S., Subbaiah, P.V.: Segmentation of vessels in fundus images using spatially weighted fuzzy c-means clustering algorithm. IJCSNS Int. J. Comput Sci. Netw. Secur. **7**(12), 102–109 (2007)
7. Sofka, M., Stewart, C.V.: Retinal vessel centerline extraction using multiscale matched filters, confidence and edge measures. IEEE Trans. Med. Imaging **25**(12), 1531–1546 (2006)
8. Lam, B.S.Y., Yan, H.: A novel vessel segmentation algorithm for pathological retina images based on the divergence of vector fields. IEEE Trans. Med. Imaging **27**(2), 237–246 (2008)
9. Staal, J., Abramoff, M.D., Niemeijer, M., Viergever, M.A., Ginneken, B.V.: Ridge-based vessel segmentation in color images of the retina. IEEE Trans. Med. Imaging **23**(4), 501–509 (2004)
10. Soares, J.V.B., Jorge, J., Leandro, G., Roberto, M., Herbert, F., Jelinek, F., Cree, M.J.: Retinal vessel segmentation using the 2-D morlet wavelet and supervised classification. IEEE Trans. Med. Imaging **25**(9), 1214–1222 (2006)

11. Haralick, R.M.: Statistical and structural approaches to texture. IEEE Trans. Syst. Man Cybern. **67**, 86–804 (1979)
12. Freeman, J.A., Skapura, D.M.: Neural Networks. Algorithms, Applications and Programming Techniques, Pearson education, New York (2004)
13. MATLAB: User's Guide, The Math Works Inc., Natick (1994–2002).

An Efficient Training Dataset Generation Method for Extractive Text Summarization

Esther Hannah and Saswati Mukherjee

Abstract The work presents a method to automatically generate a training dataset for the purpose of summarizing text documents with the help of feature extraction technique. The goal of this approach is to design a dataset which will help to perform the task of summarization very much like a human. A document summary is a text that is produced from one or more texts that conveys important information in the original texts. The proposed system consists of methods such as pre-processing, feature extraction, and generation of training dataset. For implementing the system, 50 test documents from DUC2002 is used. Each document is cleaned by pre-processing techniques such as sentence segmentation, tokenization, removing stop word, and word stemming. Eight important features are extracted for each sentence, and are converted as attributes for the training dataset. A high quality, proper training dataset is needed for achieving good quality in document summarization, and the proposed system aims in generating a well-defined training dataset that is sufficiently large enough and noise free for performing text summarization. The training dataset utilizes a set of features which are common that can be used for all subtasks of data mining. Primary subjective evaluation shows that our training is effective, efficient, and the performance of the system is promising.

Keywords Feature extraction · Dataset · Summarization · Pre-processing

E. Hannah (✉)
Department of Computer Science, Anna University, Chennai, India
e-mail: hanmoses@yahoo.com

S. Mukherjee
Department of Information Science and Technology, College of Engineering,
Anna University, Chennai, India

B. V. Babu et al. (eds.), *Proceedings of the Second International Conference on Soft Computing for Problem Solving (SocProS 2012), December 28–30, 2012,* Advances in Intelligent Systems and Computing 236, DOI: 10.1007/978-81-322-1602-5_101, © Springer India 2014

1 Introduction

The digital revolution has made digitized information easy to capture and fairly inexpensive to store. With the development of computer hardware and software, huge amount of data have been collected and stored in databases. The rate at which such data is stored is growing at a phenomenal rate. Several domains where large volumes of data are stored include the following: Financial investment, Health care, Manufacturing and Production, Telecommunication network, Scientific Domain and World Wide Web (WWW). The increasing availability of online information has necessitated intensive research in the area of automatic text summarization within the Natural Language Processing Community. Text summarization is a text reduction process, and a summary is a text that is produced from one or more texts, that convey important information in the original texts, and that is no longer than half of the original text.

2 Summarization Approaches

One of the very first works in automatic text summarization was done by Luhn et al. in 1958, demonstrates research work done in IBM, focused on technical documents [1]. Luhn proposed that the 'frequency of word' proves to be a useful measure in determining the significance factor of sentences. Words were stemmed to their roots having the stop words removed. Luhn generated a set of content words that helped to calculate the significance factor of sentences, which were then scored, and the top ranking sentences become the candidates to be part of the generated summary. In the same year, Baxendate et al. proved that the sentence position plays an important role in determining the significance factor of sentences [2]. He examined 200 paragraphs to find that in 85 % of the paragraphs the topic sentence came as the first one and in 7 % of the paragraphs it was the last sentence.

Edmundson in 1969 proposed a method for obtaining document extracts by using a linear combination of features such as cue words, keywords, title or heading, and sentence location [3]. Kupiec et al. in 1995, used human-generated abstracts as training corpus, from which he produced extracts, the feature set included sentence length, fixed phrases, sentence position in paragraph, thematic words, and uppercase words [4]. In 2004, Khosrow Kaikhah et al. proposed a new technique for summarizing news articles using a neural network that is trained to learn characteristics of sentences that should be included in the summary of the article. Rasim. M. Alguliev et al., in 2005 proposed a text summarization method that creates text summary by definition of the relevance score of each sentence by extracting sentences from the original documents [5].

Hsun-Hui Huang, Yau-Hwang Kuo, Horng-Chang Yang et al. in 2006, proposed to extract key sentences of a document as its summary by estimating the relevance of sentences through the use of fuzzy-rough sets [6]. The approach removes the problem that sentences of the same or similar semantic meaning but written in synonyms being treated differently.

Huantong Geng, Peng Zhao, Enhong Chen, Qingsheng Cai in the year 2006, described summarization based on subject information from term co-occurrence graph and linkage information of different subjects [7]. In 2007, Amini and Usunier et al. came forward with a Contextual Query Expansion Approach that makes use of Term Clustering [8]. Ladda Suanmali et al. [9] in 2009 developed a system that generates extractive summaries based on a set of features that represent the sentences in a text.. A variation to this is brought out in a sentence-oriented approach [11]. Other approaches such as multivariate [10], classification-based approaches are currently exploited for various summarization purposes.

3 Proposed Work

The overall data flow of the system is shown in Fig. 1, various processes through which data is processed. The overall system can be split into three major processes namely:

a. **Pre-processing**: Pre processing is done as a means of cleaning the document by removing words that do not contain any information that uniquely identifies a sentence.
b. **Feature Extraction**: The Feature Extraction includes 7 important features: Title Feature, Sentence Length, Term Weight, Sentence to sentence similarity, Proper Noun, Thematic Word, and Numerical Data.
c. **Generate Training Dataset**: Training datasets help to train models that will perform tasks like classification, clustering, summarization, etc. The feature 'class' identifies whether the chosen sentence is part of the summary document or not. The 'class' is marked as 'INT' or 'NOT_INT' based on its presence or absence in the summary document.

Fig. 1 Proposed system

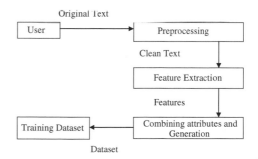

The Fig. 1 shows the overall view of the system. The input text document is pre-processed, its features are extracted and the training dataset is generated from them. The following subsections discuss each of these subprocesses.

3.1 Dataset and Pre-processing

The system made use of 50 documents from DUC2002 to generate the training dataset for document summarization. Each document consists of 7–33 sentences with an average of 21 sentences. Each document in DUC2002 collection is supplied with a set of human generated summaries provided by two different experts.

There are four main activities performed as pre-processing : Sentence Segmentation, Tokenization, Removing Stop Word, and Word Stemming. Sentence segmentation is boundary detection and separating source text into sentence. Tokenization is separating the input document into individual words. Next, Removing Stop Words, stop words are the words which appear frequently in document but provide less meaning in identifying the important content of the document such as 'a', 'an', 'the', etc. The last step for pre-processing is Word Stemming; Word stemming is the process of removing prefixes and suffixes of each word.

3.2 Sentence Features

For any task of text mining, features play an important role. The features are attributes that attempt to represent the data used for the task. The proposed approach focuses on seven features for each sentence. Each feature is given a value between '0' and '1'. Therefore, one can extract the appropriate number of sentences according to 30 % compression rate. The features and the way they are extracted are as follows:

3.2.1 Title Feature

The number of title words in the sentence contributes to title feature. Titles contain group of words that give important clues about the subjects contained in the document. Therefore, if a sentence has higher intersection with the title words, the sentence is more important than others. Eq. (1) exhibits how the value is calculated.

$$\text{Score } (Si) = \text{Number of Title words in } Si \text{ / Number of Words in Title} \quad (1)$$

3.2.2 Sentence Length

The number of words in sentence gives good idea about the importance of the sentence. This feature is very useful to filter out short sentences such as datelines and author names commonly found in articles. The short sentences are not expected to belong to the summary and the sentence length feature is calculated by the formula given in Eq. (2).

$$\text{Score } (Si) = \text{Numberof Words in } Si / \text{Numberof Words in longest sentence} \quad (2)$$

3.2.3 Term Weight

The term weight feature score is obtained by calculating the average of the Term frequency, Inverse sentence frequency (TF–ISF). Inverse term frequency helps to identify important sentences that represent the document. Term weight feature is calculated by the formula given in Eq. (3).

$$\text{Score } (Si) = \text{Sum of TF} - \text{ISF in } Si / \text{Max (Sum of TF} - \text{ISF)} \quad (3)$$

3.2.4 Sentence to Sentence Similarity

The score of this feature for a sentence is obtained by computing the ratio of the summation of sentence similarity of a sentence s with each of the other sentences over the maximum value sentence similarity as given in Eq. (4).

$$\text{Score } (Si) = \text{Sum of Sentence Similarity for } Si / \text{Max (Sum of Sentence Similarity)} \quad (4)$$

3.2.5 Proper Noun

The proper noun feature gives the score based on the number of proper nouns present in a sentence, or presence of named entity in the sentence. Usually sentences that contain proper nouns are considered to be important and these should be included in the document summary and is calculated by the formula given in Eq. (5).

$$\text{Score } (Si) = \text{Number of Proper nouns in } Si / \text{Length } (Si) \quad (5)$$

3.2.6 Thematic Word

The number of thematic word in sentence is an important feature because terms that occur frequently in a document are probably related to topic. Thematic words are words that capture main topics discussed in a given document. We used the top 10 most frequent content word for consideration as thematic. The score for this feature is calculated by the formula given in Eq. (6).

$$\text{Score (Si)} = \text{Number of Thematic Words in Si/Length (Si)} \qquad (6)$$

3.2.7 Numerical Data

This feature gives a score for the number of numerical data in a sentence. The contribution made by this feature to the weight given to a sentence is significant since a sentence that contains numerical data essentially contains important information. The score for this feature is calculated by the formula given in Eq. (7).

$$\text{Score (Si)} = \text{Number of Numerical Data in Si/Length (Si)} \qquad (7)$$

3.3 Generation of Training Dataset

The dataset that can be used as a training dataset consists of seven attributes namely: Title Feature, Sentence Length, Term Weight, Sentence to Sentence Similarity, Proper Noun, Thematic Word, and Numerical Data. The 'class' identifies whether the chosen sentence is part of the summary document or not. The 'class' is marked as 'INT' or 'NOT_INT' based on its presence or absence in the summary document. INT indicates that the sentence is *interesting* and hence is present in the model summary. NOT_INT indicates that the sentence is *not_so_interesting* and hence not present in the model summary. Figure 2 shows a sample input document and Fig. 3 shows the corresponding model summary as provided by DUC 2002.

3.3.1 Corpus Used

The TIPSTER program with its two main evaluation style conference series TREC & Document Understanding Conference-DUC (now called as Text Analysis Conference-TAC) have shaped the scientific community in terms of performance, research paradigm, and approaches. 50 documents from DUC 2002 were used for generating the training dataset. The main features of the document corpus are: Each document has minimum of seven sentences and maximum of 33 sentences. The total number of sentences in the corpus is 980.

Yasser Arafat On Tuesday Accused The United States Of Threatening To Kill Plo Officials If Palestinian Guerrillas Attack American Targets. The United States Denied The Accusation. The State Department Said In Washington That It Had Received Reports The Plo Might Target Americans Because Of Alleged Us Involvement In The Assassination Of Khalil Wazir, The Plo's Second In Command. Wazir Was Slain April 16 During A Raid On His House Near Tunis, Tunisia. Israeli Officials Who Spoke On Condition They Not Be Identified Said An Israeli Squad Carried Out The Assassination. There Have Been Accusations By The Plo That The United States Knew About And Approved Plans For Slaying Wazir. Arafat, The Palestine Liberation Organization Leader, Claimed The Threat To Kill Plo Officials Was Made In A Us Government Document The Plo Obtained From An Arab Government. He Refused To Identify The Government. In Washington, Assistant Secretary Of State Richard Murphy Denied Arafat's Accusation That The United States Threatened Plo Officials. State Department Spokesman Charles Redman Said The United States Has Been In Touch With A Number Of Middle Eastern Countries About Possible Plo Attacks Against American Citizens And Facilities. He Added That Arafat's Interpretation Of Those Contacts Was Entirely Without Foundation . Arafat Spoke At A News Conference In His Heavily Guarded Villa In Baghdad, Where Extra Security Guards Have Been Deployed. He Said Security Also Was Being Augmented At Plo Offices Around The Arab World Following The Alleged Threat. He Produced A Photocopy Of The Alleged Document. It Appeared To Be Part Of A Longer Document With The Word Confidential Stamped At The Bottom. The Document, Which Was Typewritten In English, Referred To Wazir By His Code Name, Abu Jihad. It Read:You May Be Aware Of Charges In Several Middle Eastern And Particularly Palestinian Circles That The U.S. Knew Of And Approved Abu Jihad's Assassination. On April 18th (A) State Department Spokesman Said That The United States `Condemns This Act Of Political Assassination, `Had No Knowledge Of And `Was Not Involved In Any Way In This Assassination. It Has Come To Our Attention That The Plo Leader Yasser Arafat May Have Personally Approved A Series Of Terrorist Attacks Against American Citizens And Facilities Abroad, Possibly In Retaliation For Last Month's Assassination Of Abu Jihad. Any Possible Targeting Of American Personnel And Facilities In Retaliation For Abu Jihad's Assassination Would Be Totally Reprehensible And Unjustified. We Would Hold The Plo Responsible For Any Such Attacks . Arafat Said The Document Reveals The Us Administration Is Planning, In Full Cooperation With The Israel Is, To Conduct A Crusade Of Terrorist Attacks And Then To Blame The Plo For Them. These Attacks Will Then Be Used To Justify The Assassination Of Plo Leaders . He Strongly Denied That The Plo Planned Any Such Attacks.

Fig. 2 Sample input document

Yasser Arafat On Tuesday Accused The United States Of Threatening To Kill PLO Officials If Palestinian Guerrillas Attack American Targets. The United States Denied The Accusation. The State Department Said In Washington That It Had Received Reports The PLO Might Target Americans Because Of Alleged Us Involvement In The Assassination Of Khalil Wazir, The PLO's Second In Command. Wazir Was Slain April 16 During A Raid On His House Near Tunis, Tunisia. Any Possible Targeting Of American Personnel And Facilities In Retaliation For Abu Jihad's Assassination Would Be Totally Reprehensible And Unjustified. We Would Hold The PLO Responsible For Any Such Attacks.

Fig. 3 Model summary document

Table 1 Feature categories

Feature _score	Class	Notation
0–0.2	Very low	VL
0.21–0.4	Low	L
0.41–0.6	Average	A
0.61–0.8	High	H
0.81–1.0	Very high	VH

3.3.2 Categories of Features

The feature are extracted and are categorized to five distinct classes as shown below: (Table 1).

A feature score of 0–0.2 is given a class name 'very low' a notation 'VL', score having values 0.21 till 0.4 is placed under the class low with a notation 'L'. A score of 0.41–0.6 is given the class 'average' under the notation 'A' while a score of 0.61–0.8

is given a value 'high' and a notation 'H'. The final delimit 0.8–1.0 having class name 'very high' is having a notation 'VH'.

3.3.3 Sample Training Dataset

The system generated a dataset that can be used for training any system for the task of text summarization. The dataset consists of attributes F1, F2,...F7 that represents the feature values for a given sentence and the 'class' attribute is marked as 'INT' or 'NOT_INT' based on the presence or absence of the sentence in the model summary document. INT indicates an *interesting* sentence and NOT_INT indicates a *not_so_interesting* sentence. The following table shows a sample of the training dataset generated (Table 2).

In the above table, F1 represents the 'title feature', F2 represents the 'sentence length', F3 represents the 'term weight', F4 represents the 'sentence to sentence similarity', F5 represents the 'proper noun', F6 represents the 'thematic word', and F7 represents the 'numerical data' for a given sentence.

4 Conclusions and Future Work

The system titled 'Generation of training dataset for document summarization' has successfully generated training dataset for the task of summarization. The input text document was pre-processed and the important features for each sentence of the document such as title feature, sentence length, term weight, sentence to sentence similarity, proper noun, thematic word and numerical data was extracted, and the presence or absence of the sentence in the model summary given by DUC was utilized

The system is designed only for the English language. The System can be used only with well formatted documents with respect to the grammatical rules of the English language. System cannot understand an exhaustive set of mathematical and

Table 2 Sample training dataset

Sentence #	F1	F2	F3	F4	F5	F6	F7	CLASS
1	VH	H	H	VL	L	VH	VL	NOT_INT
2	L	A	H	L	VL	H	VL	INT
3	VL	A	VH	VL	VL	VL	L	NOT_INT
4	VL	H	H	VL	L	VH	VL	NOT_INT
5	L	H	H	VL	VL	H	VL	NOT_INT
6	L	H	H	L	VL	H	H	INT
7	L	VH	H	VL	VL	VH	VL	NOT_INT

special symbols and considers them as individual strings; however the system accepts large text documents.

References

1. Luhn, H.P.: The automatic creation of literature abstract. IBM J. Res. Dev. **2**, 159–165 (1958)
2. Kupiec, J., Pedersen, J., Chen, F.: "A Trainable document summarizer". In Proceedings of the Eighteenth Annual International ACM Conference on Research and Development in Information Retrieval (SIGIR), pp. 68–73. Seattle (1995).
3. Edmundson, H.P.: New methods in automatic extracting. J. Assoc. Comput. Mach. 16(2), 264–285 (1969).
4. Baxendale, P.: Machine-made index for technical literature–an experiment. IBM J. Res. Dev. **2**, 354–361 (1958)
5. Rasim, M.: Alguliev, Effective summarization method of text documents. Proceedings of IEEE International Conference on Web Intelligence, In (2005)
6. Hsun-Hui, H., Yau-Hwang, K., Horng-Chang, Y.: Fuzzy-rough set aided sentence extraction summarization. Proceedings of the first International Conference on Innovative Computing, Information and Control, In (2006)
7. Huantong, G., Peng, Z., Enhong, C., Qingsheng, C.: A novel automatic text summarization study based on term co-occurrence. Proceedings of ICCI, In (2006)
8. Massih, R.: Amini and nicolas usunier, a contextual query expansion approach by term clustering for robust text summarization. Proceedings of DUC, In (2007)
9. Suanmali, L., Mohammed Salem, B., Salim, N.: Sentence features fusion for text summarization using fuzzy logic. In: Proceedings of HIS 2009, 142–146 (2009).
10. Esther, H., Saswati, M., Kumar, G.: An extractive text summarization based on multivariate approach. ICACTE **3**, 157 (2010)
11. Esther, H., Geetha T.V., Saswati M.: Automatic extractive text summarization based on fuzzy logic: a sentence oriented approach. LNCS 2011.

Online Identification of English Plain Text Using Artificial Neural Network

Aditi Bhateja, Ashok K. Bhateja and Maiya Din

Abstract In online communication, most of the time plain English characters are transmitted, while a few are encrypted. Thus there is a need for an automatic recognizer of plain English text (based on the characteristics of the English Language) without using a dictionary. It works for continuous text without word break-up (text without blank spaces between words). We propose a very efficient artificial neural network-based technique by selecting relevant or important features using Joint Mutual Information for online recognition of English plain text which can recognize English text from English like or random data.

Keywords Feature vector · Joint mutual information · Artificial neural network · Back-propagation

1 Introduction

The identification of text whether it is plain or not is the first requirement for the analysis of online communication. The non-plain text may be garbage or an encrypted text, which appears as random. Identification of the text can be carried out by applying a suitable pattern recognition technique using appropriate features. Every language has some inherent characteristics in terms of occurrences of letters out of the alphabet set in words, which in turn constitutes sentences. Each language has its own grammar,

A. Bhateja (✉)
Ambedkar Institute of Advanced communication Technologies and Research, Delhi, India
e-mail: aditibhateja89@gmail.com

A. K. Bhateja · M. Din
Defence Research and Development Organization, Scientific Analysis Group, Delhi, India
e-mail: akbhateja@gmail.com

M. Din
e-mail: anuragimd@gmail.com

B. V. Babu et al. (eds.), *Proceedings of the Second International Conference on Soft Computing for Problem Solving (SocProS 2012), December 28–30, 2012*, Advances in Intelligent Systems and Computing 236, DOI: 10.1007/978-81-322-1602-5_102, © Springer India 2014

which defines the constraints on word spellings and syntax of sentences. In each language there are some "vowel" like letters, whereas others are "consonant" like, which are basically characterized by the phonetic properties. The English language has five vowels ('Y' also treated sometimes as 6th vowel) and 21 consonants. Some letters such as E, T, O, A, I occur more frequently, whereas others such as B, J, X, W, Z occur occasionally in any given English text [1]. There are certain letter combinations such as WX, ZX, BX (bigrams), which are very rare whereas combinations such as TH, ER, HE occur more frequently. Similarly certain combinations of three letters like THE, ING, AND, HER, DTH (trigrams) occur more frequently. Such characteristics and affinity of certain letters with some specific letters makes one language distinguishable from the other and also enables one to check if the given text is meaningful valid plain text of the language or it is a random text (sequence of letters occurring randomly). When random text is very close to English, it becomes difficult to recognize whether it is English text or not. Therefore it is required to extract the features which are relevant for recognizing the English text from a random text. The irrelevant features create problems in recognition. Thus our goal is to filter out random data and extract English data online. Joint Mutual Information (JMI) [2] is used in ordering the features according to their decreasing order of relevance. The optimal number of relevant (distinguishing) features is selected by cumulative insertion of the features in the input of a feed-forward network, and a reduced feature set is formed. Feed-forward back-propagation Artificial Neural Network [3] with the reduced feature vector as input, is trained for the classification of English plain text and a non-English text (which look like English or random text).

In this paper, we propose an efficient scheme for the identification of English text in online communication. This scheme performs well, with very high accuracy. Data preparation and feature extraction is explained in Sect. 2. Back-propagation learning based ANN is detailed in Sect. 3. Experimental results are presented in Sect. 4. Finally, Sect. 5 concludes the paper.

2 Data Preparation and Feature Extraction

We collected 2,000 texts of English each of size 500 characters after removing special characters, numerals, and blanks, i.e., the text has only 26 upper case alphabets. We then created 1,000 English-like texts of length 500 each, by replacing some bigram/trigram, in standard plain English text, with some random bigram/trigram (e.g. THE is replaced with CMU). Thousand random texts using true random number generator and 1,000 pseudo random sequences of alphabets (by using simple substitution, Playfair [4], Vigenere [5] encryption schemes and pseudorandom generator, 250 each) each of length 500 were generated. Features considered: 26 alphabets, 48 high frequent bigrams, and 36 high frequent trigrams [6] of plain English. We selected 48 bigrams because after 48 bigrams there is a sharp decrease in frequency of the bigrams. Similarly, in case of trigram there is a sharp decrease of frequency after 36 trigrams. Feature vectors (size 110) with these features were created for each of the 5,000 texts (2,000 English text, 1,000 very close to English, 1,000 pseudo random, and 1,000 true random). For extracting the features which are relevant or important for identification of English text from non-English, JMI technique is used.

2.1 Joint Mutual Information

Filter techniques [7] are defined by a criterion 'Relevance Index' denoted by J which is intended to measure how potentially useful a feature may be when used in a classifier. An intuitive J would be some measure of correlation between the feature and the class label, i.e., the intuition being that a stronger correlation between these should imply a greater predictive ability when using the feature. For a class label Y, the 'Mutual Information' index for a feature X_k is as follows:

$$J_{\text{mim}}(X_k) = I\,(X_k; Y)$$

'mim', standing for *Mutual Information Maximization*. An important limitation is that this assumes that each feature is independent of all other features and effectively ranks the features in descending order of their individual mutual information content. However, where features may be interdependent, this is known to be suboptimal. In general, it is widely accepted that a useful and parsimonious set of features should not only be individually *relevant*, but also should not be *redundant* with respect to each other—features should not be highly correlated.

Battiti (1994) presented the 'Mutual Information Feature Selection' (MIFS) criterion:

$$J_{\text{mifs}}(X_k) = I(X_k, Y) - \beta \sum_{X_j \in S} I(X_k, X_j)$$

where S is the set of currently selected features. This includes the $I(X_k; Y)$ term to ensure feature *relevance*, but introduces a penalty to enforce low correlations with features already selected in S. The β in the MIFS criterion is a configurable parameter, which must be set experimentally. Using $\beta = 0$ would be equivalent to $J_{\text{mim}}(X_k)$, selecting features independently, while a larger value will place more emphasis on reducing inter-feature dependencies. In experiments, Battiti found that $\beta = 1$ is often optimal, though with no strong theory to explain why. The MIFS criterion focuses on reducing *redundancy*; an alternative approach was proposed by Yang and Moody (1999), and also later by Meyer et al. (2008) using the *Joint Mutual Information* (JMI), to focus on increasing *complementary* information between features.

The JMI index for feature X_k is defined as:

$$J_{\text{jmi}}(X_k) = \sum_{X_j \in S} I(X_k X_j; Y)$$

This is the information between the targets and a joint random variable $X_k X_j$, defined by pairing the candidate X_k with each previously selected feature. The idea is that if the candidate feature is complementary with existing features, it should be included.

2.2 Formation of Reduced Feature Vector

JMI algorithm arranges all the features according to decreasing order of their relevance. We used cumulative insertion technique to find the number of important features. The method used for finding the number of relevant features is given below:

Algorithm: To extract the important/relevant features

Create a feature vector consisting of the frequencies of 26 alphabets, 48 high frequent bigrams, and 36 high frequent trigrams of plain English.

Apply JMI technique to find the rank of all the 110 features according to decreasing order of their relevance.

List of relevant features $= \phi$.

For $k = 1$ to total number of features.

Select the most important feature from the list prioritized by JMI and delete it from the list.

Add this feature to the list of relevant features.

Train a three-layer ANN consisting of k number of neurons in input layer, 6 to 12 neurons in the hidden layer and 2 neurons in the output layer.

Test the performance of the network on test data.

If the performance decreases, then stop.

In this way we found that out of 110 only 20 features are important, which are used for actual classification.

---

3 Back-Propagation Learning-Based ANN

An Artificial Neural Network (ANN) is a mathematical model or computational model that is inspired by the structure and/or functional aspects of biological neural networks. A neural network consists of an interconnected group of artificial neurons, and it processes information using a connectionist approach to computation [8] (Fig. 1).

Standard back-propagation is a gradient descent algorithm, in which the network weights are moved along the negative of the gradient of the performance function. The term back-propagation refers to the manner in which the gradient is computed for nonlinear multilayer networks. Properly trained back-propagation networks tend to give reasonable answers when presented with inputs that they have never seen. Typically, a new input leads to an output similar to the correct output for input vectors used in training that are similar to the new input being presented. This generalization property makes it possible to train a network on a representative set of input/target pairs and get good results without training the network on all possible input/output pairs.

The reduced feature vectors (feature vector consisting of relevant/important features) ranked by JMI algorithm are used as input for training and weights were stored which are used for classification of unknown text.

4 Results and Analysis

2,000 texts of English, 1,000 texts of close to English, 1,000 texts of pseudo Random, and 1,000 texts of true random, each of length 500 characters, were created. Out of these 5,000 text samples 70 % of each category was used for training and the remaining 30 % was used for testing the neural network.

For each of the texts the feature vector of size 110 consisting of frequencies of 26 alphabets (A to Z), frequencies of 48 high frequent English bigrams (TH, IN, ER, RE, AN, HE, AR, EN, TI, TE, AT, ON, HA, OU, IT, ES, ST, OR, NT, HI, EA, VE, CO, DE, RA, RO, LI, RI, IO, LE, ND, MA, SE, AL, IC, FO, IL, NE, LA, TA, EL, ME, EC, IS, DI, SI, CA, UN) and frequencies of 36 high frequent English trigrams (THE, ING, AND, ION, ENT, FOR, TIO, ERE, HER, ATE, VER, TER, THA, ATI HAT, ERS, IIIS, RES, ILL, ARE, CON, NCE, ALL, EVE, ITH, TED, AIN, EST, MAN, RED, THI, IVE, REA, WIT, ONS, ESS), were formed.

JMI algorithm ranked these features in decreasing order of their relevance. A reduced list of features (list containing only the relevant features) is formed by adding one by one the most important feature from the ranked relevance list of features, and used as input vector to train a three-layer neural network. Performance of the network with increase in number of features is shown in Fig. 2.

It is clear from Fig. 2 that out of 110 features only 20 are the relevant features. Therefore a reduced feature vector of size 20 was created for all the texts. A three-layer neural network with 20 input neurons, 10 hidden neurons, and 2 output neurons was trained with the training data. The weights after training were stored in a file.

Fig. 1 Feed-forward neural network

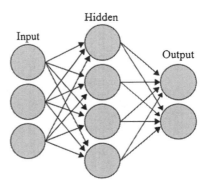

Fig. 2 Number of features versus identification performance

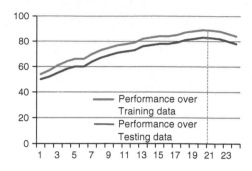

Table 1 Performance analysis

Text type	Classification score	
	Training data (%)	Testing data (%)
Plain english	96.6	93.4
Close to english	88.9	82.7
Pseudo random	97.1	96.2
True random	99.3	98.7

Network of the same architecture and stored weights was used to test all the test samples. The performance of the network is shown in Table 1.

After a certain value of mean square error (over training) the performance of the network degrades. The neural network is more than 96 % classification score achiever when it is used to classify plain English text versus pseudo random or true random text.

5 Conclusion

Online identification of plain text is very important for the analysis of communication. Relevant features improve the performance of a classifier. Joint Mutual Information-based approach for ranking the features is very effective. We have used cumulative addition approach to extract the optimal number of relevant features. Using the relevant features, back-propagation neural network recognize English plain text very accurately and efficiently.

References

1. Saxena, P.K., Pratibha, Y.: Girish, M: Index of garbledness for automatic recognition of plain english texts. Defence Sci. J. **60**(4), 415–419 (2010)
2. Yang, Howard Hua, Moody, John: Feature Selection based on Joint Mutual Information. J Comput Intell Methods Appl. Int. Comput. Sci. Convention 13, 1–8 (1999).

3. Haykin, S: Neural Networks- A Comprehensive Foundation, 2nd edn. Macmillan, New York.
4. http://en.wikipedia.org/wiki/Playfair_cipher
5. http://en.wikipedia.org/wiki/Vigen%C3%A8re_cipher
6. http://jnicholl.org/Cryptanalysis/Data/EnglishData.php
7. Brown, Gavin, Pocock, Adam, Jhao, M.J., et al.: Conditional likelihood maximization: a unifying framework for information theoretic feature selection. J. Mach. Lear. Res. 13, 27–66 (2012).
8. http://en.wikipedia.org/wiki/Artificial_neural_network

Novel Approach to Predict Promoter Region Based on Short Range Interaction Between DNA Sequences

Arul Mugilan and Abraham Nartey

Abstract Genomic studies have become one of the useful aspects of Bioinformatics since it provides important information about an organism's genome once it has been sequenced. Gene finding and promoter predictions are common strategies used in modern Bioinformatics which helps in the provision of an organism's genomic information. Many works has been carried out on promoter prediction by various scientists and therefore many prediction tools are available. However, there is a high demand for novel prediction tools due to low level of prediction accuracy and sensitivity which are the important features of a good prediction tool. In this paper, we have developed the new algorithm Novel Approach to Promoter Prediction (NAPPR) to predict eukaryotic promoter region using the python programming, which can meet today's demand to some extent. We have developed the parameters for Singlet (4^1) to nanoplets (4^9) in analyzing short range interactions between the four nucleotide bases in DNA sequences. Using this parameters NAPPR tool was developed to predict promoters with high level of Accuracy, Sensitivity and Specificity after comparing it with other known prediction tools. An Accuracy of 74 % and Specificity of 78 % was achieved after testing it on test sequences from the EPD database. The length of DNA sequence used as input has no limit and can therefore be used to predict promoters even in the whole human genome. At the end, it was found out that NAPPR can predict eukaryotic promoter with high level of accuracy and sensitivity.

A. Mugilan (✉)
Department of Bioinformatics, School of Health Science and Biotechnology,
Karunya University, Coimbatore, India
e-mail: bioinformaticsmugil@gmail.com

A. Nartey
Department of Theoretical and Applied Biology,
Kwame Nkrumah University of Science and Technology,
College of Science, Kumasi, Ghana
e-mail: abrahamnart@gmail.com

B. V. Babu et al. (eds.), *Proceedings of the Second International Conference on Soft Computing for Problem Solving (SocProS 2012), December 28–30, 2012*, Advances in Intelligent Systems and Computing 236, DOI: 10.1007/978-81-322-1602-5_103, © Springer India 2014

Keywords Positional score value · Nanoplets · Short-range-interactions · Expected promoter region

1 Introduction

Promoters are responsible for the transcription regulation of the gene downstream of it. This is because, specific conserved pattern of nucleotides known as motifs are present in this promoter regions. Transcriptional factors therefore recognize these motifs and bind it to initiate the transcription of genes. Three distinguishing parts of the promoter region can be identified, namely; the Core promoter, the Proximal promoter and the Distal promoter. The part which contains the sequence necessary to bind transcription factors is the Core promoter [1]. This part mainly contains the transcription start site, the binding site for RNA polymerase and a conserved sequence region which binds with transcription factors. Different organisms have different genomic DNA contents and there is also a variation between the coding and non-coding genomes among them. Only small portion of the human genome for instance is coding whereas larger percentage (98 %) is non-coding [2]. The coding region has two boundaries at each end, it is bounded at the 5' end by a start codon and 3' end by a stop codon [3]. Non-coding region describes the components of an organism's DNA sequence that do not encode protein sequences. Though some DNA sequences are coded or transcribed but by the virtue of the fact that they do not encode protein, they classified as non-coding sequence. Transcriptional and translational regulations of the coding sequences are carried out by these non-coding DNA sequences. Human Genome Project has the list of all human genes fully annotated to improve biomedical research [4]. The outcome of this project has served a tremendous improvement in the field of biology [5]. The result of this project was made available to the public through databases and publications of which scientists can easily get access to it to carry out further projects. These databases come in different categories, from genes, Nucleotide sequences, promoters and through coding and non-coding regions of the genomic DNA.

The Frequency Distribution Analysed Feature Selection Algorithm (FDAFSA), in which the frequency of hexamers (adjacent triplet pairs) in a dataset are considered. The second approach is the Random Triplet Pair Feature Selecting Genetic Algorithm (RTPFSGA), where the genetic algorithm to find random triplet pairs (RTPs), that is non-adjacent triplets [6]. This method was used because it has been proven in previous researches that the distributions of triplet frequencies are very useful in serving as codons which play an important role in the protein's biological functionality [7, 8]. Based on four scoring criteria, genetic algorithm was developed for predicting the operon [9]. Furthermore, there are Neural Network Promoter Prediction tool and Promoter Scan. NNPP promoter prediction tool is a promoter prediction tool that uses artificial neural network in predicting promoters. It uses a time-delay architecture method to analyze the compositional and structural properties of promoter sequence [10]. Another prediction tool, Promoter scan (PROSCAN v1.7) is also showing a low level of accuracy and specificity. It is a program that uses a weighted matrix for scoring a TATA box related sequence. It was built by using the densities of a transcription binding sites to derive a ratio between promoter and non-promoter sequence [11]. Even though a lot of works have been done on promoter predictions, there is a need to improve upon the accuracy and sensitivity of these prediction tools. There are numerous prediction tools available but their level of accuracy and specificity is low. In this research, a new promoter prediction algorithm has been introduced for eukaryotic promoter sequence known as the Novel Approach to Promoter Prediction (NAPPR).

2 Method of Promoter Prediction

2.1 Description of the Databases

We have collected 9710 (Last updated on 02 Mar. 2012) Eukaryotic promoter sequences from the Eukaryotic Promoter Database [12]. We have used 9700 promoters region (training set) for the preparation of parameters and randomly selected 10 promoter regions (test set) as the input sequences (not included in the training set). Python programming was used to count numbers of singlet (4^1), doublets (4^2), triplets (4^3), tetraplets (4^4), pentaplets (4^5), hexaplets (4^6), heptaplets (4^7), octaplets (4^8) and nanoplets (4^9) from the above dataset (9700).

2.2 Statistical Analysis

The observed frequency values from the various observed count (singlet to nanoplets) were computed using the formula below [13, 14].

$$P(x) = \frac{\sum Ni(x)}{\sum Yi}$$

where X = individual nucleotide bases (acgt) for singlet, two consecutive bases (aa, ac....gg) for doublet .... nine consecutive bases (aaaaaaaaa, aaaaaaaac......... gggggggggg) for nanoplets.

Ni = Number of count for x in the ith promoter sequence
Yi = Total number of nucleotide bases.
i varies from 1 to 9700.

Theoretical estimate for the nucleotide bases were calculated based on the singlet frequency of occurrence.

$$P(xy) = P(x) \times P(y)$$

$P(x)$ is the frequency of occurrence of singlet x

$P(y)$ is the frequency of occurrence of singlet y.

Based on the above formula, we have calculated theoretical values up to nanoplets. The chi square value for nucleotide sequence singlet to nanoplets is calculated as follows

$$\chi^2(\mathbf{x}) = (\mathbf{P}_{observed}(\mathbf{x}) - \mathbf{P}_{theoretical}(\mathbf{x}))^2 / \mathbf{P}_{theoretical}(\mathbf{x})$$

If the observed value is greater than theoretical value, chi square value is assigned as positive (preferential selection) else it is negative (non-preferential selection) [13]. The chi square value are normalized by the below formula

$$\textbf{Normalized value } (N) = \frac{Xi - \sigma}{\omega - \sigma}.$$

where Xi is the chi square value for ith DNA base ($i = a$ to g for singlet, aa to gg for doublets ..... aaaaaaaaa to gggggggg for nanoplet) is to be normalized,

σ is the minimum value occurring within the group

ω is the maximum value occurring within the group [15].

Based on the above normalized value, we have developed the Normalized prediction parameter (NPP) for singlet, doublets, triplets, tetraplets, pentaplets, hexaplets, heptaplets, octaplets and nanoplets (4^1–4^9).

2.3 Positional Score Value

We have collected the test set (10) nucleotide sequence which has 2600 nucleotide length. The above test set was calculated starting from -1491 to 1092 from the promoter region. We have subtracted 992 nucleotides from each side, the expected promoter region is $-499(991)$–$100(1591)$. Positional score value (PSV) has been determined for each nucleotide base in a test set promoter sequence and the region that shows the highest PSV's were predicted as the promoter region. A python program was written to associate the possible sequence with their respective frequency parameters. The PSV was counted along the entire length of the promoter sequence starting from the 9th position to the 2,590th position. Taking one nucleotide into account, the positional score value was calculated for all possible short-range interaction from singlet (4^1) to nanoplets (4^9). This is represented in the Table 1. An average score value of 0.5 was used as the minimum threshold value. Any PSV which is below this average value is considered as not predicted and values greater or equal to it is considered as predicted.

Table 1 A table showing manual calculation of PSV along a promoter sequence

TGTTATCTTTGCTTTTCT		
	SRI(T)	F
Singlet	T	0.16166
Doublet	TT	0.32332
	TT	0.32332
Triplet	CTT	0.31486
	TTT	0.39708
	TTG	0.20395
Tetraplet	TCTT	0.28753
	CTTT	0.35425
	TTTG	0.27641
	TTGC	0.21244
Pentaplet	ATCTT	0.09707
	TCTTT	0.22329
	CTTTG	0.18753
	TTTGC	0.16021
	TTGCT	0.15619
Hexaplet	TATCTT	0.04444
	ATCTTT	0.08481
	TCTTTG	0.11645
	CTTTGC	0.12775
	TTTGCT	0.12302
	TTGCTT	0.10790
Heptaplet	TTATCTT	0.03169
	TATCTTT	0.03334
	ATCTTTG	0.04186
	TCTTTGC	0.05681
	CTTTGCT	0.06638
	TTTGCTT	0.07236
	TTGCTTT	0.07879
Octaplet	GTTATCTT	0.00645
	TTATCTTT	0.01752
	TATCTTTG	0.01188
	ATCTTTGC	0.01390
	TCTTTGCT	0.02075
	CTTTGCTT	0.02256
	TTTGCTTT	0.04471
	TTGCTTTT	0.04250
Nanoplet	TGTTATCTT	0.00324
	GTTATCTTT	0.00324
	TTATCTTTG	0.00424
	TATCTTTGC	0.00349
	ATCTTTGCT	0.00474
	TCTTTGCTT	0.00648
	CTTTGCTTT	0.01122
	TTTGCTTTT	0.02592
	TTGCTTTTC	0.01745
	PSV(T)	4.05256

Fig. 1 A sketch of graph
showing the positions of TP,
FP, TN and FN for a prediction
tool

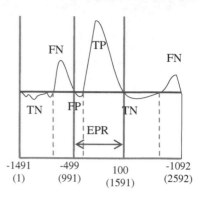

2.4 *Method to Calculate Accuracy of Prediction*

We have considered whether this predicted nucleotide is found in the Expected Promoter Region (EPR) or outside the region. The expected promoter region is the known
promoter region as in the Eukaryotic Promoter Database. Any PSV found within the
EPR which is greater or equal to the average PSV for the entire sequence was counted
and recorded as number of True Positive (TP) whiles the PSV less than the average
PSV for the entire sequence is also counted and recorded as False Positives (FP).
Also, any PSV found outside the EPR which is greater or equal to the average PSV
for the sequence is counted and recorded as False Negative (FN) whereas the PSV
outside the EPR which are less than the average PSV for the sequence were also
counted and recorded as True Negative (TN). These counting values were used to
calculate for the accuracy (ACC), sensitivity (SN) and specificity (SP) for the NAPPR
prediction tool. This is shown in Fig. 1.

The sequences used in our test set were submitted to the NNPP version 2.2. The
minimum promoter score was set to 0.5 and all the 10 sequences were predicted.
The results were then computed for true-positive and false-negative analysis in order
to obtain the sensitivity, specificity and accuracy. For the comparative analysis we
have submitted our test sequences to the web server promoter scan version 1.7 for
prediction. It was able to predict only 7 out of 10 test sequences submitted. The
results were then analyzed.

3 Prediction of Promoter Region

The test set DNA sequences were run on the following promoter sequences from
the EPD; [A2LD1], [H3F3AP1], [N6AMT2], [S100PBP], [SAMD12], [C10orf188],

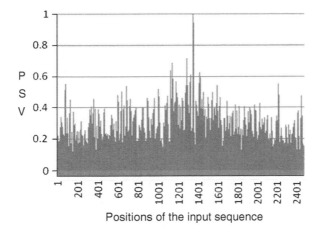

Fig. 2 Prediction of promoter region using NAPPR of H3F3AP1 gene

[CCDC51], [RAB11B], [P2RX69] and [WBP11]. Our NAPPR method is used to predict the promoter region within the above named promoter sequences. This is to test whether the proposed NAPPR method will be able to identify the promoter region with high level of accuracy and specificity. Figure 2 shows the outcomes of the predicted result of H3F3AP1. Position number 991–1591 has the experimental predicted area for promoter region. Figure 2 show that our method (NAPPR) also prepares the above promoter region.

The evaluation is then made of the number of true positives (TP), where the length and end sequence positions are correctly predicted, and the number of over-predicted positioned prediction or false positives (FP), True Negative (TN), under-predicted residues as misses or false-negative (FN) predictions. The following Accuracy calculations [16] are used:

1. The sensitivity of the method SN is given by true positives/actual Positives = TP/ (TP + FN).
2. The specificity by SP = true negatives/actual negatives which is; TN/(TN + FP).
3. Accuracy ACC = (TP+TN)/ (TP+TN+FP+FN).

4 Analysis of Promoter Prediction

According to our method, good prediction accuracy was achieved. Looking at it individually from Table 2, test sequence 2 (H3F3AP1) recorded the highest accuracy of 79 % followed by test set 5, 6 and 7 (SAMD12,C10orf18&CCDC51) of 77 % accuracy. However, test set 9 (P2RX69), predicted with the least accuracy of 61 % of which on the average prediction was lowered to 74 % accuracy.

Table 2 A table showing true-positive and true-negative calculations of the NAPPR method

SN	Promoter ID	TP	FP	TN	FN	SN	SP	ACC
1	A2LD1	41	559	1941	51	0.45	0.78	0.76
2	H3F3AP1	56	544	1979	13	0.81	0.78	0.79
3	N6AMT2	173	427	1788	204	0.46	0.81	0.76
4	S100PBP	126	474	1765	227	0.36	0.79	0.73
5	SAMD12	67	533	1919	73	0.48	0.78	0.77
6	C10orf188	48	552	1945	47	0.51	0.78	0.77
7	CCDC51	22	578	1964	28	0.44	0.77	0.77
8	RAB11B	13	587	1917	75	0.15	0.77	0.74
9	P2RX69	175	425	1403	589	0.23	0.77	0.61
10	WBP11	76	524	1782	210	0.27	0.77	0.72
**	AVERAGE	80	520	1840	152	0.41	0.78	0.74

Furthermore, looking at the number of TP in test set 2 (H3F3AP1), predicted only 56 out 600 PP values and also it predicted the highest number of true negatives leading to the least false positive prediction (13) which all contributed to the high accuracy level for that sequence.

High numbers of false negatives predicted in test set 9 were due to single sequence repeat (SSR) of the nucleotide Adenine. This is because the stability of the DNA sequence is affected by SSRs depending of the length of repeat. There is also high mutation (deletion or insertion) in GC rich repeats [17, 18]. It consist of (13, 16) nt repeats from the position of (229–241, 541–555) respectively which are all outside the EPR therefore biasing the result. This bias was due to the fact that, our program was trained to predict short-range interactions between four different nucleotides which were not present in this region and hence it has nothing to predict. Also high PSV recorded can be interpreted as over prediction of SSRs of a given type of nucleotide and length since this type of repeats might have been over represented [19]. This kind of finding may be helpful in understanding the influence of SSRs and their effects on gene or promoter prediction. From this it was found out that when a particular region of the DNA sequence is made up of more than 10 nt SSR, the program cannot predict any useful short-range-interaction between the nucleotides. Notwithstanding, high number of TPs recorded were due to the high number of C and G nucleotides present in the EPR. This was confirmed in the normalized frequency values used in the parameters. The frequency for A, C, G and T were 0.2032, 0.2019, 0.2942 and 0.3007 respectively which confirm the high percentage of CpG islands in the promoter region [20, 21]. Based on this we expected the PSV to be very high within the EPR.

Figure 3 shows the comparison between PROMOTER SCAN, NNPP and NAPPR prediction tools. Now looking at the figure, promoter scan showed the highest level of sensitivity among the three prediction tools followed by our method, the NAPPR and the least sensitive is the NNPP prediction tool. Also comparing in terms of the rate of negative prediction, the NNPP showed the highest rate of negative predictions. Finally

Fig. 3 Comparison between
PROMOTER SCAN, NNPP
and NAPPR

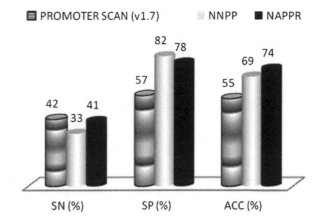

looking at the accuracy which is the most important feature of a good prediction tool, our method (NAPPR) has the highest level of accuracy followed by the NNPP and the least is the Promoter scan. Therefore despite the fact that promoter scan is more sensitive in prediction the level of accuracy is very low. Furthermore, the NNPP also is less sensitive and has the highest rate of negative prediction with medium level of accuracy and this does not qualify it as a good prediction tool.

5 Conclusion

Finally, NAPPR, the newly introduced prediction tool has the highest level of accuracy and medium level of both sensitivity and specificity and these qualities makes it an outstanding among the three methods compared. It can therefore be highly recommended to biologist as the most efficient and most reliable prediction tool. There has been a major achievement in this project that, the NAPPR method can predict with high level of accuracy and sensitivity.

Acknowledgments Glory be to God almighty for making it possible for us to come out with this kind of project. Much appreciation is also rendered to Karunya University for the support and provision of facilities toward this research project.

References

1. Smale, S., Kadonaga, J.T.: The RNA polymerase II core promoter. Ann. Rev. Biochem. **72**, 449–479 (2003)
2. Elgar, G., Vavouri, T.: Tuning in to the signals, non-coding sequence conservation in vertebrate genomes. Trends Ganet. **24**(7), 344–352 (2008)
3. Gene Structure.[http://genome.wellcome.ac.uk/doc_WTD020755.html]

4. Lander, E.S.: The new genomics, global views of biology. Science **3**, 536–539 (1996)
5. Lander, E.S.: Initial sequencing and analysis of the human genome. Nature **409**, 860–921 (2001)
6. Azad, A.K.M., Saima, S., Nasimul, N., Hyunju, L.: Prediction of plant promoters based on hexamers and random triplet pair analysis. Algorithms Mol. Biol. **6**, 19 (2011)
7. Kornev, A.P., Taylor, S.S., Ten, E.L.F.: A helix scaffold for the assembly of active protein kinases. Proc. Natl. Acad. Sci. **105**(38), 14377–14382 (2008)
8. Ten, E.L.F., Taylor, S.S., Kornev, A.P.: Conserved spatial patterns across the protein kinase family. Biochim. Biophys. Acta **1784**(1), 238–243 (2008)
9. Shuqin, W., Yan, W., Wei, D., Fangxun, S., Xiumei, W., Yanchun, L., Chunguang, Z.: Proceedings of the 8th international conference on Adaptive and Natural Computing Algorithms ICANNGA '07, Pp. 296–305 (2007)
10. Resse, M.G.: Application of a time-dalay neural network to promoter annotation in the Drosophila melanogaster genome. Comput. Chem. **26**, 51–56 (2001)
11. Prestridge, D.S.: Predicting Pol II promoter sequences using transcription factor binding sites. J. Mol. Biol. **249**, 923–932 (1995)
12. Christoph, D., Schmid, Viviane, P., Mauro, D., Rouaïda, P., Philipp, B.: Nucl. Acids Res. 32 (suppl 1), D82–D85. (2004). doi:10.1093/nar/gkh122
13. Arul, M.S.: Sequence, structure and conformational analysis of protein databases. J. Adv. Bioinform. Appl. Res. **2**, 183–192 (2011)
14. Mugilan, S.A., Veluraja, K.: Generation of deviation parameters for amino acid singlets, doublets and triplets from three-dimensional structures of proteins and its implications in secondary structure prediction from amino acid sequences. J. Bioscience. **5**, 81–91 (2000)
15. Doherty, K., Adams, R., Davey, N.: Non-Euclidean norms and data normalization. Verleysen. **6**, 181–186 (2004)
16. Óscar, B., Santiago, B.: CNN-PROMOTER, new consensus promoter prediction program based on neural networks. Revista EIA **15**, 153–164 (2011)
17. Callahan, J.L., Andrews, K.J., Zakian, V.A., Freudenreich, C.H.: Mutations in yeast replication proteins that increase CAG/CTG expansions also increase repeat fragility. Mol. Cell. Biol. **23**(21), 7849–7860 (2003)
18. Wang, G., Vasquez, K.M.: Models for chromosomal replication-independent non-B DNA structure-induced genetic instability. Mol. Carcinog. **48**(4), 286–298 (2009)
19. Kiran, J.A., Veeraraghavulu, P.C., Yellapu, N.K., Somesula, S.R., Srinivasan, S.K., Matcha, B.: Comparison and correlation of simple sequence repeats. Bioinformation **6**(5), 179–182 (2011)
20. Gardiner, G.M., Frommer, M.: CpG islands in vertebrate genomes. J. Mol. Biol. **196**, 261–287 (1987)
21. Ioshikhes, I.P., Zhang, M.Q.: Large-scale human promoter mapping using CpG islands. Nat. Benet. **26**, 61–63 (2000)

"Eco-Computation": A Step Towards More Friendly Green Computing

Ashish Joshi, Kanak Tewari, Bhaskar Pant and R. H. Goudar

Abstract There has been continuously arising demand for more computational power and is increasing day by day. A lot of new and adaptable technologies are coming forth to meet the user requirements, e.g., include clusters, grids, clouds and so on each having advantages and disadvantages providing computation, resources, or services. The problem arising is that for large computation many participating entities in the decision system are required and thereby more power is consumed. This paper focuses on minimizing the computation power required by the task (computation task or problem statement) especially in homogeneous systems here we are calling it as a Decision System which require some form of data (may be structured) along with the problem statement and thus proving to be a green computing environment (reducing use of hazardous things and maximizing the energy efficiency).

Keywords Green computing · Computation · Homogeneous systems · Dimensionality reduction · Bio-Hazards · Rough set theory

A. Joshi (✉) · K. Tewari
Research Scholar, Graphic Era University, Dehradun, India
e-mail: a.joshicse1986@gmail.com

K. Tewari
e-mail: kanak.tewari@gmail.com

B. Pant · R. H. Gounder
Department Of CSE, Graphic Era University, Dehradun, India
e-mail: rhgoudar@gmail.com

B. Pant
e-mail: pantbhaskar2@gmail.com

B. V. Babu et al. (eds.), *Proceedings of the Second International Conference on Soft Computing for Problem Solving (SocProS 2012), December 28–30, 2012*, Advances in Intelligent Systems and Computing 236, DOI: 10.1007/978-81-322-1602-5_104, © Springer India 2014

1 Introduction

As we all know the demand for computing is increasing day by day, i.e., because the data is becoming more and more complex along with its availability. To achieve the task may require more and more computational power along with other resources. These days the scenario is quite common and more and more technologies have been adopted so far.

Major focus is on adopting Green IT or green computing methodologies in order to make our planet green. The main aim of Green Computing is to utilize the resources effectively and efficiently while maintaining the overall performance of the system. Intending to achieve the 3R's- (reduce, recycle, and reuse) goal of Green Computing measures are being taken. Recent studies have shown that the usage of compute-power exceeds 50 % of that of total energy cost of the organization. The unnecessary usage of the computation power leads to energy wastage. For utilizing energy more efficiently and effectively, its wastage needs to be minimized.

Apart from implementation in homogeneous system, current technologies like Cloud Computing and Grid Computing are continuously on the search of new measures and techniques that can be worked upon the heterogeneous platforms and architectures too. Eco-friendly methodologies in relation with Green IT are on the verge to bring upon new technologies that can lesser the hazardous effect made to the planet and thus enforcing them onto the current scenario. The compute power reduction certainly reduces the unnecessary energy consumption leading its usage in other important tasks, such that energy can be utilized effectively without its wastage.

2 Related Work

Prior work has been done toward enhancing the overall performance in publish/subscribe systems. These approaches, however, were further manipulated and improved with the proposal of a 3-phase scheme in order to reconfigure the publish/subscribe systems along with the experimentation on a cluster test bed and a high performance computing platform. The algorithm phase 1 gathers performance information from the network followed by phase 2 in which the subscriptions were allocated using the information gathered and the last phase place them onto the newly built broker overlay with the relocation algorithm GRAPE [1–3].

Another contribution related to grid project, focused on the energy efficient aspect of desktop-grid computing: the Green Desktop grid, as a task of the EUFP7 DEGISCO project, the primary motivation was to optimize the compute unit energy specific consumption while avoiding the usage of energy intensive air-condition [4, 5].

Irrespective of the earlier searches made to reduce the compute-power wastage; steps are being taken to develop such systems that can operate so efficiently leading to utilization of the energy in the environment. As with respect to the earlier work done, the paper focuses on homogenous system providing measures so that the computation power usage can be minimized to such an extent to optimize the energy effectively and efficiently within the environment. Its implementation includes approach via Rough Set Theory following certain data mining techniques leading to the algorithm implementation

3 Proposed Work

In this paper, we proposed an adaptive functionality to facilitate green computing in the homogenous network. To accomplish this, present techniques of Data Mining are utilized: Classification algorithms, Discretization, Rough Set Theory concepts, etc., that proved to be quite effective in implementation.

3.1 Problem Description

Suppose a large complex problem need to be solved requiring huge computation, the problem statement consists of a large database or the input in any other form may be in the form of structured data, etc. A brief module of the system is pictorially represented below:

Figure 1 show how a task is given to the system along with the data values. We know generally whenever a large and complex task is to be achieved it requires very large processing on the data and data sets. Afterward, the processing on the data is performed. This requires high powered system to work continuously for several hours and can carry for moreover days. This certainly increases the cost of electric consumption and makes temperature of the room or the organization high.

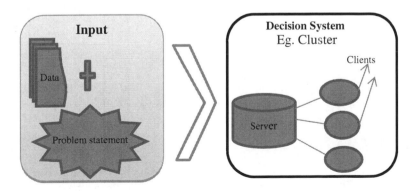

Fig. 1 A brief overview of the system

Sometimes, it happens that if there are 150 participating systems in our complex environment, it might be possible that 30 or 40 among all finishes the task within an hour of the given assignment, other 20 or 30 in the next 1 h, and so on. In general, these may have variation among them. However, if a task is assigned to 150 systems, they are all busy according to the server. But, actually the systems have finished the task, produced the output and are free. The objective of this paper focuses on reducing the amount of hours required on the processing time first, afterwards on saving the electricity measures.

The scenario taken in Fig. 1 is that of a generalized client-server approach which generally follows TCP protocol for the communication. The server may be any entity that holds the responsibilities of the clients and handling of the data with the problem statement in the system. In general, it is the first interaction to the system of the server along with performing general duties like load balancing, etc., in that specified region or portion. When the server has the task and the data, it processes the following steps:

The approach followed to simplify the problem is normally applied in relevance to the Rough Set Theory-based technique combining feature extraction and mining process through classification, Dimensionality Reduction, and Discretization. The tables in Rough set theory are termed as "Information Systems", naming rows as "objects" and columns as "attributes" shown as: $\grave{A} = (U, A)$, where U is a non-empty finite set of objects called the "universe" and A is a non-empty finite set of "attributes" such that a : $U \rightarrow V_a$ for every a CA. The set V_a is called the value set of a.

The data provided is minimized applying methods and techniques such that the resulted data set, i.e., "Reduct set" is having the same meaning as that of the original data. Applications of Rough Set Theory include Decision Support systems, Knowledge Discovery, medicine, finance, data mining, etc., as described in [5] that leads to the less power consumption and efficient usage.

In Discretization, various methods are applied on the continuous data to get the discretized data. The approach of rough set theory is mainly unable to tackle the continuous data. Thus, new techniques of Discretization were implemented. Some of them are: Boolean Reasoning, Equal Frequency Binning, and Entropy, etc. Now-a-days, it is widely used for obtaining the minimal set of attributes from the dataset following the supervised learning techniques [5].

In Data Mining, Dimensionality Reduction can be further divided into two parts: Feature Selection and Feature Extraction. The dataset available is highly dimensional. Principal Component Analysis (PCA) is one of the techniques of Feature Extraction. Thus, implementing any of these techniques on the given input dataset (structured) results into the reduced dataset that enable green computing facilities in the homogeneous environment which when followed by the trigger leads to the saving of computational power. Here, combination of all these techniques has been applied.

A generalized approach is shown in the flowchart in Fig. 2. It may be applied recursively to the system if there are certain changes in between the processing. The Fig. 2 shows the process of creation of reduct sets with relevance to the attribute selection process. The reduct set attributes when created are processed inside the system. After the sets and all attributes have been clarified from the mining techniques and the processing has been done on the new refined attribute sets, reduced and refined data is obtained. Hence, the computation task is shrinked to a very large extent and thus less time is consumed in producing the desired output.

An algorithmic approach to the Fig. 2 is shown below:
Steps with explanation to algorithm

0: Consider a decision system of a server and associated entities or clients in the network. Where s = server and $C = \{c_1, c_2, \ldots\ldots\ldots, c_n\}$ are the clients on the decision system, n stands for total number. Input is the problem statement P and the data values D.
1: Server performs steps 2–5, 7 , 8
2: Scan the Attributes of the data set in the decision system. And the problem statement, Total data set $D = \{x_1, x_2, x_3, x_4\ldots\ldots\ldots\ldots x_n\}$ where n stands for total no of attribute in the system.

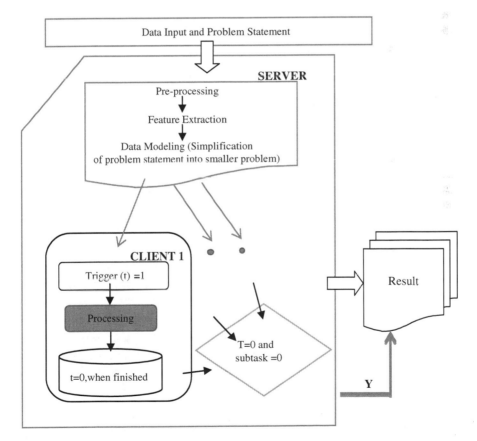

Fig. 2 Processing approach followed by task reduction step

Step 0: $C\{c_1, c_2,\ldots\ldots\ldots c_n\}, t = 0$
Step 2: $D\{x_1, x_2, x_3, x_4\ldots\ldots\ldots x_n\}$
Step 3: $R\{x1, x2, x3\ldots\ldots\ldots xm\}$ where $m < n$
Step 4: $P = \{p_1, p_2, p_3, p_4\ldots\ldots\ldots p_n\}$
Step 6: $while(i = 1$ to $n)$
{
While$(t = 1)${
$C_i[i] = P_i$
If$(P_i?D_j)${
$C_i[i] = P_i + D_j$}
If$(P_i = 0)$ \\ *refers to client finished its task*
{
Step 6: $T = 0$;
Step 7: Shutdown();
}}
Step 8: if$(p_{1\ldots\ldots n} \neq 0)$
For$(c = 1$ to $n)$
{ if (state= shutdown or $p_i = 0$)
{ wakeup();
$C_i = P_i$
Goto 6; }}

3: Apply the rough set-based approach to calculate the reduct set $R = \{x_1, x_2, x_3$
   $\ldots\ldots\ldots\ldots x_m\}$ where $m < n$.
4: Divide the problem into sub problems $P = \{p_1, p_2, p_3, p_4\ldots\ldots\ldots p_n\}$ where
   $p_1, p_2, p_3, p_4\ldots\ldots\ldots p_n$ are the sub problem of the main problem.
5: Distribute $p_1, p_2, p_3, p_4\ldots\ldots\ldots p_n$ among clients $c_1, c_2,\ldots\ldots\ldots c_n$ with the
   data set $R = \{x_1, x_2, x_3\ldots\ldots\ldots x_m\}$, and each client trigger value is set to 0
   $(t = 0)$on which task has been assigned. If any data value is required by the
   processing side it is given as value of the data D_j means at particular position j
   in data set.
6: after the clients finishes its operation it puts the trigger value as 1, $t = 0$, (Done
   by the client).
7: Server on seeing the trigger value as 1 take results from clients and initiates the
   remote shutdown procedure for the client and puts trigger value as $t = 1$.
8: If the server needs the client again it initiates the remote wakeup procedure puts
   trigger $t = 1$ and assigns the task.

A. Description:

After the new reduced task is received by the server it is then further subdivided
into smaller subtasks and is distributed among the server subentities in the system
and leads to a more precise decision system. After the subtasks have been created by
the server side, it just follows either of the mechanisms given.

Mechanism 1: if it follows a general TCP for the communication, the general
packet format as described in the RFC 793.

For this type of environment, we can modify TCP protocol. As we know that,
some bits after Data offset values, i.e., from 3 to 9 bits are reserved for future use,

We can take 1 bit for the notification purpose for notifying the server and the clients in general. Thus, we can use 1 if the work is been assigned to the client by the server. After the processing, the client put the value back to 0. So, in this way, the server come to know whether the client or a particular entity has finished its task. After the server has the information, it initiates remote shut down procedure and the entity or the client move to the shutdown mode. Suppose, after the server has finished all the tasks, if it gets a new task and if client is needed to initiate the remote wakeup procedure, it again come to live and perform several activities. Similarly, in IP-based communication we can further add a trigger bit at the end of the datagram packet. Adding 1 bit does not modify the whole packet format.

Mechanism 2: A trigger is a form of the small code segment appended with the data at the end. When it starts processing its value is set to 1 and when the client finishes the task it puts the trigger value to 0. The results and the trigger value is sent to the server. On receiving the trigger and results value, server initiates the remote shutdown of the client system. If further action is requested by the same client again the trigger value for the intended client is set to 0 and the remote wakeup procedure is initiated for the intended client. Thus, leads to the preservation of overall computational power and electricity.

A scenario illustrates the processing problem, suppose an organization wishes to set up a new Enterprise between Delhi and New York and want to build six offices in between them, also connecting these offices. For this, they need to set up research and development laboratories. Thus, enquiring about the whole situation at different locations is required. The input to the problem statement in terms of data is quite huge. They have to gather information regarding every perspective including the policies, the climate, man power, cost, competition, etc., at different regions.

A simple program illustrates the problem further and provides the remedial solution to the problem. Consider a tcp connection established between the client and server the server and the clients are interacting through concurrent mechanism which defines that several clients can establish the connection to a single server. We are now writing the full code here the server takes the input from the client means whatever we write on the client is to be written on the server side too.

In this set of programs, we can call as much clients as we can; provided the port number should be same. We use a general approach that whatever we write

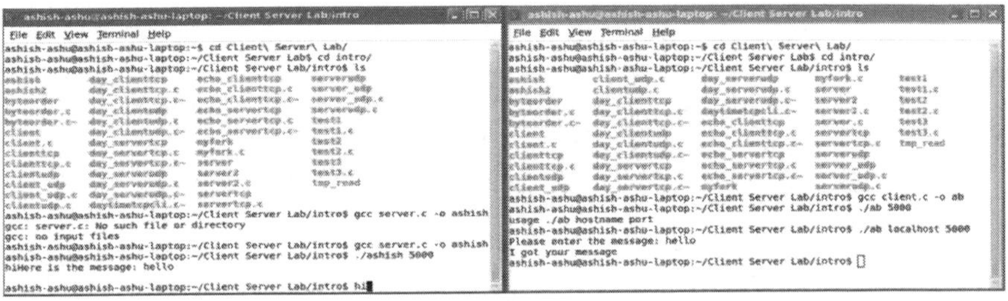

Fig. 3 Showing implementation part

Server code:

```
...newsockfd = accept(sockfd,(struct sockaddr *) &cli_addr,&clilen);
if (newsockfd < 0)
error("ERROR on accept");
bzero(buffer,256);
n = read(newsockfd,buffer,255);
if (n < 0) error("ERROR reading from socket");
printf("Here is the message: %s\n",buffer);
n = write(newsockfd,"I got your message",18);
if (n < 0) error("ERROR writing to socket");
close(newsockfd);
close(sockfd);
return 0;
}
```

Client's code:

```
....if (connect(sockfd,(struct sockaddr *) &serv_addr,sizeof(serv_addr)) < 0)
error("ERROR connecting");
printf("Please enter the message: ");
bzero(buffer,256);
fgets(buffer,255,stdin);
n= write(sockfd,buffer,strlen(buffer));
if (n < 0)
error("ERROR writing to socket");
bzero(buffer,256);
n = read(sockfd,buffer,255);
if (n < 0)
error("ERROR reading from socket");
printf("%s\n",buffer);
close(sockfd);
return 0;
}
```

on particular client is wrote remotely on the server side uses buffer concept for writing values. This Buffer will act as a semaphore or trigger in this case. But for large processing, a client will wait for the response from user side. In this type of programs, e.g., chat server for eco computing we need to save energy to much extent so whenever the client remains idle for few minutes it is automatically shut down and wakeup whenever required we can use a trigger as in case of this concept. Although we take example of client server here but the same applies to the large computation task whose data sets is first been reduced and afterwards the Eco computation facility is been provided in the system.

A screenshot of the applied mechanism is as shown below (Fig. 3).

4 Conclusion

This paper presented focuses on providing various methodologies to facilitate the green computing scenario using some of the known techniques of Data Mining on the provided dataset. Through this we are able to produce the desired results with a really good rate and make our environment quite green and reduce the environmental hazards caused to it by the unnecessary wastage of computational power with extra consumption of electricity too.

References

1. Lo, C.T.D., Qian, K.: Green computing methodology for next generation computing scientist in Computer Software and Applications Conference (COMPSAC), 2010 IEEE 34th Annual, pp. 250–251, IEEE (2010)
2. Cheung, A. K. Y., & Jacobsen, H. A.: Dynamic load balancing in distributed content-based publish/subscribe pp. 141–161, Springer Berlin Heidelberg, (2006)
3. Cheung, A.K.Y., Jacobsen, H.-A.: Publisher placement Algorithm in content-based publish/subscribe, in ICDCS'10 (2005)
4. Schott, B., Emmen, A.: Green Methodologies in Desktop-Grid, in Proceedings of the 2010 International Multi conference on Computer Science and Information Technology (IMCSIT), pp. 671–676, IEEE (2010)
5. Komorowski, J., et al. Rough sets: A tutorial. Rough fuzzy hybridization: A new trend in decision-making, 3–98 (1999)
6. Cheung, A. K. Y., & Jacobsen, H. A.: Green resource allocation algorithms for publish/subscribe systems, in proceedings of 31st (ICDCS) International Conference on Distributed Computing Systems, pp. 812–823, IEEE (2011)
7. Gruber, R., Keller, V.: HPC@ Green it: Green high performance computing methods,Springer Publications, (2010)
8. Comer, Douglas E.: Internetworking con TCP/IP. Vol. 1. Pearson, (2006)

A Comparative Study of Performance of Different Window Functions for Speech Enhancement

A. R. Verma, R. K. Singh and A. Kumar

Abstract In this paper, a speech enhancement technique proposed by Soon and Koh is examined and improved by exploiting different window functions for preprocessing of speech signals. In this method, instead of using two-dimensional (2-D) discrete Fourier transform (DFT), discrete cosine transform (DCT) is employed with a hybrid filter based on one-dimensional (1-D) Wiener filter with the 2-D Wiener filter. A comparative study of performance of different window functions such as Hanning, Hamming, Blackman, Kaiser, Cosh, and Exponential windows has been made. When compared, Cosh window gives the best performance than all other known window functions.

Keywords Cosh · Kaiser · Exponential · Hybrid filter · Wiener filter

1 Introduction

An explosive advances in recent years in the field of digital computing have provided a remarkable progress to the field of speech processing such as voice communication and voice recognition [1]. The presence of background noise in speech significantly reduces the intelligibility of speech as degrades the performance of the speech processing systems [1–3]. In early stage of research, many algorithms [4–7] were developed to suppress background noise and improve the perceptual quality

A. R. Verma (✉) · R. K. Singh · A. Kumar
PDPM Indian Institute of Information Technology Design and Manufacturing
Jabalpur, Jablpur, M.P 482005, India
e-mail: agyaram06ei03@gmail.com

R. K. Singh
e-mail: rahul09088@iiitdmj.ac.in

A. Kumar
e-mail: anilkdee@gmail.com

B. V. Babu et al. (eds.), *Proceedings of the Second International Conference on Soft Computing for Problem Solving (SocProS 2012), December 28–30, 2012*, Advances in Intelligent Systems and Computing 236, DOI: 10.1007/978-81-322-1602-5_105, © Springer India 2014

of speech. These methods were developed with some perceptual constraints placed on noise and speech signals. They involve simple signal processing and some of the speech signal is distorted during enhancement process. Hence, these techniques have a tradeoff between amount of noise removal and speech distortions which was introduced because of processing of speech signal. In the past two decades, a substantial progress has been made in the field of speech processing. So far, several efficient speech enhancement techniques such as spectral subtraction, minimum mean-square error (MMSE) estimation, Wiener filtering (linear MMSE), and Kalman filtering and subspace methods have been reported in literature [6–8]. Most commonly exploited technique is the spectral subtraction method as it is relatively easy to understand and implement [8, 9]. This technique requires only DFT of noisy signal, application of gain function, the inverse DFT and based on minimal assumption of signal and noise. Therefore, many researchers [5, 6] have used this basic spectral subtraction technique and have developed several algorithms for speech enhancement. In all these algorithms, the basic technique varies subtraction of the noise spectrum over the entire speech spectrum. Then, wiener gain function or Wiener filtering has been employed for speech enhancement [8, 9]. Thereon, many techniques based on MMSE estimation have been proposed for the speech enhancement. In these techniques, an optimal MMSE and short-time spectral amplitude estimator (STSA) for speech enhancement was introduced [5]. To estimate clean speech, the short time Fourier coefficients for continuous frame which is independent Gaussian variables of noisy speech are statistically modeled. The authors in [10] have proposed a speech enhancement technique based on adaptive filtering such as Kalman filtering, while in [11, 12], the subspace method has been used for speech enhancement. Recently, several new techniques [12–15] have been developed for speech enhancement.

Speech signal is quasi-stationary in nature and possible to observe it is by in segments. Therefore, speech signal is divided into frame block. To maintain order of continuity of first and last frame, every frame is multiply by a window function to avoid discontinuities between frames. Hamming window is most commonly used for windowing of the speech signal [5–8]. This window function gives fairly good frequency resolution as compared to other window functions. However, in speech processing, one more important spectral parameter, called side lobe roll-off ratio (S) is more essential as compared to other spectral characteristics of window function [16–19]. For the speech applications, higher side lobe ratio is required. Therefore, to improve the spectral characteristics, many window functions have been developed such as Kaiser, Saramaki, Cosh, and Exponential [16–19]. In above context, therefore, in this paper, an improved method based on DCT for speech enhancement using different window functions is presented from slight modification in Soon and Koh algorithm [7].

2 Overview of Different Windows

Window functions are widely used in digital signal processing for various applications of signal analysis [16–19]. A window is time-domain weighting function. In speech processing, because of quasi-periodic in nature and possible to observe this, a speech signal is segmented into different frames. To maintain order of continuity of first and last frame, every frame is multiplied by a window function [5, 7]. The performance of window function in different applications relies on the spectral characteristics of a window. For example, in digital filter design and image processing, the ripple ratio or main lobe to side lobe energy ratio (MSR) are more essential as compared to the side lobe roll-off ratio, while in speech processing, the side lobe roll-off ratio is more important [16–19].

In literature, a window function has been categorized in two ways: fixed length window and adjustable length window. In fixed length window, only window length controls the window's main lobe width. In the adjustable windows, in addition to the window length, the window shape parameter is also exploited for controlling the spectral characteristics such as the main lobe width and ripple ratio [18]. The authors in [18, 19] have developed new window functions, named as exponential and Cosh windows which give better the side lobe roll-off ratio and also computationally efficient. In this research, a comparative study of performance of different window functions for speech enhancement is made. In this work, following window functions are used:

Hamming Window: In 2-D, Hamming window is defined as

$$h(j,\ k) = (0.54 - 0.46\cos(\frac{2\pi j}{255})) \, (0.54 - 0.46\cos(\frac{2\pi k}{15})), \tag{1}$$

$$\text{for } 0 \le j \le 255,\ 0 \le k \le 15)$$

Hanning Window: It is defined as

$$h(j,\ k) = 0.5(1 - \cos(\frac{2\pi j}{255}))0.5(1 - \cos(\frac{2\pi k}{15})), \text{ for } 0 \le j \le 255,\ 0 \le k \le 15 \tag{2}$$

Blackman window: In 2-D, Blackman window is defined as

$$w(n) = (a_0 - a_1\cos(\frac{2\pi j}{255}) + a_2\cos(\frac{4\pi j}{255})) \, (a_0 - a_1\cos(\frac{2\pi k}{15}) - a_2\cos(\frac{4\pi k}{15})) \tag{3}$$

$$\text{for } 0 \le j \le 255,\ 0 \le k \le 15$$

where,

$$a_0 = \frac{1 - \alpha}{2},\ a_1 = \frac{1}{2} \text{ and } a_2 = \frac{\alpha}{2}. \tag{4}$$

In Eq. (3), $\alpha = 0.16$.

Kaiser window: In DT domain, Kaiser Window is defined by [16]:

$$W(n) = \begin{cases} \dfrac{I_0(\alpha\sqrt{1-(\frac{2j}{255}-1)^2})}{I_0(\alpha)} \times \dfrac{I_0(\alpha\sqrt{1-(\frac{2k}{16}-1)^2})}{I_0(\alpha)} & \text{for } 0 \le j \le 255 \text{ and } 0 \le k \le 15 \\ 0 & otherwise \end{cases}$$
(5)

where α is the adjustable parameter and $I_0(x)$ is the modified Bessel function of the first kind of order zero which can be described by the power series expansion as

$$I_0(x) = 1 + \sum_{k=1}^{\infty} \left[\frac{1}{k!}(\frac{x}{2})^k\right]^2$$
(6)

Exponential window: The authors in [17, 19] have used exponential function instead of Bessel function, and introduced a new window, called Exponential window, defined as

$$w(n) = \frac{e^{\left[\alpha\sqrt{1-(1-(2j/(255)))^2}\right]}}{e^\alpha} \times \frac{e^{\left[\alpha\sqrt{1-(1-(2k/(15)))^2}\right]}}{e^\alpha} \quad \text{for } 0 \le j \le 255 \text{ and } 0 \le k \le 15$$
(7)

where α is window shape parameter which controls ripple ratio. Exponential window provides better side lobe roll-off ratio.

Cosh window: Cosh window is defined as [18, 20]

$$w_c(n) = \begin{cases} \dfrac{\cosh(\alpha_c\sqrt{1-\left(\frac{2j}{255}\right)^2})}{\cosh(\alpha_c)} \times \dfrac{\cosh(\alpha_c\sqrt{1-\left(\frac{2k}{16}\right)^2})}{\cosh(\alpha_c)} & \text{for } 0 \le j \le 255 \text{ and } 0 \le k \le 15 \\ 0 & \text{otherwise} \end{cases}$$
(8)

where, α_c and N are the adjustable parameters to control the window spectrum in terms of the ripple ratio and main lobe width.

3 DCT-Based Methodology for Speech Enhancement

The noise taken for speech enhancement is additive noise to model background noise and both are uncorrelated. The additive noise model for speech is defined as

$$y(t) = x(t) + n(t)$$
(9)

where, $y(t)$ is the observed noisy speech, $x(t)$ is the clean speech and $n(t)$ additive background noise. The observed speech signal is divided into overlapping frame of 256 samples. The overlap taken between two consecutive frames is 75%, which means each frame is shifted by previous frame by 64 samples [7]. To maintain the order of continuity of first and last frame, every frame is multiplied by a window function. The product of window and block is called as windowed speech block. Then, 2-D DCT can be applied onto each windowed speech block. The forward 2-D DCT of windowed speech block is defined as [21].

$$W(u, v) = \alpha_u \alpha_v \sum_{j=0}^{255} \sum_{k=0}^{15} w(j, k) \cos(\frac{\pi(2m+1)u}{2M}) \cos(\frac{\pi(2n+1)v}{2N}), \quad (10)$$

$$\text{For } 0 \leq u \leq M - 1 \text{ and } 0 \leq v \leq N - 1$$

where,

$$\alpha_u = \begin{cases} \frac{1}{\sqrt{M}}, & u = 0 \\ \sqrt{\frac{2}{M}}, & 1 \leq u \leq M - 1 \end{cases} \quad \text{and} \quad \alpha_V = \begin{cases} \frac{1}{\sqrt{M}}, & v = 0 \\ \sqrt{\frac{2}{M}}, & 1 \leq v \leq M - 1 \end{cases} \quad (11)$$

In Eq. (13), M and N are the row and column size. The inverse 2-D DCT can be obtained with

$$w(j, k) = \sum_{u=0}^{M-1} \sum_{v=0}^{N-1} \alpha_j \alpha_k W(u, v) \cos(\frac{\pi(2m+1)}{2M}) \cos(\frac{\pi(2n+1)}{2N}) \quad (12)$$

where, $0 \leq j \leq M - 1$ and $0 \leq k \leq N - 1$. α_j and α_k are defined as

$$\alpha_j = \begin{cases} \frac{1}{\sqrt{M}}, & j = 0 \\ \sqrt{\frac{2}{M}}, & 1 \leq j \leq M - 1 \end{cases} \quad \text{and} \quad \alpha_k = \begin{cases} \frac{1}{\sqrt{M}}, & k = 0 \\ \sqrt{\frac{2}{M}}, & 1 \leq k \leq M - 1 \end{cases} \quad (13)$$

4 Spectral Magnitude Subtraction

In this paper, 2-D speech enhancement technique given in [7] is modified using DCT and different window functions. In Soon and Koh algorithm [7], 2-D DFT is employed with magnitude spectral subtraction. In this method, magnitude of cosine transform coefficients which is less than particular threshold is used. These coefficients depend upon the expected noise magnitude [7]. Let "Y" represents as transformed noisy speech block and $\hat{X}$ represents enhanced speech block which is defined as

$$\left| X(\hat{u}, v) \right| = \max(|Y(u, v)| - E[|N(u, v)|], 0) \quad (14)$$

where, $E[|N(u, v)|]$ is the expected mean of noise and noisy phase $\theta_{Y(u, v)}$ is represented by

$$X(\hat{u},\ v) = \left|X(\hat{u},\ v)\right| \exp(i\ \theta_Y(u,v)) \tag{15}$$

Before taking inverse DCT, weaker coefficients can be combined with the noisy phase. By applying inverse 2-D DCT, the enhanced speech block in the time domain is obtained this equation.

$$x(j,\ k) = \sum_{u=0}^{M-1}\sum_{v=0}^{N-1} \alpha_j\alpha_k\ \hat{X}(u,\ v)\ \cos(\frac{\pi(2m+1)}{2M})\ \cos(\frac{\pi(2n+1)}{2N}) \tag{16}$$

Finally, enhanced speech is obtained by reversing the blocking and framing process and is represented by this equation [12].

$$\hat{x}(t) = 0.25(f_L(j) + f_{L-1}(j+64) + f_{L-2}(j+128) + f_{L-3}(j+192)) \tag{17}$$

where,

$$L = \left\lfloor \frac{t}{64} \right\rfloor \quad \text{and} \quad j = t - 64L \tag{18}$$

The hybrid filter not only shows 2-D wiener filter effect but also shows properties of 1-D filter to remove additive audible noise. By averaging the coefficients within the surrounding frame, the hybrid wiener filter shows correlation. Figure 1 shows a block diagram of a hybrid filter for speech enhancement using DCT [7]. Equation 17 shows standard wiener filter equation. The power and noise in the cleaned speech is being known for the construction of 1-D wiener filter and it is defined as.

$$W_{1D}(u,\ v) = \frac{\xi(u,v)}{\xi(u,v)+1} \tag{19}$$

where, a-priori SNR is represented by

$$\xi(u,\ v) = \frac{E\left[X(u,v)^2\right]}{\lambda_N E[\cdot]} \tag{20}$$

and $\lambda_N = E\left[N(u,v)^2\right]$ is the noisy expected mean. Priori SNR is nothing, but combination of posterior SNR and previous frame enhanced spectrum. In this work, value of λ_N is assumed to be known because during the pauses period in the speech signal, the background noise is easily estimated. The value of $E\left[N(u,v)^2\right]$ is not known and should be estimated.

$$\xi(u,\ v) = \alpha\frac{\hat{X}(u-1,\ v)^2}{\lambda_N} + (1-\alpha)F\ \gamma(u,\ v) - 1 \tag{21}$$

The resulting speech filtered by 1-D Wiener filter is

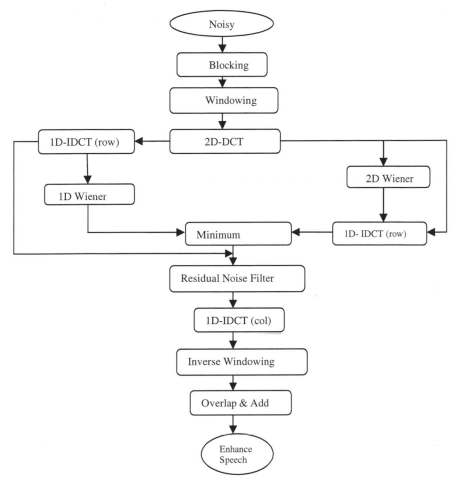

Fig. 1 Block diagram of 1-D and 2-D hybrid filter for speech enhancement [7]

$$\hat{X}_{1-D}(u, v) = \frac{\hat{\xi}(u, v)'}{\hat{X}(u, v)' + 1} Y(u, v) \tag{22}$$

where, $\hat{X}(u-1, v)$ is the estimated speech spectrum value of $X(u, v)$ in the previous frame, $\gamma(u, v) = Y(u, v)^2 / \lambda_N$ is the a-posteriori SNR, E[·] denotes the half-wave rectification function. α controls the behavior of the SNR estimator and value is normally set to 0.98. Basic method spectral magnitude subtraction does not show the 2-D properties in the 2D DFT [7]. By applying 2-D Wiener filter to speech signal, remove the mean noise magnitude level and reduce the instantaneous AC noise variation by coefficients to coefficients [7]. In this, 2-D Wiener filter is applied to the DCT coefficient. Implementation part is accomplished using 2-D spectrum filtering; therefore at the outset 2-D noise model is introduced. There are two main parts in this model, namely, the DC component and the AC component:

$$N (u, v) = N_{AC}(u, v) + N_{DC}(u, v) \tag{23}$$

where, the DC component could be expressed as $N_{DC}(u, v) = E[N(u, v)]$. The variance of $N_{AC}(u, v)$ is denoted as $\sigma_{NAC}(u, v)$ and can be obtained by Eq. (24):

$$\sigma_{NAC}(u, v)^2 = E\left[N(u, v)^2\right] - N_{DC}(u, v)^2. \tag{24}$$

The 2-D Wiener filter is based on the maximum likelihood approach. if we know the noisy speech and the additive noise, then the 2-D noise model is used:

$$\hat{X}_{2-D}(u, v) = \frac{\sigma_Y(u, v)^2 - \sigma_{NAC}(u, v)^2}{\sigma_Y(u, v)^2}$$
$$\times (Y(u, v) - \overline{Y(u, v)}) + \overline{Y(u, v)} - N_{DC}(u, v) \tag{25}$$

where, $\overline{X(u, v)}$ and $\overline{Y(u, v)}$ are the local means of $X(u, v)$ and $Y(u, v)$. "σ" indicates the local variance. Hybrid filter is the combination of 1-D and 2-D Wiener filters, Fig. 1 describe the processing of noise speech spectrogram through the hybrid Wiener filter via two channels: the 1-D Wiener filter and the 2-D Wiener filter. The result of the hybrid Wiener filtering in Fig. 1 is by taking the minimum of Eqs. (24) and (25):

$$\hat{X}(u, v) = \min(\hat{X}_{1-D}(u, v), \hat{X}_{2D}(u, v)) \tag{26}$$

Based on the listening test, the enhanced speech obtained is compared with either the 1-D Wiener filtered speech or 2-D Wiener filtered speech.

5 Result and Discussion

Results included in Table 1 illustrate the performance of different windows based on DFT and Table 2 represents the performance of different windows in the speech enhancement based on DCT. For performance analysis, several fidelity parameters such as signal-to-noise ratio (SNR), maximum error (ME), and mean square error (MSE) given in [6] are computed. As it is evident from Tables 1 and 2, Cosh window gives better performance as compared to other windows in terms of SNR and other fidelity parameters.

Table 1 Performance of different windows based on DFT

Windows	Input SNR (dB)	Output SNR (dB)	ME	MSE
Hamm	4.14	15.18	0.1045	0.2118
Hanning	4.14	9.11	15.45	0.5639
Kaiser	4.14	33.42	1.6225	0.0044
Blackman	4.14	21.26	2.5847	0.0022
Exponential	4.14	36.45\	0.1806	1.09×10^{-4}
Cosh	4.14	39.45	0.1806	1.43×10^{-4}

Table 2 Comparative performance of different windows base on DCT

Windows	Input SNR (dB)	Output SNR (dB)	ME	MSE
Hamm	4.14	18.22	0.0306	0.0631
Hanning	4.14	15.12	6.857	0.0639
Kaiser	4.14	37.66	0.0825	0.0214
Blackman	4.14	24.30	1.5821	0.0032
Exponential	4.14	41.31	0.0481	2.09×10^{-5}
Cosh	4.14	45.56	0.0806	3.43×10^{-6}

6 Conclusions

In this paper, an improved method based on DCT is presented for the speech enhancement. A comparative study of performance of different window functions has been made. The simulation results included in this paper clearly shows better performance as compared to other window functions. The most effective result is obtained using Cosh window in the 2-D DCT domain for speech enhancement. The Cosh window gives signal to noise ratio (SNR) at output is 45.56 db. It is also evident that discrete cosine transform gives better results as compared to 2-DFT, and it is computationally efficient.

References

1. Virag, N.: Single channel speech enhancement based on masking properties of the human auditory system. IEEE Trans. Speech Audio Process. **7**, 126–137 (1999)
2. Yang, L., Shuangtian, L.: DCT speech enhancement based on masking properties of human auditory System. In: Proceeding Institute of Acoustics Chinese Academy of Science, pp. 450–453 (2010)
3. Lei, S.F., Tung, Y.K.: Speech enhancement for nonstationary noises by wavelet packet transform and adaptive noise estimation. In: Proceedings of Internation Symposium on Intelligent Signal Processing and Communication Systems, pp. 41–44 (2005)
4. Chen, H., Smith, C.H., Fralick, S.: A fast computation algorithm for the discrete cosine transform. IEEE Trans. Commun. **25**, 1004–1009 (1997)

5. Ephraim, Y., Malah, D.: Speech enhancement using a minimum mean-square error short-time spectral amplitude estimator. IEEE Trans. Acoust. Speech Signal Process. (ASSP) **32**, 1109–1121 (1984)
6. Boll, S.F.: Suppression of acoustic noise in speech using spectral subtraction. IEEE Trans. Acoust. Speech Signal Process. **27**, 113–120 (1979)
7. Soon, I.Y., Koh, S.N.: Speech enhancement using 2D Fourier transform. IEEE Trans. Speech Audio process. **11**, 717–724 (2003)
8. Yuan, Z.X., Koh, S.N., Soon, I.Y.: Speech enhancement based on a hybrid algorithm. Electron. Lett. **35**, 1710–1712 (1999)
9. Vaseghi, S.V.: Advanced Digital Signal Processing and Noise Reduction, 2nd edn., pp. 270–290. Wiley, Hoboken (2000)
10. Paliwal, K.K., Basu, A.: A speech enhancement method based on Kalman filtering. In Proceeding of Internationl Conference Acoustics Speech, Signal Process. pp. 177–180 (1987)
11. Ephraim, Y., Harry, L., Trees, V.: A signal subspace approaches for speech enhancement. IEEE Trans. Speech Audio Process. **3**(4), 251–266 (1995)
12. Huang, J., Zhao, Y.: A DCT-based fast signal subspace technique for robust speech recognition. IEEE Trans. Speech Audio Process. **8**(6), 747–751 (2000)
13. Oberle, S., Kaelin, A.: HMM-based speech enhancement using pitch period information in voiced speech segments. In: Intemational Symposium on Circuits and Systems ISCAS vol. 27, pp. 114–120 (1997)
14. Makhoul, J.: A fast cosine transform in one and two dimensions. IEEE Trans. Acoust. Speech Signal Process. **28**(1), 27–34 (1980)
15. Mittal, U., Phamdo, N.: Signal/noise KLT based approach for enhancing speech degraded by colored noise. IEEE Trans. Speech Audio Process. **8**, 159–167 (2000)
16. Kaiser, J.F., Schafer, R.W.: On the use of the IO-sinh window for spectrum analysis. IEEE Trans. Acoust. Speech Signal Process. **28**, 105–107 (1980)
17. Avci, K., Nacaroglu, A.: A New Window Based on Exponential Function. In: IEEE Conference on Research in Microelectronics and Electronics pp. 69–72 (2008)
18. Kumar, A., Kuldeep, B.: Design of M-channel cosine modulated filter bank using modified exponential window. J. Franklin Inst. **349**, 1304–1315 (2012)
19. Avci, K., Nacaroglu, A.: Cosine hyperbolic window family with its application to FIR filter design. In: 3rd International Conference on Information and Communication Technologies: from Theory to Applications, ICTTA'08, Damascus, Syria (2008)
20. Kumar, A., Singh, G.K., Anand, R.S.: An improved closed form design method for the cosine modulated filter banks using windowing technique. Appl. Soft Comput. **11**(3), 3209–3217 (2011)
21. Garello, R. (ed.): Two-dimensional Signal Analysis

Methods for Estimation of Structural State of Alkali Feldspars

T. N. Jowhar

Abstract There is much interest in characterizing the variations in feldspar structures because of the abundance and importance of feldspars in petrologic processes and also due to their general significance in mineralogical studies of exsolution and polymorphism, especially order-disorder. With the appearance of new analytical and rapid methods of X-ray crystallographic study and computational techniques, the significance of feldspars in igneous and metamorphic rocks has increased tremendously. In this paper methods for estimation of structrural state of alkali feldspars is reviewed and discussed.

Keywords Alkali feldspars · X-ray crystallography · Lattice parameters · Structural state · Al-Si distribution

1 Introduction

The alkali feldspars are major components of granites and granite pegmatites. The study of their structural state and composition has always been one of interesting researches in petrology and mineralogy [1–3]. Pioneering works by Barth (1969) [4] and Laves (1952) [5], where alkali feldspar composition and structure were correlated with the formation conditions, and the effect of ordering factors on crystal structure was demonstrated.

With the appearance of new analytical and rapid methods of X-ray crystallographic study and computational techniques, the significance of feldspars in igneous and metamorphic rocks has increased [1–3, 6–13]. In this paper methods for estimation of structrural state of alkali feldspars is reviewed and discussed.

T. N. Jowhar (✉)
Wadia Institute of Himalayan Geology, Dehradun 248001, India
e-mail: tnjowhar@rediffmail.com; tnjowhar@gmail.com

B. V. Babu et al. (eds.), *Proceedings of the Second International Conference on Soft Computing for Problem Solving (SocProS 2012), December 28–30, 2012,* Advances in Intelligent Systems and Computing 236, DOI: 10.1007/978-81-322-1602-5_106, © Springer India 2014

2 Feldspar Structure

In the feldspar structure, all the four oxygen atoms of SiO_4 tetrahedron are shared with neighbouring tetrahedra, forming a three dimensional framework, giving a silicon oxygen ratio of 1:2. The ideal formula of a feldspar can be expressed as MT_4O_8. The M site is occupied by large atoms like K, Na and Ca with 9–10 coordination. T is a tetrahedral site which is occupied by small trivalent or tetravalent atoms (Al, Si). In the feldspar structure four, four membered rings of tetrahedra are joined with each other by sharing apical oxygens designated A_1 and A_2 to form eight membered rings of elliptical cross section. Each four membered ring consists of four tetrahedra, in which the central cation of two tetrahedra are of type T_1 and two of type T_2. The four membered rings of tetrahedra are joined by oxygen atoms of type A_2 on mirror plane and oxygens of type A_1 on 2-fold axes, to form eight membered ring of elliptical cross section. The M atom occupies a four fold site on mirror planes.

The structure of the $KAlSi_3O_8$ polymorphs can be described as the variants of the high-temperature sanidine structure. Each unit cell of sanidine contains $4(KAlSi_3O_8)$ formula units and sanidine has space group symmetry C2/m. The (12 Si + 4 Al) atoms in the unit cell of sanidine are distributed randomly over two crystallographically distinct tetrahedral sites T_1 and T_2. The four potassium atoms occupy special positions on mirror planes perpendicular to b-axis. The 32 oxygen atoms are located on both special positions (four on two fold axes; four on mirror planes) and on general positions.

Structure of the lower-temperatures polymorphs (maximum microcline, orthoclase) differ from that of sanidine due to the order-disorder relations of Si and Al atoms in tetrahedral sites. Ordering of Al in T site results in loss of a 2-fold axis of rotation and a mirror plane of symmetry and only the center of symmetry at the lattice plane persists, with a concomitant transition from monoclinic to triclinic symmetry. The two tetrahedral sites T_1 and T_2 of the monoclinic phase become four symmetrically distinct tetrahedral sites T_1O, T_1m, T_2O and T_2m in triclinic phase. In the structure of ideal orthoclase $(KAlSi_3O_8)$ space group C2/m aluminium is concentrated in the T_1 sites and depleted from T_2 and the probability being 50% that each T_1 site will contain aluminium. Structural refinements of maximum microcline (Triclinic) space group $C\bar{1}$ have shown that Al is strongly ordered into T_1O site [1, 3].

3 Structural State of Alkali Feldspars

The difference in symmetry of sainidine and microcline is due to order-disorder relations of Si and Al atoms over tetrahedral sites. The distribution of Al and Si in tetrahedral sites in maximum microcline (Low microcline) is of the highest degree of order and microcline has a lowest structural state. On the other hand, sanidine, with the highest structural state shows random Al-Si distribution in tetrahedral sites. In between these two end members maximum microcline and sanidine, several intermediate structural states e.g. orthoclase and intermediate microcline exist.

3.1 Determination of Al, Si Distribution in the Tetrahedral Sites of Alkali Feldspars

The Si-Al proportion in each tetrahedral site T_10, T_1m, T_20 and T_2m of alkali feldspar is difficult to determine directly from X-ray diffraction methods because the scattering factors of Al and Si are very similar. But, the ionic radius of Al is ~0.14Å larger than that of Si, therefore, it is possible to make correlations between mean T-O distances and Al content of tetrahedral sites. Multiple linear regression analyses of monoclinic K-feldspars by Phillips and Ribbe (1973) [14] have shown that the Al content of a tetrahedral site is the principle factor influencing individual T-O distances. Smith and Bailey (1963) [15], Jones (1968) [16] and Ribbe and Gibbs (1969) [17] has suggested a linear model relating the mean T-O bond length of a tetrahedron to its Al content. In a detailed study of Al-O and Si-O tetrahedral distances Smith and Bailey (1963) [15] found that the mean tetrahedral bond length in feldspars varies linearly with percentage of aluminium from 1.61 Å for Si-O to 1.75 Å for Al-O. They estimated that the average Al content of a tetrahedron can be predicted to ±5 %. This relationship determined from two and three dimensional refinements of feldspar structures has since been used to predict Al-Si distribution in feldspars and other framework structures, thereby permitting an estimate of the degree of long-range order or structural state.

3.2 X-ray Methods for Determining the Structural State of Alkali Feldspars

Three peak method of Wright: Wright (1968) [18] introduced a procedure for estimating composition and structural state of alkali feldspars from the 2θ positions of three reflections that depend principally on a single cell parameter. The reflections and the cell parameters upon which their 2θ value depends are $\bar{2}01 - a$, $060 - b$, and $\bar{2}04 - c$. The reflections $\bar{2}01$, 060 and $\bar{2}04$ are of sufficient intensity and their 2θ values can be accurately measured. Curves relating the 2θ values to the appropriate cell dimensions are essentially linear and permit easy estimation of starting cell parameters for a complete refinement of the unit cell.

130, 131 **and** 111 **splittings:** In monoclinic feldspars, 130, 131 and 111 diffractions are represented by a single peak. However, in triclinic feldspars each of these (130,131 and 111) splits into two diffractions. These splittings are useful for measuring the deviation from monoclinic geometry.

$$\text{Let } Q = \frac{1}{d^2} \quad \text{Then} \quad Q(130) - Q(1\bar{3}0) = 6a^*b^* \cos\gamma^* \tag{1}$$

$$Q(131) - (Q1\bar{3}1) = 6a^*b^* \cos\gamma^* + 6b^*c^* \cos\alpha^* \tag{2}$$

Therefore, an estimate of α^* and γ^* can be obtained from splitting of the 130 and 131 diffractions, since a*, b* and c^* can be easily obtained.

Mackenzie (1954) [19] used the splitting of the 130 diffraction, whereas Goldsmith and Laves (1954) [20] used the splitting of the 131 diffraction for microcline. Goldsmith and Laves (1954) [20] defined triclinicty Δ as:

$$\Delta = 12.5 \left[d\,(131) - d\left(1\bar{3}1\right) \right] \tag{3}$$

The value of 12.5 is chosen so that Δ reaches unity for maximum microcline.

When $\Delta = 1$ feldspar is fully ordered (Maximum microcline)

When $\Delta = 0$ feldspar is monoclinic.

For microcline, the 130 triclinic indicator is given by (Smith 1974a) [1]:

$$\Delta = 7.8 \left[d\,(130) - d\left(1\bar{3}0\right) \right] \tag{4}$$

Mackenzie (1954) [19] and Goldsmith and Laves (1954) [20] have shown that many natural k-feldspars yield blurred 130 and 131 peaks that can be interpreted as the superposition of K-feldspars with different triclinic indicators. The concept of random disorder of a multitude of very small volumes each with a different degree of Si-Al order was introduced by Christie (1962) [21].

Ragland parameters: Ragland (1970) [22] defined parameters δ and δ' based upon the data of Wright (1968) [18]:

$$\delta = \frac{9.063 + 2\theta\,(060) - 2\theta\left(\bar{2}04\right)}{0.340} \tag{5}$$

$$\text{and} \quad \delta' = \frac{9.063 + 2\theta\,(060) - 2\theta\left(\bar{2}04\right)}{0.205} \tag{6}$$

δ describes variations in structural state between the maximum microcline-low albite series and the orthoclase series. δ varies from 1.00 for maximum microcline-low albite series to 0.00 for orthoclase series. δ' describes the variations between the orthoclase and the sanidine-high albite series. δ' varies from 0.00 for orthoclase series to -1.00 for high sanidine-high albite series. Thus a unique value from -1.00 to $+1.00$ can be calculated to define the structural state of any alkali feldspar. Ragland (1970) [22] also related parameters δ and δ' to estimate Al in the T_1 sites by the following equations:

$$\text{Al } T_1 = \frac{\delta + 2.331}{3.33}; \quad \text{Al } T_1 = \frac{\delta' + 3.500}{5.0} \tag{7}$$

The above equations are based upon (i) that feldspars of the maximum microcline-low albite series are perfectly ordered ($\delta = 1$; Al $T_1 = 1$), (ii) that feldspars of the sanidine-high albite series are perfectly disordered ($\delta' = -1$; Al $T_1 = 0.5$).

Phillips and Ribbe (1973) [14] suggested the use of the following equations in estimating Al contents in monoclinic feldspars:

$$Al\ T_1 = 2.360(c - 0.4b) - 4.369 \tag{8}$$
$$Al\ T_2 = 2.256(c - 0.4b) - 4.658 \tag{9}$$

Kroll method: Kroll (1971, 1973, 1980) [23–25] has proposed a method to estimate Al, Si distribution of alkali feldspars from two lattice translations, called Tr [110] and Tr $\left[1\bar{1}0\right]$, which are the repeat distances in the [110] and $\left[1\bar{1}0\right]$ directions. He has shown that the repeat distances in the [110] and $\left[1\bar{1}0\right]$ directions were controlled by the distribution of Al atoms among the tetrahedral sites. The repeating sequence of the tetrahedra is along [110] : $T_1 0$, $T_2 0$, $T_2 m$ and along $\left[1\bar{1}0\right]$: $T_1 m$, $T_2 0$, $T_2 m$.

The differences in the repeat distance [110] and $\left[1\bar{1}0\right]$ estimates the transfer of Al from $T_1 m$ to $T_1 0$ sites, while the mean of the repeat distances measures the combined transfer of Al from T_2 to T_1 sites. This assumes that there is no interaction with the M site. However, there is an effect from the M site and Kroll prepared separate calibration curves for Na and K feldspar.

The translational distances in the two directions [110] and $\left[1\bar{1}0\right]$ are designated by Tr [110] and Tr $\left[1\bar{1}0\right]$. They are related to the cell dimensions by:

$$Tr\,[110] = \frac{1}{2}\left(a^2 + b^2 + 2ab\cos\gamma\right)^{1/2} \tag{10}$$

$$Tr\left[1\bar{1}0\right] = \frac{1}{2}\left(a^2 + b^2 - 2ab\cos\gamma\right)^{1/2} \tag{11}$$

The changing site occupancies cause the repeat distances to change their lengths: Tr [110] increases, whereas Tr $\left[1\bar{1}0\right]$ decreases during ordering of triclinic alkali feldspars.

For the K-rich and Na-rich alkali feldspars the following equations have been derived by Kroll (1980) [25] to estimate Al site occupancies:

For K-rich alkali feldspar ($Or_{70} - Or_{100}$):

$$T_1 0 = 29.424 - 3.7962\ Tr\left[1\bar{1}0\right] + 0.5355\ n_{or} \tag{12}$$
$$T_1 m = 22.838 - 2.9362\ Tr\,[110] + 0.4265\ n_{or} \tag{13}$$

For Na-rich alkali feldspars ($Or_0 - Or_{30}$):

$$T_1 0 = 28.961 - 3.7594\ Tr\left[1\bar{1}0\right] + 0.8396\ n_{or} \tag{14}$$
$$T_1 m = 17.778 - 2.3049\ Tr\,[110] + 0.7356\ n_{or} \tag{15}$$

where n_{or} = mole fraction $KAlSi_3O_8$ (or). The n_{or} content can be determined from the cell volume using Stewart and Wright (1974) [26] method:

$$n_{or} = \left(0.2962 - \sqrt{(0.953151 - 0.0013V)}\right)0.18062 \tag{16}$$

or using the following equation of Kroll and Ribbe (1983) [27] for low-albite-maximum microcline series:

$$n_{or} = -1227.8023 + 5.35958\,V - 7.81518 \times 10^{-3}V^2 + 3.80771 \times 10^{-6}V^3 \quad (17)$$

Structural indicators $\Delta(\alpha^*\gamma^*)$, $\Delta(b^*c^*)$, $\Delta(bc)$ **and** $\Delta(bc^*)$:
The $\alpha^*\gamma^*$ plot was first introduced in alkali feldspar studies by Mackenzie and Smith (1955) [28]. Subsequently, Smith (1968) [29] showed that from the quadrilateral of the end-members MF (Monoclinic Feldspars), LM (Low Microcline), LA (Low Albite) and HA (High Albite) plotted on the $\alpha^*\gamma^*$ diagram, one can derive the $\Delta(\alpha^*\gamma^*)$ indicator which measures the difference in Al content between the $T_1 0$ and $T_1 m$ tetrahedral sites. All experimental data so far obtained by number of structure analyses confirm that the $\alpha^*\gamma^*$ plot can be used for this purpose [1, 3, 8]. From the values of unit cell parameters b^* and c^*, or b and c, it is possible to obtain indicator $\Delta(b^*c^*)$ [1, 29] or the structural indicator $\Delta(bc)$ or $\Delta(bc^*)$ [26, 30, 31].

Kroll and Ribbe (1987) [32] proposed the following linear equations for determining $\Delta(bc^*)$ and $\Delta(\alpha^*\gamma^*)$:

$$\Delta(bc^*) = Al(T_1 0 + T_1 m) = \frac{b - 21.5398 + 53.8405c^*}{2.1567 - 15.8583c^*} \quad (18)$$

$$\Delta(\alpha^*\gamma^*) = Al(T_1 0 - T_1 m) = \frac{\gamma^* - 44.778 - 0.50246\alpha^*}{6.646 - 0.05061\alpha^*} \quad (19)$$

The $b - c^*$ equations produce nearly linear plots for alkali-exchange series, a substantial improvement over the relationships involving b and c cell parameters.

The distribution of Al over the four tetrahedral sites in the alkali feldspar of any structural state can now be estimated using the following equations from Stewart and Ribbe (1969) [30]:

$$Al\ T_1 0 = \frac{\Delta(bc) + \Delta(\alpha^*\gamma^*)}{2} \quad (20)$$

$$Al\ T_1 m = Al\ T_1 0 - \Delta(\alpha^*\gamma^*) \quad (21)$$

The Al in T_2 sites can be divided equally between $T_2 0$ and $T_2 m$ sites because all available feldspar structure analyses show nearly equal T-O bond lengths for these sites.

$$Al\ T_2 0 = Al\ T_2 m = \frac{[1.0 - (AlT_1 0 + Al\ T_1 m)]}{2} \quad (22)$$

Jowhar Method (1981, 1989) [6, 7]: Jowhar (1981) [6] described a procedure and made a computer program in FORTRAN IV for calculating lattice parameters and distribution of aluminium in tetrahedral sites $T_1 0$, $T_1 m$, $T_2 0$ and $T_2 m$ sites of alkali feldspars by using 4 X-ray reflections $(\bar{2}01)$, (002), (060) and $(\bar{2}04)$ and by making use of equations described by Wright (1968) [18], Wright and Stewart (1968) [33] and Orville (1967) [34]. This method however sometimes gives negative values of Al in

T sites, which is theoretically not possible. This is because of propagation of errors in estimation of lattice parameters α, β and γ. Jowhar (1989) [7] presented another rapid method for calculating lattice parameters and distribution of aluminium in tetrahedral sites $T_1 0$, $T_1 m$, $T_2 0$ and $T_2 m$ sites of alkali feldspars. This method is based on 2θ values of 6 hkl planes $(\bar{2}01)$, (130), (002), $(\bar{1}13)$, (060) and $(\bar{2}04)$ which can be easily obtained on X-ray powder diffractometer. A computer program in FORTRAN 77 on WIPRO 286 computer system was also prepared for computations involved in the determination of lattice parameters and structural state of alkali feldspars.

Coefficient of order: Ribbe and Gibbs (1967) [35] defined coefficient of order as:

$$\text{Coefficient of order} = \frac{P - 0.25}{0.75} \tag{23}$$

where P is the fractional occupancy of Al in $T_1 0$ site. Coefficient of order varies from 0 for sanidine to 0.33 for orthoclase to 1.0 for maximum microcline.

Obliquity: Obliquity, ϕ, defined by Thompson and Hovis (1978) [36] is a direct measure of the departure of a triclinic feldspar from the dimensional requirements for monoclinic symmetry. The quantity $1 - \cos\phi$ must vanish for monoclinic feldspars. In a monoclinic crystal it is necessary (but not sufficient) that the b axis and the pole b*, to (010) coincide, and that:

$$b = \frac{1}{b^*} = d\,(010) \tag{24}$$

On the other hand, in a triclinic crystal b and b* do not coincide. Thompson and Hovis (1978) [36] designated the angle between b and b* as ϕ and called it obliquity. Therefore,

$$d\,(010) = \frac{1}{b^*} = b\cos\phi \tag{25}$$

It is necessary for monoclinic symmetry that $\phi = 0$ or $\cos\phi = 1$. However, it is possible that a triclinic crystal may also have $\phi = 0$. Therefore, the vanishing of the obliquity thus ensures that only the geometric form of the unit cell meets the requirements for monoclinic symmetry, and does not prove that the crystal is monoclinic. The quantity $[d\,(131) - d\,(1\bar{3}1)]$ must vanish when ϕ vanishes, but the reverse is not true.

Thompson ordering parameters: Thompson (1969) [37] described ordering parameters X, Y and Z as:

$$X = N_{Al(T_2 0)} - N_{Al(T_2 m)} \tag{26}$$
$$Y = N_{Al(T_1 0)} - N_{Al(T_1 m)} \tag{27}$$
$$Z = \left[N_{Al((T_1 0)} + N_{Al(T_1 m)}\right] - \left[N_{Al(T_2 0)} + N_{Al(T_2 m)}\right] \tag{28}$$

where N is the fraction of each T site occupied by Al. The three ordering parameters X, Y, and Z, respectively show the deviation in the Al content of the T_2 site from monoclinic symmetry, the deviation from monoclinic symmetry of the T_1 site, and finally the amount of Al transferred from the T_2 sites into the T_1 sites. For completely disordered feldspars, all three ordering parameters are zero. For low microcline and low albite, Y and Z are both 1 but X is zero. Smith (1974a) [1] argued that alkali feldspar under equilibrium conditions would follow a one-step trend and that coherency between domains is responsible for the monoclinic symmetry at the early stages of ordering. In ideal one-step ordering path Al moves to the $T_1 0$ site at an equal rate from $T_1 m$, $T_2 0$ and $T_2 m$ sites, whereas in the ideal two-step ordering path, the Al atoms first segregate into T_1 and T_2 sites preserving monoclinic symmetry by equal occupancy of 0 and m sub-sites, and the second stage of ordering proceeds with triclinic symmetry as Al moves from $T_1 m$ to $T_1 0$. In addition, there is an infinite number of ordering paths between the extreme ideal one-step and two-step ordering path [8, 12, 13].

4 Homogeneous Equilibria in Alkali Feldspars

Estimates of maximum P-T conditions attained during metamorphism can be made relatively easily, but, the determination of the temperatures at which minerals re-equilibrate, follwing cooling, is more difficult to make (Jowhar 1994, 2005, 2012) [8, 38, 39].

Methods of thermodynamic analysis have been successful in dealing with the problem of homogeneous equilibria in crystals [37]. The intensive properties of an alkali feldspar are functions of P, T, N_{or} N_{An} and of the ordering parameters X, Y and Z. For fully equilibrated feldspars, conditions of equilibrium are (Thompson 1969) [37]:

$$\left(\frac{\delta \bar{G}}{\delta X}\right)_{P,T,N_{or},N_{An}Y,Z} = \left(\frac{\delta \bar{G}}{\delta Y}\right)_{P,T,N_{or},N_{An},X,Z} = \left(\frac{\delta \bar{G}}{\delta Z}\right)_{P,T,N_{or},N_{An},X,Y} = 0 \quad (29)$$

where $N_{or} = \frac{n_K}{n_K + n_{Na} + n_{Ca}}$ and $N_{An} = \frac{n_{Ca}}{n_K + n_{Na} + n_{Ca}}$

For monoclinic alkali feldspar the ion exchange of Al and Si between T_1 and T_2 sites may be expressed by the following reaction:

$$Al(T_2) + Si(T_1) \Leftrightarrow Al(T_1) + Si(T_2) \quad (30)$$

Thompson (1969) [37] defined the quantity K'_Z as:

$$K'_Z = \frac{N_{Al(T_1)}N_{Si(T_2)}}{N_{Si(T_1)}N_{Al(T_2)}} = \frac{(1+Z)(3+Z)}{(1-Z)(3-Z)} \quad (31)$$

K'_Z has the form of an apparent equilibrium constant.

5 Determination of Lattice Parameters of Alkali Feldspars

Determination of direct lattice parameters a, b, c, α, β, γ of alkali feldspars is carried out by solving the following Eq. 32 for triclinic crystal system (a $\neq$ b $\neq$ c $\neq$ α $\neq$ β $\neq$ γ $\neq$ 90^0) for at least 6 hkl planes [7, 8, 13, 40–42]:

$$\frac{1}{d_{hkl}^2} = \frac{\frac{h^2}{a^2}\sin^2\alpha + \frac{k^2}{b^2}\sin^2\beta + \frac{l^2}{c^2}\sin^2\gamma + \frac{2hk}{ab}(\cos\alpha\cos\beta - \cos\gamma) + \frac{2hl}{ac}(\cos\alpha\cos\gamma - \cos\beta) + \frac{2kl}{bc}(\cos\beta\cos\gamma - \cos\alpha)}{1 - \cos^2\alpha - \cos^2\beta - \cos^2\gamma + 2\cos\alpha\cos\beta\cos\gamma}$$

(32)

d spacing as a function of reciprocal unit cell parameters is:

$$\frac{1}{d_{hkl}^2} = \sigma_{hkl}^2 = \left[h^2 a^{*2} + k^2 b^{*2} + l^2 c^{*2} + 2hka^* b^* \cos\gamma^* + 2klb^* c^* \cos\alpha^* + 2lhc^* a^* \cos\beta^*\right]$$

(33)

Unit cell volume is obtained by using the relation:

$$V = abc\sqrt{1 - \cos^2\alpha - \cos^2\beta - \cos^2\gamma + 2\cos\alpha\cos\beta\cos\gamma} \tag{34}$$

$$V^* = a^* b^* c^* \sqrt{1 - \cos^2\alpha^* - \cos^2\beta^* - \cos^2\gamma^* + 2\cos\alpha^*\cos\beta^*\cos\gamma^*} \tag{35}$$

6 Conclusions

The difference in symmetry of sainidine and microcline is due to order-disorder relations of Si and Al atoms over tetrahedral sites. The distribution of Al and Si in tetrahedral sites in maximum microcline (Low microcline) is of the highest degree of order and microcline has a lowest structural state. On the other hand, sanidine, with the highest structural state, shows random Al-Si distribution in tetrahedral sites. In between these two end members maximum microcline and sanidine, several intermediate structural states, e.g. orthoclase and intermediate microcline exist. In this paper methods for estimation of structrural state of alkali feldspars is reviewed and discussed. Various ordering parameters have been used to designate the state of Al-Si distribution in alkali feldspars. Although the coefficient of order, the triclinicity, the Ragland parameter and obliquity indicate the structural state in quantitative manner, we can only draw broad generalizations with their help as regard to the actual physical environment during the formation of the alkali feldspar. On the other hand, Thompson X, Y and Z ordering parameters are more useful because it relates Gibbs function, entropy, enthalpy, etc., with the Al-Si distribution.

Acknowledgments Professor Ram S. Sharma and Professor V. K. S. Dave provided tremendous inspiration for this research work. The facilities and encouragement provided by Professor Anil K. Gupta, Director, Wadia Institute of Himalayan Geology, Dehradun, to carry out this research work is thankfully acknowledged.

References

1. Smith, J.V.: Feldspar minerals. Crystal Structure and Physical Properties, vol. 1, p. 627. Springer, Berlin (1974a)
2. Smith, J.V.: Feldspar minerals. Chemical and Textural Properties, vol. 2, p. 690. Springer, Berlin (1974b)
3. Smith, J.V., Brown, W.L.: Feldspar minerals. Crystal Structures, Physical, Chemical and Microtextural Properties,vol. 1, Second revised and extended edition, p. 828. Springer, Berlin (1988)
4. Barth, T.F.W.: Feldspars, p. 261. John Wiley, New York (1969)
5. Laves, F.: Phase relations of the alkali feldspars I and II. J. Geol. **60**(436–450), 549–574 (1952)
6. Jowhar, T.N.: AFEL, a fortran IV computer program for calculating lattice parameters and distribution of aluminium in tetrahedral sites of alkali feldspars. Comput. Geosci. **7**, 407–413 (1981)
7. Jowhar, T.N.: Determination of lattice parameters and structural state of alkali feldspars—a rapid X-ray diffraction method. Indian J. Earth Sci. **16**, 173–177 (1989)
8. Jowhar, T.N.: Crystal parameters of K-feldspar and geothermometry of Badrinath crystalline complex, Himalaya, India. Neues Jahrb. Mineral. Abh. **166**, 325–342 (1994)
9. Jowhar, T.N.: Refinement of the alkali feldspar-muscovite geothermometer and its application to the Badrinath crystalline complex, Himalaya, India. Indian Mineral. **32**, 7–20 (1998)
10. Jowhar, T.N.: Geobarometric constraints on the depth of emplacement of granite from the Ladakh batholith, northwest Himalaya, India. J. Mineral. Petrol. Sci. **96**, 256–264 (2001)
11. Jowhar, T.N., Verma, P.K.: Alkali feldspars from the Badrinath crystalline complex and their bearing on the Himalayan metamorphism. Indian Mineral. **29**, 1–12 (1995)
12. Kaur, M., Chamyal, L.S., Sharma, N., Jowhar, T.N.: Geothermometry of the granitoids of eastern higher kumaun Himalaya, India. J. Geolog. Soc. India **53**, 211–217 (1999)
13. Pandey, P., Rawat, R.S., Jowhar, T.N.: Structural state transformation in alkali feldspar: evidence for post-crystallisation deformation from a proterozoic granite kumaun Himalaya, India. J. Asian Earth Sci. **25**, 611–620 (2005)
14. Phillips, M.W., Ribbe, P.H.: The structures of monoclinic potassium-rich feldspars. Am. Mineral. **58**, 263–270 (1973)
15. Smith, J.V., Bailey, S.W.: Second review of Al-O and Si-O tetrahedral distances. Acta Crystallogr. **16**, 801–811 (1963)
16. Jones, J.B.: Al-O and Si-O tetrahedral distances in alumino-silicate framework structures. Acta Crystallogr. **B24**, 355–358 (1968)
17. Ribbe, P.H., Gibbs, G.V.: Statistical analysis and discussion of mean Al/Si-O bond distances and the aluminium content of tetrahedra in feldspars. Am. Mineral. **54**, 85–94 (1969)
18. Wright, T.L.: X-ray and optical study of alkali feldspar: II. An X-ray method for determining the composition and structural state from measurement of 2θ values for three reflections. Am. Mineral. **53**, 88–104 (1968)
19. Mackenzie, W.S.: The orthoclase-microcline inversion. Mineral. Mag. **30**, 354–366 (1954)
20. Goldsmith, J.R., Laves, F.: The microcline-sanidine stability relations. Geochim. Cosmochim. Acta **5**, 1–19 (1954)
21. Christie, O.H.J.: Feldspar structure and equilibrium between plagioclase and epidote. Am. J. Sci. **260**, 149–153 (1962)
22. Ragland, P.C.: Composition and structural state of the potassic phase in perthites as related to petrogenesis of a granitic pluton. Lithos **3**, 167–189 (1970)
23. Kroll, H.: Determination of Al. Si distribution in alkali feldspars from X-ray powder data. Neues Jahrb. Mineral. Monatsh. 91–94 (1971)
24. Kroll, H.: Estimation of the Al, Si distribution of feldspars from the lattice translations Tr [110] and Tr $\left[1\bar{1}0\right]$. I. alkali feldspars. Contrib. Miner. Petrol. **39**, 141–156 (1973)
25. Kroll, H.: Estimation of the Al, Si distribution of alkali feldspars from the lattice translations Tr [110] and Tr $\left[1\bar{1}0\right]$. Revised diagrams. Neues Jahrb. Mineral. Monatsh., 31–36 (1980)

26. Stewart, D.B., Wright, T.L.: Al/Si order and symmetry of natural alkali feldspars and the relationship of strained cell parameters to bulk composition. Bull. Soc. Franc. Miner. Cristall. **97**, 356–377 (1974)
27. Kroll, H., Ribbe, P. H.: Lattice parameters, composition and Al, Si order in alkali feldspars. In: Ribbe, P. H. (ed.) Feldspar Mineralogy, Mineralogical Society of America, Reviews in Mineralogy, vol. 2, Second Edition, pp. 57–99 (1983)
28. Mackenzie, W.S., Smith, J.V.: The alkali feldspars I. orthoclase-microperthites. Am. Mineral. **40**, 707–732 (1955)
29. Smith, J. V.: Cell dimensions b^*, c^*, α^*, γ^* of alkali feldspar permit qualitative estimates of Si, Al ordering: albite ordering process (Abstract), Geological Society of America Meeting, Mexico, p. 283 (1968)
30. Stewart, D.B., Ribbe, P.H.: Structural explanation for variations in cell parameters of alkali feldspar with Al/Si ordering. Am. J. Sci. **267–A**, 444–462 (1969)
31. Blasi, A.: Different behaviour of $\Delta(bc)$ and $\Delta(b^*c^*)$ in alkali feldspar. Neues Jahrb. Mineral. Abh. **138**, 109–121 (1980)
32. Kroll, H., Ribbe, P.H.: Determining (Al, Si) distribution and strain in alkali feldspars using lattice parameters and diffraction peak positions: a review. Am. Mineral. **72**, 491–506 (1987)
33. Wright, T.L., Stewart, D.B.: X-ray and optical study of alkali feldspar: I. Determination of composition and structural state from refined unit-cell parameters and 2V. Am. Mineral. **53**, 38–87 (1968)
34. Orville, P.M.: Unit-cell parameters of the microcline-low albite and the sanidine-high albite solid solution series. Am. Mineral. **52**, 55–86 (1967)
35. Ribbe, P.H., Gibbs, G.V.: Statistical analysis of Al/Si distribution in feldspars. Trans. Am. Geophys. Union **11**, 229–230 (1967)
36. Thompson Jr, J.B., Hovis, G.L.: Triclinic feldspars: angular relations and the representation of feldspar Series. Am. Mineral. **63**, 981–990 (1978)
37. Thompson Jr, J.B.: Chemical reactions in crystals. Am. Mineral. **54**, 341–375 (1969)
38. Jowhar, T.N.: Computer programs for P-T calculations and construction of phase diagrams: Use of TWQ, WEBINVEQ and THERMOCALC. In: Rajan, S., Pandey, P.C. (eds.) Antarctic Geoscience, Ocean-Atmosphere Interaction and Paleoclimatology, National Centre for Antarctic and Ocean Research, Goa, pp.248 262 (2005)
39. Jowhar, T.N.: Computer programs for P-T history of metamorphic rocks using pseudosection approach. Int. J. Comput. Appl. Technol. **41**, 18–25 (2012)
40. Azaroff, L.V.: Elements of X-ray Crystallography. McGraw Hill Book Company, New York (1968)
41. Appleman, D.E., Evans, H.T., Jr.: Job 9214: Indexing and least-squares refinement of powder diffraction data. Document PB 216 188, National Technical Information Service, U. S. Department of Commerce, Springfield, Virginia (1973)
42. Benoit, P.H.: Adaptation to microcomputer of the appleman-evans program for indexing and least squares refinement of powder diffraction data for unit cell dimensions. Am. Mineral. **72**, 1018–1019 (1987)

Iris Recognition System Using Local Features Matching Technique

Alamdeep Singh and Amandeep Kaur

Abstract Iris is one of the most trustworthy biometric traits due to its stability and randomness. In this paper, the Iris Recognition System is developed with the intention of verifying both the uniqueness and performance of the human iris, as it is a briskly escalating way of biometric authentication of an individual. The proposed algorithm consists of an automatic segmentation system that is based on the Hough transform, and can localize the circular iris and pupil region, occluding eyelids and eyelashes, and reflections. The extracted iris region is then normalized into a rectangular block. Further, the texture features of normalized image are extracted using LBP (Local Binary Patterns). Finally, the Euclidean distance is employed for the matching process. In this thesis, the proposed system is tested with the co-operative database such as CASIA. With CASIA database, the recognition rate of proposed method is almost 91 %, which shows the iris recognition system is reliable and accurate biometric technology.

Keywords Iris pre-processing · Normalization · Circular hough transform · LBP · Euclidean distance

1 Introduction

A biometric system provides automatic recognition of an individual based on some sort of unique attributes or traits (such as fingerprints, facial features, voice, hand geometry, handwriting, the retina, and the iris) possessed by the individual. Biometric authentication has evolved from the demerits of traditional means of authentication.

A. Singh (✉) · A. Kaur
Department of Computer Science, Punjabi University, Patiala, India
e-mail: bhinder.alam@gmail.com

A. Kaur
e-mail: aman_k2007@hotmail.com

B. V. Babu et al. (eds.), *Proceedings of the Second International Conference on Soft Computing for Problem Solving (SocProS 2012), December 28–30, 2012*, Advances in Intelligent Systems and Computing 236, DOI: 10.1007/978-81-322-1602-5_107, © Springer India 2014

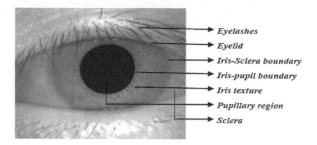

Fig. 1 Classification of human eye

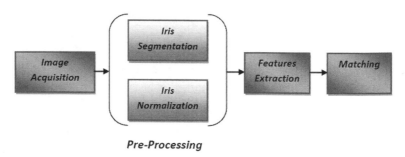

Fig. 2 Stages of iris recognition system

Biometric authentication is divided into two rudimentary categories such as: Physiological (or passive) and Behavioral (or active) biometrics. Physiological biometrics are based on parts of body like face, fingerprints palm print geometry, retina and iris recognition. While Behavioral biometrics are based on the actions taken by a person such as Voice recognition, keystroke dynamics and online/offline signatures. Iris, due to its permanence and ease of acquiring, plays a significant role among all the biometric traits. In these days Iris recognition is used as living passports, driving licenses and credit cards authentication, automobile ignition and unlocking, et cetera. Iris recognition is the process of automatically segregating people on the basis of information of an individual's iris image for biometric authentication system.

Pupil is the black circular shaped area in the eye image that controls the amount of light entering the eye by dilation and contraction. Iris is the circular shaped muscle that differentiates pupil from the sclera region. The Fig. 1 depicts the structure of the human eye.

Generally, Iris Recognition is mainly performed in four steps, as illustrated in Fig. 2: Image Acquisition, Image Pre-processing includes image segmentation and image normalization, Feature Extraction and Matching.

2 Literature Review

Since 2012, following are the main approaches for iris recognition:

At the outset, Flom and Safir [1] proposed the first mechanized biometrics system beneath extremely controlled conditions in 1987. They recommended the concept of practicing pattern recognition tools, including difference operators, edge detection algorithms, and the Hough transform, to extract iris descriptors, but no experimental results were scrutinized.

The first operational iris biometric system has been developed at University of Cambridge by Daugman [2–4]. Daugman put forwarded the concept of integro-differential operator with 2-dimensional Gabor filters and hamming distance. Contrariwise, Wildes [5], method involved computing a binary edge map followed by a Hough transform and then applied a Laplacian of Gaussian filter at multiple scales to generate a template and computed the normalized correlation as a similarity measure. Basit et al. [6] implemented an algorithm by representing eigen-irises after determining the centre of each iris and at last the recognition of irises was based on Euclidean distances. Chou, Chen and Weng [7] suggested a non-orthogonal view iris recognition system and proposed a circle rectification method to match iris images acquired at different off-axis angles. Hollingsworth et al. [8] offered a method of fusion of hamming distance and fragile bit distance. An independent component analysis method was used by Huang et al. [7] to condense the size of the iris feature without sacrificing the accuracy of iris recognition. Monro et al. [9] advised an iris coding method using differences of the discrete cosine transform coefficients that achieved highly accurate recognition results. Araghi et al. [10] proposed an Iris Recognition System based on covariance of discrete wavelet using Competitive Neural Network which discriminated noisy images very well. Bharadwaj et al. [11] proposed the Periocular biometrics as an alternative to iris recognition if the iris images were captured at a distance. They used GIST (global matcher) and complex local binary patterns (CLBP) for feature extraction with UBIRIS v2 database for recognition process.

Vatsa et al. [12] proposed a concept of curve evolution to effectively segment a non ideal iris image using the modified Mumford–Shah functional. An elastic iris blob matching algorithm was implemented by Zhenan et al. [13] to overcome the limitations of local feature based classifiers (LFC). Zhou ct al. [14] used methods such as Two-dimensional Gabor wavelet method with Log–Polar coordinates and 1-D Log–Gabor wavelet method with Polar coordinates, automatically eliminated the poor-quality images, evaluated the segmentation accuracy, and measured if the iris image had sufficient feature information for recognition. Ojala et al. [15] presented a theoretically very simple, yet efficient, multi-resolution approach to gray-scale and rotation invariant texture classification based on local binary patterns and nonparametric discrimination of sample and prototype distributions with excellent experimental results.

The database surveyed was CASIA database version 1.0 [16] includes 756 iris images from 108 eyes.

3 Brief Background of Proposed Technique

3.1 Image Acquisition

Images are acquired through a standard database—CASIA dataset (The Chinese Academy of Sciences—Institute of Automation), contains 756 grey scale eye images with 108 unique eyes or classes and 7 different images of each unique eye.

3.2 Image Pre-processing

3.2.1 Image Segmentation

The Circular Hough transform [17] is used to detect the iris/sclera boundary and iris/pupil boundary for making the circle detection process more efficient and accurate. After the completion of this process, six parameters are stored, the radius, and x (row) and y (column) centre coordinates for both circles according to the equation:

$$X_c^2 + Y_c^2 - r^2 = 0 \tag{1}$$

Afterwards, the Linear Hough Transform is used for isolation of eyelids. For segmentation of Iris, consider some general values according to *Masek's* approach for CASIA database as:

Lower pupil radius = 28,   Upper pupil radius = 75,
Lower iris radius = 80,   Upper iris radius = 150

Further, Canny edge detection is used to create an edge map, and only horizontal gradient information $|G_x|$ is taken.

$$|G| = \sqrt{G_x{}^2 + G_y{}^2} \tag{2}$$

$$|G| = |G_x| + |G_y| \tag{3}$$

Where, G_x and G_y are the gradients in x and y direction. The direction of the edge can be determined and stored by using:

$$\theta = \arctan(\frac{|G_y|}{|G_x|}) \tag{4}$$

Then the Final edges are determined by suppressing all edges that are not connected to a very certain (strong) edge (Fig. 3).

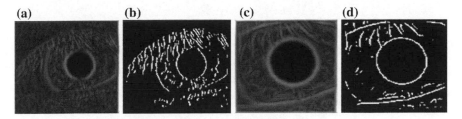

Fig. 3 Preprocessing stages: **a** Iris image after canny edge detection, **b** Iris image after hysteresis, **c** Pupil image after canny edge detection, **d** Pupil image after hysteresis

Now find circle by using circular Hough Transform (CHT). The mathematical equation for CHT is:

$$(x - a)^2 + (y - b)^2 = r^2 \qquad (5)$$

Where, a and b are the centers of the circle in the x and y direction respectively and where r is the radius. The parametric representation of the circle is:

$$x = a + r\cos(\theta) \qquad (6)$$
$$y - b + r\sin(\theta) \qquad (7)$$

For each edge point, the algorithm draw circles of different radii. From these it finds the maximum in the Hough space that is considered as the parameters of the circle.

3.2.2 Image Normalization

The segmented area is normalized to unwrap the iris region into polar coordinates by using Daugman Rubber Sheet Model [17]. The extracted iris region was then normalized into a rectangular block. This is given by:

$$r' = \sqrt{\alpha}\beta^2 - \alpha - r^2 \qquad (8)$$
$$\alpha - \alpha x^2 + \alpha y^2 \qquad (9)$$
$$\beta = \cos\left(\pi - \arctan\left(\frac{O_y}{O_x}\right) - \theta\right) \qquad (10)$$

Where, displacement of the centre of the pupil relative to the centre of the iris is given by O_x, O_y, and r' is the distance between the edge of the pupil and edge of the iris at an angle, θ around the region, and r_I is the radius of the iris (Fig. 4).

(a)   **(b)**

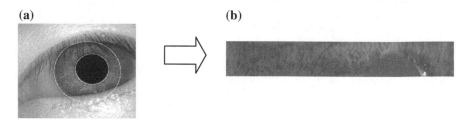

Fig. 4 Image normalization: **a** Image after segmentation, **b** after normalization

3.3 Iris Features Extraction

In this phase, LBP (Local Binary Pattern) is used to extract the features of normalized iris image. LBP is a simple yet very efficient texture operator, which labels the pixels of an image by thresholding the neighborhood of each pixel and considers the result as a binary number. The formula to compute the local texture around (x_c, y_c) for normalized iris image obtained in the Normalization stage is:

$$\text{LBP}_{p,r}(x_c, y_c) = \sum_{p=0}^{p-1} s(g_p - g_c)2^p \tag{11}$$

Where coordinates of individual neighbor, g_p is given by:

$$x_c + R\cos\frac{2\pi p}{p}, \; y_c - R\sin\frac{2\pi p}{p}, \tag{12}$$

Where, (x_c, y_c) are coordinates of the center pixel. In the next step, a set of the histogram features is computed together for LBP to construct the feature vector. The feature vector is saved for matching process (Fig. 5).

Fig. 5 Image after applying LBP on normalized iris image

3.4 Iris Matching

Finally, the Euclidean distance is employed for the matching process. An incredible feature of iris is the arbitrarily distributed irregular texture details in it. After getting histograms from both test (H_1) and training image (H_2), calculate the Euclidean distance as:

$$H_3 = \text{sqrt}\left(\sum (H_1 - H_2)^2\right) \tag{13}$$

The shortest distance gives the nearest match.

4 Proposed Algorithm

The algorithm used is as follows:

- Acquire images from CASIA database,
- Iris localization using Circular Hough Transform,
- Apply Daugman's Rubber Sheet Model to normalize the extracted region into rectangular block.
- Apply LBP to the normalized image to find the texture features which are, later on, saved in histogram,
- Finally, the Euclidean distance is used to recognize the applicable image.

5 Experimental Results

The CASIA database provided good segmentation, which undoubtedly tells apart the boundaries of iris pupil and sclera. The segmentation technique extracted the iris region of 624 out of 756 eye images, which connotes a success rate of around 83 %. The normalization process is not capable of flawlessly restructuring the similar pattern from images with the variation of pupil dilation. Then after LBP features are extracted. LBP is faster and easy to implement. The classification results are given below in the Table 1:

The accuracy of the proposed algorithms is less than Daugman's method that is a benchmark for iris recognition, but the computation speed is faster than their method. The accuracy of the proposed method is better than Masek's approach.

Table 1 Comparison of proposed method with other popular methods

Algorithm	Database	Recognition rate (%)
Daugman	CASIA	98.58
Masek	CASIA	83.92
Proposed method	CASIA	91.42

6 Conclusion

To conclude, the most complicated phase of the system is the preprocessing stage where the result of this stage is completely dependent upon the quality of the image. CHT is a popular and easy technique used to segment the iris pattern successfully from the eye image. The drawback of this method is that it still consumes enough time. After that, the iris image is normalized by the Daugman's Rubber Sheet model, which does not give the high-quality results due to the variation of pupil dilation. Then after, LBP and Euclidean Distance are used for features extraction and matching phase, which are easy to implement and gives first-rate results with 91.42 % accuracy.

References

1. Flom, L., Safir, A.: Iris recognition system. U.S. Patent IEEE, PP. 4,641,394, ©(1987)
2. Daugman, J.: High confidence visual recognition of persons by a test of statistical independence. IEEE Trans. Pattern Anal. Mach. Intell. **15**(11), 1148–1161 (1993)
3. Daugman, J.: The importance of being random: statistical principles of iris recognition. Pattern Recogn. **36**(2), 79–291 (2003)
4. Daugman, J.: Biometric personal identification system based on iris analysis. U.S. Patent Number 5,291,560 1 March 1994
5. Wildes, R.: Iris recognition: an emerging biometric technology. Proc. IEEE **85**, 1348–1363 (1997)
6. Basit, A., Javed, M., Anjum, M.: Efficient iris recognition method for human identification. Proc. WEC **2**, 24–26 (2005)
7. Huang, Y., Luo, S., Chen, E.: An efficient iris recognition system. In: International Conference on Machine Learning and Cybernetics, vol. 1, pp. 450–454. Beijing (2002)
8. Hollingsworth, K.P., Bowyer, K.W., Flynn, P.J.: Improved iris recognition through fusion of hamming distance and fragile bit distance. IEEE Trans. Pattern Anal. Mach. Intell. **33**(12), (2011), [Epub ahead of print] www.ncbi.ntm.nih.gov/pubmed/21576740
9. Monro, D., Rakshit, S., Zhang, Y. et al.: DCT-based iris recognition. IEEE Trans. Pattern Anal. Mach. Intell 29(4), 586–596 (2007)
10. Araghi, L., Shahhosseini, H., Setoudeh, F.: IRIS recognition using neural network. In: Proceedings of The International Multiconference of Engineers and Computer Scientists 2010, vol. 1, pp. 338–340. IMECS, Hong kong (2010)
11. Bharadwaj, S., Bhatt, H., Vatsa, M., Singh, R.: Periocular biometrics: when iris recognition fails. In: Proceedings of International Conference on Biometrics Theory, Applications and Systems, Washington, DC, 978–1–4244–7580–3/10/\$26.00 © 2010 IEEE
12. Vatsa, M., Singh, R., Noore, A.: Improving iris recognition performance using segmentation, quality enhancement, match score fusion, and indexing. IEEE Trans. Syst. Man Cybern. **38**(4), 896–897 (2009)
13. Sun, Z., Wang, Y., Tan, T., Cui, J.: Improving iris recognition accuracy via cascaded classifiers. IEEE Trans. Syst. Man Cybern. **35**(3), 435–441 (2005)
14. Zhou, Z., Du, Y., Belcher, C.: Transforming traditional iris recognition systems to work in nonideal situations. IEEE Trans. Ind. Electron. **56**(8), 3203–3213 (2009)
15. Ojala, T., Pietikäinen, M., MaÈenpaÈa, T.: Multiresolution gray-scale and rotation invariant texture classification with local binary patterns. IEEE Trans. Pattern Anal. Mach. Intell. **24**(7), 971–987 (2002)
16. CASIA iris image database. See http://www.cbsr.ia.ac.cn/Databases.htm

17. Masek, L., Kovesi P.: MATLAB Source Code for a Biometric Identification System Based on Iris Patterns. The School of Computer Science and Software Engineering, The University of Western Australia. http://www.csse.uwa.edu.au/_pk/studentprojects/libor/sourcecode.html (2003).

Part X
Soft Computing for Image Analysis (SCIA)

Fast and Accurate Face Recognition Using SVM and DCT

Deepti Sisodia, Lokesh Singh and Sheetal Sisodia

Abstract The main problem of face recognition is large variability of the recorded images due to pose, illumination conditions, facial expressions, use of cosmetics, different hairstyle, presence of glasses, beard, etc., especially the case of twins' faces. Images of the same individual taken at different times, different places, different postures, different lighting, may sometimes exhibit more variability due to the aforementioned factors, than images of different individuals due to gender, age, and individual variations. So a robust recognition system is implemented to recognize an individual even from a large amount of databases within a few minutes. So in order to handle this problem we have used SVM for face recognition. Using this technique an accurate face recognition system is developed and tested and the performance found is efficient. The procedure is tested on ORL face database. Results have proved that SVM approach not only gives higher classification accuracy but also proved to be efficient in dealing with the large dataset as well as efficient in recognition time. Results have proved that not only the training performance, the recognition performance but also the recognition rate raises to 100 % using SVM.

Keywords Machine learning · Support vector machine · ORL face database

D. Sisodia · L. Singh
Technocrats Institute Of Technology, Bhopal, M.P, India
e-mail: chanchal.sisodia@gmail.com

L.Singh
e-mail: lokesingh@gmail.com

S. Sisodia (✉)
Samrat Ashok Technical Institute, Vidisha, M.P, India
e-mail: sheetal.sisodia@gmail.com

B. V. Babu et al. (eds.), *Proceedings of the Second International Conference on Soft Computing for Problem Solving (SocProS 2012), December 28–30, 2012*, Advances in Intelligent Systems and Computing 236, DOI: 10.1007/978-81-322-1602-5_108, © Springer India 2014

1 Introduction

Learning theory helps give a researcher applying machine learning algorithms some rules of the thumb that tell how to best apply the algorithms. "Dr. Andrew Ng Likens" knowing machine learning algorithms to a carpenter acquiring a set of tools. However, the difference between a good carpenter and not so good one is the skill in using those tools. In choosing which one to use and how. In the same way Learning Theory gives a "machine-learnist" some intuitions about how an ML algorithm would work and helps in applying them better. A common problem that can be observed in many A.I. engineering applications is pattern recognition. The problem is as follows: given a training set of vectors, each belonging to some known category, the machine must learn, based on the information implicitly contained in this set, how to classify vectors of unknown type into one of the specified categories. SVMs provide one means of tackling this problem [1].

Therefore, using Support Vector Machine (SVM) for recognition/authentication of facial images in this paper. Face authentication is a two-class problem. As face recognition system is presented all about with a claimed identity and it is making decision whether the claimant is really that person or not.

Due to daily changes in facial images such as variations in gestures, pose, facial expressions, etc., face recognition becomes a difficult problem to be solved, especially the case of twins' faces. It is difficult to recognize the difference between both faces. It is also a problem for detection of faces in images than detection of salient features within that facial image and finally which classification model to be used that identifies clearly whether the resultant image is the desired image or not. For solution to this obstacle if we use large training image set then it might increase the complexity and if we keep training set small then the performance of facial recognition system degrades. So the only option remains to solve this problem is to reduce the dimensionality [2].

2 Structure of Support Vector Machine

A support vector machine is a machine in the sense that it is given inputs that are processed by the machine to produce outputs. Support vector machines are general algorithms based on guaranteed risk bounds of SLT, i.e., "statistical learning theory". Support vector machines are a type of state-of-the-art pattern recognition technique whose foundations stem from statistical learning theory. They have found numerous applications, such as in face recognition, character recognition, face detection, and so on [3]. The basic principle of SVM is to find an optimal separating hyperplane so as to separate two classed of patterns with maximal margin. It tries to find the optimal hyperplane making expected errors minimized to the unknown test data, while the location of the separating hyperplane is specified via only data that lie close to the decision boundary between the two classes, which are support vectors.

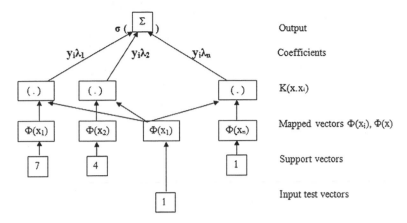

Fig. 1 SVM architecture [4]

SVM is a two-class classifier system. The SVM can be trained to classify both linearly separable and nonlinearly separable data. The SVM locates the most influential samples called SV, i.e., "support vectors". These support vectors are samples (from both classes under consideration) found closest to the decision surface being constructed [4]. The idea behind the SVM is to create a hyperplane decision surface located equidistance between two decision boundaries. Each boundary is specified by the location of support vectors that satisfy.

$$y_j[w^T x_j + b] = 1; j = 1, N_{SV} \tag{1}$$

where N_{SV} is the number of support vectors. Locating the decision surface in such a way produces an optimal hyperplane decision surface between the samples of the two classes, and therefore minimum error in classification [4].

The SVM is trained to classify samples of only one specific class-of-interest at a time. All other samples which are not classified as the class-of-interest are considered to be outside the class-of-interest. Fig. 1 shows the architecture of an SVM used to realize the linear SVM decision boundaries of Eq. 1 during construction of a hyperplane decision function [4].

3 SVMs and Statistical Learning Theory

SVMs are trained as function approximators that can be used for classification or interpretation (i.e., regression) when one wants to understand the structure of an underlying system [4].

SVMs are based on statistical learning theory where learning machine models are constructed with the goal of finding a function approximator that estimates the

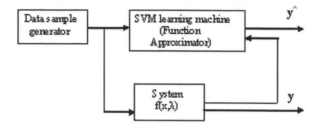

Fig. 2 SVM—learning machine model

unknown input-output dependency of a system based on a set of observable samples as shown in Fig. 2. Once an SVM has been trained to capture the underlying system structure, it may be used to classify new unknown samples [4]. Multi-class classification is possible by specifying a different class as the class-of interest each time the technique is applied. In our face detection application there are only two classes to be discriminated, i.e., faces versus non-faces [4].

Large margin classifiers are popular approaches to solve the supervised learning problems. Founded on Vapniks statistical learning theory, SVM is a representative large margin classifier that has played an important role in many areas due to its salient properties such as margin maximization and kernel substitution for classifying the data in a high-dimensional feature space [5]. As SVM considers training data as uncorrelated points, and thus is insensitive to the data distribution information, there is still room for further improvement in generalization ability. Inspired by the fact that linear discriminant analysis (LDA) and mini-max probability machine (MPM) discriminate between classes and employ the class structure into determining the classification boundary, Huang, proposed a large margin learning model: min-max margin machine (M4) that improves SVM by considering class structures into decision boundary calculation via utilizing Mahalanobis distance as the distance metric [5]. M4 does show better generalization ability in some applications, but as the true data structure is not captured, its performance sometimes is not as good as SVM [5].

It is the aim of support vector machine learning to classify data sets where the number of training data is small and where traditional use of statistics of large numbers cannot guarantee an optimal solution [6]. For example consider Fig. 3. Here two decision boundaries on the same data are seen. Both decision boundaries correctly classify the data with zero error. However, subjectively, the decision boundary on the figure on the right is a better choice [6]. The reason that this decision boundary is a better choice is because this decision boundary classifies the two classes with the maximum possible distance (the margin) between the nearest points of each class.

The goal of SVM learning applied to a classification problem is to find the maximum margin decision boundary. This is termed a maximal margin classifier. The formulation of SVM is based on the structural risk minimization (SRM) principle, whereas the empirical risk minimization (ERM) approach is commonly used in

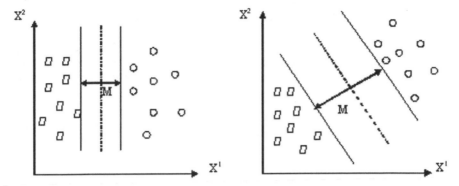

Fig. 3 *Left* Plot of two classes showing a small margin between classes defined by the decision boundary. *Right* The same plot but with the largest possible distance between opposite classes. SVM maximizes the margin M [6]

statistical learning (for example the classical batch learning for radial basis networks). This approach to SVMs gives a greater potential to generalize [6].

4 Approach Used

Discrete Cosine Transform

The DCT has properties which make it suitable to SVM learning. The discrete cosine transform (DCT) helps separate the image into parts. It transforms a signal or image from the spatial domain to the frequency domain.

When the DCT is applied on large images, the rounding effects when floating point numbers are stored in a computer system result in the DCT coefficients being stored with insufficient accuracy. The result is a deterioration in image quality. As the size of the image is increased, the number of computations increases disproportionately. For these reasons an image is subdivided into 8*8 blocks. Where an image is not an integral number of 8*8 blocks, the image can be padded with white pixels (i.e., extra pixels are added so that the image can be divided into an integral number of 8*8 blocks. The two-dimensional DCT is applied to each block so that an 8*8 matrix of DCT coefficients is produced for each block. This is termed the "DCT Matrix". The top left component of the DCT matrix is termed the "DC" coefficient and can be interpreted as the component responsible for the average background colour of the block (analogous to a steady DC current in electrical engineering). The remaining 63 components of the DCT matrix are termed the "AC" components as they are frequency components analogous to an electrical AC signal. The DC coefficient is often much higher in magnitude than the AC components in the DCT matrix.

Each component of the DCT matrix represents a frequency in the image. The further an AC component from the DC component the higher the frequency represented.

Since an image comprises hundreds or even thousands of 8*8 block of pixels so for an 8*8 block it results in this matrix. Now let us start with a block of image-pixel values. This particular block was chosen from a very upper left-hand corner of an image.

Quantization

Our 8*8 block of DCT coefficients is now ready for compression by quantization. After the quantization obtained matrix is ready for the final step of compression. Quantizing involves reassigning the value of the weight to one of limited number of values. To quantize the weights the maximum and minimum weight values (for the whole image) are found and the number of quantization levels are pre-defined. The 2-dimensional discrete cosine transform is applied to each block. After the DCT each block comprises 64 discrete cosine coefficients. The coefficients are comprised of one DC coefficients and 63 AC coefficients. The DC coefficients are separated and treated differently to the AC coefficients. Before the AC and DC coefficients are separated, the matrix of discrete cosine coefficients is divided by a quantization. The components of the quantization table are largest in the bottom-right corner. This will produce smaller AC coefficients for higher frequency components in the image.

The quantization table may be multiplied by an arbitrary number. This number is a user defined parameter which defines compression ratio. A larger number results in larger components in the quantization table which in turn results in smaller discrete cosine coefficients. The DC coefficients are Huffman encoded. The AC coefficients are mapped into a row of number using the zigzag mapping. The row of AC coefficients will tend to have nonzero coefficients at the beginning of the row while toward the end of the row, the coefficients are usually zero. Increasing the number by which the quantization matrix is multiplied will further increase the number of zeros. The row of AC coefficients is Huffman coded.

The elements of the matrix in Fig. 4 are mapped using the zig-zag sequence shown in Fig. 5 to produce a single row of numbers. That is a single row of numbers is collected as the zig- zag trail is followed in the DCT matrix. This will produce a row of 64 numbers where the magnitude tends to decrease traveling down the row of numbers [5].

In this research work the original size of each image was 92*112 pixels, with 256 grey levels per pixel. But since we have divided images into 8*8 block so we have scaled or resize the image size into 96*112 pixels. After dividing the image into 8*8 blocks, the total blocks of image becomes 168.

Following are the steps involved in the research work:

Step-1 The image is broken into 8*8 blocks of pixels.
Step-2 Working from left to right and top to bottom, DCT is applied to each block.

$$
\text{Original} = \begin{pmatrix}
125 & 124 & 124 & 126 & 123 & 124 & 127 & 124 \\
124 & 125 & 126 & 125 & 127 & 123 & 127 & 124 \\
121 & 128 & 121 & 126 & 125 & 124 & 127 & 124 \\
129 & 126 & 124 & 125 & 127 & 124 & 124 & 124 \\
127 & 124 & 126 & 122 & 129 & 123 & 127 & 123 \\
125 & 128 & 126 & 126 & 124 & 124 & 127 & 124 \\
125 & 126 & 128 & 123 & 126 & 124 & 124 & 129 \\
127 & 128 & 126 & 124 & 128 & 123 & 126 & 130
\end{pmatrix}
$$

DCT ⬇

$$
\begin{pmatrix}
1.0024 & 0.0009 & 0.0017 & -0.0008 & 0.0004 & -0.0013 & -0.0046 & 0.0039- \\
0.0039 & -0.0008 & -0.0029 & 0.0021 & -0.0011 & 0.0038 & -0.0026 & -0.0022 \\
0.0012 & -0.0022 & 0.0016 & -0.0036 & 0.0002 & -0.0003 & 0.0011 & -0.0009 \\
-0.0009 & 0.0010 & -0.0004 & -0.0000 & -0.0019 & 0.0033 & 0.0019 & 0.0011 \\
0.0009 & 0.0014 & 0.0015 & 0.0028 & 0.0029 & 0.0008 & 0.0014 & 0.0022 \\
-0.0008 & 0.0005 & 0.0011 & -0.0005 & 0.0006 & -0.0003 & 0.0018 & 0.0034 \\
-0.0006 & -0.0007 & 0.0001 & -0.0002 & -0.0018 & 0.0010 & -0.0039 & 0.0045 \\
-0.0016 & -0.0026 & -0.0004 & -0.0000 & -0.0008 & -0.0010 & -0.0004 & 0.0012
\end{pmatrix}
$$

Fig. 4 DCT maps a block of pixel color values to the frequency domain

Fig. 5 The zig-zag pattern applied to a block of DCT coefficients to produce a row of 64 coefficients

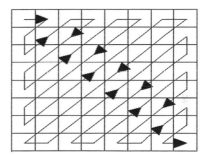

Step-3 The zigzag pattern applied to a block of DCT Coefficients to produce a row of 64 coefficients.

Step-4 DCT coefficients-Top left 3*3 block, all other components discarded.

Step-5 Support vector machine learning is applied to the absolute values of each row of AC terms.

The block diagram of proposed procedure is depicted in Fig. 7.

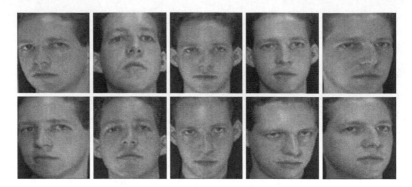

Fig. 6 Sample images from the ORL face database [7]

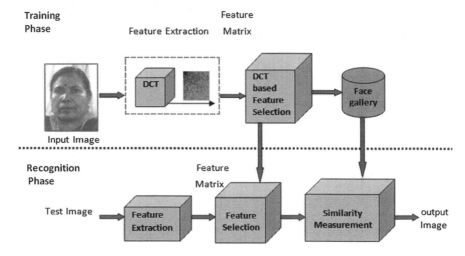

Fig. 7 Block diagram of the proposed face recognition system

5 Experiments and Results

Motivation

In order to test the recognition system, lots of face images were required. There are so many standard face databases for testing and rating a face recognition system. A standard database of face imagery was essential to supply standard imagery to the recognition system and to supply a sufficient number of images to allow testing.

Dataset

All the experiments here have been executed mainly on the face images provided by the ORL face database. The ORL Database of faces contains a set of face images taken between April 1992 and 1994 [7] as shown in Fig. 6.

There are ten different images of each of 40 distinct subjects. For some subjects, the images were taken at different times, varying the lighting, facial expressions (open/closed eyes, smiling/not smiling) and facial details (glasses/no glasses). The image files are in PGM (portable gray map) format. The size of each image is 92*112 pixels, with 256 grey levels per pixel. But since images are divided into 8*8 block so images are scaled or resized into 96*112 pixels. The images are organized in 40 directories (one for each subject), which have names of the form sX, where X indicates the subject number (between 1 and 40) i.e. s1–s40. In each of these directories, there are ten different images of that subject, which have names of the form Y.pgm, where Y is the image number for that subject (between 1 and 10) i.e. 1.pgm. The following figure shows image of one person from ORL face database.

For training, from the database 6 images of each subject is taken while the remaining 4 images is left for testing, means 240 images are used to training the database and the remaining 160 images are used for testing from the same database while some other images of the same format are also used for testing (not from the same database). Results after the experiments are shown in the Table.

Image data base	Images for training	Images for testing	SVM					
			Training time (sec)			Matching time(sec)		
			Iterations			Iterations		
			I	II	III	I	II	III
400	240	160	12.31	12.27	12.22	0.36	0.30	0.28
			18.23	18.41	18.35	1.13	1.10	1.06
			28.77	28.55	28.68	1.81	1.66	1.72
			39.75	39.58	39.69	2.43	2.50	2.54

Recognition Rate

The closer a system's measurements to the accepted value, the more accurate the system is considered to be. Recognition rate means a rate which a face recognition system recognizes an individual by matching the input image against images of all users in a database and finding the best match.

$$\text{Recognition rate} = \frac{\text{total no. of correct matches} * 100}{\text{Total no. of faces}}$$
$$= \frac{400 * 100}{400}$$
$$= 100\%$$

The following graph shows 100% recognititon rate (Fig. 8).

Training Performance

The training performance of recognition system varies according to the system configuration. This recognition system has been implemented and tested in MATLAB version 7.5 under Microsoft Windows XP operating system. The training dataset is image database taken from ORL face database [7].

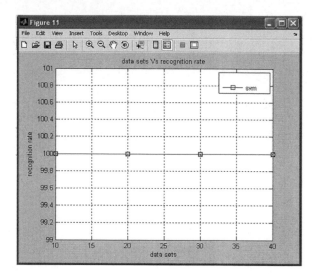

Fig. 8 Graph of recognition rate

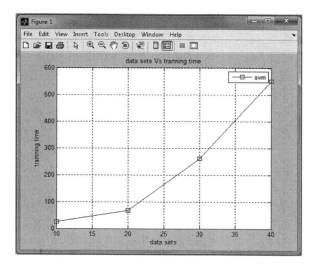

Fig. 9 Graph of training time

The following graph shows the training performance of the recognition system (Fig. 9).

Matching Time Performance

Likewise the training performance, the matching performance of the recognition system varies according to the system configuration. The following graph shows the matching performance of the recognition system (Fig. 10).

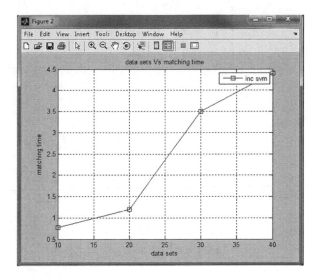

Fig. 10 Graph of matching time

6 Conclusion

By using the SVM in the DCT domain, not only the storage requirement is largely relaxed, but also the computational complexity is simplified. Some redundant information is removed by truncating the DCT coefficients so that the dimensionality of the coefficient vectors can be reduced. When the powerful features of the SVM, such as margin maximization and kernel substitution for classifying data in a high dimensional kernel space, are combined with the low dimensional DCT feature vector as described, a good compromise between the computational efficiency and performance accuracy is obtained upto 100 %.

7 Future Work

In this paper we have used SVM and DCT for face recognition. Using this technique fast and accurate face recognition system is developed and tested and the performance found is 100 %. Of course in future there can be some other ways to improve the overall performance of the face recognition such as "Incremental SVM" [8] and "Decremental SVM" [9] along with DCT [10].

References

1. Palaniswami, M., Shilton, A., Ralph, D., Owen, B.D.: Machine learning using support vector machine. The University of Melbourne, Victoria -3101, Australia and University of Cambridge Trumpington, UK
2. Amine, A., Ghouzali, S., Rziza, M., Aboutajdine, D.: Investigation of feature dimension reduction based DCT/SVM for face recognition. In: IEEE Symposium on Computers and Communications Program, pp. 188–193. IEEE, Marrakech (2008)
3. Chen, Q.-Y., Yang, Q.: Segmentation of images using support vector machine. In: Proceedings of the Third International Conference on Machine Learning and Cybernetics, pp. 3304–3306. IEEE, Shanghai (2004)
4. Shavers, C., Li, R., Lebby, G.: An SVM based approach to face detection. In: Proceedings of the 38th Southeastern Symposium on System Theory Tennessee Technological University Cookeville, pp. 362–366. IEEE, TN (2006)
5. Wang, D.F., Yeung, D.S., Tsang, E.C.C., Wang, X.Z.: Structured large margin learning. In: Proceedings of the Fourth International Conference on Machine Learning and Cybernetics, pp. 4242–4248. IEEE, Guangzhou (2005)
6. Robinson, J.: The application of support vector machines to compression of digital images, Doctor of Philosophy, the University of Auckland, Feb (2004)
7. The database can be retrieved from the following link. http://www.cl.cam.ac.uk/Research/DTG/attarchive:pub/data/att_faces.zip as a ZIP file of similar size
8. Yang, J., Li, Z.-W., Zhang, J.-P.: A training algorithm of incremental support vector machine with recombining method. In: Proceedings of the Fourth International Conference on Machine Learning and Cybernetics, pp. 4285–4288. IEEE, Guangzhou (2005)
9. Galmeanu, H., Andonie, R.: Incremental / decremental SVM for function approximation, optimization of electrical and electronic equipment. In: 11th International Conference on Digital Object Identifier, pp. 155–160. IEEE, Brasov (2008)
10. Khayam, S.A.: The Discrete cosine transform (DCT): theory and application, Department of Electrical and Computer Engineering Michigan State University, ECE 802–602 Information Theory and Coding, 10 Mar 2003

Multi-Temporal Satellite Image Analysis Using Gene Expression Programming

J. Senthilnath, S. N. Omkar, V. Mani, Ashoka Vanjare and P. G. Diwakar

Abstract This paper discusses an approach for river mapping and flood evaluation to aid multi-temporal time series analysis of satellite images utilizing pixel spectral information for image classification and region-based segmentation to extract water covered region. Analysis of Moderate Resolution Imaging Spectroradiometer (MODIS) satellite images is applied in two stages: before flood and during flood. For these images the extraction of water region utilizes spectral information for image classification and spatial information for image segmentation. Multi-temporal MODIS images from "normal" (non-flood) and flood time-periods are processed in two steps. In the first step, image classifiers such as artificial neural networks and gene expression programming to separate the image pixels into water and non-water groups based on their spectral features. The classified image is then segmented using spatial features of the water pixels to remove the misclassified water region. From the results obtained, we evaluate the performance of the method and conclude that the use of image classification and region-based segmentation is an accurate and reliable for the extraction of water-covered region.

Keywords MODIS satellite image · Gene expression programming · Artificial neural network

J. Senthilnath (✉) · S. N. Omkar · V. Mani · A. Vanjare
Department of Aerospace Engineering, Indian Institute of Science, Bangalore, India
e-mail: snrj@aero.iisc.ernet.in

S. N. Omkar
e-mail: omkar@aero.iisc.ernet.in

V. Mani
e-mail: mani@aero.iisc.ernet.in

A. Vanjare
e-mail: ashokav@aero.iisc.ernet.in

P. G. Diwakar
Earth Observation System, ISRO Head Quarters, Bangalore, Karnataka, India
e-mail: diwakar@isro.gov.in

B. V. Babu et al. (eds.), *Proceedings of the Second International Conference on Soft Computing for Problem Solving (SocProS 2012), December 28–30, 2012*, Advances in Intelligent Systems and Computing 236, DOI: 10.1007/978-81-322-1602-5_109, © Springer India 2014

1 Introduction

Multi-temporal time series analysis of satellite images plays an important role in discriminating areas of land surface changes between imaging dates. The data from NASA's MODIS satellite sensor have considerable potential for multi-temporal image analysis. The results of multi-temporal image analysis are very useful in hydrological applications such as flood detection and damage assessment. Although MODIS is a moderate resolution image sensor it does provide excellent between land and water discrimination [1]. Because of the moderate resolution of MODIS, some features such as river courses, canals, and roads appear in the images as linear segments, without additional details that would further complicate the automatic extraction of related features. Hence the complexity in extracting river networks and evaluating floods from MODIS image has attracted many researchers [1, 2]. In literature, many extraction techniques are devised based on pixel, region, and knowledge for a given image [1–6].

In this paper, we consider a combination of the pixel-based, region-based, and knowledge-based classification and segmentation methods that were used in earlier studies for the extraction of roads [3]. The same approach is used with Artificial Neural Network (ANN) and Gene Expression Programming (GEP) classifiers for this study. Multi-temporal MODIS images for the automatic extraction of river networks (before flood) and for evaluating floods (during flood) is presented. ANN and GEP classifiers are used to classify the similar spectral features of an image to differentiate water and non-water image pixels. As the classification uses only the spectral features, some misclassified water image pixels, which have spectral reflectance similar to the water, are also classified into the water group. This discrepancy between true water and misclassified water image pixels is resolved by using region-growing image segmentation and knowledge-based techniques to extract the shape and density of water-covered regions, based on the spatial features of the image. The extracted features are evaluated to differentiate water regions from non-water regions. Finally, the performance of these methods are evaluated and compared.

2 Image Classification

Optical satellite image data are often adversely affected by image "noise" due to clouds. It is therefore important to eliminate or minimize such noise before subsequent image classification and segmentation. In this study, we use a median-based switching filter, called Progressive Switching Median (PSM) filter, where both the impulse detector and the noise filter are applied progressively in iterative manner [7]. This technique is applied to remove all the cloud noise from MODIS images. In satellite imagery, the appearance of water strongly depends on the sensor's spectral sensitivity and spatial resolution. In this study, moderate-resolution gray-scale MODIS images (250 m per pixel) are used. The gray-scale intensity of any pixel in an image depends on the reflectance properties of the object it depicts. Each pixels can therefore be classified into one of the two groups based on its specific intensity level of the pixels, i.e., water and non-water. The two supervised classification techniques such as Artificial Neural Network (ANN) [8–10] and Gene Expression Programming (GEP) [11] are used to classify water image pixels. In GEP unlike ANN, provides an efficient method for obtaining classification rules in the form of a mathematical expression for a given image. Initially, the water pixels and non-water pixels from two images (before and during flood) are picked randomly for training. For the given training data set, ANN generates weights where as GEP provides mathematical expression. Using these weights and mathematical expression all pixels of the image are extracted and evaluated.

3 Image Segmentation

The objective of image segmentation is to eliminate those pixels that are wrongly classified as water. However, as some non-water pixels in the image will have reflectance properties similar to water and hence these pixels may be misclassified into the water group. Because of this misclassified water pixels get classified into true-water pixels which are a hindrance in the classification. This hindrance is removed by segmentation using spatial features of the image. The image is first divided into regions and the geometrical parameters for every region are calculated. Then these regions are segmented as true-water and false-water regions. For this we use Shape Index (SI) and Density Index (DI) defined as [3]:

$$SI = \frac{P}{4 \cdot \sqrt{A}} \tag{1}$$

$$DI = \frac{\sqrt{A}}{1 + \sqrt{VAR(X) + VAR(Y)}} \tag{2}$$

where P represents the perimeter of the region (the number of pixels on the boundary of the region), A represents the area of the region (the number of pixels of the region), $VAR(X)$ represents variance of x-coordinates of all the pixels in the region, $VAR(Y)$ represents variance of y-coordinates of all the pixels in the region, and the term $\sqrt{VAR(X)} + \sqrt{VAR(Y)}$ gives approximate radius of the region [3]. The river pixels have a high SI and a low DI, whereas non-river pixels have a low SI and a high DI. Thus, for segmentation purpose, threshold values are set for both indices: SI and DI. Regions having a SI greater than the SI threshold value and a DI less than the DI threshold value are segmented as water and non-water regions. Those regions which do not qualify as water are segmented as non-water. We see that the non-flooded regions are more than the flooded region in an image. The perimeter value of a non-flood region is greater than that of a flood region. This makes the ratio of perimeter to the area, a high value for non-flood region and a low value for flood region. Thus, non-flood regions have a higher value of SI than flood regions. Thereby, the DI for flooded regions is higher than that of non-flooded regions. Thus, for segmenting purpose, threshold values can be set for both the indexes: SI and DI. Regions having a SI lower than the threshold value and DI higher than the threshold value are segmented as flooded regions.

4 Results and Discussion

In this section, two classifier methods are used to extract the course of the Krishna river. Figure 1 shows the ground truth information of the flooded and non-flooded regions. All the flooded and non-flooded regions (cities) are shown in the map. The cities (regions) which were not affected by flood are represented by white dots whereas the cities which were affected by flood are represented by white dots with black dots in the center. The images cover an area of 18888.37 km^2 and districts like Kurnool, Mahaboobnagar, Bellary, Gulbarga, and Raichur. Figures 2a and 3a show the March 2009 image (i.e., before the flood) and September 2009 image (i.e., during the flood) MODIS satellite images. For both images, we have applied classification and segmentation technique to extract the river network (before flood) and flooded region (during flood).

Two MODIS images [12] of the same region are used to classify and validate. The classifiers were trained using 20 randomly picked samples of two classes (water

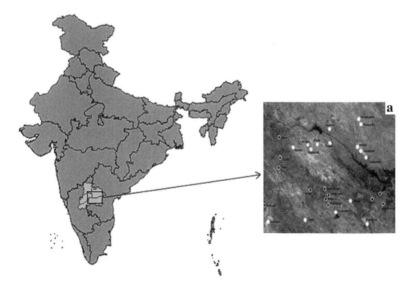

Fig. 1 Study area with flooded and non-flooded cites marked on MODIS image

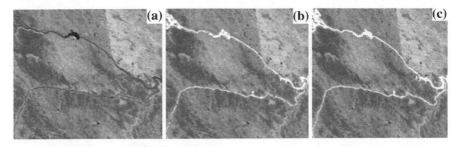

Fig. 2 **a** MODIS image of Krishna River before flood, **b** segmented image using ANN, **c** segmented Image using GEP

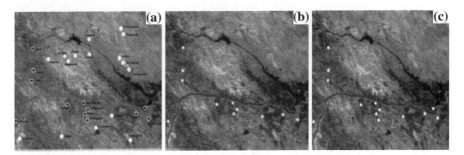

Fig. 3 **a** MODIS during flood image with ground truth information: flooded and unflooded cities; **b** segmented Image using ANN (*white points* are identified as flooded cities); **c** segmented image using GEP (*white points* are identified as flooded cities)

Table 1 Evaluating features based on ROC parameters during flood image

Methods	True positive rate	True negative rate	False positive rate	Accuracy
ANN	0.83	0.94	0.06	0.89
GEP	1	0.94	0.06	0.96

and non-water) [12]. Then the MODIS Band 2 image was tested using the trained classifiers.

The GEP expression tree generated for the training sample using before and during flood images are given in Eqs. (3) and (4) respectively.

$$M = (2 * B2) - 4.5 \qquad (3)$$

$$N = 3.3 - (11.3 * B2) \qquad (4)$$

where $B2$ represent the Band 2 of the MODIS image. Each pixel in the image is classified as water or non-water image pixel using the above equations. If the value of M and N for a pixel is >0.5 then this pixel is classified as water; otherwise this pixel is classified as non-water pixel. However, their might be some misclassified pixels.

These misclassified water pixels are properly classified as non-water pixels using region-based segmentation. For segmentation purpose, we use two indices based on geometrical features—Shape Index (SI) and Density Index (DI), in order to differentiate the water pixels from non-water pixels, by thresholding the index values. The results of ANN and GEP method are compared in terms of the Root Mean Square Error (RMSE) [13] and Receiver Operating Characteristic (ROC) parameters [14].

Figure 2a and b shows river extracted image using ANN and GEP respectively. RMSE value is analyzed for before flood image. RMSE values for ANN and GEP are 0.126 and 0.1205, respectively. We can observe that, based on the RMSE values for before flood image, GEP results in less error in comparison with ANN.

Figure 3a and b shows the list of flooded and unflooded regions (cities) using ANN and GEP respectively. From Table 1, we can observe that true positive rate and accuracy during flood image is better in GEP. The performance of GEP is better than ANN for before and during flood images.

5 Conclusions

The tasks of river mapping and flood extraction are accomplished successfully by the procedure of pixel-based spectral information for classification, and shape information for segmentation, as discussed above. This has been found to be an efficient way to extract water regions from before and during flood satellite images. In the classification stage of extracting water and non-water groups, the gene expression programming classifier proved better than the artificial neural network classifier. The results of classification using spectral information are improved through region-growing image segmentation (based on spatial feature) using similarity criteria emphasizing shape information, resulting in an effective extraction of water-covered regions.

Acknowledgments This work is supported by the Space Technology Cell, Indian Institute of Science, Bangalore and Indian Space Research Organization (ISRO). We also acknowledge the MODIS mission scientists and associated NASA personnel for the production of the remote sensing data which is used in this paper.

References

1. Brakenridge, R., Anderson, E.: MODIS-based flood detection, mapping and measurement: the potential for operational hydrological applications. Transboundary Floods: Reducing Risks through Flood Management. Springer-Verlag, pp. 1–12 (2006)
2. Khan, S.I., Hong, Y., Wang, J., Yilmaz, K.K., Gourley, J.J., Adler, R.F., Brakenridge, G.R., Policelli, F., Habib, S., Irwin, D.: Satellite remote sensing and hydrologic modeling for flood inundation mapping in Lake Victoria Basin: implications for hydrologic prediction in ungauged basins. IEEE Trans. Geosci. Remote Sens. **49**, 85–95 (2011)
3. Mingjun, S., Daniel, C.: Road extraction using SVM and image segmentation. Photogram. Eng. Remote Sens. **70**(12), 1365–1371 (2004)

4. Senthilnath, J., Rajeswari, M., Omkar, S.N.: Automatic road extraction using high resolution satellite image based on texture progressive analysis and normalized cut method. J Indian Soc. Remote Sens. **37**(3), 351–361 (2009)
5. Omkar, S.N., Senthilnath, J., Mudigere, D., Manoj Kumar, M.: Crop classification using biologically inspired techniques with high resolution satellite image. J. Indian Soc. Remote Sens. **36**(2), 172–182 (2008)
6. Senthilnath, J., Omkar, S.N., Mani, V., Tejovanth, N., Diwakar, P.G., Archana Shenoy, B.: Hierarchical clustering algorithm for land cover mapping using satellite images. IEEE J. Sel. Topics Appl. Earth Obs. Remote Sens. **5**(3), 762–768, (2012)
7. Wang, Z., Zhang, D.: Progressive switching median filter for the removal of impulse noise from highly corrupted images. IEEE Trans. Circuits Syst. **II**(46), 78–80 (1999)
8. Haykin, S.: Neural Networks—A Comprehensive Foundation, 2nd edn. Pearson Prentice Hall Publication, New Jersey (1994)
9. Omkar, S.N., Sivaranjani, V., Senthilnath, J., Mukherjee, S.: Dimensionality reduction and classification of hyperspectral data. Int. J. Aerosp. Innov. **2**(3), 157–163 (2010)
10. Omkar, S.N., Senthilnath, J.: Integration of Swarm Intelligence and Artificial Neutral Network, Neural Network and Swarm Intelligence for Data Mining, Chapter 2. In: Dehuri, S., et al. World Scientific Press, Singapore, pp. 23–65 (2011)
11. Ferreira, C.: Gene expression programming: a new adaptive algorithm for solving problems. Complex Syst. **13**(2), 87–129 (2001)
12. Senthilnath, J., Shivesh, B., Omkar, S.N., Diwakar, P.G., Mani, V.: An approach to multi-temporal MODIS image analysis using image classification and segmentation. Adv. Space Res. **50**(9), 1274–1287 (2012)
13. Arvind, C.S.: Ashoka Vanjare, Omkar, S.N., Senthilnath, J., Mani, V., Diwakar, P.G.: Multi-temporal satellite image analysis using unsupervised techniques. Adv. Comput. Inf. Technol. Adv. Intell. Syst. Comput. **177**, 757–765 (2012)
14. Fawcett, T.: An introduction to ROC analysis. Pattern Recogn. Lett. **27**(8), 861–874 (2006)

An Advanced Approach of Face Alignment for Gender Recognition Using PCA

Abhishek Kumar, Deepak Gupta and Karan Rawat

Abstract In this paper we have used principal component analysis (PCA) tool by adding mathematical rigor to provide explicit solution for gender recognition by extracting feature vector. We will implement face recognition system using PCA algorithm along with the application of kernel support vector machine for error minimization. In addition by using face-rec database. This is an Eigen face approach motivated by information theory using an images database of 545 images of male and female for improved efficiency. sometimes PCA mixes data points which lead to classification error. We are improving principal component analysis (PCA) by taking vector corresponding to k minimum error unlike conventional PCA.

Keywords Gender Recognition · Principal component analysis · Eigen faces · SVM · Euclidian distance

1 Introduction

The face is an important biometric feature of human beings. Faces are accessible 'windows' into the mechanisms that govern our emotional and social lives. A successful gender classification method has many potential applications such as human identification, smart human computer interface, computer vision approaches for monitoring people, passive demographic data collection, etc. This paper deals with gender classification based on frontal facial images we will deal with significant variation between faces that significant features are known as Eigen faces. A Principal Compo-

A. Kumar (✉) · D. Gupta
Government Engineering College Ajmer, Ajmer, Rajasthan, India
e-mail: abhishekkmr812@gmail.com

D. Gupta
e-mail: gupta_de@rediffmail.com

K. Rawat
IIIT Allahabad, Allahabad, Uttar Pradesh, India
e-mail: rawatkaran4@hotmail.com

B. V. Babu et al. (eds.), *Proceedings of the Second International Conference on Soft Computing for Problem Solving (SocProS 2012), December 28–30, 2012*, Advances in Intelligent Systems and Computing 236, DOI: 10.1007/978-81-322-1602-5_110, © Springer India 2014

nent Analysis (PCA)-based image representation [1, 2] was used along with radial basis functions and preceptor networks investigated the use of SVMs for gender classification we will discuss the learning method back propagation as well for more efficient recognition rate. But a higher recognition rate is probably achieved due to excluding of negative feature. In general, all existing method or technique is appearance based, we have shown mathematical approach for feature analysis followed by structural risk minimization with the application of SVM [3]. As in order to get more accuracy a learning algorithm is needed so we have used back propagation. that is, they learn the decision boundary between male and female classes from training images, without extracting any geometrical features such as distances, face width, face length, etc. Almost all the modern face recognition algorithms use the PCA approach as the starting point for dimensionality reduction. We need to install image processing tool box, specifically we may use imread, imresize, etc., function in implementation phase. Principal Component Analysis proves to be the most robust and novel algorithm for face recognition and this can be verified by the fact that almost every other face algorithm such as the Linear Discriminant Analysis [4] and the Gabor filter approach [3, 5] make use of the PCA for dimensionality reduction. This technique classifies images in form of Eigen faces, explained further.

2 Principal Component Analysis

In this paper, we will implement a face recognition system using the PCA algorithm [6]. Automatic face recognition systems try to find the identity of a given face image according to their memory. The memory of a face recognizer is generally simulated by a training set. In this paper, our training set consists of the features extracted from known face images of different persons. Thus, the task of the face recognizer is to find the most similar feature vector among the training set to the feature vector of a given test image. Here, we want to recognize the identity of a person where an image of that person (test image) is given to the system. We will use PCA as a feature extraction algorithm in this paper. In the training phase, we should extract feature vectors for each image in the training set. Let Ω A be a training image of person A which has a pixel resolution of $M * N$(Mrows, Ncolumns). In order to extract PCA features of Ω Awe will first convert the image into a pixel vector © A by concatenating each of the Mrows into a single vector. The length (or, dimensionality) of the ©A vector will be $M * N$. In this project, we will use the PCA algorithm as dimensionality reduction technique which transforms the vector ©A to a vector $\mathbf{w}_a$, which has a dimensionality d where $d >> M * N$. For each training image $\mathbf{a}_i$, we should calculate and store these feature vectors !i. In the recognition phase (or, testing phase), we will be given a test image Ω_j of a known person. Let αj be the identity (name) of this person. As in the training phase, we should compute the feature vector of this person using PCA and obtain !j. In order to identify, $\mathbf{a}_j$ we should compute the similarities between $\mathbf{w}_j$ and all of the feature vectors $\mathbf{w}_{is}$ in the training set. The similarity between feature vectors can be computed using Euclidean distance. The identity of the most similar $\mathbf{w}_i$ will be the output of our face recognizer. If $i = j$, it means that we have correctly identified the person j, otherwise if $i \neq j$, it means that we have misclassified the person j. of computational resources. Thus, to sum up, the jobs which the PCA technique can do are prediction redundancy removal, feature extraction and data compression.

3 Mathematical Approach of the PCA

We are considering 2-D image for working so first of all it will get converted in to 1-D vector by concatenating rows and columns T Suppose we have M vectors each of size N (rows * columns) representing a set of sampled images [7].

Let '$=p_j$' represent the values of the pixels.

$$X_i = [p1,\ p2,\ p3,\ p4,\ldots,\ pN];\ i = 1,\ 2,\ 3,\ 4,\ldots M. \tag{1}$$

then the images are mean centered when we subtract the mean image from each image vector. Let us suppose m as the mean image:

$$m = (1/M)^* (\Sigma X_i) \tag{2}$$

Let W_i be the mean centered image:

$$W_i = X_i - m \tag{3}$$

Ultimately, we have to find the values of e_i's which have the largest possible projection onto each of the w_i's. The purpose is to get M orthogonal vectors e_i for which the quantity

$$\Lambda i = (1/M)\ \Sigma(e_{iT} * w_n)2 \tag{4}$$

is normalized with the orthogonality constraint:

$$e_{iT} * e_k = \delta_{lk} \tag{5}$$

The values of e_i's and λ_i's are calculated from the Eigen vectors and the Eigen values of the covariance matrix:

$$C = W * W_T \tag{6}$$

W is a matrix formed by the column vectors wi places side by side. The size of the covariance matrix is enormous ($N * N$). It is not possible to solve for eigenvectors directly. In mathematics, there are areas where one needs to find the numbers λ and the vectors v that satisfy the equation where A is the square matrix:

$$Av = \lambda v \tag{7}$$

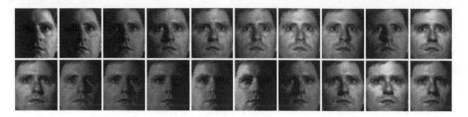

Fig. 1 An example face under a fixed view and varying illumination

Any λ satisfying the above equation is the Eigen value of A. The vector v is called the eigenvector of A. The Eigen values and eigenvectors are obtained by solving the equation:

$$[A - \lambda I] = 0 \tag{8}$$

For every λ, we calculate the corresponding eigenvectors and then normalize them. The eigenvectors are then sorted in the ascending order. This gives us the final KL Transform matrix. The covariance matrix of the final transformed image will have the Eigen values as their diagonal elements. More over the mean of final image will be zero.

4 Statistical Projection Methods

The first step is to create a database of images of different people. The database considered in this case is the—Face Recognition Database, MIT, USA‖. The images will exhibit different variations in the positioning of the head, the hair, the light content, the contrast, the skin color, and the expressions of the people (Fig. 1).

The variation of head pose or, in other words, the viewing angle from which the image of the face [8] was taken is another difficulty and essentially impacts the performance of automatic face analysis methods. For this reason, many applications limit themselves to more or less frontal face images or otherwise perform a pose-specific processing that requires a preceding estimation of the pose, like in multiview face recognition approaches. 2-D pose estimation approaches that have been presented in the literature. If the rotation of the head coincides with the image plane the pose can be normalized by estimating the rotation angle and turning the image such that the face is in an upright position. This type of normalization is part of a procedure called face alignment or face registration Fig. 2 shows some example face images with varying head pose. Fig. 3 is showing original images and their PCA pattern map.

Fig. 2 An example face under fixed illumination and varying pose

(a)

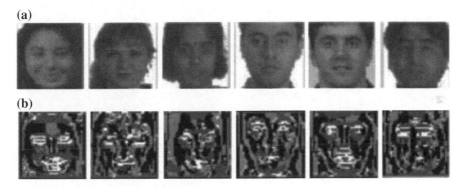

(b)

Fig. 3 Set of original images and PCA pattern maps

5 Support Vector Machines

The classification technique called Support Vector Machine (SVM) is based on the principle of Structural Risk Minimization. One of the basic ideas of this theory is that the test error rate, or structural risk $R(\alpha)$, is upper bounded by the training error rate, or empirical risk Remp and an additional term called VC-confidence which depends on the so-called Vapnik–Chervonenkis (VC)-dimension h of the classification function. More precisely, with the probability $1 - \eta$, the following holds

$$R(\alpha) < R_{\text{emp}}(\alpha) + \sqrt{h(\log(2l/h) + 1) - \log(\eta/4)/L} \qquad (9)$$

where α are the parameters of the function to learn and l is the number of training examples. The VC-dimension h of a class of functions describes its—capacity‖ to classify a set of training data points. For example, in the case of a two-class classification problem, if a function f has a VC-dimension of h there exists at least one set of h data points that can be correctly classified by f, i.e., assigned the label -1 or $+1$ to it. If the VC-dimension is too high the learning machine will over fit and show poor generalization. If it is too low, the function will not sufficiently

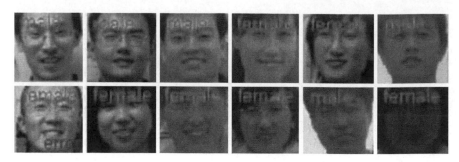

Fig. 4 Mean image for men and women created from the respective database images

approximate the distribution of the data and the empirical error will be too high. Thus, the goal of SRM is to find a h that minimizes the structural risk $R(\alpha)$, which is supposed to lead to maximum generalization (Fig. 4).

6 The Back Propagation Algorithm

We will focus on the Back propagation algorithm since it is the most common and maybe most universal training algorithm. In the context of NNs, the Back propagation (BP) algorithm has initially been presented by Rumelhart. It is a supervised learning algorithm defining an error function E and applying the gradient descent technique in the weight space in order to minimize E. The combination of weights leading to a minimum of E is considered to be a solution of the learning problem. In order to calculate the gradient of E, at each iteration, the error function has to be continuous and differentiable. Thus, the activation function of each individual perceptron. Mostly, a sigmoid or hyperbolic tangent activation function is employed, depending on the range of desired output values, i.e. $[0, 1]$ or $[-1, +1]$. Note that BP can be performed in online or offline mode, i.e., E represents either the error of one training example or the sum of errors produced by all training examples. In the following, we will explain the standard online BP algorithm, also known as Stochastic Gradient Descent, applied to MLPs [9]. There are two phases of the algorithm:

- the forward pass, where a training example is presented to the network and the activations of the respective neurons is propagated layer by layer until the output neurons.
- the backward pass, where at each neuron the respective error is calculated starting from the output neurons and, layer by layer, propagating the error back until the input neurons.

Now, let us define the error function as:

$$E = 1/2\Sigma_{pp=1}||o_p - t_p||_2 \tag{10}$$

where P is the number of training examples, o_p are the output values produced by the NN having presented example p, and t_p are the respective target values. The goal is to minimize E by adjusting the weights of the NN. With online learning we calculate the error and try to minimize it after presenting each training example.

Thus,

$$E_p = 1/2\Sigma \left\| o_p - t_p \right\|_2 = 1/2\Sigma_{kk=1}(o_{pk} - t_{pk})2 \tag{11}$$

where k is the number of output unit when minimizing this function by gradient descent, we calculate the steepest descent of the error surface in the weight space, i.e. the opposite direction of the gradient. In order to ensure convergence; the weights are only updated by a proportion of the gradient.

Thus,

$$\blacktriangledown E_p = (\partial E_p/\partial W_1, \ldots \partial E_p//\partial W_k)$$

In order to ensure convergence, the weights are only updated by a proportion of the gradient.

Thus,

$$\Delta W_k = -\lambda \partial E_p / W_k \tag{12}$$

Once we have analyze the images fixed illumination and varying pose now we will have to Eigen faces by conventional methods in order to get the corresponding Eigen vector as explained in section of PCA in above section. For the reconstruction of the test image we will need that the value of each Eigen faces. When we consider the entire training set for reconstructing the test image in Fig. 5 the reconstructed image resembles a lot with the test image shown in Fig. 5. We will analyze the performance using fuzzy logic, neural network, back propagation, and SVM. After these above step get completed, we will do the gender classification (Fig. 6).

Figures 7, 8, and 9 are showing the performance of different gender classification system on different dataset with different set of values for better analysis every time.

Fig. 5 Test input image having same number of pixels as the database images

Fig. 6 Reconstructed image using men's database

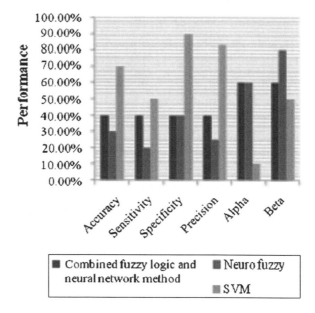

Fig. 7 Performance based on SVM, neuro fuzzy, and neural network method

On the basis of different values of threshold we have done the analysis. The different dataset is showing how at different value of threshold the accuracy is getting better each time we are altering the threshold values. The different dataset is showing not only the accuracy but sensitivity at the same time followed by specificity and precision. We have considered the value 0.1, 0.5, and 0.8 for the above dataset, respectively. So at the threshold value 0.1 SVM has achieved the accuracy of 70 % and the sensitivity of 45 % the specificity is much more than the sensitivity it is 88.72 % as shown in dataset 1f in Fig. 7. Now for the data set 2 we have the threshold value 0.5 in this case SVM has achieved the accuracy of 40 % and 25 % of sensitivity and specificity of 59.3 % similarly for the dataset 1, neuro fuzzy has achieved some different values as shown in respective figure. So through above given graph we could manage to calculate the performance of SVM, combined fuzzy logic, and neuro fuzzy in more analytical manner. As the different dataset are showing the percentage of accuracy precision, sensitivity, specificity, etc.

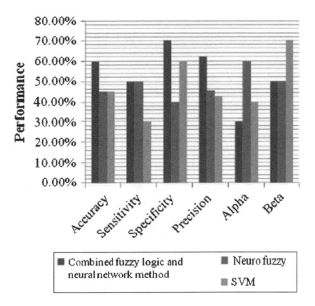

Fig. 8 Dataset 2

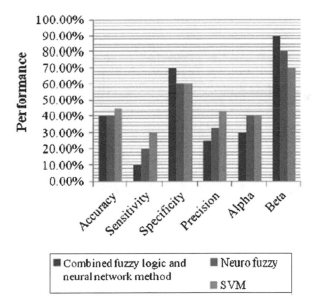

Fig. 9 Dataset 3

7 Analysis of Facial Feature With the Evaluation of Euclidian Distance

The above Fig. 10 is showing the analysis of facial feature in which it is very obvious to have different areas for the gender classification. So it is necessary to analyze facial feature in which area having more contribution for gender recognition. If we are processing on selective area the beneficial part will be the less bulky calculation and less complexity. So when we have the final test set and training set then we will go for the evaluations of Euclidian distance of that selective area of the face on which we were processing. In a mean while it evaluates how close the image from the test set is that from the training set. The consequences will be like the smaller the Euclidian distance the similarity of the face will be greater in that respect. At the same time higher recognition rate can be achieved due to excluding of the negative feature. We will divide the facial image in to smaller region as shown in the Fig. 10 in order to consider the shape information of the image. So the region around the moustache will be the specific area to go with further as the area we are going to consider is very sensitive as far as gender recognition is concerned using facial image. Women will have less intensity than men so we can calculate the brightness further to distinguish. This is how it plays an important role in gender classification as the feature analysis fig shows the frequencies of each region we are working on.

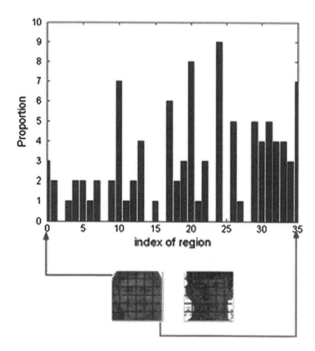

Fig. 10 Feature analysis

Table 1 Illumination changes: principal component analysis + neural network and PCA + support vector machine

Training	Testing	Correct nearest neighbour (%)	Correct kernal SVM (%)	Correct radial basis SVM (%)
4, 5, 6, 7, 8, 21	10	97.34	72.05	100
4, 5, 6, 7, 8, 21	11	100	79.41	100
4, 5, 6, 7, 8, 21	12	58.82	48.17	98.52
4, 5, 6, 7, 8, 21	20	96.34	64.70	100

Table 2 Pose changes: principal component analysis + neural network and PCA + support vector machine

Training	Testing	Correct nearest neighbor (%)	Correct kernel SVM (%)	Correct radial basis SVM (%)
1, 2, 3, 4, 6	8	83.82	33.82	76.47
1, 2, 3, 4, 6	9	97.05	98.52	100
2, 3, 4, 9	1	50	60.20	55.88
2, 3, 4, 9	6	27.94	30.88	33.82

In the next part we are going to show our experimental results with the neuro and SVM application for pose changed and illumination changed. We will show that how SVM approach along with PCA will show better classification results over conventional PCA. We are showing the experimental results by taking 60 images of men and women at one time then repeating the process for further results.

8 Experimental Results

Training samples have only uniform illumination. Testing samples have sharp illumination changes. Both training and testing samples have moderate and sharp illumination changes. We have the experimental results in Tables 1 and 2 which shows better classification above conventional PCA Use of SVM on both experiments gives more accuracy over conventional PCA in illumination changes and pose changed condition, respectively. Our aim is to reduce the classification error with advance approach of PCA by taking vector corresponding to k minimum classification error, unlike conventional PCA.

9 Conclusion

The PCA approach can effectively be used for the purpose of Face Recognition and Gender Recognition. We have used image basis function. The EIGEN FACE approach to gender recognition was motivated by information theory. It is fast relative simple but we have compared the performance using neuron network and SVM for purpose such as human computer interaction and security system. The characteristics of feature are comprehensively analyzed through the application to gender recognition performed on face image with different pose using MIT.CBCL DATA-BASE. Improved (PCA)-based gender detection with the application of (KERNAL SVM) shows robustness against different types of noises and external influences that face images can undergo in real world setting and result are more superior and efficient than other approaches we have given experimental results in Tables 1 and 2 with illumination change and pose change which is showing the improvement of this advance approach over conventional PCA. Further work can conspire improvement in accuracy and minimization of classification error as compare to classical and conventional approach of PCA.

References

1. Jain, A. K. Ross, A. Prabhakar, S.: An introduction to biometric recognition. IEEE Trans. Circ. Syst. Video Technol. **14**(1), 4–20 (2004)
2. Yan, S., Xu, D., Zhang, B., Zhang, H.J.: Graph embedding and extensions: a general framework for dimensionality reduction. IEEE Trans. Pattern Anal. Mach. Intell. **29**(1), 40–51 (2007)
3. Mohammed, A.A., Minhas, R., Jonathan Wu, Q.M., Sid-Ahmed, M.A.: Evaluation of face recognition technique using PCA, wavelets and SVM. Pattern Recognit. (Elsevier) **44**(10–11), 6404–6408 (2011)
4. Zhao, D., Liu, Z., Xiao, R., Tang, X.: Linear laplacian discrimination for feature extraction. In: Proceedings of the IEEE Conference on Computer Vision and Pattern Recognition (2007)
5. Dagher, I., Nachar, R.: Face recognition using IPCA-ICA algorithm?. IEEE Trans. Pattern Anal. Mach. Intell. **28**, 996–1000 (2006)
6. Kekre, H.B., et al.: Face and gender recognition using principal component analysis. Int. J. Comput. Sci. Eng. (IJCSE) **02**(04), 959–964 (2010)
7. Anton, H.: Elementary Linear Algebra 5e?. Wiley, Hoboken. ISBN 0-471-85223-6 (1987)
8. Utah State University—Spring 2012 STAT 5570: Statistical Bioinformatics, Notes 2.4
9. Yang, W.: Laplacian bidirectional PCA for face recognition. Neurocomputing (Elsevier) **74**, 487–493 (2010)

Analysis of Pattern Storage Network with Simulated Annealing for Storage and Recalling of Compressed Image Using SOM

Manu Pratap Singh and Rinku Sharma Dixit

Abstract In this paper, we are analyzing the SOM-HNN for storage and recalling of fingerprint images. The feature extraction of these images is performed with FFT, DWT, and SOM. These feature vectors are stored as associative memory in Hopfield Neural Network with Hebbian learning and Pseudoinverse learning rules. The objective of this study is to determine the optimal weight matrix for efficient recalling of the memorized pattern for the presented noisy or distorted and incomplete prototype patterns from the Hopfield network using Simulated Annealing. This process minimizes the effect of false minima in the recalling process.

Keywords Pattern storage network · SOM · FFT · DWT · Simulated annealing

1 Introduction

Pattern recognition has been implemented most commonly using the feedback neural networks. The key idea in pattern storage by feedback networks is the formation of basins of attraction in the energy landscape of the output state space. The number of basins of attraction depends only on the network, i.e., the number of processing units and their interconnection strengths. When the number of patterns to be stored is less than the number of basins of attraction, i.e., stable states then spurious stable states will exist, which do not correspond to any desired pattern. Determination of

M. Pratap Singh (✉)
Department of Computer Science, Institute of Engineering and Technology, Dr. B.R.Ambedkar University, Agra (Khandari), Uttar Pradesh, India
e-mail: manu_p_singh@hotmail.com

R. Sharma Dixit
Manav Rachna College of Engineering, Sector-43, Aravali Hills, Surajkund—Badkal RoadFaridabad, Haryana, India
e-mail: rinkudixit.mrce@mrei.ac.in

B. V. Babu et al. (eds.), *Proceedings of the Second International Conference on Soft Computing for Problem Solving (SocProS 2012), December 28–30, 2012*, Advances in Intelligent Systems and Computing 236, DOI: 10.1007/978-81-322-1602-5_111, © Springer India 2014

the number and locations of the basins of attraction in the network is not normally possible, but it is possible to estimate the capacity of the network and the average probability of error in recall.

The standard Hopfield model uses a very simple weight matrix formed with one-shot hebbian learning that produces a network with relatively poor capacity and performance [1]. This capacity and performance can be enhanced by adopting the modified Pseudoinverse rule. Pattern storage for continuum features as the input data can be characterized with the self organizing map for dimension reduction and feature extraction with Hopfield energy function analysis [2]. Iterations of the competitive learning between the input and the feedback output layer reduce the neighboring region in the processing elements of feedback layer and at each learning iteration, the used activation dynamics of the feedback layer leads the network toward the stable state. The feedback layer, which behaves as the HNN, at the equilibrium stable state reflects the stored pattern at the minimum energy state. It reflects that we may explore the possibilities of mapping of the features from the pattern space to the feature space and simultaneously encode the presented patterns. This SOM-HNN combination can be used to enhance the capabilities of the HNN. However, the stable states may not necessarily correspond to the memorized pattern and this can be one of the major pitfalls of the SOM-HNN formulation. Since the energy surface comprises several basins of attraction, there may be many local minima and on presenting the memorized pattern there is an equal probability that the activation dynamics converges to some false minima instead of the memorized pattern. Error in pattern recall due to false minima can be reduced by using a *stochastic update* [3] of the state for each unit, instead of the deterministic update. This paper explores this strategy wherein noise is introduced into the network dynamics. In effect, the network will reach states of lower energy but will also occasionally jump to higher energy states. This extended dynamics can help the network to skip out of local minima of the energy function. This approach will be implemented through the process of simulated annealing controlled by the temperature parameter to check the recall of the memorized patterns. The appearance of false minima in the SOM-HNN networks can be formulated as an Optimization problem for optimizing the activation dynamics. Such network when presented a prototype input pattern or its noisy variant should move stochastically through the activation dynamics in such a way that the network settles only into the energy basin corresponding to the memorized pattern thus avoiding the convergence to some false minima. This has been implemented in the simulations in this paper.

2 Preprocessing for Feature Extraction

The feature extraction algorithms extract unique information from the images [4]. The pattern set, i.e., scanned fingerprint images of multiple individuals are pre-processed before converting them to suitable patterns. The scanned RGB images are converted to *Grayscale*, enhanced and made sharper using *histogram equalization* and finally converted into a binary image by thresholding about the histogram peak. The efficiency of the feature extraction method decides to a great extent the quality of the image. Therefore, we are employing FFT and DWT for feature extraction to consider the pattern for storage.

The two-dimensional X-by-Y FFT and inverse X-by-Y FFT relationships are respectively given as

$$F(p,q) = \sum_{x=0}^{X-1} \sum_{y=0}^{Y-1} f(x,y) e^{-j2 \prod px/X} e^{-j2 \prod qy/Y} \begin{matrix} p = 0, 1, \ldots, X-1 \\ q = 0, 1, \ldots, Y-1 \end{matrix} \quad (1)$$

$$f(x,y) = \frac{1}{XY} \sum_{p=0}^{X-1} \sum_{q=0}^{Y-1} F(p,q) e^{j2\Pi px/X} e^{j2\Pi qy/Y} \begin{matrix} x = 0, 1, \ldots, X-1 \\ y = 0, 1, \ldots, Y-1 \end{matrix} \quad (2)$$

The values $F(p,q)$ are the FFT coefficients of $f(x,y)$.

Wavelet analysis expresses the original image in terms of a sum of basis functions, which are the shifted and scaled versions of the original (or mother) wavelet [4] and can often compress or de-noise a signal without appreciable degradation. Each DWT is characterized by a transform function pair or set of parameters that define the pair. Transform functions also called the wavelets are obtained from a single prototype wavelet called mother wavelet by dilations and shifting as:

$$\psi_{a,b}(t) = \frac{1}{\sqrt{a}} \psi \frac{t-b}{a} \quad (3)$$

where, a is the scaling parameter and b is the shifting parameter. The transform functions can be represented by three separable 2-D wavelets, i.e., $\psi^H(x,y)$, $\psi^V(x,y)$, $\psi^D(x,y)$ and one separable 2-D scaling function, i.e., $\varphi(x,y)$. The digitized binary images are subjected to FFT and DWT filtering and the refined and filtered images are obtained by Inverse FFT and DWT, respectively.

The filtered images are then converted to bipolar patterns. The general form of the lth pattern vector is $x_l = [x_{l1}, x_{l2}, x_{l3}, \ldots, x_{lN}]^T$ where $N = 1$ to 900. All the L image pattern vectors are presented to the feedback network for storage in the form of a comprehensive matrix P of order N x L.

3 Self Organizing Maps

Self-organizing maps can be used for feature mapping. The feature map can often be effectively used for the feature extraction from the input data for their recognition or, if the neural network is a regular two-dimensional array, to project and visualize high dimensional signal spaces on such a one or two dimensional display. It is used to deal for the patterns which represent the continuity in the feature space.

The features extracted from the SOM can be used as patterns for storing or encoding in the feedback neural network of Hopfield type. The associative memory feature of HNN for pattern storage and their recalling can be accomplished to incorporate symmetric feedback synaptic interconnection between the processing units with bipolar state of the output layer in self organizing map. Obviously, it is quite interesting to incorporate feedback connections among the processing units of grid in SOM for pattern recognition [2, 5, 6]. In this network, the processing elements of the feedback layer, i.e., grid of SOM are fully interconnected with the symmetric connection strength represented with weight vector M. The processing elements of the input layers are connected to each of the processing element of the feedback layer with connection strength represented with weight vector W. The presented input pattern to the network is K-dimensional with continuum features, applied one at a time. The network trained to map the similarities in the set of input patterns and at any given time only few of the input may be turned on. That is, only the corresponding links are activated to accomplish the aim of capturing the features in the space of input pattern and the connections are like soft wiring dictated by the unsupervised learning mode in which the network is expected to work [5]. There are several ways of implementing the feature mapping process. In one of the method, output layer is organized into predefined receptive fields, and the unsupervised learning should perform the feature mapping by activating the appropriate connections [7, 8]. Another method of implementing the feature mapping process is to use the architecture of a competitive learning network with on center off surround type of connections among units, but at each stage the weights are updated not only for the winning units, but also for the units in its neighborhood [9]. This neighborhood region may be progressively reduced during learning. The input pattern vector $X = \{x_1, x_2, x_3, \ldots, x_k\} \in R^n$ is applied to the processing elements of the input layer. The processing elements of the input layer are connected with each element in the SOM grid. This grid contains the feedback layer region. We associate connection strength $W_i = [w_{i1}, w_{i2}, \ldots, w_{in}]^T \in R^n$ to every processing element of the feedback layer. The initial value of W^T is selected randomly. Now the input pattern vector X is applied on the processing units of the input layer. The linear output of these processing units feed the weighted input through feed forward connection to the SOM grid. The activation of the jth process unit of the feed back layer can be represented as:

$$y_j = \sum_{i=1}^{K} w_{ij} x_i \qquad (4)$$

where, $j = 1$ to N (Number of units in the feed back layer.)

A winning unit, say P is selected among all the processing units of the feedback layer as:

$$y_p = \max_i (y_j)$$
$$W_P^T X \geq W_j^T X \text{ for all j}$$

Geometrically, the above relationship means that the input vector X is closest to the weight vector w_p among all w_i i.e.,

$$\sum_{i=1}^{k} (x_i - w_{Pi}) \leq \sum_{i=1}^{k} (x_i - w_{ji}) \quad \text{for all } j \tag{5}$$

Hence, during learning, the nodes that are topographically close up to certain geometric distance will activate each other to learn from the same input vector X and the weights associated with the winning unit P and its neighboring units r are updated as:

$$w_{iP}(t+1) = w_{iP}(t) + \lambda(P,r)[x_i(t) - w_{ij}(t)] \tag{6}$$

For $i = 1$ to K and $j = 1$ to N, here the $\lambda(P,r)$ is the neighborhood function. The neighborhood region reduces in successive iterations of the training process. Therefore, with this mechanism of competitive learning, the SOM is able to learn in unsupervised mode the feature mapping of the input pattern with continuum feature space.

4 Hopfield Neural Network

The proposed Hopfield Model to store L of patterns each of order $N \times 1$ consists of N processing units and $N * N$ connection strengths. The state of the processing unit is considered bipolar with symmetric connection strength between the processing units. Each neuron can be in one of the two stable states, i.e., ± 1. Storage as patterns is accomplished with Hebbian rule and the Pseudoinverse Rule. The Hebbian Rule to store L patterns is given by the summation of correlation matrices for each pattern as:

$$W_{ij} = \frac{1}{N} \sum_{l=1}^{L} x_{li} * l_j \quad \text{for } i \neq j$$
$$= 0 \qquad \text{for } i = j, 1 \leq i \leq N \tag{7}$$

where N is the number of units/neurons in the network.

The standard pseudoinverse rule is known to be better than the hebbian rule in terms of the capacity (N), recall efficiency and pattern corrections [10]. The Pseudoinverse Rule to store the pattern set P of Eq. (5) is given by

$$W = PP^{\dagger} \tag{8}$$

where $P^{\dagger}$ is its pseudoinverse of P [11, 12]. But pseudoinverse rule is neither local nor incremental as compared to the hebbian rule. These problems can be solved by modifying the rule in such a way that some characteristics of hebbian learning are also incorporated such that locality and incrementality is ensured. Hence the weight matrix is first calculated using hebbian rule stated in Eq. (7) then the pseudoinverse of the weight matrix can be obtained as:

$$W_{\text{pinv}}^L = (W^L)^\dagger (W^L * (W^L)^\dagger)^{-1} \tag{9}$$

where $(W^L)^\dagger$ is the transpose of the weight matrix W^L and $(W^L*(W^L)^\dagger)^{-1}$ is the inverse of the product of W^L and its transpose. After storing pattern set P in the HNN using either the Hebbian or the Pseudoinverse rule, the performance of the network needs be tested for the memorized patterns, their noisy variants and also for incomplete pattern information. For this, the process of recalling is considered, whereby a test pattern, which can be the memorized pattern or its noisy form, is input into the network and the network is allowed to evolve through its activation dynamics. The output state of the network is then tested for resemblance with one of the expected stable states.

5 Stochastic Recall using Simulated Annealing

In associative memory architectures, the memorized patterns act as attractors in the state space of the network. In effect as the memorized pattern is presented it moves to the attractor that it most closely resembles to. However there exist additional attractors that do not correspond to any memorized pattern and thus are unhelpful. These spurious attractors may lead to error in pattern recall. The addition of noise to the deterministic HNN can be beneficial in the elimination of the spurious minima because as the free energy landscape is changed, the spurious local minima of the energy function may no longer be stable and therefore the probability that the network converges to a memorized pattern is increased [13].

In the stochastic case the update rule is probabilistic instead of being deterministic. Whenever a unit is selected for updating, the energy either decreases or remains the same. It is possible to realize a transition to a higher energy state from a lower energy state by using a stochastic update [3] in each unit instead of the deterministic update of the output function. In this process the probability of firing for an activation value of y can be expressed as

$$P(s = 1/y) = 1/1 + e^{-(y-\theta)/T} \tag{10}$$

where T is the temperature of the network. At $T = 0$ the stochastic update reduces to a deterministic update. As the temperature is increased, the uncertainty in making the update according to $f(y)$ increases, giving thus a chance for the network to go to a higher energy state. Therefore the result of Hopfield energy analysis, i.e. $\Delta V \leq 0$, will be no longer true for nonzero temperatures. Finally when $T = \infty$, then the update of the unit does not depend on the activation value y any more. The state of the network changes randomly from 1 to -1 or vice versa.

The training and recall of patterns in a stochastic network use the process of annealing controlled by the temperature parameter. Simulated Annealing has been widely applied to solve optimization problems [14, 15]. The recall of the memorized patterns corresponding to the presented noisy prototype patterns from the HNN can also be formulated as an optimization problem. Here the criterion is to minimize the network energy in such a way that the network settles into the global minima corresponding to the memorized pattern and thus avoids getting trapped into any false minima. The annealing schedule, becomes critical in realizing the desired probability of states near $T = 0$. Thus, for recall, when a prototype input pattern or its noisy variant is presented, the network is allowed to evolve through its stochastic dynamics following an annealing schedule and reach an equilibrium state near $T = 0$. This will reduce the effects of local minima and thus reduces the probability of error in the recall. The temperature is initialized to a high value and then a neuron is randomly selected from the network for updating. The energy of the network for the state s is computed via the Hopfield energy function as

$$E(s) = -\frac{1}{2} \sum_i \sum_j w_{ij} s_i s_j \qquad (11)$$

The state of the chosen neuron is flipped to generate a new configuration. The energy of the new configuration is again computed using Eq. (11). Change in energy for change of state in the unit k from 0 to 1, for example, is given by:

$$\Delta E = E(s_k = 1) - E(s_k = 0) = -\sum_j w_{kj} s_s = -x_k \qquad (12)$$

If ΔE is negative then the new configuration obtained as a result of flipping of the chosen neuron is considered as a better state and the new state is accepted. But if the change in energy is positive then the new state is accepted with a probability calculated as:

$$P(s_i) = \frac{1}{1 + \exp^{(\Delta E_i / T)}} \qquad (13)$$

The above procedure continues several times until a thermal equilibrium is reached. At this point the temperature is lowered and the procedure is repeated. This continues until the temperature reaches a very small value. A critical aspect here is the choice of initial temperature and the annealing schedule, specified as $T_{k+1} = cT_k$, where T_k is the current temperature, T_{k+1} is the new temperature and c is a constant with value ranging between 0 and 1, with actual working range 0.8–0.9 for gradual reduction of temperature and skipping of false minima.

6 Simulation Design and Results

In the simulation design and implementation the proposed HNN is designed with 900 processing units. This HNN is trained for pattern storage with two learning rules i.e. Hebbian and Pseudoinverse Rules. The patterns for storage have been preprocessed and filtered through FFT and DWT separately. In one implementation the preprocessed patterns are directly entered into the HNN while in other the preprocessed patterns are first fed into the SOM and the feature spaces generated thereafter are stored into the HNN. Thereafter the process of recalling is initiated for prototype input patterns and for the noisy prototype input patterns of already memorized patterns. The recall is evaluated with the standard Hopfield recall algorithm and with Simulated Annealing. The results show that the SOM-HNN when trained with FFT filtered patterns shows poor results as compared to the HNN. The storage capacity with hebbian rule is reduced to almost half i.e. $\frac{1}{2}\left(N/2\log_2 N\right)$ while that with Pseudoinverse rule is reduced to one-ninth i.e $N/9$. But on the other hand the same network when trained with DWT filtered patterns shows enhanced storage capacity with both the learning rules. With hebbian rule the capacity increases to $0.28N$ while with pseudoinverse rule the capacity is maintained at the maximum i.e. N. The Pattern recall ability of the SOM-HNN for the noisy variants of the FFT and DWT feature vectors is better with DWT filtered patterns as compared to the FFT filtered patterns. But it is also observed from the results that the occurrence of false minima is increased in this network as compared to a stand-alone HNN. But this drawback can also be taken care of by stochastic recalling of the patterns using Simulated Annealing, whereby the occurrence of false minima is been completely eliminated.

7 Conclusions

Efficiency of the Pattern Storage networks is dependent on preprocessing techniques, the learning methods adopted and the activation dynamics of the network. Modified Pseudoinverse learning rule can be used to enhance the network capabilities. The continuum feature spaces created by SOM when stored in the Hopfield network can lead to substantial enhancement in the capacity and recall efficiency but this mapping amplifies the occurrence of false minima, which can be eliminated by using stochastic update instead of deterministic update of the network units in the recall mechanism.

References

1. Davey, N., Hunt, S.P., Adams, R.: High capacity recurrent associative memories. Neurocomputing - IJON **62**, 459–491 (2008). doi:10.1016/j.neucom.2004.02.007
2. Gill, S., Sharma, N.K., Singh, M.P.: Study of pattern storage techniques in self organizing map using hopfield energy function analysis. In: Proceedings of ADCOM–2006, pp. 640–641. (1–4244-0716-8/06/2006/IEEE).

3. Stroeve, S., Kappen, B., Gielen, S.: Stimulus segmentation in a stochastic neural network with exogenous signals. In: Ninth International Conference on Artificial Neural Networks, 1999, ICANN 99, vol. 2, pp. 732–737 (1999)
4. Sandirasegaram, N., English, R.: Comparative analysis of feature extraction (2D FFT & wavelet) and classification (Lp metric distances, MLP NN & HNet) algorithms for SAR imagery. In. Proceedings Of SPIE **5808**, 314–325 (2005)
5. Kaski, S., Kangas, J., Kohonen, T.: Bibliography of self-organizing map (SOM) papers: 1981–1997. Neural Comput. Surv. **1**(3&4), 1–176 (1998)
6. Kohonen, T.: Self-organized formation of topologically correct feature maps. Biol. Cybernet. **43**, 59–69 (1982b)
7. Kohonen, T.: Self-organizing formation of topologically correct feature maps. Biol. Cybern. 43, 59–69 (1982)
8. Hubel, D.H., Wiesel, T.N.: Receptive fields, binocular interaction, and functional architecture in the cat's visual cortex. J. Physiol. London **160**, 106–154 (1962)
9. Colin, M., Utku, S., Hugues, B.: Introduction of a Hebbian unsupervised learning algorithm to boost the encoding capacity of Hopfield networks, neural networks. In: Proceedings of the IEEE International Joint Conference on IJCNN '05. 2005, vol. 3, pp. 1552–1557 (2005)
10. Gorodnichy, D.O.: The influence of self connection on the performance of pseudoinverse autoassociative networks. In: Proceedings of CVPR Workshop on Face Processing in Vodeo (FPIV '04) (2004)
11. Emmert-Streib, F.: Active learning in recurrent neural networks facilitated by a hebb-like learning rule with memory. Neural Inf. Process. Lett. Rev. 9(2), 31–40 (2005)
12. Labiouse, C.L., Salah, A.A., Starikova, I.: The impact of connectivity on the memory capacity and the retrieval dynamics of hopfield-type networks. In: Proceedings of the Santa Fe Complex Systems Summer School, pp. 77–84 (2002)
13. Davey, N., Adams, R.G.: Stochastic dynamics and high capacity associative memories. In: Proceedings of the Ninth International Conference on Neural Information Processing, pp. 1666–1671 (2002)
14. Jeffrey, W., Rosner, R.: Optimization algorithms: simulated annealing and neural network processing. Astrophys. J. **310**(1), 473–481 (1986)
15. Salcedo Sanz, S.: A hybrid hopfield network-simulated annealing approach for frequency assignment in satellite communication systems. IEEE Trans. Syst. Man Cybern. B: Cybern. 34(2), 1108–1116 (2004)

Nonrigid Image Registration of Brain MR Images Using Normalized Mutual Information

Smita Pradhan and Dipti Patra

Abstract Registration is an advanced technique which maps two images spatially and can produce an informative image. Intensity-based similarity measures are increasingly used for medical image registration that helps clinicians for faster and more effective diagnosis. Recently, mutual information (MI)-based image registration techniques have become popular for multimodal brain images. In this chapter, normalized mutual information (NMI) method has been employed for brain MR image registration. Here, the intensity patterns are encoded through similarity measure technique. NMI is an entropy-based measure that is invariant to the overlapped regions of the two images. To take care of the deformations, transformation of the floating image is performed using B-spline method. NMI-based image registration is performed for similarity measure between the reference and floating image. Optimal evaluation of joint probability distribution of the two images is performed using parzen window interpolation method. The hierarchical approach to nonrigid registration based on NMI is presented in which the images are locally registered and nonrigidly interpolated. The proposed method for nonrigid registration is validated with both clinical and artificial brain MR images. The obtained results show that the images could be successfully registered with 95 % of correctness.

Keywords Medical image registration · Mutual information · Normalized mutual information · B-spline method

S. Pradhan
IPCV Laboratory, Department of Electrical Engineering, NIT, Rourkela, India
e-mail: ssmita.pradhan@gmail.com

D. Patra (✉)
Deptartment of Electrical Engineering, NIT, Rourkela, India
e-mail: DPATRA@nitrkl.ac.in

B. V. Babu et al. (eds.), *Proceedings of the Second International Conference on Soft Computing for Problem Solving (SocProS 2012), December 28–30, 2012*, Advances in Intelligent Systems and Computing 236, DOI: 10.1007/978-81-322-1602-5_112, © Springer India 2014

1 Introduction

Image registration is the process of transforming different sets of data into one coordinate system. Registration methods can be classified by a number of characteristics, which include the type of transformation and the registration measure. Although it has applications in many fields, medical image registration is important among them. Medical image registration has a wide range of potential applications, but the emphasis is on radiological imaging. Modern three-dimensional treatment radiation planning is based on sequences of tomographic images. Computed tomography (CT) has the potential to quantitatively characterize the physical properties of heterogeneous tissue in terms of electron densities. Magnetic resonance (MR), is very often superior to CT, especially for the task of diferentiating between healthy tissue and tumor tissue. Positron emission tomography (PET), single photo emission tomography (SPECT), and MRS (magnetic resonance spectroscopy) imaging have the potential to include information on tumor metabolism. They have specific properties and deliver complementary information. The images supply important information for delineation of tumor and target volume, and for therapy monitoring.

A widespread survey of image registration methods have been published by Brown and Zitova et.al. They have classified the image registration techniques as intensity-based methods and feature-based methods [1, 2]. In intensity-based methods, similarity measure has an important role, which quantifies the relationship of transformation between the images. The most commonly used similarity measures are based on intensity differences, intensity cross correlation, and information theory. Among them, mutual information (MI) has gained wide interest in the medical image registration field [3]. When the assumptions of corresponding intensities are not one-to-one related, maximization of mutual information (MMI) is widely applicable [4]. Pluim et al. proposed to combine MI with an image gradient-based term that favors similar orientation of edges in both images [5].

As MI is computed on voxel-by-voxel basis, it does not take the spatial information inherent to the original image. To overcome this drawback, variations of MI have been proposed. Pluim et al., suggested interpolation artifacts in similarity measures [6]. Likar et al. developed a hierarchical image subdivisions strategy, that decomposes the nonrigid matching problem into an elastic interpolation of numerous local rigid registration [7]. For overlapping sub regions of the image, a nonrigid registration scheme was proposed by extending the intensity joint histogram with a third channel representing a spatial label by Chen et al. [8]. Rueckert et al. employed for transformation modeling the multiresolution scheme [9]. Studholme et al. employed normalized mutual information (NMI) for nonrigid registration of serial brain magnetic resonance imaging (MRI) due to the local intensity changes, which are mainly caused by imaging distortions and biological changes of the brain tissue [10]. The optimization algorithm for nonrigid medical image registration based on cubic b-spline and maximization of MI is derived in [11].

MI faces difficulties for registration of small-sized images. To overcome this limitation, Andronache et al. used the MI for global registration and cross correlation to register the small image patches [12]. Besides the Shanon's entropy, other divergence measures have been used such as Tsallis entropy by Khader et al. [13].

In image registration, finding homologous landmarks is a challenging task due to lack of redundancy in anatomical information, in different modalities. Intensity-based techniques circumvent these problems as they do not require any geometrical landmarks. Their basic principle is to search, the transformation that maximizes a criterion measuring the intensity similarity of corresponding voxels. In this chapter, the registration scheme has been proposed using similarity measure-based method by incorporating NMI. The NMI approach is a robust similarity measure technique used for multimodal medical image registration. Moreover, the NMI-based registration is less sensitive to the changes in the overlap of two images. Here, we propose a non-rigid image registration approach by optimizing the NMI as similarity measure and B-spline method for modeling the transformation of the deformation field between the reference and floating image pairs.

2 Problem Statement

Here the NMI-based similarity measure is used as a matching criterion to solve the image alignment problem.

Let A and B be two misaligned images to be registered where A is the reference image and B is the floating image. The floating image B is a deformed image with a deformation field. The deformation field is described by a transformation function T(r, μ) where μ is the set of transformation parameter to be determined. The image registration problem may be formulated as an optimization problem

$$\hat{\mu} = \arg\max_{\mu} NMI(A(r), B(T(r; \mu))) \tag{1}$$

For alignment of the transformed target image $B(T(r; \mu))$ with the reference image A, we need the set of transformation parameters ⁻ that maximizes the image cost function $NMI(A(r), B(T(r; \mu)))$.

3 Problem Formulation

3.1 Registration by Normalized Mutual Information

The notion of image registration based on MI has been proposed by Maintz et al. [3]. The MI of two images is a combination of the entropy values of the images. The entropy of an image can be computed by estimation of the probability distribution of the image intensities [5].

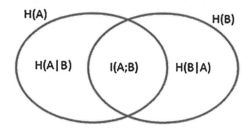

Fig. 1 Mutual Information of two images

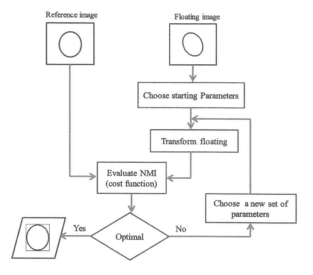

Fig. 2 Block diagram of NMI

If A and B are two random variables, then the amount of information that one variable contains about another is evaluated by MI (Fig. 1) which is given as

$$I_{(A:B)} = \sum_{a,b} p_{A,B}(a,b) \log\left(\frac{p_{A,B}(a,b)}{p_A(a)p_B(b)}\right) \tag{2}$$

where $P_A(a)$ and $P_B(b)$ are the marginal probability mass function and $P_{AB}(a,b)$ is the joint probability mass function. MI is related to the entropies by

$$I(A,B) = H(A) + H(B) - H(A,B) \tag{3}$$

where H(A) and H(B) are the entropies of A and B and H(A, B) is the joint entropy defined as (Fig. 2)

$$H(A) = -\sum_a p_A(a).\log p_A(a)$$

$$H(B) = -\sum_b p_B(b).\log p_B(b)$$

$$H(A, B) = -\sum_{a,b} p_{a,b}(a, b).\log p_{a,b}(a, b) \tag{4}$$

The MI of two images determines the uncertainty of one of the images when the other is known. MI is assumed to be maximum when the images are registered. According to Studholme et al., the registration quality might decrease despite an increasing MI value whenever overlap between voxel occurs. At that time, MI is maximum but the quality of the registration is not optimal [10]. To counter the effect of increasing MI with decreasing registration quality NMI is considered. The transformation yielding the highest NMI value is assumed to be the optimal registration of two images. An estimate of the intensity distribution of the images is necessary to compute the NMI value. NMI is a well-established registration quality measure which can be defined in terms of image entropies,

$$NMI(A, B) = \frac{H(A) + H(B) - H(A, B)}{H(A, B)} \tag{5}$$

3.2 Optimization Using Parzen Window

To calculate the NMI between the reference image A and the floating image B using a transformation $T(r; \mu)$, the joint histogram $H(a, b; \mu)$ has been taken. The popular Parzen window interpolation method is formulated to obtain the joint histogram. Parzen window places a kernel at a particular bin r and updates all the bins falling under the kernel with the corresponding kernel value. The parzen window joint histogram is given by;

$$H(a, b; \mu) = \sum_{r \in A} w_a(I_A(r_a) - a)w_b(I_B(T(r_a; \mu)) - b) \tag{6}$$

where $I_A(r)$: intensity of A, $I_B(r)$: intensity of B, $T(r; \mu)$, maps every reference position r_a to the corresponding floating position $r_b = T(r_a; \mu)$ with a given set of parameters μ. w_a and w_b are the Parzen window kernels used to distribute an intensity over the neighboring bins.

3.3 Transformation Model

The transformation model defines how one image can be deformed to match another; it characterizes the type and number of possible deformations. Several transformation models have been proposed for nonrigid image registration. In this chapter, we adopt the second order B-splines model for transformation of deformed image. By contrast, B-splines are only defined in the vicinity of each control point; perturbing the position of one control point only affects the transformation in the neighborhood of the point. Because of this property, B-splines are often referred to as having "local support". B-spline based nonrigid registration techniques are popular due to their general applicability, transparency, and computational efficiency. The B-spline model is situated between a global rigid registration model and a local nonrigid model at voxel scale. Its locality or nonrigidity can be adapted to a specific registration problem by varying the mesh spacing and thus the number of degrees of freedom.

Let Φ denote a $rx \times ry$ mesh of control points $\Phi_{i,j}$ with a uniform spacing Δ. Then, the 2D transformation at any point $r = [x, y]^T$ in the target image is interpolated using a linear combination of a B-spline convolution kernel as follows:

$$T(r; \mu) = \sum_{ij} \eta_{ij} \beta^2 \left(\frac{r - \phi_{ij}}{\Delta} \right) \tag{7}$$

where $\beta^{(2)}(r) = \beta^{(2)}(x)\beta^{(2)}(y)$ is a separable B-spline convolution kernel, and η_{ij} are the deformation coefficients associated to the control points $\Phi_{i,j}$.

3.4 Simulation and Results

The optimization of nonrigid transformation using B-spline interpolation method leads to align the images. The time required for optimization steps to reach registration can be used as a measure of computational speed. The interpretation of NMI function is based on dispersion of the joint histogram, meaning the less dispersed the joint histogram, the better the two images are assumed to be registered. Under this interpretation, minimization of the dispersion of the joint histogram is related to maximization of NMI value.

The registrations were performed for a set of deformed brain MR images in our simulation. In the first set, a T1 weighted brain MR image and a deformed image that are considered as reference and floating images and, are shown in Fig. 3a, b respectively. The deformation field is shown in Fig. 3c. The registered image is shown in Fig. 3d. The joint histograms of the two images before registration and after registration are shown in Fig. 3e and f respectively. It can be seen that the histogram is not dispersed, rather focused after registration using NMI-based similarity measure. The dispersion shows the misalignment between the images with a

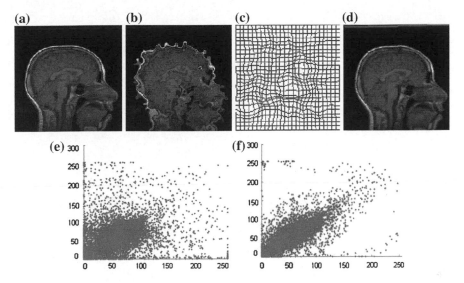

Fig. 3 **a** Floating image **b** Reference image **c** Deformation field **d** Registered image **e** Joint histogram before registration **f** Joint histogram after registration

percentage of 4.009. The optimization timing or computational timing is recorded as 41.14 s.

Another set of MR images are taken, where Fig. 4a is the floating image which is deformed and rotated with an angle of 30 %. The reference image is shown in Fig. 4b. The floating image is first rotated for best alignment with the reference one shown in Fig. 4c. The deformation field is presented in Fig. 4d. The joint histogram of Fig. 4a, b is shown in Fig. 4f, where it can be observed that the histogram is dispersed due to misalignment between the two images. But in case of Fig. 4g it can been seen that the histogram is very much focused after the registration process. The computation time for optimal registration is recorded as 18.51 s.

The optimized NMI value, the percentage of misregistration error (MRE), and the computational time for optimization metrics for both experiments are presented in Table 1. The normalized value signifies the alignment between the two images after registration process. The higher the value of NMI means more alignment of the images. MRE is misregistration error calculated by considering the registered image with respect to the reference images. Less value of MRE means best alignment of the images. Computational time is the time required for optimal registration. From

Table 1 Calculated NMI value with % age of Mis-reg Error and comp. time

Image	NMI value	MRE (%)	Comp. timing (s)
Figure 3	0.2880	4.0090	18.51
Figure 4	0.3960	7.5041	41.14

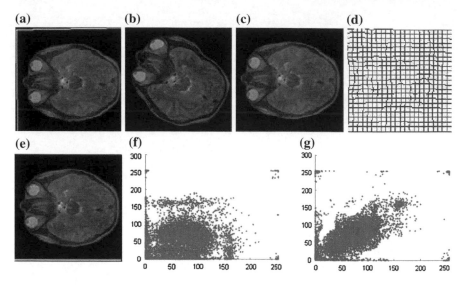

Fig. 4 **a** Floating image **b** Reference image **c** 30° rotated floating image **d** Deformation field **e** Registered image **f** Joint histogram before registration **g** Joint histogram after registration

the table, it can be concluded that the optimization process for aligning the images takes more time with a low percentage of misalignment and vice versa.

4 Conclusion

In this paper, intensity-based technique is applied for alignment problem as we can work directly with the volumetric data. We propose the registration scheme by optimizing the NMI as similarity measure. This measure produces accurate registration results on both artificial and clinical brain images that we have tested. B-spline method is used for modeling the nonrigid deformation field of the target image with respect to the reference image. Parzen window interpolation method is employed for joint histogram plot. Current attempts are made to register the image with multimodal images. This work can be extended towards multimodal brain image registration.

References

1. Brown, L.G.R.: A survey of image registration techniques, ACM Comput. Surv. **24**(4), 325–376 (1992)
2. Zitova, Barbara, Flusser, Jan: Image registration methods: a survey. Image Vis. Comput. **21**, 977–1000 (2003)

3. Pluim, J., Maintz, J., Viergever, M.: Mutual-information-based registration of medical images: a survey. IEEE Trans. Med. Imag. **22**(8), 986–1004 (2003)
4. Viola, P., Wells, W.M.: Alignment by maximization of mutual information. In: ICCV '95 Proceedings of the Fifth International Conference on Computer Vision, IEEE Computer Society (1995)
5. Pluim, J., Maintz, J., Viergever, M.: Image registration by maximization of combined mutual information and gradient information. IEEE Trans. Med. Imag. **19**(8), 809–814 (2000)
6. Pluim, J.P.W., Maintz, J.B.A., Viergever, M.A.: Interpolation artefacts in mutual information-based image registration. Comput. Vis. Image Understand. **77**(2), 211–232 (2000)
7. Likar, B., Pernu, F.: A hierarchical approach to elastic registration based on mutual information. Image Vis. Comput. **19**(1–2), 33–44 (2001)
8. Chen, H.M., Varshney, P.K.: Mutual information-based CT-MR brain image registration using generalized partial volume joint histogram estimation. IEEE Trans. Med. Imag. **22**(9), 1111–1119 (2003)
9. Rueckert, D., Aljabar, P., Heckemann, R.A., Hajnal, J.V., Hammers, A.: Diffeomorphic registration using b-splines. Medical Image Computing and Computer-Assisted Intervention, Lecture Notes Computer Science, **4191**, 702–709. Springer, New york (2006)
10. Studholme, C., Drapaca, C., Iordanova, B., Cardenas, V.: Deformation-based mapping of volume change from serial brain MRI in the presence of local tissue contrast change. IEEE Trans. Med. Imag. **25**(5), 626–639 (2006)
11. Klein, S., Staring, M., Pluim, J.P.W.: Evaluation of optimization methods for nonrigid medical image registration using mutual information and B-splines. IEEE Trans. Image Process. **16**(12), 2879–2890 (2007)
12. Andronache, A., Siebenthal, M.v., Szekely, G., Cattin, P.: Nonrigid registration of multimodal images using both mutual information and cross-correlation. Med. Image Anal. **12**, 3–15 (2008)
13. Khader, M., Hamza, A.B., An entropy-based technique for nonrigid medical image alignment. In: Proceedings of the 14th International Workshop Combinatorial Image, Analysis, pp. 444–455 May 2011

Part XI
Soft Computing for Communication, Signals and Networks (CSN)

A Proposal for Deployment of Wireless Sensor Network in Day-to-Day Home and Industrial Appliances for a Greener Environment

Rajat Arora, Sukhdeep Singh Sandhu and Paridhi Agarwal

Abstract Wireless automation sensor communication network (WASCN) is the promising tool for energy conservation [1] according to the research work of this paper. This paper considers the two aspects of analyzing an appliance, i.e., at manufacturer level and user condition level of that appliance. At manufacturer level, the manufacturer company of the particular appliance will itself provide the concerned data of appliance and its effect on the environment, whereas at user level, these data may depend on the infrastructure or the environment of that particular appliance where it is being used. Now to analyze the 2nd impact we have environment and infrastructure manager (EIM). EIM will deploy WASCN [2] between various appliances to measure the effects of the appliance according to user, infrastructure as well as the environment in which it is being used. These data recorded by the EIM will be shared to analyze the effect of a particular appliance in various particular conditions which in turn will help the customers or an industry to choose the environment friendly products considering the cost too.

Keywords WASCN · EIM · Energy conservation

R. Arora
Mechanical Engineering Department, IIT Kanpur, Kanpur, UP, India
e-mail: rajat.arora9464@gmail.com

S. S. Sandhu (✉) · P. Agarwal
Computer Science and Engineering Department, Gyan Vihar University, Jaipur, Rajasthan, India
e-mail: sukhdeepsingh90@gmail.com

P. Agarwal
e-mail: paridhi.agarwal1990@gmail.com

B. V. Babu et al. (eds.), *Proceedings of the Second International Conference on Soft Computing for Problem Solving (SocProS 2012), December 28–30, 2012*, Advances in Intelligent Systems and Computing 236, DOI: 10.1007/978-81-322-1602-5_113, © Springer India 2014

1 Introduction

One of the most common and of utmost concern in today's world is energy conservation, that too, environmental friendly. Today various companies are coming up with eco friendly appliances. The need of hour is to change the old devices with newer ones but for large-scale industries it is not possible to change them all one at a time. So, in this paper we discuss the alternative method for energy conservation in an environment friendly way. This paper argues on the fact that WASCN has all the potential to improve the energy efficiency of such devices allowing EIM to target more and more green purchases [3].

WASCN are low cost wireless sensors [4] that ensure the consumption of low power of radio technologies. WASCN promises the monitoring of appliances via a common EIM interface. Simultaneous monitoring of appliances via WASCN provides the path of identification of appliances requiring maintenance or replacement.

Refreshed data being generated after regular intervals creates an opportunity for a manufacturer to develop energy as well as environment friendly products considering infrastructure as well as the environment in which appliance will be used via detailed power profile being generated by EIM in this regard.

2 Proposed Architecture for Energy Conservation and a Step Ahead to Greener Environment

This section of our paper describes the components of our proposed architecture and how the same components will help in conserving energy [5] and ENVIRONMENT. The main thing is that to achieve our desired goal all these components should be entangled with each other via Internet services so as to share their data.

2.1 Executive Roles

We have identified different executive roles at different levels that are as follows:

  i> EIM
 ii> Manufacturers
iii> End users

EIMs are the major proposed executive role in this system. EIMs are responsible for maintaining a power profile of a particular appliance working under some environmental conditions as well as space and area constraints. This power profile may be different from the ideal power profile of an appliance. Also, we may note here that power profile in one particular environment and area of the same appliance may differ from other power profile in some other area size and environment. This feature of EIM helps the customer to choose the most appropriate appliance according to his environment and area size which in turn helps customer saving energy and environment by choosing the appropriate appliance.

The second Executive role is that of the manufacturers. Manufacturers create the power profile of an appliance in ideal conditions which is compared to EIM's power profiles. Now as we have pointed earlier, all the executive role players are connected to each other in a network so a manufacturer can create one appliance for two different environments for, e.g., homemade appliances may have different environmental impacts and energy consumption than industries.

End users are the ones who help EIM to make a power profile of their appliances by deploying WASCN with each appliance of theirs. It is also the responsibility of the end user to check the power profile of appliances and choose the best appliance which will help in conserving energy and environment according to their available conditions.

2.2 WASCN and Its Features

WASCNs are the major proposed technology in our paper. According to our proposed architecture WASCN are small chips that are entangled with each appliance which in turn help in making the power profile. WASCN records the effect of appliance on the environment and vice versa and the energy consumed by that appliance in a particular area and environment. The sensed data is then sent to EIM. EIM is connected to WASCN which will record the data of all the appliances in a particular home or industry which may be helpful for the new customer to compare the profiles of one appliance of the same company being used by different end users situated at different areas having two different environmental conditions and select the eco friendly appliance.

Various features expected to be possessed by WASCN are:

- Low power: WASCN are small chips or sensors that consume very less operating power [6].
- Reliability: As the same WASCN will be applied in different areas having different environment therefore, it should work in any of the conditions without any failure or it must be reliable enough to be used.
- Low cost: There may be thousands of appliances at some places being used so the operating cost and the maintenance cost must be low and within the budget of the end user [7].
- Submission of manufacturer data: It is also expected that WASCN may be used by manufacturers also to record their profiles and share data. So, WASCN should fit into any conditions and may be used by any user.

A. Power profiles.

Power profile [8] refers to the power consumed by the device in a particular area and environment and the worst-case will always be considered. Some areas may be hot while some areas may be cold so the behaviour of one particular appliance will be different in both the conditions. Also, some areas may be compact and some areas may be wide so the heat dissipated by the device in compact area will be more which will disturb the environmental condition of that area. Thus here we choose the appliance which will dissipate the minimum heat. So, we conclude that both area as well as environmental condition somewhere has effect on appliance which may again have an adverse effect on our environment itself and may consume more energy. Therefore, this power profile helps in comparing data of different appliances [9] in same conditions, different appliances in different conditions, same appliances in same conditions and same appliances in different conditions.

B. Assembling of various power profiles by different executive roles

Now the need of the hour says to gather all power profiles recorded by EIM and the manufacturers. All these profiles are shared online and data of different appliances in same conditions, different appliances in different conditions, same appliances in same conditions, and same appliances in different conditions is compared so that we can work upon the existing appliances which are a threat to the environment or it may also help in the case where we intend to replace the existing appliances with new ones or it may also work when we go on for buying new appliances.

C. Power profile repository

Power profile repository [10] is like a database which stores various power profiles generated by EIMs and the manufacturers. This repository is not maintained at end user side, rather we make EIM and manufacturer responsible for holding the same, but at the same time we should also keep in mind that this repository should be accessible [11] by every executive role (Fig. 1).

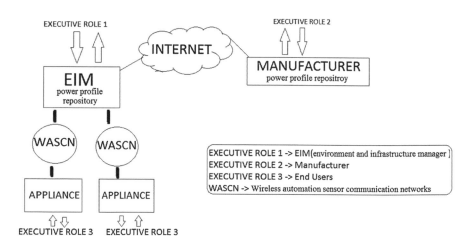

Fig. 1 Architecture of the proposed model

3 Working of our Proposed System

In the following section we explain the working of our proposed architecture:-

First, executive role 2, i.e., the manufacturer will develop an appliance and will make the power profile of the appliance in the ideal conditions. The power profile is then stored in the power profile repository which will in turn be shared on the Internet. Then comes the role of the executive role 3, i.e., end user of an appliance. Now in our proposed system executive role 3 is assigned the duty to deploy WASCN with every appliance they are using. Now according to the surroundings for, e.g., area, temperature, room space, weather conditions, etc., the sensors will record the effect an appliance on the environment and vice versa. Its power consumption also depends on the surrounding. So, according to the conditions its power profile will be recorded and will be stored in the repository of executive role 1, i.e., EIM. Now EIM is responsible for storing data of different appliances in same conditions, different appliances in different conditions, same appliances in same conditions, and same appliances in different conditions & is compared so that we can work upon the existing appliances which are threat to the environment or it may also help in the case where we intend to replace the existing appliances with new ones or it may also work when we go for buying new appliances. Likewise, it will be beneficial for end users to search for the best suitable appliance according to their surroundings which in turn will help us saving in the environment and reduce the electricity consumption hence, saving the energy too.

4 Conclusion

Through this paper we have tried to present an energy conservation model which will also help in making the present environment greener. Wireless sensors [12] introduced itself has a low power consumption and is environmental friendly which helps customer to go for an efficient purchase of an appliance which also cuts short the maintenance cost keeping in mind the green environment as its main aim.

5 Review

A Review by
  Mr DEEPAK SACHAN
  Alumni IIT Roorkie Electronics and Communication Engineering.
  MBA SungKyunKwan University Seoul, South Korea.
  Manager Samsung Electronics Gurgaon
  "There is hope if people will begin to awaken that spiritual part of them, that heartfelt knowledge that we are caretakers of this planet."—Brooke Medicine Eagle

According to the proposed idea if the research scientists really implement it in the practical life then it would change the whole current scenario and mind sets of people towards efficient use of home appliances leading to a greener environment. Till now whenever we talk about purchasing an appliance it's just the deal between an end user and manufacturer but both manufacturer as well as end user are least bothered about environmental aspects If Executive role 1 and WASCN (according to the proposed idea) comes into picture then there is hope to save environment from the dangerous aspects of appliances. I'm moved by the idea and as a manager at Samsung Electronics Gurgaon we'll try to conduct research on the idea in our industry too.

References

1. http://info.iet.unipi.it/anastasi/papers/adhoc08.pdf
2. http://arri.uta.edu/acs/networks/WirelessSensorNetChap04.pdf
3. http://www.deepgreenrobot.org/using-green-technology-under-the-threat.html
4. http://www.sciencedaily.com/releases/2011/04/110412143123.htm
5. http://www.sciencedirect.com/science/article/pii/S1570870508000954
6. Daabaj, K., Dixon, M.W. and Koziniec, T.: Adaptive transmission power control scheme for wireless sensor networks. In: 2011 MUPSA Multidisciplinary Conference, 29 Sept, Murdoch University, Murdoch (2011)
7. http://cache.freescale.com/files/microcontrollers/doc/ref_manual/DRM094.pdf
8. http://www.ict-aim.eu/fileadmin/user_files/deliverables/AIM-D2-3v2-0.pdf
9. http://www.ecnmag.com/articles/2012/08/designing-low-cost-wireless-sensor-networks-real-world-applications
10. http://researchrepository.murdoch.edu.au/4460/
11. http://doc.oracle.com/cd/E23507_01/Platform.20073/ATGPersProgGuide/html/s0401settingupacompositeprofilereposi01.html
12. http://www.sensor-networks.org/

Energy-Aware Mathematical Modeling of Packet Transmission Through Cluster Head from Unequal Clusters in WSN

Raju Dutta, Shishir Gupta and Mukul Kumar Das

Abstract Clustering techniques in wireless sensor networks (WSNs) compare to random selection techniques is less costly due to the saving of time in journeys, reduction in number of transmissions and receptions at each node, identification, contacts, etc., which are valuable for increasing the overall network life, scalability of WSNs. Clustering sensor nodes is an effective and efficient technique for achieving the requirement. The maximizing lifetime of network by minimizing energy consumption poses a challenge in design of protocols. Therefore, proper organization of clustering and orientation of nodes within the cluster becomes one of the important issues to extend the lifetime of the whole sensor network through Cluster Head (CH). We investigate the problem of energy consumption in CH rotation in WSNs. In this paper, CH selection algorithm has been proposed from an unequal cluster. The total energy and expected number of packet retransmissions in delivering a packet from the sensor node to other nodes have been mathematically derived. In this paper, we applied the approach for producing energy-aware unequal clusters with optimal selection of CH and discussed several aspects of the network mathematically and statistically. The simulation results demonstrate that our approach of re-clustering in terms of energy consumption and lifetime parameters.

Keywords WSN · Packet transmission · Unequal cluster · Cluster head · Lifetime

R. Dutta (✉)
Narula Institute of Technology, Kolkata, India
e-mail: rdutta80@gmail.com

S. Gupta · M. K. Das
Indian School of Mines, Dhanbad, India
e-mail: shishir_ism@yahoo.com

M. K. Das
e-mail: mkdas12@gmail.com

B. V. Babu et al. (eds.), *Proceedings of the Second International Conference on Soft Computing for Problem Solving (SocProS 2012), December 28–30, 2012*, Advances in Intelligent Systems and Computing 236, DOI: 10.1007/978-81-322-1602-5_114, © Springer India 2014

1 Introduction

Cluster schemes are hierarchical. Selection of sensor node as a CH in WSN greatly affects the power consumption efficiency of the network. In conventional clustering, the network is divided and sensor nodes are accumulated into groups, known as clusters and then one sensor node from cluster is selected to lead as clusterhead. The distribution of the clusterhead may be uneven. In unequal clusters size, the selection strategy of CH effects in uneven energy consumption of the network. For a given number of sampling units, cluster sampling is more convenient and less costly. The advantages of cluster sampling are

1. Within a cluster, all normal nodes send their data to the CH. The resulting absence of flooding scheme, multiple route which is energy saving.
2. The backbone network consists only of the CHs, which are fewer in number then all nodes in the entire network and therefore simpler.
3. the change of nodes within the cluster affects only that cluster but not the entire network.
4. collection of data for neighbouring elements is easier, cheaper, faster and operationally more convenient then observing units spread over a region.
5. It is less costly then simple random sampling due to the saving of time in journeys, identification, contacts, etc.
6. When a sampling frame of elements may not be readily available.

Energy conservation is always a challenging issue in wireless sensor networks. Lifetime of wireless network based on battery power, which has limited energy source. For increasing the lifetime of such network and nodes, it is important that one has to find the techniques either increasing the battery power or an alternative source of energy for the nodes. One of the methods for increasing the lifetime of nodes is by adjusting the transmission power of sensor node during transmission. However, adjusting the transmission power is not always sufficient to improve the battery power of sensor nodes and optimize the energy consumption. Now-a-days to increase in the network life time interference plays an important role for minimizing the energy consumption. Due to interference, the quality of service of wireless network to a great extent causing collision of packets, packet loss, and retransmission frequency. Decreasing interference level may save the node power by minimizing collision, retransmission and congestion. In this paper, we consider transmission and reception power of nodes for a good quality routing path for delivering the packets from source to destination. Our protocol minimizes the total energy consumption of routing path from source to destination and balances load among the nodes. We propose our protocol with shortest path routing.

2 Related Work

In this paper [1] proposed a distributed, randomized clustering techniques to find optimum cluster size and cost to organize the sensors in a wireless sensor network within clusters. In [2] it describes mathematical modeling of packet transmission through CH from unequal clusters in WSN. Low Energy Adaptive Clustering Hierarchy (LEACH) described in [3] the rotation of CH among the nodes and a cluster based data gathering in WSN. MECH [4] explains a communication range and sensor nodes select themselves as CH if satisfying the fixed node degree criterion. ACE-C [5] pointed out for even distribution of sensor nodes and avoid re-clustering in each round. Reconfiguration of the clusterhead explained in [6] and also avoid re-clustering technique power efficient routing protocol is given in [7] for selecting CH in power efficient at even distribution of CH. Power consumption and maximizing network lifetime during communication of sensor node in WSN has been discussed in [8]. In [9] efficient clustering techniques to optimize the system lifetime in wireless sensor network. In [10] topology control of wireless networks is achieved using link level interference. In [11] sender computes the potential interference of receivers and adjusts their transmission ranges to reduce the receiver interference using global information. The energy consumption pattern in an integrated interference-aware and confidentiality-enhanced multipath routing scheme for continuous data streams on Wi-Fi based multi-hop wireless ad hoc networks has also been proposed [11]. In [13, 14] dynamic virtual carrier sensing and interference aware routing protocol to select the optimum path based on two criteria, shortest path and interference of nodes, have been proposed. The sensor network is a wireless collection of portable devices that offers are a variety of services, including area and wildlife monitoration, etc. [15]. A cluster-based approach is introduced in [16] using PSO, which was found to function better than Low-Energy Adaptive Clustering Hierarchy-Centralized (LEACH-C) [17] protocol in terms of energy efficiency. Another improved PSO has been proposed in [18] for improving the performance of the optimization technique. In [19] the research work evaluates a routing optimization method on the basis of graph theory and PSO.

3 Network Model and Assumptions

The assumptions made to describe our network scenario (Fig. 1).

- The sensor network is assumed to be a circular geographic region with the sink S, positioned at coordinate $(x_0 y_0)$, and radius R_s.
- The sensors are uniformly deployed in the sensing area A_S. Moreover, the number of sensor nodes is distributed according to 2-dimensional Poisson point process with ρ as the expected density of A_{ca}.
- The cluster covers circular area with its CH at the center o with radius R_{ca}.

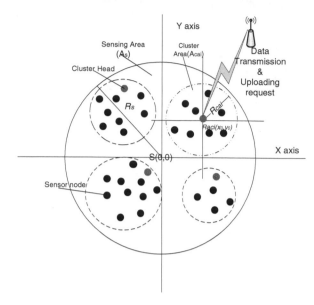

Fig. 1 Network model with data transmission from sensor node and uploading request from BS

- There are total k clusters in the sensor network. Further, owing to the uniform node deployment strategy, we can compute an approximation for the cluster radius, R_{ca}:

$$K \times A_{ca} = A_s \Rightarrow K \times \pi \times R_{ca}^2 = \pi \times R_s^2 \Rightarrow R_{ca} = \sqrt{\frac{R_s}{K}}$$

$$\text{where} \quad K = \sum_{i=1}^{k} K_i \quad \text{and} \quad R_{ca} = \sum_{i=1}^{k} R_{cai}$$

- The base station (or sink) periodically sends a request to the CH of unequal cluster size to upload samples collected by the sensors (Fig. 1). On receiving the request, the CH broadcasts a data-gathering-signal to all its cluster members.
- In our contribution we have applied unequal clustered sensor network, where the nodes are stationery. The basic aim is to find optimized position for CH from the base station, i.e., the distance is as close as possible to the base station. Such localization for CH would ultimately minimize the average distance covered by the sensors to transmit data to the CH and to the base station.

3.1 Power Consumption Schemes

The normal nodes in a cluster only transmit and relay their data to their CH. In addition to transmitting their data, the CHs are also receiving data from the normal nodes and transferring these data. The CHs therefore consume more power then other normal nodes and when the CHs run out of energy the cluster will break down. To avoid this situation and keep network healthy, we have implemented a Cluster Selection Efficient Protocol (CSEP), where clusters will be selected in the basis of few assumption.

- Here we consider a dynamic clustered network where clusters are unequal.
- There may not be even distribution of CH.
- In each cluster, the CH is selected by CH routing protocol.
- In each cluster, a node will be selected as a CH whose energy level is high then others node.
- The CH will be selected from the cluster if the variance within the cluster is less.
- The CH will be selected from the cluster if the distance from the base station is less. But if the closer cluster is unable to find CH then next closer distance cluster have opportunity to select CH.
- The sensor nodes which are in the boundary of the cluster consumes more energy then other node.
- CH rotation is associated with Re-clustering technique.

3.2 Cluster Head Selection Algorithm

In this section, CHs are selected. If $E_{residual} > E_{threshold}$ satisfy then a sensor become KEY NODE then it broadcast the MESSAGE. Then each KEY NODE receives the MESSAGE within the range of the sensor node. If a node satisfies all the condition $E_{residual} > E_{threshold}$ and $d\left(n_j, BS\right) < d(n_K, BS)$ and $E_{residual} > E_{any\ node}$ simultaneously then it becomes CH otherwise re-clustering will be carried out.

Lemma *In every cluster, only one CH exists at a $Timer_{Phase}$ within the radio range (R_{range}) of the cluster.*

Proof Let us consider S be the set of all sensor node in a cluster of size M_i. Let us consider $S = \{ x,\ if\ d\ (x,\ y) < R_{range}\}$. Let $p, q \in S$ then if we consider p is a CH, then by our algorithm of selecting of CH we can say p. $E_{residual} > q\ q.\ E_{residual}$. Again if we take q as a CH then according to the algorithm we have p. $E_{residual} > q$ q. $E_{residual}$. Which is only possible if p = q. So In every cluster only one CH exists at a $Timer_{Phase}$ within the radio range (R_{range}) of the cluster.

Algorithm 1 Clusterhead selection algorithm

```
/*
∀n_j ∈ M_i
Timer_Phase, each node gets sufficient time to complete operation.
1. state ← NODE;
2. if (E_residual > E_threshold) then
3.           state ← KEY NODE;
4.     broadcast MESSAGE;
5.     while (Timer phase not expired) do
6.               msg ← receive MESSAGE;
7.               if ((E_residual > E_threshold) & d(nj, BS)< d(nK, BS)
                    &(E_residual > E_any node))then
8.                        state ← CH;
9.               else if
10.                       state ← NODE;
11.                       break;
12.      end if
13.      end while
14. end if
15. if (state = KEY NODE) then
16.      state ← CH;
17 end if
```

4 Unequal Cluster Sampling and Sampling for Proportion

Estimator of Mean and its Variance. In many practical situations, cluster size vary. Suppose there are N clusters. Let the ith cluster consists of Mi elements (i =1, 2,...,N) and $\sum_{i}^{n} M_i = M_0$ The cluster population mean per element $\bar{Y}$ is defined by

$$\bar{Y} = \frac{\sum_{i=1}^{N} \sum_{j=1}^{M_i} y_{ij}}{\sum_{i=1}^{N} M_i} = \frac{\sum_{i=1}^{N} M_i \bar{y}_{i.}}{\sum_{i=1}^{N} M_i} = \frac{\sum_{i=1}^{N} M_i \bar{y}_{i.}}{M_0}$$ where $\bar{y}_i$ is the mean per element of the ith

cluster. We may also defined the pooled mean $\bar{Y}_N = \frac{\sum_{i=1}^{N} \bar{y}_i}{N}$

Suppose it is required to estimate the proportions of elements belongs to a specified classes when the population consists N clusters, each of size M and a random sample of n clusters is selected. Suppose the M elements in any cluster can be classified into two classes. Assuming that $y_{ij} = 1$ if the jth elements of the ith cluster belongs to the class = 0, otherwise.

It can be easily seen that $p_i = \frac{a_i}{M}$ is the true proportion in the ith cluster, a_i being the number of elements in the ith cluster belongs to the specified class. An unbiased estimator of the population proportion $P = \frac{\sum_{i}^{N} a_i}{NM} = \frac{\sum_{i}^{N} p_i}{N}$ is given by $\widehat{P_c} =$

$\frac{\sum_{i}^{n} p_i}{n} = \bar{p}$ where p_i, the proportion of elements is belongs to the specified class in the ith cluster of the sample. The sampling variance of $\widehat{P_c}$ is given by [4] $V(\widehat{P_c}) = \frac{N-n}{n} \frac{\sum_{i=1}^{N}(p_i-P)^2}{N-1} = \frac{(1-f)}{n}S_b^2$ is the variance between cluster proportions and is given by $MS_b^2 = \sum_i^N \frac{(p_i-P)^2}{N-1} = \frac{N}{N-1}PQ$, where $Q = 1 - P$. For large N, we have $S^2 \cong S_b^2 + S_w^2 = PQ$ and the within variance S_w^2 is given by $\sum_i^N \frac{P_iQ_i}{N}$. Hence the intracluster correlation coefficient ρ can be written as $\rho = 1 - \frac{\sum_i^N MP_iQ_i}{(M-1)PQ}$ and the range of ρ is given by $-\frac{1}{M-1}(-\frac{M_0-1}{M_0(M-1)} \le \rho \le \frac{(N-1)S_b^2}{NS^2}$ where $S_b^2 = \sum_i^N \frac{M_i(M_i-1)(\bar{y}_i - \bar{Y})^2}{M(M-1)(N-1)}$ therefore, the sampling variance, in terms of the intracluster correlation coefficient can be expressed as

$$V(\widehat{P_c}) = \frac{(1 - f)\,NPQ}{(N - 1)\,nM}[1 + (M - 1)\,\rho] \tag{1}$$

An estimation of total number of units belonging to the specified class, can be obtained by multiplying by $\widehat{P_c}$ by NM and the expression for its sampling variance is N^2M^2 times that given by Eq. (1). If Simple Random Sampling (SRS) of nM elements could be taken, the variance of the sampling proportions $\widehat{P}$ would be obtained by binomial theory and is given by

$$V_{bin}(\widehat{P}) = (1 - f)\frac{NPQ}{nM(NM - 1)}$$

The efficiency of cluster sampling as compared to SRS, without replacement, can be obtained as

$$\frac{V(\widehat{P_c})}{V_{bin}(\widehat{P})} = \frac{(MN - 1)[1 + (M - 1)\rho]}{N - 1} \tag{2}$$

And for large N

$$\frac{V(\widehat{P_c})}{V_{bin}(\widehat{P})} = M[1 + (M - 1)\,\rho]$$

If the cluster sizes M_i are variable, the estimate $\widehat{P_c} = \frac{\sum a_i}{\sum M_i}$ is a ratio estimate. Its variance is given by approximately by the formula (1), $V(\widehat{P_c}) = \frac{N-n}{nM}\frac{\sum_{i=1}^{N} M_i^2(p_i-P)^2}{N-1}$ if this sample is compared with a simple random sample of nM elements, we find as a generalization of Eq. (2), by

$$\frac{V(\widehat{P_c})}{V_{bin}(\widehat{P})} = \frac{(\bar{M}N - 1)\left[1 + (\bar{M} - 1)\rho\right]}{N - 1} \text{ where } \bar{M} = \frac{\sum M_i}{N} \tag{3}$$

5 Varying Probability Cluster Sampling

In many practical situations, cluster size is positively correlated with the variable under study. In these cases, it is advisable to select the clusters with probability proportional to the number of elements in the cluster of their sizes M_i. There are several applications in which size is not the number of elements in the cluster and some other measure is considered for probability selection. In this paper, we shall confine ourselves to discussion of the case where the probability of selection is proportional to cluster size M_i. Let p_i ($0 < p < 1$) be the probability of selecting the ith cluster if size M_i ($i = 1, 2,...,N$) at each draw with $\sum_i^N p_i = 1$. Suppose that $z_{ij} = \frac{M_i y_{ij}}{M_0 p_i}$ for $j = 1, 2, \ldots\ldots, M_i$ and $i = 1, 2, \ldots\ldots N$, further suppose that n clusters are selected by pps with replacement so that $z_{ij} = \frac{M_i \bar{y}_{i.}}{M_0 p_i}$ for $i = 1, 2, \ldots N$. If clusters are selected with probabilities proportional to size $p_i = \frac{M_i}{M_0}$ and with replacement, then an unbiased estimator of $\bar{Y}$ is given by $\bar{z}_n = \sum_i^n \frac{\bar{y}_{i.}}{n}$ with variance

$$V(\bar{z}_n) = \sum_i^n \frac{M_i}{nM_0} \left(\bar{y}_{i.} - \bar{Y}\right)^2 \tag{4}$$

and unbiased estimator of the variance $V(\bar{z}_n)$ is given by $V(\bar{z}_n) = \frac{\sum_i^n (\bar{y}_{i.} - y_n)^2}{n(n-1)}$.

The efficiency of sampling n unequal clusters with pps, without replacement as compared to Simple random sampling by taking $n\bar{M}$ elements, can be obtained by Eq. (4) with the variance $V_{SR}(\bar{y}) = \frac{S^2}{n\bar{M}}$ which can be expanded by introducing $\bar{y}_{i.}$ in the above relation. Thus,

$$V_{SR}(\bar{y}) = \frac{1}{n\bar{M}} \frac{1}{N\bar{M}} \left[\sum_i^N \sum_j^{M_i} \left(y_{ij} - \bar{y}_{i.}\right)^2 + \sum_i^N M_i \left(\bar{y}_{i.} - \bar{Y}\right)^2 \right]$$

$$= \frac{\sigma_w^2}{n\bar{M}} + V(z_n) \text{ where } \sigma_w^2 = \frac{1}{N\bar{M}} \sum_i^N \sum_j^{M_i} \left(y_{ij} - \bar{y}_{i.}\right)^2 = \sum_i^N \frac{M_i \sigma_i^2}{N\bar{M}}$$

Hence, the efficiency of cluster sampling is given by

$$E = \frac{V_{SR}}{V_{pps}} = \left[\bar{M} \left(1 - \frac{\sigma_i^2}{\sigma^2} \right) \right]^{-1} \qquad (5)$$

It can be seen that E is always greater than 1.

6 Expected Number of Retransmission Attempts

Initially, we assume that an aggregation tree exists in every cluster with the CH as the root. Moreover, within the cluster, transmission of data to the CH follows the path in the aggregation tree. We assume that there are h hops or links between the source node and CH. Also, every link (between two sensors) possesses link failure probability (p_{LFP}). Clearly, for a tree of m links, the number of transmissions required for one successful sending of data to CH is also m. Now, the probability of m successful transmission for one successful end-to-end data delivery (towards CH) is $(1 - p_{LFP})^m$. Also, the probability of at least one failed transmission, leading to failed data delivery, is $[1 - (1 - p_{LFP})^m]$. Let random variable Y which denotes the number of successful data delivery attempts. Further, $(\mu - 1)$ failures followed by one successful attempt satisfy geometric distribution and is given by: $p[Y = \mu] = [1 - (1 - p_{LFP})^m]^{(\mu-1)} \times (1 - p_{LFP})^m$

Therefore, the expected number of attempts leading to a successful delivery of data is:

$$E[\mu] = \sum_{\mu-1}^{\infty} \mu \times p[Y = \mu] = \sum_{\mu=1}^{\infty} \mu \times [1 - (1 - p_{LFP})^m]^{(\mu-1)} \times (1 - p_{LFP})^m = \frac{1}{(1 - p_{LFP})^m} \qquad (6)$$

7 Energy Consumption Model

The radio energy dissipated by a sensor node is mainly in form of electronics and amplifier energy. Here E_{elec} is the energy dissipated per bit to run the radio electronics (E_{elec}) largely depends on how efficiently the signal is encoded, modulated, and filtered and ξamp is the energy expended to run the power amplifier for transmitting a bit over unit distance. However, the energy dissipation rate in the radio amplifier (ε_{amp}) is directly proportional to d^λ, where d is the distance between the source (member of the cluster) and destination node CH and λ is the path loss component. Therefore, the expected value of d^2, represented by $E[d^2]$ is obtained as following:

$$E\left[d^2\right] = \int \int [(x - x_0)^2 + (y - y_0)^2] \times \rho\,(x, y)dx\,dy$$

$$= \rho \int_{x_0}^{p} \int_{y_0}^{\sqrt{p^2 - x_0^2}} \left[(x - x_0)^2 + (y - y_0)^2\right] dx\,dy$$

$$= \frac{\rho}{3}\,(p - x_0) \left[\begin{array}{l} 2p^2\left(\sqrt{p^2 - x_0^2} - 2y_0\right) + \\ 2px_0\left(y_0^2 - \sqrt{p^2 - x_0^2}\right) + y_0\left(2x_0^2 + y_0\left(3\sqrt{p^2 - x_0^2} - y_0\right)\right) \end{array}\right]$$

Where $x \le p \le R_s$, $\rho = \frac{\sum_{i=1}^{k} \rho_i}{K}$ and $\rho_i = \frac{K_i}{\pi R_{cai}^2}$

To transmit an *m-bit* packet over a distance d the energy used by a sensor node is given by:

$$e_T = \begin{cases} m \times E_{elec} + m \times \varepsilon_{amp} \times & d_i^2 & d_i \langle v_0 \\ m \times E_{elec} + m \times \varepsilon_{amp} \times & D^4 & D \ge v_0 \end{cases} \qquad (7)$$

In Eq. (7) v_0 is the threshold distance and d_i denotes the distance between the $(i+1)$th node and ith node in the cluster and D is the beyond threshold distance where the signal strength is affected by multi-path fading between the leader and the base station. In our approach, we have used both the free space (distance 2 power loss) and the multipath fading (distance 4 power loss) channel modes. In our model, we assume that inter-nodal distances are small compared to distance between the nodes and the Base Station (BS). Thus for communication among sensors, we take $n = 2$, and that between the leader and BS, we take $n = 4$, in Eq. (7). The energy spent in the receiving the packet is, $e_R = m \times E_{elec}$. Finally the total energy expanded (E_{NODE}) by a sensor node to support transmission—reception operation as well as in other energy consumption (E_{other}) can be determined by the following equation:

$$E_{NODE} = \lambda \times e_T + (1 - \lambda) \times e_R + E_{other}$$

$$= \lambda \times \left[mE_{elec} + m\,\varepsilon_{amp} \times \frac{\rho}{3}\,(p - x_0) \left[\begin{array}{l} 2p^2\left(\sqrt{p^2 - x_0^2} - 2y_0\right) + \\ 2px_0\left(y_0^2 - \sqrt{p^2 - x_0^2}\right) + \\ y_0\left(2x_0^2 + y_0\left(3\sqrt{p - x_{02} - y_0}\right)\right) \end{array}\right] \right]$$

$$+ (1 - \lambda)\,mE_{elec} + E_{other}$$

where other energy consumption due to the environmental noise. Finally, the total energy in delivering a packet can be expressed as following:

$$E_{\text{TOTAL}} = h \times E_{\text{NODE}} \times E[\mu]$$

$$= h \times \left[\left[\lambda \times \left[mE_{\text{elec}} + m\, \varepsilon_{\text{amp}} \times \tfrac{\rho}{3}\, (p - x_0) \left[\begin{array}{l} 2p^2 \left(\sqrt{p^2 - x_0^2} - 2y_0 \right) + \\ 2px_0 \left(y_0^2 - \sqrt{p^2 - x_0^2} \right) + \\ y_0 \left(2x_0^2 + y_0 \left(3\sqrt{p^2 - x_0^2} - y_0 \right) \right) \end{array} \right] \right] \\ + (1 - \lambda)\, mE_{\text{elec}} + E_{\text{other}} \right] \right]$$

$$\times \frac{1}{(1 - p_{\text{LFP}})^m}$$

(8)

where h hops or links between the source node and CH.

8 Simulation Results

Extensive simulation carried out, and the results of the distribution power consumption of sensors are shown in Fig. 2.

9 Conclusion

In this paper, we have considered unequal clustering techniques to minimize interference, transmission power, and reception power of nodes to derive a good quality routing path for delivering the packets from source to destination based on our pro-

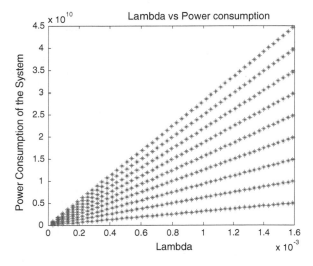

Fig. 2 Power consumption reduces according to the increase of lambda

posed algorithm. For a given source to destination pair in the network, there may exist more than one routing paths. But according to our proposed model only the optimum route will be considered based on our algorithm for cluster and CH selection. Figure 2 shows the power consumption according to the different values of lambda. We compared our protocol with other sampling and Eq. (5) shows that the efficiency of cluster sampling is always better then other sampling. Equation 3 conclude that the sampling variance $\widehat{P}_c$ of unequal clustering gives better result if this sample is compared with a simple random sample of nM elements and if intracluster correlation coefficient ρ is small. Equation 8 gives the total energy consumption for delivering a data packet. At present, we are carrying further studies in this area.

References

1. Dutta, R., Gupta, S., Das, M.K.: Mathematical modelling of packet transmission through cluster head from unequal clusters in WSN. In: IEEE International Conference Parallel, Distributed and Grid, Computing (PDGC-2012), University of Jaypee, Guna, 515–520 (2012)
2. Dutta, R, Gupta, S., Das, M.K.: Efficient statistical clustering techniques for optimizing cluster size in wireless sensor network. In: International Conference on Modeling, Optimization and Computing (ICMOC-2012) (Procedia Eng. J.,), vol. 38, pp. 1501–1507. Elsevier, London (2012)
3. Heinzelman, W., Chandrakasan, A., Balakrishnan, H.: Energy-efficient communication protocol for wireless microsensor network. In: Proceedings of 33rd HICS **2**, 1–10 (2000)
4. Chang, R., Kuo, C.: An energy efficient routing mechanism for wireless sensor networks. In: Proceedings of International Conference in Advanced Information Networking and Applications (AINA '06), **5**(2) (2006)
5. Liu, C., Lee, C., Chun Wang, L.: Distributed clustering algorithms for datagathering in wireless mobile sensor networks, Elsevier science. J. Parallel Distrib. Comput. **67**, 1187–1200 (2007)
6. Kim, N., Heo, J., Kim, H., Kwon, W.: Reconfiguration of clusterheads for load balancing in wireless sensor networks, Elsevier science. J. Compos. Commun. **31**, 153–159 (2008)
7. Zhang, W., Linag, Z., Hou, Z., Tan, M.: A power efficient routing protocol for wireless sensor network. In: Proceedings of IEEE International Conference in Networking, Sensing and Control, London, 20–25 (2007)
8. Dutta, R., Gupta, S., Das, M.K.: Power consumption and maximizing network lifetime during communication of sensor node in WSN. Procedia Technology, vol. 4, pp. 158–162. Elsevier, London (2012)
9. Dutta, R., Saha, S., Mukhopadhyay, A.K.: Efficient clustering techniques to optimize the system lifetime in wireless sensor network. In: IEEE - International Conference on Advances in Engineering, Science and Management, IEEE Explore pp. 679–683.(2012)
10. Burkhart, M., Rickenbach, P., Wattenhofer, R., Zollinger, A.: Does topology control reduce interference?. In: Proceedings of ACM MobiHoc. (2004)
11. Krunz, M., Muqattash, A., Lee, S.: Transmission power control in wireless ad hoc networks: challenges, solutions, and open Issues. J. IEEE Netw. 8–14 (2004)
12. Lee, C.W.: Wireless transmission energy analysis for interference-aware and confidentiality-enhanced multipath routing. In: Proceedings of 8th IEEE International Symposium on Network Computing and Applications, (2009)
13. Lou, W., Liu, W., Fang, Y.: SPREAD: enhancing data confidentiality in mobile ad hoc networks. In: Proceedings of IEEE INFOCOM, 2404–2413 (2004)
14. Hou, K., Qin, Y.: A dynamic virtual carrier sensing and interference aware routing protocols in wireless mesh networks. In: Proceedings of IEEE ICCS, (2008)

15. Akyildiz, I.F., Su, W., Sankarasubramaniam, Y., Cayirci, E.: Wireless sensor networks: a survey. J. Comput. Netw. **384**, 393–422 (2002)
16. Latiff, N.M.A., Tsemenidis, C.C., Sheriff, B.S.: Energy-aware clustering for wireless sensor networks using particle swarm optimization. In: The 18th Annual IEEE International Symposium on Personal, Indoor and Mobile Radio, Communications (PIMRC'07), 1–5 (2007)
17. Heinzelman, W.B., Chandrakasan, A.P., Balakrishnan, H.: An application-specific protocol architecture for wireless microsensor networks. IEEE Trans. Wirel. Commun. **14**, 660–670 (2002)
18. Yang, E., Erdogan, A.T., Arslan, T., Barton, N.: An improved particle swarm optimization algorithm for power-efficient wireless sensor networks. In: ECSIS Symposium on Bio-inspired, Learning, and Intelligent Systems for Security (BLISS), 76–82 (2007)
19. Cao, X., Zhang, H., Shi, J., Cui, G.: Cluster heads election analysis for multi-hop wireless sensor networks based on weighted graph and particle swarm optimization. In: Fourth International Conference on Natural Computation, 599–603 (2008)

Evaluation of Various Battery Models for Bellman Ford Ad hoc Routing Protocol in VANET Using Qualnet

Manish Sharma

Abstract The present era in communication is the era of VANET which simply represents the communication between fast moving vehicles. The VANET is expanding day by day from planes to higher altitude, from metropolitan to towns etc. At higher altitude the performance of VANET becomes prime area of consideration. Because power consumption is more due to average QoS (Quality of Service). Everyday days new functions like gaming, internet, audios, videos, credit card functions etc. are being introduced leading to fast CPU clock speed hence more battery consumption. Since energy conservation is main focus now days therefore in this paper we studied and compared various battery models for residual battery capacity taking Bellman Ford ad hoc routing protocol in to consideration along with various VANET parameters like nodes, speed, altitude, area etc. in real traffic scenario. The battery models Duracell AA(MX-1500), Duracell AAA(MN-2400), Duracell AAA(MX-2400), Duracell C-MN(MN-1400) are compared in hilly scenario using Qualnet as a simulation tool taking nearly equal real scenarios after a frequent study of that region. The performance of various battery models for residual battery capacity is compared so that right battery model shall be chosen in hilly VANET. Varying parameters of VANET shows that in the real traffic scenarios battery models AA (MX-1500) and C-MN (MN-1400) performs more efficiently for energy conservation.

Keywords VANET · Bellman ford · Duracell · Qualnet

1 Introduction

Vehicular Ad hoc Network (VANET) is a new communication mode by which fast moving vehicles communicate. The concept of VANETs is quite simple: by incorporating the wireless communication and data sharing capabilities, the vehicles can

M. Sharma (✉)
Department of Physics, Government College, Dhaliara, Himachal Pradesh 177103, India
e-mail: manikambli@rediffmail.com

B. V. Babu et al. (eds.), *Proceedings of the Second International Conference on Soft Computing for Problem Solving (SocProS 2012), December 28–30, 2012*, Advances in Intelligent Systems and Computing 236, DOI: 10.1007/978-81-322-1602-5_115, © Springer India 2014

be turned into a network providing similar services like the ones with which we are used to in our offices or homes. VANETs are direct offshoot of Mobile Ad hoc Networks (MANETs) [1] but with distinguishing characteristics like, movement at high speeds, constrained mobility, sufficient storage and processing power, unpredictable node density and difficult communication environment with short link lifetime etc. It has been found that at higher altitude the mobile terminals in fast moving vehicles like cars, buses and trains are frequent signal breakdowns as compared to pedestrians leading to more battery consumption. Since we can't change geographical and physical conditions hence in order to improve QoS and energy conservation in fast moving vehicles various battery models and light weight routing protocols needed to be studied. So that right selection of the battery model along with routing protocol can be made. Now days more and more functions like gaming, internet, audios, videos, credit card functions etc. are being introduced leading to fast CPU clock speed hence more battery consumption. Therefore performance of battery models counts a lot in any scenario. Moreover, transmission speed of cellular networks is increasing exponentially. It has increased from few kbps to 10 several mbps, a third generation system in last 10 years. Unfortunately, battery capacity has been tripled in the last 10 years [2, 3] throughout the world. As the dependence on mobile is increasing day by day therefore offsetting the energy conservations of the mobile terminals is the formidable challenge and hence need immediate attention.In this paper we studied and compared various battery models for Bellman Ford [4] ad hoc routing protocol taking in to consideration various VANET parameters like nodes, speed, altitude, area etc. in real traffic scenario. The battery models [5] are compared in hilly scenario using Qualnet as a Simulation tool [6] taking almost real scenarios after a frequent study of that region and Potentials for optimizing the deployed transport and routing Protocols is investigated. These battery models are different in impedance, weight, ANSI and IEC. Special care is taken in to provide realistic scenarios of both road traffic and network usage. A micro simulation environment for road traffic supplied vehicle movement information, which is then fed in to an event-driven network simulation that congaed and managed a VANET model based on this mobility data. The protocols and their various parameters of the transport, network, data link, and physical layers is provided by well-tested implementations for the networks simulation tool, while VANET mobility is performed by our own implementation.

2 Ad hoc Routing Protocols

Routing protocol [7] is a standard that controls how nodes decide how incoming packets are routed between devices in a wireless domain and further distinguished in many types. There are mainly three types of routing protocols: Proactive [7], Reactive [8] and Hybrid [9]. These protocols are having different criteria for designing and classifying routing protocols for wireless ad hoc network. The factors which must be taken under considerations while choosing a protocol are multicasting, loop free, multiple routes, unidirectional link support, power consumption etc. The selection of right protocol is the target for the fast moving vehicles so that QoS and power saving shall be improved [10, 11]. Routing protocols are always challenging in the fast moving nodes as their performance degrades and such type of network is difficult to manage as fast handoff, signal quality, Interference maximizes along with other geographical factors. In addition to mobility, the power availability to the mobile node is also a serious consideration. Unlike typical wired-link routers, the power source of mobile nodes comes from non-permanent power sources such as batteries. There are mainly three types of routing protocols [12]. In our work the chosen protocols is Bellman Ford ad hoc routing protocol which is proactive in nature and the simple reason behind chosing this protocol because it uses distance-vector algorithm.

2.1 Bellman Ford Routing Protocol

The term vector–distance, Ford–Fulkerson, Bellman–Ford, and Bellman are synonymous with distance-vector [4, 13], the last two are taken from the names of researchers who published the idea. The term distance-vector refers to a algorithms class where routers use to propagate routing information. The idea behind distance-vector algorithms is quite simple. The router keeps a list of all known routes in a table. When it boots, a router initializes its routing table to contain an entry for each directly connected network. Each entry in the table identifies a destination network and gives the distance to that network, usually measured in hops. For example, Fig. 1 shows an existing table in a router, K, and an update message from another router, J. Periodically, each router sends a copy of its routing table to any other router it can reach directly. When a report arrives at router K from router J, K examines the set of destinations reported and the distance to each. If J knows a shorter way to reach a destination, or if J lists a destination that K does not have in its table, or if K currently routes to a destination through J and J's distance to that destination changes, K replaces its table entry. For example, Fig. 1 shows an existing table in a router, K, and an update message from another router, J.

Note that if J reports distance N, an updated entry in K will have distance N+1 (the distance to reach the destination from J plus the distance to reach J). Of course, the routing table entries contain a third column that specifies a next hop. The next hop entry in each initial route is marked *direct delivery*. When router K adds or updates an entry in response to a message from router J, it assigns router J as the next hop for that entry.

3 Battery Models

Batteries are continued to power an increasing number of electronic systems, their life becomes a primary design consideration. By understanding both the source of energy

Destination	Distance	Route		Destination	Distance
Net1	0	Direct		Net1	2
Net2	0	Direct	→	Net4	3
Net4	8	Router L		Net17	6
Net17	5	Router M	→	Net21	4
Net24	6	Router J		Net24	5
Net30	2	Router Q		Net30	10
Net42	2	Router J	→	Net42	3

(a)   (b)

Fig. 1 **a** An existing route table for a router K, and **b** an incoming routing update message from router J. The marked entries will be used to update existing entries or add new entries to K's table

and the system that consumes it, the battery life can be maximized. There are basically two types of battery models alkaline or lechlanche cells. The zinc/potassium hydroxide/manganese dioxide cells, commonly called alkaline [5] or alkaline-manganese dioxide cells and the zinc–carbon cells are called as Lechlanche cells. The alkaline cells have a higher energy output than zinc–carbon (Lechlanche) cells. The use of an alkaline electrolyte, electrolytic ally prepared manganese dioxide, and a more reactive zinc powder contributes to a higher initial cost than zinc–carbon cells. However, due to the longer service life, the alkaline cell is actually more cost-effective based upon cost-per-hour usage, particularly with high drains and continuous discharge. The high-grade, energy-rich materials composing the anode and cathode, in conjunction with the more conductive alkaline electrolyte, produce more energy than could be stored in standard zinc carbon cell sizes. The product information and test data included in this section represent Duracell's newest alkaline battery products. Note that both these battery models have some common battery parameters Table 1.

Table 1 Parameters of Duracell AA (MX-1500), Duracell AAA (MX-2400), Duracell AAA (MN-2400), Duracell C-MN (MN-1400)

Parameters	AA (MX-1500)	AAA (MX-2400)	AAA (MN-2400)	C-MN (MN-1400)
Nominal voltage	1.5 V	1.5 V	1.5 V	1.5 V
Operating voltage	1.6–0.75 V	1.6–0.75 V	1.6–0.75 V	1.6–0.75 V
Impedance	81 m-ohm @ 1 kHz	114 m-ohm @ 1 kHz	114 m-ohm @ 1 kHz	136 m-ohm @ 1 kHz
Typical weight	24 g (0.8 oz.)	11 g (0.4 oz.)	11 g (0.4 oz.)	139 g (0.4 oz.)
Typical volume	$8.4\,cm^3 (0.5\,in^3)$	$3.5\,cm^3 (0.2\,in^3)$	$3.5\,cm^3 (0.2\,in^3)$	$3.5\,cm^3 (0.2\,in^3)$
Storage temperature	-20–$35\,°C$	-20–$35\,°C$	-20–$35\,°C$	-20–$35\,°C$
Operating temperature	-20–$54\,°C$	-20–$54\,°C$	-20–$54\,°C$	-20–$54\,°C$
Terminals	Flat	Flat	Flat	Flat
ANSI	15 A	24 A	24 A	13 A
IEC	LR06	LR03	LR03	LR20

4 Simulation Tool

Network simulation [1] methodology is often used to verify analytical models, generalize the measurement results, evaluate the performance of new protocols that are being developed, as well as to compare the existing protocols. Actually Qualnet [6] is a comprehensive suite of tools for modelling large wired and wireless networks. It uses simulation and emulation to predict the behaviour and performance of networks to improve their design, operation and management. Qualnet is a commercial simulator that grew out of GloMoSim, which was developed at the University of California, Los Angeles, UCLA, and is distributed by Scalable Network Technologies [6]. The QualNet simulator is C++ based. All protocols are implemented in a series of C++ files and are called by the simulation kernel. QualNet comes with a java based graphical user interface (GUI). Here while creating scenario the mobility model selected is Random Waypoint Mobility Model [14] in which a mobile node begins the simulation by waiting a specified pause-time Table 2.

Table 2 Simulation parameters

Parameters	Values
Simulator	Qualnet version 5.o.1
Terrain size	1500×1500
Simulation time	3000 s
Number of nodes	15
Mobility	Random way point, Pause time = 0 s
Speed of vehicles	Minimum = 5 m/s Maximum = 20 m/s
Routing protocols	Bellman Ford
Medium access protocol	802.11 MAC, 802.11 DCF
T× power	150 dbm
Data size	512 bytes
Data interval	250 ms
Number of sessions	5
Altitude	1500 m
Weather mobility	100 ms
Battery models	Duracell AA (MX-1500) Duracell AAA (MN-2400) Duracell AAA (MX-2400) Duracell C-MN (MN-1400)

5 Designing of Scenario

The scenario is designed in such a way that it undertakes the real traffic conditions as shown in Fig. 2. We have chosen 15 fast moving vehicles in the region of 1500×1500 with the random way point mobility model. There is also well defined path for some of the vehicles, so that real traffic conditions can also be taken care of. It also shows wireless node connectivity of few vehicles using CBR application. The area for simulation is hilly area with altitude of 1500 meters. Weather mobility intervals is 100 ms. Pathloss model is two ray with max prop distance of 100 m.The data size is 512 bytes and the data interval is 250 ms. The speed of the vehicles varies from 5 m/s to 20 m/s and for precision the numbers of sessions are five. Battery models are Duracell AA(MX-1500), Duracell AAA(MN-2400), Duracell AAA(MX-2400), and Duracell C-MN(MN-1400). The simulation is performed with different node mobility speed and CBR (Constant bit rate) traffic flow. CBR traffic flows with 512 bytes are applied. Simulations is made in different speed utilization with IEEE 802.11 Medium access control (MAC) and Distributed Coordination Function (DCF) ad hoc mode and the channel frequency is 2.4 GHz and the data rate 2 mbps.

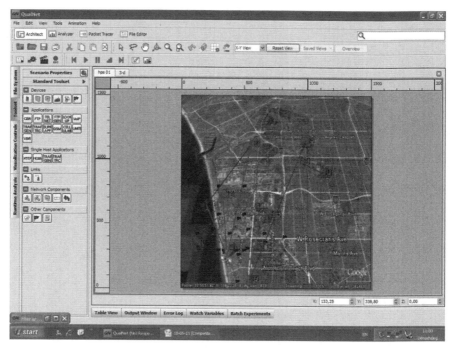

Fig. 2 Qualnet VANET scenario

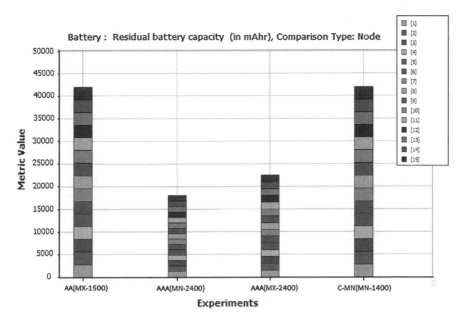

Fig. 3 Residual battery capacity for Bellman Ford routing protocol

6 Results and Discussions

The simulation result brings out some important characteristic differences between the various battery models as shown in Fig. 3. It has been found that the residual battery capacity for Duracell AA (MX-1500) and Duracell C-MN (MN-1400) is 4200 mA/h is quite high as compared to other two. This is because they offer low impedance of 81 m-ohm @ 1 kHz and 136 m-ohm @ 1 kHz as compared to 250 m-ohm @ 1 kHz for Duracell AAA (MN-2400). Moreover ANSI values are 15 A and 13 A as compared to 24 A for both Duracell AAA (MN-2400) and Duracell AAA (MX-2400). The IEC values are LR6 and LR20 for Duracell AA (MX-1500) and Duracell C-MN (MN-1400) as compared to LR03 for other two. The typical weight values are also meant for considerations in hilly areas.

7 Conclusion

Evaluation of the feasibility and the expected quality of VANETs operated as per various battery models, shows significant results. It has been found that the battery models depending upon their impedance, volume, weight, various temperatures operating range show significant variations in performance. Since energy conservations is most required now days. Hence its study leads to solutions to many problems.

Note that performance of these battery models also depends upon routing protocols and other parameters like, number of times link broke, signal transmitted and received, throughput, IEEE 802.11 (MAC), IEEE 802.11 (DCF), IP, FIFO, broadcast received, time difference (Real time and simulation time) and power saving etc. Result shows that for comfort and safety applications in a V2V communication environment the significant requirements are high and reliable quality of service packet. In general, position-based routing is more promising than other routing protocols for VANETs because of the geographical constrains. These routing protocols would improve the traffic control management and provide the information in timely manner to the concern authority and drivers. From the result and analysis, we now look into further improving the use battery models and ad hoc routing protocols in the VANET communication of the vehicular environment [14]. The scope of this research on VANET would be a very vast for ad hoc routing protocol that is used for comfort and safety applications.

References

1. Raw, R.S., Das, S.: Performance comparison of position based routing protocols in vehicle-to-vehicle (V2V) communication. Intl. J. Eng. Sci. Technol. 3(1), 435–444 (2011)
2. Toor, Y., Laouiti, A., Muhlethaler, P.: Vehicular ad hoc network applications and related technical issues. IEEE. Commun. Surv. Tutorials 10(3), 74–88 (2008)
3. Paradiso, J.A., Starner, T.: Energy scavenging for mobile and wireless electronics. IEEE. Pervasive. Comput. 4(1), 18–27 (2005)
4. Comer, D.,E.: Internetworking with TCP/IP: Principles,Protocols and Architecture, Vol. I, 4th edn. Prentice Hall, Upper Saddle River (2000)
5. http://www.duracell.com/
6. Qualnet simulator version 5.0.1. Scalable Network Technologies. http://www.scalable-networks.com/
7. Lasson, T., Hedman, N.: Routing Protocols in Wireless Ad Hoc Network. Lulea University of Technology, Stockholm (1998)
8. Sommer,C., Dressler, F.: The DYMO Routing Protocol in VANET Scenarios. University of Erlangen-Nuremberg, Germany (2009)
9. Mustafa, B., Waqas Raja, U.: Issues of routing in VANET. Thesis MCS-2010-20, School of Computing, Blekinge Institute of Technology (2010)
10. IEEE, wireless LAN medium access control (MAC) and physical layer (PHY) specifications. IEEE Std. 802.11-1997
11. Forouzan, B., A.: Data Communications and Networking. Networking Series, Tata Mcgraw-Hill, India (2005)
12. Perkins, C., Royer, E., Das, S., Marina, K.: Performance comparison of two on-demand routing protocols for ad hoc network. IEEE. Pers. Commun, 16–28 (2001)
13. Kozierok C.M.: The TCP/IP Guide Version 3, No Starch Press, San Francisco (2005)
14. Khan, I.: Performance evaluation of ad hoc routing protocols for vehicular ad hoc networks. Thesis, Mohammad Ali Jinnah University

A Comparative Analysis of Emotion Recognition from Stimulated EEG Signals

Garima Singh, Arindam Jati, Anwesha Khasnobish, Saugat Bhattacharyya, Amit Konar, D. N Tibarewala and R Janarthanan

Abstract This paper proposes a scheme to utilize the unaltered direct outcome of brain's activity viz. EEG signals for emotion detection that is a prerequisite for the development of an emotionally intelligent system. The aim of this work is to classify the emotional states experimentally elicited in different subjects, by extracting their features for the alpha, beta, and theta frequency bands of the acquired EEG data using PSD, EMD, wavelet transforms, statistical parameters, and Hjorth parameters and then classifying the same using LSVM, LDA, and kNN as classifiers for the purpose of categorizing the elicited emotions into the emotional states of neutral, happy, sad, and disgust. The experimental results being a comparative analysis of the different classifier performances equip us with the best accurate means of emotion recognition from the EEG signals. For all the eight subjects, neutral emotional state is classified with an average classification accuracy of 81.65 %, highest among the other three emotions. The negative emotions including sad and disgust have better average

G. Singh · A. Jati · S. Bhattacharyya · A. Konar · R. Janarthanan
Department of Electronics and Telecommunication Engineering, Jadavpur University,
Kolkata, India
e-mail: garima201290@gmail.com

A. Jati
e-mail: arindamjati@gmail.com

S. Bhattacharyya
e-mail: saugatbhattacharyya@gmail.com

A. Konar
e-mail: konaramit@yahoo.co.in

R. Janarthanan
e-mail: srmjana_73@yahoo.com

A. Khasnobish (✉) · D.N. Tibarewala
School of Bioscience and Engineering, Jadavpur University, Kolkata, India
e-mail: anweshakhasno@gmail.com

D.N. Tibarewala
e-mail: biomed.ju@gmail.com

B. V. Babu et al. (eds.), *Proceedings of the Second International Conference on Soft Computing*   1109
for Problem Solving (SocProS 2012), December 28–30, 2012, Advances in Intelligent Systems
and Computing 236, DOI: 10.1007/978-81-322-1602-5_116, © Springer India 2014

classification accuracy of 76.20 and 74.96 % as opposed to the positive emotion i.e., happy emotional state, the average classification accuracy of which turns out to be 73.42 %.

Keywords Emotion recognition · Electroencephalogram (EEG) · Power spectral density (PSD) · Wavelet transform (WT) · Empirical mode decomposition (EMD) · Statistical parameters (STAT) · Hjorth parameters · Linear discriminant analysis (LDA) · Linear support vector machine (LSVM) · K-nearest neighbor (kNN)

1 Introduction

In today's era, when no realm of science remains untouched by the miracles of artificial intelligence; 'Emotion,' no longer remains a terminology that solely defines human acts and performances, but is being incorporated to enhance the performance of emotionally challenged computers/machines as well [1, 2]. However, development of emotionally intelligent systems and devices, demand the recognition of the emotion as a vital step forward in this regard. The quality of performance of a system would improve with its ability to atone the negative emotional effects by the use of positive emotions accordingly. The act of emotion recognition may be achieved either through physiological signals or through external gestures and facial expressions of the individual. EEG signals, being the unmodified direct outcome of one's brain activity [3, 4] and independent of the hemodynamic of the brain, rather rely on the electrical potentials obtained from scalp due to various activities in brain and henceforth are not susceptible to voluntary suppression or modification. The procedure involves sending information regarding emotional changes, consciousness, and thinking to the frontal and prefrontal lobes of the cerebral cortex, which are then recorded by electrodes placed on the scalp above these regions. Raw EEG signals are needed to be processed and classified into different emotional categories, using different features and intelligent classifier algorithms.

Various strategies used for the purpose of feature extraction are power spectral densities, statistical parameters, wavelet transforms, Hjorth parameters, Fourier transforms, short time Fourier transforms (STFTs), empirical mode decompositions, higher order crossings, etc. [5–8] while the next step of emotion classification is done using Bayes' classifiers, support vector machines, fuzzy classifiers, genetic algorithms, K-means [9–14].

The aim of this research is to classify the emotional states experimentally elicited in the different subjects under consideration, by extracting the features from the alpha, beta, and theta frequency bands using power spectral density (PSD), empirical mode decomposition (EMD), Hjorth parameters, and wavelet transforms and then classifying the same using linear support vector machine (LSVM), linear discriminant analysis (LDA), and k-nearest neighbor (kNN) as classifiers. Thus categorizing the elicited emotions into the emotional states of neutral, happy, sad, and disgust are achieved in this work (Fig. 1).

Fig. 1 The Approach

The paper is divided into six sections. The experimental setup and data organization is explained in Sect. 2. The feature extraction techniques and classifiers are mentioned in Sects. 3 and 4, respectively. Performance analysis of the classifiers is given in Sect. 5. Experimental results and conclusions are listed in Sect. 6.

2 Experimental Data Description

It has been noted that among various modes to stimulate emotions in subjects under a controlled environment, audio-visual cues have the potential to elicit discrete emotions more effectively [15]. Keeping in mind these findings and the age group and cultural background of the subjects under study, after extensive search of movie clips, 20 movie clips are selected for eliciting happy, sad, and disgust emotions, 5 movie clips of each category. The movie clips were first shown to ten subjects, and based on their manual feedback which was recorded in the consent and feedback form, the final eight movie clips (two clips for each emotion) were selected for these discrete emotions. But these 10 subjects did not take part in the actual experiments.

The stimulus consisted of a blank screen for 10 seconds, followed by few soothing pictures of total 30 seconds duration. This was given to relax the subjects so that their neutral EEG baseline could be detected. This relaxing clip was followed by a movie clip first of the corresponding emotion, followed by the relaxing clip in the end (Fig. 2).

The experimental phase involved collection of EEG data from eight different subjects who were subjected to audio-visual stimulus relating to the four emotional states viz. neutral, happy, sad, and disgust. The subjects were given a brief introduction of the experimental procedure followed by filling up of consent form. The EEG signals were recorded using the NEUROWIN EEG amplifier of Nasan Medicals with a sampling rate of 250 Hz from electrodes positioned at F3, F4, Fp1 and Fp2, since they lie over the frontal and prefrontal lobe [16, 17]. Common reference montage was used for the recording purpose which represents the EEG signal as a difference of potentials of each electrode to one reference point that is placed on either ear lobe

To extract the EEG signals free from these artifacts, the raw signals were preprocessed in the MATLAB environment using an elliptical band pass filter of order 10 and bandwidth 4–32 Hz. The bandwidth of the filter is chosen as mentioned above since the required frequency bands are theta (4–7 Hz), alpha (8–12 Hz) and beta (13–32 Hz) that lie well within the chosen range.

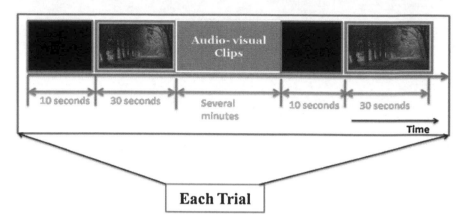

Fig. 2 Schematic representation of the stimulus

3 Feature Extraction

Five feature extraction techniques namely, power spectral density (PSD) [18, 19], Hjorth parameters, empirical mode decomposition (EMD), wavelet transform (WT) [20, 21], and the statistical parameters (STAT) [22, 23] of the wavelet coefficients have been employed to extract features from the preprocessed EEG signal.

4 Classification

Our ultimate goal is to classify the features extracted from the EEG signal related to various emotions into their respective classes with maximum attainable classification accuracy. For this job we have used three different classifiers (1) Linear Discriminant Analysis (LDA) [19, 22, 23], (2) Linear Support Vector Machine (LSVM) [22, 23] and (3) k-nearest neighbor (kNN) [22, 23].

5 Performance Analysis

Features were extracted from the preprocessed stimulated EEG signals that had been recorded from the subjects, using power spectral density (PSD), empirical mode decomposition (EMD), Hjorth parameters, statistical parameters, and wavelet coefficients. These features were then classified using LSVM (linear support vector machine), LDA (linear discriminant analysis), and kNN (k-nearest neighbor) classifies into the respective emotional states of neutral, happy, sad, and disgust. The overall classification results among all eight subjects for the four emotions using the

Table 1 Classification accuracies of emotion classification

Emotions	Classifiers	Subjects								Classifier mean	Emotion mean
		01	02	03	04	05	06	07	08		
Neutral	LDA	84.5	80.23	76.7	77.43	80.6	77.07	73.0	74.61	78.05	81.65
	LSVM	83.18	84.88	78.5	78.56	88.4	80.07	77.9	81.86	81.67	
	kNN	84.87	86.77	82.6	82.87	92.4	81.67	81.7	88.56	85.2	
Happy	LDA	69.6	67.15	71.1	66.43	75.9	65.59	62.7	71.13	68.71	73.42
	LSVM	75.59	71.77	76.4	71.32	80.5	74.17	70.1	78.86	74.85	
	kNN	75.29	72.03	73.4	72.08	91.6	73.2	70.0	85.77	76.69	
Sad	LDA	75.31	70.93	71.2	69.44	77.0	63.59	66.4	75.51	71.18	76.20
	LSVM	82.03	78.34	77.3	73.72	81.7	74.74	72.5	81.02	77.68	
	kNN	80.71	79.63	72.9	74.45	91.7	75.53	73.6	89.37	79.75	
Disgust	LDA	67.51	67.97	69.8	65.21	84.0	72.5	64.6	79.60	71.42	74.96
	LSVM	74.95	72.47	77.0	69.55	83.1	75.91	70.7	80.71	75.56	
	kNN	75.92	71.39	75.5	70.52	96.1	74.43	71.2	88.15	77.91	

three classifiers and all the five features are enlisted in Table 1. For all the eight subjects, neutral emotional state is classified with an average classification accuracy of 81.65 %, highest among the four emotions under consideration. The negative emotions including sad and disgust have better average classification accuracy of 76.20 and 74.96 % than the positive emotion i.e., happy emotional state whose average classification accuracy is 73.42 %, being the lowest among the four emotions considered in this work. It is seen that with increase in average classification accuracy, the variance decreased. kNN proved to be classifying the emotional states from EEG signal better that LSVM followed by LDA when all the five features are considered together. Subject 05 showed the highest CAs with all the four emotions in all the different classifiers used. kNN was found out to be the most suitable classifier for emotion classification with the four emotions being classified from the stimulated EEG signals with an average classification accuracy of 76.6 %. Conclusively, A combination of PSD, Statistical features and kNN classifier is found to perform most efficiently for emotion classification.

6 Conclusion

This work presented a comparative study of classifying the different emotional states from the stimulated EEG signals. Recognition of the emotions elicited in different subjects with the usage of audio-visual stimuli has been a prerequisite for achieving our final goal of developing a complete emotionally adaptive intelligent system. Four emotions (viz. neutral, happy, sad, and disgust) were classified suitably by three classifiers (viz. LDA, LSVM, and kNN) from five extracted features (viz. PSD, EMD, wavelet transforms, statistical parameters, and Hjorth parameters) of stimulated EEG signals. Performances of all the classifiers were compared and an overall classification accuracy of 76.6 % was found. PSD and statistical features were found to be better feature extractors among the other techniques used. kNN classified emotions with higher classification accuracy than the other classifiers when all the features are considered together.

With the utility of other modalities like speech, facial expressions etc., developing a multimodal emotion recognition system, that employs the emotional correlates of other physiological signals, research in the area can be extended further to develop an emotionally intelligent system lies.

References

1. Chakraborty, A., Konar, A.: Emotional intelligence: a cybernetic approach, studies in computational intelligence. 1st (edn.), Springer, Hiedelberg (2009)
2. Cornelius, R.R.: Theoretical approaches to emotion. In: Proceedings of the ISCA Workshop on Speech and Emotion, Belfast (2000)
3. Pantic, M., Rothkrantz, L.J.M.: Toward an affect-sensitive multimodal human-computer interaction. In: Proceedings of the IEEE Invited Speaker, vol. 91, no. 9, September (2003)
4. Chanel, G., Kronegg, J., Grandjean, D., Pun, T.: Emotion assessment: arousal evaluation using EEG's and peripheral physiologicalsignals. Lect. Notes Comput. Sci. vol. 4105, pp. 530 (2006)
5. Picard, R.W., Vyzas, E., Healey, J.: Toward machine emotional intelligence: analysis of affective physiological state. IEEE Trans. Pattern Anal. Mach. Intell. 23(10), 1175–1191 (2001)
6. Jung, T.: Removing electroencephalographic artifacts by blind source separation. J. Psychophysiol. 37, 163–178 (2000)
7. Gott, P.S., Hughes, E.C., Whipple, K.: Voluntary control of two lateralized conscious states: validation by electrical and behavioral studies. Neuropsychologia 22, 65–72 (1984)
8. Murugappan, M., Rizon, M., Nagarajan, R., Yaacob, S., Zunaidi, I., Hazry, D.: EEG feature extraction for classifying emotions using FCM and FKM. J. Comput. Commun. 1, 21–25 (2007)
9. Das S., Halder A., Bhowmik P., Chakraborty A., Konar A., Janarthan R.: A support vector machine classifier of emotion from voice and facial expression data. In: Proceedings of the IEEE 2009 World Congress on Nature and Biologically Inspired Computing (NaBIC 2009), Coimbatore, pp. 1010–1015 (2009)
10. Srinivasa, K.G., Venugopal, K.R., Patnaik, K.R.: Feature extraction using fuzzy C-means clustering for data mining systems. Int. J. Comput. Scie. Netw. Secur. 6, 230–236 (2006)
11. Michael, S., Chambers J.A.: Brain computer interfacing. In: Proceedings of the EEG Signal Processing, pp. 239–265, Wiley, NJ (2007)
12. Lotte, F. et al.: A review of classification algorithms for EEG-based Brain-computer interfaces. J. Neural. Eng. 4(2), (2007)
13. Rezaei, S., Tavakolian, K., Nasrabadi, A.M., Setarehdan, S.K.: Different classification techniques considering brain computer interface applications. J. Neural. Eng. 3(2), 139–144 (Jun 2006)
14. Xu, W., Guan, C., Siong, C.E., Ranganatha, S., Thulasidas, M., Wu, J.: High accuracy classification of EEG signal. In: Proceedings of the 17th International Conference on Pattern Recognition (ICPR), vol. 2, 2004, pp. 391–394. Cambridge (2004)
15. Ledoux, J.: Brain mechanisms of emotion and emotional learning. Curr. Opin. Neurobiol. 2, 191–197 (1992)
16. Davidson, R.I., Jackson, D.C., Kahn, N.H.: Emotion, plasticity, context, and regulation: perspectives from affective neuroscience. Psychol. Bull. 126(6), 89–909 (2000)

17. Niemic, C.P.: Studies of emotion: a theoretical and empirical review of psychophysiological studies of emotion. J. Undergrad. Res. pp. 15–18 (2002)
18. Sanei, S., Chambers J.A.: Brain computer interfacing. In: Proceedings of the EEG Signal Processing, pp. 239–265, Wiley, NJ (2007)
19. Stoica, P., Moses, R.: Introduction to spectral analysis. Prentice Hall, NJ, USA (1997)
20. Proakis, J.G., Malonakis, D.G.: Digital signal processing: principles. algorithm and applications, 3rd (edn.), Prentice Hall, NJ, USA (1996)
21. Oppenheim, A., Schafer, R.: Digital signal processing. Prentice Hall, NJ, USA (1975)
22. Alpaydin, E.: Introduction to achine earning. MIT Press, Cambridge (2004)
23. Webb, A.R.: Statistical pattern recognition, 2nd edn. Wiley, Reprint (2004)

Propagation Delay Analysis for Bundled Multi-Walled CNT in Global VLSI Interconnects

Pankaj Kumar Das, Manoj Kumar Majumder, B. K. Kaushik and S. K. Manhas

Abstract Multi-walled carbon nanotube (MWCNT) bundle potentially provided attractive solution in current nanoscale VLSI interconnects. This research paper introduces an equivalent single conductor (ESC) model for bundled MWCNT that contains a number of MWCNTs with different number of shells. A driver-interconnect-load (DIL) system employing CMOS driver is used to calculate the propagation delay. Using DIL system, propagation delay is compared for bundled CNT structures containing different number of MWCNTs. At global interconnect lengths, delay is significantly reduced for the bundled CNT containing more number of MWCNTs with lesser number of shells. It is observed that compared to the bundles containing lesser number of MWCNTs, the overall delay is improved by 9.89 % for the bundle that has more number of MWCNTs.

Keywords Carbon nanotube (CNT) · Multi-walled CNT (MWCNT) · Propagation delay · Interconnect · Nanotechnology · VLSI

P. K. Das (✉) · M. K. Majumder · B. K. Kaushik · S. K. Manhas
Microelectronics and VLSI Group, Department of Electronics and Computer Engineering,
Indian Institute of Technology Roorkee, Roorkee 247667, India
e-mail: pankaj_jkd@yahoo.co.in

M. K. Majumder
e-mail: manojbesu@gmail.com

B. K. Kaushik
e-mail: bkk23fec@iitr.ernet.in

S. K. Manhas
e-mail: samanfec@iitr.ernet.in

B. V. Babu et al. (eds.), *Proceedings of the Second International Conference on Soft Computing for Problem Solving (SocProS 2012), December 28–30, 2012*, Advances in Intelligent Systems and Computing 236, DOI: 10.1007/978-81-322-1602-5_117, © Springer India 2014

1 Introduction

In recent years, carbon nanotubes (CNT) have aroused a lot of research interest for their applicability as VLSI interconnects in future high speed electronics due to their extremely desirable electrical and thermal properties [1, 2]. CNTs have long mean free paths (MFPs) in the order of several micrometers that provide low resistivity and possible ballistic transport. CNTs are formally known as allotropes of carbon. They are made by rolling up graphene sheets in cylindrical form. Graphene is a monolayer sheet of graphite with sp^2 bonding of carbon atoms arranged in honeycomb lattice structure. The sp^2 bonding in graphene is stronger than sp^3 bonds in diamond [2], making graphene the strongest material. CNTs have unique atomic arrangement as well as interesting physical properties that include large current carrying capability, long ballistic transport length, high thermal conductivity, and mechanical strength [1–3]. These remarkable properties make CNTs one of the most promising research materials for the future nanoscale technology. These extraordinary physical properties of CNTs make them exciting prospects for a variety of applications in the areas of microelectronics/ nanoelectronics, spintronics, optics, material science, mechanical, and biological fields [3, 4]. Particularly, in the area of nanoelectronics, CNTs and graphene nanoribbons (GNRs) show a lot of research interest in their applicability as energy storage (such as supercapacitors) devices, energy conversion devices (including thermoelectric and photovoltaic devices), field emission displays and radiation sources, nanometer semiconductor transistors, nano-electromechanical systems (NEMS), electrostatic discharge (ESD) protection, interconnects, and passives [1–4]. The unique physical properties are mainly due to the structure of CNTs that depends on the rolling up direction of graphene sheets. Depending on the number of rolled up graphene sheets, CNTs can be categorized as single-walled CNTs (SWCNTs) and multi-walled CNTs (MWCNTs).

MWCNTs have similar current carrying capabilities compared to metallic SW-CNTs but are easier to fabricate than SWCNTs due to easier control of the growth process. In fact most of the CNT interconnect fabrication efforts have targeted multi-walled CNTs (MWCNTs) [5]. The typical diameter of SWCNT is about 1 nm and depending on its chirality, it can be either metallic or semiconducting [5, 6] in nature. MWCNT consists of several coaxial CNT shells and each shell can have different chirality depending on their rolled up direction. MWCNTs may have diameters in the range of a few to hundreds of nanometers. Initially, most experiments have indicated that only the outer shell in an MWCNT conducts. Recently, it has been confirmed that all shells can conduct if they are properly connected to contacts [6, 7]. Earlier experiments made contacts to outer shells only and the impact of inner shells is very small on overall conduction. This research paper analyzes the propagation delay for bundled MWCNT structures containing different number of shells. An equivalent single conductor (ESC) model is introduced to estimate the delay using driver-interconnect-load (DIL) system [7].

The paper is organized into four different sections. Section 1 provides details of the current research scenario and briefs about the works carried out. Section 2 describes the geometry and ESC model of bundled MWCNT structure. Details of simulation setup and comparison of delay for different bundled MWCNT structures is discussed in Sect. 3. Finally, Sect. 4 concludes the paper.

2 Geometry and ESC Model of Bundled MWCNT

Consider the geometry of single and bundled MWCNT over ground plane as shown in Fig. 1. The diameters of MWCNT are considered as $D_1, D_2, D_3, \ldots\ldots, D_n$, where D_1 and D_n are the innershell and outershell diameters respectively. l is the nanotube length, S_i is intershell spacing, and H represents the distance between center of MWCNT and ground plane. It is assumed that the spacing between each shell is fixed, while diameter of outermost nanotube can change over a fixed interval. For current fabrication technology, intershell spacing S_i can be formulated as [6–8]

$$S_i = \frac{D_n - D_{n-1}}{2} \approx 0.34 \, \text{nm} \qquad (1)$$

Bundled MWCNT with width w and height h contains a number of MWCNTs with spacing of S_p as shown in Fig. 1b. The total number of metallic MWCNTs in bundle can be formulated as

$$N_{MWCNT} = \beta \left[N_h N_w - (N_h/2) \right] \qquad (2)$$

where N_w and N_h are the number of MWCNTs in bundle at horizontal and vertical directions respectively. The center to center distance between adjacent MWCNTs in bundle can be expressed using the cross-section of bundle (Fig. 1b)

$$S_p = S_i + D_n \qquad (3)$$

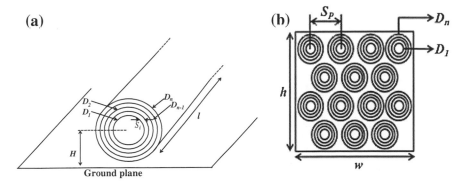

Fig. 1 Geometry of **a** single and **b** bundled MWCNT

where $S_i = 0.34$ nm [7] is the spacing between MWCNTs in bundle that depends on the Vander Waals gap [8, 9]. Based on the geometry, ESC model of bundled MWCNT is presented in Fig. 2b that is simply developed on the basis of multi-conductor transmission line as shown in Fig. 2a.

For the ESC model, it is assumed that all the shells of MWCNTs are properly contacted. Assuming the metallic to semiconducting tube ratio as β, the approximate number of conducting channels per shell is given by [10]

$$N_{channel/shell}(d) \approx (ad + b)\beta, d > 4\,\text{nm}$$

$$\approx 2\beta, d < 4\,\text{nm} \qquad (4)$$

where $a = 0.1836\,\text{nm}^{-1}$ and $b = 1.275$ [10]. Typical value of β can be defined as 1/3 for an MWCNT bundle [10]. The number of conducting channels per shell in any given shell j varies with respect to the shell diameter.

The ESC model of Fig. 2 exhibits $p.u.l.$ resistance, capacitance, and inductances that can be expressed as [8–10]

$$\hat{R}' = \left[\sum_{j=1}^{s} (2n_j/R') \right]^{-1} = R'/2n_{tot} \qquad (5)$$

$$\hat{L}'_e = L'^{s,\,s}_e \text{ and } \hat{L}'_k = \left[\sum_{j=1}^{s} \left(1/L'^{j,j}_k\right) \right]^{-1} = L'_k/2n_{tot} \qquad (6)$$

$$C'_e = C'^{s,\,s}_e \text{ and } \hat{C}'_q \approx 2n_{tot}C'_q \qquad (7)$$

Fig. 2 a Basic transmission line and **b** ESC model of MWCNT bundle intercon-nects

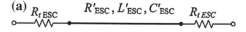

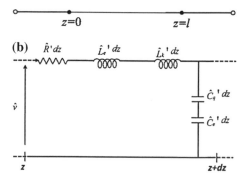

where n_{tot} is the total number of conducting channels in MWCNT that can be presented as

$$n_{tot} = n_1 + n_2 + \ldots + n_j + \ldots + n_s \qquad (8)$$

Using the equations from (5–8), it is observed that the interconnect parasitics
such as resistance, inductance and capacitance primarily depends on the number of
MWCNTs in bundle and can be expressed as [8, 11]

$$L'_k = h/2e^2 v_F, \, C'_q = 2e^2/hv_F, \, R' = h/2e^2 v_F \tau \qquad (9)$$

where $v_F = 8 \times 10^5$ m/s is the Fermi velocity and τ is the diffusion time of CNT.
Apart from this, the ESC model considers terminal contact resistance R_{cj} of the
jth shell of MWCNT [8]. This resistance accounts the effect of voltage drop at the
metal-nanotube interface.

3 Performance Analysis

Propagation delay is compared using DIL system as shown in Fig. 3. A CMOS driver
with supply voltage $V_{dd} = 1$V is used for accurate estimation of delay. The ESC
model of bundled MWCNT represents interconnects line in the bus architecture.
The interconnect line ranging from length 100 to 1, 000 μm is terminated by a load
capacitance C_L of 10 aF [7].

Using the above-mentioned setup, propagation delay is compared for different
bundled MWCNT structures. Figure 4 presents the variation of delay at different
interconnect lengths. It is observed that the delay increases with interconnect lengths,
whereas it reduces for the bundle containing more number of MWCNTs. The reason
is that the delay primarily depends on interconnect parasitic such as resistance and
capacitance that reduces for higher number of MWCNTs in bundle. With smallest
value of parasitic associated, the bundle containing MWCNTs of four shells results
in smaller delay as compared to other bundled MWCNTs that are presented in Fig.
5. Finally, Table 1 summarizes the percentage improvement in delay for the bundle
containing MWCNTs of four shells that shows an overall improvement of 7.93 %
compared to other bundles.

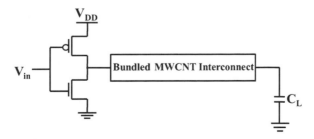

Fig. 3 Simulation setup for bundled MWCNT interconnects using DIL system

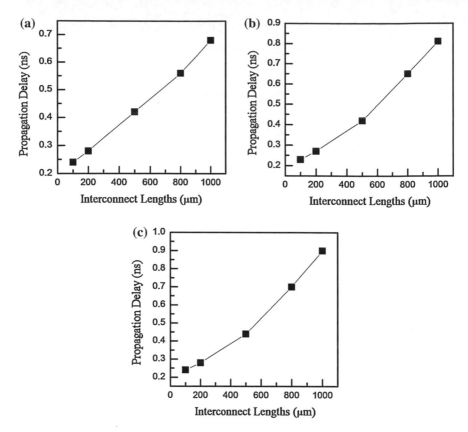

Fig. 4 Propagation delays for bundled CNTs containing **a** 4, **b** 6, **c** 8-shell MWCNTs

Fig. 5 Propagation delay
with varying interconnect
lengths for different bundled
MWCNTs

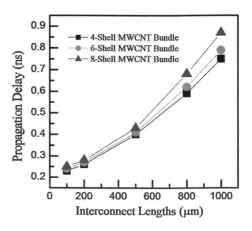

Table 1 Percentage improvement in delay *w.r.t.* 8-shell bundled MWCNT

Interconnect lengths (μm)	Bundle with MWCNT shell = 4	Bundle with MWCNT shell = 6
100	5.00	4.00
200	6.98	5.57
500	8.14	7.65
800	13.23	8.82
1,000	15.79	9.19

4 Conclusion

This research paper presents a comparative analysis between bundled CNT containing MWCNTs with different number of shells. Based on the number of shells and conducting channels, a generalized ESC model has been developed that primarily follows basic transmission line theory. Using the ESC model, propagation delay has been evaluated with the help of DIL system at different global interconnect lengths. It has been observed that the bundle containing more number of MWCNTs exhibits significant improvement in delay compared to other bundle structures. Therefore, from fabrication point of view, it is desirable to use a bundle containing lesser number of shells for future global VLSI interconnects.

5 Biography

Pankaj Kumar Das received his B. Tech degree in Electronics and Communication Engineering from SLIET, Longowal in the year of 2006. Currently, he is

working towards his M.Tech in Microelectronics and VLSI group at Indian Institute of Technology, Roorkee.

Since 2006, he has been served as Assistant Professor at Department of Electronics and Communication Engineering, Sant Longowal Institute of Engineering and Technology, Longowal. He has received best faculty and research award during his services. He has published more than five papers in national and international conferences. His current research area includes microelectronics-based VLSI circuits, circuit modeling, and analysis of carbon naotubes and nanowires.

Manoj Kumar Majumder (SM'11) received his B. Tech and M. Tech in the year of 2007 and 2009, respectively. Currently, he is working towards his Ph.D. in Microelectronics and VLSI group at Indian Institute of Technology, Roorkee.

From 2009 to 2010, he was associated with academic activities in electronics department in Durgapur Institute of Advanced Technology and Management (DIATM), West Bengal, India. His current research interest is Carbon Nanotube-based VLSI interconnects and circuit modeling. He has published more than 35 research papers in various international journals and conferences.

Mr. Majumder has obtained Graduate Aptitude Test Engineering (GATE) fellowship in 2007 and MHRD fellowship in 2010.

Brajesh Kumar Kaushik (M'10) received his Ph.D. in 2007 from Indian Institute of Technology, Roorkee.

In 1998, he joined as a lecturer at G.B.Pant Engineering College Pauri Garhwal, where he served as Assistant Professor till 2009. Since 2009, he has been working as Assistant Professor in VLSI group at IIT Roorkee. His current research area includes Electronic simulation, Low power VLSI Design. He has published more than 150 research papers in various international journals and conferences.

Dr. Kaushik has received many awards and recognitions from IBC, Cambridge such as top 100 scientists in World-2008 and International Educator of 2008.

Sanjeev Kumar Manhas (M'00) received his Ph.D. in 2003 from De Montfort University, Leicester UK.

In August, 2007 he joined the Indian Institute of Technology, Roorkee where he has served as Assistant Professor till now. His current research area includes silicon nanowire-based circuit design, parasitic evaluation and fabrication technologies, MOSFET modeling and reliability, DRAM leakage mechanism and refresh, VLSI technologies and OTFTs. He has both academic and industrial experience.

Dr. Manhas has received many academic awards such as Microelectronics best paper award in 2002, graduate merit award in 1991, and Indian National Research Scholarship award in 1993.

References

1. Li, H., Xu, C., Srivastava, N., Banerjee, K.: Carbon nanomaterials for next-generation interconnects and passives: physics, status and prospects. IEEE Trans. Electron. Devices **56**(9), 1799–1821 (2009)
2. Javey, A., Kong, J.: Carbon Nanotube Electronics. Springer, New York (2009)
3. Majumder, M.K., Pandya, N.D., Kaushik, B.K., Manhas, S., K.: Analysis of crosstalk delay and area for MWNT and bundled SWNT for global VLSI interconnects. In: Proceedings of the 13th IEEE International Symposium on Quality Electronic Design (ISQED 2012), pp. 291–294, Santa Clara (2012)
4. Avorious, P., Chen, Z., Perebeions, V.: Carbon-based electronics. Nat. Nanotechnol. **2**(10), 605–613 (2007)

5. Yu, M.F., Lourie, O., Dyer, M.J., Moloni, K., Kelly, T.F., Ruoff, R.S.: Strength and breaking mechanism of multiwalled carbon nanotubes under tensile load. Science **287**(5453), 637–640 (2000)
6. Naeemi, A., Meindl, J.D.: Compact physical models for multiwall carbon nanotube interconnects. IEEE Electron. Device Lett. **27**(5), 338–340 (2006)
7. Majumder, M.K., Pandya, N.D., Kaushik, B.K., Manhas, S.K.: Analysis of MWCNT and bundled SWCNT interconnects: impact on crosstalk and area. IEEE Electron. Device Lett. **33**(8), 1180–1182 (2012)
8. Sarto, M.S., Tamburrano, A.: Single-conductor transmission-line model of multiwall carbon nanotubes. IEEE Trans. Nanotechnol. **9**(1), 82–92 (2010)
9. Amore, M.D., Sarto, M.S., Tamburrano, A.: Fast transient analysis of next-generation interconnects based on carbon nanotubes. IEEE Trans. Electromagn. Compat. **52**(2), 496–503 (2010)
10. Subash, S., Kolar, J., Chowdhury, M.H.: A new spatially rearranged bundle of mixed carbon nanotube as VLSI interconnection. IEEE Trans. Nanotechnolo. **7**, 1–10 (2011)
11. Burke, P.J.: Luttinger liquid theory as a model of the gigahertz electrical properties of carbon nanotubes. IEEE Trans. Nanotechnol. **1**(3), 129–144 (2002)

Improvement in Radiation Parameters Using Single Slot Circular Microstrip Patch Antenna

Monika Kiroriwal and Sanyog Rawat

Abstract This paper investigates a new geometry of circular microstrip patch antenna using rectangular slot which can be used for WLAN and Wi-Max application. This geometry obtained bandwidth enhancement upto 9.58 % in comparison with conventional design and there is also improvement in other radiation parameters like gain, efficiency and return loss. For entire bandwidth the radiation pattern is stable and uniform.

Keywords Bandwidth · Circular microstrip patch antenna · Return loss · Coaxial probe feed · Radiation pattern

1 Introduction

Microstrip antennas are the most rapidly developing field in the last 20 years [1]. At present use of microstrip antennas are increasing day by day in wireless communication due to their low profile structure and also other advantage of low cost, light weight, ease of fabrication [2], conformable to mounting surface and being integrated in active devices [3]. Microstrip patch antenna may be square, rectangular, circular, elliptical, triangular and other desired configuration [4]. Coaxial probe fed microstrip antennas provide excellent isolation between the feed network and the radiating elements and yield very good front to back ratios [5]. Among the conventional microstrip antenna geometries, circular microstrip antenna is most widely analyzed antenna due to easy modeling and applicable boundary conditions [6]. The major drawback of circular microstrip patch antenna is narrow bandwidth and low gain especially at lower microwave frequencies [7]. Drawbacks which are responsible for the limited application in consumer world and it was realized that after doing

M. Kiroriwal (✉) · S. Rawat
ASET, Amity University Rajasthan, Rajasthan, India
e-mail: m.kiroriwal@rediffmail.com

B. V. Babu et al. (eds.), *Proceedings of the Second International Conference on Soft Computing for Problem Solving (SocProS 2012), December 28–30, 2012*, Advances in Intelligent Systems and Computing 236, DOI: 10.1007/978-81-322-1602-5_118, © Springer India 2014

some changes in the geometry of the conventional patch, these limitations may be removed [8]. To overcome this limitation, one of the methods is to cut slots in various shapes. For example in circular patch antenna bandwidth was increased up to 4.13 % of a skimmer shaped circular microstrip patch antenna [9] and in another example bandwidth is improved by 2.3 times of the conventional antenna bandwidth (1.4 %) using slots in circular microstrip patch antenna [10].

In this paper geometry is simulated using electromagnetic simulator, Zealand IE3D software [11]. This software works on method of moment. This circular microstrip patch antenna has good thermal, emissivity and mechanical properties: light weight, low profile, robust and it is easily constructed and it comes under the second band (5.15–5.825GHz) of IEEE802.11 WLAN and high band (5.25–5.85 GHz) of Wi-Max application. The Sect. 2 comprises of antenna geometry and in the Sect. 3 of this paper simulated and measured results are discussed followed by conclusion in the Sect. 4.

2 Antenna Configuration

Here conventional circular patch microstrip antenna is considered the reference antenna to compare the results of that simulated from single slot circular patch antenna.

2.1 Antenna Design

The geometry of the proposed antenna structure is shown in Fig. 1a. The patch has the diameter of 14 mm. A 50 Ω coaxial probe is used to connect the microstrip patch at coordinates and it is made fixed for both the conventional and the single slot circular microstrip patch antenna. The dimension of single slot is 0, −4 mm with 8 mm length and 2.75 mm width. Impedance bandwidth of about 9.38 % can be obtained from the above geometry.

2.2 Antenna Prototype

For validation of the proposed antenna geometry is fabricated as shown in Fig. 1b, on a supporting substrate FR4 with dielectric constant, $\varepsilon_r = 4.4$ and the thickness of the substrate, h = 1.59 mm. Many simulations are done for optimizing the length; width and location of the slot and best result are obtained with defined length and width of the slot. Due to existence of the slot, the current distribution changed and another mode is excited, each mode has its own cut off frequency, so modified geometry has a new resonant frequency, which is less than the conventional patch resonant frequency.

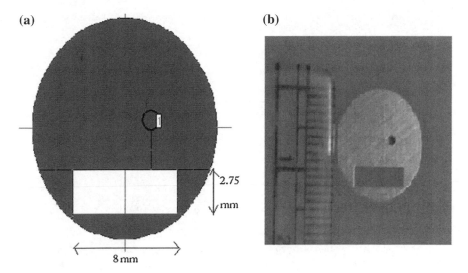

Fig. 1 **a** Proposed geometry of *circular* microstrip patch antenna. **b** Fabricated antenna prototype

3 Simulated and Measured Results

3.1 Return Loss and Bandwidth

Return Loss is a measure of how much power is delivered from the source to a load and measured by S_{11} parameters. Bandwidth is the range of frequencies over which the antenna can operate effectively. Bandwidth can be calculated by going 10 dB down in return loss.

As shown in Fig. 2 simulated return loss of the single slot circular microstrip patch antenna is −38.3067 dBi at resonating frequency 5.32 GHz and from the return loss curve the simulated bandwidth obtained is 9.38 % and measured bandwidth is 9.58 % with −35 dBi return loss.

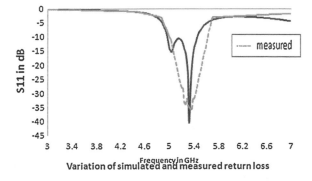

Fig. 2 Measured and simulated return loss for proposed *circular* x microstrip patch antenna

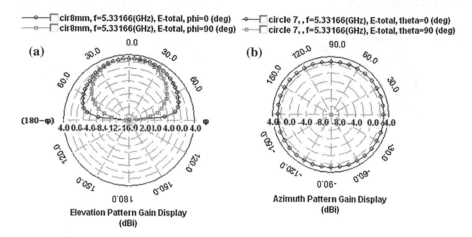

Fig. 3 **a** Computed elevation radiation pattern. **b** Azimuth radiation pattern for proposed *circular* microstrip patch antenna

3.2 Radiation Pattern

A plot through which it is visualizes where the antenna transmits or receives power. The microstrip antenna radiates normal to its patch surface. So, the elevation pattern for $\phi = 0$ and $\phi = 90$ degrees are important for the simulation. The simulated elevation and azimuth radiation pattern of proposed antenna is illustrated in Fig. 3a and b. Elevation radiation pattern is smooth and uniform over the band of frequencies and azimuth radiation pattern is omni directional.

3.3 Smith Chart

Smith Chart provides the information about polarization and the impedance match of the radiating patch. Figure 4 shows the smith chart for proposed antenna and the input impedance of 49.68 $\Omega - j2.129$ at resonant frequency 5.32 GHz. This smith chart shows that the antenna is circularly polarized with some impurity.

3.4 Gain

Gain is basically measure of the effectiveness of a directional antenna as compared to a standard non-directional antenna. The gain observed for the proposed circular patch antenna is shown in Fig. 5. The gain is found to be uniform over the whole the frequency band Tables 1 & 2.

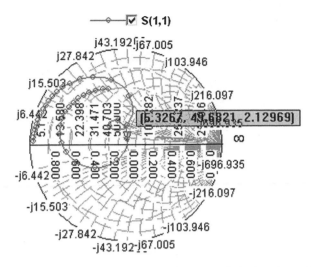

Fig. 4 Variation of input impedance with frequency for proposed circular microstrip patch antenna

Fig. 5 Computed variation of gain with frequency for proposed circular microstrip patch antenna

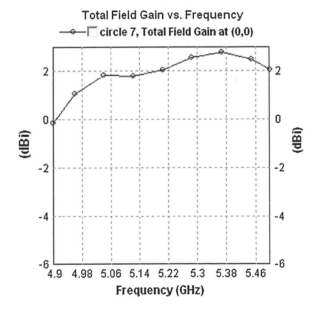

Table 1 Variation in antenna parameters with slot area

Sr. no.	Slot area (mm)^2	Resonant frequency (GHz)	Return loss (dBi)	Gain (dBi)	Band-width (%)	Efficiency (%)
1.	20	5.45	−32.01	2.85	7.9	42.26
2.	20.3	5.30	−31.61	2.67	8.1	40.66
3.	20.54	5.35	−35.77	2.81	9.1	41.91
4.	22	5.32	−38.30	2.75	9.38	41.45
5.	24	5.31	−35.98	2.73	8.8	41.21

Table 2 Comparison of conventional and proposed antenna

Sr.no.	Patch antenna	Resonant frequency (GHz)	Return loss (dBi)	Gain (dBi)	Band-width (%)	Efficiency (%)
1.	Conventional patch antenna	5.48	-27.43	1.28	2.53	29.99
2.	Proposed antenna	5.32	-35	2.75	9.58	41.45

4 Conclusion

In this paper, the radiation performance of single slot circular microstrip patch antenna is simulated by applying IE3D full-wave electromagnetic simulator and results are compared with conventional circular patch antenna. Using optimization it is observed that as we increase the slot area there is a decrement in the resonant frequency. Measured results indicate that the antenna exhibits bandwidth up to 9.56 % by optimizing the length and width of the slot in antenna geometry. There is also improvement in radiation characteristics like gain 2.75 dBi and efficiency 41.45 %. In radiation pattern, the direction of maximum radiation is normal to the patch geometry and also found to be stable over the entire bandwidth.

References

1. El Aziz D.A., Hamad, R.: Wideband circular microstrip antenna for wireless communication Systems. 24th National Radio Science Conference (NRSC), Egypt, pp. 1–8 (2007)
2. sfahlani, S.H.S., Tavakoli, A., Dehkhoda, P.: A compact single layer dual-band microstrip antenna for satellite application. IEEE Antennas Wirel. Propag. Lett. **10**, 931–934 Piscataway, USA (2011)
3. Constantine A. B.: Antenna Theory: Analysis and Design. 3rd edn, Wiley, (ed.) New York May (2005)
4. Garg, R., Bhartia, P., Bhal, I.J., Ittipiboon, A.: Microstrip antenna design book. Artech House, New York (2001)
5. Pozar, D.M.: Microstrip antennas. In: IEEE Proceedings **80**, 79–91 (1992)

6. Surmeli, K., Turetken, B.: U-Slot stacked patch antenna using high and low dielectric constant material combination in S- Band. General Assembly and Scientific Symposium, URSI, Hiroshima pp. 1–4 (2011)
7. Singh, J., Singh, A.P., Kamal, T.S.: Design of circular microstrip antenna using artificial neural networks. In: Proceedings of the World Congress on Engineering, Vol 2, U.K. (2011)
8. Wong, K.L.: Compact and broadband microstrip antenna. John Wiley & Sons, New York (2002)
9. Ryu, M.R., Woo, J. M., Hu, J.: Skimmer shaped linear polarized microstrip antennas for miniaturization. International Conference on Advance Communication Technology, Vol. (1) pp. 758 –762 (2006)
10. Wu, J. W., Jui H. L.: Slotted circular microstrip patch antenna for bandwidth enhancement. IEEE Proceeding Microwave Antennas Propagation, Vol. 2, pp. 272–275 (2003)
11. Huang, C. Y., Wu, J. Y., Wong, K.: LIE3D Software, Release 14.65 (Zeland Software Inc. Freemont, USA), April (2010)

A Π- Slot Microstrip Antenna with Band Rejection Characteristics for Ultra Wideband Applications

Mukesh Arora, Abha Sharma and Kanad Ray

Abstract Federal Communications Commission (FCC) revealed that a bandwidth of 7.5 GHz from 3.1 to 10.6 GHz is for Ultra Wideband (UWB) wireless communication. UWB is a rapidly advancing technology for high data rate wireless communication. The main challenge in UWB antenna design is achieving the very broad impedance bandwidth with compactness in size. The proposed antenna has the capability of operating between 1.1 and 11.8 GHz. In this paper, a rectangular patch antenna is designed with truncated corners at the ground as well as at the patch. A $\prod$-shaped slot is cut out from the patch to get the complete UWB. After that two equal size of slits on sides of the $\prod$ shape are cut out from the patch which dispends the WLAN. The proposed antenna uses Rogers RT/duroid substrate with a thickness of 1.6 mm and relative permittivity of 2.2. The aperture coupled feed is used for excitation. The proposed antenna is simulated using HFSS 11 software.

Keywords Microstrip patch antenna · Ultra wide band · Dispend band · Aperture coupled feed · I slot · WLAN

1 Introduction

The UWB technology is one of the most demanding technologies which provides high data rate transmission. The FCC approved rules for the commercial use of UWB in 2002. UWB uses the unlicensed spectrum from 3.1 to 10.6 GHz allocated

M. Arora
Department of Electronics and Communication, SKIT, Jaipur, Rajasthan, India

A. Sharma
Department of Electronic and Communication, GCT, Jaipur, Rajasthan, India

K. Ray (✉)
Institute of Engineering and Technology, JKLakshmipat University, Jaipur, Rajasthan, India
e-mail: kanadray@jklu.edu.in

B. V. Babu et al. (eds.), *Proceedings of the Second International Conference on Soft Computing for Problem Solving (SocProS 2012), December 28–30, 2012*, Advances in Intelligent Systems and Computing 236, DOI: 10.1007/978-81-322-1602-5_119, © Springer India 2014

by FCC [1, 2]. The FCC allocated a bandwidth of 7.5 GHz from 3.1 to 10.6 GHz for UWB applications. Low-cost UWB antennas are required for various applications such as wireless communications, medical imaging, parking radars, and other military applications and indoor positioning. It has many advantages over conventional wireless communication technology such as low power consumption, high speed transmission, and large impedance bandwidth. But within this range some other licensed narrowband systems exist for which the UWB causes the potential interference. Systems like WLAN for IEEE 802.11/a/b/g operate in 5.15–5.35 GHz and 5.725–5.825 GHz. This band interferes with the UWB system, so it is desirable to remove this band from ultra wideband [3]. In order to improve coexistence of UWB with other wireless standards, research has been done to devise mechanisms to reject certain bands within the pass band of the UWB [4].

One common and simple way to achieve this function is cutting a slot on the patch [5]. This slot adds an extra path to the antenna's current, eventually introducing, an additional resonance circuit. As the resonance frequency of the slot depends on its position and dimensions one can optimize them so that the rejected band can be adjusted to cover the WLAN band. But often it is difficult to achieve sharp and narrow stop band while the interference signals have narrow bandwidth usually.

According to Shannon-Nyquist criterion,

$$C = \mathcal{CC} \; B \; \log_2 \left[1 + \text{SNR} \right] \text{bits/sec.}$$

where C is the maximum transmission data rate, B denotes bandwidth of the channel, and SNR stands for the Signal-to-Noise ratio. The maximum achievable data rate or capacity for the ideal band limited additive white Gaussian noise (AWGN) channel is related to the bandwidth and the signal-to-noise ratio [6]. From above principle, the transmit data rate can be increased either by enhancing the bandwidth or transmission power. Since many portable devices are battery operated, the power cannot be easily increased, so a large frequency bandwidth is a better solution to achieve a high data rate.

Aperture coupled microstrip antenna couples patch to the feed line through a slot. In this technique the radiating microstrip patch element is etched on the top of the antenna substrate, and the microstrip feed line is etched on the bottom of the feed substrate [7] (Fig. 1).

In this paper, a compact microstrip patch antenna with band-rejection characteristics for UWB application is proposed. First, the ground as well as patch is truncated at all its four corners, which reduced the size of the antenna. Second, a ∏-shaped slot is cut out from the patch which gives the continuous frequency band from 1.12 to 10.86 GHz covering the entire ultra wideband. To improve upon the sharpness of stop band we explore the combination of slots with the insertion of slits on the patch (another conventional method). After that two equal-sized rectangular slits are cut out on the sides of the ∏ shape in the patch which notches the band from 5.07 to 5.84 GHz covering the WLAN band. By removing this band from UWB, a major problem of interference associated with UWB band is resolved. The performance characteristics of the proposed antenna are simulated using HFSS 11 software.

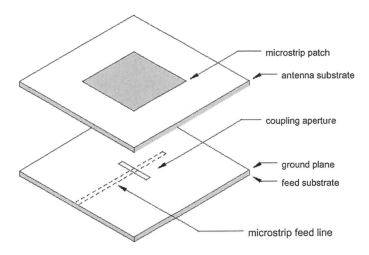

Fig. 1 Microstrip antenna with aperture coupled feed [8]

Table 1 Antenna dimension of Fig. 2

Length of trunk corner in ground (L1)	5.3 mm
Width of trunk corner in ground (W1)	5.7 mm
Length of trunk corner in ground (L2)	6.65 mm
Width of trunk corner in ground (W2)	7.4 mm
Length of trunk corner in patch (L3)	3 mm
Width of trunk corner in patch (W3)	3 mm
Length of trunk corner in patch (L4)	3 mm
Width of trunk corner in patch (W4)	4 mm

2 Antenna Geometry and Design

In this design, a Substrate of Rogers RT/duroid of relative permittivity 2.2 is used to design the antenna. The substrate dimension is 22.8 × 26.6 × 1.6 mm. A rectangular patch is used with dimensions of 13.2 × 17 mm. The ground and patch is truncated at all its four corners. The aperture coupled feed is used for excitation. The geometry of proposed antenna is discussed in Table 1. The designed patch antenna is shown in Fig. 2.

3 Measured Results

The measured return loss is shown in Fig. 3 while the VSWR plot and radiation pattern are shown in Figs. 4 and 5. The measured return loss plot shows that there one resonant frequency band from 1.08 to 8.42 GHz with return loss of −43.72 dB. The measured VSWR is also less than 2.

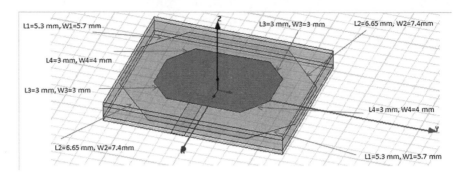

Fig. 2 Geometry of the antenna

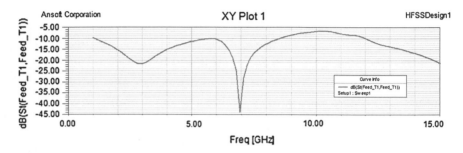

Fig. 3 Simulated return loss S_{11} V/s frequency

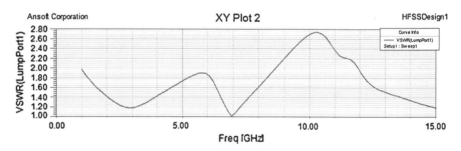

Fig. 4 The measured VSWR

4 UWB Antenna Geometry and Design

In this design, a $\prod$-shaped slot is cut out from the truncated patch. This $\prod$-shaped slot gives frequency band from 1.12 to 10.86 GHz which covers the entire UWB band. The dimensions of substrate and patch are the same as that used in the previous design. The geometry is shown in Fig. 6. The dimensions of I slot are shown in Table 2.

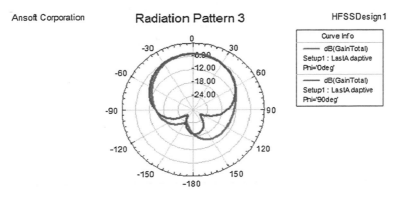

Fig. 5 Radiation pattern

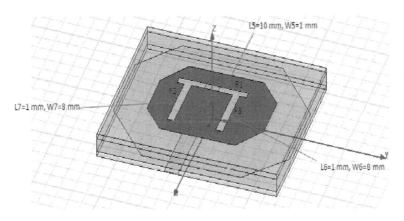

Fig. 6 Geometry of the antenna

Table 2 Antenna dimension of Fig. 6

Length of R1	10 mm
Width of R1	1 mm
Length of R2	1 mm
Width of R2	8 mm
Length of R3	1 mm
Width of R3	8 mm

5 Measured Results

The return loss plot gives the impedance bandwidth of 162.57 % from 1.12 to 10.86 GHz which is less than -10 dB as shown in Figs. 7, 8 and 9 show the VSWR plot and radiation pattern for the proposed antenna.

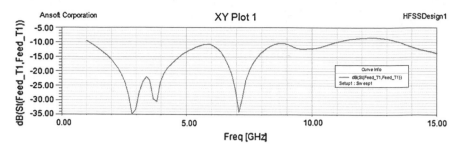

Fig. 7 Simulated return loss S_{11} V/sfsrequency

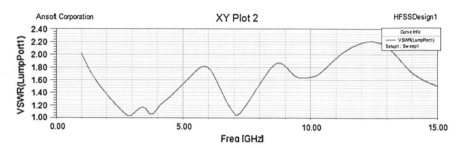

Fig. 8 The measured VSWR

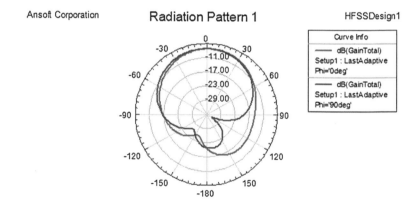

Fig. 9 Radiation pattern for UWB antenna

Figures 10, 11 and 12 show the current distributions at 3, 7, and 10 GHz frequencies respectively.

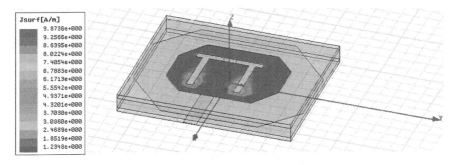

Fig. 10 Current distribution at 3 GHz

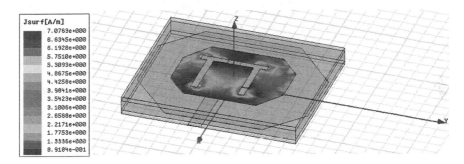

Fig. 11 Current distribution at 7 GHz

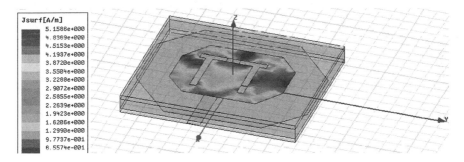

Fig. 12 Current distribution at 10 GHz

6 UWB Antenna with Band Notch

We have simulated and observed earlier that a ∏-shaped slot cut out from the truncated patch gives a continuous band from 1.12 to 10.86 GHz covering the complete UWB. In this design we introduced two symmetric-sized rectangular slits P1 and P2. It dispends a band from 5.07 to 5.84 GHz which covers WLAN. By removing

Table 3 Antenna dimension of Fig. 13

Length of Rectangular Slit P1	1.5 mm
Width of Rectangular Slit P1	7 mm
Length of Rectangular Slit P2	1.5 mm
Width of Rectangular Slit P2	7 mm

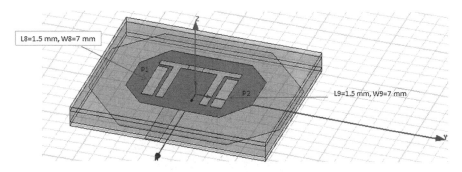

Fig. 13 Geometry of Notched UWB antenna

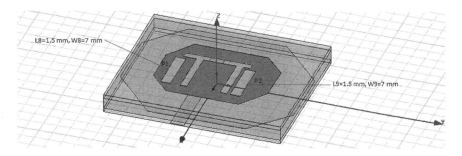

Fig. 14 Geometry of completely WLAN dispended Notched UWB antenna

WLAN band from UWB band, interference problem associated with UWB band is resolved. The dimensions of rectangular slits are shown in Table 3.

In Fig. 13, two rectangular slits equidistant from ⊓ slot are introduced. These rectangular slits at the present positions do not remove the band which covers the complete WLAN band, by properly optimizing them we have obtained a complete UWB —without WLAN band which resolves the interference problem. The completely WLAN dispended microstrip antenna is shown in Fig. 14.

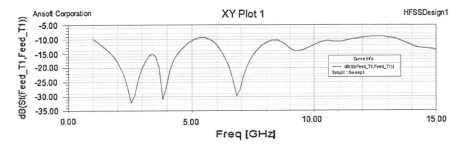

Fig. 15 Simulated return loss S_{11} V/S frequency

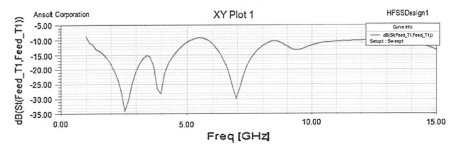

Fig. 16 Simulated return loss S_{11} V/S frequency

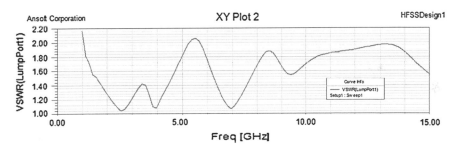

Fig. 17 Measured VSWR for UWB antenna

7 Measured Results

In the return loss plot shown in Fig. 15, two resonant bands exist. First, frequency band existing from 1.02 to 5.05 GHz resulted in impedance bandwidth of 132.81 % and another band— from 5.72 to 11.42 GHz gives the bandwidth of 66.54 %. From the above analysis it is observed that it does not remove the complete WLAN.

From Fig. 16, it is observed that it covers the complete UWB along with the removal of the WLAN band. The first resonant band is from 1.08 to 5.07 GHz and the second band is from 5.84 to 11.82 GHz (Figs. 17 and 18).

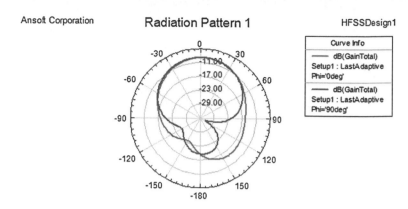

Fig. 18 Radiation pattern for completely WLAN dispended Notched UWB antenna

8 Conclusion

A small-sized ultra wideband microstrip patch antenna with band rejection character-istics has been obtained and the results are simulated. These results are encouraging and it is expected to produce a suitable structure for ultra wideband applications. The proposed antenna resulted in wide bandwidth performance from 1.1 to 11.8 GHz. This designed antenna can be used in any compact handheld devices as well as in military applications, radar, etc., involving UWB applications without interference with WLAN.

In most of the antennas, designed incorporating slots, it is fairly difficult to create multiple frequency notches. We hope that the method will be helpful for creating multiple frequency notches as well as achieving sharp stop band.

References

1. Federal Communications Commission, Washington, DC: FCC report and order on ultra wide-band technology (2002)
2. Schantz, H.G., Wolenec, G., Myszka, E.M.III.: Frequency notched UWB antennas. In: Proceedings of IEEE Conferences Ultra Wideband System Technology, 214–218 November 2003
3. Lim, E.G., Wang, Z., Lei, C.-U., Wang, Y., Man, K.L.: Ultra wideband antennas—past and present. IAENG Int. J. Comput. Sci. **37**:3 (2010)
4. Zhang, J., Xu, Y., Wang, W.: Microstrip-fed semi-elliptical dipole antennas for ultrawideband communications. IEEE Trans. Antennas Propag. **56**(1), 241–244 (2008)
5. Jang, J.W., Hwang, H.Y.: An improved band-rejection UWB antenna with resonant patches and a slot. IEEE Antennas Wirel Propag. Lett. **8**, 299–302 (2009)
6. Taub's Principles of Communication systems, Herberttaub, DonaldL: Schilling, GoutamSaha, TMH, 3rd edn. ISBN 0-07-064811-5 (2009)
7. Balanis, C.:Antenna Theory, 2nd edn. Chapter 14. ISBN 0-471-59268-4 Wiley (1997)
8. A Review of Aperture Coupled Microstrip Antennas: History, Operation, Development, and Applications. http://www.ecs.umass.edu/ece/pozar/aperture.pdf

A Technique to Minimize the Effect On Resonance Frequency Due to Fabrication Errors of MS Antenna by Operating Dielectric Constant

Sandeep Kumar Toshniwal and Kanad Ray

Abstract This paper presents a method to minimize the effect on resonance frequency due to fabrication error of microstrip patch antenna. When a patch antenna is fabricated, dimension of the patch may be slightly differentfrom its calculated value due to error in the fabrication operations, which causes into variation of its resonance frequency. To overcome this problem this paper presents a new technique to minimize the effect on resonance frequency due to fabrication error of MS antenna by operating dielectric constant. Effective dielectric constant of substrate is changed in such a way that the resonant frequency is set back to the calculated value.

Keywords Micro strip Antenna · Resonance frequency · Composite dielectric constant · Multilayer structure.

1 Introduction

The rapid growth of communication system developed a great demand of small antennas. The most commonly and popularly used small or miniature antenna is microstrip patch antenna [1].

Nowadaysmicrostrip antennas are most popular antennas for HF applications. The microstrip patch antenna posseses a number of important advantages such as low profile, low weight, low manufacturing cost and having conformability. The main advantage of this antenna is that we can easily change the frequency, bandwidth and

S. K. Toshniwal (✉)
Department of Electronics and Communication, Kautilya Institute of Technology
and Engineering and School of Management, Jaipur, Rajasthan, India
e-mail: toshniwal.sandeep@gmail.com

K. Ray
Institute of Engineering and Technology, JK Lakshmipat University, Jaipur, Rajasthan, India
e-mail: kanadray@jklu.edu.in

B. V. Babu et al. (eds.), *Proceedings of the Second International Conference on Soft Computing for Problem Solving (SocProS 2012), December 28–30, 2012,* Advances in Intelligent Systems and Computing 236, DOI: 10.1007/978-81-322-1602-5_120, © Springer India 2014

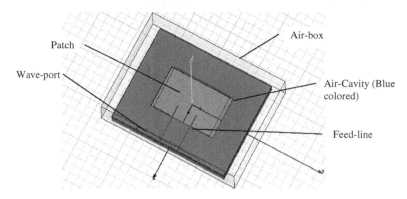

Fig. 1 Layout of MS Antenna with air-cavity drown by HFSS

gain by changing the parameters of MS antenna [2]. We can modify the frequency, bandwidth and gain by changing either the patch or dielectric width and length. This paper investigates the effect of frequency variation with respect to the change in dielectric constant [3]. For that we simply cut the material below the patch and consider it as an air-cavity [4]. The new effective dielectric constant gives the new value of resonance frequency. All analysis has been carried out using High Frequency Structural Simulator (HFSS). Change in resonance frequency has been plotted against area of cavity to take a broad view of the results. The paper is mainly divided into two sections. Section 2, describes the design methodology explained in brief. Section 3, describes results and discussion of the calculated parameters (Fig. 1).

2 Antenna Design and Simulation

Dimensions of antenna patch were calculated using standard equations. The essential parameters required for designing of MS antenna are $f_o = 2.4\text{GHz}$, $\varepsilon_r = 4.4$, $h = 6.4\,\text{mm}$.

The length, width and other parameters related to designing of the MS antenna are calculated by EM Calculator.

Here are the calculated parameters:

$$W = 38.036\,\text{mm}$$

$$\varepsilon_{\text{reff}} = (\varepsilon_r + 1/2) + (\varepsilon_r - 1/2)[1 + 12h/w]^{-1/2}$$

$$\text{Leff} = 32.5\,\text{mm}, \Delta L = 2.8215\,\text{mm}, L = 27\,\text{mm}$$

Simulations have been done for two values of resonance frequency (f_o)–2 GHz and 2.4 GHz. For each value of resonance frequency (f_o) two values of substrate

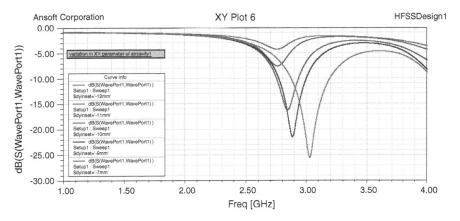

Fig. 2 Variation of resonance frequency (f_o) with various values of area of air cavity1

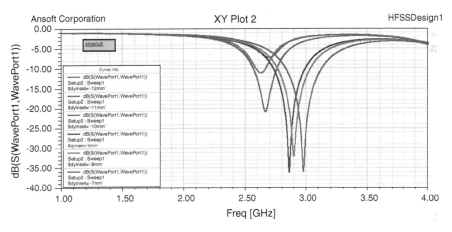

Fig. 3 Variation of resonance frequency (f_o) with various values of area of air cavity2

height (h)–1.2 mm and 6.4 mm have been investigated. A cavity has been created in the antenna dielectric just below the patch. Various values of cavity area have been considered. HFSS simulations yielded the resonance frequency (f_o) of the structure. Figures 2 and 3 are the simulation results which show variation in f_o with area of the air cavity 1 and air cavity 2 respectively.

3 Results and Discussion

Figures 4 and 5 shows dependence of resonance frequency with the area air cavity 1 **(When we inserted air-cavity in only one layer of dielectric)** and air cavity 2 **(When we inserted air-cavity in two layers of dielectric)**.

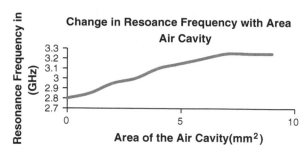

Fig. 4 Change in dimension of air cavity1 (vary in area)

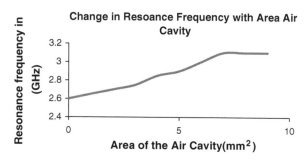

Fig. 5 Change in dimension of air cavity2 (vary in area)

Here we see the effects of change of area of cavity with respect to frequency, which is as in Table 1.

The result we are getting through inserting air-cavities in two layers is different from when we are inserting air-cavity in one layer because there is a variation in thickness of the air-cavity. Difference is only that the initial frequency is different in both of them. But both the frequencies linearly increased with area below 6 mm and after that frequency is constant.

Above results indicate that when a cavity is made in the antenna dielectric just below the patch, the resonance frequency f_o of the structure increases.

	Area	Frequency
Table 1 Effect of area of cavity with respect to frequency	$< 6mm^2$	Linearly increase
	$>= 6mm^2$	Constant

4 Conclusion

Through our simulation we conclude that, to enhance the range of frequencies, there is no need to change in the basic structure of the MS antenna, instead of that we can directly cut the material equal to the dimension of the air-cavity and find the change in the frequency. Effective dielectric constant of the substrate is changed in such a way that the resonance frequency is set back to the calculated value if change occurs due to error in fabrication process. Results indicate that when a cavity is made in the antenna dielectric just below the patch, the resonance frequency f_o of the structure increases.

These results are found here only for FR-4 epoxy but are equally valid for other dielectric materials (i.e. duroid, Rogers) also.

Acknowledgments The authors are thankful to the JKLU, Jaipur and KITE, Jaipur for encouragement to pursue this work and for providing the supporting materials.

References

1. Garg, R.., Bhartia, P., IndorBahl and Ittipiboon, A.. Microstrip Antenna Design Handbook, pp. 1–68, 253–316. Artech House Inc., Norwood, MA (2001)
2. Kumar, G and Ray, K.P.: Broadband Microstrip Antenna. Artech House, London 2003.
3. Hoorfar, A., Perrotta, A.: An experimental study of microstrip antennas on very high permittivity ceramic substrates and very small ground planes. IEEE Trans. Antennas Propag. **49**(4), 838 (2001)
4. Lee, B. and Harackiewicz, F.J.: Miniature microstrip antenna with a partially filled high-permittivity substrate. IEEE Trans. Antennas Propag. **50**(8), 1160 (2002)

Part XII
Soft Computing for Industrial Applications (SCI)

Artificial Neural Network Model for Forecasting the Stock Price of Indian IT Company

Joydeep Sen and Arup K. Das

Abstract The central issue of the study is to model the movement of stock price for Indian Information Technology (IT) companies. It has been observed that IT industry has some promising role in Indian economy. We apply the artificial neural networks (ANNs) for modeling purpose. ANNs are flexible computing frameworks and its universal approximations applied to a wide range with desired accuracy. In the study, multilayer perceptron (MLP) models, which are basically feed-forward artificial neural network models, are used for forecasting the stock values of an Indian IT company. On the basis of various features of the network models, an optimal model is being proposed for the purpose of forecasting. Performance measures like R^2, standard error of estimates, mean absolute error, mean absolute percentage error indicate that the model is adequate with respect to acceptable accuracy.

Keywords Artificial neural network · Financial forecasting · Stock price · Indian IT companies · Multilayer perceptron.

1 Introduction

The study considers modeling the movement of stock price for Indian Information Technology (IT) companies. It has been observed that IT industry has some promising role in Indian economy. A number of profitable Indian companies today belong to the IT sector and a great deal of investment interest is now being focused. Companies in BSE IT index are those that have more than 50 % of their turnover from IT

J. Sen (✉)
Population Studies Unit, Indian Statistical Institute, Kolkata, India
e-mail: joydp.sen@gmail.com

A. K. Das
SQC and OR Division, Indian Statistical Institute, Kolkata, India
e-mail: akdas@isical.ac.in

B. V. Babu et al. (eds.), *Proceedings of the Second International Conference on Soft Computing for Problem Solving (SocProS 2012), December 28–30, 2012*, Advances in Intelligent Systems and Computing 236, DOI: 10.1007/978-81-322-1602-5_121, © Springer India 2014

related activities like IT infrastructure, IT education, software training, telecommunication services, networking infrastructure, software development, hardware manufacture, vending, support and maintenance. BSE IT index constituents represent about 10.22 % of the free float market capitalization.

In the literature, Box-Jenkins ARMA and ARIMA models [1] have respectively been adopted for financial forecasting. These traditional models do not learn immediately as the arrival of new data, instead they must be re-estimated periodically.

De Groot and Wurtz [2] analyzed univariate time series forecasting using feedforward neural networks. Due to the inherently noisy, non-stationary and chaotic nature of financial time series, it is often difficult to forecast based on the restricted statistical models. Therefore, artificial neural networks (ANNs) have received an increasing attention in time series forecasting in recent years.

El-Hammday and Abo-Rizka [3] developed a recurrent neural network model trained by ARIMA analysis for forecasting stock price in Egyptian Stock Market. Zhang and Wu [15] proposed an improved bacterial chemotaxis optimization (IBCO), which is later integrated into back-propagation (BP) artificial neural network to develop an efficient forecasting model for predicting various stock indices. Panahian [9] used a multilayer perceptron (MLP) neural network to determine the relationship between some variables as independent factors and the level of stock price index (TEPIX) as a dependent element in Iranian stock market. Kara et al. [6] observed that the average performance of ANN model in predicting the daily Istanbul Stock Exchange (ISE) National 100 Index was found significantly better than that of SVM model.

Several attempts [7, 8, 13, 14] have been done to implement artificial neural network to model Indian financial indexes like BSE SENSEX and NIFTY. But only a few studies have been done for the different Indian IT companies. The purpose of the study is to model the stock price of Indian IT companies using artificial neural network and forecast the same.

We organise the rest of this paper in the following way. In Sect. 2, a brief exposition of the methodology used in the study is discussed. Section 3 provides the analysis and results along with discussion. A conclusion of the study is presented in Sect. 4.

2 Methodology

Artificial neural networks are computational structures modeled on the gross structure of human brain. Neural network can model the behavior of known systems without being given any rule or models. Moreover neural networks may be considered as flexible nonlinear non-parameterized models where the parameters may be adopted according to the available data.

As the design is based on the human brain, they are made to obtain knowledge through learning. The process of learning for a neural network consists of regulating the weights of each of its node considering the input of the neural network and its expected output.

Fig. 1 Multilayer Perceptron

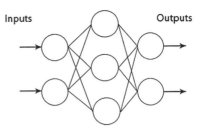

The model is characterized by a network of three layers (viz. input, hidden and output) of simple processing units connected by acyclic links. The relationship between the output (y_t) and the inputs $(y_{t-1}, \ldots, y_{t-p})$ has the following mathematical representation :

$$y_t = w_0 + \sum_{j=1}^{q} w_j g \left(w_{oj} + \sum_{i=1}^{p} w_{i,j} y_{t-i} \right) + e_t$$

where $w_{i,j} (i = 0, 1, 2, \ldots, p; j = 1, 2, \ldots, q)$ and $w_j (j = 0, 1, 2, \ldots, q)$ are model parameters often called connection weights, p is the number of input nodes, q is the number of hidden nodes. Identity, sigmoid, tanh, logistic, exponential functions are used as the hidden layer activation function and output activation function. The choice of q is data dependent and no systematic, well-accepted rule exists in deciding q. Experiments are conducted to find p.

MLP is one of the most popular network architectures used today, due to Rumelhart and McClelland [11]. Each performs a biased weighted sum of inputs and passes this activation level through a transfer function to produce the outputs, and the units are arranged in a layered feed-forward topology. The network thus has a simple interpretation in a form of input–output model with weights and thresholds (biases) as the free parameters of the model. Such networks can model functions of almost arbitrary complexity. MLPs are variable tools when one has less knowledge about the form of the relationship between inputs and outputs. The architecture of a MLP is shown in Fig. 1.

To adjust weights properly, we apply a general method for non-linear optimization, called Broyden–Fletcher–Goldfarb–Shanno (BFGS) algorithm. Sum of squares (SOS) is used as error function.

The accuracy of the prediction for each ANN model has been compared by the indexes of training performance and the testing perfection. The efficiency of ANN models varies with the number of input layers (p) and hidden layers (q). The residual of the proposed ANN model is tested for the presence of autocorrelation and white noise property.

Table 1 Summary of the selected ANN model for INFOSYS limited

Network name	MLP 6-7-1
Training perfection	0.99762
Training error	0.00013
Testing perfection	0.99772
Testing error	0.00014
Algorithm	BFGS 26
ERROR function	SOS
Hidden activation function	Tanh
Output activation function	Identity

3 Analysis and Results

To analyze the stock market price, we consider the daily data of consecutive five financial years from 2006 to 2011 of the stock-values of INFOSYS Limited. The data is collected at a particular time interval on every working day. On Saturday, Sunday and other national holidays, Bombay Stock Exchange remains closed and also there is no transaction in stock exchange market.

The closing price represents the most up-to-date valuation of a security until trading commences again on the next trading day. Closing prices provide a useful marker for investors to assess changes in stock prices over time. The closing price of a day can be compared to the previous closing price in order to measure market sentiment for a given security over a trading day. So we consider only closing price of the stock for our analysis.

The accuracy of prediction for each ANN model has been compared by the indexes of training performance and the testing performance. The efficiency of ANN models vary with the number of input layers (p) and hidden layers (q). p and q are chosen based on experimentation.

The whole data set is being divided into three groups as 80, 15 and 5 % of the total observations for the purpose of training, testing and validation to develop the ANN model. For having the least training performance and testing performance, the multilayer perceptron model with six input nodes, seven hidden nodes and one output node (i.e. MLP 6-7-1 model) is chosen to model and forecast the daily stock price of INFOSYS Limited. Summary of the selected ANN model is depicted in Table 1.

The actual close-price and the predicted close-price obtained by using the selected ANN model are shown in Fig. 2. The weight of the network parameters of MLP 6-7-1 is presented in Table 2.

The Var1 indicates the actual close-price and 2.MLP 6-7-1 indicates the predicted close price based on the ANN model of MLP 6-7-1.

Following the similar methodology discussed earlier, we apply the multilayer perceptron model for forecasting the daily stock prices of two other Indian IT companies: Tata Consultancy Services Ltd. (TCS Ltd.) and WIPRO Ltd. A performance summary of the optimum multilayer perceptron model for each of these three Indian IT companies is given in Table 3.

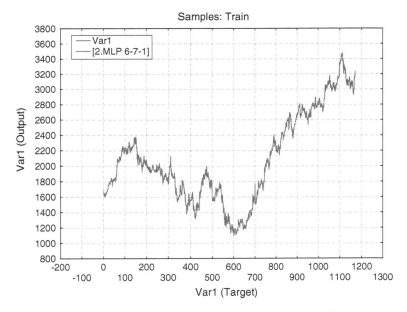

Fig. 2 Actual and predicted (by MLP 6-7-1) closing stock price of INFOSYS Limited

Table 2 Network weights of MLP 6-7-1

Serial number	Connections (MLP 6-7-1)	Weight values (MLP 6-7-1)
1	Input bias –> hidden neuron 1	0.137006
2	Input bias –> hidden neuron 2	0.144430
3	Input bias –> hidden neuron 3	−0.178096
4	Input bias –> hidden neuron 4	−0.340784
5	Input bias –> hidden neuron 5	−0.058491
6	Input bias –> hidden neuron 6	−0.264130
7	Input bias –> hidden neuron 7	−0.072084
8	Hidden neuron 1 –> Var1	−0.323277
9	Hidden neuron 2 –> Var1	0.327605
10	Hidden neuron 3 –> Var1	0.290812
11	Hidden neuron 4 –> Var1	0.710852
12	Hidden neuron 5 –> Var1	0.191150
13	Hidden neuron 6 –> Var1	0.151419
14	Hidden neuron 7 –> Var1	−0.610038
15	Hidden bias –> Var1	0.292043

Fisher's Kappa $= 7.352384$ (for Infosys Ltd.) indicates that the MLP residual of Infosys Ltd. is white noise. We obtain similar results for the TCS Ltd. and WIPRO Ltd. Performance measures presented in Table 3 show that MLP models provide excellent prediction and these models are adequate for modeling the stock price.

Table 3 Performance summary of INFOSYS Ltd., TCS Ltd. and WIPRO Ltd

Indian IT company	Infosys Ltd.	TCS Ltd.	WIPRO Ltd.
Fitted model	MLP 6-7-1	MLP 5-6-1	MLP 6-6-1
Standard error of the fitted model	39.55929	22.57515	28.46415
R^2 of the fitted model	0.99531	0.99135	0.99395
MAE of the fitted model	29.91343	14.67768	18.653321
MAPE of the fitted model	0.01547	0.01831	0.02303

4 Conclusion

The present study is designed to model the stock price for the Indian IT companies using artificial neural network for the purpose of prediction. We observe that artificial neural network performs well to model stock price. This phenomenon is true in the sense that ANN may capture the non-linear and chaotic behavior as compared to other conventional time series models. Our experiment has shown that the multilayer perceptron model is capable of predicting the stock price with acceptable accuracy. In this study it has been observed that multilayer perceptron model has ability to explain the time series pattern of the stock market data.

References

1. Box, G.E.P., Jenkins, G.M.: Time series analysis: forecasting and control. Holden-day Inc., San Francisco (1976)
2. De Groot, C., Wurtz, D.: Analysis of univariate time series with connectionist nets: a case study of two classical examples. Neurocomputing **3**, 177–192 (1991)
3. El-Hammady, A.H., Abo-Rizka, M.: Neural network based stock market forecasting. IJCSNS **11**(8), 204–207 (2011)
4. Freeman, J.A., Skapura, D.M.: Neural network algorithms, application and programming techniques. Addision Wesley (1991)
5. Hornick, K., Stinchcombe, M., White, H.: Universal approximation of an unknown mapping and its derivatives using multilayer feedforward networks. Neural Netw. **3**, 551–560 (1990)
6. Kara, Y., Boyacioglu, M.A., Baykan, O.K.: Predicting direction of stock price index movement using artificial neural networks and support vector machines: the sample of the Istanbul stock exchange. Expert Syst. Appl. **38**, 5311–5319 (2011)
7. Manjula, B., Sarma, S.S.V.N., Naik, R.L., Shruthi, G.: Stock Prediction using Neural Network. IJAEST. **10**(1), 13–18 (2011)
8. Merh, N., Saxena, V.P., Pardasani, K.R.: A comparison between hybrid approaches of ANN and ARIMA for Indian stock trend forecasting. Bus. Int. J. **3**(2), 23–43 (2010)
9. Panahian, H.: Stock market index forecasting by neural networks models and nonlinear multiple regression modeling: study of Iran's capital market. Am. J. Sci. Res. **18**, 35–51 (2011)
10. Qi, M., Zhang, G.P.: An investigation of model selection criteria for neural network time series forecasting. Eur. J. Operat. Res. **132**(3), 666–680 (2001)
11. Rumelhart, D.E., McClelland, J.L., The PDP Research Group: Parallel distributed processing: explorations in the microstructure of cognition, vols. 1–2. MIT Press, Cambridge (1986)

12. Schoneburg, E.: Stock price prediction using neural network: a project report. Neurocomputing **2**(1), 17–27 (1990)
13. Thenmozhi, M.: Forecasting stock index returns using neural networks. DBR **7**(2), 59–69 (2006)
14. Vashisth, R., Chandra, A.: Predicting stcok returns in nifty index: an application of artificial neural network. IRJFE **49**, 15–23 (2010)
15. Virli, F., Freisleben, B.: Neural network model selection for financial time series prediction. Comput. Stat. **16**(3), 451–463 (2001)
16. Zhang, Y., Wu, L.: Stock market prediction of S&P 500 via combination of improved BCO approach and BP neural network. Expert. Syst Appl. **36**(5), 8849–8854 (2009)

V/f-Based Speed Controllers for an Induction Motor Using AI Techniques: A Comparative Analysis

Awadhesh Gupta, Lini Mathew and S. Chatterji

Abstract This paper presents a comparative analysis of speed controllers for three-phase induction-motor scalar speed control. The control strategy consists in keeping constant the voltage and frequency ratio of the induction-motor supply source. First, a conventional control loop including a PI controller is realized. Then a fuzzy-control system is built on a MATLAB platform, which uses speed and difference in speed variation to change both the fundamental voltage amplitude and frequency of a sinusoidal pulse width modulated inverter. An alternative optimized method using Genetic Algorithm is also proposed. The controller performance, in relation to reference speed and step change of speed variations with constant load-torque, is evaluated by simulating it on a MATLAB platform. A comparative analysis of these with conventional proportional-integral controller is also presented.

Keywords Artificial intelligence (AI) · Adjustable speed drives (ASDs) · Proportional-integral (PI) · Fuzzy Logic Controller (FLC) · Genetic Algorithm (GA)

1 Introduction

Adjustable speed drives (ASDs) are the essential and endless demand of industries and researchers. These are widely used in industries to control the speed of conveyor systems, blower speeds, machine tool speeds, and other applications that require adjustable speeds. In many industrial applications, traditionally, dc motors were the workhorses for the ASDs due to their excellent speed and torque response [1, 2]. These however have the inherent disadvantage of commutator and mechanical

A. Gupta (✉)
EED, MNNIT, Allahabad, India
e-mail: awadhesh.g@rediffmail.com

L. Mathew · S. Chatterji
EED, NITTTR, Chandigarh, India
e-mail: lenimathew@yahoo.com

B. V. Babu et al. (eds.), *Proceedings of the Second International Conference on Soft Computing for Problem Solving (SocProS 2012), December 28–30, 2012*, Advances in Intelligent Systems and Computing 236, DOI: 10.1007/978-81-322-1602-5_122, © Springer India 2014

brushes, which undergo wear and tear with the passage of time [3]. An induction machine is the most widely used motor in the industry because of its high robustness, reliability, low cost, high efficiency, and good self-starting capability. In spite of these capabilities induction motors suffer from two disadvantages. First, the standard motor is not a true constant-speed machine; its full load slip varies from less than 1 % (in high horse power motors) to more than 5 % (in fractional horse power motors). Second, it is not inherently capable of providing variable-speed operation. These limitations can be solved through the use of adjustable speed controllers [4]. The basic action involved in adjustable speed control of induction motor is to apply a variable voltage magnitude and variable frequency to the motor so as to obtain variable speed operation. Voltage source inverter can be used to meet the aforesaid purpose. Induction motors offer a very challenging control problem due to its nonlinearity of dynamical system, motor parameter variation, and difficulty in measuring rotor variables. Moreover, the widely used PID controller does not offer satisfactory results when adaptive algorithms are required [5–8]. This problem can be solved by using artificial intelligent control techniques such as neural network, fuzzy logic, genetic algorithms, and particle swarm optimization techniques which can enhance the performance of the system to a great extent without requiring an exact model of induction motor.

2 Controller Design

The conventional closed loop PI control scheme is developed using the MATLAB/SIMULINK for V/f-based speed control of induction motor which is shown in Fig. 1. The various parameters required for the model of Fig. 1 saved in an m-file is

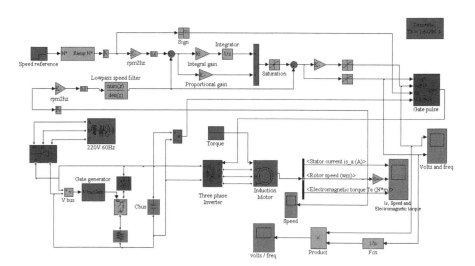

Fig. 1 SIMULINK model of IMD with PI controller

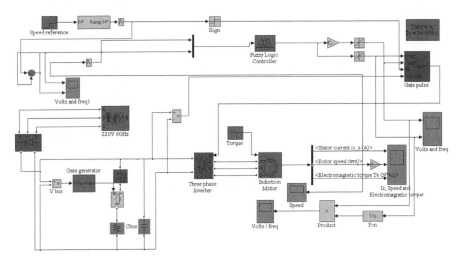

Fig. 2 SIMULINK model of IMD with Fuzzy Logic Controller

run first. Then SIMULINK model of Fig. 1 is run to get the response of conventional P-I controller. Another approach namely Fuzzy Logic Controller design is shown in Fig. 2. The actual speed of the induction motor is sensed and is compared with the reference speed. The error so obtained is processed in a P-I controller and its output sets the inverter frequency. The synchronous speed, obtained by adding actual speed ω_m and the slip speed ω_{slip} determines the inverter frequency. The reference signal for closed loop control of machine terminal voltage is generated from the frequency.

3 Results and Discussions

The induction motor drive system has been simulated with P-I controller and then a fuzzy controller is employed in place of P-I controller. After that the P-I controller parameters have been optimized using genetic algorithm. The optimum value of K_p and K_i have been chosen for the controller. All the controllers are tested for speed tracking with constant load torque conditions. Results obtained by employing the different controllers have been finally compared. The drive is subjected to operate at different speeds with constant load torque to test the robustness of the different controllers. The rating and parameters of the three-phase induction motor which has been considered as the test case is given in Table 1.

Different cases under which the simulation tests have been carried out are

(i) Step change in reference speed.
(ii) Tracking of reference speed.
(iii) Robustness test against constant load torque.

Table 1 Induction motor parameters

Rated value	Power (P)	3,730 VA
	Voltage (V)	460 V
	Frequency (f)	60 Hz
	Speed (N)	1,760 rpm
	Pole pairs	2
Constants	Stator resistance (Rs)	1.115 Ω
	Rotor resistance (Rr)	1.083 Ω
	Stator inductance (Ls)	0.005974 H
	Rotor inductance (Lr)	0.005974 H
	Mutual inductance	0.2037 H
	Inertia constant (J)	0.02 kg m^2
	Damping constant (B)	0.005752 Nm s

The simulation responses of the drive system with P-I controller, Fuzzy controller and Genetic Algorithm (GA) optimized P-I controller are shown in Figs. 3, 4 and 5, respectively. The reference speed is changed from 1,000 to 1,200 rpm at time, $t = 2$ s, and again from 1,200 to 1,500 rpm at time, $t = 4$ s and after that the motor is continued to run at this speed till, $t = 6$ s. The load torque is kept constant. The responses of stator current, speed, and electromagnetic torque are also shown in Figs. 3, 4 and 5. It can be seen from the response of Fig. 3 that the current ripples and

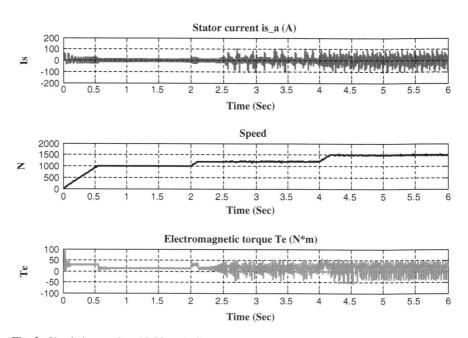

Fig. 3 Simulation results with PI controller

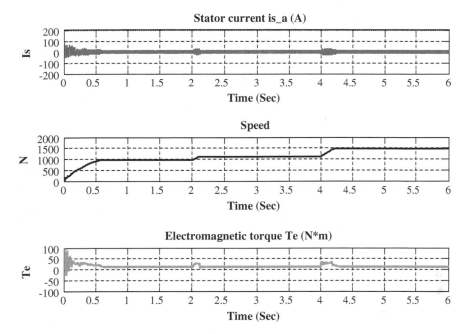

Fig. 4 Simulation results with fuzzy controller

torque ripples at 1,200 and 1,500 rpm are quite high while at 1,000 rpm the ripples are less. From Figs. 3 and 4, it is clear that, in case of fuzzy controller, the speed response is smoother than P-I controller. Also, the torque ripples and current ripples are minimized with fuzzy controller. From Figs. 3, 4 and 5, it is clear that, in case of GA optimized P-I controller the speed response is further smoother than P-I controller and fuzzy controller. The torque ripples and current ripples are also greatly minimized.

The simulation response of the drive system with P-I and GA optimized P-I controller at different set points are shown in Figs. 6 and 7. In case of P-I controller, it is observed that the reference speed is changed from 1,500 to 1,000 rpm at time, $t = 2$ s, and again from 1,000 to 1,200 rpm at time, $t = 4$ s and after that the speed reference of motor is changed from 1,200 to 900 rpm at $t = 5$ s. In case of GA optimized P-I controller, it is observed that the reference speed is changed from 1,500 to 1,200 rpm at time, $t = 2$ s and again from 1,200 to 1,000 rpm at time, $t = 3$ s and after that the speed reference of motor is changed from 1,000 to 600 rpm at $t = 5$ s. The load torque is kept constant in both the cases. The response of stator current, speed and electromagnetic torque are also shown in Figs. 6 and 7. From Figs. 6, 4 and 7, it is clear that the induction motor has run up to 900 rpm with P-I controller, up to 1,000 rpm with fuzzy controller and up to 600 rpm with GA optimized P-I controller. Thus the speed control range has been increased with GA optimized P-I controller.

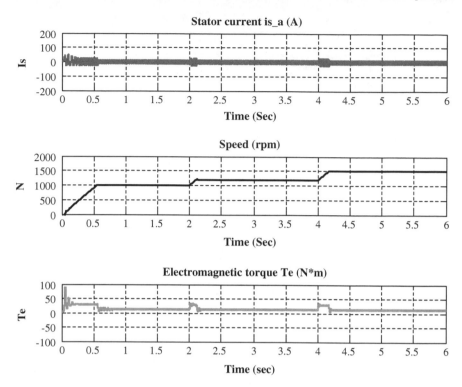

Fig. 5 Simulation results with GA optimized controller

4 Comparative Analysis of Results

The different controllers were tested for their speed response at different set points. The results obtained for the drive with P-I controller, Fuzzy controller, and GA optimized P-I controller are shown in Figs. 8, 9 and 10, respectively. For all the three controllers the reference speed is changed from 1,000 to 1,200 rpm at time, $t = 2$ s, and again from 1,200 to 1,500 rpm at time, $t = 4$ s and after that the motor is continued to run at this speed till, $t = 6$ s.

The speed response of the drive with conventional controller shows the rise time of 0.6 s as shown in Fig. 8. The overshoot value at 1,000 rpm is within reasonable limits but at 1,200 and 1,500 rpm, there exists sustained oscillations. The steady-state response of the drive is found to be poor. This could be due to the improper choice of integral gain. Table 2 shows the parameters of the P-I Controller.

The speed response of the drive with fuzzy controller as shown in Fig. 8 shows smoother response as compared to the conventional controller. Comparing Figs. 8 and 9, it is observed that the speed ripples are suppressed to a great extent but the steady-state error of the drive is found to be poor. Although the drive shows smoother response it generates appreciable offset. The speed ripples in the steady-state as well

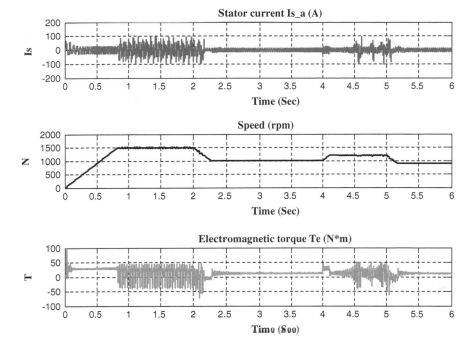

Fig. 6 Simulation results with PI controller with different set points

as in dynamic conditions are found to be reduced. This offset error can be resolved by making the fuzzy rule base more exhaustive.

The speed response of the drive with GA optimized controller as shown in Fig. 10 shows much smoother response as compared to the conventional controller and fuzzy controller. The speed ripples are greatly suppressed with respect to the other controllers' response mentioned earlier. The steady-state error of the drive at speeds 1,000, 1,200, and 1,500 rpm is almost negligible, depicted in Fig. 10. In case of GA optimized controller, the steady-state response as well as dynamic response is considerably improved with no torque ripples, as compared to the other controllers, thus, making the drive robust. Table 3 shows the parameters of the P-I controller which is obtained by the optimization process with the help of GA toolbox of MATLAB software with IATE considered as objective function.

5 Comparison Among Controllers

Table 4 shows the summary of comparative assessment of all the controllers considered here.

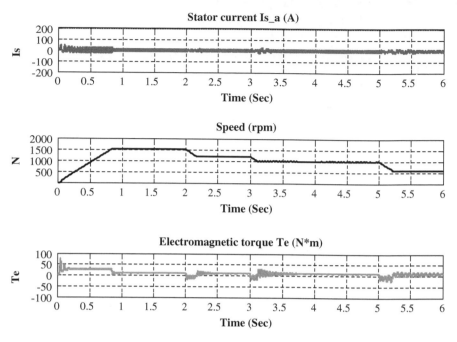

Fig. 7 Simulation results with GA optimized controller with different set points

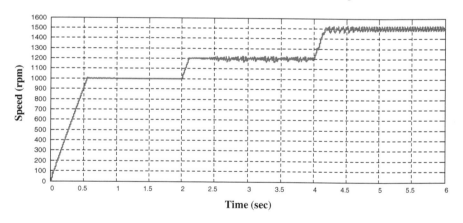

Fig. 8 Speed response with P-I controller

In view of the above discussions, it is clear that the transient as well as dynamic performance of the GA Optimized P-I controller is far superior to the conventional and fuzzy controllers. The speed control range obtained is also more as compared to the other controllers mentioned. The torque ripples and current ripples have also been minimized. Thus, GA-Optimized P-I controller is inferred as the best choice.

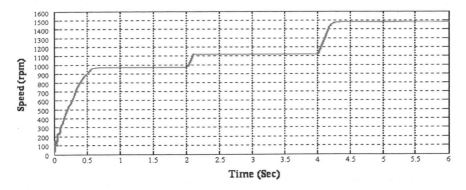

Fig. 9 Speed response with fuzzy controller

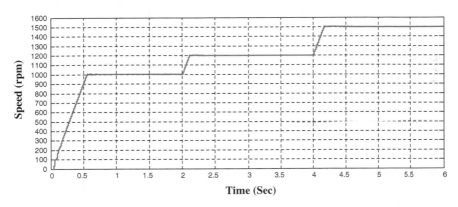

Fig. 10 Speed response with GA optimized controller

Table 2 P-I controller parameters

Sr. No.	Parameter	Value
1.	K_p	39
2.	K_i	50

Table 3 GA optimized P-I controller parameter

Sr. No.	Parameters	Value
1.	K_p	09
2.	K_i	10

Table 4 Comparative assessment of the controllers

Property	Conventional P-I	Fuzzy controller	GA-optimized P-I
Starting transient performance	Poor	Good	Very Good
Steady-state performance	Poor	Poor	Very Good
Speed-control range	Up to 900 rpm	Up to 1,000 rpm	Up to 600 rpm
Robustness	Average	Good	Very Good

6 Conclusion

In this work, artificial intelligence-based scalar speed control for an induction motor has been examined. First, a PI algorithm was thoroughly explored. After this, fuzzy and GA control system have been investigated to cope better with the dynamic conditions. The main conclusions are listed below:

(i)   Conventionally, P-I controller is suited for the speed control of induction motor.
(ii) The unique characteristics of the proposed GA-based tuning technique are found to be completely model independent.
(iii) This study demonstrates that GA technique can solve tuning problem of the controller more efficiently.
(iv) The conventional PID control mechanism performs poorly. If the system has to be controlled without assumption and linearization, the heuristic search algorithm, especially GA is found to be the best choice.
(v)   Unlike the other methods, the GA technique is simple, fast, and easy to implement in a variety of control loops and yields much better results as compared to the currently available conventional tuning methods.
(vi) The simulation results have shown that the performance of the system under dynamic conditions have improved significantly which is the main drawback of the scalar speed control.
(vii) The simulation results have shown that the speed control range is considerably increased by using the GA optimization technique to the P-I controller.
(viii) The dynamic response with the Fuzzy controller shows much more improvement but at the same time the offset error is not completely eliminated.
(ix) In the simulation response with the GA optimization, the offset error is eliminated considerably.

The investigator attempts to modernize the V/f control strategy, which already exists in various industries, by incorporating artificial intelligent control techniques such as Fuzzy Logic and GA, so as to give good dynamic performance along with the steady-state performance.

References

1. Javadi, S.: Induction motor drive using fuzzy logic. In: Proceedings of the 7th WSEAS International Conference on System Theory and Scientific Computation, Athens, Greece, August 2007
2. Bose, B.K.: Modern Power Electronics and AC Drives, A Text Book, 4th edn. Pearson Education, New York (2004)
3. Singh, B., Ghatak Choudhuri, S.: DSP based implementation of vector controlled induction motor drive using fuzzy pre-compensated proportional integral speed controllers. IEEMA J. 17(4), 77–84 (2007)

4. Koreboina, V.B., Magajikondi, S.J., Raju, A.B.: Modeling, simulation and PC based implementation of a closed loop speed control of VSI fed induction motor drive. In: Proceedings of the IEEE International Conference on Power Electronics, Banglore, January 2011
5. Muthuselvan, N.B., Dash, S.S., Somasundaram P.: A high performance induction motor drive system using fuzzy logic controller. In: Proceedings of the IEEE International Conference on Power Electronics, Chennai, November 2006
6. Yang, L., Li, Y., Chen, Y., Li, Z.: A novel fuzzy logic controller for indirect vector control induction motor drive. In: Proceedings of the 7th World Congress on Intelligent Control and Automation, pp. 24–28, Beijing, China, June 2008
7. Zerikat, M., Mechernene, A., Chekroun, S.: High-performance sensorless vector control of induction motor drives using artificial intelligent technique. In: Proceedings of the International Conference on Methods and Models in Automation and Robotics, pp. 67–75, Oran, Algeria, August 2010
8. Chaudhary, P.S., Patil, P.M., Patil, S.S., Kulkarni, P.P., Holmukhe, R.M.: Comparison of performance characteristic of a squirrel cage induction motor by three phase sinusoidal and PWM inverter supply using MATLAB digital simulation. In: Proceedings of the Third International Conference on Emerging Trends in Engineering and Technology, pp. 362–367, Pune, India, November 2010

Part XIII
Soft Computing for Information Management (SCIM)

Enhanced Technique to Identify Higher Level Clones in Software

S. Mythili and S. Sarala

Abstract Code copy and reuse are the most common way of programming practice. Code duplication occurs in every software program. A function, a module, or a file is duplicated for various reasons. The copied part of the source code with or without modification is called a code clone. Several tools have been designed to detect duplicated code fragments. These simple code clones assists to identify the design level similarities. Recurring patterns of simple clones indicate the presence of design level similarities called higher level clones. In this work we describe a new technique using fingerprinting to find higher level clones in software. Initially the simple clones are found, and then using LSH, we compare the fingerprints to find recurring patterns of method level, file level, and directory level clones. Finally, experiments and results shows that the proposed method finds all higher level clones in the software.

Keywords Software clones · Clone detection · Similarity hashing · Fingerprinting

1 Introduction

Recent research suggests that most of the codes in large software system are duplicated. Programmers normally clone the software by simple copy and paste method. After copying, the original code is modified according to the programmer's need.

S. Mythili (✉)
Research Scholar, Department of Information Technology,
Bharathiar University, Coimbatore 46, India
e-mail: smythili78@gmail.com

S. Sarala
Department of Information Technology, School of Computer Science and Engineering,
Bharathiar University, Coimbatore 46, India
e-mail: sriohmau@yahoo.co.in

B. V. Babu et al. (eds.), *Proceedings of the Second International Conference on Soft Computing for Problem Solving (SocProS 2012), December 28–30, 2012,* Advances in Intelligent Systems and Computing 236, DOI: 10.1007/978-81-322-1602-5_123, © Springer India 2014

Fig. 1 Higher level clones

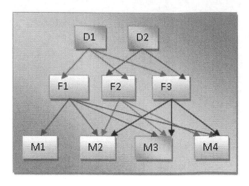

This copy and paste programming practice often creates exactly matched or similar portion of the code, which are called as code clones. Several tools are available to find these code clones. DECARD [1], Cp-Miner [2], CCFinder [3] and CloneDR [4] are some of the best clone detection techniques to find clones in the software and the Web applications. The drawback of most of the clone detection tools are that they concentrate on finding simple clones and do not concentrate on a wider picture looking for design level similarities. Simple clones are lower level clones with similar code fragments. These lower level clones when combined to form design level similarities are called as Higher Level Clones. Examples of higher level clones are shown in Fig. 1.

2 Literature Review

This section describes the various categorizations of code clones that have been proposed by different authors.

One of the first categorizations of clones in software was proposed by Mayrand et al. [5]. This categorization classifies candidate code clones according to the types and degrees of differences between code segments. The type of differences considered are function names, layout, expressions, and control flow. Using these attributes, clones are categorized in to eight types.

Balazinska et al. [6] created a schema for classifying various cloned methods based on the differences between the two functions that are cloned The differences are accounted in five groups. These categories are used by Balazinska et al. to produce software-aided reengineering systems for code clone elimination.

Major limitations of the above categorizations are that they concentrate only on function clones. These function clones will cover only 30–50 % of the cloning activity. As an improvement to this a new taxonomy was designed by Cory Kapser [7] and the new scheme is based on different attributes of the clone including similarity, locality within the software system, scope of the cloned code, degree of similarity of regions the clones exist in, and the type of region the clone occurs in. This taxonomy is a hierarchical classification of code clones which helps in finding Same File Clones, Same Region Clones, Function to Function Clones, Structure Clones, Macro Clones. Heterogeneous Clones, and Misc. Clones.

Koschke et al. [8] define another clone classification using only the types of differences between clones. In this classification only three types of clones are considered, as described above for its use in the Bellon benchmark. Basit et al. [9] have produced a clone classification scheme that groups clones according to higher level structures such as files or classes. The data mining techniques "market basket analysis" is used for searching frequently co-occurring clones.

3 Proposed Model for Finding Higher Level Clones

Figure 2 represents an idea of the new technique for finding the higher level clone patterns in the software system. The software system taken into consideration is a sample java system. The first step in the above technique is the preprocessing and formattings. It accepts the source code to be compared and preprocesses. This involves *white space* removal, *comment* removal and converting the lines of code into general format. In the general format conversion the source code is traversed for data types and variable names. Then the data types and the variable names are converted into a general format for, e.g., void sum(int a, int b) → void sum($v, $v);

The next step of the above technique is to split the source code considered into tokens. After tokenizing the tokens of the arbitrary length are converted into fingerprints using the *Message Digest Algorithm.* Most of the clone detection methods uses various algorithms which compares the literals and the identifiers for finding the clones. The proposed technique finds the code clones using fingerprinting technique.

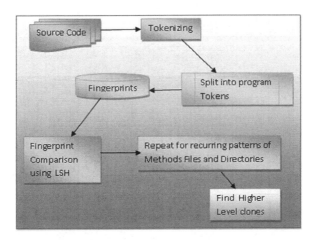

Fig. 2 Proposed technique

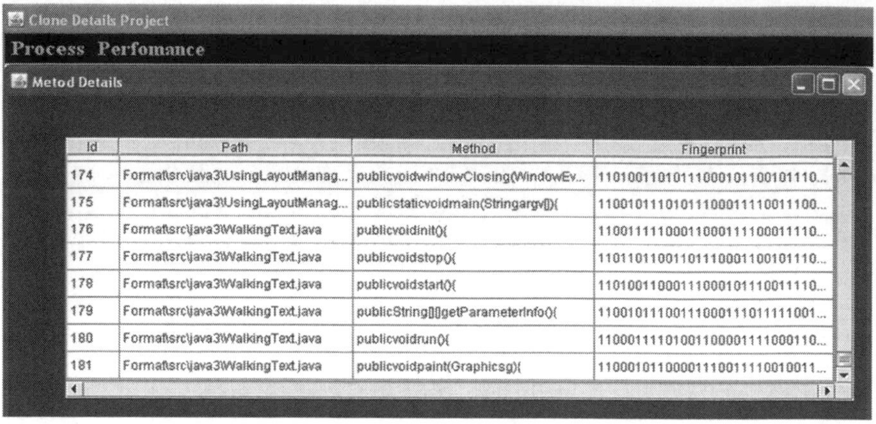

Fig. 3 Fingerprints for the methods

Figure 3, shows the fingerprints generated for the methods. In the same way fingerprints are generated for files and directories.

The major idea of our proposed technique is to find the similarity using fingerprints. One of the advantages of this technique is to map a large dataset of arbitrary length into a same bit sequence. We use the *Message Digest Algorithm* [10] for this purpose. Fingerprints are computed for every token at the method level, file level, and the directory level. The *Message Digest Algorithm* generates unique fingerprints for every token. At the method level the fingerprints are calculated for the tokens and the similar tokens will have the similar fingerprint. At the file level fingerprints are calculated for the sequence of statements and similar files will have similar fingerprints which gives the File Level clones. And at the directory level fingerprints are calculated for the group of file in the same directory and such similar directories will have similar fingerprints.

The next step in the above process is to find the similarity between the fingerprints. The similarity is calculated using a hash function called Locality Sensitive Hashing. A LSH function is a hash function which hashes for vectors such that the probability that two vectors having the same hash value is strictly decreasing function of their corresponding distance. In other words we can say that two vectors having the smaller distance will have the higher probability of having the same hash code. Let a hashing family be defined as

$$hi = (pi) \text{ where pi} = \text{ith bit of } p.$$

$$\text{Pr}_H[h(p) \neq h(q)] = \frac{||p, q||H}{d}. \tag{1}$$

$$\text{Pr}_H[h(p) = h(q)] = 1 - \frac{||p, q||H}{d}. \tag{2}$$

When similar points collide Pr must be $\geq 1 - \left(1 - \frac{1}{p1}\right)^K$

When dissimilar points collide Pr must be $\leq P_2^K$

The LSH function [11] is used to hash the fragments into smaller sets called buckets based on their hash code. The cloned fragments having similar hash codes are mapped into same buckets. LSH similarity comparison is computed using the formula Fig. 2. After finding the similarity the values are compared with a threshold value to classify them as *Exactly Similar Clones, Most Similar Clones* and *Least Similar Clones*. Under each threshold value method level, file level, and directory level are identified. The last process is to form clone pairs and form the clone set. The clones which come under the same classification are compared to form clone pairs. From the clone pairs method level, file level and directories level clone sets are formed.

3.1 Algorithm for Finding Higher Level Clones

4 Implementation and Evaluation

Experiments were carried over on the Sample Java System given in Table 2. The system includes the following contents.

Table 1 Detection algorithm

Algorithm for finding Higher Level Clones
1. For every file in the software
a. Find for files
b. Find for directories
2. For every file in the directory
a. Collect all the methods
3. Generate the Fingerprints for methods, files, and directories
4. Find the similarity using LSH
5. Find the ClonePair and CloneSets for recurring patters of methods, files, and directories.

The algorithm for finding higher level clone is given in Table 1

Our main aim is to find the cloned code in terms of higher level patterns, i.e., methods, files and directories. In the method level two similar methods of the same content will have the same fingerprints and will be identified as method level clones. Similarly, the file level and the Directory level clones are identified. The new technique proposed by us enables to find the Clone Pairs and their corresponding CloneSets for a particular threshold value. Figure 4. shows the Clone Pairs identified for the exactly similar clones at the method level.

To find the exactly similar clones the threshold (Th) values are varied from 1.0, 0.9, and 0.8. To find the *Most Similar Clones* the Th values are 0.6, 0.7, and 0.75.

Table 2 Sample Java system

Sample Java System	Contents
No of files	129
No of methods	181
No of directories	3

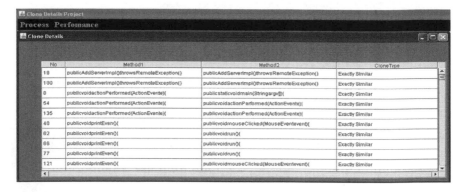

Fig. 4 Clone pairs for exactly similar clones

Table 3 Exactly and most similar clone pairs

Threshold value		Methods		Files		Directory	
ES	MS	ES	MS	ES	MS	ES	MS
0.8	0.6	221	814	76	1	4	0
0.9	0.7	89	451	70	2	4	0
1.00	0.75	57	170	62	2	4	0

Table 4 Exactly and most similar clone sets

Threshold value		Methods		Files		Directory	
ES	MS	ES	MS	ES	MS	ES	MS
0.8	0.6	57	57	110	0	2	0
0.9	0.7	54	42	99	3	2	0
1.00	0.75	53	43	90	3	2	0

The same process is continued to find the *Least Similar Clones*. By changing the threshold values the following Clone Pairs and the Clone Sets are detected and the results are shown in Tables 3 and 4. The graph for the *Exactly Similar Clone* pairs and clone sets are shown in Figs. 5, 6.

5 Conclusion and Future Work

In this work, a technique has been presented to detect the higher level clones in the softwares such as File Level, Method level, and the Directory level clones. This technique uses Message Digest Algorithm to find out the fingerprints and uses LSH

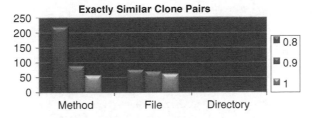

Fig. 5 Exactly similar clone pairs

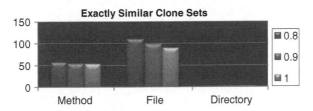

Fig. 6 Exactly similar clone sets

to compare it. The output produced by this system is a group of Clone pairs and Clone sets for Files, Methods and Directories. From the clone sets the number of Methods, Files, and directories which belong to the *Exactly Similar, Most Similar* and the *Least Similar* category are identified. The algorithm is tested using a sample java system. According to our sample system 53 out of 181 methods, 90 out of 129 files, and 2 out of 3 directories are cloned. In future, clone pairs can be clustered using the clustering techniques and the structural similarities can be found. Further the algorithm can be extended to detect clones in other data structures.

References

1. Jiang, L., Misherghi, G., Su, Z., Glondu, S.:Deckard: scalable and accurate tree-based detection of code clones. In: Proceedings of International Conference on Software Engineering, pp. 96–105, (2007)
2. Li, Z., Lu, S., Myagmar, S., Zhou, Y.: CP-miner: a tool for finding copy-paste and related bugs in operating systemcode. In: Proceedings of the Symposium Operating System Design and Implementation, pp. 289–302, (2004)
3. Kamiya, T., Kusumoto, S., Inoue, K.: CCFinder: a multilinguistic token-based code clone detection system for large scale source code. In: IEEE Transaction on Software Engineering, pp. 654–670, (2002)
4. Baxter, I.D., Yahin, A., Moura, L., Sant'Anna, M., Bier, L.:Clone detection using abstract syntax trees. In: International Conference on Software Maintenance, pp. 368-378, (1998)
5. Mayrand, J., Leblanc, C., Merlo, E.: Experiment on the automatic detection of function clones in a software system using metrics. In: International Conference on Software Maintenance, pp. 244–253, (1996)

6. Balazinska, M., Merlo, E., Dagenais, M., Lague, B., Kontogiannis K.: Measuring clone based reengineering opportunities. In: IEEE Symposium on Software Metrices, pp. 292–303, (1999)
7. Kapser, C.: Toward a taxonomy of clones in source code: a case study. In: Evolution of Large Scale Industrial Software Architectures, pp. 67-78, (2003)
8. Koschke, R., Falke, R., Frenzel, P.: Clone detection using abstract syntax suffix trees. In: Working Conference on Reverse Engineering, pp. 253–262, (2006)
9. Basit, H.A., Jarzabek,S.: Detecting higher-level similarity patterns in programs. In: Proceedings of the 10th European Software Engineering Conference held Jointly with 13th ACM SIGSOFT, International Symposium on Foundations of Software Engineering, pp. 156–165, (2005)
10. Rivest, R.: The MD5 message digest algorithm. RFC1321, Network Working Group. www.ietf.org/rfc/rfc1321.txt. (1992)
11. Gionis, A., Indyk P., Motwani, R.: Similarity search in high dimensions via hashing. In: Proceedings of the 25th International Conference on Very Large Data Base, 518–529, (1999)

Privacy Protected Mining Using Heuristic Based Inherent Voting Spatial Cluster Ensembles

R. J. Anandhi and S. Natarajan

Abstract Spatial data mining i.e., discovery of implicit knowledge in spatial databases, is very crucial for effective use of spatial data. Clustering is an important task, mostly used in preprocessing phase of data analysis. It is widely recognized that combining multiple models typically provides superior results compared to using a single, well-tuned model. The idea of combining object partitions without accessing the original objects' features leads us to knowledge reuse termed as cluster ensembles. The most important advantage is that ensembles provide a platform where vertical slices of data can be fused. This approach provides an easy and effective solution for the most haunted issue of preserving privacy and dimensionality curse in data mining applications. We have designed four approaches to implement spatial cluster ensembles and have used these for merging vertical slices of attribute data. In our approach, we have brought out that by using a guided approach in combining the outputs of the various clusterers, we can reduce the intensive distance matrix computations and also generate robust clusters. We have proposed hybrid and layered cluster merging approach for fusion of spatial clusterings and used it in our three-phase clustering combination technique. The major challenge in fusion of ensembles is creation and manipulation of voting matrix or proximity matrix of order n^2, where n is the number of data points. This is very expensive both in time and space factors, with respect to spatial data sets. We have eliminated the computation of such expensive voting matrix. Compatible clusterers are identified for the partially fused clusterers, so that this acquired knowledge will be used for further fusion. The apparent advantage is that we can prune the data sets after every $(m-1)/2$ layers. Privacy preserving has become a very important aspect as data sharing between organizations is also difficult. We have tried to provide a solution for this problem. We have obtained clusters

R. J. Anandhi (✉)
Department of Computer Science and Engg, The Oxford College of Engineering, Bangalore, India
e-mail: rjanandhi@hotmail.com

S. Natarajan
Department of Information Science and Engineering, PESIT, Bangalore, India
e-mail: snatarajan_44@gmail.com

B. V. Babu et al. (eds.), *Proceedings of the Second International Conference on Soft Computing for Problem Solving (SocProS 2012), December 28–30, 2012*, Advances in Intelligent Systems and Computing 236, DOI: 10.1007/978-81-322-1602-5_124, © Springer India 2014

from the partial datasets and then without access to the original data, we have used the clusters to help us in merging similar clusters obtained from other partial datasets. Our ensemble fusion models are tested extensively with both intrinsic and extrinsic metrics.

Keywords Cluster ensembles · Degree of agreement · Performance metrics · Spatial attribute data

1 Introduction

Clustering ensembles can be applied in various application environments and can be summarized based on the final clusters formed. The foremost utility of cluster ensembles is its ability to cluster categorical data. A noteworthy capability of clustering ensembles is that it provides a very natural and elegant way to cluster categorical data without any external labeling and uses the underlying data structure. Let us consider a data set with tuples $t_1, ..t_n$ over a set of categorical attributes $A_{1...}A_m$. The basic idea here is to view each attribute A_j as a way of producing a simple clustering of the data. If A_j contains k_j distinct values, then it is considered as A_j partitions the data into k_j clusters, one cluster for each value. Then, clustering fusion considers all those m clusterings produced by the m attributes and tries to find a consensus clustering that agrees as much as possible with all of them. Cluster ensemble approaches for fusing clusterings can also be extended to incorporate domain knowledge along with human guidance, when available. The clustering ensembles framework provides several ways for dealing with missing values in categorical data. One of the most important features of the usage of clustering ensembles is that there is no need to specify the number of clusters in the result. The formulation of clustering ensembles or aggregation [1] gives one inherent way of automatically selecting the number of clusters. If majority of the input clusterings place two objects in the same cluster, then it will not be beneficial for a clustering fusion solution to split these two objects.

Privacy-preserving data mining is a most studied subject in the last decade. There can be a common situation where a database table is vertically split and different attributes are maintained in different sites. For such cases, ensembles offer a natural model for clustering. the data maintained in all sites as a whole and also in a privacy-preserving manner. The only information revealed is which tuples are clustered together and no information is revealed about data values of any individual tuples. This becomes a welcome boom for the privacy preserving related issues.

The rest of the paper is organized as follows. The related work is in Sect. 2. The proposed knowledge guided fusion ensemble technique and its application in vertical slices merging, thereby favoring privacy protected data mining, is discussed in Sect. 3. In Sect. 4, we present experimental test platform and results with discussion. Finally, we conclude with a summary and our planned future work in this area of research.

2 Literature Survey

2.1 Work Done in Cluster Ensembles

The goal of cluster ensemble is to combine the clustering results of multiple clustering algorithms to obtain better quality and robust clustering results. Even though many clustering algorithms have been developed, not much work is done in cluster ensembles in data mining and machine learning community. There are few special categories of consensus function such as Voting Approach, Hyper graph Partitioning, Mutual Information Algorithm, Finite Mixture model and Co-association based functions, all of which we have studied and consolidated during our literature survey.

Strehl and Ghosh [2] have considered three different consensus functions for ensemble clustering. The Cluster based Similarity Partitioning Algorithm (CSPA) and HyperGraph Partitioning Algorithm (HGPA) proposed by them has a computing complexity of $O(kn^2H)$ and $O(knH)$. Their proposed algorithms improved the quality and robustness of the solution, but their greedy approach is the slowest and often is intractable for large n. Moreover all their algorithms are based on static approaches, in the sense they do not reuse the gained knowledge during fusion.

Azimi et al. [5], proposed a new clustering ensemble method, that generates a new feature space from initial clustering outputs. Multiple runs of an initial clustering algorithm like k-means generate a new feature space, which is significantly better than pure or normalized feature space and also reduces the local maxima effect. Then, applying a simple clustering algorithm on this generated feature space can obtain the final partition significantly better than pure data or the initial clusterings.Fischer and Buhmann [6], and also Dudoit and Fridlyand [4], have implemented a combination of partitions by re-labeling and voting. Their works practiced direct re-labeling approaches to the correspondence problem. A re-labeling can be done optimally between two clusterings using the Hungarian algorithm as in basic voting approaches. Topchy et al. [7] have developed a different consensus function based on information theoretic principles, using generalized Mutual Information (MI). Fred [3] proposed to summarize various clustering results using a co-association matrix. The rational of the his approach is to weigh associations between sample pairs by the number of times they co-occur or repeat in a cluster from the set of data partitions produced by independent runs of clustering algorithms.

2.2 Motivation for the Current Work in Ensembles

Many clustering algorithms are capable of producing different partitions of the same data because they capture various distinct aspects of the data. Due to the lowest computational complexity in mutual information based clustering ensembles technique and simplicity in voting technique we have a hybrid combination of both of these, resulting in a new heuristic algorithm called Hybrid Inherent Voting Ensemble

Fusion technique [8–10]. We have tried to find answers for which clusterers to merge first, how to merge and how to reuse the gained intelligence during merge in an effective way [12].

3 Proposed Hybrid Inherent Voting Spatial Cluster Ensembles

3.1 Motivation and Challenges

To achieve compromise between individual yet conflicting clusterings is an important aim of cluster ensembles. The problem of combination of multiple clusterings brings its own new and tough challenges. The major difficulty is in arriving at a consensus partition from the various output partitions of different clustering algorithms. Unlike supervised classification and the classifier ensembles, here the patterns are unlabeled. This implies that there is no explicit correspondence of label information that is available from different clusterings. An extra complexity arises when different partitions contain different numbers of clusters, often resulting in an intractable label correspondence problem. This results in either under fitting or over fitting the data. Most of the existing methods consider all generated partitions, for deciding the concluding clusters. This may not be the optimal solution because some ensemble members are less accurate than others and these in turn may have detrimental effects on the final performance. It is to be noted that technique we have used are fundamentally different from the existing methods because we aim to design and employ heuristics for selection of ensemble members, without considering the characteristics of the data sets and the clusterings. Our goal is to select ensembles based on the underlying implicit clusters of the data set and also to make use of the various approaches available in the clustering algorithm themselves.

3.2 The Inherent Voting Approach in Spatial Cluster Ensembles

The hybrid inherent Voting approach has two phases, ensemble initialization process and the fusion process. Basically, there are two basic possibilities for fusion namely static and dynamic approaches. In the static approach, all the compatible pairs for fusion are found using methods like co-association and then the actual merge takes place. The second method does not have that kind of clear cut demarcation between the merging pairs ahead of the fusion process. This dynamic approach follows the simultaneous select and fuse process, so that any added knowledge gained during the fusion process can further guide in selection of better fusion points. Our approach is based on the dynamic fusion method. The layer based merging uses the clusterers as layers and selects two layers at a time for the fusion. This technique avoids the need for generation of the costly distance matrix, thereby exploiting the inherent voting advantages. We felt that the knowledge gained during the intermediate merge stages, is not at all considered nor utilized till now in any of the existing work. None of the existing approaches use the merged knowledge to guide the fusion of the remaining clusterers. But, we had an intuition that this knowledge can actually help in guiding the fusion process and will be very useful in an unsupervised clustering scenario. Combination of approaches like voting model, mutual information model and co-association model might lead to overcome the weakness of each one of them. Hence we have tried a hybrid combination of all these approaches, resulting in a new heuristic algorithm called Hybrid Inherent Voting Ensemble Fusion technique.

We have used Shannon entropy measure to calculate the entropy between all the clusterers and form a **Clusterer Compatibility Matrix (CCM)** of the order m x m, where 'm' is the number of clusterers. The corresponding i and j values for the maximum 'CCM(i,j)'th value indicates that these two clusterers (i and j), if selected as initial clusterers for fusion, will provide a stronger base information for guiding further fusions. We have built two fusion models using the above mentioned CCM technique termed as **S**lice and **D**ice **E**nsemble **M**odel [8] and other without using CCM technique, as in Inherent Voting Depth First Merge and Cyclic Merge [11].

Selecting the best clusterer pair with whom the fusion can commence, based on the clusterer compatibility has given the initial head start for the whole fusion process resulting in more accurate clusters. Extra knowledge about the underlying data structure is already available in the form of label vectors. Hence highly computational intensive methods like normalized mutual information computation or counting the number of all the co associations or checking for complete majority votes are not needed for selection of fusion joints. The merge and purge technique, where we keep purging the input data set once their confidence levels are above a threshold, along with the process of merging the data points, helps in fast convergence of the algorithm as well as leads to the effective use of the memory resources. Validation of the ensemble fusion results of clusters, generated in this unsupervised manner has been done using both intrinsic and extrinsic metrics.

We have formulated four greedy and heuristics approaches [8–12] to exploit the hidden knowledge in the spatial data clusterings.

During the second phase of fusion, we have framed four models for fusing the clusterings formed in first phase, using the dynamic approach. This phase comprises of activities like **(i)** computation of clusterer compatibility matrix, **(ii)** finding the initial fusion seed from clusterer compatibility matrix using partition entropy. Then, the fusion joints are identified, for those clusterings of the clusterers, selected by the fusion seed component, using the proposed similarity measures. The main component of the ensemble fusion model is the consensus functions based on the various heuristic models like Inherent Voting Depth First merge model, Cyclic Merge model, Slice and Dice Ensemble Merge model and the Nebulous Pool merge model. Framing different similarity measures for finding the fusion joints, elegant but effortless approximation base methods for identifying the prime fusion seed, i.e., clusterer compatibility identification, usage of different consensus techniques are the key techniques in this work. We have also categorized the experimental data sets into small, medium and large based on the number of instances, number of dimensions and number of underlying classes.

3.3 Parameters and Definitions for the Hybrid Fusion

Fusion Joint Set—FJ_{ij}: Set of probable matching pairs of ith clusterer with the jth clusterer, based on the maximum entropy factor or maximum degree of overshadow factor. The cardinality of this fusion set is equal to the maximum number of clusters in i or j. This set will be used for deciding where the fusion, if happens, will most likely yield optimal information gain in the resultant clusters.

Clusterer Compatibility Matrix—CCM(m x m): Symmetric matrix where m is the total number of clusterers considered for fusion. **CCM(i, j):** An integer value representing the Shannon entropy between ith clusterer and jth clusterer. The maximum value in a cell of the matrix, say CCM(i,j), indicates that there is maximum partition entropy between clusterer 'i' and 'j', indicative of probable profitable fusion. This value forms the seed for the fusion.

DegreeofAgreement(DoAfactor): Cumulative ratio of the index of the merging level to the total number of clusterers.

3.4 Proposed Models for the Hybrid Fusion

Inherent Voting Depth First Merge (IVDFM) algorithm is framed with the notion that if normally a clustering with maximum number of data points is handled properly then the convergence of such fusion will happen faster. This is seen even during manually clustering objects, where we tend to handle the bigger components first by intuition. And also the clusterings with comparatively lesser number of data points will tend to disturb the bigger clusterings' co-occurrence measure. Another feature of this model is that we will not be explicitly building the complete static voting matrix though we will rely on majority opinions. The IVDFM fusion model starts with the first clusterer's clustering which has maximum data points (L_{max1}). The selection of the order of the clusterer goes with the way input is provided. That is, we have considered first come first serve basics during selection of clusterers. The IVDFM model uses both the similarity measures i.e., MeCos co-efficient based on, Minimum enclosing circle's over shadow and the measure based on partition entropy, separately, for finding the compatible clustering fusion joint. In Cyclic Merge Model (CycM), the clustering ensembles are combined in sliced pair, called Matching Groups set, $MG_{mk,}$ using clusterer compatibility matrix. The main difference between this model and the IVFDM model is that equal and fair chances are given to all the clusterings in the first layer to choose their fusion joints. The layered Slice and Dice Ensemble Merge model (SDEM) is an enhanced model from the earlier IVDFM and CycM, in the sense that it incorporates heuristics in selection of clusterers as well as clusterings for the fusion. And also it tries to make the model more efficient both in the terms of run time as well as space complexity. During the merge phase, fusion joint sets are formed by using either MeCos coefficient or the partition entropy between the data elements of the clusterings. The usage of similarity between core points is a normally used approach, which is very sensitive to the presence of outliers. The co-association based approach is computationally intensive with regard to spatial data. Instead we have found that the usage of cardinality of set intersection in association with partition entropy performs better in respect to identifying the concluding cluster partitions as well as resolves the label naming issues very easily and elegantly.

During the merging phase, initially, for each data point in the clusterer in focus, the **Degree of Agreement** (DoA) is calculated. There are two types of merge that we have used in the fusion step. Normal merge takes place between two clusterers when the DoA value is less than the threshold. Pruned Merge occurs when the DoA value is equal to or greater than the threshold DoA value. This ensures that once, certain data points have the marginal majority, they can be placed in the corresponding concluding clusters. This type of pruned merge helps to drastically reduce the data points half way through the fusion process. Spatial data with spatial auto correlated elements tends to benefit from this pruned merging approach.

4 Experimental Platform, Results and Inferences

4.1 Metrics Used in the Study

In the study of the hybrid inherent voting ensembles, we have used intra cluster density, inter cluster density and Dunn indices from the group of intrinsic measures as metrics. Dunn's index is a metric of, how well a set of clusters represent compact separated clusters. The Dunn index defines the ratio between the minimal inter cluster distances to maximal intra cluster distance. The Dunn index is limited to the interval [0, 1] and should be maximized. Dunn's index (α) for a partition U is defined as in Eq. (1)

$$\alpha(c, U) = \frac{\min_{1 \leq q \leq c} \min_{1 \leq r \leq c, r \neq q} \text{dist}(C_q, C_r)}{\max_{1 \leq p \leq c} \text{diam}(C_p)} \tag{1}$$

These metrics measure the goodness of a clustering structure without referring to any external information and depends on the under lying data structure itself. Inter cluster density, an intrinsic metric is based on how distant are the data elements in one cluster from elements in other clusters. This metric comes under More-The-Better category of factor values in the performance study. Intra cluster density, an intrinsic metric is based on proximity of elements are within one cluster to each other.

4.2 The Experimental Test Platform

In the test platform, both homogeneous as well as heterogeneous ensembles are considered. For the clusterings in homogeneous ensembles, we have used k-means with different values of k. For the heterogeneous ensembles, we have created the ensemble clusters using k-means, CLARA, Average Link agglomerative algorithm, fuzzy clustering and Wards divisive algorithms. The entire ensembles are from different clustering techniques; hence we have made the test platform with as much diversity as possible. The resources needed for executing the above process is a PC with Pentium Dual Core processor with 4 GB RAM and the code for implementing the ensemble fusion models is written using Matlab 7.0.

4.3 Data Sets Used in the Experiments

The cluster ensemble applications were tested on both real and synthetic data-sets. The data sets were downloaded from University of California, Irvine (UCI) Machine Learning repository and http://www.Strehl.com. As an indication of the impact of the archive, it has been extensively cited, making it one of the top 100 most cited "papers" in all of Computer Science research. These data sets serve as a benchmark for most of the algorithm verification and validation in Data Mining. We have collected the experimental data sets in such way that it includes as much diversity as possible. Based on the properties of the data sets, we have categorized them into small, average and large groups. This will help us to get the insight of which ensemble fusion approach will be more suitable for a particular data set. We have also chosen data sets with more number of dimensions and instances and classes, which reflects the basic traits of spatial data.

4.4 Experimentation Results and Inferences

The input for the ensemble algorithms are from base clusterers. Each of these algorithms was executed at least 15 times to see that no discrepancy due to the base algorithm is passed on as inputs. To check how useful homogenous ensembles can be, while handling huge dimensions in data set and to verify the quality and robustness of the final clusters, metrics like Cluster Purity and Dunn Indices were used. Initially the data set is divided vertically, simulating partial attributes and clusterings are formed for ensembles, using k-means clustering algorithm. This way the availability of partial attributes is simulated and homogenous base clusterers are formed based on different disjoint attributes of the same data set and is used for base clustersers' formation. The clusterings are merged using the hybrid fusion model. A

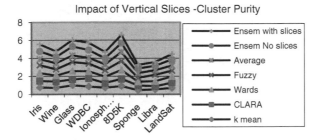

Fig. 1 Validation with cluster purity

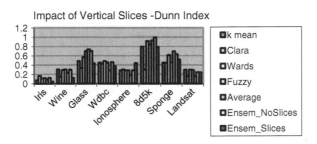

Fig. 2 Validation with dunn indices

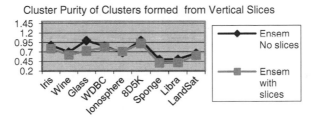

Fig. 3 Behavior of vertical slices ensemble fusion: cluster purity

comparison of the base clusterers' values against the fused clusters generated from partial view data sets is done to check for the deviations between them.

The graphs shown in the Figs. 1 and 2 bring out the deviations of the Cluster Purity and Dunn Index values between the clusters formed from the complete data set and clusters formed from partial view of data. We can see that the purity of the clusters has been reduced in the cases where the dimensions are low. In data sets like Ionosphere (34 D), Sponge (45 D), Libra (91 D) and Landsat (36 D), we have achieved almost the same purity. Similarly, the Dunn indices are 38 % better in Ionosphere data set, revealing that robust clusters can also be formed from having partial view of the data set attributes using our fusion process. This is graphically shown in Figs. 3 and 4.

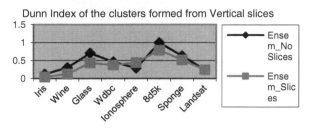

Fig. 4 Behavior of vertical slices ensemble fusion: dunn index

5 Conclusion and Future Work

The concept of fusion of clusterers derived from different clustering algorithms ensures that we have robust, novel and compact cluster outputs. And also clustering is an important functionality and also most of the times it used as pre processing step in the field of Data Mining. First, we have tested several non-spatial datasets which are normally used as bench marks for data clustering. Then we tested how our layer based methodology can work with spatial data. We have evaluated our work incorporating spatial feature spaces. The proposed automated layered merge approach was able to provide the acceptable accuracy with more efficiency both in space constraint and in computations. The results of the fusion of vertically sliced attribute data against the clusters derived from complete data was promising and proves that cluster compactness of the fused clusters derived from partial data is also in acceptable margin, when the dimensions are large. This is a promising result for data analysts who are concerned about the privacy of data attributes and who are deprived of some fields in their data due to privacy issues. However, more work should be carried out to provide support for more real life data from satellites and incomplete data. Future work in the short term will focus on how to acquire such datasets, and continue with more testing, in spite of current security concerns in distributing such data.

References

1. Gionis, A., Mannila, H., Tsaparas, P.: Clustering aggregation. J. ACM Trans. Knowl. Disc. Data (TKDD) 1(1), (2007). doi:10.1145/1217299.1217303
2. Strehl, A., Ghosh, J.: Cluster ensembles - a knowledge reuse framework for combining multiple partitions. J. Mach. Learn. Res. 3, (2002)
3. Fred, A.L.N.: Finding consistent clusters in data partitions. In: Proceedings of Second International Workshop on Multiple Classifier Systems, pp. 309–318. Springer-Verlag, London (2001)
4. Dudoit, S., Fridyand, J.: Bagging to improve the accuracy of a clustering procedure. Oxford J. Bioinform. 19(9), 1090–1099 (2003). doi:10.1093/bioinformatics/btg038

5. Azimi, J., Abdoos, M., Analoui, M.: A new efficient approach in clustering ensembles, LNCS, **4881**, pp. 395–405 (2007)
6. Fischer, B., Buhmann, J.M.: Bagging for path-based clustering. IEEE Trans. Pattern Anal. Mach. Intell. **25**(11):1411–1415 (2003)
7. Topchy, A., Jain, A.K., Punch, W.:Combining multiple weak clusterings. Proceeding of the Third IEEE International Conference on Data Mining, pp. 331–338 (2003). ISBN :0-7695-1978-4 doi:10.1109/ICDM.2003.1250937
8. Anandhi, R.J., Natarajan, S.: Efficient consensus function for spatial cluster ensembles: An heuristic layered approach. Proceedings of International Symposium on Computing, Communication and Control, Singapore (2009). ISBN 978-9-8108-3815-7
9. Anandhi, R.J., Natarajan, S.:A novel method for combining results of clusters in spatial cluster ensembles: A layered depth first merge approach with inherent voting. Int. J. Algorithms, Comp. Math. **2**(4), 53–58 (2009). ISSN 0973–8215
10. Anandhi, R.J., Natarajan, S.:An enhanced clusterer aggregation using nebulous pool. Proceedings of ACM -w International Conference of Celebration of Women in Computing, India, ACM Digital, Library (2010). 978–1-4503-0194-7
11. Anandhi, R.J., Natarajan,S.:A robust-knowledge guided fusion of clustering ensembles. Int. J. Comput. Sci. Inf. Sci. ISBN 1947-5500 **8**(4), LJS Publisher and IJCSIS Press, Pennsylvania, USA (2010)
12. Anandhi, R.J., Natarajan, S.: Efficient and effortless similarity measures for spatial cluster ensembles. CiiT Int. J. Artif. Intell. Syst. Mach. Learn. Pr. ISSN 0974–9667 and Online: ISSN 0974–9543, doi:AIML112010010, **2**(11), pp. 359–365 (2010)

Smart Relay-Based Online Estimation of Process Model Parameters

Bajarangbali and Somanath Majhi

Abstract This paper presents online estimation of unstable and integrating time delay process model parameters using a smart relay. The describing function (DF) approximation of relay not only results in simpler analytical expressions but also enables one to estimate the model parameters with significant accuracy. Measurement noise is an important issue during estimation of process model parameters. The smart relay is capable of emulating the dynamics of a conventional relay and also of rejecting the ill effects of measurement noise. Simulation results show the usefulness of the identification technique.

Keywords Relay · Hysteresis · Smart relay · Process model

1 Introduction

Industrial controllers are tuned offline or online depending upon the ways process model parameters are estimated. In online identification, relay is connected in parallel with a controller to extract the process information from the measurements made on the limit cycle output. A sustained oscillatory output can be obtained from relay feedback test. Several authors have used the describing function analysis for estimation of integrating and unstable time delay process model parameters. Since Relay is a nonlinear device it is approximated by an equivalent gain using describing function technique. Initially, Åström and Hägglund [1] proposed the use of relay feedback technique combined with describing function approximation to determine the

Bajarangbali (✉) · S. Majhi
Department of Electronics and Electrical Engineering, Indian Institute
of Technology,Guwahati 781039, India
e-mail: bajarangbali@iitg.ernet.in

S. Majhi
e-mail: smajhi@iitg.ernet.in

B. V. Babu et al. (eds.), *Proceedings of the Second International Conference on Soft Computing for Problem Solving (SocProS 2012), December 28–30, 2012*, Advances in Intelligent Systems and Computing 236, DOI: 10.1007/978-81-322-1602-5_125, © Springer India 2014

ultimate gain and frequency. Later Luyben [2] pioneered the use of relay feedback with describing function analysis for process identification. Accurate process model parameters cannot be estimated because of the measurement noise, which may even fail the test. Sources of measurement noise are many subsystems those do not possess low-pass characteristics. Using a relay with hysteresis, the effect of measurement noise can be reduced [1]. Majhi [3] proposed relay-based identification of integrating process model parameters by state space technique. Recently Panda et al. [4] proposed estimation of integrating and time delay processes using single relay feedback test. Liu and Gao [5] derived exact relay response expressions for integrating and unstable processes by using Newton-Raphson iteration method. Estimation of parameters of unstable processes with time delays is a difficult task. Because the limit cycle exists only when the ratio of time delay to unstable time constant is less than 0.693 [6, 7]. Majhi and Atherton [8] proposed online tuning of process controllers for an unstable first order plus time delay systems by using single relay feedback test and state space technique. Li et al. [9] used two relay tests for estimation of stable and unstable process model parameters. But in their method the use of an additional relay test is tedious and time consuming. Later some authors improved relay autotuning by using Fourier analysis or by exact analysis. These improvements do not get rid of the practical constraints of relay-based estimation. The conventional relay method is an offline identification method. But sometimes offline identification may be dangerous since it affects the operational process regulation which may not be acceptable for certain critical applications. Therefore, the tuning under tight continuous closed loop control often known as online tuning is preferred. Marchetti et al. [10] developed identification method for open-loop unstable processes by using two relay tests. But the above-mentioned methods are not simple and straightforward like the describing function method, and also they are time consuming. It becomes extremely difficult to obtain simple explicit expressions for unstable and integrating process model parameters by exact analysis methods.

This paper, proposes a smart relay and a straightforward method to estimate the process model parameters. The mathematical analyses of the authors' recent publication [11] have been extended here to estimate online parameters of unstable and integrating time delay process models. This paper is organized as follows. Section 2 describes the proposed identification methods and estimation of process model parameters. Simulation examples are presented in Sect. 3. And conclusions are given in Sect. 4.

2 Proposed Identification Methods

This section presents the procedure for online estimation of unstable and integrating process model parameters by relay-based closed loop tests.

The conventional online identification scheme is shown in Fig. 1. Here it can be seen that relay is connected in parallel with a PID controller $G_c(s)$ in the loop. Figure 2 shows equivalent representation of Fig. 1. For the purpose of identification,

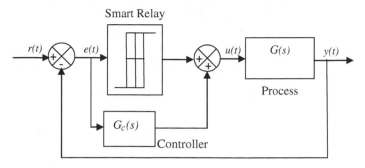

Fig. 1 Online identification scheme

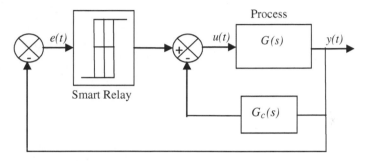

Fig. 2 Equivalent representation of Fig. 1

reference input in Fig. 2 is considered to be zero. The process gets stabilizing signal from the inner feedback controller $G_c(s)$ thereby improving its stability during identification.

Switching function such as ideal symmetrical relay encounters practical problems under real-time implementation. Multiple switching of relay may occur due especially to the measurement noise. Rustamov et al. [12] have proposed a method for synthesis of a relay regulator with fuzzy switching times on the basis of a macrovariable. A smart relay to overcome the problems associated with multiple switching is implemented with the help of fuzzy rules. The smart relay is assumed to have a nonlinear gain, M, expressed as

$$M = \frac{4h(\sqrt{A^2 - \varepsilon^2} - j\varepsilon)}{\pi A^2} \tag{1}$$

where h and ε are parameters of the relay and A is the amplitude of the relay input signal.

2.1 Unstable FOPDT Process Model

Let the dynamics of $G(s)$ be represented by the unstable FOPDT transfer function model

$$G_m(s) = \frac{K_1 e^{-\theta_1 s}}{T_1 s - 1} \tag{2}$$

Here, the model has three unknown parameters K_1, T_1 and θ_1. where K_1, T_1 and θ_1 are the steady-state gain, time constant, and the time delay respectively. Let the form of the PID controller be

$$G_c(s) = K_p \left(1 + \frac{1}{T_i s} + \frac{T_d s}{\gamma T_d s + 1} \right) \tag{3}$$

where K_p, T_i, T_d and γ are proportional gain, integral time constant, derivative time constant and derivative filter constant respectively. Since γ is very small, the derivative filter term in Eq. (3) is neglected in the following analysis.

2.1.1 Estimation of T_1 and θ_1

Here T_1 and θ_1 of the process model are estimated online from the measurements of peak amplitude (A) and time period (P) of the limit cycle output signal, for any non-zero settings of the relay height and hysteresis, i.e., $h \neq 0$ and $\varepsilon \neq 0$. The following condition should be satisfied for a periodic solution to correspond to a stable limit cycle,

$$M \bar{G}_m(j\omega) = -1 \tag{4}$$

where $\omega = \frac{2\pi}{P}$ and

$$\bar{G}_m(j\omega) = \frac{G_m(j\omega)}{1 + G_c(j\omega) G_m(j\omega)} \tag{5}$$

Then, Eq. (4) can be written as

$$G_m(j\omega) [M + G_c(j\omega)] = -1 \tag{6}$$

Substitution of $G_m(j\omega)$, M and $G_c(j\omega)$ in Eq. (6) and solving gives

$$\frac{K_1 e^{-j\omega\theta_1}}{j\omega T_1 - 1} (a + jb) = -1 \tag{7}$$

where

$$a = \frac{4h\sqrt{A^2 - \varepsilon^2}}{\pi A^2} + K_p$$

$$b = K_p \omega T_d - \frac{K_p}{\omega T_i} - \frac{4h\varepsilon}{\pi A^2}$$

The steady-state gain K_1 can be estimated by several methods. Equating the magnitude and phase angle of both sides of Eq. (7), gives

$$T_1 = \frac{\sqrt{K_1^2(a^2 + b^2) - 1}}{\omega} \tag{8}$$

$$\theta_1 = \frac{\tan^{-1}\left(\frac{b}{a}\right) + \tan^{-1}(\omega T_1)}{\omega} \tag{9}$$

T_1 and θ_1 are estimated using Eqs. (8) and (9), respectively.

2.2 Integrating SOPDT Process Model

Next, let the dynamics $G(s)$ be represented by an integrating SOPDT process model transfer function as

$$G_m(s) = \frac{K_2 e^{-\theta_2 s}}{s(T_2 s + 1)} \tag{10}$$

The model parameters are K_2, T_2 and θ_2.

2.2.1 Estimation of T_2 and θ_2

The steady-state gain K_2 can be estimated by several methods. Measurements of peak amplitude (A) and time period (P) made on the limit cycle output signal for any non-zero settings of the relay height and hysteresis, i.e., $h \neq 0$ and $\varepsilon \neq 0$ are used to estimate the model parameters. Repeating the procedure given in Sect. 2.1 the loop gain is expressed as

$$\frac{K_2 e^{-j\omega\theta_2}}{j\omega(j\omega T_2 + 1)}(a + jb) = -1 \tag{11}$$

where

$$a = \frac{4h\sqrt{A^2 - \varepsilon^2}}{\pi A^2} + K_p$$

$$b = K_p \omega T_d - \frac{K_p}{\omega T_i} - \frac{4h\varepsilon}{\pi A^2}$$

Equating the magnitude and phase angle of both sides of Eq. (11), gives

$$T_2 = \frac{\sqrt{\frac{K_2^2(a^2+b^2)}{\omega^2} - 1}}{\omega} \qquad (12)$$

$$\theta_2 = \frac{\frac{\pi}{2} + \tan^{-1}\left(\frac{b}{a}\right) - \tan^{-1}(\omega T_2)}{\omega} \qquad (13)$$

Thus the parameters T_2 and θ_2 are estimated from Eqs. (12) and (13) respectively.

2.3 Initial PID Controller Parameters

The relay induces stable limit cycle output when the fictitious process model $\bar{G}_m(s)$ satisfies the condition in Eq. (3). Extensive simulation studies show that the choice of initial controller parameters $K_p = 0.01$, $T_i = 1$ and $T_d = 0.25$, results in a stable limit cycle output. The relation $T_i = 4T_d$ is maintained for good transient response of the closed loop system.

3 Simulation Results

Example 1
Consider the unstable FOPDT process transfer function [10]

$$G(s) = \frac{e^{-0.4s}}{s - 1}$$

The smart relay with the parameters $(h, \varepsilon) = (1, 0.1)$ induces the limit cycle output as shown in Fig. 3. Substituting the peak amplitude and time period of the output signal in Eqs. (8) and (9), the model parameters are estimated as $T_1 = 0.8293$ and $\theta_1 = 0.4285$. Initial PID controller parameters chosen are $K_p = 0.01$, $T_i = 1$ and $T_d = 0.25$. To validate the proposed method, Nyquist plots of actual process transfer function, and the identified model are shown in Fig. 4. It is evident from the plots that

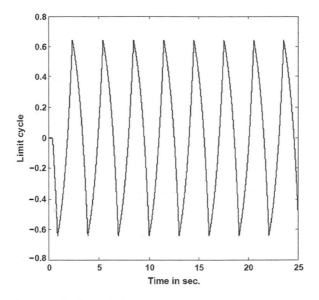

Fig. 3 Limit cycle output for Example 1

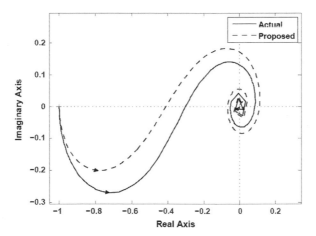

Fig. 4 Nyquist plots for Example 1

the proposed method can be used to estimate online the process model parameters of unstable FOPDT systems.

Example 2

Consider the integrating SOPDT process transfer function [3]

$$G(s) = \frac{e^{-1.5s}}{s(5s+1)}$$

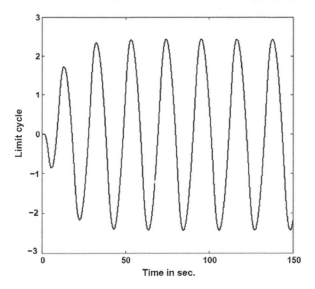

Fig. 5 Limit cycle output for Example 2

The peak amplitude and time period are measured from the limit cycle output shown in Fig. 5 resulting from the autotuning test with the relay parameters $(h, \varepsilon) = (1, 0.1)$. Substituting the measured values in Eqs. (12) and (13), the model parameters $T_2 = 5.0344$ and $\theta_2 = 1.636$ are estimated. K_2 is assumed to be known apriori. Initial PID controller parameters chosen are $K_p = 0.01$, $T_i = 1$, and $T_d = 0.25$. The Nyquist plots of actual process transfer function, and that of the process model are compared in Fig. 6. It can be concluded from the proximity of the plots that the proposed method can be used to estimate the process model parameters of integrating SOPDT systems online.

4 Conclusion

Online estimation technique for process model parameters of integrating and unstable processes with time delay is presented. A smart relay is used to overcome the difficulties associated with the conventional symmetrical relay. Two examples, one each for an integrating process and unstable process, are considered to validate the proposed method and results are compared using Nyquist plots. Automatic tuning of the controllers, as and when required, can be done employing the smart relay that induces sustained oscillatory output.

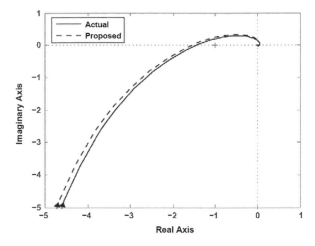

Fig. 6 Nyquist plots for Example 2

References

1. Åström, K.J., Hägglund, T.: Automatic tuning of simple regulators with specifications on phase and amplitude margins. Automatica **20**, 645–651 (1984)
2. Luyben, W.L.: Derivations of transfer functions for highly non-linear distillation columns. Ind. Eng. Chem. Res. **26**, 2490–2495 (1987)
3. Majhi, S.: Relay based identification of processes with time delay. J. Process Control **17**, 93–101 (2006)
4. Panda, R.C., Vijayan, V., Sujatha, V., Deepa, P., Manamali, D., Mandal, A.B.: Parameter estimation of integrating and time delay processes using single relay feedback test. ISA Trans. **50**, 529–537 (2011)
5. Liu, T., Gao, F.: Identification of integrating and unstable processes from relay feedback. Comput. Chem. Eng. **38**, 3038–3056 (2008)
6. Atherton, D.P.: Improving accuracy of autotuning parameter estimation. In: Proceedings of the IEEE International Conference on Control Applications, pp. 51–56, Hartford (1997)
7. Atherton, D.P., Majhi, S.: Plant parameters identification under relay control. In: Proceedings of the IEEE Conference on Decision and Control, pp. 1272–1277, Tampa (1998)
8. Majhi, S., Atherton, D.P.: Online tuning of process controllers for an unstable FOPDT. IEEE Proc. Control Theory Appl. **147**(4), 421–427 (2000)
9. Li, W., Eskinat, E., Luyben, W.L.: An improved auto tune identification method. Ind. Eng. Chem. Res. **30**, 1530–1541 (1991)
10. Marchetti, G., Scali, C., Lewin, D.R.: Identification and control of open-loop unstable processes by relay methods. Automatica **37**, 2049–2055 (2001)
11. Bajarangbali, Majhi, S.: Relay based identification of systems. IJSER 3(6)ISSN. 2229–5518 (2012)
12. Rustamov, G.A., Namazov, M.B., Misrikhanov, L.M.: Synthesis of a relay regulator with fuzzy switching times on the basis of a macro-variable. Autom. Control Comput. Sci. **41**(3), 158–163 (2007)

Software Cost Estimation Using Similarity Difference Between Software Attributes

Divya Kashyap and A.K. Misra

Abstract The apt estimate of the software cost in advance is one of the most challenging, difficult and mandatory task for every project manager. Software development is a critical activity which requires various considerable resources and time. A prior assessment of software cost directly depends on the expanse of these resources and time, which in turn depends in the software attributes and its characteristics. Since there are many precarious and dynamic attributes attached to every software project, the accuracy in prediction of the cost will rely on the prudential treatment of these attributes. This paper deals with the methods of selection, quantification and comparison of different attributes related to different projects. We have tried to find the similarity difference between project attributes and then consequently used this difference measurement for creating the initial cost proposals of any software project that has some degree of correspondence with the formerly completed projects whose total cost is fairly established and well known.

Keywords Software development cost · Software attributes · Cost estimation · k-nearest neighbor classifier · Analogy and similarity difference

1 Introduction

Accurate software cost approximation is a widely researched issue related to every industry engaged in software design and development. Despite many estimation models that have been proposed till date, there is a continuous requirement of new

D. Kashyap (✉)
Department of Computer Science and Engineering, MNNIT, Allahabad, India
e-mail: div.kashyap@gmail.com

A. K. Misra
Department of Computer Science and Engineering, MNNIT, Allahabad, India
e-mail: akm@mnnit.ac.in

B. V. Babu et al. (eds.), *Proceedings of the Second International Conference on Soft Computing for Problem Solving (SocProS 2012), December 28–30, 2012*, Advances in Intelligent Systems and Computing 236, DOI: 10.1007/978-81-322-1602-5_126, © Springer India 2014

systems to forecast the software cost because of continuous and swift changes occurring in the sphere of information technology and software development life cycle.

A prototype is always needed that span through all the factors or attributes that are consequent with the project and consequently forms precise development cost estimate. "Cost estimation is a critical and arduous job which is mandatorily required to silhouette the future software development activities" [1]. It is also required for software overall project management, contract negotiations, scheduling, resource allocation and project planning [2, 3]. Thus it may be informed that the timely production of fully-functional and quality software directly depends on the initial estimates of the required cost, effort and resources.

All of the cost prediction models [4], proposed till date, fall in one of the two categories:

(a) Parametric or Algorithmic models.
(b) Non-Algorithmic methods.

Tables 1 and 2 lists various models/methods belonging to these categories.

Algorithmic models [4] are based on the mathematical formulas and depend on the measurements and processing of certain project attributes. Most of the algorithmic models are of the form of:

Table 1 Parametric or algorithmic models

1	COCOMO–II [4–6]
2	Function points [7, 8]
3	Putnam slim [9]
4	SEER SEM [10]
5	SLOC [11]
6	The Doty model [12]
7	Price-S model [13]
8	Walston-Felix multiplicative model [14]
9	Estimacs [15, 16]
10	Checkpoint [17]

Table 2 Non-algorithmic methods

1	Analogy costing [18–20]
2	Expert judgment [21, 22, 24, 25]
3	Parkinson method [26–28]
4	Price-to-win [29]
5	Bottom-up approach [30]
6	Top-down approach [30]
7	Delphi [31]
8	Machine learning [32]

$$\text{Size (KLOC)} = f(\text{cost affecting factors})$$
$$\text{Effort (person months)} = f(\text{size})$$
$$\text{Software cost (INR)} = f(\text{Effort})$$

Cost factors are the variables or project attributes, numerous in numbers that simultaneously effect the software development and the form of function f varies from model to model.

Non-Algorithmic models are the proposed heuristics that requires a considerable reasoning, logic and a large knowledge base. These methods are generally based on the terminologies like *"learning by experience"* or *"trial by case studies"*. Non-algorithmic methods are based on discovery while algorithmic methods focus on calculation.

Algorithmic methods are good because they have fixed and well defined step by step procedure to deliver out the estimate.

Non-Algorithmic models are good because they do not require extraneous parameters that require tuning or calibration according to the measurement environment. In this paper combined both of these approaches in order to enjoy the benefits of both strategies.

A previously completed software project may shed some light on the cost estimation of the current project if both of them have similar attributes and characteristics. Although there exists many factors that describe software project; tried to sort out some the essential nominal, linguistic and numerical attributes [33–35] that prominently affect the software development effort. Then, to find the cost of new software, we find the previously completed projects that have a least similarity difference with the new project in terms of these attributes. To collect these similar projects we have used clustering algorithm to create a cluster of similar projects. Then, we have tried to deliver out the possible cost range of the new project based on the costs of these clustered projects. The flow of our paper goes in the following manner: In section two we have discussed various types of attributes and the operation that can be applied on them. In section three we have proposed and algorithm to create a cluster of projects that has a least similarity difference. Later on in section four we have tried to figure out the performance of our algorithm on some predefined project datasets.

2 Software Attributes

Obviously, the similarity between the bygone projects and the current project is gauged on the basis of some parameters or project characteristics. To find out these cost affecting attributes is a separate area of research in itself and researchers have proposed various factors that potentially affect the software development cost. For more information on software attributes we redirect the reader to [36–38].

To establish a correspondence between two projects we have divided various attributes into three categories, nominal attributes, linguistic attributes and numerical attributes. The following sub-section describes these categories and the attributes of interest covered under each category.

2.1 Nominal Attributes (No)

A nominal scale is the lowest level of measurement scale. The attributes covered under this category are categorical. This scale is used to classify a measurement into one of the predetermined class [34, 39]. The attributes covered under this category are:

Name: Implementation programming language.
Values: C, C++, Java, .Net, Scripting, PL/SQL or PHP

Name: Development platform.
Values: Mainframe, mid-range, personal computer or mobile.

Name: Type of software.
Values: System software, application software, programming software, operating system or embedded system.

Name: Existence of overall schedule or plan.
Values: Yes or no.

Name: Use of modern programming practices.
Values: Yes or no.

Name: Software development model in practice.
Values: Waterfall model, prototype model, spiral model, agile development or iterative model.

Name: Project priority.
Values: High or low.

Name: Risks associated with the project.
Values: Low, medium or high.

Name: Type of software release.
Values: Firsthand release or next version.

Similarity difference D_{no} between two projects $P1$ and $P2$ in the nominal attributes space on any nominal attribute (No_i) is calculated as below:

$$D_{no}(P1 \rightarrow No_i \cdot P2 \rightarrow No_i) = \begin{cases} \text{if}(P1 \rightarrow No_i \cdot \text{value} = P2 \rightarrow No_i \cdot \text{value}) \\ \qquad\qquad\qquad \text{then} \quad 0 \\ \qquad\qquad\qquad \text{else} \quad 1 \end{cases}$$

In this category each attribute can have a predefined fixed value or we may say that it belong to a fix class. So to calculate the similarity difference on the basis of some attribute belonging to this category, we directly compare their values. If their values are identical i.e. they can be mapped to same class, the similarity difference between the projects is 0 otherwise the similarity difference would be 1.

2.2 Linguistic Attributes (Li)

Some software attributes neither have a concrete class, into which they can be classified nor have a fix numerical value. These attributes are known as linguistic attributes and these can be measured on a relative scale, like: *(very low, low, medium, high, and very high)* [40, 41]. Although this scale is partial and imprecise but it is quick, easy and less efforts are made to optimize it. For maintaining standards we have taken all the COCOMO-II [29] cost drivers under this category. It uses 17 cost drivers [5, 30, 42] named as: *RELY, CPLX,DOCU,DATA,RUSE,TIME,PVOL,STOR,ACAP,PCON, PCAP,PEXP, AEXP, LTEX,TOOL,SCED,SITE and five scaling factors named as*: *Precedentedness (PREC), Development Flexibility (FLEX), Risk Resolution (RESL), Team Cohesion (TEAM), Process maturity (PMAT)*. Out of these five scale factors four will be assumed nominal, while the remaining fifth (PMAT) rating will based on the development process maturity level, and hence, will vary according to the rating of process maturity level.

Similarity difference D_{ls} between two projects $P1$ and $P2$ in the linguistic attributes space on any linguistic attribute (Li_i) is calculated as below:

$$D_{Li}(P1 \rightarrow Li_i, P2 \rightarrow Li_i) = \begin{cases} \text{if } (P1 \rightarrow Li_i, \text{value} = P2 \rightarrow Li_i, \text{value}) \\ \qquad \text{then} \quad 0 \\ \text{elseif } (P1 \rightarrow Li_i, \text{value} = P2 \rightarrow Li_i \text{ value}, \pm 1) \\ \qquad \text{then} \quad 0.5 \\ \qquad \text{else} \quad 1 \end{cases}$$

In this category each attribute is measured on a relative scale varying from 1 to 5. If the compared attribute have the same value then the similarity difference between projects is 0. Or if according to attribute values, they belong to neighbor classes (*very low and low* or *low and medium* or *medium and high* or *high and very high*) their similarity difference is 0.5. Otherwise the similarity difference is 1.

2.3 Numerical Attributes (Nu)

Software attributes for which we can associate a fix and precise numerical value, can be termed as numerical attributes. Some of the software numerical attributes that we have considered are as follows:

Name: Functions points [7, 8]
Name: Quality index [43, 44]
Name: Project duration (in months)
Name: Team size
Name: Number of consultants.
Name: Number of intended users.

Similarity difference D_{nu} between two projects $P1$ and $P2$ in the numerical attributes space on any numeric attribute (Nu_i) is calculated as below:

$$D_{nu}(P1 \rightarrow Nu_i, P2 \rightarrow Nu_i) = \begin{cases} \text{if} (P1 \rightarrow Nu_i, \text{value} = P2 \rightarrow Nu_i, \text{value}) \\ \qquad \text{then} \quad 0 \\ \text{elseif} (P1 \rightarrow Nu_i, \text{value} = P2 \rightarrow Nu_i, \text{value} \pm 5\%) \\ \qquad \text{then} \quad 0.25 \\ \text{elseif} (P1 \rightarrow Nu_i, \text{value} = P2 \rightarrow Nu_i, \text{value} \pm 10\%) \\ \qquad \text{then} \quad 0.5 \\ \text{elseif} (P1 \rightarrow Nu_i, \text{value} = P2 \rightarrow Nu_i, \text{value} \pm 20\%) \\ \qquad \text{then} \quad 0.75 \\ \qquad \text{else} \quad 1 \end{cases}$$

In this category each attribute can have any numerical value. If the value of the attribute of the source project and the target project is same then the similarity difference between the projects would be 0 otherwise the similarity difference would be 0.25 or 0.5 or 0.75 if the target projects attribute value lies in the range of $\pm 5\%$ or $\pm 10\%$ or $\pm 20\%$ target projects attribute value respectively. In the absence of both of the above cases the similarity difference would be 1.

3 Similarity Difference Measurement

Hitherto, we have the target software described in the form of various attributes and we have a set of bygone software projects whose cost is properly established and these projects also have the required set of attributes. Now we have to find the projects from the database that are similar to target software on the basis of these attributes. Apparently we are estimating the cost using analogy. Although human intelligence is the best live example of analogy based reasoning, lot of software tools like ANGEL [45], ESTOR [46], DD-EbA [47], ANALOGY-X [48] and Fuzzy classification [34] are also present in the market for software cost estimation. By combining the results from various algorithms that are using analogy and some latest researches on improving these results in presence of imprecisions, noise and outliers [49], we are in a position to say that these methods have outperformed many of the parametric and non-algorithmic methods we quote in Sect. 1.

Our algorithm is a further improvement over the other methods because all the existing methods have not explored various categories of attributes. So, they didn't have separate methods to map attributes belonging to different categories. Also, we have used k-nearest neighbor [50] classifier [49] to form a cluster of k-similar projects in the attribute space. These projects form a set of k closest projects which have the least similarity difference. Finally the cost of target project is proposed according to the cost of these k similar projects. The following paragraphs we have discussed the mathematical model of the problem.

Any entity SP $(No_{1...i}, Li_{1...j}, Nu_{1...p})$ is defined as a software project SP, described on the basis of three categories of attributes i.e. Nominal attributes (No), Linguistic attributes (Li) and Numerical attributes (Nu). No_x, Li_y and Nu_z, specifies the value of xth, yth and zth attribute in the respective category. In context to the above statement the following algorithm tends to find out the cost estimate of a target software project SP^T according to the cost of k most similar projects from the database of n old projects $SP_{1...n}$. The value D_{tn} is the similarity difference between the target software project and the nth old project present in the database and it is calculated according to the following algorithm.

Algorithm 1: Finding k-similar projects.

After execution of Algorithm 1, we get a cluster of k existing projects in the similarity difference space. This cluster contains the projects that are most similar to the target project when the projects are compared against the set of attributes describes in the previous section. And the cost of target project SP^T $(No_{1...i}, Li_{1...j}, Nu_{1...p})$ can be estimated cost as the average cost of a project that belongs to the formed cluster, and this relationship can be mathematically expressed as:

$$\text{Cost}\left(SP^T\right) = \frac{1}{k}\sum\nolimits_{i=1}^{k}\text{Cost}\,(SP_i) \tag{1}$$

4 Performance Analysis

To scrutinize the performance of our model, we have used Mean Magnitude Relative Error (MMRE) [51], as a benchmark.

$$\text{MMRE} = \frac{1}{I}\sum\nolimits_{i=1}^{I}\frac{|\text{Actual\_cost}_i - \text{Etimated\_cost}_i|}{\text{actual\_cost}_i} \tag{2}$$

i is the total number of fitness cases. Actual_cost$_i$ and Estimated_cost$_i$ represent the value of actual cost and model calculated cost respectively for the ith project. To set up the test-bed we have used 120 sample projects from ISBSG data repository [52]. Out of these 120 projects, 20 (the value for I is 20) are used as the target projects for which we only know the attributes and rest 100 are used as the data repository of old projects for which the cost as well as the value of software attributes is well known. To recognize importance of number of old projects, we increase the number of old projects gradually in our test from 50 to full 100 with an interval of 5. And from the results we can conclude that MMRE of the similarity difference method is very less and it decreases with an increase in the number of old projects (Fig. 1). It can also be clinched that given a very large database of old projects, MMRE may

Parameters:
Inputs:
i: Total number of Nominal Attributes *(No)*.
j: Total number of Linguistic Attributes *(Li)*.
p: Total number of Numerical Attributes *(Nu)*.
No_x: Value of xth Nominal Attributes; $1 \leq x \leq i$.
Li_y: Value of yth Linguistic Attributes; $1 \leq y \leq j$.
Nu_z: Value of zth Numerical Attributes; $1 \leq z \leq p$.
$SP^T(No_{1...i}, Li_{1...j}, Nu_{1...p})$: Target Software Project.
$SP_{1...n}(No_{1...i}, Li_{1...j}, Nu_{1...p})$: n old projects.
k: Variable; $k << n$.
Output:
Set SP^S of k most similar projects.
Local variables:
D_{no}: Similarity difference in nominal attributes space
D_{li}: Similarity difference in linguistics attributes space
D_{nu}: Similarity difference in numerical attributes space
D_{tn}: Total Similarity difference between SP^T and SP_n
Procedure:

Step 1. set $SP^S = \{\emptyset\}$
Step 2. for each project pr in database $(1 \ldots n)$

        set $D_{no} = 0$
        set $D_{li} = 0$
        set $D_{nu} = 0$
        for each nominal attribute at $(1 \ldots i)$
        set $D_{no}+ = D_{no}(SP^T \rightarrow No_{at}, SP_{pr} \rightarrow No_{at})$
        end for
        for each linguist attribute at $(1 \ldots j)$
        set $D_{li}+ = D_{li}(SP^T \rightarrow Li_{at}, SP_{pr} \rightarrow Li_{at})$
        end for
        for each numerical attribute at $(1 \ldots p)$
        set $D_{li}+ = D_{li}(SP^T \rightarrow Nu_{at}, SP_{pr} \rightarrow Nu_{at})$
        end for
        set $D_{tn} = D_{no} + D_{li} + D_{nu}$
        end for

Step 3. sort SP_n on the values of D_{tn} in increasing order
Step 4. move the top k elements into set SP^S
Step 5. return SP^S

cease to trifle value. In our executions we have taken the value of parameter k as five i.e. we have estimated the cost of a project from the cost of its five nearest neighbors.

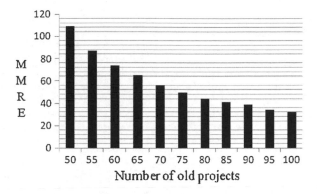

Fig. 1 Performance of cost estimation algorithm based on similarity difference measurement

5 Conclusion and Future Work

Despite of the presence of various software estimation models, software project managers are still encountering the complications and challenges in predicting the cost of forth coming software. It is so because of inherent vitality in the software attributes and wide dimensions of software development methodology. In our present paper we have described a method based on 'similarity difference measure' to estimate the cost of a software project. To calculate the similarity difference between softwares we have defined any software on the basis of three aspects, namely Nominal Attributes, Linguistic Attributes and Numerical Attributes. We have described the methods to calculate the similarity difference for each category. Then we have used this difference to find out k most similar projects or nearest neighbors in the similarity difference space. We have also tried to validate this procedure using MMRE as the error benchmark.

To extend this model in future we are trying the find out more cost affecting attributes in each category, to increase the precision of the model. Secondly we are also trying to reframe the similarity difference measurement procedure by assigning the weights to attributes

References

1. Gray, R., MacDonell, S.G., Shepperd, M.J.: Factors systematically associated with errors in subjective estimates of software development effort: the stability of expert judgment, IEEE 6th Metrics Symposium, IEEE Computer Society, pp. 216–227, Los Alamitos (1999)
2. Nolan, A., Abrahao, S.: Dealing with cost estimation in software product lines: experiences and future directions. Software Product Lines: Going Beyond, pp. 121–135. LNCS, Springer, Berlin (2010)
3. Leung, H., Fan, Z.: Software Cost Estimation. Hong Kong Polytechnic University, Handbook of Software Engineering (2002)

4. Ma, J., and Mu, L.: Comparison Study on methods of software cost estimation, supported by Hebei Provincial Construction of Science and Technology Research Program, 2010 IEEE
5. Boehm, B.W.: Software engineering economics. IEEE Trans. Softw. Eng. **10**(1), 4–21 (1981)
6. Boehm, B. W., Valerdi, R.: Achievements and challenges in cocomo-based software resource estimation, IEEE Softw 25(5), 74–83. doi:10.1109/MS, 2008
7. Zheng, Y., Wang, B.: Estimation of software projects effort based on function point, Proceedings of 4th International Conference on Computer, Science and Education 2009
8. Fu, Y. -F., Liu, X.-D., Yang, R. -N., Du, Y. -L., Li, Y. -J.: A software size estimation method based on improved FPA. Second WRI World Congress on, Software Engineering (2010)
9. Kemerer, C.F.: An empirical validation of software cost estimation models. Commun. ACM **30**(5), 416–429 (1987)
10. GalorathandM, D.D., Evans, W.: Software Sizing, Estimation, and Risk Management: When Performance is Measured Performance Improves. Auerbach, Boca Raton (2006)
11. Pressman, R.S.: Software Engineering: A Practioner's Approach, 6th edn. McGraw-Hill, New York (2005). ISBN 13: 9780073019338
12. Herd, J.R., Postak, J.N., Russell, W.E., Steward, K.R.: Software cost estimation study–Study results, Final technical report, RADC-TR77-220, vol. I. Doty Associates Inc., Rockville (1977)
13. The PRICE software model user's manual. Moorestown, PRICE Systems (1993)
14. Walston, C.E., Felix, C.P.: A method of programming measurement and estimation. IBM Syst. J. **16**(1), 54–73 (1977)
15. Rubin, H.A.: Macroestimation of software development parameters: the ESTIMACS system, In: SOFTFAIR Conference on Software Development Tools, Techniques and Alternatives (Arlington, July 25–28) , pp. 109–118. IEEE Press, New York (1983)
16. Robin, H.A.: Using ESTIMACS E. Management and Computer Services, Valley Forge (1984)
17. Checkpoint for Windows User's Guide, version 2.3.1. Burlington, Software Productivity Research (1996)
18. Walkerden, F., Jeffery, R.: An empirical study of analogy-based software effort estimation. Empirical Softw. Eng. 4, 135–158 (1999). Kluwer Academic Publishers, Boston
19. Kocaguneli, E., Bener, A.B.: Exploiting the essential assumptions of analogy-based effort estimation. IEEE Trans. Softw. Eng. **38**(2), 425–438 (2012)
20. Keung, J.: Software development cost estimation using analogy: A review. Australian, software engineering conference, 2009
21. Cuadrado-Gallego, J. J., Rodríguez-Soria, P., Martín-Herrera, B.: Analogies and differences between machine learning and expert based software project effort estimation. 11th ACIS international conference on software engineering, artificial intelligence, networking and par-allel/distributed, computing (2010)
22. Jorgensen, M.: Practical guidelines for expert-judgment-based software effort estimation. IEEE Softw. **22**(3), 57–63 (2005)
23. Jorgenson, M., Shepperd, M.: A Systematic review of software development cost estimation studies. IEEE Trans. Softw. Eng. **33**(1), 33–53 (2007)
24. Hughes, R.T.: Expert judgment as an estimating method. Inf. Softw. Technol. **38**, 67–75 (1996)
25. Parkinson, G.N.: Parkinson's Law and Other Studies in Administration. Houghton-Miffin, Boston (1957)
26. Leung, H.: Fan . Software Cost Estimation, IEEE Transactions on Software Engineering, Z. (1984)
27. Heemstra, F.J.: Software cost estimation. Inf. Softw. Technol. **34**(10), 627–639 (1992)
28. Putnam, L. H., Fitzsimmons, A.: Estimating software cost. Datamation (1979)
29. Boehm, B.W.: Software Cost Estimation Using COCOMO II. Prentice-Hall, Englewood Cliffs (2000)
30. Albrecht, A.J., Gaffney, J.E.: Software function, source lines of codes and development effort prediction: a software science validation. IEEE Trans. Softw. Eng. **9**, 639–648 (1983)
31. Alkoffash, M.S., Bawanehand, M.J.: Al Rabea, Ai: Which software cost estimation model to choose in a particular project. J. Comput. Sci. **4**(7), 606–612 (2008)

32. Khatibi, V., Jawawi, D.N.A.: Software cost estimation methods: a review. J. Emerg. Trends Comput. Inf. Sci. **2**(1), 21–29 (2011)
33. Idri, A., Abran, A., Khoshgoftaar, T. M.: Estimating software project effort by analogy based on linguistic values metrics. Eighth IEEE international symposium on software metrics (METRICS'02), pp. 21, 2002
34. Idri, A., Abran, A., Khoshgoftaar, T.M.: Fuzzy case-based reasoning models for software cost estimation. Soft Computing in Software Engineering, pp. 64–96. Springer-Verlag, Berlin (2004)
35. Morasca, S., Briand, L. C.: Towards a theoretical framework for measuring software attributes. In: Proceedings of the Fourth International Symposium on Software Metrics, Albuquerque, November 1997, pp. 119–126. IEEE Computer Society (1997)
36. Sommerville, I.: Software Engineering, 7th edn. Addison-Wesley, Boston (2004)
37. Clarke, P., O'Connor, R.V.: The situational factors that affect the software development process: Towards a comprehensive reference framework. J. Inf. Softw. Technol. **54**(5), 433–447 (2012)
38. Lagerström, R., von Würtemberg, L.V., Holm, H., Luczak, O.: Identifying factors affecting software development cost. Proceedings of the Fourth International Workshop on Software Quality and Maintainability (SQM), March, In (2010)
39. Raschia, G., Mouaddib, N.: SAINTETIQ: a fuzzy set-based approach to database summarization. Fuzzy Sets Syst. **129**, 137–162 (2002)
40. Zadeh, L.A.: Fuzzy set. Inf. Control **8**, 338–353 (1965)
41. Zadeh, L.A.: A computational approach to fuzzy quantifiers in natural languages. Comput. Math. **9**, 149–184 (1983)
42. Khatibi, V., Jawawi, D.N.A.: Software cost estimation methods: A review. CIS J. **2**, 21–29 (2011)
43. McCall, J. A., Richards, P. K., Walters, G. F.: Factors in software quality. Technical report RADC-TR-77-369. U.S. Department of Commerce, Washington, DC (1977)
44. Pressman, R.S.: Software Engineering: A Practitioner's Approach, 5th edn. McGraw-Hill series in computer science, New York (2001)
45. Shepperd, M., Schofield, C.: Estimating software project effort using analogies. IEEE Trans. Softw. Eng. **23**(12), 736–743 (1997)
46. Mukhopadhyay, T., Vicinanza, S., Prietula, M.J.: Examining the feasibility of a case-based reasoning model for software effort estimation. MIS Quart. **16**(2), 155–171 (1992)
47. Kosti, M. V., Mittas, N., Angelis, L.: DD-EbA: An algorithm for determining the number of neighbors in cost estimation by analogy using distance distributions, 3rd artificial intelligence techniques in software engineering workshop, Larnaca, 7 October 2010
48. Keung, J.W., Kitchenham, B.A., Jeffery, D.R.: Analogy-X: Providing statistical inference to analogy-based software cost estimation. IEEE Trans. Softw. Eng. **34**(4), 471–484 (2008)
49. Shepperd, M., Kadoda, G.: Using simulation to evaluate predictions systems. In: Proceedings of the 7th International Symposium on Software Metrics, England, pp. 349–358. IEEE Computer Society (2001)
50. Coomans, D., Massart, D.L.: Alternative k-nearest neighbour rules in supervised pattern recognition: Part 1. k-nearest neighbour classification by using alternative voting rules. Anal. Chim. Acta **136**, 15–27 (1982). doi:10.1016/S0003-2670(01)95359-0
51. Shin, M., Goel, A.L.: Emprirical data modeling in software engineering using radial basis functions. IEEE Trans. Softw. Eng. **26**, 567–576 (2000)
52. ISBG: Online data repository. http://www.isbsg.org/. Accessed 28 Feb 2012

Mining Knowledge from Engineering Materials Database for Data Analysis

Doreswamy and K. S. Hemanth

Abstract With growing science and technology in manufacturing industry, an electronic database as grown in a diverse manner. In order to maintain, organizing and analyzing application-driven databases, a systematic approach of data analysis is essential. The most succeeded approach for handling these problems is through advanced database technologies and data mining approach. Building the database with advance technology and incorporating data mining aspect to mine the hidden knowledge for a specific application is the recent and advanced data mining application in the computer application domain. Here in this article, association rule analysis of data mining concepts is investigated on engineering materials database built with UML data modeling technology to extract application-driven knowledge useful for decision making in different design domain applications.

Keywords Data mining · Object-oriented · Advanced engineering materials · Association rule

1 Introduction

Advancement in sensing and digital storage technologies and their dramatic growth in the applications ranging from market analysis to scientific data explorations have created many high-volume and high- dimensional data sets. Most of the data stored in electronic media have influenced the development of efficient mechanisms of maintaining and automatic information retrieval for data analysis and data summarize.

Doreswamy · K. S. Hemanth (✉)
Department of Computer Science, Mangalore Univeristy, Mangalagangotri,
Mangalore, Karnataka 574199, India
e-mail: doreswamyh@yahoo.com

K. S. Hemanth
e-mail: reachhemanthmca@gmail.com

B. V. Babu et al. (eds.), *Proceedings of the Second International Conference on Soft Computing for Problem Solving (SocProS 2012), December 28–30, 2012*, Advances in Intelligent Systems and Computing 236, DOI: 10.1007/978-81-322-1602-5_127, © Springer India 2014

In order to maintain and efficiently analyze the high-dimensional data, the advanced DBMS and the data analysis methods respectively contribute the data organizing methods and knowledge discovery from large database. To represent the concept of both technologies a growing advance engineering materials property database is considered. Increasing engineering technology in civil construction, automobiles, and aerospace, etc., emerges with new engineering materials with new features. This leads complexity in maintaining the database through relational data model. This motivates for the construction of an advanced object-oriented database model (OODM) for engineering materials data [1]. By constructing the OODM for materials data, it is possible to maintain and extract the information from the large database in an efficient manner. Extracting knowledge from the OODM database for the selection of materials for different applications required an efficient and modern technology. Most popular and effective method for extracting hidden knowledge from the complicated database is by using data mining technique [2]. Construction of highly effective data mining rules for the selection of materials is investigated. Quantitative association rules are built on performance of mechanical behavior of materials. These rules filter the materials according to the application and support to process of selection of materials.

The remaining sections in the paper are organized as follows: in Sect. 2 a brief review of advanced engineering materials database and construction of a database is discussed. Extraction of relevant information through mining concept is presented in Sect. 3. Implementation and experimental results are discussed in the Sect. 4. And the conclusion is discussed in the Sect. 5.

2 Advanced Engineering Materials' Database

An engineering materials database is a database used to store empirical data of materials in such a way that they can be retrieved efficiently by humans or computer programs. Engineering materials databases are the basis of materials informatics [3], manufacturing industries [4], and the related disciplines [3]. Design and development of materials databases are done to enrich the availability and accessibility of materials data to materials scientist, researcher, and design engineers in manufacturing industries [5, 6].

Materials database consist of a variety of information about materials behaviour and manufacturing process [7]. Currently, in the market a few major online materials' databases provide rich information about material properties and manufacturing processes in an ad hac manner. The major advantages of materials database are to provide information about materials behavior to engineer while selecting the material for engineering applications.

Thus, it is an important task in engineering materials field (Material Informatics), especially in informatics to investigate searchability and comparability in existing knowledge. Developing an advanced database, which supports for knowledge system, is the core task of the research in computer application.

One of the most promising approaches to advanced database support for engineering applications is the concept of advanced object-oriented database management systems (AOODBMS) [8]. Due to the object-oriented nature of the database model, it is much simpler to approach a problem with these needs in terms of objects. Code can be directly applied to a database, and thus saves time and money in development and maintenance [9].

The main intention of developing an object-oriented database models for advance engineering materials data is to (1) increase searchable and comparable of existing knowledge about materials data in large databases, (2) increase availability of information and expandability of database for future, and (3) expanding the knowledge on proper application of estimation methods of material behavior and mainly for mining prospective.

An object-oriented database model for advanced engineering materials is implemented in C# .Net technology. A GUI model is designed for accessing and analyzing the data without interacting the back end database, the database can be retrieved; updated and new data can be inserted from the GUI module. Complete process of construction of object-oriented database is explained in the form of UML class diagram. The class diagram consists of 12 classes (Engineering materials (), Reinforcement Materials (), Matrix Materials (), Thermal_Prop (), Environment _Prop (), Mechanical_Prop (), Electrical _Prop (), Chemical_Prop (), Optical_Prop (), Magnetic_Prop (), Manufacturing_Prop (), and Acoustical_Prop (). Engineering materials class is the main class where the information of requested materials is being retrieved. Engineering materials class is associated with metric and reinforcement class, therefore objects of reinforcement and matrix class can be described through engineering materials class. Engineering materials class is also associated with Thermal_Prop, Environment_Prop, Mechanical_Prop, Electrical_Prop, Chemical_Prop, Optical_ Prop, Magnetic_Prop, Manufureturing_Prop, and Acoustical_Prop classes. Similar to the other class, certain objects can be desirable through engineering materials class. The class diagram of the object-oriented data model for engineering materials is shown in Fig. 1.

3 Extraction of Relevant Information Through Quantitative Association Rules Mining

Selection of materials for application requirement is purely depending on the performance and the cost of materials. Here, we consider the performance-based selection through quantitative association rule mining. Usually materials are analyzed by their performance like strength, toughness, rigidity, durability, and formability [4, 6]. This performance is analyzed by their experimental results stored in the database as different features(Table 1). The availability of detailed information on materials experimental values has led to the development of techniques that automatically look for association between each behavior that are quantitatively stored in the database [10, 11].

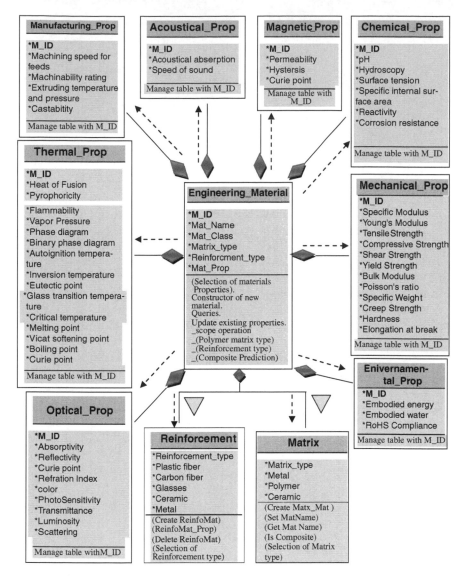

Fig. 1 The static structure of classes in the system with its relationship through UML notation

4 Implementation and Experimental Results

In this article an advanced object-oriented engineering material database is implemented in C# .Net technology. Materials data are collected from website and handbooks of engineering materials and materials science books and blogs [6, 12–14]. Mainly engineering materials data supporting for Civil engineering, Automobiles,

Table 1 The database samples

Mat_type	bM	cS	eB	sS	sM	tS	yS	yM	Hrds
Polymer	0	0	2	0.004	0.015	0.9	11.24	0.016	1,165
Polymer	0	0	8	1.2	0.147	11	12.3	0.14	1,035
Ceramic	15	23.47	0	1.7	3	421.1	0	10.2	905
Ceramic	23	42.11	0	5.6	5.95	463.1	0	32.1	775
Metal	0	0	4.3	5.1	1.6	19.6	1.3	3.1	645
Composite	0.0018	0	255	2	41.6	200	90	20	385
Composite	235	950	0	10	45.3	1500	0	250	255

Minimum support = 50 % and Minimum Confidence = 60 %

Where *bM* Bulk Modulus, *cS* Compressive Strength, *eB* Elongation at Break, *sS* Shear strength, *sM* Specific Modulus, *tS* Tensile Strength, *yS* Yield Strength, *yM* Young Modulus Strength, *Hrds* Hardness

Table 2 Representation of qualitative association rules constructed for selecting the materials for different application requirements

Rules	Support (%)	Confidence (%)
Strength => <Material(), tS :1000..1500> And <Material(), yS:500..1000> And <Materials(), sS:0.1..100> And <Materials(),cS:18..200>	50	100
Strength=> <Material(), tS:...> And <Materials(),yS:...>	50	70
Rigidity=><Materials(),yM:10..100> And <Materials(),sM:0.1..100)	50	100
Rigidity=><Materials(),yM:10..100>	50	50
Formability=><Materials(),eB:100..500> And <Materials(),Dsty;0.1..3>	50	100
Formability=><Materials(), Dsty:0.1..3>	50	50
Toughness=><Materials(),eB:100..500> And <Materials(),yS:0.85..70>	50	100
Toughness=> <Materials(),eB:100..500>	50	50
Durability=> <Materials(),Hrds:1000..2000> And <Materials(),tS:0.85..200>	50	100
Durability=> <Materials(),tS:0.85..200>	50	50

and Aerospace applications are collected. The database consists of 1,299 materials dataset with eight properties tables. Total of 32 behaviors are considered and organized in eight property tables, which are related to Engineering applications. In this experiment performance criteria are depend on mechanical properties. In mechanical property table, nine behaviors (attribute) are used to construct the rules. Each time different application is selected through the GUI and the corresponding materials are analyzed. The Table 3 shows the percentage of materials selection under different classes for different applications.s

Experimental results are shown in Table 2 which represent materials which satisfy the qualitative association rules that supports the required performance by a different

Table 3 The percentage of materials satisfying QA rules under different classes' database for different applications

Materials class	Civil engineering materials (%)	Aerospace engineering materials (%)	Automobiles engineering materials (%)
Polymer	5	68	30
Ceramic	30	2.5	00
Metal	58	20	35
Composite	10	70	62

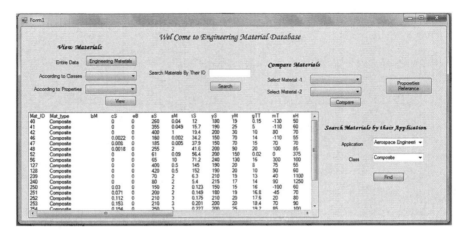

Fig. 2 A GUI model for QA rule analysis

application. A GUI interface for selecting materials for engineering application is represented in Fig. 2.

5 Conclusion

In this paper advanced object-oriented data model for advanced engineering materials data is constructed in order to maintain and summarizing the materials data effectively and user friendly. Through data mining approach, quantitative association rules are investigated on materials datasets for selecting the materials based on their performance for different engineering applications. The model is compatible with effective data analysis rules. The proposed quantitative association rules can be used for the retrieval of effective information from the large application-driven engineering materials database.

References

1. Walls, M.D.: Data Modeling, 2nd Revised edn. URISA, Park Ridge (2007)
2. Han, J., Kamber, M., Pie, J.: Data Mining: Concepts and Techniques, 3rd edn. Margan Kaufmann, Burlington (2012)
3. Rajan, K.: Materials informatics. Mater. Today **8**(10), 38–45 (2005)
4. Ashby, M.F.: Materials Selection in Mechanical Design, 3rd edn. Pergamon Press, Oxford (2005)
5. Doreswamy, Manohar, M.G., Hemanth, K.S.: Object-oriented database model for effective mining of advanced engineering materials data sets. In: The Second International Conference on Computer Science Engineering and Applications (CCSEA-2012), pp. 129–137 (2012)
6. Budinski, K.G.: Engineering Materials Properties and Selection, 5th edn. Prentice Hall Publishing, New York (2000)
7. Callister, W.D Jr.: Materials Science and Engineering, 5th edn. Wiley, New York (2000)
8. Umoh, U.A., Nwachukwu, E.O., Eyoh, I.J., Umoh, A.A.: Object oriented database management system: a UML design approach. Pacific J. Sci. Technol. **10**(2), 355–365 (2009)
9. Satheesh, A., Patel, R.: Use of object-oriented concepts in databases for effective mining. Int. J. Comput. Sci. Eng. **1**(3), 206–216 (2009)
10. Srikant, R., Agrawal, R.: Mining quantitative association rules in large relational tables. In: Proceedings of the ACM-SIGMOD: Conference on Management of Data, Montreal, Canada, June 1996
11. Watanabe, T., Takahashi, H.: A study on quantitative association rules mining algorithm based on clustering algorithm. Biomed. Soft Comput. Hum. Sci. **16**(2), 59–67 (2011)
12. Online database for materials' properties. http://www.makeitfrom.com/ (2012)
13. Online material data obtained from literature research and from experiments performed during work on projects and doctoral thesis. http://www.matdat.com/ (2012)
14. Online materials properties database. http://www.matweb.com/ (2012)

Rule Based Architecture for Medical Question Answering System

Sonal Jain and Tripti Dodiya

Abstract As the wealth of online information is increasing tremendously, the need for question-answering systems is evident. Current search engines return ranked lists of documents to the users query, but they do not deliver the precise answer to the queries. The goal of a question-answering system is to retrieve answers to questions rather than full documents or best-matching passages, as most information retrieval systems currently do. Patients/Medical students have many queries related to the medical terms, diseases, and its symptoms. They are inquisitive to find these answers using search engines. But due to keyword search used by search engines it becomes quite difficult for them to find the correct answers for the search item. This paper proposes the architecture of question-answering system for medical domain and discusses the rule-based question processing and answers retrieval. Rule formation for retrieval of Answers has also been discussed in the paper.

Keywords Question answering · Question processing · Document processing · Answer processing

1 Introduction

Internet has made a tremendous wealth of information available which is accessible through information retrieval (IR) search engines. The search engines return ranked lists of documents to the users query, but they do not deliver the precise answer to the queries. Question-answering (QA) systems are a step ahead of

S. Jain (✉)
J K Laksmipat University, Jaipur, India
e-mail: drsonalamitjain@gmail.com

T. Dodiya
GLS Institute of Computer Applications, Ahmedabad, India
e-mail: triptidodiya@glsica.org

B. V. Babu et al. (eds.), *Proceedings of the Second International Conference on Soft Computing for Problem Solving (SocProS 2012), December 28–30, 2012*, Advances in Intelligent Systems and Computing 236, DOI: 10.1007/978-81-322-1602-5_128, © Springer India 2014

Information retrieval systems [3]. The goal of a question-answering system is to retrieve answers to questions rather than full documents or best-matching passages, as most information retrieval systems currently do. They take as input a natural language question (e.g., *"what is the capital of India?"*) and return a short passage or the precise text (e.g. *"New Delhi"*) that provide the answer. They combine techniques from the Information Retrieval (IR), Information Extraction (IE), and more broadly Natural Language Processing (NLP) fields [3].

Research and development of QA systems have been evaluated since 1999 in the yearly TREC (text retrieval conference) evaluations conducted by NIST and have been supported in the U.S by AQUAINT program. QA systems are also evaluated by two other workshops namely CLEF and NTCIR.

Patients/Medical students have many questions related to the medical terms, diseases, and its symptoms. They frequently need to seek answers for their questions. Medical information retrieval systems (e.g., PubMed) are used by physicians and medical students to retrieve information [2]. But studies indicate that as the average time taken to find the answer is more, the question is likely to be abandoned. Also, it has been surveyed in [8] that most of the current QA systems though found to be useful to obtain information, showed some limitations in various aspects which can be resolved for the user satisfaction. QA systems in the medical field can bring significant time saving to the patients as well as students who are often unhappy with the complexity of the IR systems.

2 Process Model

Most of the QA systems contain three process modules namely question processing, document processing, and answer processing [2, 4–7] as shown in the Fig. 1.

2.1 Question Processing

Question processing is performed to understand the question posed by the user. It can be further classified into (a) Question Analysis and Classification and (b) Query Reformulation.

Question analysis and classification require several steps such as parse the question, classify the question on one of the following types such as who, what, when,

Fig. 1 Three main process models of QA systems

which, why, and where [4]. It is used to find the type of question and the expected answer which further helps in searching the right passages from a large number of documents. For answer extraction from large documents, patterns should be provided by the system to find the correct answer and then send the query for further document processing [5].

Query reformulation transforms input question to a set of queries which acts as input to a document retrieval engine. Thus syntax and the semantic relations between the words in the question can be used [5]. A very important part of the system "usage knowledge" database saves the previously asked questions and their answers. If a new question matches a previously asked question then the answer can be extracted from the usage knowledge and presented to the user. But if there is some difference between the user question and the saved question in terms of name, adverb, adjective, or preposition then the system searches the ontology to find the similarity between the words. Usage knowledge helps in reducing the time to search the answer.

2.2 Document Processing

Document processing relies on the extraction of the documents keeping in focus the question. The query generated from the question processing module acts as an input to a document retrieval engine. A set of relevant documents are selected from which the candidate answer passages are extracted containing the relevant text. These candidate answers will acts as input to the answer processing module.

2.3 Answer Processing

The answer processing module consists of (a) answer matching and ranking (b) user answer voting.

In answer matching the candidate answers extracted from the document processing modules are matched against the type of expected answer generated in the question processing module. Some of the passages will be removed from the collected candidate answers and the most relevant ones will be further send for ranking. The ranking component classifies the answers and gives each of them a priority number. The answers are ranked based on the distance of the keywords, answer type and the answer repetition in the candidate answers [5]. The answers ranked as highest to lowest will be displayed as the answer list to the users for further voting.

In answer voting, the users will be shown the list of answers generated from the answer matching and ranking component. The users will vote the best answer generated by the system. If the users votes for the topmost answer, then the question-answer pair will be saved in the usage knowledge for further use. A validation grade will be assigned to the [q, a] pair. The next time [q, a] pair is selected by the user, the system increments the validation grade. If the user does not find any of the answers relevant to be voted from the list, then the system asks for a new question with some additional information and sends it back to question processing module.

A detailed architecture of the medical QA system is described in the next section. This architecture has used the process model outlined in this section.

3 Medical QA System Architecture

Figure 2 shows the architecture of our system. Each module in the system is described in further section.

1. *Query Interface* In this part the user will input the question using the interface. A list of questions will be presented in the interface to give idea on how to formulate questions. If the user question matches the one in the list he can select it and the answer is retrieved from the usage knowledge. Also, if the question is similar to the one that is saved in the usage knowledge, the highest ranked answers are extracted and presented. If the user is not satisfied with the answer he/she can reformulate the question.
2. *Query Analyser* It segments the question into subjects, verb, adjectives, prepositional phrases, and objects. The type of words and their synonyms are searched in the WordNet lexical database and domain specific ontology. The subjects form as the keywords and are used in query reformulation.

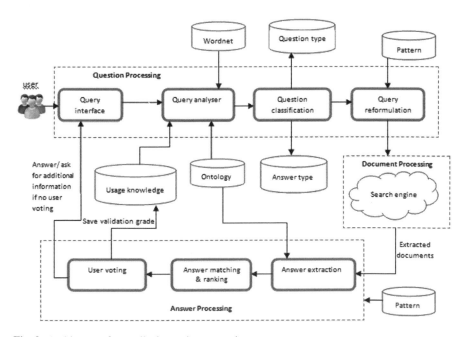

Fig. 2 Architecture for medical question answering system

3. *WordNet* This part acts like a dictionary containing all the words that are in related domains and used to search the type of words and their synonyms. It will be used as lexical database whose building block is synset, a set of synonym words representing the underlying lexical concept [9].

4. *Ontology* This part helps in surveying the questions and the answers semantically. Ontology for medical domain will be created for semantic analysis of the questions and answers.

5. *Usage knowledge* It acts like FAQ and is domain specific. First, the user question will be searched in usage knowledge. If the question structure is the same as the one in usage knowledge, the answer is presented to the user. If there are differences between the words, then domain ontology will be searched to find the synonyms. For example, the two questions will seem to be different if the new question uses the word "operation" and the saved question uses "surgery". But if in the domain ontology they both are saved as synonyms, then the system consider the two questions and the type of their answers as same.

6. *Question classification* This part will be used to find the type of question and answers. The question is classified according to the type like*who, what, when, which, why, and where.* In some cases the type of question does not clearly indicate the expected answer. For example, the type of answer for questions starting with *"when"* will be "date" but for questions starting with *"what"* the expected answer can be a definition, number or procedure. Table 1 shows the question classification and the answer type.

7. *Question reformulation* The user question (Q) will be reformulated to (Q') by using the rules. The input question will be transformed to a set of keyword queries which acts as input to a document retrieval engine. The keywords from the question can be words that are noun, adverbs, objects, subject, or words in quotation. To find the correct answer, the system uses patterns. Also use of syntax relations, semantic relations, and usage knowledge are important for question reformulation.

8. *Search engine* It searches for the documents based on the set of keywords in the reformulated query and the answer patterns.

9. *Answer extraction* The documents from the search engine will be sent to answer extraction module which will extract the information satisfying the reformulated question (Q') from the collected documents.

10. *Answer matching and ranking* In this part the candidate answers extracted will be matched against the answer type and the co-occurrence patterns. Rule-base will be designed for the answer retrieval as shown in Table 2. Irrelevant passages will be removed and the most relevant ones will be ranked. Answer ranking can be based on keywords distance and rate of keywords in the answer. The answers will be listed and shown to the user from the highest to the lowest rank.

11. *User voting* Here the user will check the listed answers and give a validation answer grade to them. This will be further saved in the usage knowledge. If the user does not find the answer to be relevant and gives no validation grade, then the system will ask for a new question or to give more information about the question asked which will be again passed to the question processing module.

Table 1 Question classification and answer type

Question classification	Sub classification	Answer type	Example
What	With preposition	Procedure	What is the procedure/process for kidney stone removal?
		Definition	What is the meaning of pancreas?
		Advantages/ Disadvantages	What are advantages of chemoprophylaxis?
		Digits	What is length of large intestine?
			What is normal blood pressure in adult?
	Without preposition	Definition	What is obesity?
When		Date	When was penicillin invented?
Which	Which-when	Date	Which date is celebrated as world cancer day?
	Which-what	Methods	Which are the methods for kidney stone treatment?
	Which	List	Which are the reproductive organs of the body?
Who		Person/ Authority	Who received noble prize for Prostate cancer treatment?
			Who did first kidney transplant?
How		Procedure	How to calculate BMI?
Where		Place	Where is thyroid gland?
			Where is mayo clinic?
How many		Digits	How many bones are there in human body?
			How many teeth does an adult have?
Other than 'Wh' questions	With or without preposition	Procedure/Process	Procedure/Process for kidney stone removal
		Date	World cancer day
		Define	Define obesity
		List	List of salivary glands
		Rules	Rules of murphy
		Benefits	Benefits of exercise
		Hazards	Hazards of smoking
		Difference	Difference between constipation and diarrhea
		Describe	Describe peripheral smear

Table 2 Rules for answer retrieval

Pattern type picked up from question	Answer retrieval	Examples of answer

Definition/
Define/
Defining/
Mean/
Meaning
	a. Subject is………
	   i. Obesity is……
	b. Subject is defined as
	   i. Obesity is defined as …………..
	c. Subject called ………..
	   i.Obesity called
	d. Subject known as ……….
	   i. Obesity is known as…….
	e. Subject means……
	   i.Obesity means …
	f. ………….is called *subject* …………..is called obesity
	Sentence containing the said pattern will be displayed. In addition sentence in continuation will be checked, if it starts with pronoun, it is presumed that it is related with the sentence containing pattern and hence will be displayed. The process of displaying sentence would be stopped once a sentence starting with noun is fetched
		Obesity is defined as the body mass index greater than 28

Benefits/ Advantages/ Merits and Demerits/ Disadvantages	Advantages *a*. Advantages of *subject*…. (include further) *b. It/Subject helps* in…. *c. It/Subject benefits* in…. *d. It/Subject can/would help to* …. *e. It/Subject use in*…….. Disadvantages *a*.Disadvantages of *subject*…. *b. It/Subject* can't help in…. *c. It/Subject* can't use in…. *d. It Restricts* to……….. *e. Itcan't apply* to…… Sentence containing the said pattern will be displayed. In addition sentence in continuation will be checked, if it starts with pronoun, it is presumed that it is related with the sentence containing pattern and hence will be displayed. The process of displaying sentence would be stopped once a sentence starting with noun is fetched. Here "It" in the sentence is considered as subject	Advantages of regular exercise 1. Stay fit 2. Strengthen muscles and bones 3. Improves digestion

(continued)

Table 2 (continued)

Pattern type picked up from question	Answer retrieval	Examples of answer
Date	Sentence should contain subject and any combination of below: -Day/Day of month -Month -Year	Date of penicillin invention
Describe	Apply All the rules for definition, characteristics, process, advantages and disadvantages	1.Definition 2.Characteristics 3.Process 4.Advantages and disadvantages
Key phrase	Follow the rules as discussed in describe	

4 Conclusion and Future Work

The medical QA system addresses patients' and medical students' information needs. Corresponding to the growth of medical information there is a growing need of QA systems that can utilize the ever-accumulating information. In this paper, we have described the architecture for medical QA system. The question classification and the rule-base for answer retrieval are designed. Answers to medical questions should be searched in the most reliable data available. Also, answers to complex medical questions often need to span more than one sentence. The extraction of the answers can be performed from single documents or from multiple documents. Building the usage knowledge improves the system performance to a great extent. Also, continued research on using semantic knowledge in the QA systems is required.

The medical domain poses a particular challenge for QA with its highly complex terminology. As the first step toward building the proposed system, we have constructed the rules for question type and expected answer. We envisage the following tasks ahead for the development of the medical QA system.

- Construction of medical ontology.
- Development of more sophisticated techniques for natural language question analysis and classification.
- Development of effective methods for answer generation.
- Extensive utilization of semantic knowledge throughout the QA process.
- Incorporation of logic and reasoning mechanism.

References

1. Demner-Fushman, Dina, Lin, Jimmy: Answering clinical questions with knowledge-based and statistical techniques. Comput. Linguist. **33**(1), 63–103 (March 2007)
2. Hong, Yu., Lee, Minsuk, Kaufman, David, Ely, John, Osheroff, Jerome A., Hripcsak, George, Cimino, James: Development, implementation, and a cognitive evaluation of a definitional question answering system for physicians. J. Biomed. Inform. **40**, 236–251 (2007)
3. Pierre, J., Zweigenbaum, P.: Towards a medical question-answering system: a feasibility study. In: Medical Informatics Europe, Studies in Health Technology and Informatics, vol. 95, pp. 463–468. IOS Press, Amsterdam (2003)
4. Vargas-Vera, Maria, Motta, Enrico, Domingue, John: AQUA: an ontology-driven question answering system. In: Maybury, M. (ed.) New directions in question answering. AAAI Press, Menlo Park (2003)
5. Kangavari, M.R., Samira, G., Manak, G.: Information retrieval: improving question answering systems by query reformulation and answer validation. World Acad. Sci. Eng. Technol. **48**, 303–310 (2008)
6. Minsuk, L., James, C., Hai Ran, Z., Carl, S., Vijay, S., John, E., Hong Y.: Beyond information retrieval–medical question answering, AMIA Annu. Symp Proc. 469–473 (2006)
7. Athenikosa, S.J., Hanb, H.: Biomedical question answering: a survey. Comput. Methods Programs Biomed. **99**, 1–24 (2010)
8. Dodiya, T., Jain, S.: Comparison of question answering systems. Adv. Intell. Syst. Comput. ISBN 978-3-642- 32063-7. vol. 182, 99–107 Sprimger
9. http://wordnet.princeton.edu/

Part XIV
Soft Computing for Clustering (SCCL)

Adaptive Mutation-Driven Search for Global Minima in 3D Coulomb Clusters: A New Method with Preliminary Applications

S. P. Bhattacharyya and Kanchan Sarkar

Abstract A single-string-based evolutionary algorithm that adaptively learns to control the mutation probability (p_m) and mutation intensity (Δ_m) has been developed and used to investigate the ground-state configurations and energetics of 3D clusters of a finite number (N) of 'point-like' charged particles. The particles are confined by a harmonic potential that is either isotropic or anisotropic. The energy per particle (E_N/N) and its first and second differences are analyzed as functions of confinement anisotropy, to understand the nature of structural transition in these systems.

Keywords 3D Coulomb clusters · Genetic algorithm · Simulated annealing · Mutation probability · Mutation intensity

1 Introduction

Coulomb clusters, having concentric multi-shell structures, are observed in laser-cooled trapped ion systems. Three-dimensional (3D) multi-shell structures of dust particles under spherically symmetric potential have been extensively investigated [1–4] partly because of their analogy to other physically interesting systems, e.g., cold ionic systems, classical artificial atoms, and one or multi-component plasmas. Dust grains, or solid particles of μm to sub-μm sizes, are observed in various low-temperature laboratory plasmas such as process plasmas, dust plasma crystals, space plasmas (e.g., planetary rings, comets, noctilucent clouds), 'plasma crystals,' or

S. P. Bhattacharyya (✉)
IIT Bombay, Mumbai 400076, India
e-mail: pcspb@chem.iitb.ac.in

K. Sarkar
Indian Association for the Cultivation of Science, Jadavpur, Kolkata 700032, India
e-mail: pcks@iacs.res.in

B. V. Babu et al. (eds.), *Proceedings of the Second International Conference on Soft Computing for Problem Solving (SocProS 2012), December 28–30, 2012*, Advances in Intelligent Systems and Computing 236, DOI: 10.1007/978-81-322-1602-5_129, © Springer India 2014

Coulomb lattices of charged dust grains in a plasma. The massive dust grains are generally highly charged. Dust grains that are immersed in plasma and/or radiative environments are electrically charged owing to processes such as plasma current collection, photoemission, or secondary emission. The first experimental realization of a spherical 3D cloud of monodisperse dust particles was achieved [2] by clever manipulation of thermophoretic forces and the plasma-induced field to counter the action of gravity, and the application of lateral external potential resulting in parabolic confinement.

It is important to understand how the global (local) minimum energy configurations emerge in such systems, how their energetics, stabilities, and other attributes change as functions of varying strength and symmetry of the confining potential, number and types of particles trapped, etc. Some theoretical investigations of structural properties and melting behavior in 3D Coulomb clusters have been very recently reported in [5]. The minimum energy structure of an assembly of cold charged particles depends on the delicate balance between inter-ionic Coulomb energy (two body) and the confining potential (one body). Since many local minima are possible, it is always challenging to locate the lowest energy structure. It is possible to invoke a population (n_{pop}) based method like the genetic algorithm (GA) [6] to handle the problem. GA has been very successful and popular among the cluster physics community [7, 8]. The use of a population of strings makes the search thorough, although the necessity of evaluating n_{pop} number of configurations in every generation could make the search costly relative to a single-string method (e.g., simulated annealing method) unless the GA hits the solution at least n_{pop} times faster than any method using a single string. Parallelization of the string evaluations step can make the GA more efficient [9]. For problems with rugged energy landscapes, simulated annealing (SA) [10] may encounter what has been known as the 'freezing' problem [11], because the escape rate from local minimum diverges with decreasing temperature (can be amended at the expense of introducing additional parameters). It could be profitable to develop a single-string-based evolutionary strategy [12–15] that could, in principle, outperform a GA-based search through the complex potential energy landscape. We propose an adaptive mutation-only algorithm and evaluate its performance vis-a-vis that of GA and SA in the context of search for the global minimum on the potential energy surfaces of Coulomb clusters.

2 The Method

The configuration of the system of N–confined charges is completely defined by a geometry string S containing the Cartesian coordinates (x_i, y_i, z_i) of the N charges (q_i):

$$S_k = s(x_1^k, y_1^k, z_1^k; x_2^k, y_2^k, z_2^k; \ldots; x_N^k, y_N^k, z_N^k) \equiv (\xi_1^k, \xi_2^k, \ldots, \xi_{3N-1}^k, \xi_{3N}^k) \quad (1)$$

The initial coordinate values are generated randomly within a specified range. The fitness (f_k) of S_k is determined by computing the potential energy $V_k(\xi_1, \ldots, \xi_{3N})$ for the kth configuration of charges encoded by S_k as follows:

$$f_k = \frac{1}{(V_k - V_L^k)^2 + \delta} \tag{2}$$

where,

$$V_k = \sum_{i=1}^{N-1} \sum_{j>i}^{N} \frac{q_i q_j}{(r_{ij})_k} + \frac{1}{2} K \sum_{i=1}^{N} (r_i^2)_k \tag{3}$$

and

$$
\begin{aligned}
(r_{ij})_k &= \sqrt{(x_i^k - x_j^k)^2 + (y_i^k - y_j^k)^2 + (z_i^k - z_j^k)^2} \\
(r_i)_k &= \sqrt{(x_i^k)^2 + (y_i^k)^2 + (z_i^k)^2}
\end{aligned}
\tag{4}
$$

V_L^k is an updatable lower bound to the current potential energy V_k. δ is a small constant ($\approx 10^{-12}$) that takes care of exponential overflow. We start with a randomly chosen geometry string S_0 ($k = 0$) and allow it to undergo a mutation in geometry with a current mutation probability P_m^k and intensity Δ_m^k. Supposing that the pth coordinate (ξ_p) in S_k has been selected for mutation, the mutated coordinate becomes

$$\xi_p^{k+1} = \xi_p^k + (-1)^M \Delta_m^k r \tag{5}$$

where M is a random integer, and r is a uniformly distributed random number in the range $(0, 1)$. Δ_m^k is the current mutation intensity. The mutated-geometry string S_{k+1} is used to compute V_{k+1} and the fitness f_{k+1}. If $f_{k+1} \geq f_k$ the mutation is retained; if not, it is rejected and another mutation is attempted. The current mutation intensity is automatically determined from the experience so far by the following heuristics. If more than 20 % of the mutations were accepted in the last 100 trials, we set $\Delta_m^l = \Delta_m^{l-1} \times (1 + r)$, where r is a random number in the range $(0, 1)$. If less than 10 % of the trials were accepted during the same span, we set $\Delta_m^l = \Delta_m^{l-1} \times \frac{1}{1+r}$. In all other cases, $\Delta_m^l = \Delta_m^{l-1}$, Δ_m^0 being set equal to 0.5.

The adaptive adjustment of P_m is also done following a similar logic:

$$P_m^l = P_m^{l-1} \times \frac{1}{1+r} \tag{6}$$

if less than 10 % of the trials led to successful mutation during the last 100 generation, else

$$P_m^l = P_m^{l-1} \times (1 + r) \tag{7}$$

if more than 30 % of the trials produced successful mutations. Otherwise, P_m^l is set equal to P_m^{l-1}. P_m^0 is chosen to have an initial value between $\frac{1}{3}$ and $\frac{1}{4}$ randomly. The value of P_m^k is kept fixed once it is $\leq \frac{1}{3N}$, N being the number of charges in the cluster. The logic of adaptive control of Δ_m^k and P_m^k can be found it the fact that if a large number of trials are producing successful mutations (fitness-increasing mutations), then it would be reasonable to allocate higher values of probability and intensity of mutation. If too few trials are leading beneficial mutations, both Δ_m^k and P_m^k should be reduced from their current values.

3 Results and Discussion

We consider a 3D cluster of 200 unit charges randomly distributed in space. Values of Δ_m^0 and P_m^0 are chosen to be 0.5 and 0.25, respectively. Initially, V_L is approximately chosen and lowered if $(V_k - V_L) \leq \varepsilon$, $(\varepsilon = 10^{-2})$. The confinement is harmonic and isotropic with $k = 1.0$.

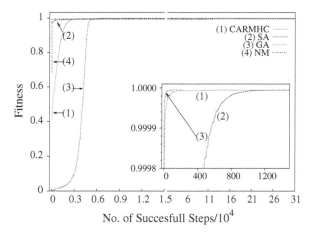

Fig. 1 Comparison of fitness evolution profiles for 200 unit positive charges confined in a harmonic trap obtained in CARMHC, NM, SA, and GA. The evolution in the GM region in an expanded scale has been shown in the inset. Compared to the CARMHC, SA requires much longer run (with multiple cooling and heating) in order to reach the desired optimum. NM ends up in a relatively higher energy structure. GA reaches the GM in a fewer number of 'generations,' each generation involving 20–30 fitness evaluation

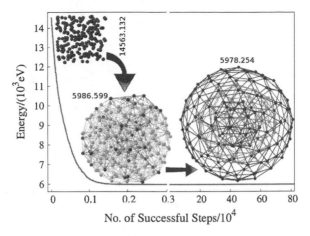

Fig. 2 Energy and structural evolution in CARMHC as the search proceeds to locate the GM

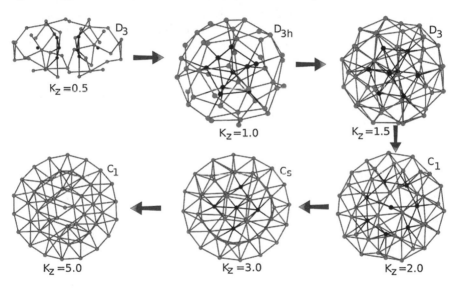

Fig. 3 Structural evolution for a system consisting of 50 charged particles with varying confinement strength along the z-axis (K_z). The point symmetry of each structure is also displayed

Figure 1 shows the growth of fitness as a function of the number of the successful trials (generations), while Fig. 2 displays the energy evolution profile along with snapshots of the mutated structures at different stages of evolution. For comparison, we have also included the GA energy evolution profile for the best string in a population of size (30) using tournament selection, arithmetic mutation, and BLX-α crossover. The crossover probability was kept fixed at $p_c = 0.7$, while the mutation probability p_m was set at 0.05. The initial mutation intensity used is $\Delta_m = 0.02$. It is clear that our single-string adaptive random mutation hill climbing method performs

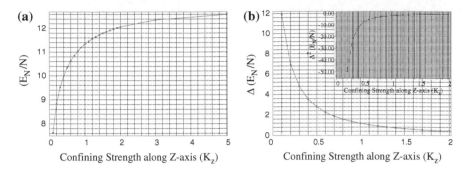

Fig. 4 a Energy evolution profile for a system consisting of 50 charged particles with varying confinement strength along the z-axis (K_z) and **b** first and second differences in energy

better than the GA as far as the rate of convergence to the global minimum is concerned. We have also compared the performance of our CARMHC method with that of simulated annealing (SA) applied to the same problem. SA as usual takes longer time as well as several heating and cooling schedules to reach the final structure, often requiring multiple starts.

Figure 3 displays the structural evolution in a Coulomb cluster with $N = 200$ when $k_x = k_y \neq k_z$. The anisotropy destroys the symmetry of the structures that were produced in isotropic confinement. The optimal energy (E_n) and the first and second differences of energy have been analyzed as functions of k_z (Fig. 4a, b). Apparently, there is a continuous structural phase transition as K_z attains a high value.

We are extending these studies to monodisperse clusters of much larger sizes and looking into the effects of charge anisotropy in polydisperse clusters on the possibility of structural phase transition.

Acknowledgments K. Sarkar thanks the CSIR, Government of India, New Delhi, for the award of senior research fellowship, and S.P.B. thanks the DAE, Government of India, for the award of Raja Ramanna Fellowship.

References

1. Zuzic, M., et al.: Phys. Rev. Lett. **85**, 4064 (2000)
2. Arp, O., Block, D., Piel, A., Melzer, A.: Phys. Rev. Lett. **93**, 165004 (2004)
3. Kählert, H., Bonitz, M.: Phys. Rev. Lett. **104**, 015001 (2010)
4. Buluta, I.M., Hasegawa, S.: Phys. Rev. A **78**, 042340 (2008)
5. Apolinario, S.W.S., Peeters, F.M.: Phys. Rev. E **83**, 041136 (2011)
6. Holland, J.H.: Adaptation in Natural and Artificial Systems: An Introductory Analysis with Applications to Biology, Control and Artificial Intelligence. MIT Press, Cambridge, MA, USA (1992)
7. Sobczak, P., Kucharski, L., Kamieniarz, G.: Comput. Phys. Commun. **182**, 1900 (2011)
8. Ali, M., Smith, R., Hobday, S.: Comput. Phys. Commun. **175**, 451 (2006)

9. Nandy, S., Sharma, R., Bhattacharyya, S.P.: Appl. Soft Comput. **11**, 3946 (2011)
10. Kirkpatrick, S., Gelatt, C.D., Vecchi, M.P.: Science **220**, 671 (1983)
11. Li, Z., Scheraga, H.A.: Proc. Nat. Acad. Sci. **84**, 6611 (1987)
12. Mitchell, M., Holland, J.H., Forrest, S.: When will a genetic algorithm outperform hill climbing?. In: Advances in Neural Information Processing Systems, vol. 6, pp. 51–58, Morgan Kaufmann, (1993)
13. Sharma, R., Bhattacharyya, S.P.: Direct search for wave operator by a genetic algorithm (ga): route to few eigenvalues of a hamiltonian. In: IEEE Congress on, Evolutionary Computation, pp. 3812–3817, (2007)
14. Sarkar, K., Sharma, R., Bhattacharyya, S.P.: J. Chem. Theory Comput. **6**, 718 (2010)
15. Sarkar, K., Sharma, R., Bhattacharyya, S.P.: Int. J. Quantum Chem. **112**, 1547 (2012)

A New Rough-Fuzzy Clustering Algorithm and its Applications

Sushmita Paul and Pradipta Maji

Abstract Cluster analysis is a technique that divides a given data set into a set of clusters in such a way that two objects from the same cluster are as similar as possible and the objects from different clusters are as dissimilar as possible. A robust rough-fuzzy c-means clustering algorithm is applied here to identify clusters having similar objects. Each cluster of the robust rough-fuzzy clustering algorithm is represented by a set of three parameters, namely, cluster prototype, a possibilistic fuzzy lower approximation, and a probabilistic fuzzy boundary. The possibilistic lower approximation helps in discovering clusters of various shapes. The cluster prototype depends on the weighting average of the possibilistic lower approximation and probabilistic boundary. The reported algorithm is robust in the sense that it can find overlapping and vaguely defined clusters with arbitrary shapes in noisy environment. The effectiveness of the clustering algorithm, along with a comparison with other clustering algorithms, is demonstrated on synthetic as well as coding and non-coding RNA expression data sets using some cluster validity indices.

Keywords Cluster analysis · Robust rough-fuzzy · C-means clustering algorithm

1 Introduction

Cluster analysis is one of the important problems related to a wide range of engineering and scientific disciplines such as pattern recognition, machine learning, psychology, biology, medicine, computer vision, web intelligence, communications, and remote sensing. It finds natural groups present in a data set by dividing the data set

S. Paul (✉) · P. Maji
Machine Intelligence Unit, Indian Statistical Institute, Kolkata, India
e-mail: sushmita_t@isical.ac.in

P. Maji
e-mail: pmaji@isical.ac.in

B. V. Babu et al. (eds.), *Proceedings of the Second International Conference on Soft Computing for Problem Solving (SocProS 2012), December 28–30, 2012*, Advances in Intelligent Systems and Computing 236, DOI: 10.1007/978-81-322-1602-5_130, © Springer India 2014

into a set of clusters in such a way that two objects from the same cluster are as similar as possible and the objects from different clusters are as dissimilar as possible. Hence, it tries to mimic the human ability to group similar objects into classes and categories [5].

A number of clustering algorithms have been proposed to suit different requirements [5, 6]. One of the most widely used prototype based partitional clustering algorithms is Hard C-Means (HCM) [10]. The hard clustering algorithms generate crisp clusters by assigning each object to exactly one cluster. When the clusters are not well defined, that is, when they are overlapping, one may desire fuzzy clusters. In this regard, the problem of pattern classification is formulated as the problem of interpolation of the membership function of a fuzzy set in [2], and thereby a link with the basic problem of system identification is established. A seminal contribution to cluster analysis is Ruspini's concept of a fuzzy partition [12]. The application of fuzzy set theory to cluster analysis was initiated by Dunn and Bezdek by developing fuzzy ISODATA [4] and Fuzzy C-Means algorithms (FCM) [3].

The FCM relaxes the requirement of the HCM by allowing gradual memberships [3]. In effect, it offers the opportunity to deal with the data that belong to more than one cluster at the same time. It assigns memberships to an object those are inversely related to the relative distance of the object to cluster prototypes. Also, it can deal with the uncertainties arising from overlapping cluster boundaries. Although the FCM is a very useful clustering method, the resulting membership values do not always correspond well to the degrees of belonging of the data, and it may be inaccurate in a noisy environment [7]. However, in real data analysis, noise and outliers are unavoidable. To reduce this weakness of the FCM, and to produce memberships that have a good explanation of the degrees of belonging for the data, Krishnapuram and Keller [7] proposed Possibilistic C-Means (PCM) algorithm, which uses a possibilistic type of membership function to describe the degree of belonging. However, the PCM sometimes generates coincident clusters [1].

Rough set theory is a new paradigm to deal with uncertainty, vagueness, and incompleteness. It is proposed for indiscernibility in classification or clustering according to some similarity [11]. In this regard, a rough-fuzzy clustering algorithm, termed as *robust Rough-Fuzzy C-Means* (rRFCM) [9], is applied in the current study. It integrates judiciously the merits of rough sets, and probabilistic and possibilistic memberships of fuzzy sets. While the integration of both membership functions of fuzzy sets enables efficient handling of overlapping partitions in noisy environment, the concept of lower and upper approximations of rough sets deals with uncertainty, vagueness, and incompleteness in cluster definition. Each cluster is represented by a set of three parameters, namely, a cluster prototype or centroid, a possibilistic lower approximation, and a probabilistic boundary. The cluster prototype depends on the weighting average of the possibilistic lower approximation and probabilistic boundary. Integration of probabilistic and possibilistic membership functions avoids the problems of noise sensitivity of the FCM and the coincident clusters of the PCM. The reported algorithm is robust in the sense that it can find overlapping and vaguely defined clusters with arbitrary shapes in noisy environment.The effectiveness of the reported algorithm, along with a comparison with other clustering algorithms, is demonstrated on synthetic as well as coding and non-coding RNA expression time-series data sets using some standard cluster validity indices.

2 Robust RFCM Algorithm

This section reports a new c-means algorithm, termed as rRFCM. Let $X = \{x_1, \cdots, x_j, \cdots, x_n\}$ be the set of n objects and $V = \{v_1, \cdots, v_i, \cdots, v_c\}$ be the set of c centroids, where $x_j \in \Re^m$ and $v_i \in \Re^m$. Each of the clusters β_i is represented by a cluster center v_i, a lower approximation $\underline{A}(\beta_i)$ and a boundary region $B(\beta_i) = \{\overline{A}(\beta_i) \setminus \underline{A}(\beta_i)\}$, where $\overline{A}(\beta_i)$ denotes the upper approximation of cluster β_i. According to the definitions of lower approximation and boundary of rough sets, if an object $x_j \in \underline{A}(\beta_i)$, then $x_j \notin \underline{A}(\beta_k), \forall k \neq i$, and $x_j \notin B(\beta_i), \forall i$. That is, the object x_j is contained in β_i definitely. Hence, the memberships of the objects in lower approximation of a cluster should be independent of other centroids and clusters. Also, the objects in lower approximation should have different influence on the corresponding centroid and cluster. From the standpoint of "compatibility with the cluster prototype", the membership of an object in the lower approximation of a cluster should be determined solely by how far it is from the prototype of the cluster, and should not be coupled with its location with respect to other clusters. As the possibilistic membership v_{ij} depends only on the distance of object x_j from cluster β_i, it allows optimal membership solutions to lie in the entire unit hypercube rather than restricting them to the hyperplane given by FCM. Whereas, if $x_j \in B(\beta_i)$, then the object x_j possibly belongs to cluster β_i and potentially belongs to another cluster. Hence, the objects in boundary regions should have different influence on the centroids and clusters, and their memberships should depend on the positions of all cluster centroids. So, in the rRFCM, the membership values of objects in lower approximation are identical to the PCM, while those in boundary region are the same as the FCM, and are as follows:

$$\mu_{ij} = \left[\sum_{k=1}^{c} \left(\frac{d_{ij}^2}{d_{kj}^2} \right)^{\frac{1}{m_1-1}} \right]^{-1} \; ; \; v_{ij} = \left[1 + \left\{ \frac{d_{ij}^2}{\eta_i} \right\}^{\frac{1}{(m_2-1)}} \right]^{-1}$$

$$\text{subject to} \quad \sum_{i=1}^{c} \mu_{ij} = 1, \forall j, \text{ and} 0 < \sum_{j=1}^{n} \mu_{ij} < n, \forall i,$$

$$\text{also } 0 < \sum_{j=1}^{n} v_{ij} \leq n, \forall i; \text{ and } \max_{i} v_{ij} > 0, \forall j;$$

where η_i is the scale parameter. The centroid for the rRFCM is computed as:

$$v_i = \begin{cases} w\mathscr{C}_1 + (1-w)\mathscr{D}_1 & \text{if } \underline{A}(\beta_i) \neq \emptyset, B(\beta_i) \neq \emptyset \\ \mathscr{C}_1 & \text{if } \underline{A}(\beta_i) \neq \emptyset, B(\beta_i) = \emptyset \\ \mathscr{D}_1 & \text{if } \underline{A}(\beta_i) = \emptyset, B(\beta_i) \neq \emptyset \end{cases} \quad (1)$$

$$\text{where } \mathscr{C}_1 = \frac{\sum\limits_{x_j \in \underline{A}(\beta_i)} (v_{ij})^{\acute{m}_2} x_j}{\sum\limits_{x_j \in \underline{A}(\beta_i)} (v_{ij})^{\acute{m}_2}}; \quad \mathscr{D}_1 = \frac{\sum\limits_{x_j \in B(\beta_i)} (\mu_{ij})^{\acute{m}_1} x_j}{\sum\limits_{x_j \in B(\beta_i)} (\mu_{ij})^{\acute{m}_1}}.$$

The process starts by choosing c objects as the initial centroids of the c clusters. The possibilistic memberships of all the objects are calculated. Let $v_i = (v_{i1}, \cdots, v_{ij}, \cdots, v_{in})$ represents the possibilistic cluster β_i associated with the centroid v_i. After computing v_{ij} for c clusters and n objects, the values of v_{ij} for each object x_j are sorted and the difference of two highest memberships of x_j is compared with a threshold value δ_1. Let v_{ij} and v_{kj} be the highest and second highest memberships of x_j. If $(v_{ij} - v_{kj}) > \delta_1$, then $x_j \in \underline{A}(\beta_i)$, otherwise $x_j \in B(\beta_i)$ and $x_j \in B(\beta_k)$ if $v_{ij} > \delta_2$. After assigning each object in lower approximations or boundary regions of different clusters based on the thresholds δ_1 and δ_2, the probabilistic memberships μ_{ij} for the objects lying in the boundary regions are computed. The new centroids of different clusters are computed as per (1). The thresholds δ_1 and δ_2 control the size of granules of rough-fuzzy clustering. In practice, the following definitions work well:

$$\delta_1 = \frac{1}{n} \sum_{j=1}^{n} (v_{ij} - v_{kj}) \quad \delta_2 = \frac{1}{\acute{n}} \sum_{j=1}^{\acute{n}} v_{ij} \quad (2)$$

where n is the total number of objects, v_{ij} and v_{kj} are the highest and second highest memberships of object x_j. On the other hand, the objects with $(v_{ij} - v_{kj}) \leq \delta_1$ are used to calculate the threshold δ_2; where $\acute{n}$ is the number of objects those do not belong to lower approximations of any cluster and v_{ij} is the highest membership of object x_j.

3 Experimental Results and Discussions

The performance of the rRFCM algorithm is compared extensively with that of different c-means algorithms on 1 synthetic ($X32$), 1 mRNA, and 1 miRNA expression data sets, which are downloaded from *Gene Expression Omnibus* (www.ncbi.nlm.nih.gov/geo/) with accession numbers GDS1013, and GSE9449. The synthetic data set $X32$ consists of $n = 32$ objects in $\Re^2$ with two clusters. The object x_{30} is outlier or noise, and the object x_7 is the so called inlier or bridge. The objects belong to either of two groups and both the clusters are overlapping in nature.

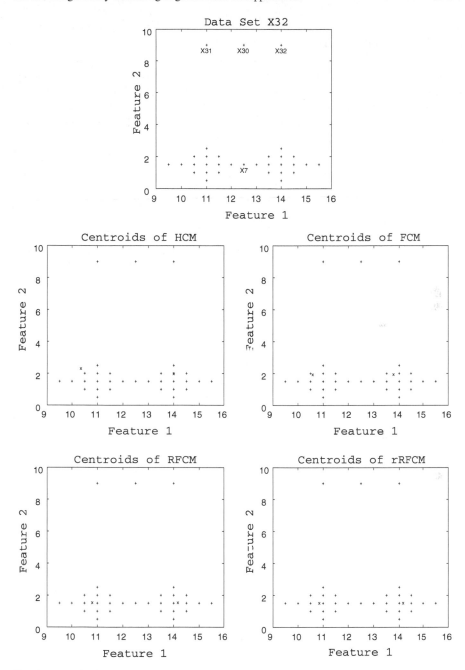

Fig. 1 Example data set $X32$ and clusters prototypes of different c-means algorithms

The algorithms compared are HCM [5, 10], FCM [3], and rough-fuzzy c-means (RFCM) [8]. The input parameters used, which are held constant across all runs, are values of fuzzifiers $\acute{m}_1 = 2.0$ and $\acute{m}_2 = 2.0$, value of w, used in the rRFCM algorithm is set to 0.99. For expression data sets the value of c for each data set is decided by using the cluster identification via connectivity kernels (CLICK) algorithm [13].

Two randomly generated initial centroids, along with two scale parameters and the final prototypes of different c-means, are reported in Table 1 for the synthetic data set. Fig. 1 depicts the scatter plot of the $X32$ synthetic data set. Fig. 1 also represents the scatter plots of $X32$ synthetic data set along with the clusters prototypes obtained using different c-means algorithms. The objects of synthetic data sets are represented by $+$, while $\times$ depicts the positions of cluster prototypes.

Finally, Table 2 presents the performance of different c-means algorithms for optimum values of λ. The results are presented for one synthetic data set, one mRNA and one miRNA expression data sets with respect to Silhouette index, DB index, Dunn index, and β index. All the results reported in this table establish the fact that the rRFCM algorithm is superior to other c-means clustering algorithms, irrespective of the cluster validity indices and data sets used.

Table 1 Cluster prototypes of different c-means for data set $X32$

Different algorithms	$X32$	
	Centroid 1	Centroid 2
Initial	11.088 2.382	14.100 2.000
Scale	$\eta_1 = 7.090$	$\eta_2 = 7.090$
HCM	10.353 2.294	14.000 1.969
FCM	10.585 1.919	13.770 1.924
RFCM	10.798 1.525	14.168 1.530
rRFCM	10.844 1.500	14.156 1.500

Table 2 Performance of different c-means algorithms on different data sets

Different data sets	Different algorithms	Cluster validity indices			
		Silhouette index	DB index	Dunn index	β index
Synthetic	HCM	0.408	1.306004	0.993485	1.376322
	FCM	0.416	1.779841	0.907058	1.366696
	RFCM	0.416	1.667268	0.987276	1.301975
	rRFCM	0.628	0.131575	13.862039	19.259300
mRNA	HCM	0.240	0.771935	0.001268	22.379993
	FCM	0.250	1.432223	0.002458	22.454958
	RFCM	0.255	0.769563	0.001619	22.666529
	rRFCM	0.580	0.364647	0.154374	263.586884
miRNA	HCM	0.256	0.907291	0.199289	2.745963
	FCM	0.192	4.979097	0.005345	2.007121
	RFCM	0.345	0.816723	0.103278	2.427899
	rRFCM	0.427	0.595101	1.030297	6.110056

4 Conclusion

In this paper, the application of the rRFCM algorithm on three different types of data has been demonstrated. Integration of the merits of rough sets, fuzzy sets, and c-means algorithm generates better results as compared to other c-means algorithms. The effectiveness of the rRFCM algorithm, along with a comparison with other algorithms, is demonstrated on one synthetic data set, one mRNA, and one miRNA microarray data sets.

Acknowledgments This work is partially supported by the Indian National Science Academy, New Delhi (Grant No. SP/YSP/68/2012). The work was done when one of the authors, S. Paul, was a Senior Research Fellow of Council of Scientific and Industrial Research, Government of India.

References

1. Barni, M., Cappellini, V., Mecocci, A.: Comments on a possibilistic approach to clustering. IEEE Trans. Fuzzy Syst. **4**(3), 393–396 (1996)
2. Bellman, R.E., Kalaba, R.E., Zadeh, L.A.: Abstraction and pattern classification. J. Math. Anal. Appl. **13**, 1–7 (1966)
3. Bezdek, J.C.: Pattern Recognition with Fuzzy Objective Function Algorithm. Plenum, New York (1981)
4. Dunn, J.C.: A fuzzy relative of the ISODATA process and its use in detecting compact, Well-separated clusters. J. Cybern. **3**, 32–57 (1974)
5. Jain, A.K., Dubes, R.C.: Algorithms for Clustering Data. Prentice Hall, Englewood Cliffs (1988)
6. Jain, A.K., Murty, M.N., Flynn, P.J.: Data clustering: a review. ACM Comput. Surv. **31**(3), 264–323 (1999)
7. Krishnapuram, R., Keller, J.M.: A possibilistic approach to clustering. IEEE Trans. Fuzzy Syst. **1**(2), 98–110 (1993)
8. Maji, P., Pal, S.K.: RFCM: a hybrid clustering algorithm using rough and fuzzy sets. Fundamenta Informaticae **80**(4), 475–496 (2007)
9. Maji, P., Paul, S.: Rough-fuzzy clustering for grouping functionally similar genes from microarray data. In: Proceedings of the 10th Asia Pacific Bioinformatics Conference, pp. 307–320, Australia (2012)
10. McQueen, J.: Some methods for classification and analysis of multivariate observations. In: Proceedings of the 5th Berkeley Symposium on Mathematics, Statistics and Probability, 281–297 1967
11. Pawlak, Z.: Rough Sets: Theoretical Aspects of Resoning About Data. Kluwer, Dordrecht (1991)
12. Ruspini, E.H.: Numerical methods for fuzzy clustering. Info. Sci. **2**, 319–350 (1970)
13. Shamir, R., Sharan, R.: CLICK: a clustering algorithm for gene expression analysis. In: Proceedings of the 8th International Conference on Intelligent Systems for Molecular Biology, 2000

A Novel Rough Set Based Clustering Approach for Streaming Data

Yogita and Durga Toshniwal

Abstract Clustering is a very important data mining task. Clustering of streaming data is very challenging because streaming data cannot be scanned multiple times and also new concepts may keep evolving in data over time. Inherent uncertainty involved in real world data stream further magnifies the challenge of working with streaming data. Rough set is a soft computing technique which can be used to deal with uncertainty involved in cluster analysis. In this paper, we propose a novel rough set based clustering method for streaming data. It describes a cluster as a pair of lower approximation and an upper approximation. Lower approximation comprises of the data objects that can be assigned with certainty to the respective cluster, whereas upper approximation contains those data objects whose belongingness to the various clusters in not crisp along with the elements of lower approximation. Uncertainty in assigning a data object to a cluster is captured by allowing overlapping in upper approximation. Proposed method generates soft-cluster. Keeping in view the challenges of streaming data, the proposed method is incremental and adaptive to evolving concept. Experimental results on synthetic and real world data sets show that our proposed approach outperforms Leader clustering algorithm in terms of classification accuracy. Proposed method generates more natural clusters as compare to k-means clustering and it is robust to outliers. Performance of proposed method is also analyzed in terms of correctness and accuracy of rough clustering.

Keywords Clustering · Streaming data · Cluster approximation · Rough set

Yogita (✉) · D. Toshniwal
Indian Institute of Technology, Roorkee, India
e-mail: thakranyogita@gmail.com

D. Toshniwal
e-mail: durgafec@iitr.ernet.in

B. V. Babu et al. (eds.), *Proceedings of the Second International Conference on Soft Computing for Problem Solving (SocProS 2012), December 28–30, 2012*, Advances in Intelligent Systems and Computing 236, DOI: 10.1007/978-81-322-1602-5_131, © Springer India 2014

1 Introduction

Digital data is increasing enormously in the present times. Applying data analysis techniques to such huge data repositories is thus very important currently. Data mining aims to extract useful information and patterns from huge quantities of data and can be applied on a variety of data types.

Nowadays many applications are generating streaming data for an example real-time surveillance, medical systems, internet traffic, online transactions and remote sensors. Data streams and streaming data are synonymous. Data streams are temporally ordered, fast changing, massive, and potentially infinite sequence of data objects [6]. Unlike traditional data sets, it is impossible to store an entire data stream or to scan through it multiple times due to its tremendous volume. New concepts may keep evolving in data streams over time. Evolving concepts require data stream processing algorithms to continuously update their models to adapt to the changes.

Clustering is an important data stream mining technique. Inherent uncertainty involved in real worlds streaming data increases the challenge of cluster analysis of data streams.

Rough set theory was introduced by Pawlak [10]. Rough sets have its application across a wide range of fields such as: data mining, medical data processing, information retrieval, machine learning, knowledge based systems etc. Rough set can be used to deal with uncertainty and vagueness involved in real world cluster analysis [5, 11]. It describes a cluster by a pair of two crisp set one is lower approximation and another is upper approximation. Lower approximation comprises of the data objects that are the sure members of a cluster. Upper approximation contains those data objects whose belongingness to clusters is uncertain along with the elements of lower approximation. In Rough Set based clustering [5, 8] an object must satisfy the following properties:

1. An object can be part of at most one lower approximation.
2. For a cluster C and object x, if x belongs to lower approximation of cluster C, then x also belongs to upper approximation of C.
3. If an object is not part of any lower approximation, then it belongs to two or more upper approximations.

There are many algorithms in literature for clustering of static and stored data sets which are based on a variety of approaches like Partitioning approach, hierarchical approach, density based approach, model based approach. Most of existing clustering methods require multiple scanning. Such methods cannot be used for clustering streaming data. Though some research work has been done on clustering of streaming data [1, 2, 4] but most of these methods do not deal with the uncertainty involved due to the lake of information.

In this paper we have proposed a novel rough set based clustering method for streaming data. This method uses rough sets for handling the uncertainty involved in clustering. It describes a cluster as a pair of lower approximation and an upper approximation. As oppose to most of existing clustering methods, keeping in view the streaming data environment proposed clustering method is incremental and adaptive to evolving concepts of data.

The rest of this paper is organized as follows: Section 2 discusses related work. Section 3 describes proposed method. Section 4 presents experimental results. Section 5 concludes the paper.

2 Related Work

Several clustering methods for streaming data have been proposed recently. A brief description of some of them is given here. STREAM [9] is very first algorithm for data stream clustering. It scans data once and organise them in buckets. Median of each bucket is taken and weighted by number of objects in a bucket. In next step theses weighted medians are clustered. CluStream is proposed by Aggarwal et al. [2]. It divides the clustering process in online phase and offline phase. Summary statics are stored in online phase and based on that clustering is done in offline phase. HPStream is proposed in [1] it uses a fading function and dimention projection for high dimensional data. It performs better then CluStream and STREAM. Cao et al. has proposed DenStream [4]. It is a density-based clustering algorithm. DenStream also involves two phase clustering. It uses the concept of micro-cluster to store an approximate representation of the data points. DenStream is able to identify clusters with arbitrary shapes. E-Stream is introduced by Udommanetanakit in [13] for data stream clustering. It supports the concept evolution of data over time.

Pawlak has proposed the rough sets [10]. Rough-K-means is proposed by Lingras in [8]. It is rough set based extension of K-means clustering algorithm. Represent a cluster by lower and upper approximation. Zhou et al. has suggested an approach for adapting the parameters of rough-K-means based on the number of iterations [15]. Wang in [5] has proposed a new type of adaptive weights based on the rough accuracy relating to different rough clusters and a new hybrid threshold by combining the difference and distance thresholds, refine the algorithm for assigning objects into lower and upper approximations. Asharaf et al. in [3] proposed a incremental approach for clustering interval data by employing rough set to deal with uncertainty involved in clustering. Yogita et al. in [14] has proposed a rough set based framework for exception discovery. Joshi et al. in [7] has presented interval-K-mean and compared the Rough-K-mean, Fuzzy-K-mean and interval-K-means.

3 Proposed Method Rough Set Based Clustering Approach for Streaming Data

In this section preliminary concepts and proposed scheme is described. Abstract pictorial representation of proposed technique is given in Fig. 1.

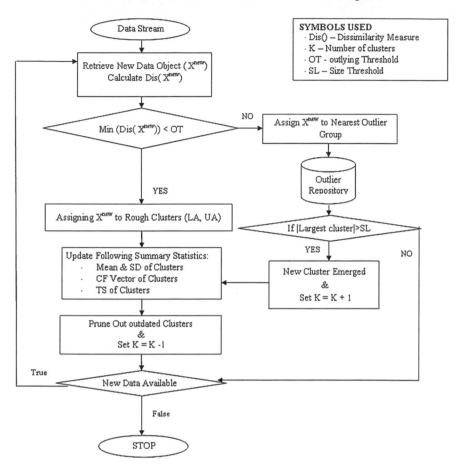

Fig. 1 Block diagram–rough set based data stream clustering

3.1 Preliminary Concepts

- Data Stream–A DataStream DS $= x_1, x_2, \ldots\ldots\ldots, x_n$ is a unbounded sequence of data objects. Object $x_i = (x_{1i}, x_{2i}, \ldots\ldots\ldots, x_{mi})$ is characterize by a set of m attributes.
- Rough Cluster It is represented by a pair of lower approximation and upper approximation.
- Lower Approximation $(LA(C_i))$ Lower approximation of a cluster comprises of the data objects that definitely belongs to that cluster.
- Upper Approximation $(UA(C_i))$ UA of a cluster contains those data objects whose belongingness to that cluster is not crisp along with the elements of LA.
- Mean of LA Mean of the LA of a cluster C_i is represented by $MLA(C_i)$. Its value is the average of the all elements of $LA(C_i)$

- Mean of UA Mean of the UA of a cluster C_i is represented by $MUA(C_i)$. Its value is the average of the all elements of $UA(C_i)$
- Mean of Cluster Mean of a cluster C_i is the weighted sum of $MLA(C_i)$ and $MUA(C_i)$ as given by following Eq. [15]

$$Mean\ of\ Cluster(C_i) = W_1 \times MLA(C_i) + W_u \times MUA(C_i)$$

with $W_1 + W_u = 1.W_1$ and W_u specifies the importance of LA and UA in cluster mean calculation.

- Standard Deviation of Cluster $(SD(C_i))$- It is the standard deviation of LA of a cluster. We have considered only definite members of cluster for calculation of value of $(SD(C_i))$.
- Cluster Feature Vector (CF-Vector) It is a six tuple vector defined as $< nl, sl, ssl, nu, su, ssu >$ where nl is the number of data objects in LA, sl is the sum of all the data objects of LA in each dimension, ssl is the sum of square of all the data objects in each dimension, nu is the number of data objects in UA, su is the sum of all the data objects of UA in each dimension, ssu is the sum of square of all the data objects in each dimension U.
- Time Stamp of Cluster $(TS(C_i))$ It represent the last time a data object is assigned to the lower approximation of a cluster.

3.2 Proposed Technique Phases

Proposed method comprises of the following steps:
    Initial cluster can be generated by any static clustering algorithm on a small fraction of data. At that time LA is equal to UA and CF vector is stored and then following steps starts.

- Step1: Data stream is input to this phase. It retrieves the new data object from data stream and calculates the dissimilarity of new object to all the clusters using Euclidean measure.
- Step 2: In this phase, minimum dissimilarity of new data object out of all clusters is found and compared to a threshold (OT). If minimum dissimilarity is to nearest cluster is smaller than OT then go to Step 3 otherwise go to Step 4.
- Step 3: In this phase, new data object is assigned to LA and UA of cluster based on following conditions: Let $d(x, Ci)$ be the dissimilarity between new object x and the cluster Ci. The difference between $d(x, Ci) - d(x, Cj), 1 \leq i, j \leq K$, k is the number of clusters, is used to determine the membership of x as follows:
  - If $d(x, Ci) - d(x, Cj) \leq$ threshold, then $x \in UA\ (Ci)$ and $x \in UA\ (Cj)$. Furthermore, x will not be a part of any LA.
  - Otherwise, $x \in LA(Ci4)$,such that $d(x, Ci)$ is the minimum for $1 \leq i \leq K$. In addition, $x \in UA(Ci)$.

Go to Step 7 by skipping Steps 4, 5 and 6.

- Step 4: In Step 2 if minimum dissimilarity of new object is to nearest cluster is larger than threshold (OT) than control flow directly jump to this step. Because such new object can be an outlier or it may be a member of new emerging cluster that appear as outlier currently. Outlying objects are stored as groups in outlier repository. New object is assigned to most similar outlier group. Go to Step 5.
- Step 5: In this phase, size of largest group is compared to user specified size if it is greater then got to Step 6 otherwise go to Step 9.
- Step 6: Largest group from outlier repository is considered as new emerged cluster [13] and number of clusters is increased by one. Let K is the number of clusters the $K = K + 1$.
- Step 7: It is the updation phase. It updated in following manner:

  - If new object is assigned to LA(Ci) then update MLA, MUA, complete CF vector, Time stamp, Mean and SD of corresponding cluster.
  - If new object is assigned to UA of two or more cluster then update MUA, only <nu, su, ssu> portion of CF vector, Mean of all clusters.
  - If new cluster has emerged then initialise its summary statistics that are MLA, MUA, CF vector, Time stamp, Mean and SD. Her LA = UA is assumed.

- Step 8: In streaming data some clusters may disappear over the time as no new data is assigned to them. Such clusters represent the disappearing concepts of data [13]. To keep a track of such clusters a time stamp is associated with cluster and updated whenever a data object is assigned to LA of that cluster. If time stamp of a cluster is older than an allowed time then that cluster is considered as obsolete cluster and it discarded and number of clusters decreased.
- Step 9: If more streaming data is available then for processing that go to Step 1 otherwise stop.

3.3 Difference Between Proposed Clustering Technique and Rough K-Means

- Proposed clustering technique is incremental in nature as oppose to Rough K-means which is iterative in nature.
- Proposed clustering technique performs well on streaming data as proved by results in Sect. 4 and it requires only single scan of data. But Rough K-means requires multiple scanning of data which is not possible in case of streaming data.
- Proposed clustering technique is adaptive to concept evolution as it focuses on new upcoming clustering and discards the obsolete clusters. But there is no such feature in Rough K-means as it is meant for static data.
- Rough K-means is sensitive to outliers that can worsen its performance as oppose to this Proposed clustering technique is robust to outliers.

4 Experimental Results

We have done all implementation in matlab R2010a. We have compared our results with Leader clustering algorithm which is an incremental method of clustering. In next coming subsection we will focus on performance analysis of proposed method.

4.1 Data Sets

Experiments are conducted on synthetic as well as real data sets (Table 1). Real data sets were taken from UCI machine learning repository [12].

There are total 10 classes in yeast data that are following: CYT, NUC, MIT, ME3 ME2, ME1, EXC, VAC, POX, and ERL. These classes represent the localization site of protein. Out of these classes we have not considered ERL as a cluster instead consider as outliers because its size is very small. Sequence number attribute of yeast dataset is not used as it is not relevant. In abalone data set there are in all 29 classes that corresponds to the age of abalone. In our experiments only large sized 20 classes are taken as clusters others are of very small size assumed as outliers. In synthetic data there are five clusters each having 2,000 samples and 20 outlier. This data is generated by a mixture of normal distribution and then based on statistical characteristics like mean, standard deviation, class distribution outliers are planted. In all experiments $Wl = 0.75$, $Wu = 0.25$, $OT = 2 * SD$, $ST = 5$ for yeast data set, $ST = 9$ for abalone and $ST = 10$ for synthetic data set.

4.2 Comparison of Proposed Method with K-Means Clustering

In this section, we will analyse the result of proposed clustering method and K-means clustering. For visual representation, we have used a synthetic data consisting of total 150 samples. Original data set is shown in Fig. 2.

There are clusters in the data seta. From Figs. 3 and 4. It is clear the proposed method gives move natural clusters as compare to K-means method because proposed

Table 1 Characteristics of the data sets

Data set name	Number of instances	Number of attributes	Number of clusters	Outliers
Yeast	1,484	8	9	5
Abalone	4,177	8	20	24
Synthetic dataset	10,020	6	5	20

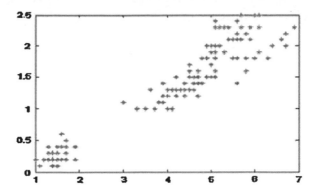

Fig. 2 Original data

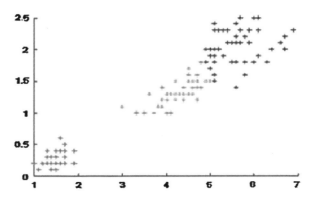

Fig. 3 *K*-means clustering results (*Red, Blue, yellow* colored are clusters and it does not give any overlapping points

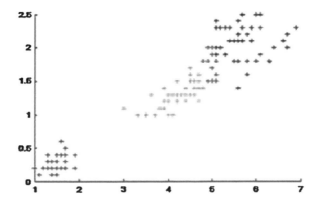

Fig. 4 Proposed method clustering results (*Red, Blue, yellow* colored are clusters and *green* color represent the overlap of *yellow and red* cluster)

method can handle uncertainty of belongingness of a data objects by allowing over-lapping cluster(see Fig. 4) which is not allowed in K-means.

4.3 Effect of Outlier on Proposed Method and K-Means Clustering

In this section, we will analyse the effect of outlier on proposed clustering method and K-means clustering. For visual representation, we have used a synthetic data consisting of total 158 samples. Out of which 8 are outliers other 150 represent 3 clusters. Original data set is shown in Fig. 5.

From Fig. 6, it is concluded that due to the presence of outliers K-means clusters centers are get distorted because K-mean clustering algorithm cannot handle outliers and assign them to clusters. Proposed method is robust to outliers so it can be seen from Fig. 7 that cluster centers are more natural.

It is concluded the proposed method generates more intuitive and better quality clusters as compare to K-means.

4.4 Performance Comparison in terms of Classification Accuracy

In this section classification accuracy of proposed and leader clustering algorithm is compared on all three data sets. For this datasets are divided in a ration of 70 % for cluster generation training and 30 % for testing. Cluster prototypes (Center of clusters) are used for nearest neighbour classifier.

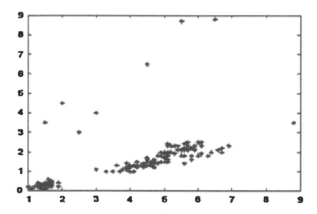

Fig. 5 Original data with outliers

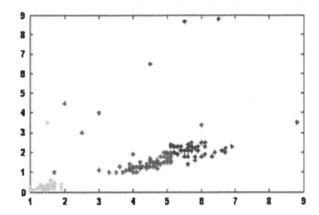

Fig. 6 *K*-means clustering results in presence of outliers (*Red, Blue, yellow* colored are clusters and *magenta* color represent the centers of clusters)

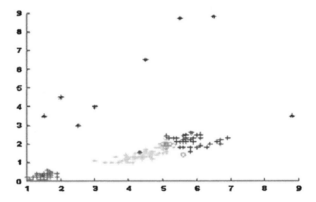

Fig. 7 Proposed method clustering results in presence of outliers (*Red, Blue, yellow* colored are clusters and *green* color represent the overlap of *yellow* and *blue* cluster, *black* color are outliers, *magenta* color cluster centers)

Figure 8 shows that classification accuracy of proposed method is much better than the leader clustering algorithm. It is because proposed scheme allow the overlap and more importance is given to sure members in finding the prototype of clusters.

4.5 Performance Analysis of Proposed Method

In this section, the performance of proposed method is analysed in the terms of correctness and accuracy. The correctness of rough clustering is defined as the correct rate of lower approximations, because the objects in the lower approximation of a rough cluster are the representative members of this cluster [5].

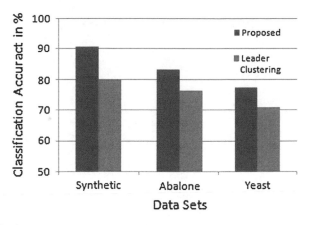

Fig. 8 Classification accuracy

Table 2 Performance of proposed method in terms of accuracy and correctness

Data set name	Correctness in %	Accuracy in %
Yeast	76.11	70.67
Abalone	79.84	68.89
Synthetic dataset	88.91	91.3

The correctness of rough clustering:

$$Correctness = \frac{\sum\limits_{j=1}^{K} |LA(Cj) \cap A(Cj)|}{\sum\limits_{j=1}^{K} |LA(Cj)|} \times 100$$

The accuracy of rough clustering describes the classification ability of rough clustering. A rough clustering with higher correctness and accuracy is better than that with lower values.

The accuracy of rough clustering:

$$Accuracy = \frac{\sum\limits_{j=1}^{K} |LA(Cj)|}{\sum\limits_{j=1}^{K} |UA(Cj)|} \times 100$$

In above equations K number of clusters and A(Cj) is the set of objects of that are actually labelled as class Cj.

The higher the value of correctness and accuracy better is the performance of rough clustering. It can be analysed from Table 2 that the performance of proposed method is good on all three data sets. Correctness and accuracy both are highest for synthetic data sets.

5 Conclusion

In this paper, we have presented a rough set based clustering technique for streaming data. Rough set is employed to capture the inherent uncertainty involved in cluster analysis. To face the challenges of data stream processing our proposed scheme is incremental and dynamic in nature. Proposed clustering method is robust to outliers and generates soft-clusters.

Experimental results show that the proposed method generates more meaningful clusters as compare to K-mean. It performance is better than leader clustering in terms of classification accuracy on all three datasets. Performance of proposed method is also good in terms of correctness and accuracy of rough clustering on all used data sets.

Thus the results show that the proposed technique is very promising for clustering streaming data. Further analysis of the performance of the proposed method is being further examined by using more real world and larger datasets.

References

1. Aggarwal, C., Han J., Wang J., Yu P.S.: A framework for projected clustering of high dimensional data streams. In: Proceedings of the 30th VLDB Conference (2004)
2. Aggarwal, C., Han, J., Wang, J., Yu, P.S.: A framework for clustering evolving data streams. In: Proceedings of 2003 International Conference on Very Large Data Bases (VLDB03), Berlin (2004)
3. Asharaf, S.: Narasimha Murty, M., Shevade, S.K.: Rough set based incremental clustering of interval data. Pattern Recogn. Lett. **27**(6), 515–519 (2006)
4. Cao, F., Ester, M., Qian, W., Zhou, A.: Density-based clustering over evolving data tream with noise. In: Proceedings of the 6th SIAM International Conference on Data Mining (SIAM 2006), pp. 326–337 (2006)
5. Hailiang, W., Mingtian, Z.: A refined rough k-means clustering with hybrid threshold. Rough Sets and Current Trends in Computing. Lecture Notes in Computer Science, vol. 7413, pp. 26–35 (2012)
6. Jiawei, H., Mieheline, K.: Data Mining, Concepts and Techniques. 2nd edn. Morgan Kaufmann, Massachusetts (2006)
7. Joshi, M., Yiyu, Y., Lingras, P., Virendrakumar, C.B.: Rough, fuzzy, interval clustering for web usage mining. In: Proceedings of 10th International Conference on Intelligent Systems Design and Applications (ISDA), pp. 397–402 (2010)
8. Lingras, P.: Rough set clustering for web mining. In: Proceedings of 2002 IEEE International Conference on Fuzzy Systems, pp. 1039–1044 (2002)
9. OCallaghan, L., Mishra, N., Meyerson, A., Guha, S.: Streaming data algorithms for high-quality clustering. In: Proceedings of ICDE Conference, pp. 685–704 (2000)
10. Pawlak, Z.: Rough Sets. Int. J. Inf. Commun. Comp.Sci. **11**, 145–172 (1982)
11. Pawlak, Z.: Some Issues on rough sets. Trans. Rough. Sets. **3100**, 1–58 (2004)

12. UCI Machine Learning Repository Irvine: CA University of California, School of Information and Computer, Irvine, CA (2010)
13. Udommanetanakit, K., Rakthanmanon, T., Waiyamai, K.: E-Stream. In: Evolution-Based Technique for Stream Clustering, pp. 605–615. Springer, Heidelberg (2007)
14. Yogita, Saroj, Kumar, D., Pal, V.: Rules + Exceptions: automated discovery of comprehensible decision rules. In: Proceedings of IEEE International Advance Computing Conference, pp. 1479–1484, TIET Patiala, India (2009)
15. Zhou, T., Zhang, Y.N., Lu, H.L.: Rough k-means cluster with adaptive parameters. In: 6th International Conference Machine on Learning and Cybernetics, pp. 3063–3068 (2007)

Optimizing Number of Cluster Heads in Wireless Sensor Networks for Clustering Algorithms

Vipin Pal, Girdhari Singh and R P Yadav

Abstract Clustering of sensor nodes is an energy efficient approach to extend lifetime of wireless sensor networks. It organizes the sensor nodes in independent clusters. Clustering of sensor nodes avoids the long distance communication of nodes and hence prolongs the network functioning time. The number of cluster heads is an important aspect for energy efficient clustering of nodes because total intra-cluster communication distance and total distance of cluster heads to base station depends upon number of cluster heads. In this paper, we have used genetic algorithms for optimizing the number of cluster heads while taking trade-off between total intra-cluster distance and total distance of cluster heads to base station. Experimental results show that proposed scheme can efficiently optimize the number of cluster heads for clustering of nodes in wireless sensor networks.

1 Introduction

Wireless sensor networks [1, 2] are application specific and consist of large number of sensor nodes deployed in a harsh environment/area. Sensor nodes sense area and send information to base station located outside/inside of area via single or multi-hop. Lifetime of network depends upon limited battery power of nodes. Due to harsh working area, it is not possible to change or replace battery of nodes. Hence, energy efficiency is critical issue for wireless sensor networks.

V. Pal (✉) · G. Singh
Malaviya National Institute of Technology, Jaipur, India
e-mail: vipinrwr@yahoo.com

G. Singh
e-mail: girdharisingh@rediffmail.com

R. P. Yadav
Rajasthan Technical University,Kota, India
e-mail: rp_yadav@yahoo.com

B. V. Babu et al. (eds.), *Proceedings of the Second International Conference on Soft Computing for Problem Solving (SocProS 2012), December 28–30, 2012,* Advances in Intelligent Systems and Computing 236, DOI: 10.1007/978-81-322-1602-5_132, © Springer India 2014

Clustering of nodes is an energy efficient approach for wireless sensor networks. Clustering approaches [3, 4] increases energy efficiency of network by avoiding long distance communication of nodes. Nodes are organized in independent sets or clusters. At least one cluster head is selected for each cluster. Clustering algorithms apply data aggregation techniques [5] which reduce the collected data at cluster head in the form of significant information. Cluster heads then send the aggregated data to base station.

For energy efficiency in wireless sensor network, the number of cluster heads is an important issue for clustering algorithms. Value of total intra-cluster communication distance and total distance of cluster heads to base station depends upon number of cluster heads. If numbers of cluster heads are less, total distance of cluster heads to base station decreases, while total intra-cluster communication distance increases. When numbers of cluster heads are more, total intra-cluster communication distance decreases, but there is increase in total distance of cluster heads to base station.

In this paper, genetic algorithm is applied for finding optimal number of cluster head while taking trade-off of total intra-cluster communication distance and total distance of cluster heads to base station. Experimental results show that proposed solution is effective to optimize number of cluster heads.

Rest of paper is organized as: Sect. 2 describes related work for energy efficient clustering algorithms. Section 3 describes genetic algorithm to optimize number of cluster heads. Section 4 describes results, and Sect. 5 concludes the work of paper.

2 Related Work

Low energy adaptive cluster hierarchy (LEACH) [6] is fully distributed algorithm. In setup phase cluster heads selection, cluster formation and TDMA scheduling are performed. In steady phase, nodes send data to cluster head and cluster head aggregate the data. Aggregated data are sent to base station. After a fix round time, re-clustering is performed. Role of cluster head is rotated to all the sensor nodes to make the network load balance. LEACH scheme does not guarantee about equal number of cluster heads in each round and number of nodes in each cluster.

LEACH-C [7] is centralized algorithm to form cluster and to assign duty of cluster heads. During setup phase, nodes send information about respective location and energy level to BS. BS formulates clusters using simulated annealing algorithm [8]. Algorithm provides CHs such that nodes minimize their transmission distance and conserve energy. After the formation of clusters and cluster heads, BS broadcasts a message that contains the information of CH ID for each node. The steady phase is same as of LEACH.

Adaptive decentralized re-clustering protocol (ADRP) [9] selects a cluster head and set of next heads for upcoming few rounds based on residual energy of each nodes and average energy of cluster. A round of ADRP has two phases: initial phase and cycle phase. In the initial phase, nodes send status of their energy and location to base station. Base station partitions the network in clusters and selects a cluster head for each cluster along with a set of next heads. In the cycle phase, cluster head aggregates the data and sends to the base station. In the re-cluster stage, nodes transit to cluster head from set of next heads without any assistance from base station. If the set of next heads is empty, initial phase is executed again. Re-clustering energy consumption is avoided for few rounds, but node death from next cluster head list makes network unbalanced.

References [10–13] applies genetic algorithm to clustering algorithm to find better cluster formation. All these approaches combined effect of total intra-cluster communication distance and total distance of cluster heads to base station. Two adverse factors are combined in one factor for fitness function. So, work of this paper takes effect of these two papers independently to find optimal number of cluster heads.

3 Genetic Algorithm to Optimize Number of Cluster Head

Genetic algorithms are heuristic search-based algorithms and are useful for searching and optimizing problems. GAs are based on theory of survival of the fittest. GA has population of individuals, and each individual represents a solution. Fitness value of each individual is calculated, and best-valued individual always has better chance of survival. Survived individuals go under genetic transformations, crossover and mutation. Then, there is new population of individuals that is fitter to previous one. Procedure of GA is shown in Fig. 1.

- Population: Population consists of various individual solutions for the problem. Larger the size of population, higher is the accuracy of algorithm. Length of individual depends upon number of nodes in network as a 1 in individual represents node as cluster head, while a 0 means nodes is member node. Initial population is generated randomly.
- Fitness Function: Survivability of an individual depends upon its fitness value. Fitness value of each individual is calculated according a fitness function. In our work, fitness function consists of following three parameters.

  - Number of cluster heads (CH)
  - Total intra-cluster communication distance (IC)
  - Total distance from cluster heads to base station (BSD)

  Value of last to parameters depends upon first. Less number of cluster heads has less total distance from cluster heads to base station but has high total intra-cluster communication distance. While high number of cluster heads has less total intra-cluster communication distance but has more total distance from cluster heads to base station. After scaling the fitness function, we have fitness function as:

$$\text{Fitness} = (N - \text{CH}) + \frac{\text{IC}}{100} + \frac{\text{BSD}}{100}$$

Fig. 1 Genetic algorithm

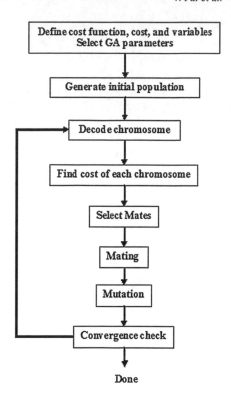

Done

where N is total number of nodes in network. Fitness function shows that there is more emphasis on decreasing total distance from cluster heads to base station.

- Selection: Selection is the process of choosing individuals from current population for new population. The purpose of the selection process in a genetic algorithm is to give more reproductive chances to those population members that are better fit. The selection procedure may be implemented in a number of ways like roulette wheel selection, tournament selection, Boltzmann selection, rank selection, and random selection. In this work, roulette wheel selection procedure is applied to select chromosomes for generating new population.
- Crossover: In this paper, one-point crossover method is used. The crossover operation takes place between two chromosomes with probability specified by crossover rate. These two chromosomes exchange portions that are separated by the crossover point. The following is an example of one-point crossover.

| Individual 1 | 0 | 1 | 1 | 1 | 0 | 0 | 1 | 1 | 1 | 0 | 1 | 0 |
| Individual 2 | 1 | 0 | 1 | 0 | 1 | 1 | 0 | 0 | 1 | 0 | 1 | 0 |

After crossover, two offsprings are created as below:

| Offspring 1 | 0 | 1 | 1 | 1 | 0 | **1** | **0** | **0** | **1** | **0** | **1** | **0** |
| Offspring 2 | 1 | 0 | 1 | 0 | 1 | **0** | **1** | **1** | **1** | **0** | **1** | **0** |

- Mutation: The mutation operator is applied to each bit of a chromosome with a probability of mutation rate. After mutation, a bit that was 0 changes to 1 and vice versa.

| Before mutation | 0 | 1 | 1 | **1** | 0 | 0 | 1 | 1 | 1 | 0 | 1 | 0 |
| After mutation | 0 | 1 | 1 | **0** | 0 | 0 | 1 | 1 | 1 | 0 | 1 | 0 |

4 Results

Genetic algorithm for optimizing number of cluster heads is first implemented in C++ language. Sensor network topologies of 50 nodes over $50 \times 50 \, \text{m}^2$ are generated. Base station is located outside the field (100, 25).

Figure 2 shows best fitness values and average fitness values for 100 simulations. The best fitness graph represents average of best fitness value of each iteration of a simulation, while average fitness graph shows average of fitness value of each chromosome of each iteration of a simulation. Random topology is generated for

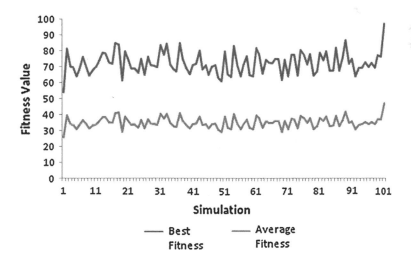

Fig. 2 Fitness for various simulations

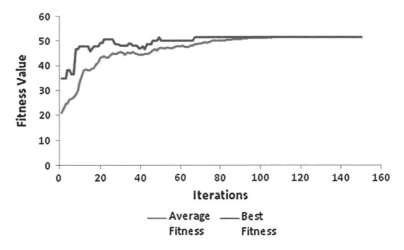

Fig. 3 Fitness with respect to generations

each simulation. There is significant difference between best fitness and average fitness values for each simulation

Figure 3 shows change in average fitness value and best fitness value over increase in generations. There is increase in both average fitness value and best fitness value as iterations increases. Both converge at one point after 120 iterations. That shows that fitness function is successful to converse the solution for finding optimal number of cluster heads in network.

Figure 4 shows number of cluster heads in network as iterations increases. At first, there are 30 cluster heads, while as iterations increased number of clusters are decreased and after 115 iterations it is constant, i.e., it optimized. So, the suggested fitness function is effective to optimize the solution.

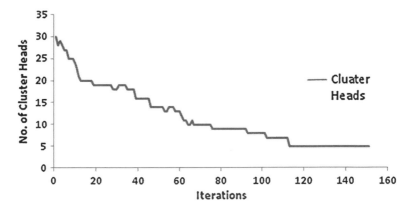

Fig. 4 Number of cluster heads over iterations

4.1 Proposed Scheme for Wireless Sensor Network

Nodes are deployed randomly in field. Nodes are considered to be location aware. Nodes send their information about location and remaining energy to base station. Genetic algorithm is applied to optimize number of cluster heads and their respective cluster formation at base station. Base station broadcasts complete information about cluster heads, member nodes of cluster heads, and data transmission/receive schedule of nodes to network. All nodes in network receive the broadcasted message and update their status; Setup phase is completed. Nodes send sensed data to cluster heads. Cluster heads performs aggregation. Reduced data are then sent to base station. After completion of current round, nodes send their updated information to base station for new cluster head selection and cluster formation.

5 Conclusion and Future Work

The number of cluster heads is a critical issue for energy efficient clustering of sensor nodes in wireless sensor networks. Work of this paper has optimized the number of cluster heads for clustering while taking trade-off of inter-cluster communication distance and cluster heads to base station distance. Results show that proposed scheme has effectively optimized the cluster heads.

Based on the work of this paper, we have also proposed a framework for centralized clustering of wireless sensor network. In future, we will implement the proposed framework and will compare the performance of proposed scheme with other existing clustering scheme.

References

1. Akyildiz, I., Su, W., Sankarasubramaniam, Y., Cayirci, E.: Wireless sensor networks: a survey. Comput. Netw. **38**(4), 393–422 (2002)
2. Estrin, D., Govindan, R., Heidemann, J.S., Kumar, S.: Next century challenges: scalable coordination in sensor networks. In: MOBICOM. pp. 263–270 (1999)
3. Younis, O., Krunz, M., Ramasubramanian, S.: Node clustering in wireless sensor networks: recent developments and deployment challenges. IEEE Network Mag. **20**, 2025 (2006)
4. Abbasi, A.A., Younis, M.: A survey on clustering algorithms for wireless sensor networks. Comput. Commun. **30**(14–15), 2826–2841 (2007)
5. Karl, H., Willig, A.: Protocols and Architectures for Wireless Sensor Networks. Wiley, England (2007)
6. Heinzelman, W.R., Chandrakasan, A., Balakrishnan, H.: Energy efficient communication protocol for wireless microsensor networks. In: Proceedings of the 33rd Hawaii International Conference on System Sciences, pp. 10–20 (2000)
7. Heinzelman, W., Chandrakasan, A., Balakrishnan, H.: An application-specific protocol architecture for wireless microsensor networks. IEEE Trans. Wireless Commun. **1**(4), 660–670 (2002)

8. Murata, T., Ishibuchi, H.: Performance evaluation of genetic algorithms for flowshop scheduling problems. In: Proceeding First IEEE Conference on IEEE World Congress on Computational Intelligence, Evolutionary Computation, pp. 812–817 (1994)
9. Bajaber, F., Awan, I.: Adaptive decentralized re-clustering protocol for wireless sensor networks. J. Comput. Syst. Sci. **77**(2), 282292 (2011)
10. Heidari, E., Movaghar, A.: An efficient method based on genetic algorithms to solve sensor network optimization problem. Int J. GRAPH-HOC **3**(1), 18–32 (2011)
11. Sajid, H., Abdul, W.M., Obidul, I.: Genetic algorithm for hierarchical wireless sensor networks. J. Netw. **2**(5), 87–97 (2007)
12. Jin, F., Parish, D.J.: Using a genetic algorithm to optimize the performance of a wireless sensor network. In: Proceeding PGNet, (2007)
13. Jenn-Long, L., Chinya, V.R.: LEACH-GA: genetic algorithm-based energy-efficient adaptive clustering protocol for wireless sensor networks. Int. J. Mach. Learn. Comput. **1**(1), 79–85 (2011)

Data Clustering Using Cuckoo Search Algorithm (CSA)

P. Manikandan and S. Selvarajan

Abstract Cluster Analysis is a popular data analysis in data mining technique. Clusters play a vital role for users to organize, summarize and navigate the data effectively. Swarm Intelligence (SI) is a relatively new subfield of artificial intelligence which studies the emergent collective intelligence of groups of simple agents. It is based on social behavior that can be observed in nature, such as ant colonies, flocks of birds, fish schools and bee hives. SI technique is integrated with clustering algorithms. This paper proposes new approaches for using Cuckoo Search Algorithm (CSA) to cluster data. It is shown how CSA can be used to find the optimally clustering N object into K clusters. The CSA is tested on various data sets, and its performance is compared with those of K-Means, Fuzzy C-Means, Fuzzy PSO and Genetic K-Means clustering. The simulation results show that the new method carries out better results than the K-Means, Fuzzy C-Means, Fuzzy PSO and Genetic K-Means.

Keywords Clustering · Swarm Intelligence (SI) · CSA · K-Means · Fuzzy C-Means · Fuzzy PSO · Genetic K-Means

1 Introduction

Clustering analysis identifies and groups the data, where each cluster consists of similar objects and dissimilar to objects of other clusters. For high quality clusters, the inter-cluster similarity is low and the intracluster similarity is high [15, 17, 24].

P. Manikandan (✉)
Paavaai Group of Institutions, Namakkal, Tamilnadu, India
e-mail: mani.p.mk@gmail.com

S. Selvarajan
Muthayammal Technical Campus,Rasipuram, Tamilnadu, India
e-mail: asselvarajan@rediffmail.com

B. V. Babu et al. (eds.), *Proceedings of the Second International Conference on Soft Computing for Problem Solving (SocProS 2012), December 28–30, 2012*, Advances in Intelligent Systems and Computing 236, DOI: 10.1007/978-81-322-1602-5_133, © Springer India 2014

Swarm Intelligence (SI) [1, 6] is the collective behaviour which has no definite organized system but it has a self-organized system, which may be natural or artificial. The expression was introduced by Gerardo Beni and Jing Wang in 1989.

A survey of the clustering algorithms can be found in [13, 17]. Genetic algorithms are used in [3, 5, 14, 19, 21, 23] while an analytical review of the use of neural networks in clustering is given in [12]. Clustering algorithms based on Ant Colony Optimization are used in [2, 4, 8, 10, 26]. While in [7, 9, 16, 18] clustering algorithms based on Particle Swarm Optimization are applied.

This paper aims to propose Cuckoo Search Algorithm (CSA) to solve the clustering problem. The proposed CSA is applied for optimally clustering N object into K clusters. The CSA is tested on various data sets, and its performance is compared with those of K-Means, Fuzzy C-Means, Fuzzy PSO and Genetic K-Means clustering. The simulation results illustrate that this algorithm not only has a better response but also converges more quickly than the ordinary evolutionary methods. The rest of the paper is organized as follows: Proposed technique of CSA for Data Clustering is described in Sect. 2. The detailed result and discussion is discussed in Sect. 3. Conclusion and Future Research is described in Sect. 4.

2 The Proposed Cuckoo Search Algorithm for Clustering

2.1 Cuckoo Breeding Behaviour

Cuckoo birds attract attention because of their unique aggressive reproduction strategy. Cuckoos engage brood parasitism in which a bird lays and discards its eggs in the nest of another species. Species lay their eggs in shared nests.

While flying, some animals and insects follow the path of long path with sudden 900 turns shared with short, random movements. This random walk is called Levy flight and it describes foraging patterns in natural systems, such as systems of ants, bees and bumbles. These flights can also be noticed in the movements of chaotic fluids. One example of Levy flight paths is depicted on Fig. 1.

Levy-style search behavior [22] and random search in general has been applied to optimization and implemented in many search algorithms [11, 20]. One of such algorithms is CSA [25]. Preliminary results show its promising capability.

2.2 Description of the Cuckoo Search Algorithm

Cuckoo Search Algorithm (CSA) is population based stochastic global search metaheuristics. It is based on the general random walk system which will be briefly discussed in this paper.

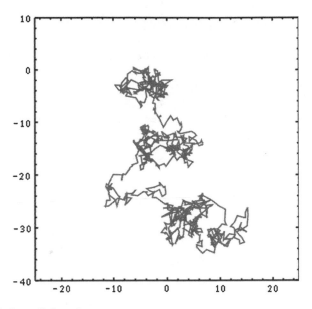

Fig. 1 Possible Levy flight path

In CSA, Nature's systems are complex and thus, they cannot be modeled by computer algorithms in its basic form. Simplification of natural systems is necessary for successful implementation in computer algorithms.

One approach is to simplify CSA through three below presented approximation rules:

- Cuckoos chose random location (nest) for laying their eggs. Artificial cuckoo can lay only one egg at a time.
- Elitist selection process is applied, so only the eggs with highest quality are passed to the next generation.
- Host nests number is not adjustable. Host bird discovers cuckoo egg with probability $pd \epsilon [0, 1]$. If cuckoo egg is disclosed by the host, it may be thrown away, or the host may abandon its own nest and commit it to the cuckoo intruder.

To make the things even simpler, the last assumption can be approximated by the fraction of pd of n nests that are replaced by new nests with new random solutions. Considering maximization problem, the quality (fitness) of a solution can simply be proportional to the value of its objective function. Other forms of fitness can be defined in a similar way; the fitness function is defined in genetic algorithms and other evolutionary computation algorithms.

A new solution $x^{(t+1)}$ for cuckoo i is generated using a Levy flight according to the following equation:

$$x_i (t + 1) = x_i (t) + \alpha^\wedge Levy (\lambda) , \tag{1}$$

Where $\alpha(\alpha > 0)$ represents a step scaling size. This parameter should be related to the scales of problem the algorithm is trying to solve. In most cases, α can be set to the value of 1 or some other constant. The product $^\wedge$ represents entry-wise multiplications.

Equation (2) states that described random walk is a *Markov chain,* whose next location depends on two elements: current location (first term in Eq. 2) and transition probability (second term in the same expression).

The random step length is drawn from a Levy distribution which has an infinite variance with an infinite mean:

$$Levy \sim u = t - \lambda \qquad (2)$$

$$\text{Where } \lambda \in [0, 3].$$

3 Computational Results

3.1 Dataset Description

The performance of the proposed methodology is tested on four benchmark instances taken from the UCI Machine Learning Repository (http://archive.ics.uci.edu/ml/datasets.html). The datasets are chosen to include a wide range of domains and their characteristics. The data sets are wine, blood-transfusion, breast cancer Wisconsin and Space.

3.2 Experimental Results

Data clustering as one of the important data mining techniques is a fundamental and widely used method to achieve useful information about data. In face of the clustering problem, clustering methods still suffer from trapping in a local optimum and cannot often find global clusters. In order to overcome the shortcoming of the available clustering methods, this paper presents efficient clustering algorithms based on CSA. The obtained experimental results such as, Fitness values of various iterations for wine, Blood-Transfusion, Breast Cancer-Wisconsin and Space dataset are shown in Tables 1, 2, 3 and 4.

3.3 Comparative Analysis

In this section, the scope is to compare the CSA clustering algorithm with several typical algorithms including K-Means, Fuzzy C-Means, Fuzzy PSO and Genetic K-Means. Four data sets have been used. The performance analysis has been made by

Table 1 Fitness values of various iterations for wine dataset

Iterations	Fitness value				
	CSA	K-Means	F C-Means	Fuzzy PSO	G K-Means
1	17958.31	16556.18	40063.98	19593.3	18318.59
2	16618.49	16556.18	39079.49	19593.3	17552.52
3	16580.31	18295.45	40098.22	18754.42	17275.87
4	16580.31	16556.18	22331.47	17951.09	17189.49
5	16580.31	18295.45	19146.47	17951.09	17055.54
Average	16721.93	16904.03	24464.64	18351.02	17058.94

Table 2 Fitness values of various iterations for blood-transfusion dataset

Iterations	Fitness				
	CSA	K-Means	F C-Means	Fuzzy PSO	G K-Means
1	297053.7	325565.5	700113.2	308098.6	340869.2
2	297053.7	366406.4	651359.1	308098.6	312919.2
3	297053.7	325565.5	440347.1	308098.6	308862.5
4	297053.7	366406.4	436916.4	307469.5	295575.1
5	297053.7	325565.5	413273.8	291531.3	295575.1
Average	295006.5	333733.7	411692.1	298095.3	303167.7

Table 3 Fitness values of various iterations for breast cancer-Wisconsin dataset

Iterations	Fitness				
	CSA	K-Means	F C-Means	Fuzzy PSO	G K-Means
1	73676602	75011185	161000000	78837480	80818812
2	73676602	72714041	141000000	78837480	74635073
3	72308694	72714041	130000000	73974520	74635073
4	72308694	72714041	76971891	73974520	74635073
5	72308694	72714041	80138269	73470934	74635073
Average	72582276	74453340	96231713	74644960	75253447

plotting the graphs. By analyzing the plotted graph, the performance of the proposed technique has significantly improved compared with other techniques. Figures 2, 3, 4 and 5 shows the graph which is plot between iteration and fitness value for our proposed technique as well as other algorithms. In Figs. 2, 3, 4 and 5, our proposed technique (plotted in orange color) shows better performance when compared to the other techniques.

Table 4 Fitness values of various iterations for space dataset

Iterations	Fitness				
	CSA	K-Means	F C-Means	Fuzzy PSO	G K-Means
1	158.2374	232.5886	1218.797954	161.3488	201.6643
2	158.2374	143.5048	436.3725051	161.3488	200.3161
3	158.2374	143.5048	268.5698966	161.3488	200.3161
4	158.2374	238.0524	260.0512394	161.3488	200.3161
5	158.2374	238.0524	143.3624333	161.3488	200.3161
Average	155.4592	208.049	318.8903314	160.3235	200.4509

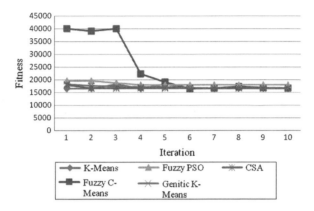

Fig. 2 Iteration versus fitness value of wine dataset.

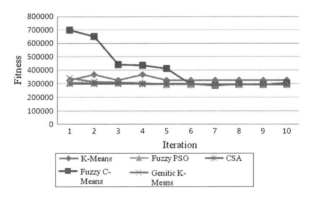

Fig. 3 Iteration versus fitness value of blood-transfusion dataset

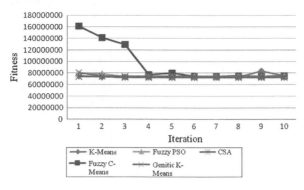

Fig. 4 Iteration versus fitness value of breast cancer Wisconsin dataset

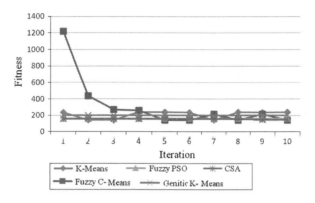

Fig. 5 Iteration versus fitness value of space dataset

4 Conclusion and Future Research

A CSA to solve clustering problems has been developed in this paper. To evaluate the performance of the CSA, it is compared with other stochastic algorithms viz. K-Means, Fuzzy C-Means, Fuzzy PSO and Genetic K-Means clustering algorithms on several well known data sets. The experimental results indicate that the proposed clustering algorithm is comparable to the other algorithms in terms of fitness function. The result illustrates that the proposed CSA algorithm can be considered as a viable and an efficient heuristic. In future CSA clustering algorithm which has been discussed can be hybrid with Swarm Intelligence or some conventional clustering algorithms for achieving optimal results.

References

1. Abraham, A., Guo, H., Liu, H.: Swarm Intelligence: Foundations, Perspectives and Applications, Swarm Intelligence in Data Mining. Springer, Germany (2006)
2. Azzag, H., Venturini, G., Oliver, A., Gu, C.: A hierarchical ant based clustering algorithm and its use in three real-world applications. J. Oper. Res. **179**, 906–922 (2007)
3. Babu, G., Murty, M.: A near-optimal initial seed value selection in k-means algorithm using a genetic algorithm. Pattern Recogn. Lett. **14**(10), 763–769 (1993)
4. Chen, L., Tu, L., Chen, H.: A novel ant clustering algorithm with digraph. In: Wang, L., Chen, K., Ong, Y.S. (eds.) LNCS, pp. 1218–1228. Springer, Berlin (2005)
5. Cowgill, M., Harvey, R., Watson, L.: A genetic algorithm approach to cluster analysis. Comput. Math. Appl. **37**, 99–108 (1999)
6. Bonabeau, E., Dorigo, M., Theraulaz, G.: Swarm Intelligence: From Natural to Artificial Systems. Oxford University Press, New York (1999)
7. Nie, F., Tu, T., Pan, M., Rong, Q., Zhou, H.: Advances in Intelligent and Soft Computing, vol. 139, pp. 67–73. Springer, Heidelberg (2012)
8. He, Y., Hui, S.C., Sim, Y., et al.: A novel ant-based clustering approach for document clustering. In: Ng, H.T. (ed.) LNCS, pp. 537–544. Springer, Berlin (2006)
9. Janson, S., Merkle, D., et al.: A new multi-objective particle swarm optimization algorithm using clustering applied to automated docking. In: Blesa, M.J. (ed.) LNCS, pp. 128–141. Springer, Berlin (2005)
10. Kao, Y., Cheng, K., et al.: An ACO-based clustering algorithm. In: Dorigo, M. (ed.) LNCS, pp. 340–347. Springer, Berlin (2006)
11. Ozdamar, L.: A dual sequence simulated annealing algorithm for constrained optimization. In: Proceedings of the 10th WSEAS International Conference on, Applied Mathematics, pp. 557–564 (2006)
12. Liao, S.-H., Wen, C.-H.: Artificial neural networks classification and clustering of methodologies and applications - literature analysis from 1995 to 2005. Expert Syst. Appl. **32**, 1–11 (2007)
13. Verma, M., Srivastava, M., Chack, N., Kumar Diswar, A., Gupta, N.: A comparative study of various clustering algorithms in data mining. Int. J. Eng. Res. Appl. (IJERA) **2** 1379–1384 (2012)
14. Meng, L., Wu, Q.H., Yong, Z.Z., et al.: A faster genetic clustering algorithm. In: Cagnoni, S. (ed.) LNCS, pp. 22–33. Springer, Berlin (2000)
15. Mirkin, B.: Mathematical Classification and Clustering. Kluwer, Dordrecht (1996)
16. Paterlini, S., Krink, T.: Differential evolution and particle swarm optimization in partitional clustering. Computational Statistics and Data Analysis **50**, 1220–1247 (2006)
17. Rokach, L., Maimon, O.: Clustering methods. In: Maimon, O., Rokach, L. (eds.) Data Mining and Knowledge Discovery Handbook, pp. 321–352. Springer, New York (2005)
18. Shen, H.-Y., Peng, X.-Q., Wang, J.-N., Hu, Z.-K.: A mountain clustering based on improved PSO algorithm. In: Wang, L., Chen, K., Ong, Y.S. (eds.) LNCS, pp. 477–481. Springer, Berlin (2005)
19. Sheng, W., Liu, X.: A genetic it k-medoids clustering algorithm. J. Heuristics **12**, 447–466 (2006)
20. Chen, T.Y., Cheng, Y.L.: Global optimization using hybrid approach. WSEAS Trans. Math. **7**(6), 254–262 (2008)
21. Tseng, L., Yang, S.: A genetic approach to the automatic clustering problem. Pattern Recogn. **34**, 415–424 (2001)
22. Viswanathan, G.M., Raposo, E.P., da Luz, M.G.E.: Lévy flights and superdiffusion in the context of biological encounters and random searches. Phys. Life Rev. **5**(3), 133–150 (2008)
23. Wu, F.-X., Zhang, W.J., Kusalik, A.J.: A genetic k-means clustering algorithm applied to gene expression data. In: Xiang, Y., Chaib-draa, B. (eds.) LNAI, pp. 520–526. Springer, Berlin (2003)

24. Xu, R., Wunsch II, D.: Survey of clustering algorithms. IEEE Trans. Neural Networks **16**(3), 645–678 (2005)
25. Yang, X.S., Deb, S.: Engineering Optimisation by Cuckoo Search. Int. J. Math. Model. Numer. Optim. **1**(4), 330–343 (2010)
26. Yang, Y., Kamel, M.S.: An aggregated clustering approach using multi-ant colonies algorithms. Pattern Recogn. **39**, 1278–1289 (2006)

Search Result Clustering Through Expectation Maximization Based Pruning of Terms

K. Hima Bindu and C. Raghavendra Rao

Abstract Search Results Clustering (SRC) is a well-known approach to address the lexical ambiguity issue that all search engines suffer from. This paper develops an Expectation Maximization (EM)-based adaptive term pruning method for enhancing search result analysis. Knowledge preserving capabilities of this approach are demonstrated on the AMBIENT dataset using Snowball clustering method.

Keywords Information retrieval · Clustering · FPtree · Expectation maximization

1 Introduction

Due to enormous growth of information available on the Web, search engine users are swamped with millions of search results, especially in case of broad and ambiguous queries. Ambiguity arises from the low number of query words [7]. Users typically give short queries with an average query length of three words per query [12]. Many search engines use diversification techniques to avoid duplicate results on the first pages of the results. This approach enables quick retrieval of one relevant result per subtopic, but may not facilitate retrieval of more results of user interest [13].

Search Results Clustering (SRC) is a solution to address the lexical ambiguity issue [2]. This approach is especially useful in case of polysemous queries. Search Results Clustering partitions the results obtained in response to a query into a set of labeled clusters. Each cluster corresponds to a subtopic of the query. Therefore, the user need not accurately predict the words used in the documents that best satisfy his

K. Hima Bindu (✉) · C. Raghavendra Rao
Department of Computer and Information Sciences, University of Hyderabad,
Hyderabad, Andhra Pradesh, India
e-mail: himagopal@gmail.com

C. Raghavendra Rao
e-mail: crrcs@uohyd.ernet.in

B. V. Babu et al. (eds.), *Proceedings of the Second International Conference on Soft Computing for Problem Solving (SocProS 2012), December 28–30, 2012*, Advances in Intelligent Systems and Computing 236, DOI: 10.1007/978-81-322-1602-5_134, © Springer India 2014

or her information needs. Instead, the user can start with a generic or broad query and quickly navigate to relevant subtopic. Further, user can grasp the semantic structure in the search results or refine the queries by using the cluster labels.

Search Results Clustering has specific challenges in contrast to traditional clustering methods. These include quick response time (as this is an online activity), ephemeral clustering (clustering happens when the user issues a query), meaningful cluster labels (must be human readable), and no fixed number of clusters.

A Frequent Itemset-based document clustering technique [4] can generate meaningful cluster labels and does not require the number of clusters as an input parameter. However, identification of frequent itemsets is a complex and time-consuming task and also involves subjectivity. Hence, we used Expectation Maximization (EM) [3] to prune the irrelevant terms to make our approach meet search result analysis challenges, in particular the SRC requirements. Snowball clustering proposed by [6] develops clusters starting with highest score (fitness) to least score. Hence, we have chosen to build a clustering method based on these approaches.

The outline of this paper is as follows. Section 2 briefly discusses the search results acquisition procedure and the preprocessing methods. The objective term pruning approach based on EM is presented in Sect. 3. Section 4 briefs the clustering and labeling algorithm. Section 5 shows our experimental results and the comparison with other popular data centric algorithms used for SRC. We conclude the paper in Sect. 6.

2 Search Results Acquisition and Preprocessing

The search results of a search engine are acquired by using the search engine's API by sending HTTP requests. All major search engines provide APIs with restrictions on the number of queries per day as a free service and as a paid service without such limitations. By sending a RESTful (Representational State Transfer) request to the public search engine APIs, the results are available in either JSON or XML format. Usually, the first 100 results are considered.

The title of each search result along with one or two lines summary of the web page (called as snippet) forms a search result document. These search results are usually preprocessed by tokenization, stemming (Porter Stemmer) and stopword removal. Stop words are frequently occurring words, which do not carry semantics. We used the stop word list of 571 words from SMART[1] system. Then, the words are stemmed by Porter's suffix stripping algorithm[2]. By using Vector space model, each result is represented as a TF-IDF vector [8]. The terms of a result's title and snippet after these steps form the features.

---

[1] http://www.lextek.com/manuals/onix/stopwords2.html

[2] http://tartarus.org/~martin/PorterStemmer/

3 Term Pruning with Expectation Maximization

Expectation Maximization is frequently used for data clustering in Machine learning, as it can handle latent variables in the data. Expectation Maximization method is developed for characterizing the population characteristics, which is a mixture of finite fundamental factors. Each factor will have its own uncertainty model characterized by associated probability distribution. The characteristics of the population will obey an uncertainty model characterized by mixed distribution, which is generalization of fundamental factors. This mixture model will have set of parameters, namely mixing proportions, and the parameters of fundamental factor distributions. If one assumes that the population is a mixture of k fundamental factors, and each factor obeys normal distribution, when the characteristic under study is one-dimensional, then the mixture model will have parameter's mixing proportions $\pi_1, \pi_2, \ldots \pi_k; \left(\pi_i > 0 \text{ and } \sum_{i=1}^{k} \pi_i = 1 \right)$ and (π_i, σ_i) as the mean and standard deviation of ith fundamental factor. Estimation process of these parameters is addressed by EM [3] based on the sample.

The terms occurring in each preprocessed search result constitute the population for the analysis. However, some of the terms are relevant and some are not relevant for building the knowledge. Thus, the distribution of the TF-IDF values of the terms can be viewed as a mixture of relevant and irrelevant terms (i.e., as a mixture of two distributions). The simpler way of representing it is by using Gaussian distribution, $f(x) = \pi_1 f_1 (x|\mu_1, \sigma_1) + \pi_2 f_2 (x|\mu_2, \sigma_2)$, where f_1 is the probability distribution function of TF-IDF values of relevant terms with mean μ_1 and standard deviation σ_1, f_2 is the probability distribution function of TF-IDF values of non-relevant terms with mean μ_2 and standard deviation σ_2; π_1 and π_2 are the mixing proportions. These parameter estimates must have the property that $\mu_1 > \mu_2$. Based on the parameters obtained for the mixture model, the threshold on TF-IDF values is $\frac{\pi_1\mu_1 + \pi_2\mu_2}{\pi_1 + \pi_2}$, from [14]. Any term with TF-IDF value more than this threshold is classified as relevant term, otherwise irrelevant term.

4 The Clustering and Labeling Algorithm

Search Results Clustering algorithms must be faster to have minimum latency between query submission and cluster presentation. Our algorithm performs clustering and labeling on the fly by processing only the search results without using external knowledge, hence, it is a lightweight approach. It labels the clusters automatically by using only the text concepts (frequent termsets) available in the search results.

The frequent termsets are identified by FPGrowth [5], by considering the relevant terms retained after the EM based term pruning, snowball clustering method [6] has been applied to arrive at the clusters along with the labels. The size of the relevant terms set due to EM makes the frequent itemset analysis manageable, though it is a tedious task.

5 Results and Analysis

An illustration of EM-based term pruning is provided here, by considering the first ten results of the query "Beagle," from the AMBIENT[3]. dataset.

5.1 Terms Extraction

The first 10 results after tokenization (using white space as delimiter) resulted in 189 terms, are shown in Sect. 5.1.1. With the preprocessing by stemming and stop-word removal as discussed in the Sect. 3, 118 terms are retained, these are shown in Sect. 5.1.2. The TF-IDF vectors are constructed for these terms. When EM algorithm is run, the threshold on TF-IDF values is obtained as 1.17, based on the method described in Sect. 4. It is observed that 28 terms' TF-IDF satisfy this threshold. These are treated as relevant terms, which are given in Sect. 5.1.3.

5.1.1 Terms Obtained After Tokenization

(EC), (software), -, –, ..., 2, 3, 5, :, A, ANSI/ISO, American, Apple, Architecture, BEAGLE, Beagle, Beagle, Beagle-type, Beagle, Beagles, Breed, Britannica, British, C++, C++, CenterÂ®, Club, Computation, Consortium, Debian, Desktop, Dog, Download, Earth, Encyclopaedia, English, Evolutionary, Files, GNOME, God's, Google, Group, Guide, Head, Hound, Information, Information, Kennel, Linux, Main, Mars, Open, Owner's, PSSRI, Package, Page, Pillinger, Professor, Profile:, Search, Size, Spotlight, Standard, Storage, The, There's, University-based, University, Unix, W3, Wikipedia, Windows, a, all, also, among, an, and, any, architectures, are, article, as, at, available, be, beagle, beagle, blog-post, breed, built, but, by, can, centuries, child, coat, code, coded, color, come, compliant, critters, cutest, data, dog, dog, dogs, easy-care, encyclopedia, enhance, entirely, existed, experience, exploration, fairly, for, framework, free, get, good, gotten, green, group, has, have, head, heavy, hound, idea, in, indexing, industrial, is, its, led, long, medium, member, modern, not, of, on, or, other, over, overwhelmed, pack, page, partners, people, personal, pet, point, post, power, product, puppies, puppy, reference, researchers, search, short, should, similar, simplicity, single, sized, skull, sleek, small, solidly, sporty, standard, system, the, their, this, to, together, tool, website, where, which, will, with, you, your, â€d.

---

[3] AMBIguous ENTries : http://credo.fub.it/ambient.

5.1.2 Terms After Pre Processing by Stemming and Stop-Word Removal

american, ansiiso, appl, architectur, articl, avail, beagl, beagletyp, blogpost, breed, britannica, british, built, center, centuri, child, club, coat, code, color, come, compliant, comput, consortium, critter, cutest, data, debian, desktop, dog, download, earth, easycar, ec, encyclopaedia, encyclopedia, english, enhanc, entir, evolutionari, exist, experi, explor, fairli, file, framework, free, gnome, god, good, googl, gotten, green, group, guid, head, heavi, hound, idea, index, industri, inform, kennel, led, linux, long, main, mar, medium, member, modern, open, over, overwhelm, owner, pack, packag, page, partner, peopl, person, pet, pilling, point, post, power, product, professor, profil, pssri, puppi, refer, research, search, short, similar, simplic, singl, size, skull, sleek, small, softwar, solidli, sporti, spotlight, standard, storag, system, togeth, tool, univers, universitybas, unix, w3, websit, wikipedia, window.

5.1.3 Terms After EM-Based Term Pruning

american, architectur, britannica, british, club, code, debian, desktop, download, encyclopaedia, encyclopedia, explor, free, guid, inform, kennel, led, main, mar, open, owner, page, profil, puppi, search, softwar, w3, wikipedia.

5.2 Analysis

For the query "Beagle", by considering the first 100 search results, the number of terms after stemming and stop-word removal is 840. With term pruning, the number of terms is 230. The tokenized termset size is not reported as it not the interest of researchers.

The average terms set size for all the 44 queries of AMBIENT dataset, resulted without term pruning as 865.18 (with standard deviation 74.06), and with term pruning as 274.22 (with standard deviation 72.22). The percentage of reduction in terms set size is 68.30.

The associated FP Trees of pre and post term pruning have been constructed and the corresponding estimates of the size (obtained by multiplying the maximum depth and maximum width) are derived and reported in Fig. 1. For the 44 datasets of AMBIENT, the Mean (Standard Deviation) of FPTree sizes before and after term pruning are 2486.48 (SD = 286.12) and 962.82 (SD = 243.29), respectively. The percentage of reduction in FP tree sizes is 61.28.

The "degree of knowledge representation" (Kappa) developed in the Rough Set theory [10] which quantifies the knowledge representation in considered features is employed, treating the terms as features. It is observed that pre pruning produced mean Kappa as 1 (SD = 0), indicating that the pre pruned termset possess hundred percent knowledge about the ground truth. Whereas post pruning term set has mean Kappa 0.95 (SD = 0.04) indicates that, the knowledge representation by post pruned

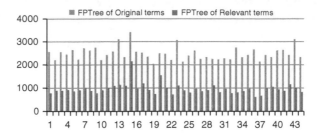

Fig. 1 FP Tree sizes of the AMBIENT dataset

term set will possess the knowledge about the ground truth from 0.8 to 1 (with mean 0.95). The relevant terms set may lead at the most 20% loss in knowledge representation, 68.02% gain by reduction in term set size and with 61.28 gain by FPTrees size reduction. This suggest that the relevant term set obtained by EM-based term set pruning can be employed in search result analysis in Web Mining.

The search result clustering method discussed in Sect. 4 has been considered for demonstrating the effectiveness of the recommended method. The distribution of the no of clusters with and without pruning is given in Fig. 2. The term pruning approach resulted in more specific clusters, so the number of clusters has increased.

RandIndex (RI) [11] is used as the evaluation measure, to compare our approach against the ground truth. The Rand Index values of each query when the clustering method is run with and without term pruning, are reported in Fig. 3. With term pruning, every query's RI value improved (statistically significant from our experiments).

When the Snowball clustering algorithm is run on the AMBIENT datasets, without term pruning, the average Rand Index has a low value of 58.83. When the term pruning is performed, RI value raised to 61.72. Comparison of our approach against some of the data centric SRC approaches, which do not use external resources is presented in Table 1. Lingo [9] uses Singular Value Decomposition, STC [15] uses Suffix trees and KeySRC [1] is based on Key Phrases.

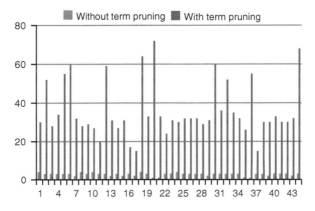

Fig. 2 Number of clusters with and without term pruning

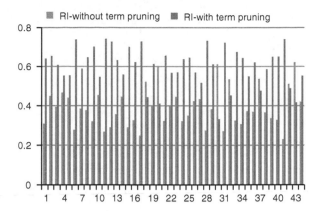

Fig. 3 RI values with and without term pruning

Table 1 Average rand index

Clustering method	Rand index
Lingo	62.75
STC	61.48
KeySRC	66.49
Snowball without term pruning	58.83
Snowball with EM based term pruning	61.72

6 Conclusion

The data centric SRC is effective due to the pruning of terms beyond well-defined stop words, through the EM algorithm. The search results are clustered and labeled with Snowball clustering method, without any human interaction or even external knowledge. As the threshold computed by EM algorithm is adaptive, the recommended method is adaptive machine learning technique. EM based term pruning has shown two benefits—overcome the curse of dimensionality and less runtime memory (FP Tree sizes are low).

References

1. Bernardini, A., Carpineto, C., D'Amico, M.: Full-subtopic retrieval with keyphrase-based search results clustering. In: Proceedings of Web Intelligence 2009, IEEE Computer Society, pp. 206–213 Milan (2009)
2. Carpineto, C., Osinski, S., Romano, G., Weiss, D.: A survey of web clustering engines. ACM. Comput. Surv. (CSUR) **41**(3) (2009). Article No. 17. ISSN:0360–0300
3. Dempster, A.P., Laird, N.M., Rubin, D.R.: Maximum likelihood from incomplete data via the EM algorithm. J. Roy. Stat. Soc.: Ser. B (Methodol.) **39**(1), 1–38 (1977)
4. Fung, B., Wang, K., Ester, M.: Hierarchical document clustering using frequent itemsets. In: Proceedings of SIAM International Conference on Data Mining, pp. 59–70 (2003)

5. Han, J., Pei, J., Yin, Y.: Mining frequent patterns without candidate generations. In: Proceedings of ACM SIGMOD International Conference on Management of Data, Dallas, TX, USA (2000)
6. Hima Bindu K., Raghavendra Rao C.: Association rule centric clustering of web search results. In: Proceedings of MIWAI'11, Vol. 7080/2011, pp. 159–168, Hyderabad (2011). doi:10.1007/978-3-642-25725-4_14
7. Kamvar, M., Baluja, S.: A large scale study of wireless search behavior: google mobile search. In: Proceedings of CHI'06, pp. 701–709, NewYork (2006)
8. Manning, C. D., Raghavan, P., Schutze, H.: Introduction to Information Retrieval. Cambridge University press, New York, USA (2008)
9. Osinski, S., Stefanowski, J., Weiss, D.: Lingo: Search results clustering algorithm based on singular value decomposition. In: Proceedings of the International Intelligent Information Processing and Web Mining Conference. Advances in Soft Computing, Springer, 359–368 (2004)
10. Pawlak, Z.: Rough Sets—Theoretical Aspects of Reasoning about Data. Kluwer, Boston (1991)
11. Rand, W. M.: Objective criteria for the evaluation of clustering methods. J. Am. Stat. Assoc. **66**(336), 846–850 (1971)
12. Scaiella, U., Ferragina, P., Marino, A., Ciaramita, M.: Topical clustering of search results. In: Proceedings of WSDM 2012, pp. 223–232. ACM, New York (2012)
13. Taghavi, M., et al.: An analysis of web proxy logs with query distribution pattern approach for search engines. Comput. Stand. Interfaces. 162–170 (2011). doi:10.1016/j.csi.2011.07.001
14. Wajahat Ali, M. S., Raghavendra Rao, C., Bhagvati, C., Deekshatulu, B. L.: EMTrain++: EM based incremental training algorithm for high accuracy printed character recognition system. Int. J. Comput. Intell. Res. **5**(46), 365–371 (2010)
15. Zamir, O. And Etzioni, O.: Web document clustering: A feasibility demonstration. In: Proceedings of the 21st International ACM SIGIR Conference on Research and Development in Information Retrieval. ACM Press, 46–54 (1998)

Intensity-Based Detection of Microcalcification Clusters in Digital Mammograms using Fractal Dimension

P. Shanmugavadivu and V. Sivakumar

Abstract This paper presents a novel method to locate and segment the microcalcification clusters in mammogram images, using the principle of fractal dimension. This proposed technique detects the edges using the intensities of the regions/objects in the image, the Fractal dimension of the image, which is image-dependent in such a way that leads to the segmentation of microcalcification clusters in the image. Hence this fractal dimension based detection of microcalcifations is proved to produce excellent results and the location of the detected microcalcifications clusters complies with the specifications of dataset of the mini-MIAS database accurately, which substantiate the merit of the proposed technique.

Keywords Fractal dimension · Edge detection · Mammogram · Microcalcifications · Image segmentation

1 Introduction

The recent advancements in Digital Image Processing (DIP) have broadened its application horizons in Medical Image Processing, and thus have added a new dimension in the diagnosis of medical images. Image segmentation, an element of DIP, deals with subdividing an image into its constituent regions/objects and the level of segmentation is determined by the requirements of the problem domain. This technique has gained popularity in many areas including medical imaging, satellite imaging and under-water imaging, as it provides a platform to segment and retain the vital

P. Shanmugavadivu (✉) · V. Sivakumar
Department of Computer Science and Applications, Gandhigram Rural Institute–Deemed University, Gandhigram, Tamil Nadu 624302, India
e-mail: psvadivu67@gmail.com

V. Sivakumar
e-mail: sivakumar.vengusamy@gmail.com

B. V. Babu et al. (eds.), *Proceedings of the Second International Conference on Soft Computing for Problem Solving (SocProS 2012), December 28–30, 2012*, Advances in Intelligent Systems and Computing 236, DOI: 10.1007/978-81-322-1602-5_135, © Springer India 2014

objects/regions of interest (RoI) and ignore the insignificant details of an image, which has to pay-off on reduced computation time and storage space, during the subsequent phases like image analysis and image registration. Edge detection is a subprocess of image segmentation using which the prospective pixels that constitute contour/boundary of RoI are identified [1, 2].

Breast cancer is one of the primary causes of early mortality among women in both developed and developing countries [3]. Recently mammography has emerged as the most popular and reliable source for the early detection of breast cancer and other abnormalities in the breast tissues. The texture, shape and dimension of breast tissues play a significant role in the detection of breast cancer which is made up of lobules, the glands for milk production and the ducts that connect lobules to the nipple. According to the radiologists' version, a patient is confirmed to have breast cancer if certain types of masses or calcifications which appear as small white specks are identified in the mammogram [4]. Generally, a mass is explained as a 'spot' or 'density' that may appear in the mammogram and calcifications are described as deposits of calcium into various tissues within the breast [5, 6]. As the breast tissues very much possess self-similar property of fractal objects, fractal-based techniques can be extended to mammograms for the diagnosis and analysis of mammograms [6–10].

The rough geometric shapes which can be subdivided into discrete parts, each of which if replicates the structure of whole object, are termed as Fractals. Every fractal object is characterized by its fractal dimension, which is an important parameter that measures the fractal property of an object. It has got wider applications in the fields of image segmentation and shape recognition [6, 7, 10–14]. This paper projects the potential of combining the potential of edge intensities and fractal dimension, using which, the microcalcifications present in the input images are precisely segmented.

In this paper, Sect. 2 describes the basics of edge detection as well as fractal dimension and the computational methodology of the segmentation of microcalcifications clusters presented in Sect. 3. The results and discussion and the conclusions are presented in Sects. 4 and 5 respectively.

2 Basics of Techniques Used in the Proposed Work

2.1 Edge Detection

Due to the limitations in the image acquisition mechanism and its dependent attributes, the edges may get blurred, and thus the pixels constituting the edges may not appear distinct with respect to neighbours, which in turn may misguide the process of edge detection. Moreover, the rate of sampling and quantization also adds a new dimension to the problem. Hence, it is important to enhance the quality of any input image, prior to the application of image segmentation [2]. An edge is a vector variable with two components, magnitude and direction. The edge magnitude is the magnitude of the gradient and the edge direction is rotated with respect to the gradient direction by $-90°$. The gradient direction gives the direction of maximum growth of the function. The first-order derivatives in an image are computed using the Gradient and the second-order derivatives are obtained using the Laplacian.

Sobel is a powerful edge detection technique that uses the doubled center coefficient of mask to give more importance for the center point. For computation of the gradient of an image, sobel operators provide both a differencing and a smoothing effect [1, 2]:

$$G_x = (z_7 + 2z_8 + z_9) - (z_1 + 2z_2 + z_3)$$
$$G_y = (z_3 + 2z_6 + z_9) - (z_1 + 2z_4 + z_7) \tag{1}$$

The sobel mask for computing x-derivative is anti-symmetry with respect to the y-axis. It has the positive sign on the right side and negative sign on the left side. It should be noted that the sum of all coefficients is equal to zero to make sure that the response of a constant intensity area is zero. The sobel operator is often used as a simple detector of horizontality and verticality of edges [1, 2].

2.2 Fractal Dimension

Fractals are the geometric primitives that are self-similar and irregular in nature. Self-similarity is defined as a property where a subset is indistinguishable from the whole, when magnified to the size of the whole. In 1982, Mandelbrot introduced the *Fractal geometry* to the world of research. Fractals are of rough geometric shapes which can be subdivided in parts, each of which is reduced to similar of the whole [11, 12].

Fractal objects are characterized by their fractal dimension (defined as D) which is an important characteristic of fractals as it has got information about their geometric structure. In the fractal world, the fractal dimension of an object, need not be an integer number and is normally greater than its topological dimension (i.e. $D \geq d$) [11–13].

In Euclidean n-space, the bounded set X is said to be self-similar when X is the union of N_r distinct non-overlapping copies of itself, each of which is similar to X scaled down by a ratio r. Fractal dimension D of X can be derived from the relation [11], as

$$D = \frac{\log(N_r)}{\log(\frac{1}{r})} \tag{2}$$

3 Methodology

The proposed work initially finds the fractal dimension, D for a given input mammogram image by box counting method using Eq. (2). Then, the input image undergoes image enhancement process that maps the intensity values of the input grayscale image A to new values in B by compressing the low end and expanding the high end such that 1% of data is saturated at low and high intensities of A. This increases the contrast of the input image.

Further, edges are detected for enhanced input image B using the sobel operator and those edge intensities are noted. Now those edge intensities are compared with the fractal dimension of the image and the pixels for which the fractal dimension value is greater than those edge intensities are noted and marked separately in the image that leads to the segmentation of microcalcifications clusters in the given input mammogram image. As, in the images of mini-Mammographic Image Analysis Society (MIAS) database several types of noises and imaging artifacts are present, by default the detected image consists of the entire frame that the image acquires which contains even the chest wall behind portion, high intensity rectangular labels and low intensity labels too. The proposed methodology aims to exclude such high intensity noises in the detected image. To omit such unwanted noise areas and to acquire the required portion of the frame, the concept of region of interest is considered to set the desired region. This gives the projection of microcalcifications part alone that results in exact segmented image. The algorithm for the above methodology is given as follows.

Algorithm : Detection of microcalcification clusters from a mammogram image
Aim : To detect microcalcification clusters
Input : A 2-dimensional mammogram image, A
Output : Segmented portions of microcalcifications in A
Step 1: Read a 2-dimensional input mammogram image A
Step 2: Enhance the input image A to acquire B
Step 3: $[M, N] \leftarrow$ IMSIZE [B]
Step 4: If $M > N$ then $r \leftarrow M$
                    Else $r \leftarrow N$
Step 5: Compute fractal dimension D using Eq. (2).
Step 6: Detect the edges of objects in B to C using sobel operator
Step 7: Note the edge intensities b(i, j) of C
Step 8: Compare D with b(i, j) of C
Step 9: Plot the pixels for which $D > b(i, j)$ of B in D
Step 10: Considering only the region of interest in D that results in the segmented image.
Step 11: Stop.

---

4 Results and Discussion

The proposed procedure was implemented using Matlab 7.8. to ascertain the merit of the proposed method, different mammogram images were collected from the mini-mammographic database of the MIAS from the Pilot European Image Processing Archive (PEIPA) at the University of Essex (ftp://peipa.essex.ac.uk/ipa/pix/mias/). As per the principle of the proposed methodology, the input image has been enhanced

Fig. 1 **a–c** Original images. **d–f** Enhanced images of (**a–c**). **g–i** Intensity values of edges of (**d–f**). **j–l** Detected images of (**g–i**). **m–o** Artifacts cleared images of (**j–l**). **p–r** Segmented mass of (**m–o**)

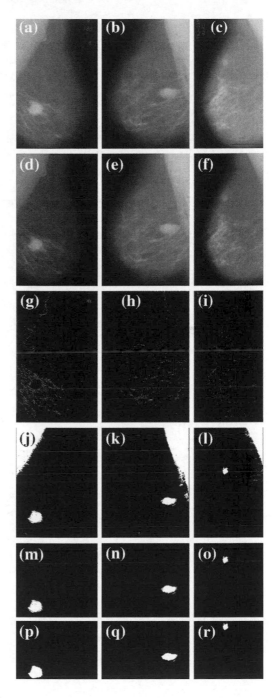

by adjusting the contrast values. Then the edges are detected for the enhanced image using sobel operator and those edge intensities are recorded in another image. For illustrative purpose, the results of three mammograms (mdb028.pgm (image1), mdb023.pgm (image2), mdb025.pgm (image3)) are depicted in Fig.1a–r. The original images and the corresponding enhanced images for the input mammograms are depicted in Fig.1a–c and Fig.1d–f respectively.

The recorded intensity values of edges that are detected using sobel operator are depicted in Fig.1g–i. After finding the fractal dimension of the input images, it is compared with those edge intensities and the pixels whose Fractal dimension value is greater than those edge intensities are recorded and marked separately in an image that leads to the detection of microcalcifications which is depicted in Fig.1j–l.

Fig.1j–l shows that it contains the pectoral muscle portion of the breast and such unwanted areas are removed by considering only the region of interest so that the desired region is acquired which is depicted in Fig.1m–o. Dilation and erosion are done on the artifacts cleared images as post-processing methods to smoothen the image so that the interior gaps are filled which results in the segmentation of microcalcification clusters which are shown in Fig.1p–r. As the validation process, finally the location of the coordinates of the detected mass in the final segmented image is verified and very much complies with the location specified in the mini-MIAS dataset. Hence, it is clearly understood that the proposed work using fractal dimension is proved to be precise and robust in detection of microcalcification clusters in a digital mammogram.

5 Conclusions

This method proposes an edge intensity based novel method for the detection of microcalcifications in the mammogram images, using fractal dimension, that enables the early diagnosis of breast cancer. This paper provides a framework that seems to make segmentation easier by considering the optimal RoI in the detected masses in order to avoid the inclusion of pectoral muscle portion. As per the mini-MIAS database, the location of the (x,y) image-coordinates of centre of circumscribed mass abnormalities for the input images mdb023, mdb025 and mdb028 are (538, 681), (674, 443) and (338, 314) respectively which substantiates the microcalcifications detected by the proposed method that vouches its merit. The future scope of the study covers the classification of the detected masses, either as benign or malignant, using texture analysis.

References

1. Gonzalez, R.C., Woods, R.E.: Digital Image Processing, Second edition. Prentice Hall, New york (2009)
2. Sonka, M., Hlavac, V., Boyle, R.: Digital Image Processing and Computer Vision. Cengage Learning, Stamford (2008)
3. Kom, G., Tiedeu, A., Kom, M.: Automated detection of masses in mammogram by local adaptive thresholding. Comput. Biol. Med. **37**, 37–48 (2007)
4. Maitra, I.K. et al.: Technique for preprocessing of digital mammogram. Comp. Meth. Programs Biomed. **107**, 175–188 (2012)
5. Stojic, Tomislav, et al.: Adaptation of multifractal analysis to segmentation of microcalcifications in digital mammograms. Physica A **367**, 494–508 (2006)
6. Shanmugavadivu, P., Sivakumar, V.: Fractal approach in digital mammograms: a survey, NCSIP-2012, pp. 141–143, ISBN: 93-81361-90-8, (2012)
7. Shanmugavadivu, P., Sivakumar, V.: Fractal dimension-based texture analysis of digital images. Procedia Eng. **38**, 2981–2986, ISSN: 1877–7058, (Elsevier-Scidirect) (2012)
8. Cheng, H.D., Cai, X., Chen, S., Hu, L., Lou, X.: Computer-aided detection and classification of microcalcifications in mammograms: a survey. Pattern Recogni. **36**, 2967–2991 (2003)
9. Claudio, M., Mario, M., D'Elia, C., Tortorella, F.: A computer-aided detection system for clustered microcalcifications. Artif. Intell. Med. **50**, 23–32 (2010)
10. Mohamed, W.A., Alolfe, M.A., Kadah, Y.M.: Fast fractal modeling of mammograms for microcalcifications detection. In: 26th National Radio Science Conference, Egypt (2009)
11. Addison, P.S.: Fractals and Chaos. IOP Publishing, Bristol (2005)
12. Welstead, S.: Fractal and wavelet image compression techniques. In: Tutorial Texts in Optical Engineering, Vol.TT40. SPIE Press (http://www.spie.org/bookstore/tt40/) (1999)
13. Lopes, R., Betrouni, N.: Fractal and multifractal analysis: a review. Med. Image Anal. **13**, 634–649 (2009)
14. Vuduc, R.: Image segmentation using fractal dimension. Report on GEOL 634, Cornell University (1997)

Part XV
General Soft Computing
Approaches and Applications

Palmprint Recognition Using Geometrical and Statistical Constraints

Aditya Nigam and Phalguni Gupta

Abstract This paper proposes an efficient biometrics system based on palmprint. Palmprint ROI is transformed using proposed local edge pattern (LEP). Corner like features are extracted from the enhanced palmprint images as they are stable and highly discriminative. It has also proposed a distance measure that uses some geometrical and statistical constraints to track corner feature points between two palmprint ROI's. The performance of the proposed system is tested on publicly available PolyU database consisting of 7,752 and CASIA database consisting of 5,239 hand images. The feature extraction as well as matching capabilities of the proposed system are optimized and it is found to perform with CRR of 99.97 % with ERR of 0.66 % for PolyU and CRR of 100 % with ERR of 0.24 % on CASIA databases respectively.

Keywords Biometrics · Palmprint · Local Binary Pattern (LBP) · Phase only Correlation (POC) · Optical Flow

1 Introduction

Reliable human identification and authentication in real time is a huge challenge and one of the most desirable tasks of our society. Hence researchers got motivated to design an efficient, effective, robust and low cost personal authentication systems. Biometrics can be an alternative to the token-based traditional methods as they are easier to use and harder to circumvent. Authentication systems based on biometrics are extensively used in computer security, banking, law enforcement etc. Exponential increase in the computational power and society driven business, guides biometrics

A. Nigam (✉) · P. Gupta
Indian Institute of Technology, Kanpur, India
e-mail: naditya@cse.iitk.ac.in

P. Gupta
e-mail: pg@cse.iitk.ac.in

B. V. Babu et al. (eds.), *Proceedings of the Second International Conference on Soft Computing for Problem Solving (SocProS 2012), December 28–30, 2012*, Advances in Intelligent Systems and Computing 236, DOI: 10.1007/978-81-322-1602-5_136, © Springer India 2014

Table 1 Trait-wise
challenges and issues

	Challenges	Issues
Face	Pose, expression, illumination, aging, rotation, translation and background	Too many challenges
Finger	Rotation and translation	Acceptance
Iris	Segmentation and illumination	Acquisition, coopera-tion, accep-tance
Palm	Illumination, rotation and translation	–

research so as to meet the practical real field challenges. Behavioral as well as physiological biometrics based characteristics (such as face [12, 15, 16], fingerprint [20], iris [8], knuckleprint [7, 17], gait, voice, vein patterns *etc.*) are used to develop robust, accurate and highly efficient personal authentication systems. But every biometric trait has its own challenges (like occlusion, pose, expression, illumination, affine transformation) and trait specific issues (such as user acceptance, cooperation) as shown in Table 1. Hence one cannot consider any trait to be best for all applications.

Hand based biometric recognition systems (e.g. palm print [5], fingerprint [20] and finger knuckleprint [17]) have gathered attraction over past few years because of their good performance and inexpensive acquisition sensors. The inner part of the hand is called palm and the extracted region of interest in between fingers and wrist is termed as palmprint. Pattern formation within this regions are suppose to be stable as well as unique even monozygotic twins are found to have different palmprint patterns. Hence one can consider it as a well discriminative biometrics trait.

Palmprint's prime advantages over fingerprint includes its higher acceptance socially because it is never being associated with criminals and larger ROI area as compared to fingerprint images. Larger ROI ensures abundance of structural features including principle lines, wrinkles, creases and texture pattern even in low resolution palmprint images, that enhances system's speed, accuracy and reduces the cost. Some other factors favoring palmprint includes its lesser user cooperation, non-intrusive and cheaper acquisition sensors.

Huge amount of work has been reported on palmprint based authentication systems because of the above mentioned advantages. Developed systems have adopted either structural or statistical features analysis.

- Structural Features: Morphological operations are applied on sobel edges [10] to extract line-like features from palmprints. Isolated points as well as some points along principle line are considered as features for palmprint authentication [9].
- Statistical Features: Several systems are developed that transform palmprint features into higher dimensional space where they are considered as a points using PCA, LDA, ICA and their combinations. Direction based filtering using gabor is performed to obtained *palmcode* [22], *compcode* [11] and *ordinalcode* [21]. Moments like Zernike [6] and transforms like Stockwell [4] and Fourier [5] are applied on palmprint low resolution images to achieve better results.
- Combined: System proposed in [3] uses Kernal PCA and ICA gabor filter response to extract the kernal principle components of gabor features.

This paper proposes a transformation technique as well as a distance measure which is robust against illumination and slight amount of local non-rigid distortions. The performance of the system is studied on publicly available PolyU and CASIA datasets where the hand image are obtained under varying environment. The rest of the paper is organized as follows. Section 2 discusses mathematical basis of the proposed system. Section 3 presents the proposed palmprint based recognition system. Experimental results are analyzed in Sect. 4. Conclusions are presented in the last section.

2 Mathematical Basis

2.1 LBP based Image Enhancement

In Ref. [18], LBP (Local Binary Pattern) is proposed that assumes pixel's relative gray value with respect to its eight-neighborhood pixels can be more stable than its own gray value. In Ref. [16] gt-transformation is proposed that can also preserve the distribution of gray level intensities in facial images. It also helps to address the problems like robustness against illumination variation and local non-rigid distortions.

2.2 Phase Only Correlation (POC)

It has been used for image registration and alignment purposes. Let f and g are $M \times N$ images such that $m = -M_0 \ldots M_0 (M_0 > 0)$ and $n = -N_0 \ldots N_0 (N_0 > 0)$. Also $M = 2M_0 + 1$ and $N = 2N_0 + 1$. F and G are the DFT of f and g images and F can be defined as:

$$F(u, v) = \sum_{m=-M_0}^{M_0} \sum_{n=-N_0}^{N_0} f(m, n)e^{-j2\pi(\frac{mu}{M}+\frac{nv}{N})} \tag{1}$$

$$= A_F(u, v)e^{j\theta_F(u,v)} \tag{2}$$

$$G(u, v) = \sum_{m=-M_0}^{M_0} \sum_{n=-N_0}^{N_0} g(m, n)e^{-j2\pi(\frac{mu}{M}+\frac{nv}{N})} \tag{3}$$

$$= A_G(u, v)e^{j\theta_G(u,v)} \tag{4}$$

Cross Phase Spectrum between G and F is defined as:

$$R_{GF}(u, v) = \frac{G(u, v) \times F^*(u, v)}{|G(u, v) \times F^*(u, v)|} = e^{j(\theta_G(u,v)-\theta_F(u,v))} \tag{5}$$

Phase Only Correlation is the IDFT of $R_{GF}(u, v)$ defined as:

$$P_{gf}(m, n) = \frac{1}{MN} \sum_{u=-M_0}^{M_0} \sum_{v=-N_0}^{N_0} R_{GF}(u, v)e^{j2\pi(\frac{mu}{M}+\frac{nv}{N})} \tag{6}$$

Phase Only Correlation can be obtained by taking the IDFT of $R_{GF}(u, v)$. If two images are "Same" the POC function $P_{gf}(m, n)$ becomes kronecker delta function $\delta(m, n)$. If two images are "Similar" POC function gives a distinct sharp peak otherwise the peak value drops significantly. Hence peak height of $P_{gf}(m, n)$ can be considered as a similarity measure.

2.3 Lukas Kanade Tracking

Sparse optical flow between two images can be estimated using Lukas Kanade tracking algorithm [13]. Let there be some feature at location (x, y) at time instant t with intensity $I(x, y, t)$ and this feature has moved to the location $(x + \delta x, y + \delta y)$ at time instant $t + \delta t$. LK Tracking [13] make use of three assumptions to estimate the optical flow.

- **Brightness Consistency**: Little variation in brightness for small δt.

$$I(x, y, t) \approx I(x + \delta x, y + \delta y, t + \delta t) \tag{7}$$

- **Temporal Persistence**: Small feature movement for small δt. One can estimate $I(x + \delta x, y + \delta y, t + \delta t)$ by applying Taylor series on Eq. (7) and neglecting the high order terms to get

$$\frac{\delta I}{\delta x}\delta x + \frac{\delta I}{\delta y}\delta y + \frac{\delta I}{\delta t}\delta t = 0 \tag{8}$$

Dividing both sides of Eq. (8) by δt one gets

$$I_x V_x + I_y V_y = -I_t \tag{9}$$

where V_x, V_y are the respective components of the optical flow velocity for feature at pixel $I(x, y, t)$ and I_x, I_y and I_t are the local image derivatives in the corresponding directions.

- **Spatial Coherency**: It assumes local constant flow (i.e. a patch of pixels moves coherently) to estimate the optical flow vector $\widehat{V}$ (2×1 matrix) for any corner. A non-iterative method that assume constant flow vector (V_x, V_y) within 5×5 neighborhood (i.e. 25 neighboring pixels) around the current feature point (center pixel) is used to estimate its optical flow. Hence, an overdetermined linear system of 25 equations is obtained which can be solved using least square method as

$$\underbrace{\begin{pmatrix} I_x(P_1) & I_y(P_1) \\ \vdots & \vdots \\ I_x(P_{25}) & I_y(P_{25}) \end{pmatrix}}_{C} \times \underbrace{\begin{pmatrix} V_x \\ V_y \end{pmatrix}}_{V} = -\underbrace{\begin{pmatrix} I_t(P_1) \\ \vdots \\ I_t(P_{25}) \end{pmatrix}}_{D} \tag{10}$$

where rows of the matrix C represent the derivatives of image I in x, y directions and those of D are the temporal derivative at 25 neighboring pixels. The 2×1 matrix $\widehat{V}$ is the estimated flow of the current feature point determined as

$$\widehat{V} = (C^T C)^{-1} C^T (-D) \tag{11}$$

The final location $\widehat{F}$ of any feature point can be estimated using its initial position vector $\widehat{I}$ and the estimated flow vector $\widehat{V}$ by

$$\widehat{F} = \widehat{I} + \widehat{V} \tag{12}$$

The tracking performance depends on how well these three assumptions are satisfied.

3 Proposed System

The proposed authentication system works in four phases: ROI extraction, image enhancement, feature extraction and finally feature level matching. In first phase ROI is extracted from palmprint images using the method suggested in Ref. [5] and shown in Fig. 1. In second phase extracted ROI is transformed so as to achieve better

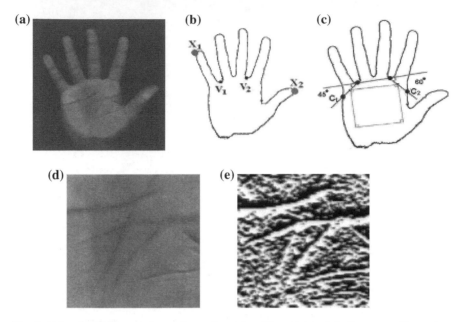

Fig. 1 Palmprint ROI extraction and transformation [5]. **a** Original. **b** Contour. **c** Key points. **d** Palmprint ROI. **e** Transformed

and robust feature representation using proposed local edge pattern (LEP) method. Corner features are extracted in third phase. Finally, matching between features is performed using proposed UTC dissimilarity measure. Tasks performed in each phase are explained as follows.

[A] **Palmprint ROI Extraction** [5]: Hand images are thresholded to obtain binarized image from which hand contour is extracted. Four key-points (X_1, X_2, V_1, V_2) are obtained on the hand contour as shown in Fig. 1b. Then two more key-points are obtained, C_1 that is the intersection point of hand contour and the line passing from V_1 with a slope of 45° and C_2 that is the intersection point of line passing from V_2 with a slope of 60° as shown in Fig.1c. Finally the midpoints of the line segment $V_1 C_1$ and $V_2 C_2$ are joined which is considered as one side of the required square ROI. The final extracted palmprint ROI is shown in Fig. 1d. The ROI extracted palmprints are normalized to 100×100 size.

[B] **Local Edge Pattern (LEP) Based Transformation:** Palmprint images have abundant texture information including wrinkles, principle line, creases, ridges, minutia that can be useful for recognition purposes. In order to achieve illumination invariance *edgemaps* are extracted and worked on. Sobel edge detector uses pure central difference operator to reduce artifacts, but it lacks perfect rotational symmetry. Scharr operators are obtained by optimized minimization of weighted mean squared angular error in Fourier domain under the condition that resulting filters are numerically consistent. Hence Scharr kernels are better derivative kernels than usual sobel kernels. To achieve robust features x-derivative of palmprint ROI is convolved with 3×3 Scharr kernel. The proposed *local edge pattern* (LEP) based transformation calculates *scharr_code* for each pixel from the obtained derivative image. Every palmprint ROI is transformed into its *lep_code* (as shown in Fig. 1e) that is robust to illumination and local non-rigid distortions.Any palmprint ROI, P is transformed by applying the x-direction scharr filter to obtain derivative image P^d. The *lepcode* can be obtain by evaluating *scharr_code* for every pixel $P_{j,k}$ using the derivative image P^d. The *scharr_code* for any pixel is a 8 bit binary number whose ith bit is defined as

$$scharr\_code_i = \begin{cases} 1 & \textbf{if } (Neigh[i] > 0) \\ \\ 0 & \textbf{otherwise} \end{cases} \tag{13}$$

where Neigh[i], $i = 1, 2, \ldots 8$ are the x-direction scharr gradient of eight neighboring pixels centered at pixel $P_{j,k}$ obtained from P^d.

In *lepcode* (as shown in Fig. 1e), every pixel is represented by its *scharr_code* which is an encoding of edge pattern in its $eight - neighborhood$. The *scharr_code* of any pixel retains only the sign of the derivative in its neighborhood hence any sudden change in the illumination cannot affect *scharr_code* much because the edge pattern near the pixel (within a neighborhood) remains to be more or less same. This property ensures robustness of the proposed system in illumination varying environments.

[C] **Feature Extraction:** Corners can be an ideal candidate for features because they are robust to illumination variations and can be tracked accurately. Corner features can provide enough information for tracking because they possess strong derivative in two orthogonal directions. In order to find corners, eigen analysis of structure tensor is done.The structure tensor of second-order is a common tool for local orientation estimation. It is defined as Cartesian product of the gradient vector $(I_x, I_y)^T$ with itself. Any image I can be seen as a function represented as a discrete array of samples $I[P]$, where P corresponds to any pixel. The 2D structure tensor at any pixel P at ith row and jth column can be defined as

$$\begin{pmatrix} I_x^2(P) & I_x(P).I_y(P) \\ I_x(P).I_y(P) & I_y^2(P) \end{pmatrix} \tag{14}$$

where $I_x(P)$ and $I_y(P)$ are the partial derivatives sampled at pixel P that are estimated using finite difference formula by considering all the pixels within $(2K + 1) \times (2K + 1)$ window (N) centered at pixel P. The value of K is empirically chosen to be five. The above matrix can have two eigen values λ_1 and λ_2 such that $\lambda_1 \geq \lambda_2$ with e_1 and e_2 as the corresponding eigenvectors. Like [19], all pixels having $\lambda_2 \geq T$ (smaller eigen value greater than a threshold) are considered as corner feature points.

[D] **Feature Matching:** Let A and B are two palmprint ROI that are to be matched and P_a, P_b are their corresponding *lepcode*. In order to make the decision on matching between A and B, LK tracking has been used (as discussed in Sect. 2). A dis-similarity measure UTC (Unsuccessfully Tracked Corners) is defined that tries to quantify the performance of LK tracking using some geometrical as well as statistical constraints. It is assumed that the tracking performance of LK algorithm will be good between features of same subject while degrades substantially for others. The tracking performance depends on how well the three assumptions brightness consistency, temporal persistence and spatial coherency (as discussed in Sect. 2) are satisfied. The geometrical and statistical constraints constraints that has to be satisfied in order to consider any feature at $a(i, j)$ to be successfully tracked are defined below.

Geometric and Statistical Constraints

- The euclidean distance between $a(i, j)$ and its estimated tracked location should be less than or equal to a pre-assigned threshold T_d.
- The tracking error defined as pixel-wise sum of absolute difference between a local patch centered at $a(i, j)$ and that of its estimated tracked location patch should be less than or equal to a pre-assigned threshold T_e.
- The phase only correlation (as discussed in Sect. 2) [14] between a local patch centered at $a(i, j)$ and that of its estimated tracked location patch should be more than or equal to a pre-assigned threshold T_p.

However all tracked corners may not be the true matches, because of noise, local non-rigid distortions in iris and less difference in inter class matching as compared with intra class matching. Hence a notion of consistent optical flow is introduced .

Algorithm 1 $UTC(P_a, P_b)$

Require: The $lepcode$ P_a, P_b of transformed palmprint ROI images.
  N_a and N_b are the number of corners in P_a, P_b respectively.
Ensure: Return the symmetric function $UTC(P_a, P_b)$.
1: Track all the corners of $lepcode$ P_a in $lepcode$ P_b using LK tracking algorithm.
2: Calculate the number of successfully tracked corners (i.e. stc_{ab}) that have their tracked spatial location with in T_d, their local patch dissimilarity under the T_e and have their phase only correlation more than T_p.
3: Similarly calculate successfully tracked corners of $lepcode$ P_b in $lepcode$ P_a (i.e. stc_{ba}).
4: Quantize optical flow directions into eight directions and find out the consistent direction that have maximum number of successfully tracked features.
5: Calculate the number of corners that are successfully tracked and have consistent optical flow (i.e. cof_{ab} and cof_{ba}).
6: $utc_{ab} = 1 - \frac{cof_{ab}}{N_a}$;
7: $utc_{ba} = 1 - \frac{cof_{ba}}{N_b}$;
8: **return** $UTC(P_a, P_b) = \frac{utc_{ab} + utc_{ba}}{2}$;

3.1 Consistent Optical Flow

The direction of optical flow at any feature at $a(i, j)$ can be calculated by determining the slope of the line joining $a(i, j)$ and its tracked location estimated using LK tracking algorithm. It can be noted that true matches have the optical flow which can be aligned with the actual affine transformation between the two images. The estimated optical flow direction is quantized into eight directions and the most consistent direction is selected as the one which has most number of successfully tracked features. Any pair of matching corners other than the most consistent direction is considered as false matching pair.

In Algorithm 1, two *lepcodes* P_a and P_b are matched. Successfully tracked corners of *lepcode* P_a in *lepcode* P_b and vice-versa are calculated using constraints as defined above. The consistent optical flow direction is evaluated and finally the number of unsuccessfully tracked corners of P_a in P_b and vice-versa is calculated. The $UTC(P_a, P_b)$ measure is made symmetric by taking the average of utc_{ab} and utc_{ba}.

Some properties of U T C distance measure

1. $UTC(A, B) = UTC(B, A)$.
2. $UTC(A, A) = 0$.
3. $UTC(A, B)$ always lies in the range $[0, 1]$.

4 Experimental Results

This section analyses the performance of the proposed system. The performance of the system is measured using correct recognition rate (CRR) in case of identification and equal error rate (EER) for verification. CRR of the system is defined by

$$CRR = \frac{N_1}{N_2} \tag{15}$$

where N_1 denotes the number of correct (Non-False) top best match of palmprint ROI and N_2 is the total number of palmprint ROI in the query set.

At a given threshold, the probability of accepting the impostor, known as false acceptance rate (FAR) and probability of rejecting the genuine user known as false rejection rate (FRR) are obtained. EER is the value of FAR for which FAR and FRR are equal.

$$EER = \{FAR|FAR = FRR\} \tag{16}$$

The proposed system is tested on two publicly available benchmark palmprint databases CASIA [1] and PolyU [2]. All possible genuine as well as imposter matching are considered to evaluate the performance of the system. The CASIA palmprint database have $5,502$ palmprint images taken from 312 subjects (i.e. 624 distinct

Table 2 Parameterized analysis

Parameters			PolyU			Casia		
T_p	T_d	T_e	d'	CRR	EER	d'	CRR	EER
0.15	150	600	2.137	100	0.7913	2.3121	99.97	0.2604
0.25	200	600	2.140	100	0.6665	*2.3166*	*99.97*	*0.2459*
0.35	200	600	*2.235*	*100*	*0.6626*	2.3168	99.97	0.2757
0.45	200	600	2.32	100	0.9035	2.224	99.97	0.3435
0.4	290	600	2.287	100	0.6834	2.3070	99.97	0.2907

Table 3 Performance analysis

Systems	CASIA CRR/ERR (%)	PolyU CRR/ERR (%)
PalmCode [22]	99.619/3.6730	99.916/0.5338
CompCode [11]	99.716/2.0130	99.964/0.3082
OrdinalCode [21]	99.843/1.7540	100.00/0.709
Zernike moments [6]	99.75/2.003	100.00/0.2939
Stockwell transform [4]	100/1.1606	100.00/0.1
Proposed	99.97/0.245	100.00/0.6626

palms). From each subject eight images are collected from left as well as right hand. Hence the total matching possible are 20,498,256 out of which 31,696 matching are genuine while rest are imposter matching. The PolyU palmprint database have 7, 752 palmprint images taken from 193 subjects (i.e. 386 distinct palms). From each subject 20 images are collected from left as well as right hand palm in two sessions (10 images per session). Hence the total matching possible are 59,590,680 out of which 146,680 matching are genuine while rest are imposter matching. There are some palms in both databases with incomplete or missing data such palms are discarded for this experiment.

4.1 Parameterized Analysis

The proposed UTC dissimilarity measure is primarily parameterized by three parameters T_e, T_p and T_d. Our system is tested using these parameters as input and their values are fixed so as to maximize the performance of the system. The parametrized analysis for some parameter sets on both databases are shown in Table 2 and the corresponding Receiver Operating Characteristics curves are drawn in Fig. 2a, b.

The proposed system has been compared with state of the art palmprint systems [11, 21, 22]. It is found that the CRR of the proposed system is more than 99.95 % for both databases. The CRR/ERR of other systems are shown in Table 3. Further, EER of the proposed verification system is 0.245 % for CASIA database which

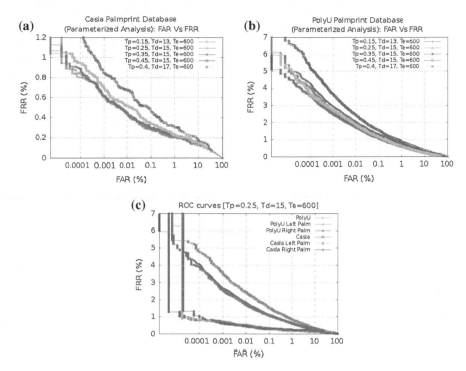

Fig. 2 The ROC curves for various experiments. **a** Parameterized analysis—CASIA. **b** Parametrized analysis—polyU. **c** ROC curves for all databases

is better than the reported systems and that for PolyU is 0.6626 % which is also comparable to the reported systems.

For each database, ROC curves which plots FRR against FAR is shown in Fig. 2c for best parameters. In another experiment both databases are divided into left and right palms. It has been found as also shown in Fig. 2c that right palm's performance is marginally better than the left palm for both databases. Also, by classifying query palm into left or right, matching will only be performed with either left or right training set. This will reduce the computational cost hugely with out loosing much on performance.

5 Conclusion

This paper has proposed a dissimilarity measure termed as Unsuccessfully Tracked Corners (UTC) to compare structural features in palmprint ROI's. Further, a palmprint based user authentication system is designed using it. In order to address the issue of varying illumination conditions edge-maps are extracted and worked on. The proposed palmprint system first transformed ROI, extracted from palmprint images using proposed LEP method then extract the corner features. Finally, proposed UTC measure is used for matching to achieve the appearance based comparison on palmprint ROI's. Rigorous experimentation shows that UTC can work effectively and efficiently under environments, having slight amount of variations in illumination, rotation and translation. The system has been tested on publicly available PolyU as well as CASIA databases of palmprint images. It has achieved CRR of more than 99.9 % for the top best match, in case of identification for both databases and EER of 0.24 and 0.66 % for CASIA and PolyU respectively, in case of verification. The proposed system has been compared with well known palmprint based systems [11, 21, 22] and is found to perform better.

References

1. Casia database. http://www.cbsr.ia.ac.cn
2. Polyu database. http://www.comp.polyu.edu.hk/biometrics
3. Aykut, M., Ekinci, M.: Kernel principal component analysis of gabor features for palmprint recognition. In: ICB, pp. 685–694 (2009)
4. Badrinath, G.S., Gupta, P.: Stockwell transform based palm-print recognition. Appl. Soft Comput. **11**(7), 4267–4281 (2011)
5. Badrinath, G.S., Gupta, P.: Palmprint base phase-difference information. Future Gen. Comp. Syst. **28**(1), 287–305 (2012)
6. Badrinath, G.S., Kachhi, N.K., Gupta, P.: Verification system robust to occlusion using low-order zernike moments of palmprint sub-images. Telecommun. Sys. **47**(3–4), 275–290 (2011)
7. Badrinath, G.S., Nigam, A., Gupta, P.: An efficient finger-knuckle-print based recognition system fusing sift and surf matching scores. In: International Conference on Information and Communications Security, ICICS, pp. 374–387 (2011)
8. Bendale, A., Nigam, A., Prakash, S., Gupta, P.: Iris segmentation using improved hough transform. In: International Conference on Intelligent Computing, ICIC, pp. 408–415 (2012)
9. Duta, N., Jain, A.K., Mardia, K.V.: Matching of palmprints. Pattern Recogn. Lett. **23**(4), 477–485 (2002)
10. Han, C.C., Cheng, H.L., Lin, C.L., Fan, K.C.: Personal authentication using palm-print features. Pattern Recogn. **36**(2), 371–381 (2003)
11. Kong, A.W.K., Zhang, D.: Competitive coding scheme for palmprint verification. In: ICPR(1), pp. 520–523 (2004)
12. Kumar, J., Nigam, A., Prakash, S., Gupta, P.: An efficient pose invariant face recognition system. In: International Conference on Soft Computing for Problem Solving, SocProS (2), pp. 145–152 (2011)
13. Lucas, B.D., Kanade, T.: An iterative image registration technique with an application to stereo vision. In: IJCAI, pp. 674–679 (1981)
14. Miyazawa, K., Ito, K., Aoki, T., Kobayashi, K., Nakajima, H.: An effective approach for iris recognition using phase-based image matching. IEEE Trans. Pattern Anal. Mach. Intell. **30**(10), 1741–1756 (2008)
15. Nigam, A., Gupta, P.: A new distance measure for face recognition system. In: International Conference on Image and Graphics, ICIG, pp. 696–701 (2009)
16. Nigam, A., Gupta, P.: Comparing human faces using edge weighted dissimilarity measure. In: International Conference on Control, Automation, Robotics and Vision, ICARCV, pp. 1831–1836 (2010)
17. Nigam, A., Gupta, P.: Finger knuckleprint based recognition system using feature tracking. In: Chinese Conference on Biometric Recognition, CCBR, pp. 125–132 (2011)

18. Ojala, T., Pietikäinen, M., Mäenpää, T.: Multiresolution gray-scale and rotation invariant texture classification with local binary patterns. IEEE Trans. Pattern Anal. Mach. Intell. **24**(7), 971–987 (2002)
19. Shi, J., Tomasi: Good features to track. In: Computer Vision and, Pattern Recognition, pp. 593-600 (1994). doi:10.1109/CVPR.1994.323794
20. Singh, N., Nigam, A., Gupta, P., Gupta, P.: Four slap fingerprint segmentation. In: International Conference on Intelligent Computing, ICIC, pp. 664–671 (2012)
21. Sun, Z., Tan, T., Wang, Y., Li, S.Z.: Ordinal palmprint represention for personal identification. In: CVPR (1), pp. 279–284 (2005)
22. Zhang, D., Kong, W.K., You, J., Wong, M.: Online palmprint identification. Pattern Anal. Mach. Intell. **25**(9), 104–1050 (2003). doi:10.1109/TPAMI.2003.1227981

A Diversity-Based Comparative Study for Advance Variants of Differential Evolution

Prashant Singh Rana, Kavita Sharma, Mahua Bhattacharya, Anupam Shukla and Harish Sharma

Abstract Differential evolution (DE) is a vector population-based stochastic search optimization algorithm. DE converges faster, finds the global optimum independent to initial parameters, and uses few control parameters. The exploration and exploitation are the two important diversity characteristics of population-based stochastic search optimization algorithms. Exploration and exploitation are compliment to each other, i.e., a better exploration results in worse exploitation and vice versa. The objective of an efficient algorithm is to maintain the proper balance between exploration and exploitation. This paper focuses on a comparative study based on diversity measures for DE and its prominent variants, namely JADE, jDE, OBDE, and SaDE.

Keywords Diversity measures · Stochastic search · Exploration–exploitation · Differential evolution · Comparative study

1 Introduction

Differential evolution (DE), proposed by Rainer Storn and Ken Price in 1995, is a popular evolutionary algorithm (EA) and exhibits good results in a wide variety of problems from diverse fields [20]. Like other EAs, DE uses mutation, crossover, and selection operators at each generation to move its population toward the global optimum. The performance of DE mainly depends on two components: One is its offspring vector generation strategy (i.e., mutation and crossover operators) and the

P. S. Rana (✉) · M. Bhattacharya · A. Shukla · H. Sharma
ABV—Indian Institute of Information Technology and Management, Gwalior,
Madhya Pradesh, India
e-mail: psrana@gmail.com

K. Sharma
Government Polytechnic College, Kota, Rajasthan, India
e-mail: k_sharma28@rediffmail.com

B. V. Babu et al. (eds.), *Proceedings of the Second International Conference on Soft Computing for Problem Solving (SocProS 2012), December 28–30, 2012*, Advances in Intelligent Systems and Computing 236, DOI: 10.1007/978-81-322-1602-5_137, © Springer India 2014

other is its control parameters (i.e., population size NP, scaling factor F, and crossover control rate CR).

Researchers are continuously working to improve the performance of DE. Some of the recently developed variants of DE with appropriate applications can be found in [1, 4, 6, 17, 27]. Experiments over several numerical benchmark problems [29] show that DE performs better than the genetic algorithm (GA) [11] and particle swarm optimization (PSO) [12]. DE has successfully been applied to various areas of science and technology, such as chemical engineering [15], signal processing [5], mechanical engineering design [26], machine intelligence, and pattern recognition [19]. DE outperforms in nonlinear, non-convex, multi-model, constrained, and discrete optimization problems. The results indicate that DE has the advantage of fast convergence and low computational consumption of function evaluations. But it is found that DE easily trapped in local optimum [14, 16, 21] due to its greedy updating process and intrinsic differential property results in a premature convergence.

Diversity has a significant effect on the performance of an algorithm [9]. It shows the behavior of the algorithm during the global optimum search process. A large value of diversity implies more exploration, while low implies more exploitation. It is expected that an optimization algorithm retains high diversity in the early stage of the search process and proportionally decreases diversity as search progresses. The performance of an algorithm can be ranked on the basis of their diversity measures. There are many diversity measures in the literature [2, 9, 10, 13, 24, 28]. A good study on diversity measures for PSO process is given in [18], and the presented study is based on this study.

In this paper, seven important diversity measures (Listed in Table 1) have been used to quantify the diversity of DE and its prominent variants such as JADE [30], jDE [3], OBDE [23], and SaDE [22]. The Table 2 lists the benchmark problems that are taken into consideration for diversity measures to quantify the dispersion of individuals in the population. Further, effect of outliers has been analyzed over the diversity measures.

Rest of the paper is organized as follows: Sect. 2 describes various diversity measures. In Sect. 3, importance and behavior of diversity measures are discussed. Experimental results are shown in Sect. 4. Finally, in Sect. 5, paper is concluded.

Table 1 Diversity measure

S.no.	Diversity measure
1	Population diameter
2	Population radius
3	Average distance around the population center
4	Geometric average distance around the population center
5	Normalized average distance around the population center
6	Average of the average distance around all individuals in the population
7	Population coherence

Table 2 Test problems

Test problem	Objective function	Search range	I	Acceptable error
Ackley	$f_1(x) = -20 + e + exp\left(-\frac{0.2}{I}\sqrt{\sum_{i=1}^{I} x_i{}^3}\right)$ $-exp(\frac{1}{I}\sum_{i=1}^{I} \cos(2\pi x_i)x_i)$	$[-1, 1]$	30	$1.0E-15$
Griewank	$f_2(x) = 1 + \frac{1}{4000}\sum_{i=1}^{I} x_i^2 - \prod_{i=1}^{I} \cos\left(\frac{x_i}{\sqrt{i}}\right)$	$[-600, 600]$	30	$1.0E-15$
Rosenbrock	$f_3(x) = \sum_{i=1}^{I}(100(x_{i+1} - x_i^2)^2 + (x_i - 1)^2)$	$[-30, 30]$	30	$1.0E-15$
Sphere	$f_4(x) = \sum_{i=1}^{I} x_i^2$	$[-5.12, 5.12]$	30	$1.0E-15$

2 Diversity Measures

There are many diversity measuring strategies available in the literature [18]. Most of the diversity measures are based on the distance metric of individuals. Diversity measures are also differed in terms of normalization of parameters or distance metric. Further, the measures are differed based on the choice of population center which may be global best solution found so far or may be spatial. In this section, seven different diversity measures (Listed in Table 1) based on the Euclidean distance metric are discussed. Further, global best population center is used in this paper wherever required opposed to a spatial population center. Generally, spatial population center and global best population center can be considered equivalent where the global best is not necessarily centered position in the population. Further, for normalization of parameters, the population diameter is used, instead of the radius of population.

2.1 Population Diameter

The population diameter is defined as the distance between two farthest individuals, along any axis [25], of the population as shown in Fig. 1.

The diameter D is calculated using Eq. (1):

$$D = \max_{(i \neq j) \in [1, N_p]} \left(\sqrt{\sum_{k=1}^{I}(x_{ik} - x_{jk})^2}\right) \qquad (1)$$

Fig. 1 Population diameter

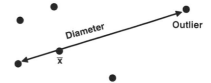

where N_p is the population size, I is the dimensionality of the problem, and x_{ik} is the kth dimension of the ith individual position.

In Fig. 1, an outlier individual is also shown. In a population, a significantly deviated individual from the remaining individuals is often termed as *outlier*. From Fig. 1, it can be seen that the presence of an outlier can significantly affect the diameter of a population.

2.2 Population Radius

The radius of a population is defined as the distance between the population center and the individual in the population which is farthest away from it [25], as shown in Fig. 2.

The population radius is calculated using Eq. (2):

$$R = \max_{i \in [1, N_p]} \left(\sqrt{\sum_{k=1}^{I} (x_{ik} - \bar{x}_k)^2} \right) \qquad (2)$$

where the parameters have same meaning as for the population diameter. $\bar{x}$ is the position of population center, and $\bar{x}_k$ represents the kth dimension of $\bar{x}$.

Now, it is evident that the population diameter (D) and radius (R) are two important diversity measures. A large value of D or R exhibits exploration of the search region, while low value exhibits exploitation. However, both are badly affected with outliers.

2.3 Average Distance around the Population Center

The average distance from the population center, D_A, can be defined as the average of distances of all individuals from the population center. This measure is given in [13] and defined in Eq. (3)

Fig. 2 Population radius

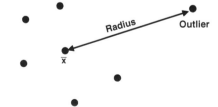

$$D_A = \frac{1}{N_p} \sum_{i=1}^{N_p} \left(\sqrt{\sum_{k=1}^{I} (x_{ik} - \bar{x}_k)^2} \right) \tag{3}$$

where the notations have their usual meaning. A low value of this measure shows population convergence around the population center, while a high value shows large dispersion of individuals from the population center.

2.4 Geometric Average Distance around the Population Center

Geometric average is not significantly affected by outliers in the population on the high end. The geometric average distance around the population center is defined in Eq. (4).

$$D_{GM} = \left(\prod_{i=1}^{N_p} \sqrt{\sum_{k=1}^{I} (x_{ik} - \bar{x}_k)^2} \right)^{\frac{1}{N_p}} \tag{4}$$

2.5 Normalized Average Distance around the Population Center

This diversity measure is almost the same as the average distance of all individuals from the population center. The only difference is that the average distance is normalized using the population diameter. This normalization can also be done by the radius of the population. This diversity measure is given in [25] and described by Eq. (5):

$$D_N = \frac{1}{N_p \times D} \sum_{i=1}^{N_p} \left(\sqrt{\sum_{k=1}^{I} (x_{ik} - \bar{x}_k)^2} \right) \tag{5}$$

2.6 Average of the Average Distance around all Individuals in the Population

In this measure, first the average distances, considering each individual as a population center, are calculated, and then, an average of all these average distances is taken. It is described by Eq. (6).

$$D_{all} = \frac{1}{N_p} \sum_{i=1}^{N_p} \left(\frac{1}{N_p} \sum_{j=1}^{N_p} \sqrt{\sum_{k=1}^{I} (x_{ik} - x_{jk})^2} \right) \tag{6}$$

This diversity measure shows average dispersion of every individual in the population from every other individual in the population.

2.7 Population Coherence

This diversity measure is given in [10] and described by Eq. (7):

$$S = \frac{s_c}{\bar{s}} \tag{7}$$

where s_c represents the step size of population center which is defined in Eq. (8):

$$s_c = \frac{1}{N_p} \left\| \sum_{i=1}^{N_p} \tilde{s}_i \right\|_2 \tag{8}$$

where $\tilde{s}_i$ is the vector of step size for ith individual as indicated in Eq. (9) and $\bar{s}$ shows the average individual step size in the population and is defined by Eq. (10). $\|.\|_p$ is the Euclidean p-norm.

$$\tilde{s}_i = F \times (x_{i_1} - x_{i_2}) \tag{9}$$

Here, F is a scaling factor and x_{i_1}, x_{i_2} are randomly selected individuals from the population such that $i_1 \neq i_2$. This equation is used in the mutation process of DE.

$$\bar{s} = \frac{1}{N_p} \sum_{i=1}^{N_p} \|\tilde{s}_i\|_2 \tag{10}$$

This diversity measure is calculated by averaging the step sizes of all the individuals in a population with respect to population center.

The dispersion of the individuals in DE and its prominent variants such as SaDE, jDE, OBDE, and JADE could be quantified, using the various diversity measures described in this section. The diversity measures show a trend of exploration or exploitation of the population and helps to analyze the behavior of the population-based algorithms.

3 Discussion

The two important diversity characteristics of population-based stochastic search optimization algorithms are exploration and exploitation. Exploration capability explores the solution search space to find the possible solution region, whereas exploitation capability exploits a particular region for a better solution. Exploration and exploitation are compliment to each other, i.e., a better exploration results in worse exploitation and vice versa. The main objective of an efficient algorithm is to maintain a proper balance between exploration and exploitation in the population. Hight value of diversity measure shows the exploration, whereas low value exhibits exploitation. A decreasing diversity measure through iterations represents the transition of exploration to exploitation. On the basis of these characteristics, the following conclusions have been drawn:

- The population diameter presents a required decrease by iterations, as $(x_{ik} - x_{jk})^2$ (refer Eq. (1)) tends to zero for all the individuals as the population converges to a solution. The same behavior is shown by the population radius as the distance between each individual and population center decreases as the population converges with iterations. Further, it is clear from the Eqs. (1) and (2) that the population diameter and population radius both are very sensitive to the outlier individual. Considering Figs. 1 and 3, it can be shown that the diverse behavior of the current population shown in Figs. 1 and 3 is same, if population radius or population diameter is the diversity measure, while in Fig. 1, the individuals are more diversified than in Fig. 3.
- The diversity measure D_A, which is shown in Eq. (3), is robust measure as compared to the population diameter and population radius because it is based on the average distance of all individuals in the population from the population center. Hence, this diversity measure is considered less affected due to the outliers as compared to the population diameter and population radius. But an extreme farthest outlier may skew the individual's dispersion significantly in the population. Further, $(x_{ik} - \bar{x}_k)^2 \to 0$ (refer Eq. (3)) for all individuals in the population as population converges. The same behavior is shown by the diversity measure D_{all} given by Eq. (6) because for all individuals in the population, the component $(x_{ik} - x_{jk})^2$ also approaches to zero as population converges.
- The diversity measure D_{GM} shown in Eq. (4) is again a robust measure for measuring diversity. In statistics, geometric average is relatively less affected from outliers.
- The diversity measure D_N shown in Eq. (5) is the ratio of the average distance D_A to the population diameter D. Here, diameter is considered as a normalization parameter used to normalize the average distance around the population center. In this measure of dispersion, as population converges, D_N and D both approaches to zero with iterations. Further, in this dispersion, as the normalization is done by the population diameter or the population radius, it is significantly influenced by

Fig. 3 Population diameter and outlier

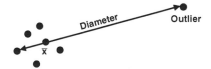

the outlier individuals. Therefore, D_A and D_{all} still may be considered as a better choice for measuring diversity of the population.

- The diversity measure S, which is shown in Eq. (7), is the ratio of the absolute step size of the population center to the average step size of all individuals in the population. A high value of the population center step size implies that all the individuals in the population are moving in the same direction. Further, a low value implies that most of the individuals are moving to opposite directions. A high value of average individual's step size in population implies that the solutions are significantly changing the positions, which implies exploration of the search space, while a low value shows the convergence in the population, i.e., exploiting the solution search space found so far. Thus, S could be used to analyze the diversity behavior of the algorithm.

Figures 4 and 5 show a large population diversity for small population coherence and large population coherence, respectively. The individuals are dispersed in the search space, whereas the step size of population center is relatively low and relatively high, respectively.

Further, by analyzing Figs. 6 and 7, it is clear that the value of S does not depend completely on the diversity of population. Therefore, it could be concluded that the population coherence is not proportional to population diversity of individuals in the DE and its prominent variants are not a true measure of diversity.

The outcome of the above discussion is that the effect of outlier is significant in most of the diversity measures and it biases the measure of dispersion. However, the effect of outliers could be minimized; it cannot be ignored completely.

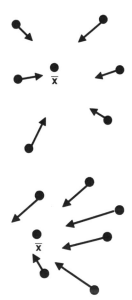

Fig. 4 High individuals' diversity and small population coherence

Fig. 5 High individuals' diversity and large population coherence

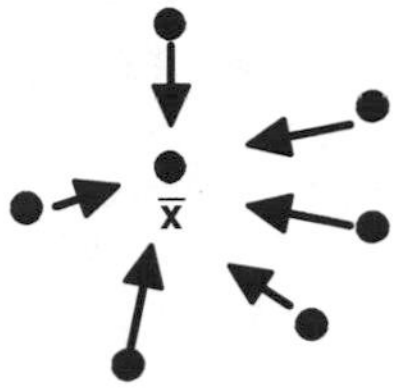

Fig. 6 Low individuals' diversity and small population coherence

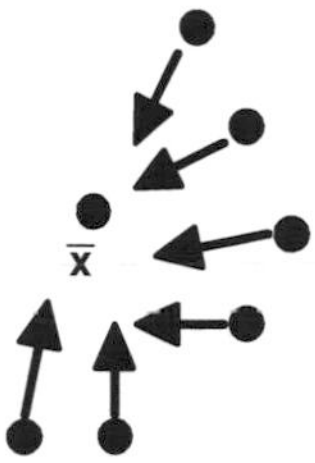

Fig. 7 Low individuals' diversity and large population coherence

4 Experimental Setting and Results

To analyze the various diversity measures for DE and its prominent variants, experiments have been carried out on four well-known benchmark test problems (Listed in Table 2). For these experiments, the following experimental setting is adopted:

- Population size NP = 50 [7, 8],
- $\Gamma = 0.5$,
- CR = 0.9,
- The stopping criterion is either the maximum number of function evaluations (which is set to be 200000) is reached or the acceptable error (mentioned in Table 2) has been achieved,
- The number of runs = 100, and graphs are plotted using the mean of each run.
- Scaling, which is used to construct the graph outputs to the interval [0, 1], is shown below:

$$\bar{y} = \frac{\bar{y} - \min(\bar{y})}{\max(\bar{y}) - \min(\bar{y})}.$$

This is done to make comparisons of all measures in the same range.

Figures 8, 9, 10, 11, 12, 13, and 14 show the diversity measures based on diameter, radius, average distance around the population center, geometric average distance, normalize average distance, average of the average distance around all individuals in the population, and population coherence for the considered advance variant of DE, respectively. It is clear from these figures that, in terms of diversity in population, jDE is better, while in terms of convergence speed, JADE is better, among considered DE variants.

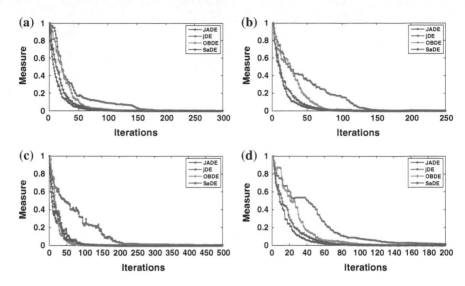

Fig. 8 Diversity measure for population diameter. **a** Diameter for Ackley function. **b** Diameter for Griewank function. **c** Diameter for Rosenbrock function. **d** Diameter for sphere function

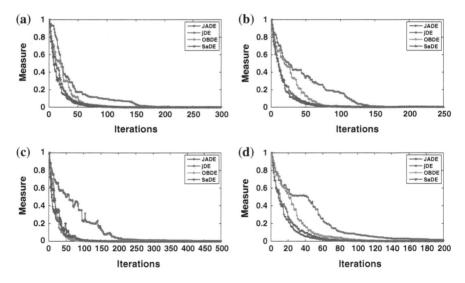

Fig. 9 Diversity measure for radius. **a** Radius for Ackley function. **b** Radius for Griewank function. **c** Radius for Rosenbrock function. **d** Radius for sphere function

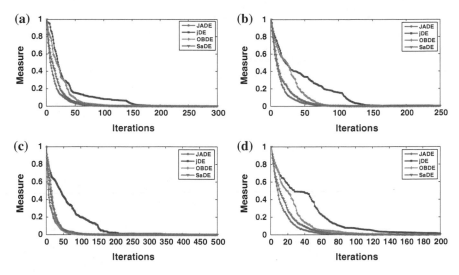

Fig. 10 Diversity measure for average distance around population center. **a** Average distance for Ackley function. **b** Average distance for Griewank function. **c** Average distance for Rosenbrock function. **d** Average distance for sphere function

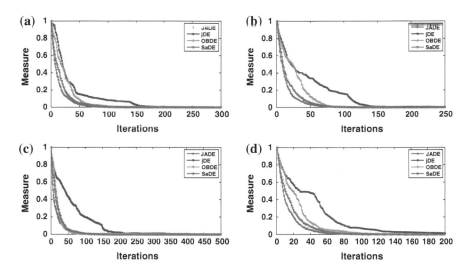

Fig. 11 Diversity measure for geometric average distance. **a** Geometric average distance for Ackley function. **b** Geometric average distance for Griewank function. **c** Geometric average distance for Rosenbrock function. **d** Geometric average distance for sphere function

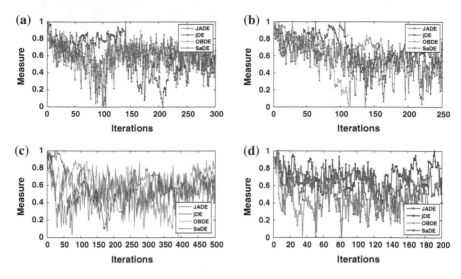

Fig. 12 Diversity measure for normalized average distance. **a** Normalized average distance for Ackley function. **b** Normalized average distance for Griewank function. **c** Normalized average distance for Rosenbrock function. **d** Normalized average distance for sphere function

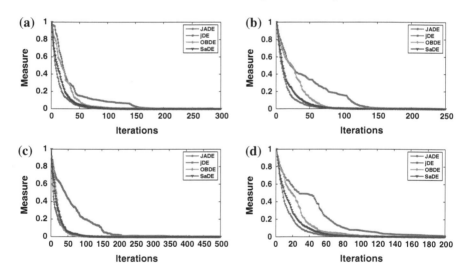

Fig. 13 Diversity measure for average of the average distance around all individuals in the population. **a** Average of average distance for Ackley function. **b** Average of average distance for Griewank function. **c** Average of average distance for Rosenbrock function. **d** Average of average distance for sphere function

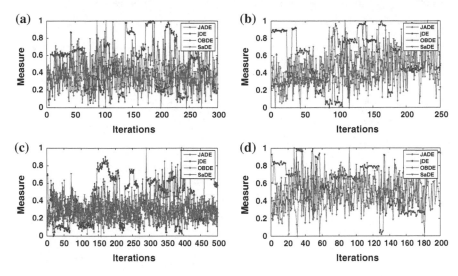

Fig. 14 Diversity measure for coherence. **a** Coherence for Ackley function. **b** Coherence for Griewank function. **c** Coherence for Rosenbrock function. **d** Coherence for sphere function

5 Conclusion

In this paper, various diversity measures are studied and analyzed to measure the dispersion in the population of prominent DE variants, namely JADE, jDE, OBDE, and SaDE. In population-based algorithms, diversity measures are used to investigate the exploration and exploitation characteristics. Further, the diversity measures are analyzed on four well-known benchmark problems. The outcome of the experiments shows that the value of diversity measures proportionally decreases with the iterations. Further, it is found that the diversity measures are more or less affected by the outliers in the population. Diversity measures like the average distance around the population center, the geometric mean average distance around the population center, and the average of average distance around the population center are less affected by the outliers and could be used for analyzing the exploration and exploitation of the solution search space in the population. The considered DE variants are compared on the basis on these diversity measure, and it is found that diversity of jDE algorithm is better, while convergence speed of JADE algorithm is better among considered algorithms.

References

1. Bansal, J.C., Sharma, H.: Cognitive learning in differential evolution and its application to model order reduction problem for single-input single-output systems. Memetic Comput.1–21, (2012)

2. Blackwell, T.M.: Particle swarms and population diversity i: Analysis. In GECCO, pp. 103–107,2003.
3. Brest, J., Greiner, S., Boskovic, B., Mernik, M., Zumer, V.: Self-adapting control parameters in differential evolution: A comparative study on numerical benchmark problems. Evolutionary Computation, IEEE Transactions on $10(6)$, 646–657 (2006)
4. Chakraborty, U.K.: Advances in differential evolution. Springer, Berlin (2008)
5. Das, S., Konar, A.: Two-dimensional IIR filter design with modern search heuristics: A comparative study. Int. J. Comput. Intell. Appl. $6(3)$, 329–355 (2006)
6. Das, S., Suganthan, P.N.: Differential evolution: A survey of the state-of-the-art. IEEE Trans. Evol. Comput. 99, 1–28 (2010)
7. Diwold, K., Aderhold, A., Scheidler, A., Middendorf, M.: Performance evaluation of artificial bee colony optimization and new selection schemes. Memetic Comput., 1–14 (2011).
8. El-Abd, M.: Performance assessment of foraging algorithms vs. evolutionary algorithms. Inf. Sci. (2011).
9. Engelbrecht, A.P.: Fundamentals of computational swarm intelligence. Recherche 67, 02 (2005)
10. Hendtlass, T., Randall, M.: A survey of ant colony and particle swarm meta-heuristics and their application to discrete optimization problems, pp. 15–25. In: Proceedings of the Inaugural Workshop on Artificial Life (2001).
11. Holland, J.H.: Adaptation in natural and artificial systems. The University of Michigan Press, Ann Arbor (1975)
12. Kennedy, J., Eberhart, R.: Particle swarm optimization. In: Proceedings of IEEE International Conference on. Neural Networks 4, 1942–1948 (1995)
13. Krink, T., VesterstrOm, J.S., Riget, J.: Particle swarm optimisation with spatial particle extension. In: Proceedings of the 2002 Congress on, Evolutionary Computation, CEC'02, pp. 1474–1479 (2002)
14. Lampinen, J., Zelinka, I.: On stagnation of the differential evolution algorithm. In: Proceedings of MENDEL, pp. 76–83. Citeseer (2000).
15. Liu, P.K., Wang, F.S.: Inverse problems of biological systems using multi-objective optimization. J. Chin. Inst. Chem. Eng. $39(5)$, 399–406 (2008)
16. Mezura-Montes, E., Velázquez-Reyes, J., Coello Coello, C.A.: A comparative study of differential evolution variants for global optimization. In: Proceedings of the 8th annual conference on Genetic and evolutionary computation, pp. 485–492. ACM (2006).
17. Neri, F., Tirronen, V.: Recent advances in differential evolution: a survey and experimental analysis. Artif. Intell. Rev. $33(1)$, 61–106 (2010)
18. Olorunda, O., Engelbrecht, A.P.: Measuring exploration/exploitation in particle swarms using swarm diversity. In: Evolutionary Computation (IEEE World Congress on Computational Intelligence), pp. 1128–1134 (2008).
19. Omran, M.G.H., Engelbrecht, A.P., Salman, A.: Differential evolution methods for unsupervised image classification. In: The 2005 IEEE Congress on. Evolutionary Computation 2, 966–973 (2005)
20. Price, K.V.: Differential evolution: A fast and simple numerical optimizer. In: Fuzzy Information Processing Society. NAFIPS, Biennial Conference of the North American, IEEE, pp. 524–527 (1996).
21. Price, K.V., Storn, R.M., Lampinen, J.A.: Differential evolution: a practical approach to global optimization. Springer, Berlin (2005)
22. Qin, A.K., Huang, V.L., Suganthan, P.N.: Differential evolution algorithm with strategy adaptation for global numerical optimization. IEEE Trans. Evol. Comput. $13(2)$, 398–417 (2009)
23. Rahnamayan, S., Tizhoosh, H.R., Salama, M.M.A.: Opposition-based differential evolution. IEEE Trans. Evol. Comput. $12(1)$, 64–79 (2008)
24. Ratnaweera, A., Halgamuge, S., Watson, H.: Particle swarm optimization with self-adaptive acceleration coefficients. In: Proceedings od 1st International Conference on Fuzzy System Knowledge. Discovery, pp. 264–268 (2003).
25. Riget, J., Vesterstrøm, J.S.: A diversity-guided particle swarm optimizer-the ARPSO. Dept. Comput. Sci., Univ. of Aarhus, Aarhus, Denmark. Tech. Rep 2, 2002 (2002)

26. Rogalsky, T., Kocabiyik, S., Derksen, R.W.: Differential evolution in aerodynamic optimization. Can. Aeronaut. Space J. **46**(4), 183–190 (2000)
27. Sharma, H., Bansal, J., Arya, K.: Dynamic scaling factor based differential evolution algorithm. In: Proceedings of the International Conference on Soft Computing for Problem Solving (SocProS 2011) Dec 20–22, 2011, pp. 73–85. Springer (2012).
28. Vesterstrom, J.S., Riget, J., Krink, T.: Division of labor in particle swarm optimisation. In: IEEE proceedings of the 2002 Congress on Evolutionary Computation, CEC'02., 2, pp. 1570–1575 (2002)
29. Vesterstrom, J., Thomsen, R.: A comparative study of differential evolution, particle swarm optimization, and evolutionary algorithms on numerical benchmark problems. In: IEEE Congress on, Evolutionary Computation, CEC2004, 2, pp. 1980–1987, 2004.
30. Zhang, J., Sanderson, A.C.: Jade: adaptive differential evolution with optional external archive. IEEE Trans. Evol. Comput. **13**(5), 945–958 (2009)

Computing Vectors Based Document Clustering and Numerical Result Analysis

Neeraj Sahu and G. S. Thakur

Abstract This paper presents new approach analytical results of document clustering for vectors. The proposed analytical results of document clustering for vectors approach is based on mean clusters. In this paper we have used six iterations I_1 to I_6 for document clustering results. The steps Document collection, Text Pre-processing, Feature Selection, Indexing, Clustering Process and Results Analysis are used. Twenty news group data sets are used in the experiments. The experimental results are evaluated using the numerical computing MATLAB 7.14 software. The experimental results show the proposed approach out performs.

Keywords Mean clusters · Clustering technique · Vectors and iterations

1 Introduction

Document clustering is an important issue in text mining. Clustering has been widely applicable in different areas of science, technology, social science, biology, economics, medicine and stock market. Clustering problem appears in other different field like pattern recognition, statistical data analysis, bio-informatics, etc. There exist many clustering methods in the literature. In last recent years lot of research work has been done on document clustering. Some contributions are as follows: Approximation Schemes for Euclidean k-median and Related Problems [1], Multi-dimensional Binary Search Trees Used for Associative Searching [2], Refining Initial Points for k-means Clustering [3], Geometric Clusterings [4], Learning Mixtures of

N. Sahu (✉)
Singhania University Rajasthan, Rajasthan, India
e-mail: neerajsahu79@gmail.com

G. S. Thakur
MANIT, Bhopal, India
e-mail: ghanshyamthakur@gmail.com

B. V. Babu et al. (eds.), *Proceedings of the Second International Conference on Soft Computing for Problem Solving (SocProS 2012), December 28–30, 2012*, Advances in Intelligent Systems and Computing 236, DOI: 10.1007/978-81-322-1602-5_138, © Springer India 2014

Gaussians [5], Centroidal Voronoi Tesselations: Applications and Algorithms[6], Exact and Approximation Algorithms for Clustering [7], Pattern Classification and Scene Analysis [8], Clustering and the Continuous k-means Algorithm [9], Approximate Range Searching, Computational Geometry: Theory and Applications [10], Cluster Analysis of Multivariate Data: Efficiency versus Interpretability of Classification [11], Advances in Knowledge Discovery and Data Mining [12], Introduction to Statistical Pattern Recognition [13], Some Fundamental Concepts and Synthesis Procedures for Pattern Recognition Pre-processors [14], A Two-Round Variant of EM for Gaussian Mixtures [15], An Introduction to Probability Theory and its Applications [16], An Efficient k-means Clustering Algorithm [17], Optimal Algorithm for Approximate Nearest Neighbour Searching [18], Convergence Properties of the k-means Algorithms, Advances in Neural Information Processing Systems 7 [19], A Spatial Filtering Approach to Texture Analysis [20], A Database Interface for Clustering in Large Spatial Databases [21], Hesitant Distance Similarity Measures for Document Clustering [22], Classification of Document Clustering Approaches [23], Architecture Based Users and Administrator Login Data Processing [24].

The paper is organized as follows. Section 1 describes the introduction and review of literature. In Sect. 2, the methodology of document clustering is described. In Sect. 3, the numerical results analysis of document clustering for computing vectors is described. In Sect. 4, experimental results are described. In Sect. 5, evaluation measurement is described. Finally, we concluded and proposed some future directions in the conclusion section.

2 Methodology

In the document clustering different the steps are used. The steps are as follows:

2.1 Document Collection

In this phase we collect relevant documents like e-mail, news, web pages etc. from various heterogeneous sources. These text documents are stored in a variety of formats depending on the nature of the data. The datasets are downloaded from the UCI KDD archive. This is an online repository of large datasets and has wide variety of data types.

2.2 Text Pre-processing

Text pre-processing means transforming documents into a suitable representation for the clustering task. The text documents have different stop words, punctuation marks, special character and digits and other characters. Algorithm 1 is removed

HTML tags and stop words from the text documents. After removing stop words, word stemming is performed. Word stemming is the process of suffix removal to general word stems. A stem is a natural group of words with similar meaning. In text-pre-processing we performed the following task: Removal of HTML tags and special character, stop words and word stemming. Algorithm 1 is proposed for text pre-processing. The proposed algorithm removes special characters and stop words from the document.

Algorithm1: This algorithm obtain to remove stop words & special characters
Input: A document Data Base D and List of Stop words L
D = {d1, d2, d3, ..., dk} ;where 1 <= k <= i
tij is the jth term in ith document
Output: All valid stem text term in D
for (all di in D) do
for (1 to j) do
Extract tij from di
If(tij in list L)
Remove tij from di
End for
End for

2.3 Dimension Reduction

High dimension is the greatest challenge of document clustering, so dimension reduction became major issue of clustering. This module performs two functions- indexing and feature selection. In indexing method we assign the value to the terms in the documents. After indexing, feature selection method is applied. Feature selection is the process of removing indiscriminate terms from the documents to improve the document clustering accuracy and reduce the computational complexity. Algorithm 2 is proposed for indexing and feature selection.

Algorithm 2: Basic algorithm obtain for feature selection
Input: A document DataSet D and y minimum threshold value, N is the counter
D = {d1, d2, d3, ..., dk} ; where 1 <= k <= i
tij is the jth term in ith document
Output: Documents Dataset D after feature selection
for (all di in D) do
for (1 to j) do
Count total occurrence of tij in document di
Assign the total occurrence of tij in N
If (N < y)
Remove tji from the document di
End for
End for

Algorithm 2 removes all low frequency terms from the documents. This improves clustering effectiveness and reduces the computational complexity. The document clustering Algorithm 3 is developed. In this algorithm the formulae from I_1 to I_6 are used to calculate the similarities.

Algorithm 3: k-mean algorithm obtains to final cluster of each vector for each Iteration.
Step 1: Input eight vectors and three initial Clusters (k=3) form of vector coming from input vectors.
Step 2: initialize x_1, y_1, z_1 for each vector and d_1, d_4, d_7 for each initial clusters vector
Step 3: Produce and compare $cos\theta_1$ one by one.
Step 4: $cos\theta_1$ maximum value from each vector final cluster of each vector for first Iteration.
Step 5: After first Iteration find new mean cluster for each Iteration.
Step 6: Repeat step 2 to step 5 for all Iteration until all final clusters vector are not same for next Iterations.

---

3 Vector Representation of Text Document

Text document representation is a challenging issue in text mining. After performing text pre-processing, feature selection and dimension reduction the text documents can be represented as a vector form. Illustrative example I: Let D be a set of documents $D = \{d1, d2, d3, d4, d5, d6, d7, d8\}$ where D is a documents set and d_i are the text documents $i \leq 1 \leq 8$. The documents d_1 to d_8 is represented in vector form. $d_1 = (2, 7, 10)$, $d_2 = (2, 3, 5)$, $d_3 = (8, 3, 5)$, $d_4 = (5, 3, 8)$, $d_5 = (7, 3, 5)$, $d_6 = (6, 4, 7)$, $d_7 = (11, 2, 3)$, $d_8 = (4, 3, 9,)$. Here the tables are Tables 1 and 2. Clustering algorithm K-mean in my algorithm we choose k=3 and centres are d_1, d_4, d_7.

Table 1 Final clusters in iteration I to VI

Iteration → Vectors ↓	I	II	III	IV	V	VI
d_1	1	1	1	1	1	1
d_2	3	1	1	1	1	1
d_3	3	3	3	3	3	3
d_4	2	2	2	3	2	2
d_5	3	2	2	2	2	2
d_6	2	3	2	2	2	2
d_7	2	3	3	3	3	3
d_8	2	2	2	2	2	2

Table 2 Document clustering results percentages

Iteration vectors	Accuracy percentages	Iteration vectors	Accuracy percentages
I_1V_a	2.37	I_3V_e	51.31
I_1V_b	4.75	I_3V_f	53.68
I_1V_c	7.12	I_3V_g	56.05
I_1V_d	9.50	I_3V_h	59.43
I_1V_e	11.87	I_4V_a	61.80
I_1V_f	14.25	I_4V_b	64.18
I_1V_g	16.62	I_4V_c	66.55
I_1V_h	19.82	I_4V_d	68.93
I_2V_a	21.37	I_4V_e	71.33
I_2V_b	23.75	I_4V_f	73.67
I_2V_c	26.12	I_4V_g	76.05
I_2V_d	28.50	I_4V_h	79.45
I_2V_e	30.87	I_5V_a	81.82
I_2V_f	33.25	I_5V_b	84.19
I_2V_g	35.62	I_5V_c	86.59
I_2V_h	39.44	I_5V_d	88.99
I_3V_a	41.81	I_5V_e	91.39
I_3V_b	44.19	I_5V_f	93.79
I_3V_c	46.56	I_5V_g	96.16
I_3V_d	48.94	I_5V_h	98.56

4 Experimental Results

In this paper the unstructured datasets are used. The datasets are downloaded from UCI KDD Archive [4]. This is an online repository of large datasets with wide variety of data types. This repository has twenty newsgroups dataset for text analysis. This data set consists of 20,000 messages taken from Usenet newsgroup. The subset of twenty newsgroups is mini newsgroup. We have done our experiments on 20 newsgroup datasets. Each category contains 1,000 documents, so there are 20,000 documents for experiments. The five categories Computer Hardware, Computer Graphics, Medical, Sports and Automobile are used in first experiment. The experimental results are shown in Figs. 1, 2, 3, 4, 5, 6, 7, 8, 9, 10, 11 and 12 with eight vectors d_1 to d_8.

The document clustering results percentages show accuracy of clustering process. Here percentages are increases iteration by iterations.

Fig. 1 Accuracy percentages
for iteration I

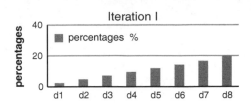

Fig. 2 Accuracy percentages
for iteration II

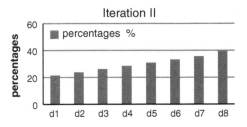

Fig. 3 Accuracy percentages
for iteration III

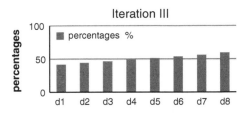

Fig. 4 Accuracy percentages
for iteration IV

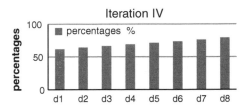

Fig. 5 Accuracy percentages
for iteration V

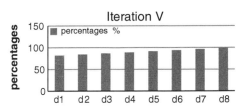

Fig. 6 Accuracy percentages
from I to VI

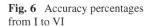
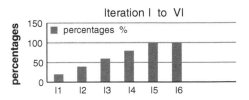

Fig. 7 final clusters for iteration I

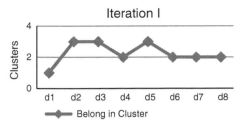

Fig. 8 final clusters for iteration II

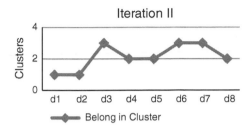

Fig. 9 final clusters for iteration III

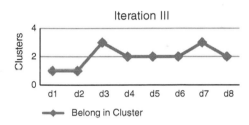

Fig. 10 final clusters for iteration IV

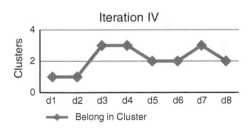

Fig. 11 final clusters for iteration V

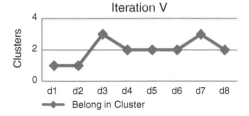

Fig. 12 final clusters for
iteration VI

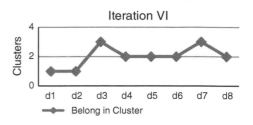

The graphical representations of the results are shown in Figs. 1, 2, 3, 4, 5, 6, 7, 8, 9, 10, 11, 12. We performed our experiments on five newsgroups—Computer graphics, Computer hardware, Automobile, Sports and Medical. In this research the 80% dataset is used as the training dataset and 20% dataset as the test dataset.

5 Conclusions

This paper analyzed document clustering using computing vectors. The experimental results of document clustering are efficient and accurate. The proposed approach works efficiently. The accuracy is best as compare to other existing algorithm of the document clustering.

Acknowledgments This work is supported by research grant from MPCST, Bhopal M.P., India, Endt.No. 2427/CST/R&D/2011 dated 22/09/2011.

References

1. Arora, S., Raghavan, P., Rao, S.: Approximation schemes for Euclidean k-median and related problems. In: Proceedings of the 30th Annual ACM Symposium on Theory of Computing, May 1998, pp. 106–113
2. Bentley, J.L.: Multidimensional binary search trees used for associative searching. Comm. ACM **18**, 509–517 (1975)
3. Bradley, P.S., Fayyad, U.: Refining initial points for k-means clustering. In: Proceedings of the 15th International Conference on Machine Learning, 1998, pp. 91–99
4. www.kdd.ics.uci.edu
5. Dasgupta, S.: Learning mixtures of Gaussians. In: Proceedings of the 40th IEEE Symposium on Foundations of Computer Science, Oct 1999, pp. 634–644
6. Du, Q., Faber, V., Gunzburger, M.: Centroidal Voronoi tesselations: Applications and algorithms. SIAM Rev. **41**, 637–676 (1999)
7. Agarwal, P.K., Procopiuc, C.M.: Exact and approximation algorithms for clustering. In: Proceedings of the Ninth Annual ACM-SIAM Symposium on Discrete Algorithms, Jan 1998, pp. 658–667
8. Duda, R.O., Hart, P.E.: Pattern Classification and Scene Analysis. Wiley, New York (1973)
9. Faber, V.: Clustering and the continuous k-means algorithm. Los Alamos Sci. **22**, 138–144 (1994)

10. Arya, S., Mount, D.M.: Approximate range searching. Comput. Geom. Theor. Appl. **17**, 135–163 (2000)
11. Forgey, E.: Cluster analysis of multivariate data: efficiency versus interpretability of classification. Biometrics **21**, 768 (1965)
12. Fayyad, U.M., Piatetsky-Shapiro, G., Smyth, P., Uthurusamy, R.: Advances in Knowledge Discovery and Data Mining. AAAI/MIT Press, Cambridge (1996)
13. Fukunaga, K.: Introduction to Statistical Pattern Recognition. Academic Press, Boston (1990)
14. Ball, G.H., Hall, D.J.: Some fundamental concepts and synthesis procedures for pattern recognition preprocessors. In: Proceedings of the International Conference on Microwaves, Circuit Theory, and Information Theory, Sept 1964
15. Dasgupta, S., Shulman, L.J.: A two-round variant of EM for Gaussian mixtures. In: Procedings of the 16th Conference on Uncertainty in Artificial Intelligence (UAI-2000), June 2000, pp. 152–159
16. Feller, W.: An Introduction to Probability Theory and Its Applications, 3rd edn. Wiley, New York (1968)
17. Alsabti, K., Ranka, S., Singh, V.: An Efficient k-means clustering algorithm. In: Proceedings of the First Workshop High Performance Data Mining, Mar 1998
18. Arya, S., Mount, D.M., Netanyahu, N.S., Silverman, R., Wu, A.Y.: An optimal algorithm for approximate nearest neighbor searching. J. ACM **45**, 891–923 (1998)
19. Bottou, L., Bengio, Y.: Convergence properties of the k-means algorithms. In: Tesauro, G., Touretzky, D. (eds.) Advances in Neural Information Processing Systems 7, pp. 585–592. MIT Press, Cambridge (1996)
20. Coggins, J.M., Jain, A.K.: A spatial filtering approach to texture analysis. Pattern Recognit. Lett. **3**, 195–203 (1985)
21. Ester, M., Kriegel, H., Xu, X.: A database interface for clustering in large spatial databases. In: Procedings of the First International Conference on Knowledge Discovery and Data Mining (KDD-95), 1995, pp. 94–99
22. Neeraj, S., Thakur, G.S.: Hesitant distance similarity measures for document clustering. In: IEEE Conference—2011 World Congress on Information and Communication Technologies Mumbai, 11–14 Dec 2011. ISBN: 978-1-4673-0125-1
23. Sahu, S.K., Sahu, N., Thakur, G.S.: Classification of document clustering approaches. Intl. J. Comput. Sci. Softw. Eng. **2**(5), 509–513 (2012). ISSN (Online): 2277 128X
24. Sahu, B., Sahu, N., Thakur, G.S.: Architecture based users and administrator login data processing. In: International Conference on Intelligent Computing and Information System (ICICIS-2012), Pachmarhi, Piparia, 27–28 Oct 2012. ISSN (Online): 2249–071X
25. Bradley, P.S., Fayyad, U., Reina, C.: Scaling clustering algorithms to large databases. In: Proceedings of the Fourth International Conference on Knowledge Discovery and Data Mining, 1998, pp. 9–15

Altered Fingerprint Identification and Classification Using SP Detection and Fuzzy Classification

Ram Kumar, Jasvinder Pal Singh and Gaurav Srivastava

Abstract Fingerprint recognition is one of the most commonly used biometric technology. Even if fingerprint temporarily changes (cuts, bruises) it reappears after the finger heals. Criminals started to be aware of this and try to fool the identification systems applying methods from ingenious to very cruel. It is possible to remove, alter, or even fake fingerprints (made of glue, latex, silicone), by burning the fingertip skin (fire, acid, other corrosive material), by using plastic surgery (changing the skin completely, causing change in pattern—portions of skin are removed from a finger and grafted back in different positions, like rotation or "Z" cuts, transplantations of an area from other parts of the body like other fingers, palms, toes, and soles). This paper presents a new algorithm for altered fingerprints detection based on fingerprint orientation field reliability. The map of the orientation field reliability has peaks in the singular point locations. These peaks are used to analyze altered fingerprints because, due to alteration, more peaks as singular points appear with lower amplitudes.

Keywords Fingerprints · Alteration · Image enhancement · Reliability · Singular points

R. Kumar
M.Tech Scholor in R.K.D.F Institute of Technology and Science, Bhopal, India
e-mail: hr.coet@gmail.com

J.P. Singh and G. Srivastava
Assistant Professor in Computer Science and Engineering in R.K.D.F Institute of Technology and Science, Bhopal, India
e-mail: jasvinder162@gmail.com

G. Srivastava
e-mail: gashr83@gmail.com

B. V. Babu et al. (eds.), *Proceedings of the Second International Conference on Soft Computing for Problem Solving (SocProS 2012), December 28–30, 2012*, Advances in Intelligent Systems and Computing 236, DOI: 10.1007/978-81-322-1602-5_139, © Springer India 2014

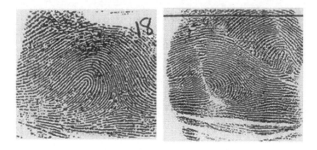

Fig. 1 A fingerprint altered by switching two parts of a 'Z' shaped cut [2]

1 Introduction

Fingerprint alteration is not a new phenomenon. As early as in 1934, John Dillinger, the infamous bank robber and a dangerous criminal, applied acid to his fingertips [1]. Since then, there has been an increase in the reported cases of fingerprint alteration. In 1995, a Criminal was found to have altered his fingerprints by making a "Z" shaped cut into the finger and switching the finger skin the two parts (see Fig. 1). In 2009, a Chinese woman successfully deceived the Japan immigration fingerprint system by performing surgery to swap fingerprints on her left and right hands [3]. Fingerprint alteration has even been performed at a much larger scale involving multiple individuals. Hundreds of asylum seekers have cut, abraded, and burned their fingertips to prevent identification by EURODAC, a European Union fingerprint system for identifying asylum seekers [2]. Additional cases of fingerprint alteration have been compiled in [2].

The primary purpose of fingerprint alteration [4] is to evade identification using techniques that vary from abrading, cutting, and burning fingers to performing plastic surgery.

Fingerprint alteration constitutes a serious "attack" against a border control fingerprint identification system since it defeats the very purpose for which the system was deployed in the first place, i.e., to identify individuals on a watch-list.

The goal of this work is to introduce the problem of fingerprint alteration and to develop methods to automatically detect and classify altered fingerprints.

2 Type of Altered Fingerprint

According to the changes made to the ridge patterns, fingerprint alterations may be categorized into three types:

   i. Obliteration
  ii. Distortion
 iii. Imitation (see Fig. 2).

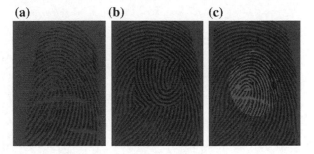

Fig. 2 Three types of altered fingerprints. **a** Obliterated fingerprint **b** distorted fingerprint, and **c** imitated fingerprint

For each type of alteration, its characteristics and possible countermeasures are described.

3 Methodology

In this section, we consider the problem of automatic detection of alterations and classification based on analyzing singular point reliability map of orientation field.

A set of features is first extracted from the ridge orientation field of an input fingerprint and then a fuzzy classifier is used to classify it into natural or altered fingerprint and its alteration type.

3.1 Fingerprint Database

Due to lack of altered fingerprint data, I use constructed database synthetically form the original fingerprints. I collect more than 100 original fingerprints of my friends to analyze the proposed system. After collecting the original fingerprint, I altered those fingerprint by the three different techniques of fingerprint alteration named obliteration, distortion, and imitation respectively. For analyzing the system, we use 49 images for all type of alteration.

3.2 Preprocessing on Fingerprint

The aim of the image preprocessing stage is to increase both the accuracy and the interpretability of the digital data during the image processing stage. The preprocess-

(a)   **(b)**   **(c)**

(d)   **(e)**   **(f)**

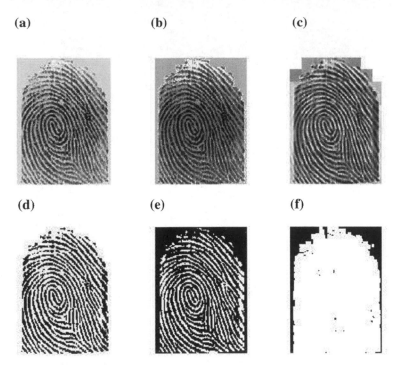

Fig. 3 Preprocessing steps: **a**. Real altered fingerprint image; **b**. Image obtain after Histogram Equalization; **c**.Image obtained after FFT; **d**. Binary image; **e**. Distance transform image; **f**. Region of interest.

ing takes place prior to any principal component analysis. The main steps for preprocessing are enhancement, binarization, distance transform, and segmentation.

3.3 Image Enhancement

The fingerprint input image is enhanced in the spatial domain by applying the Histogram Equalization technique, for a better distribution of the pixel values. Considering the real altered fingerprint from Fig. 3a, the result image is represented in Figs. 3b and 4)

3.4 Histogram Equalization

The histogram of a digital image with gray levels in the range $[0, L-1]$ is a discrete function:

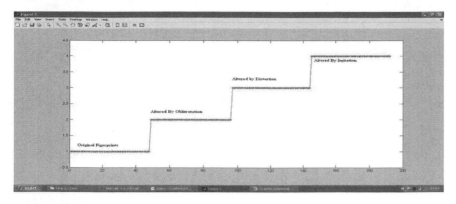

Fig. 4 Classification of fingerprint using fuzzy classification

$$p\ (r_k) = \frac{n_k}{n};\qquad\qquad\qquad (1)$$

where r_k represents the k 'th gray level, n_k is the number of pixels in the image with that gray level, n represents the total number of pixels in the image, $k = 1,1,\ldots L, L = 256$.

3.5 *Binarisation*

Most minutiae extraction algorithms operate on basically binary images where there are only two levels of interest: The black pixels represent ridges, and the white pixels represent valleys. Binarisation [4] converts a greylevel image into a binary image. This helps in improving the contrast between the ridges and valleys in a fingerprint image, and consequently facilitates the extraction of minutiae.

4 Singular Point Detection

Singular point detection is the most challenging task; it is an important process for fingerprint image alignment, fingerprint classification, and fingerprint matching. In the following subsections, we propose orientation reliability and singular point position methods.

5 Orientation Field Reliability Map

The fingerprint image is made up of pattern of ridges and valleys; they are the replica of the human fingertips. The fingerprint image represents a system of oriented texture and has very rich structural information within the image. This flow-like pattern forms an orientation field extracted from the style of valleys and ridges. In the large part of fingerprint topologies, the orientation field is quite smooth, while in some areas, the orientation appears in a discontinuous manner. These regions are called singularity or singular points, including core and delta and are defined as the centers of those areas. In addition, the reference point is defined here as the point with maximum curvature on the convex ridge. The reliability of the orientation filed describes the consistency of the local orientations in a neighborhood along the dominant orientation is used to locate the unique reference point constantly for all types of fingerprints.

6 Experiment Result

In this paper, the singular point is defined as the point with maximum curvature on the convex ridge. In natural fingerprints, the reliability orientation image generally has one or two sharp point, while in altered fingerprints more points are detected with smaller values. Starting from this observation, the altered fingerprint analysis can be done using the density and the count of the singular points.

The proposed algorithm for fingerprint analysis based on the estimation of orientation field and the computation of the reliability was tested with real altered fingerprint and simulated altered fingerprint obtained from natural fingerprint images by using synthetic method due to unavailability of altered fingerprint database.

For the classification, we capture 49 fingerprints of friends and altered them synthetically. On training, it will be classified according to the type of alteration. The graph will show the classification .

7 Conclusion and Future Scope

The proposed system design will be tested using altered fingerprints synthesized in the way typically observed in operational cases with good performance.

This paper proposes a method to consistently and precisely locate the singular points (core and delta) in fingerprint images. The method applied is based on the enhanced fingerprint image orientation reliability. Classification of three different types of altered fingerprint using fuzzy clustering method, the false acceptance rate is 16.7 % and false rejection rate is 13.8 % of 192 fingerprint database. In context with K-Mean classification on the same database, the false acceptance rate is 20.7 %

and false rejection rate is 13.4 %. Classification using fuzzy clustering is batter for false acceptance rate and 0.4 % less for false rejection rate.

The current altered fingerprint detection algorithm can be improved along the following directions:

1. Reconstruct altered fingerprints. For some types of fingerprints where the ridge patterns are damaged locally or the ridge structure is still present on the finger but possibly at a different location, reconstruction is indeed possible.
2. Match altered fingerprints to their unaltered mates. A matcher specialized for altered fingerprints can be developed to link them to unaltered mates in the database utilizing whatever information is available in the altered fingerprints.

In this paper, we use the synthetic fingerprints due to the unavailability of public databases, but this method will give a way for researcher to do in the fields of altered fingerprint recognition.

References

1. Jain, A.K., Yoon, S., Feng, J.: Altered fingerprints: analysis and detection. IEEE Trans. Pattern Anal. Mach. Intell. **34**, 451–464 (2012)
2. Singh, K.: Altered Fingerprints. http://www.interpol.int/Public/Forensic/fingerprints/research/alteredfingerprints.pdf (2008)
3. Jain, A.K.: Surgically altered fingerprints help woman evade immigration. http://abcnews.go.com (2009)
4. Cummins, H.: Attempts to alter and obliterate fingerprints. J. Am. Inst. Crim. Law Criminol. **25**, 982–991 (1935)

Optimal Advertisement Planning for Multi Products Incorporating Segment Specific and Spectrum Effect of Different Medias

Sugandha Aggarwal, Remica Aggarwal and P. C. Jha

Abstract A large part of any firm's investment goes in advertising and therefore planning of an appropriate media for advertisement is the need of today so as to achieve the best returns in terms of wider reach over potential market. In this paper, we deal with a media planning problem for multiple products of a firm in a market which is segmented geographically into various regional segments with diverse language and cultural base. As such each of these regional segments responds to regional advertising as well as national advertising which reaches them with a fixed spectrum. The objective is to plan an advertising media (national and regional media) for multiple products in such a way that maximizes the total reach which is measured through each media exclusively as well as through their combined impact. The problem is formulated as a multi-objective programming problem and solved through goal programming technique. A real life case is provided to illustrate the applicability of the proposed model.

Keywords Multiple products · Market segmentation · National advertising · Regional advertising · Spectrum effect · Media planning · Mathematical programming

S. Aggarwal (✉) · P. C. Jha
Department of Operational Research, University of Delhi, Delhi, India
e-mail: sugandha_or@yahoo.com

R. Aggarwal
Department of Management, Birla Institute of Technology and Science, Pilani, India
e-mail: remica_or@rediffmail.com

P. C. Jha
e-mail: jhapc@yahoo.com

B. V. Babu et al. (eds.), *Proceedings of the Second International Conference on Soft Computing for Problem Solving (SocProS 2012), December 28–30, 2012*, Advances in Intelligent Systems and Computing 236, DOI: 10.1007/978-81-322-1602-5_140, © Springer India 2014

Abbreviations

. Max-maximize
. Min-minimize
. s.t.-subject to
. GP-goal programming

1 Introduction

Companies ranging from large multinational corporations to small retailers rely on different forms of promotion to market their products and services. Advertising is one of the most important forms and is done through various media sources such as print media and electronic media. Planning a suitable media source includes appropriate selection of advertising media and development and allocation of advertising budget. Selection of media largely depends on the target potential market in which product needs to be advertised. The target market could be a single/sole market or may be divided into several segments based on certain criteria. These criteria are defined either geographically, psycho-graphically, sociologically, etc. Take for example geographic segmentation of target market. Here, market is segmented according to geographic criteria—nations, states, regions, countries, cities, neighborhoods, zip codes, population density, or climate. This paper considers a geographically segmented market in which two types of advertising is done, national advertising and regional advertising. National advertising simultaneously reaches all the geographically segmented regions of the country. Various sources of national advertising include national newspapers, national television channels, magazines, websites, etc. Whereas regional advertising targets a specific regional segment of the country through regional newspapers, regional radio channels, regional TV channels, etc.

Regional advertising targets a more concentrated customer base while national advertising has the potential to bring in larger amounts of customers. As national advertisement reaches masses it creates an effectiveness segment-spectrum, which is distributed over various regional segments of the country. Thus, in each region, people get affected by regional advertisements as well as national advertisements with a spectrum effect. Such concept of advertising holds well in a country like India which has a large cultural and language base. Companies advertise their products in Indian market by giving ad messages in national language as well as people belonging to a particular region are emotionally targeted by regional advertising in their native language. In this way they relate themselves with the product in a better way. For example, Hindustan Unilever Limited (HUL), India's largest consumer goods company prepares different versions of ads for its consumable products. They advertise their products in Hindi language national television channels such as Zee TV, Doordarshan, Star Plus, Life OK, etc., to create a mass demand all over the country. Advertisements floating on these channels create a spectrum effect in several regions of country as it is viewed by people all over India. Along with it, the same advertisement is dubbed in various state dominant regional languages, such as Gujarati, Telugu, Tamil, Kannada, Punjabi, Marathi, etc., and telecasted in regional channels also. In this way they try to reach maximum number of people all over the country.

In this chapter, both national as well as regional media are considered as a source of advertising multiple products in different regions. A media planning problem is formulated for allocating total budget available among several national and regional media that are found suitable for advertising the products, maximizing total reach in all the regions where each region's people get affected by regional advertisement corresponding to that region as well as national advertisement with a fixed spectrum. Goal programming technique is used to solve the formulated problem.

The chapter is organized as follows: Section 2 gives a brief literature review. Section 3 formulates the media planning problem for multiple products. Solution procedure for the formulated problem has been given in Sect. 4. A real life case problem is presented in Sect. 5 which clearly illustrates solution methodology. Concluding remarks are made in Sect. 6.

2 Literature Review

A plethora of work has been done on media planning problems. Charnes et al. [1] introduced a GP model for media selection to address problems associated with the critical advertising measurement of frequency and reach. A similar kind of problem was introduced by Lee [2] as well. De Kluyver [3] proposed the more realistic use of hard and soft constraints for linear programming model used in media selection. An integer GP model was developed by Keown and Duncan [4] to solve media selection problems and improved upon sub-optimal results produced by linear programming and non-integer GP models. Some of the prominent literature relating multi-criteria decision making with media selection includes the works by Moynihan et al. [5] and Fructhter and Kalish [6]. They contended that the mathematical requirements of the MCDM model for media selection force the media planner to create an artificial structuring of the media selection criteria. An information resource planning using an analytic hierarchy process based GP model was developed by Lee and Kwak [7]. Mihiotis and Tsakiris [8] discussed the best possible combination of placements of commercial with the goal of the highest rating subject to constrained advertisement budgets. Kwak et al. [9] presented a case study that considers two options: Industrial and consumer products. In order to resolve the strategic decision making about dual market high technology products, a mixed integer GP model is developed to facilitate the advertising media selection process. A chance constraints GP model for the advertising planning problem was proposed by U.K. Bhattacharya [10] to decide the number of advertisement in different advertising media and the optimal allocation of the budget assigned to the different media so as to maximize the total reach. Jha et al. [11, 12] formulated a media planning problem for allocating the available budget among the multiple media that are found suitable for the advertising of single/multiple products considering marketing segmentation aspect of advertising. Spectrum effect of an advertising channel has been discussed recently by Burrato et al. [13] assuming that an advertising channel has an effectiveness segment-spectrum, which is distributed over potential market segments and solutions to deterministic optimal control problems were obtained. However, the above models have not included the segment specific and spectrum effect of different media in media allocation problems. In the following section, a model has been developed which deals with optimal allocation of budget for multiple products advertised through different regional and national media in order to maximize national and regional reach considering the spectrum effect of national media in the regional segments.

3 Model Development

The model is based on an Indian company which manufactures a range of consumable products. It advertises its products through national and regional media in four geographically segmented regions. The media considered are print media (different types of newspapers) and different channels of television. The total budget (A) available is fixed. It is desired to find optimal number of advertisements of different products to be allocated to different national and regional media that would maximize the total advertising reach of all products in each segment of the market, under cost budget and frequency constraints. The advertisements is carried through a national newspaper and a national channel of television alongwith two regional newspapers and two regional channels of each region. Each of these media (regional as well as national) have many suboptions, i.e., advertisements can be given in front page (FP) or other pages (OP) of a newspaper. Similarly considering a television channel, advertising can be done in prime time (PT) as well as in other time (OT).

$p^r_{is_{c}jk_{ij}}$ $C^r_{ijk_{ij}l_{ij}}$ $X^r_{ijk_{ij}}$ $\underline{P}_{is_c jkij}$ $\underline{C}_{ijk_{ij}l_{ij}}$ $C_{ijk_{ij}l_{ij}}$ $C_{ijk_{ij}l_{ij}}$ $\forall i = 0, 1, \ldots, 4$ Let is the total circulation/average number of viewers of national (for $i = 0$) and ith regional segment's jth media, k_{ij}th media option and l_{ij}th slot respectively (where $j = 1$ represents newspaper, $j = 2$ represents television; $k_{ij} = 1, 2$ $\forall i = 1, 2, \ldots, 4$; $j = 1, 2$ as in case of regional segments, for each media two media options is considered in each region; $k_{0j} = 1$ for national media as only one type of national newspaper and TV channel is considered; $l_{ij} = 1, 2$ $\forall i = 0, 1, \ldots, 4$; $j = 1, 2$ represents slots, i.e., front page and other page in newspapers and prime time and other time in television). Let $\underline{C}_{ijk_{ij}l_{ij}}$ is circulation common to national media for ith regional segment's jth media, k_{ij}th media option, l_{ij}th slot. It is assumed that media chosen for advertisement for all the products is same. Let $C^r_{ijk_{ij}l_{ij}}$ is advertisement cost of inserting one advertisement for rth product in national/ith regional segment's jth media, k_{ij}th media option and l_{ij}th slot. It has been assumed that the cost of advertisements at any pages other than the front page is the same. The readership profile/viewership profile represents the percentage of people in different criteria who are reading or viewing the advertisement. $p^r_{is_{c}jk_{ij}}$ is percentage of people for rth product who fall under s_cth criteria and read/view national/ith regional segment's jth media, k_{ij}th media option, and l_{ij}th slot.

Let $\underline{p}^r_{is_c jk_{ij}}$ is percentage of people common to national media for rth product who fall under s_cth criteria and read/view national/ith regional segment's jth media, k_{ij}th media option, and l_{ij}th slot. Different criteria for specifying the target population to whom the advertisement should reach can be age, income, education, etc. Let w_{cj} be the weight corresponding to s_cth criteria, jth media. Let α_{ij} be the spectrum effect of jth national media on ith regional segment.

Also let $x^r_{ijk_{ij}l_{ij}}$ be decision variable corresponding number of advertisements of rth product in national/ith regional segment's jth media, k_{ij}th media option, and l_{ij}th slot. Let Z_0 be reach objective corresponding to national media and Z_i (for $i = 1, 2, \ldots 4$) be exclusive reach objective for ith segment's regional media. Exclusive as in each region, there are common people who read/view both national and regional media and these people get subtracted from the total reach of that region as they are already included in the total national reach objective. Thus, problem for finding optimal number of advertisements of different products to be allocated to different media that would maximize advertising reach at national and regional level under cost–budget and advertisement frequency constraint can be formulated as:

$$
\text{Max}
\begin{cases}
Z_0 = \displaystyle\sum_{r=1}^{2}\sum_{j=1}^{2}\sum_{l_{0j}=1}^{2} \alpha_{1j}\left[\left(\left(\sum_{c=1}^{q} w_{cj}\, p^r_{0j1l_{0j}}\right)C_{0j1l_{0j}}\right)\right]x^r_{0j1l_{0j}} \\[2ex]
\quad + \displaystyle\sum_{r=1}^{2}\sum_{j=1}^{2}\sum_{l_{0j}=1}^{2} \alpha_{2j}\left[\left(\left(\sum_{c=1}^{q} w_{ci}\, p^r_{0j1l_{0j}}\right)C_{0j1l_{0j}}\right)\right]x^r_{0j1l_{0j}} \\[2ex]
\quad + \displaystyle\sum_{r=1}^{2}\sum_{j=1}^{2}\sum_{l_{j}=1}^{2} \alpha_{3j}\left[\left(\left(\sum_{c=1}^{q} w_{cj}\, p^r_{0j1l_{0j}}\right)C_{0j1l_{0j}}\right)\right]x^r_{0j1l_{0j}} \\[2ex]
\quad + \displaystyle\sum_{r=1}^{2}\sum_{j=1}^{2}\sum_{l_{j}=1}^{2} \alpha_{4j}\left[\left(\left(\sum_{c=1}^{q} w_{cj}\, p^r_{0j1l_{0j}}\right)C_{0j1l_{0j}}\right)\right]x^r_{0j1l_{0j}} \\[2ex]
Z_1 = \displaystyle\sum_{r=1}^{2}\sum_{j=1}^{2}\sum_{k_{1j}=1}^{2}\sum_{l_{1j}=1}^{2}\left[\left(\sum_{c=1}^{q} w_{cj}\, p^r_{1s_c jk_{1j}l_{1j}}\right)C_{1jk_{1j}l_{1j}}\right. \\[2ex]
\qquad \left. -\left(\sum_{c=1}^{q} w_{cj}\,\underline{p}^r_{1s_c jk_{1j}l_{1j}}\right)\underline{C}_{1jk_{1j}l_{1j}}\right]x^r_{1jk_{1j}l_{1j}} \\[2ex]
Z_2 = \displaystyle\sum_{r=1}^{2}\sum_{j=1}^{2}\sum_{k_{2j}=1}^{2}\sum_{l_{2j}=1}^{2}\left[\left(\sum_{c=1}^{q} w_{cj}\, p^r_{2s_c jk_{2j}l_{2j}}\right)C_{2jk_{2j}l_{2j}}\right. \\[2ex]
\qquad \left.\left(\sum_{c=1}^{q} w_{cj}\,\underline{p}^r_{2s_c jk_{2j}l_{2j}}\right)\underline{C}_{2jk_{2j}l_{2j}}\right]x^r_{2jk_{2j}l_{2j}} \\[2ex]
Z_3 = \displaystyle\sum_{r=1}^{2}\sum_{j=1}^{2}\sum_{k_{3j}=1}^{2}\sum_{l_{3j}=1}^{2}\left[\left(\sum_{c=1}^{q} w_{cj}\, p^r_{3s_c jk_{3j}l_{3j}}\right)C_{3jk_{3j}l_{3j}}\right. \\[2ex]
\qquad \left. -\left(\sum_{c=1}^{q} w_{cj}\,\underline{p}^r_{3s_c jk_{3j}l_{3j}}\right)\underline{C}_{3jk_{3j}l_{3j}}\right]x^r_{3jk_{3j}l_{3j}} \\[2ex]
Z_4 = \displaystyle\sum_{r=1}^{2}\sum_{j=1}^{2}\sum_{k_{4j}=1}^{2}\sum_{l_{4j}=1}^{2}\left[\left(\sum_{c=1}^{q} w_{cj}\, p^r_{4s_c jk_{4j}l_{4j}}\right)C_{4jk_{4j}l_{4j}}\right. \\[2ex]
\qquad \left. -\left(\sum_{c=1}^{q} w_{cj}\,\underline{p}^r_{4s_c jk_{4j}l_{4j}}\right)\underline{C}_{4jk_{4j}l_{4j}}\right]x^r_{4jk_{4j}l_{4j}}
\end{cases}
$$

$$\text{s.t.} \sum_{r=1}^{2}\sum_{j=1}^{2}\sum_{l_{0j}=1}^{2} c^r_{0j1l_{0j}} x^r_{0j1l_{0j}} + \sum_{r=1}^{2}\sum_{i=1}^{4}\sum_{j=1}^{2}\sum_{k_{ij}=1}^{2}\sum_{l_{ij}=1}^{2} c^r_{ijk_{ij}l_{ij}} x^r_{ijk_{ij}l_{ij}} \le A \quad (P1)$$

$$\sum_{r=1}^{2}\sum_{j=1}^{2}\sum_{l_{0j}=1}^{2} c^r_{0j1l_{0j}} x^r_{0j1l_{0j}} \ge vA$$

$$\left.\begin{array}{l} x^r_{0j1l_{0j}} \ge t^r_{0j1l_{0j}}, x^r_{0j1l_{0j}} \le u^r_{0j1l_{0j}} \\ x^r_{0j1l_{0j}} \ge 0 \text{ and integers} \end{array}\right\} \forall r = 1,2; j = 1,2; l_{0j} = 1,2$$

$$\left.\begin{array}{l} x^r_{ijk_{ij}l_{ij}} \ge t^r_{ijk_{ij}l_{ij}}, x^r_{ijk_{ij}l_{ij}} \le u^r_{ijk_{ij}l_{ij}} \\ x^r_{ijk_{ij}l_{ij}} \ge 0 \text{ and integers} \end{array}\right\} \forall r = 1,2; i = 1,2,...,4;$$

$$j = 1,2; l_{ij} = 1,2; k_{ij} = 1,2$$

where v is the minimum proportion of the total budget required to be allocated to national media $t^r_{ijk_{ij}l_{ij}}$ is the minimum number of advertisements of rth product in national/ith regional segment's jth media, k_{ij}th media option, and l_{ij}th slot, respectively. Similarly $u^r_{ijk_{ij}l_{ij}}$ is the maximum number of advertisements of rth product in national/ith regional segment's jth media, k_{ij}th media option, and l_{ij}th slot.

The optimal value of Z_0, Z_1, Z_2, Z_3, Z_4 obtained on solving the above problem can be set as individual aspiration level to be achieved from different segments. Problem (P1) can then be written as a multi-objective programming problem.

$$\text{Max}\left\{\begin{array}{l}
Z_0 = \sum_{r=1}^{2}\sum_{j=1}^{2}\sum_{l_{0j}=1}^{2} \alpha_{1j}\left[\left(\left(\sum_{c=1}^{q} w_{cj} p^r_{0j1l_{0j}}\right) C_{0j1l_{0j}}\right)\right] x^r_{0j1l_{0j}} \\
+ \sum_{r=1}^{2}\sum_{j=1}^{2}\sum_{l_{0j}=1}^{2} \alpha_{2j}\left[\left(\left(\sum_{c=1}^{q} w_{cj} p^r_{0j1l_{0j}}\right) C_{0j1l_{0j}}\right)\right] x^r_{0j1l_{0j}} \\
+ \sum_{r=1}^{2}\sum_{j=1}^{2}\sum_{l_{j}=1}^{2} \alpha_{3j}\left[\left(\left(\sum_{c=1}^{q} w_{cj} p^r_{0j1l_{0j}}\right) C_{0j1l_{0j}}\right)\right] x^r_{0j1l_{0j}} \\
+ \sum_{r=1}^{2}\sum_{j=1}^{2}\sum_{l_{j}=1}^{2} \alpha_{4j}\left[\left(\left(\sum_{c=1}^{q} w_{cj} p^r_{0j1l_{0j}}\right) C_{0j1l_{0j}}\right)\right] x^r_{0j1l_{0j}} \\
Z_1 = \sum_{r=1}^{2}\sum_{j=1}^{2}\sum_{k_1j=1}^{2}\sum_{l_1j=1}^{2} \left[\left(\sum_{c=1}^{q} w_{cj} p^r_{1scjk_1jl_1j}\right) C_{1jk_1jl_1j} \\
- \left(\sum_{c=1}^{q} w_{cj} \underline{p}^r_{1scjk_1jl_1j}\right) \underline{C}_{1jk_1jl_1j}\right] x^r_{1jk_1jl_1j} \\
Z_2 = \sum_{r=1}^{2}\sum_{j=1}^{2}\sum_{k_2j=1}^{2}\sum_{l_2j=1}^{2} \left[\left(\sum_{c=1}^{q} w_{cj} p^r_{2scjk_2jl_2j}\right) C_{2jk_2jl_2j} \\
- \left(\sum_{c=1}^{q} w_{cj} \underline{p}^r_{2scjk_2jl_2j}\right) \underline{C}_{2jk_2jl_2j}\right] x^r_{2jk_2jl_2j} \\
Z_3 = \sum_{r=1}^{2}\sum_{j=1}^{2}\sum_{k_3j=1}^{2}\sum_{l_3j=1}^{2} \left[\left(\sum_{c=1}^{q} w_{cj} p^r_{3scjk_3jl_3j}\right) C_{3jk_3jl_3j} \\
- \left(\sum_{c=1}^{q} w_{cj} \underline{p}^r_{3scjk_3jl_3j}\right) \underline{C}_{3jk_3jl_3j}\right] x^r_{3jk_3jl_3j} \\
Z_4 = \sum_{r=1}^{2}\sum_{j=1}^{2}\sum_{k_4j=1}^{2}\sum_{l_4j=1}^{2} \left[\left(\sum_{c=1}^{q} w_{cj} p^r_{4scjk_4jl_4j}\right) C_{4jk_4jl_4j} \\
- \left(\sum_{c=1}^{q} w_{cj} \underline{p}^r_{4scjk_4jl_4j}\right) \underline{C}_{4jk_4jl_4j}\right] x^r_{4jk_4jl_4j}
\end{array}\right.$$

$$\text{s.t.} \sum_{r=1}^{2}\sum_{j=1}^{2}\sum_{l_{0j}=1}^{2} c^r_{0j1l_{0j}} x^r_{0j1l_{0j}} + \sum_{r=1}^{2}\sum_{i=1}^{4}\sum_{j=1}^{2}\sum_{k_{ij}=1}^{2}\sum_{l_{ij}=1}^{2} c^r_{ijk_{ij}l_{ij}} x^r_{ijk_{ij}l_{ij}} \leq A \quad \text{(P2)}$$

$$\sum_{r=1}^{2}\sum_{j=1}^{2}\sum_{l_{0j}=1}^{2} c^r_{0j1l_{0j}} x^r_{0j1l_{0j}} \geq vA$$

$$Z_0 \geq Z_0^*,\ Z_1 \geq Z_1^*,\ Z_2 \geq Z_2^*,\ Z_3 \geq Z_3^*,\ Z_4 \geq Z_4^*$$

$$\left.\begin{array}{l} x^r_{0j1l_{0j}} \geq t^r_{0j1l_{0j}},\ x^r_{0j1l_{0j}} \leq u^r_{0j1l_{0j}} \\[4pt] x^r_{0j1l_{0j}} \geq 0 \text{ and integers} \end{array}\right\} \forall r = 1,2;\ j = 1,2;\ l_{0j} = 1,2$$

$$\left.\begin{array}{l} x^r_{ijk_{ij}l_{ij}} \geq t^r_{ijk_{ij}l_{ij}},\ x^r_{ijk_{ij}l_{ij}} \leq u^r_{ijk_{ij}l_{ij}}, \\[4pt] x^r_{ijk_{ij}l_{ij}} \geq 0 \text{ and integers} \end{array}\right\} \forall r = 1,2;\ i = 1,2,\ldots,4;$$

$$j = 1,2;\ l_{ij} = 1,2;\ k_{ij} = 1,2$$

Problem (P2) leads to an infeasible solution. GP approach has been used to obtain a feasible compromised solution to the problem.

4 Solution Methodology: GP

In a simpler version of GP, management sets goals and relative importance (weights) for different objectives. Then an optimal solution is defined as one that minimizes both positive and negative deviations from set goals simultaneously or minimizes the amount by which each goal can be violated. First, we solve the problem using rigid constraints only and then the goals of objectives are incorporated depending upon whether priorities or relative importance of different objectives are well defined or not. The problem (P2) can be solved in two stages as follows:

Stage 1

$$\textbf{Min } g_0\ (\eta, \rho, X) = \rho_1 + \eta_2 + \sum_{r=1}^{2}\sum_{j=1}^{2}\sum_{l_{0j}=1}^{2} \eta^r_{0j1l_{0j}} + \sum_{r=1}^{2}\sum_{i=1}^{4}\sum_{j=1}^{2}\sum_{k_{ij}=1}^{2}\sum_{l_{ij}=1}^{2} \eta^r_{ijk_{ij}l_{ij}}$$

$$+ \sum_{r=1}^{2}\sum_{j=1}^{2}\sum_{l_{0j}=1}^{2} \rho''^r_{0j1l_{0j}} + \sum_{r=1}^{2}\sum_{i=1}^{4}\sum_{j=1}^{2}\sum_{k_{ij}=1}^{2}\sum_{l_{ij}=1}^{2} \rho''^r_{ijk_{ij}l_{ij}}$$

$$\text{s.t.} \sum_{r=1}^{2}\sum_{j=1}^{2}\sum_{l_{0j}=1}^{2} c^r_{0j1l_{0j}} x^r_{0j1l_{0j}} + \sum_{r=1}^{2}\sum_{i=1}^{4}\sum_{j=1}^{2}\sum_{k_{ij}=1}^{2}\sum_{l_{ij}=1}^{2} c^r_{ijk_{ij}l_{ij}} x^r_{ijk_{ij}l_{ij}} + \eta_1 - \rho_1 = A$$

$$\text{(P3)}$$

$$\sum_{r=1}^{2}\sum_{j=1}^{2}\sum_{l_{0j}=1}^{2} c_{0j1l_{0j}}^{r}x_{0j1l_{0j}}^{r} + \eta_2 - \rho_2 = vA$$

$$\left.\begin{array}{l} x_{0j1l_{0j}}^{r} + \eta_{0j1l_{0j}}^{r} - \rho_{0j1l_{0j}}^{r} = t_{0j1l_{0j}}^{r}, x_{0j1l_{j}} + \eta_{0j1l_{0j}}^{\prime r} - \rho_{0j1l_{0j}}^{\prime r} = u_{0j1l_{j}} \\ x_{0j1l_{0j}}^{r} \geq 0 \text{ and integers} \\ \eta_{0j1l_{0j}}^{r}, \rho_{0j1l_{0j}}^{r}, \eta_{0j1l_{0j}}^{\prime r}, \rho_{0j1l_{0j}}^{\prime r} \geq 0 \end{array}\right\}$$

$$\forall r = 1, 2; j = 1, 2; l_{0j} = 1, 2$$

$$\left.\begin{array}{l} x_{ijk_{ij}l_{ij}}^{r} + \eta_{ijk_{ij}l_{ij}}^{r} - \rho_{ijk_{ij}l_{ij}}^{r} = t_{ijk_{ij}l_{ij}}^{r}, x_{ijk_{ij}l_{ij}}^{r} + \eta_{ijk_{ij}l_{ij}}^{\prime r} - \rho_{ijk_{ij}l_{ij}}^{\prime r} = u_{ijk_{ij}l_{ij}}^{r}, \\ x_{ijk_{ij}l_{ij}}^{r} \geq 0 \text{ and integers} \\ \eta_{ijk_{ij}l_{ij}}^{r}, \rho_{ijk_{ij}l_{ij}}^{r}, \eta_{ijk_{ij}l_{ij}}^{\prime r}, \rho_{ijk_{ij}l_{ij}}^{\prime r} \geq 0, \end{array}\right\}$$

$$\eta_1, \rho_1, \eta_2, \rho_2 \geq 0$$

$$\forall r = 1, 2; j = 1, 2; l_{ij} = 1, 2; k_{ij} = 1, 2$$

where η and ρ are the over and under achievement (positive and negative deviational) variables of the goals for their respective objective/constraint function. $g_0(\eta, \rho, X)$ is the goal objective function corresponding to rigid constraints. Let (η^0, ρ^0, X^0) be the optimal solution for the problem (P3) and $g_0(\eta^0, \rho^0, X^0)$ be the corresponding objective function value then final problem can be formulated using the optimal solution of the problem (P3) through the problem (P2).

Stage 2

$$\textbf{Min } g(\eta, \rho, X) = \lambda_0 \eta_0 + \sum_{i=1}^{4} \lambda_{si}\eta_{si}$$

$$\textbf{s.t.}\sum_{r=1}^{2}\sum_{j=1}^{2}\sum_{l_{0j}=1}^{2} c_{0j1l_{0j}}^{r}x_{0j1l_{0j}}^{r} + \sum_{r=1}^{2}\sum_{i=1}^{4}\sum_{j=1}^{2}\sum_{k_{ij}=1}^{2}\sum_{l_{ij}=1}^{2} c_{ijk_{ij}l_{ij}}^{r}x_{ijk_{ij}l_{ij}}^{r} + \eta_1 - \rho_1 = A$$

$$\text{(P4)}$$

$$\sum_{r=1}^{2}\sum_{j=1}^{2}\sum_{l_{0j}=1}^{2} c_{0j1l_{0j}}^{r}x_{0j1l_{0j}}^{r} + \eta_2 - \rho_2 = vA$$

$$Z_0 + \eta_0 - \rho_0 = Z_0^*, \ Z_i + \eta_{si} - \rho_{si} = Z_i^* \quad \forall i = 1, 2, \ldots, 4$$

$$\left.\begin{array}{l} x_{0j1l_{0j}}^{r} + \eta_{0j1l_{0j}}^{r} - \rho_{0j1l_{0j}}^{r} = t_{0j1l_{0j}}^{r}, x_{0j1l_{j}} + \eta_{0j1l_{0j}}^{\prime r} - \rho_{0j1l_{0j}}^{\prime r} = u_{0j1l_{0j}}^{r} \\ x_{0j1l_{0j}}^{r} \geq 0 \text{ and integers} \\ \eta_{0j1l_{0j}}^{r}, \rho_{0j1l_{0j}}^{r}, \eta_{0j1l_{0j}}^{\prime r}, \rho_{0j1l_{0j}}^{\prime r} \geq 0 \end{array}\right\}$$

$$\forall r = 1, 2; j = 1, 2; l_{0j} = 1, 2$$

$$x^r_{ijk_{ij}l_{ij}} + \eta^r_{ijk_{ij}l_{ij}} - \rho^r_{ijk_{ij}l_{ij}} = t^r_{ijk_{ij}l_{ij}}, x^r_{ijk_{ij}l_{ij}} + \eta^{\prime r}_{ijk_{ij}l_{ij}} - \rho^{\prime r}_{ijk_{ij}l_{ij}} = u^r_{ijk_{ij}l_{ij}},$$

$$x^r_{ijk_{ij}l_{ij}} \geq 0 \text{ and integers}$$

$$\eta^r_{ijk_{ij}l_{ij}}, \rho^r_{ijk_{ij}l_{ij}}, \eta^{\prime r}_{ijk_{ij}l_{ij}}, \rho^{\prime r}_{ijk_{ij}l_{ij}} \geq 0,$$

$$g_0(\eta, \rho, X) = g_0\left(\eta^0, \rho^0, X^0\right)$$

$$\forall r = 1, 2; i = 1, 2, \ldots, 4; j = 1, 2; l_{ij} = 1, 2; k_{ij} = 1, 2$$

$$\eta_0, \rho_0, \eta_1, \rho_1, \eta_2, \rho_2 \geq 0$$

$$\eta_{si}, \rho_{si} \geq 0 \quad \forall i = 1, 2, \ldots, 4$$

$g(\eta, \rho, X)$ is objective function of the problem (P3). Goal programming is used to find a compromise solution.

5 Case Problem

A company is considered which has to advertise its two types of consumable products, targeting middle income group people in its four selected regions. Advertisement is done in national as well as regional newspapers and television channels. The fixed budget available for advertising is ₹ 2,50,00,000. It is desired to allocate at least a minimum of 35 % of the total budget in national media. Data corresponding to circulation figures for newspapers, average number of viewers for TV channels, advertisement costs in various newspapers, and TV channels are given below in Tables 1, 2, 3, and 4, respectively. Readership profile and Viewership profile matrix based on random sample of size 200 is listed in Tables 5 and 6. Finally, the minimum and maximum number of advertisements in various Newspapers and Television

Table 1 Circulation figure for Newspapers ('10000)

NN	Region 1				Region 2				Region 3				Region 4			
	RNP1		RNP2		RNP1		RNP2		RNP1		RNP2		RNP1		RNP2	
	T	C	T	C	T	C	T	C	T	C	T	C	T	C	T	C
180	52	17	45	15	36	12	25	6	42	15	41	13	34	13	32	9

Table 2 Circulation figure for TV channels ('100000)

	NCH	Region 1				Region 2				Region 3				Region 4			
		RCH1		RCH2		RCH1		RCH2		RCH1		RCH2		RCH1		RCH2	
		T	C	T	C	T	C	T	C	T	C	T	C	T	C	T	C
PT	600	110	40	100	30	140	60	130	40	110	40	130	50	120	50	150	70
OT	400	80	30	60	28	90	40	80	35	70	20	90	30	70	25	100	40

Table 3 Advertisement cost (per 100 column cm) in Newspapers

		NN	Region 1		Region 2		Region 3		Region 4	
			RNP1	RNP2	RNP1	RNP2	RNP1	RNP2	RNP1	RNP2
P1	FP	1900	820	750	480	450	690	650	400	410
	OP	1300	550	400	225	200	400	380	220	210
P2	FP	1800	700	630	450	410	640	580	390	350
	OP	1100	460	350	210	200	340	320	175	170

Table 4 Advertisement cost (per 10 seconds spot) in TV Channels ('1000)

		NCH	Region 1		Region 2		Region 3		Region 4	
			RCH1	RCH2	RCH1	RCH2	RCH1	RCH2	RCH1	RCH2
P1	PT	120	36	32	41	40	37	39	37	45
	OT	80	16	17	20	18	18	20	14	20
P2	PT	108	33	29	37	36	34	36	34	41
	OT	72	15	16	18	17	17	18	13	18

Table 5 Readership profile matrix for newspapers

	NN	Region 1				Region 2				Region 3				Region 4			
		RNP1		RNP2		RNP1		RNP2		RNP1		RNP2		RNP1		RNP2	
		T	C	T	C	T	C	T	C	T	C	T	C	T	C	T	C
P1 FP		0.52	0.49	0.27	0.45	0.22	0.5	0.28	0.45	0.2	0.46	0.21	0.44	0.22	.46	0.22	0.43	0.16
OP		0.34	0.33	0.2	0.31	0.14	0.33	0.21	0.32	0.13	0.33	0.16	0.33	0.16	0.34	0.14	0.29	0.1
P2 FP		0.45	0.46	0.23	0.42	0.21	0.45	0.23	0.42	0.14	0.43	0.19	0.4	0.2	0.44	0.21	0.41	0.19
OP		0.28	0.3	0.14	0.26	0.15	0.3	0.16	0.26	0.1	0.26	0.13	0.25	0.12	0.28	0.16	0.22	0.13

Table 6 Viewership profile matrix for television

	NCH	Region 1				Region 2				Region 3				Region 4			
		RCH1		RCH2		RCH1		RCH2		RCH1		RCH2		RCH1		RCH2	
		T	C	T	C	T	C	T	C	T	C	T	C	T	C	T	C
P1 PT		0.56	0.48	0.28	0.45	0.25	0.47	0.23	0.5	0.21	0.49	0.25	0.43	0.2	0.44	0.23	0.52	0.29
OT		0.35	0.32	0.16	0.29	0.15	0.31	0.14	0.34	0.15	0.33	0.17	0.29	0.14	0.29	0.16	0.35	0.2
P2 PT		0.42	0.47	0.29	0.43	0.27	0.45	0.22	0.45	0.2	0.42	0.2	0.54	0.3	0.45	0.21	0.4	0.19
OT		0.33	0.37	0.23	0.35	0.2	0.33	0.15	0.36	0.15	0.31	0.16	0.22	0.13	0.3	0.17	0.22	0.1

Channels and the spectrum effect of national media on different regional segments are given in Tables 7, 8, and 9, respectively.

Using the above given data we solve (P1) with each objective subject to constraints individually. The optimal values so obtained for each objective is then set as aspiration level to be achieved for reach corresponding to national media and each of the four segment's regional media. The multi-objective programming problem combining all the objectives and incorporating the individual aspirations obtained can be written as:

Table 7 Minimum and maximum number of advertisements in various Newspapers

		NN	Region 1		Region 2		Region 3		Region 4	
			RNP1	RNP2	RNP1	RNP2	RNP1	RNP2	RNP1	RNP2
P1	FP	[12,20]	[10,16]	[7,12]	[6,15]	[4,12]	[7,15]	[6,15]	[6,12]	[4,15]
	OP	[7,12]	[4,12]	[3,7]	[6,12]	[6,9]	[5,14]	[4,12]	[4,6]	[2,6]
P2	FP	[10,16]	[12,15]	[9,12]	[7,15]	[6,12]	[7,15]	[6,16]	[3,12]	[3,15]
	OP	[7,12]	[7,12]	[3,9]	[3,12]	[3,7]	[7,14]	[3,12]	[3,7]	[2,7]

Table 8 Minimum and maximum advertisement in various TV channels

		NCH	Region 1		Region 2		Region 3		Region 4	
			RCH1	RCH2	RCH1	RCH2	RCH1	RCH2	RCH1	RCH2
P1	PT	[3,12]	[2,8]	[1,6]	[1,9]	[1,5]	[3,8]	[2,10]	[1,8]	[2,6]
	OT	[1,6]	[2,4]	[1,3]	[1,4]	[0,2]	[2,6]	[1,5]	[0,3]	[1,4]
P2	PT	[2,10]	[2,12]	[1,3]	[2,9]	[2,5]	[1,8]	[1,7]	[1,6]	[1,6]
	OT	[1,7]	[3,7]	[1,6]	[2,7]	[0,4]	[2,5]	[0,5]	[0,4]	[0,4]

Table 9 Spectrum effect of national media on different regional segments

	Region 1	Region 2	Region 3	Region 4
National Newspaper	0.27	0.21	0.25	0.22
National Channel	0.21	0.26	0.23	0.27

$$
\text{Max} \begin{bmatrix}
\begin{aligned}
Z_0 ={} & \{0.95(936000x_{0111}^1 + 612000x_{0112}^1 + 810000x_{0111}^2 + 504000x_{0112}^2) \\
& + 0.97(3360000x_{0211}^1 + 1400000x_{0212}^1 + 2520000x_{0211}^2 + 1320000x_{0212}^2)\} \\
Z_1 ={} & \{208900x_{1111}^1 + 137600x_{1112}^1 + 169500x_{1121}^1 + 118500x_{1122}^1 \\
& + 416000x_{1211}^1 + 208000x_{1212}^1 + 375000x_{1221}^1 + 132000x_{1222}^1 \\
& + 200100x_{1111}^2 + 132200x_{1112}^2 + 157500x_{1121}^2 + 94500x_{1122}^2 \\
& + 401000x_{1211}^2 + 227000x_{1212}^2 + 349000x_{1221}^2 + 154000x_{1222}^2\} \\
Z_2 ={} & \{146400x_{2111}^1 + 93600x_{2112}^1 + 100500x_{2121}^1 + 72200x_{2122}^1 \\
& + 520000x_{2211}^1 + 223000x_{2212}^1 + 566000x_{2221}^1 + 219500x_{2222}^1 \\
& + 134400x_{2111}^2 + 88800x_{2112}^2 + 96600x_{2121}^2 + 59000x_{2122}^2 + 498000x_{2211}^2 \\
& + 237000x_{2212}^2 + 505000x_{2221}^2 + 235500x_{2222}^2\} \\
Z_3 ={} & \{161700x_{3111}^1 + 114600x_{3112}^1 + 151800x_{3121}^1 + 114500x_{3122}^1 \\
& + 439000x_{3211}^1 + 197000x_{3212}^1 + 459000x_{3221}^1 + 219000x_{3222}^1 \\
& + 152100x_{3111}^2 + 89700x_{3112}^2 + 138000x_{3121}^2 + 86900x_{3122}^2 + 382000x_{3211}^2 \\
& + 185000x_{3212}^2 + 552000x_{3221}^2 + 159000x_{3222}^2\} \\
Z_4 ={} & \{127800x_{4111}^1 + 97400x_{4112}^1 + 123200x_{4121}^1 + 83800x_{4122}^1 \\
& + 413000x_{4211}^1 + 163000x_{4212}^1 + 577000x_{4221}^1 + 270000x_{4222}^1 \\
& + 122300x_{4111}^2 + 74400x_{4112}^2 + 1141]00x_{4121}^2 + 58700x_{4122}^2 + 435000x_{4211}^2 \\
& + 167500x_{4212}^2 + 467000x_{4221}^2 + 180000x_{4222}^2\}
\end{aligned}
\end{bmatrix}
$$

s.t. $1900x_{0111}^1 + 1300x_{0112}^1 + 120000x_{0211}^1 + 80000x_{0212}^1 + 820x_{1111}^1 + 550x_{1112}^1$

$\quad + 750x_{1121}^1 + 400x_{1122}^1 + 36000x_{1211}^1 + 16000x_{1212}^1 + 32000x_{1221}^1$

$\quad + 17000x_{1222}^1 + 480x_{2111}^1 + 225x_{2112}^1 + 450x_{2121}^1 + 200x_{2122}^1 + 41000x_{2211}^1$

$\quad + 20000x_{2212}^1 + 40000x_{2221}^1 + 18000x_{2222}^1 + 690x_{3111}^1 + 400x_{3112}^1 + 650x_{3121}^1$

$\quad + 380x_{3122}^1 + 37000x_{3211}^1 + 18000x_{3212}^1 + 39000x_{3221}^1 + 20000x_{3222}^1$

$\quad + 400x_{4111}^1 + 220x_{4112}^1 + 410x_{4121}^1 + 210x_{4122}^1 + 37000x_{4211}^1 + 14000x_{4212}^1$

$\quad + 45000x_{4221}^1 + 20000x_{4222}^1 + 1800x_{0111}^2 + 1100x_{0112}^2 + 108000x_{0211}^2$

$\quad + 72000x_{0212}^2 + 700x_{1111}^2 + 460x_{1112}^2 + 630x_{1121}^2 + 350x_{1122}^2 + 33000x_{1211}^2$

$\quad + 15000x_{1212}^2 + 29000x_{1221}^2 + 16000x_{1222}^2 + 450x_{2111}^2 + 210x_{2112}^2 + 410x_{2121}^2$

$\quad + 200x_{2122}^2 + 37000x_{2211}^2 + 18000x_{2212}^2 + 36000x_{2221}^2 + 17000x_{2222}^2$

$\quad + 640x_{3111}^2 + 340x_{3112}^2 + 580x_{3121}^2 + 320x_{3122}^2 + 34000x_{3211}^2 + 17000x_{3212}^2$

$\quad + 36000x_{3221}^2 + 18000x_{3222}^2 + 390x_{4111}^2 + 175x_{4112}^2 + 350x_{4121}^2 + 170x_{4122}^2$

$\quad + 34000x_{4211}^2 + 13000x_{4212}^2 + 41000x_{4221}^2 + 18000x_{4222}^2 \leq 25000000$

$1900x_{0111}^1 + 1300x_{0112}^1 + 120000x_{0211}^1 + 80000x_{0212}^1 + 1800x_{0111}^2 + 1100x_{0112}^2$

$\quad + 108000x_{0211}^2 + 72000x_{0212}^2 \geq .35 \times 25000000$

$Z_0 \geq 71181200,\ Z_1 \geq 23458500,\ Z_2 \geq 18629800,\ Z_3 \geq 23455000,$

$Z_4 \geq 15999600$

$x_{0111}^1 \geq 12,\ x_{0112}^1 \geq 7,\ x_{0211}^1 \geq 3,\ x_{0212}^1 \geq 1,\ x_{1111}^1 \geq 10,\ x_{1112}^1 \geq 4,\ x_{1121}^1 \geq 7,$

$x_{1122}^1 \geq 3,\ x_{1211}^1 \geq 2,\ x_{1212}^1 \geq 2,\ x_{1221}^1 \geq 1,$

$x_{1222}^1 \geq 1,\ x_{2111}^1 \geq 6,\ x_{2112}^1 \geq 6,\ x_{2121}^1 \geq 4,\ x_{2122}^1 \geq 6,\ x_{2211}^1 \geq 1,\ x_{2212}^1 \geq 1,$

$x_{2221}^1 \geq 1,\ x_{2222}^1 \geq 0,\ x_{3111}^1 \geq 7,\ x_{3112}^1 \geq 5,$

$x_{3121}^1 \geq 6,\ x_{3122}^1 \geq 4,\ x_{3211}^1 \geq 3,\ x_{3212}^1 \geq 2,\ x_{3221}^1 \geq 2,\ x_{3222}^1 \geq 1,\ x_{4111}^1 \geq 6,$

$x_{4112}^1 \geq 4,\ x_{4121}^1 \geq 4,\ x_{4122}^1 \geq 2,\ x_{4211}^1 \geq 1,$

$x_{4212}^1 \geq 0,\ x_{4221}^1 \geq 2,\ x_{4222}^1 \geq 1,\ x_{0111}^2 \geq 10,\ x_{0112}^2 \geq 7,\ x_{0211}^2 \geq 2,\ x_{0212}^2 \geq 1,$

$x_{1111}^2 \geq 12,\ x_{1112}^2 \geq 7,\ x_{1121}^2 \geq 9,\ x_{1122}^2 \geq 3,$

$x_{1211}^2 \geq 2,\ x_{1212}^2 \geq 3,\ x_{1221}^2 \geq 1,\ x_{1222}^2 \geq 1,\ x_{2111}^2 \geq 7,\ x_{2112}^2 \geq 3,\ x_{2121}^2 \geq 6,$

$x_{2122}^2 \geq 3,\ x_{2211}^2 \geq 2,\ x_{2212}^2 \geq 2,\ x_{2221}^2 \geq 2,$

$x_{2222}^2 \geq 0,\ x_{3111}^2 \geq 7,\ x_{3112}^2 \geq 7,\ x_{3121}^2 \geq 6,\ x_{3122}^2 \geq 3,\ x_{3211}^2 \geq 1,\ x_{3212}^2 \geq 2,$

$x_{3221}^2 \geq 1,\ x_{3222}^2 \geq 0,\ x_{4111}^2 \geq 3,\ x_{4112}^2 \geq 3,$

$x_{4121}^2 \geq 3,\ x_{4122}^2 \geq 2,\ x_{4211}^2 \geq 1,\ x_{4212}^2 \geq 0,\ x_{4221}^2 \geq 1,\ x_{4222}^2 \geq 0,\ x_{0111}^1 \leq 20,$

$x_{0112}^1 \leq 12,\ x_{0211}^1 \leq 12,\ x_{0212}^1 \leq 6,$

$x_{1111}^1 \leq 16,\ x_{1112}^1 \leq 12,\ x_{1121}^1 \leq 12,\ x_{1122}^1 \leq 7,\ x_{1211}^1 \leq 8,\ x_{1212}^1 \leq 4,\ x_{1221}^1 \leq 6,$

$x_{1222}^1 \leq 3,\ x_{2111}^1 \leq 15,\ x_{2112}^1 \leq 12,$

$x^1_{2121} \leq 12, x^1_{2122} \leq 9, x^1_{2211} \leq 9, x^1_{2212} \leq 4, x^1_{2221} \leq 5, x^1_{2222} \leq 2, x^1_{3111} \leq 15,$

$x^1_{3112} \leq 14, x^1_{3121} \leq 15, x^1_{3122} \leq 12,$

$x^1_{3211} \leq 8, x^1_{3212} \leq 6, x^1_{3221} \leq 10, x^1_{3222} \leq 5, x^1_{4111} \leq 12, x^1_{4112} \leq 6, x^1_{4121} \leq 15,$

$x^1_{4122} \leq 6, x^1_{4211} \leq 8, x^1_{4212} \leq 3,$

$x^1_{4221} \leq 6, x^1_{4222} \leq 4, x^2_{0111} \leq 16, x^2_{0112} \leq 12, x^2_{0211} \leq 10, x^2_{0212} \leq 7, x^2_{1111} \leq 15,$

$x^2_{1112} \leq 12, x^2_{1121} \leq 12, x^2_{1122} \leq 9,$

$x^2_{1211} \leq 12, x^2_{1212} \leq 7, x^2_{1221} \leq 3, x^2_{1222} \leq 6, x^2_{2111} \leq 15, x^2_{2112} \leq 12, x^2_{2121} \leq 12,$

$x^2_{2122} \leq 7, x^2_{2211} \leq 9, x^2_{2212} \leq 7,$

$x^2_{2221} \leq 5, x^2_{2222} \leq 4, x^2_{3111} \leq 15, x^2_{3112} \leq 14, x^2_{3121} \leq 16, x^2_{3122} \leq 12, x^2_{3211} \leq 8,$

$x^2_{3212} \leq 5, x^2_{3221} \leq 7, x^2_{3222} \leq 5,$

$x^2_{4111} \leq 12, x^2_{4112} \leq 7, x^2_{4121} \leq 15, x^2_{4122} \leq 7, x^2_{4211} \leq 6, x^2_{4212} \leq 4, x^2_{4221} \leq 6,$

$x^2_{4222} \leq 4$

$x^r_{0j1l_{0j}} \geq 0$ and integers $\forall r = 1, 2; j = 1, 2; l_{0j} = 1, 2$

$x^r_{ijk_{ij}l_{ij}} \geq 0$ and integers $\forall r = 1, 2; i = 1, 2, ..., 4; j = 1, 2; l_{ij} = 1, 2; k_{ij} = 1, 2$

This problem gives an infeasible solution when solved mathematically. Hence GP technique is used to obtain a compromised solution to the problem.

5.1 Solution Procedure: GP

Stage 1

$$g_0(\eta, \rho, X) = \rho_1 + \eta_2 + \eta^1_{0111} + \eta^1_{0112} + \eta^1_{0211} + \eta^1_{0212} + \eta^2_{0111} + \eta^2_{0112}$$

Min

$$
\begin{aligned}
&+ \eta^2_{0211} + \eta^2_{0212} + \eta^1_{1111} + \eta^1_{1112} + \eta^1_{1121} + \eta^1_{1122} + \eta^1_{1211} + \eta^1_{1212} + \eta^1_{1221} \\
&+ \eta^1_{1222} + \eta^2_{1111} + \eta^2_{1112} + \eta^2_{1121} + \eta^2_{1122} + \eta^2_{1211} + \eta^2_{1212} + \eta^2_{1221} + \eta^2_{1222} \\
&+ \eta^1_{2111} + \eta^1_{2112} + \eta^1_{2121} + \eta^1_{2122} + \eta^1_{2211} + \eta^1_{2212} + \eta^1_{2221} + \eta^1_{2222} + \eta^2_{2111} \\
&+ \eta^2_{2112} + \eta^2_{2121} + \eta^2_{2122} + \eta^2_{2211} + \eta^2_{2212} + \eta^2_{2221} + \eta^2_{2222} + \eta^1_{3111} + \eta^1_{3112} \\
&+ \eta^1_{3121} + \eta^1_{3122} + \eta^1_{3211} + \eta^1_{3212} + \eta^1_{3221} + \eta^1_{3222} + \eta^2_{3111} + \eta^2_{3112} \\
&+ \eta^2_{3121} + \eta^2_{3122} + \eta^2_{3211} + \eta^2_{3212} + \eta^2_{3221} + \eta^2_{3222} + \eta^1_{4111} + \eta^1_{4112} \\
&+ \eta^1_{4121} + \eta^1_{4122} + \eta^1_{4211} + \eta^1_{4212} + \eta^1_{4221} + \eta^1_{4222} + \eta^1_{4111} \\
&+ \eta^2_{4112} + \eta^2_{4121} + \eta^2_{4122} + \eta^2_{4211} + \eta^2_{4212} + \eta^2_{4221} + \eta^2_{4222} + \rho'^1_{0111} \\
&+ \rho'^1_{0112} + \rho'^1_{0211} + \rho'^1_{0212} + \rho'^2_{0111} + \rho'^2_{0112} + \rho'^2_{0211} + \rho'^2_{0212} + \rho'^1_{1111} \\
&+ \rho'^1_{1112} + \rho'^1_{1121} + \rho'^1_{1122} + \rho'^1_{1211} + \rho'^1_{1212} + \rho'^1_{1221} + \rho'^1_{1222} + \rho'^2_{1111} \\
&+ \rho'^2_{1112} + \rho'^2_{1121} + \rho'^2_{1122} + \rho'^2_{1211} + \rho'^2_{1212} + \rho'^2_{1221} \\
&+ \rho'^2_{1222} + \rho'^1_{2111} + \rho'^1_{2112} + \rho'^1_{2121} + \rho'^1_{2122} + \rho'^1_{2211} + \rho'^1_{2212}
\end{aligned}
$$

$$+ \rho_{2221}'^1 + \rho_{2222}'^1 + \rho_{2111}'^2 + \rho_{2112}'^2 + \rho_{2121}'^2 + \rho_{2122}'^2 + \rho_{2211}'^2 + \rho_{2212}'^2 + \rho_{2221}'^2$$

$$+ \rho_{2222}'^2 + \rho_{3111}'^1 + \rho_{3112}'^1 + \rho_{3121}'^1 + \rho_{3122}'^1 + \rho_{3211}'^1 + \rho_{3212}'^1 + \rho_{3221}'^1$$

$$+ \rho_{3222}'^1 + \rho_{3111}'^2 + \rho_{3112}'^2 + \rho_{3121}'^2 + \rho_{3122}'^2 + \rho_{3211}'^2 + \rho_{3212}'^2 + \rho_{3221}'^2$$

$$+ \rho_{3222}'^2 + \rho_{4111}'^1 + \rho_{4112}'^1 + \rho_{4121}'^1 + \rho_{4122}'^1 + \rho_{4211}'^1 + \rho_{4212}'^1 + \rho_{4221}'^1$$

$$+ \rho_{4222}'^1 + \rho_{4111}'^2 + \rho_{4112}'^2 + \rho_{4121}'^2 + \rho_{4122}'^2 + \rho_{4211}'^2 + \rho_{4212}'^2 + \rho_{4221}'^2 + \rho_{4222}'^2$$

s.t. $1900x_{0111}^1 + 1300x_{0112}^1 + 120000x_{0211}^1 + 80000x_{0212}^1 + 1800x_{0111}^2$

$\qquad + 1100x_{0112}^2 + 108000x_{0211}^2 + 72000x_{0212}^2 + \eta_2 - \rho_2 = 0.35 \times 25000000$

$x_{0111}^1 + \eta_{0111}^1 - \rho_{0111}^1 = 12,\ x_{0112}^1 + \eta_{0112}^1 - \rho_{0112}^1 = 7,$

$x_{0211}^1 + \eta_{0211}^1 - \rho_{0211}^1 = 3,\ x_{0212}^1 + \eta_{0212}^1 - \rho_{0212}^1 = 1,$

$x_{1111}^1 + \eta_{1111}^1 - \rho_{1111}^1 = 10,\ x_{1112}^1 + \eta_{1112}^1 - \rho_{1112}^1 = 4,$

$x_{1121}^1 + \eta_{1121}^1 - \rho_{1121}^1 = 7,\ x_{1122}^1 + \eta_{1122}^1 - \rho_{1122}^1 = 3,$

$x_{1211}^1 + \eta_{1211}^1 - \rho_{1211}^1 = 2,\ x_{1212}^1 + \eta_{1212}^1 - \rho_{1212}^1 = 2,$

$x_{1221}^1 + \eta_{1221}^1 - \rho_{1221}^1 = 1,\ x_{1222}^1 + \eta_{1222}^1 - \rho_{1222}^1 = 1,$

$x_{2111}^1 + \eta_{2111}^1 - \rho_{2111}^1 = 6,\ x_{2112}^1 + \eta_{2112}^1 - \rho_{2112}^1 = 6,$

$x_{2121}^1 + \eta_{2121}^1 - \rho_{2121}^1 = 4,\ x_{2122}^1 + \eta_{2122}^1 - \rho_{2122}^1 = 6,$

$x_{2211}^1 + \eta_{2211}^1 - \rho_{2211}^1 = 1,\ x_{2212}^1 + \eta_{2212}^1 - \rho_{2212}^1 = 1,$

$x_{2221}^1 + \eta_{2221}^1 - \rho_{2221}^1 = 1,\ x_{2222}^1 + \eta_{2222}^1 - \rho_{2222}^1 = 0,$

$x_{3111}^1 + \eta_{3111}^1 - \rho_{3111}^1 = 7,\ x_{3112}^1 + \eta_{3112}^1 - \rho_{3112}^1 = 5,$

$x_{3121}^1 + \eta_{3121}^1 - \rho_{3121}^1 = 6,\ x_{3122}^1 + \eta_{3122}^1 - \rho_{3122}^1 = 4,$

$x_{3211}^1 + \eta_{3211}^1 - \rho_{3211}^1 = 3,\ x_{3212}^1 + \eta_{3212}^1 - \rho_{3212}^1 = 2,$

$x_{3221}^1 + \eta_{3221}^1 - \rho_{3221}^1 = 2,\ x_{3222}^1 + \eta_{3222}^1 - \rho_{3222}^1 = 1,$

$x_{4111}^1 + \eta_{4111}^1 - \rho_{4111}^1 = 6,\ x_{4112}^1 + \eta_{4112}^1 - \rho_{4112}^1 = 4,$

$x_{4121}^1 + \eta_{4121}^1 - \rho_{4121}^1 = 4,\ x_{4122}^1 + \eta_{4122}^1 - \rho_{4122}^1 = 2,$

$x_{4211}^1 + \eta_{4211}^1 - \rho_{4211}^1 = 1,\ x_{4212}^1 + \eta_{4212}^1 - \rho_{4212}^1 = 0,$

$x_{4221}^1 + \eta_{4221}^1 - \rho_{4221}^1 = 2,\ x_{4222}^1 + \eta_{4222}^1 - \rho_{4222}^1 = 1,$

$x_{0111}^2 + \eta_{0111}^2 - \rho_{0111}^2 = 10,\ x_{0112}^2 + \eta_{0112}^2 - \rho_{0112}^2 = 7,$

$x_{0211}^2 + \eta_{0211}^2 - \rho_{0211}^2 = 2,\ x_{0212}^2 + \eta_{0212}^2 - \rho_{0212}^2 = 1,$

$x_{1111}^2 + \eta_{1111}^2 - \rho_{1111}^2 = 12,\ x_{1112}^2 + \eta_{1112}^2 - \rho_{1112}^2 = 7,$

$x_{1121}^2 + \eta_{1121}^2 - \rho_{1121}^2 = 9,\ x_{1122}^2 + \eta_{1122}^2 - \rho_{1122}^2 = 3,$

$x_{1211}^2 + \eta_{1211}^2 - \rho_{1211}^2 = 2,\ x_{1212}^2 + \eta_{1212}^2 - \rho_{1212}^2 = 3,$

$x_{1221}^2 + \eta_{1221}^2 - \rho_{1221}^2 = 1,\ x_{1222}^2 + \eta_{1222}^2 - \rho_{1222}^2 = 1,$

$x_{2111}^2 + \eta_{2111}^2 - \rho_{2111}^2 = 7,\ x_{2112}^2 + \eta_{2112}^2 - \rho_{2112}^2 = 3,$

$x_{2121}^2 + \eta_{2121}^2 - \rho_{2121}^2 = 6,\ x_{2122}^2 + \eta_{2122}^2 - \rho_{2122}^2 = 3,$

$x_{2211}^2 + \eta_{2211}^2 - \rho_{2211}^2 = 2,\ x_{2212}^2 + \eta_{2212}^2 - \rho_{2212}^2 = 2,$

$$x_{2221}^2 + \eta_{2221}^2 - \rho_{2221}^2 = 2, \; x_{2222}^2 + \eta_{2222}^2 - \rho_{2222}^2 = 0,$$
$$x_{3111}^2 + \eta_{3111}^2 - \rho_{3111}^2 = 7, \; x_{3112}^2 + \eta_{3112}^2 - \rho_{3112}^2 = 7,$$
$$x_{3121}^2 + \eta_{3121}^2 - \rho_{3121}^2 = 6, \; x_{3122}^2 + \eta_{3122}^2 - \rho_{3122}^2 = 3,$$
$$x_{3211}^2 + \eta_{3211}^2 - \rho_{3211}^2 = 1, \; x_{3212}^2 + \eta_{3212}^2 - \rho_{3212}^2 = 2,$$
$$x_{3221}^2 + \eta_{3221}^2 - \rho_{3221}^2 = 1, \; x_{3222}^2 + \eta_{3222}^2 - \rho_{3222}^2 = 0,$$
$$x_{4111}^2 + \eta_{4111}^2 - \rho_{4111}^2 = 3, \; x_{4112}^2 + \eta_{4112}^2 - \rho_{4112}^2 = 3,$$
$$x_{4121}^2 + \eta_{4121}^2 - \rho_{4121}^2 = 3, \; x_{4122}^2 + \eta_{4122}^2 - \rho_{4122}^2 = 2,$$
$$x_{4211}^2 + \eta_{4211}^2 - \rho_{4211}^2 = 1, \; x_{4212}^2 + \eta_{4212}^2 - \rho_{4212}^2 = 0,$$
$$x_{4221}^2 + \eta_{4221}^2 - \rho_{4221}^2 = 1, \; x_{4222}^2 + \eta_{4222}^2 - \rho_{4222}^2 = 0,$$
$$x_{0111}^1 + \eta_{0111}'^1 - \rho_{0111}'^1 = 20, \; x_{0112}^1 + \eta_{0112}'^1 - \rho_{0112}'^1 = 12,$$
$$x_{0211}^1 + \eta_{0211}'^1 - \rho_{0211}'^1 = 12, \; x_{0212}^1 + \eta_{0212}'^1 - \rho_{0212}'^1 = 6,$$
$$x_{1111}^1 + \eta_{1111}'^1 - \rho_{1111}'^1 = 16, \; x_{1112}^1 + \eta_{1112}'^1 - \rho_{1112}'^1 = 12,$$
$$x_{1121}^1 + \eta_{1121}'^1 - \rho_{1121}'^1 = 12, \; x_{1122}^1 + \eta_{1122}'^1 - \rho_{1122}'^1 = 7,$$
$$x_{1211}^1 + \eta_{1211}'^1 - \rho_{1211}'^1 = 8, \; x_{1212}^1 + \eta_{1212}'^1 - \rho_{1212}'^1 = 4,$$
$$x_{1221}^1 + \eta_{1221}'^1 - \rho_{1221}'^1 = 6, \; x_{1222}^1 + \eta_{1222}'^1 - \rho_{1222}'^1 = 3,$$
$$x_{2111}^1 + \eta_{2111}'^1 - \rho_{2111}'^1 = 15, \; x_{2112}^1 + \eta_{2112}'^1 - \rho_{2112}'^1 = 12,$$
$$x_{2121}^1 + \eta_{2121}'^1 - \rho_{2121}'^1 = 12, \; x_{2122}^1 + \eta_{2122}'^1 - \rho_{2122}'^1 = 9,$$
$$x_{2211}^1 + \eta_{2211}'^1 - \rho_{2211}'^1 = 9, \; x_{2212}^1 + \eta_{2212}'^1 - \rho_{2212}'^1 = 4,$$
$$x_{2221}^1 + \eta_{2221}'^1 - \rho_{2221}'^1 = 5, \; x_{2222}^1 + \eta_{2222}'^1 - \rho_{2222}'^1 = 2,$$
$$x_{3111}^1 + \eta_{3111}'^1 - \rho_{3111}'^1 = 15, \; x_{3112}^1 + \eta_{3112}'^1 - \rho_{3112}'^1 = 14,$$
$$x_{3121}^1 + \eta_{3121}'^1 - \rho_{3121}'^1 = 15, \; x_{3122}^1 + \eta_{3122}'^1 - \rho_{3122}'^1 = 12,$$
$$x_{3211}^1 + \eta_{3211}'^1 - \rho_{3211}'^1 = 8, \; x_{3212}^1 + \eta_{3212}'^1 - \rho_{3212}'^1 = 6,$$
$$x_{3221}^1 + \eta_{3221}'^1 - \rho_{3221}'^1 = 10, \; x_{3222}^1 + \eta_{3222}'^1 - \rho_{3222}'^1 = 5,$$
$$x_{4111}^1 + \eta_{4111}'^1 - \rho_{4111}'^1 = 12, \; x_{4112}^1 + \eta_{4112}'^1 - \rho_{4112}'^1 = 6,$$
$$x_{4121}^1 + \eta_{4121}'^1 - \rho_{4121}'^1 = 15, \; x_{4122}^1 + \eta_{4122}'^1 - \rho_{4122}'^1 = 6,$$
$$x_{4211}^1 + \eta_{4211}'^1 - \rho_{4211}'^1 = 8, \; x_{4212}^1 + \eta_{4212}'^1 - \rho_{4212}'^1 = 3,$$
$$x_{4221}^1 + \eta_{4221}'^1 - \rho_{4221}'^1 = 6, \; x_{4222}^1 + \eta_{4222}'^1 - \rho_{4222}'^1 = 4,$$
$$x_{0111}^2 + \eta_{0111}'^2 - \rho_{0111}'^2 = 16, \; x_{0112}^2 + \eta_{0112}'^2 - \rho_{0112}'^2 = 12,$$
$$x_{0211}^2 + \eta_{0211}'^2 - \rho_{0211}'^2 = 10, \; x_{0212}^2 + \eta_{0212}'^2 - \rho_{0212}'^2 = 7,$$
$$x_{1111}^2 + \eta_{1111}'^2 - \rho_{1111}'^2 = 15, \; x_{1112}^2 + \eta_{1112}'^2 - \rho_{1112}'^2 = 12,$$
$$x_{1121}^2 + \eta_{1121}'^2 - \rho_{1121}'^2 = 12, \; x_{1122}^2 + \eta_{1122}'^2 - \rho_{1122}'^2 = 9,$$
$$x_{1211}^2 + \eta_{1211}'^2 - \rho_{1211}'^2 = 12, \; x_{1212}^2 + \eta_{1212}'^2 - \rho_{1212}'^2 = 7,$$
$$x_{1221}^2 + \eta_{1221}'^2 - \rho_{1221}'^2 = 3, \; x_{1222}^2 + \eta_{1222}'^2 - \rho_{1222}'^2 = 6,$$
$$x_{2111}^2 + \eta_{2111}'^2 - \rho_{2111}'^2 = 15, \; x_{2112}^2 + \eta_{2112}'^2 - \rho_{2112}'^2 = 12,$$

$$x_{2121}^2 + \eta_{2121}'^2 - \rho_{2121}'^2 = 12, \; x_{2122}^2 + \eta_{2122}'^2 - \rho_{2122}'^2 = 7,$$
$$x_{2211}^2 + \eta_{2211}'^2 - \rho_{2211}'^2 = 9, \; x_{2212}^2 + \eta_{2212}'^2 - \rho_{2212}'^2 = 7,$$
$$x_{2221}^2 + \eta_{2221}'^2 - \rho_{2221}'^2 = 5, \; x_{2222}^2 + \eta_{2222}'^2 - \rho_{2222}'^2 = 4,$$
$$x_{3111}^2 + \eta_{3111}'^2 - \rho_{3111}'^2 = 15, \; x_{3112}^2 + \eta_{3112}'^2 - \rho_{3112}'^2 = 14,$$
$$x_{3121}^2 + \eta_{3121}'^2 - \rho_{3121}'^2 = 16, \; x_{3122}^2 + \eta_{3122}'^2 - \rho_{3122}'^2 = 12,$$
$$x_{3211}^2 + \eta_{3211}'^2 - \rho_{3211}'^2 = 8, \; x_{3212}^2 + \eta_{3212}'^2 - \rho_{3212}'^2 = 5,$$
$$x_{3221}^2 + \eta_{3221}'^2 - \rho_{3221}'^2 = 7, \; x_{3222}^2 + \eta_{3222}'^2 - \rho_{3222}'^2 = 5,$$
$$x_{4111}^2 + \eta_{4111}'^2 - \rho_{4111}'^2 = 12, \; x_{4112}^2 + \eta_{4112}'^2 - \rho_{4112}'^2 = 7,$$
$$x_{4121}^2 + \eta_{4121}'^2 - \rho_{4121}'^2 = 15, \; x_{4122}^2 + \eta_{4122}'^2 - \rho_{4122}'^2 = 7,$$
$$x_{4211}^2 + \eta_{4211}'^2 - \rho_{4211}'^2 = 6, \; x_{4212}^2 + \eta_{4212}'^2 - \rho_{4212}'^2 = 4,$$
$$x_{4221}^2 + \eta_{4221}'^2 - \rho_{4221}'^2 = 6, \; x_{4222}^2 + \eta_{4222}'^2 - \rho_{4222}'^2 = 4,$$

$$\left. \begin{array}{l} x_{0j1l_{0j}}^r \geq 0 \text{ and integers} \\ \eta_{0j1l_{0j}}'^r, \rho_{0j1l_{0j}}'^r, \eta_{0j1l_{0j}}''^r, \rho_{0j1l_{0j}}''^r \geq 0 \end{array} \right\} \forall r = 1, 2; \, j = 1, 2; \, l_{0j} = 1, 2 \qquad (*)$$

$$\left. \begin{array}{l} x_{ijk_{ij}l_{ij}}^r \geq 0 \text{ and integers} \\ \eta_{ijk_{ij}l_{ij}}'^r, \rho_{ijk_{ij}l_{ij}}'^r, \eta_{ijk_{ij}l_{ij}}''^r, \rho_{ijk_{ij}l_{ij}}''^r \geq 0, \end{array} \right\} \forall r = 1, 2; \, j = 1, 2; \, l_{ij} = 1, 2; \, k_{ij} = 1, 2$$

$$\eta_1, \rho_1, \eta_2, \rho_2 \geq 0$$

Stage 2

$$\mathbf{Min} g(\eta, \rho, X) = \lambda_0 \eta_0 + \lambda_{s1} \eta_{s1} + \lambda_{s2} \eta_{s2} + \lambda_{s3} \eta_{s3} + \lambda_{s4} \eta_{s4}$$

$$\text{s.t.} \, Z_0 + \eta_0 - \rho_0 = 71181200, \; Z_1 + \eta_{s1} - \rho_{s1} = 23458500,$$
$$Z_2 + \eta_{s2} - \rho_{s2} = 18629800,$$
$$Z_3 + \eta_{s3} - \rho_{s3} = 23455000, \; Z_4 + \eta_{s4} - \rho_{s4} = 15999600$$
$$\eta_0, \eta_{s1}, \eta_{s2}, \eta_{s3}, \eta_{s4}, \rho_0, \rho_{s1}, \rho_{s2}, \rho_{s3}, \rho_{s4} \geq 0$$

together with constraints of problem (*).

Giving equal weightage to national and regional segments, a satisfactory advertising reach is received from both the national media as well as each of the regional media of Regions 1,2,3, and 4 given as 68271200, 18926500, 13617800, 18621000, and 11318600, respectively. The budget allocated to the national media is ₹1159600 and for Regions 1, 2, 3, and 4 is ₹ 367250, ₹ 315690, ₹ 406740, ₹ 247875 respectively.

The total reach thus generated given the total budget as ₹ 2,50,00,000 all the media is 130755100. The number of advertisements to be allocated to different media for

Table 10 Advertisement of different products allocated to different media

National Media	Region 1's Media	Region 2's Media	Region 3's Media	Region 4's Media
x^1_{0111} 20	x^1_{1111} 16	x^1_{2111} 15	x^1_{3111} 15	x^1_{4111} 12
x^1_{0112} 12	x^1_{1112} 12	x^1_{2112} 12	x^1_{3112} 14	x^1_{4112} 6
x^1_{0211} 4	x^1_{1121} 12	x^1_{2121} 12	x^1_{3121} 15	x^1_{4121} 15
x^1_{0212} 1	x^1_{1122} 7	x^1_{2122} 9	x^1_{3122} 12	x^1_{4122} 6
x^2_{0111} 16	x^1_{1211} 2	x^1_{2211} 1	x^1_{3211} 3	x^1_{4211} 1
x^2_{0112} 12	x^1_{1212} 2	x^1_{2212} 1	x^1_{3212} 2	x^1_{4212} 0
x^2_{0211} 4	x^1_{1221} 1	x^1_{2221} 1	x^1_{3221} 2	x^1_{4221} 2
x^2_{0212} 1	x^1_{1222} 1	x^1_{2222} 0	x^1_{3222} 1	x^1_{4222} 1
	x^2_{1111} 15	x^2_{2111} 15	x^2_{3111} 15	x^2_{4111} 12
	x^2_{1112} 12	x^2_{2112} 12	x^2_{3112} 14	x^2_{4112} 7
	x^2_{1121} 12	x^2_{2121} 12	x^2_{3121} 16	x^2_{4121} 15
	x^2_{1122} 9	x^2_{2122} 7	x^2_{3122} 12	x^2_{4122} 7
	x^2_{1211} 2	x^2_{2211} 2	x^2_{3211} 1	x^2_{4211} 1
	x^2_{1212} 3	x^2_{2212} 2	x^2_{3212} 2	x^2_{4212} 0
	x^2_{1221} 1	x^2_{2221} 2	x^2_{3221} 1	x^2_{4221} 1
	x^2_{1222} 1	x^2_{2222} 0	x^2_{3222} 0	x^2_{4222} 0

both the products in different slots are given below in Table 10 products in different slots are given. Problem is solved using optimization software LINGO 11.0 (Thiriez [14]).

6 Conclusion

In this paper, a media planning problem is considered for advertising multiple products of a firm in a segmented market at both national and regional level. The objective is to allocate the advertisements so as to capture maximum advertising reach from all media. Problem is formulated as a multi-objective programming problem and a compromise solution is achieved using GP approach.

References

1. Charnes, A., Cooper, W.W., DeVoe, J.K., Learner, D.B., Reinecke, W.: A goal programming model for media planning. Manag. Sci. **14**, 422–430 (1968)
2. Lee, S.M.: Goal Programming for Decision Analysis. Auerbach Publishers, Philadelphia (1972)

3. De Kluyver, C.A.: Hard and soft constraints in media scheduling. J. Advertisement Res. **18**, 27–31 (1978)
4. Keown, A.J., Duncan, C.P.: Integer goal programming in advertising media selection. Decis. Sci. **10**, 577–592 (1979)
5. Moynihan, G.P., Kumar, A., D'Souza, G., Nockols, W.G.: A decision support system for media planning. Comput. Ind. Eng. **29**(1), 383–386 (1995)
6. Fructhter, G.E., Kalish, S.: Dynamic promotional budgeting and media allocation. EJOR **111**(1), 15–27 (1998)
7. Lee, C.W., Kwak, N.K.: Information resource planning using AHP-based goal programming model. J. Oper. Res. Soc. **51**, 1–8 (1999)
8. Mihiotis, A., Tsakiris, I.: A mathematical programming study of advertising allocation problem. Appl. Math. Comput. **148**(2), 373–379 (2004)
9. Kwak, N.K., Lee, Chang Won, Kim, Ji Hee, et al.: An MCDM model for media selection in the dual consumer/industrial market. EJOR **166**, 255–265 (2005)
10. Bhattacharya, U.K.: A chance constraints goal programming model for the advertising planning problem. EJOR **192**, 382–395 (2009)
11. Jha, P. C., Aggarwal, R. and Gupta, A.: Optimal Media Selection of Advertisement Planning in a Segmented Market. Om Parkash (Ed.), Advances in Information Theory and Operations Research: interdisciplinary trends, VDM Verlag, Germany. 248–268 (2010)
12. Jha, P.C., Aggarwal, R., Gupta, A.: Optimal Media Planning for Multi-product in Segmented Market. Applied Mathematics and Computation. Elsevier. **217**(16), 6802–6818 (2011)
13. Buratto, A., Grosset, L., Viscolani, B.: Advertising a new product in a segmented market. EJOR. **175**, 1262–1267 (2006)
14. Thiriez, H.: OR software LINGO. EJOR. **12**, 655–656 (2000)

Two Storage Inventory Model for Perishable Items with Trapezoidal Type Demand Under Conditionally Permissible Delay in Payment

S R Singh and Monika Vishnoi

Abstract This article develops a two warehouse deterministic inventory model for deteriorating items with trapezoidal type demand under conditionally permissible delay in payments. A rented warehouse is used when the ordering quantity exceeds the limited capacity of the owned warehouse, and it is assumed that deterioration rates of items in the two warehouses may be different. In contrast to the traditional deterministic two-warehouse inventory model with shortages at the end of each replenishment cycle, an alternative model in which each cycle begins with shortages and ends without shortages is proposed. Deterioration rate is taken to be time-dependent. Shortages are allowed and fully backlogged. Then a solution procedure is shown to find the optimal replenishment policy of the considered problem. At last, article provides numerical example to illustrate the developed model. Sensitivity analysis is also given with respect to major parameters.

Keywords Two warehouse · Trapezoidal type demand · Shortages · Deterioration · Conditionally permissible delay in payments

1 Introduction

Demand is the most volatile of all the market forces, as it is the least controlled by management personnel. Even a slight change in the demand pattern for any particular item causes a lot of havoc with the market concerned. Overall, it means that every time the demand for any commodity goes a noticeable change, the inventory manager has to reformulate the complete logistics of management for that item. Here, one thing

S. R. Singh (✉) · M. Vishnoi
D.N. (P.G.) College, Meerut, C.C.S. University, Meerut 250002, U.P, India
e-mail: shivrajpundir@gmail.com

M. Vishnoi
e-mail: monikavishnoi11@gmail.com

B. V. Babu et al. (eds.), *Proceedings of the Second International Conference on Soft Computing for Problem Solving (SocProS 2012), December 28–30, 2012*, Advances in Intelligent Systems and Computing 236, DOI: 10.1007/978-81-322-1602-5_141, © Springer India 2014

becomes very apparent, even if the firm is able to take the jolt of changed customer's preference, it will not be able to take the sweep of formulating a new inventory management theory every time. Previously, two types of time dependent demands, i.e., linear and exponential have been studied. The main limitation in linear time-dependence of demand rate is that it implies a uniform change in the demand rate per unit time. This rarely happens in the case of any commodity in the market. On the other hand, an exponential rate of change in demand is extraordinarily high and the demand fluctuation of any commodity in the real market cannot be so high. It concludes that demand rate never been constant or an increasing function or decreasing function of time. To observe these fluctuation in demand pattern, ramp-type demand rate taken into consideration.

Trapezoidal type demand pattern is more realistic in comparison to ramp-type demand in many cases like fad or seasonal goods coming to market. The demand rate for such items increases with the time up to certain time and then ultimately stabilizes and becomes constant, and finally the demand rate approximately decreases to a constant or zero, and the begins the next replenishment cycle. Hill [4] first proposed a time dependent demand pattern by considering it as the combination of two different types of demand such as increasing demand followed by a constant demand in two successive time periods over the entire time horizon and termed it as ramp-type time dependent demand pattern. Wu [11] further investigated the inventory model with ramp-type demand rate such that the deterioration followed the Weibull distribution deterioration and partial backlogging. Giri et al. [3] extended the ramp-type demand inventory model with Weibull deterioration distribution. Manna and Chaudhuri [5] have developed a production inventory model with ramp-type two time periods classified demand pattern where the finite production rate depends on the demand.

The effect of deterioration of physical goods in stock is very realistic feature of inventory control. For more details about deteriorating items one can see the review paper of Goyal and Giri [2]. In the past few decades, inventory problems for deteriorating items have been widely studied. In general, deterioration is defined as the damage, spoilage, dryness, vaporization, etc., that result in decrease of usefulness of the original one. In 2006, the Wu et al. [12] defined a new phenomenon as non-instantaneous deteriorating and considered the problem of determining the optimal replenishment policy for such items with stock dependent demand. Soon, Ouyang et al. [6] further developed an inventory model for non-instantaneous deteriorating items with permissible delay in payment.

Generally, when suppliers provide price discounts for bulk purchases or the products are seasonal, the retailers may purchase more goods than can be stored in his own warehouses (OW). Therefore, a rented warehouse (RW) is used to store the excess units over the fixed capacity W of the own warehouse. Usually, the rented warehouse is to charge higher unit holding cost than the own warehouse, but to offer a better preserving facility resulting in a lower rate of deterioration for the goods than the own warehouse. To reduce the inventory costs, it will be economical to consume the goods of RW at the earliest. Consequently, the firm stores goods in OW before RW, but clears the stocks in RW before OW. Chung and Huang [1] proposed a two-warehouse inventory model for deteriorating items under permissible delay in payments, but they assume that the deteriorating rate of two warehouses are the same. Rong et al. [7] developed an optimization inventory policy for a deteriorating item with imprecise lead time, partially/fully backlogged shortages, and price dependent demand under two-warehouse system. Vishnoi and Shon [8] presented a two-warehouse production inventory model for deteriorating items with inflation induced demand and partial backlogging. Singh and Vishnoi [9] developed optimal replenishment policy for deteriorating items with time dependent demand under the learning effect. Vishnoi and Shon [10] explored an economic order quantity model for non-instantaneous deteriorating items with stock dependent demand, time varying partial backlogging under permissible delay in payments and two warehouses.

So far it seems none has tried to investigate this important issue with the assumptions of two warehouse, trade credit, deterioration, and shortages. The whole combination is very unique and very much practical. We think that our work will provide a solid foundation for the further study of this kind of important models with trapezoidal type demand rate. In this paper, we proposed the work as follows:

(i)   We introduce a trapezoidal type demand rate, with its variable part being linear function of time.
(ii) Variable rate of deterioration.
(iii) Shortages are allowed.
(iv) Infinite planning horizon consider in this model.
(v)   This paper divides in two sections

   (a) Model 1. When shortages occur at the end of the cycle with conditionally permissible delay in Payments
   (b) Model 2. When shortages occur at the beginning of the cycle without permissible delay in Payments

2 Assumptions and Notations

2.1 Assumptions

The mathematical model of the economic order quantity problem here is based on the following assumptions:

- The trapezoidal type demand rate, $D(t)$, which is positive and consecutive, is assumed to be a trapezoidal type function of time, that is,

$$D(t) = \begin{cases} a_1 + b_1 t, & t \leq t_W \\ D_1, & t_W \leq t \leq t_1 \\ a_2 - b_2 t, & t_1 \leq t \leq T \end{cases} \quad \text{and} \quad D(t) = \begin{cases} a_3 - b_3 t, & t \leq T' \\ D_2, & T' \leq t \leq t'_W \\ a_4 + b_4 t, & t'_W \leq t \leq t'_1 \end{cases}$$

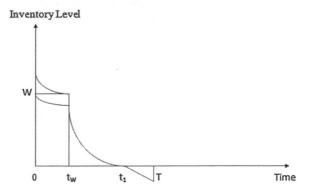

Fig. 1 Graphical representation of a two warehouse inventory system

where t_W is the time at which demand is changing from the linearly increasing demand to constant demand and t_1 is time at which demand is changing from the constant demand to the linearly deceasing demand for Model 1. (See Fig.1) and $a_1, a_2 > b_1, b_2$ and T' is the time at which demand is changing from the linearly decreasing demand to constant demand, and t_W' is time at which demand is changing from the constant demand to the linearly increasing demand for Model 2. (See Fig. 3) and $a_3, a_4 > b_3, b_4$.

- Shortages are allowed and fully backlogged.
- There is no repair or replenishment of deteriorated units during the period.
- Deterioration rate is $\theta_1(t) = \theta_1 t$, for RW and $\theta_2(t) = \theta_2 t$ for OW, which is a continuous function of time
- Deteriorated items are neither replaced nor repaired.
- A single item inventory is considered over the prescribed period.
- Planning horizon is infinite and lead-time is zero.
- The retailer can accumulate revenue and earn interest after his/her customer pays for the amount of purchasing cost to the retailer until the end of the trade credit period offered by the supplier.

2.2 Notations

The following notations are used in the proposed study:

W	Capacity of the OW
C_{h1}	The holding cost per unit per unit time in OW
C_{h2}	The holding cost per unit per unit time in RW
C_D	The purchasing cost per unit
C_S	Shortage Cost per unit per unit time
C_0	Fix amount of the replenishment costs per $ per order
p	The purchasing cost per unit.

c	The selling price per unit.
I_p	The capital opportunity cost in stock per \$ per year.
I_e	The interest earned per \$ per year.
t_W	The time at which the inventory level reaches zero in RW for Model.1
t_1	The time at which the inventory level reaches zero in OW for Model.1
T	The time at which the shortage reaches the lowest point in the replenishment cycle for Model.1
T'	The time at which the shortage occur for Model.2
t'_W	The time at which the inventory level reaches zero in RW for Model.2
t'_1	The time at which the inventory level reaches zero in OW for Model.2
$q_1(t)$	The level of positive inventory in RW at time t for Model.1
$q_2(t), q_3(t)$	The level of positive inventory in OW at time t for model.1
$q_4(t)$	The level of negative inventory at time t for Model.1
$q_1(t)$	The level of negative inventory at time t for Model.2
$q_2(t)$	The level of positive inventory in RW at time t for Model.2
$q_3(t)$	The level of positive inventory in OW at time t for Model.2
$q_4(t)$	The level of positive inventory in OW at time t for Model.2
TC_i	The present value of the total relevant cost per unit time for Model i, $i = 1, 2$

3 Formulation and Solution of the Model

There are two storage shortage models under the assumptions described in previous section, i.e., one is the traditional model and another is staring with shortage model so far can be found in literature. The traditional model is depicted graphically in Fig. 1. It starts with an instant replenishment and ends with shortages. It has been studied in several papers. In contrast, we propose the other shortage model in which the demand will be met at the end of cycle. In fact, the inventory level in our proposed model starts with shortages and ends without shortages. The proposed shortage model is depicted graphically in Fig. 3.

3.1 Shortages Occur at the End of the Cycle with Permissible Delay in Payment (Model 1)

In this section, we discussed the deterministic inventory model for deteriorating items with the traditional two warehouse model where shortage occur at the end of the cycle at time $t = 0$, a lot size of certain units enters the system from which a portion is backlogged toward previous shortages, W units are kept in OW, and the rest is stored in RW. The goods of OW are consumed only after consuming the goods kept in RW. During the interval $(0, t_W)$, the inventory in RW gradually decreases due to deterioration and linearly increasing demand and it vanishes at $t = t_W$. At OW, the inventory W remains same in the interval $(0, t_W)$. During the interval (t_W, t_1) the inventory level is depleted due to constant demand and deterioration. By the time t_1, both warehouses are empty and thereafter the shortages are allowed to occur with linearly decreasing function of time. The shortage quantity is supplied to customers at the beginning of the next cycle. By the time t_2, the replenishment cycle restarts. The objective of the traditional model is to determine the timings of t_W, t_1, and T, so that the total relevant cost (including holding, deterioration, shortage, and ordering costs) per unit time of the inventory system is minimized when in between retailer get some trade credit privilege.

Mathematically, the system can be represented by the following system of differential equations:

$$q_1'(t) + \theta_1(t)q_1(t) = -(a_1 + b_1 t), \qquad 0 \leq t \leq t_W \tag{1}$$

$$q_2'(t) + \theta_2(t)q_2(t) = 0, \qquad 0 \leq t \leq t_W \tag{2}$$

$$q_3'(t) + \theta_2(t)q_3(t) = -D_1, \qquad t_W \leq t \leq t_1 \tag{3}$$

$$q_4'(t) = -(a_2 - b_2 t), \qquad t_1 \leq t \leq T \tag{4}$$

With the boundary conditions $q_1(t_W) = 0$, $q_2(0) = W$, $q_3(t_1) = 0$, $q_4(t_1) = 0$, one can arrive the following equations

$$q_1(t) = a_1 (t_W - t) + \frac{a_1 \theta_1}{6} \left(t_W^3 - t^3 \right) + \frac{b_1}{2} \left(t_W^2 - t^2 \right) + \frac{b_1 \theta_1}{8} \left(t_W^4 - t^4 \right)$$
$$- \frac{a_1 \theta_1}{2} \left(t_W t^2 - t^3 \right) - \frac{b_1 \theta_1}{4} \left(t_W^2 t^2 - t^3 \right), \qquad 0 \leq t \leq t_W \tag{5}$$

$$q_2(t) = W e^{-\frac{\theta_2 t^2}{2}} = W - W \frac{\theta_2 t^2}{2}, \qquad 0 \leq t \leq t_W \tag{6}$$

$$q_3(t) = D \left[(t_1 - t) + \frac{\theta_2}{6} \left(t_1^3 - t^3 \right) - \frac{\theta_2}{2} \left(t_1 t^2 - t^3 \right) \right], \qquad t_W \leq t \leq t_1 \tag{7}$$

$$q_4(t) = a_2 (t_1 - t) + \frac{b_2}{2} \left(t_1^2 - t^2 \right), \qquad t_1 \leq t \leq T \tag{8}$$

i **Ordering Cost** Since replenishment is done at the start of the cycle, the ordering cost per cycle is given by

$$A = C_0 / T \tag{9}$$

ii **Holding Cost for RW** The inventory holding cost in RW per cycle can be derived as

$$HC_{RW} = \frac{C_{h2}}{T} \left[\int_0^{t_W} q_1(t) dt \right] = \frac{C_{h2}}{T} \left(\frac{a_1}{2} t_W^2 + \frac{b_1}{3} t_W^3 + \frac{a_1 \theta_1}{12} t_W^4 + \frac{b_1 \theta_1}{15} t_W^5 \right) \tag{10}$$

iii **Holding Cost for OW** The inventory holding cost in OW per cycle can be derived as

$$
\begin{aligned}
HC_{OW} &= \frac{C_{h1}}{T}\left[\int_0^{t_W} q_2(t)\mathrm{d}t + \int_{t_W}^{t_1} q_3(t)\mathrm{d}t\right] \\
&= \frac{C_{h1}}{T}\left[\left(W t_W - \frac{W\theta_2}{6}t_W^3\right) + D\left\{\left(\frac{t_1^2}{2} - t_1 t_W + \frac{t_W^2}{2}\right)\right.\right. \\
&\quad \left.\left. + \frac{\theta_2}{6}\left(\frac{3t_1^4}{4} - t_1^3 t_W + \frac{t_W^4}{4}\right) + \frac{\theta_2}{6}\left(\frac{3t_1^4}{4} - t_1^3 t_W + \frac{t_W^4}{4}\right)\right\}\right] \quad (11)
\end{aligned}
$$

iv **Deterioration Cost** The deterioration cost per cycle can be derived as

$$
\begin{aligned}
DC &= \frac{C_D}{T}\left[\int_0^{t_W}\theta_1(t)q_1(t)\mathrm{d}t + \int_0^{t_W}\theta_2(t)q_2(t)\mathrm{d}t + \int_{t_W}^{t_1}\theta_2(t)q_3(t)\mathrm{d}t\right] \\
&= \frac{C_D}{T}\left[\theta_1\left(\frac{a_1}{6}t_W^3 + \frac{b_1}{8}t_W^4 + \frac{a_1\theta_1}{120}t_W^5 + \frac{b_1\theta_1}{48}t_W^6\right) + \theta_2\left(W - W\frac{\theta_2}{8}t_W^4\right)\right. \\
&\quad \left. D\theta_2\left\{\left(\frac{t_1^3}{6} + t_1\frac{t_W^2}{2} + \frac{t_W^3}{3}\right) + \frac{\theta_2^2}{6}\left(\frac{3t_1^5}{10} - t_1^3\frac{t_W^2}{2} + \frac{t_W^5}{5}\right) - \frac{\theta_2^2}{2}\left(\frac{t_1^5}{20} - t_1\frac{t_W^4}{4} + \frac{t_W^5}{5}\right)\right\}\right] \\
&\quad\quad\quad\quad\quad\quad\quad\quad\quad\quad\quad\quad\quad\quad\quad\quad\quad\quad\quad\quad\quad\quad\quad\quad\quad\quad (12)
\end{aligned}
$$

v **Shortage Cost** The shortage cost per cycle can be derived as

$$
SC = C_S\left[\int_{t_1}^{t_2}[-q_4(t)]\,\mathrm{d}t\right] = \frac{C_S}{T}\left[a_2\left(\frac{t_1^2}{2} - t_1 T + \frac{T^2}{2}\right) + \frac{b_2}{2}\left(\frac{2t_1^3}{3} - t_1^2 T + \frac{T^3}{3}\right)\right] \quad (13)
$$

vi **The interest payable opportunity cost** There are three cases depicted as Fig. 2.

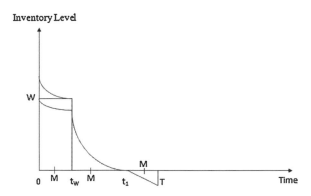

Fig. 2 Different cases of trade credit period M

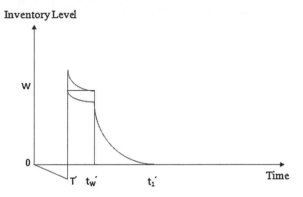

Fig. 3 Graphical representation of a two warehouse inventory system

Case 1 $M \leq t_W < T$ or $0 < M \leq t_W$ In this case, the annual interest payable is

$$IP_1 = \frac{cI_P}{T} \left[\int_M^{t_W} q_1(t)\mathrm{d}t + \int_M^{t_W} q_2(t)\mathrm{d}t + \int_{t_W}^{t_1} q_3(t)\mathrm{d}t + \int_{t_1}^{T} q_4(t)\mathrm{d}t \right]$$

$$IP_1 = \frac{cI_P}{T} \left[a_1\left(\frac{t_W^3}{2} + t_W M + \frac{M^2}{2}\right) + \frac{a_1\theta_1}{6}\left(\frac{3t_W^4}{4} + t_W^3 M + \frac{M^4}{4}\right) \right.$$

$$+ \frac{b_1}{2}\left(\frac{2t_W^3}{3} + t_W^2 M + \frac{M^3}{3}\right) + \frac{b_1\theta_1}{8}\left(\frac{4t_W^5}{5} - t_W^4 M + \frac{M^5}{5}\right)$$

$$- \frac{a_1\theta_1}{2}\left(\frac{t_W^4}{12} - t_W \frac{M^3}{3} + \frac{M^4}{4}\right) - \frac{b_1\theta_1}{4}\left(\frac{2t_W^5}{5} - t_W^2 \frac{M^3}{3} + \frac{M^5}{5}\right)$$

$$+ \left(Wt_W - \frac{W\theta_2}{6}t_W^3 - WM + \frac{W\theta_2}{6}M^3\right) + D\left\{\left(\frac{t_1^2}{2} - t_1 t_W + \frac{t_W^2}{2}\right)\right.$$

$$+ \frac{\theta_2}{6}\left(\frac{3t_1^4}{4} - t_1^3 t_W + \frac{t_W^4}{4}\right) - \frac{\theta_2}{2}\left(\frac{t_1^4}{12} - t_1 \frac{t_W^3}{3} + \frac{t_W^4}{4}\right)\right\}$$

$$\left. + a_2\left(\frac{t_1^2}{2} - t_1 T + \frac{T^2}{2}\right) + \frac{b_2}{2}\left(\frac{2t_1^3}{3} - t_1^2 T + \frac{T^3}{3}\right) \right] \qquad (14)$$

Case 2 $t_W < M \leq T$ In this case, the annual interest payable is

$$IP_2 = \frac{cI_P}{T}\left[\int_M^{t_1} q_3(t)\mathrm{d}t + \int_{t_1}^{T} q_4(t)\mathrm{d}t\right]$$

$$= \frac{cI_P}{T}\left(\frac{t_1^2}{2} - t_1 M + \frac{M^2}{2}\right) + \frac{\theta_2}{6}\left(\frac{3t_1^4}{4} - t_1^3 M + \frac{M^4}{4}\right)$$

$$- \frac{\theta_2}{2} \left(\frac{t_1^4}{12} - t_1 \frac{M^3}{3} + \frac{M^4}{4} \right) \right\} + a_2 \left(\frac{t_1^2}{2} - t_1 T + \frac{T^2}{2} \right)$$
$$+ \frac{b_2}{2} \left(\frac{2t_1^3}{3} - t_1^2 T + \frac{T^3}{3} \right) \right]$$

$$(15)$$

Case 3 $M > t_1$ In this case, no interest charges are paid for the items.

vii **The opportunity interest earned** There are two cases as follows:

Case 1 $0 < M \leq t_W$ or $M \leq T$ In this case, the annual interest earned is

$$IE_1 = \frac{pI_e}{T} \left[\int_0^M (a_1 + b_1 t) t \, dt \right] = \frac{pI_e}{T} \left[a_1 \frac{M^2}{2} + b_1 \frac{M^3}{3} \right]$$

$$(16)$$

Case 2 $t_W < M \leq t_1$ In this case, the annual interest earned is

$$IE_2 = \frac{pI_e}{T} \left[\int_0^{t_W} (a_1 + b_1 t) t \, dt + \int_{t_W}^M D_1 t \, dt \right]$$
$$= \frac{pI_e}{T} \left[a_1 \frac{t_W^2}{2} + b_1 \frac{t_W^3}{3} + D_1 \frac{M^2}{2} - D_1 \frac{t_W^2}{2} \right]$$

$$(17)$$

Case 3 $M > t_1$ In this case, the annual interest earned is

$$IE_3 = \frac{pI_e}{T} \left[\int_0^{t_W} (a_1 + b_1 t) t \, dt + \int_{t_W}^{t_1} D_1 t \, dt \right]$$
$$= \frac{pI_e}{T} \left[\left\{ \left(a_1 \frac{t_W^2}{2} + b_1 \frac{t_W^3}{3} \right) + \frac{D_1}{2} \left(t_1^2 - t_W^2 \right) + \left(a_2 \frac{M^2}{2} - b_2 \frac{M^3}{3} - a_2 \frac{t_1^2}{2} + b_2 \frac{t_1^3}{3} \right) \right\} \right.$$
$$\left. + (M - t_1) \left\{ \left(a_1 \frac{t_W^2}{2} + b_1 \frac{t_W^3}{3} \right) + \frac{D_1}{2} \left(t_1^2 - t_W^2 \right) + \left(a_2 \frac{M^2}{2} - b_2 \frac{M^3}{3} - a_2 \frac{t_1^2}{2} + b_2 \frac{t_1^3}{3} \right) \right\} \right]$$

$$(18)$$

viii **Total Cost**

Therefore, the annual total relevant cost for the retailer can be expressed as:

$$
\begin{aligned}
\text{TC}\,(t_W, t_1, T) = \ & \text{ordering cost} + \text{Inventory holding cost in RW} \\
& + \text{Inventory holding cost in OW} + \text{deteriorating cost} \\
& + \text{shortage cost} + \text{interest payable opportunity cost} \\
& - \text{opportunity interest earned.}
\end{aligned}
$$

$$(19)$$

3.2 Shortage Occurs at the Beginning of the Cycle Without Permissible Delay in Payments (Model 2)

In this section, we discussed the deterministic inventory model for deteriorating items with the two warehouse model where shortage occurs at the beginning of the cycle is discussed. At time $t = 0$, it starts with zero inventory. At the time interval $(0, T')$ shortages are allowed to occur with linearly decreasing function of time. At time $t = T'$, excess quantity of lot size enters to remove the previous shortages. It is assumed that the management owns a warehouse with fixed capacity and any quantity exceeding this should be stored in RW, which is assumed to be available with abundant space. Deterioration starts by the time T' and also it is the time for changing the demand pattern from decreasing trend to constant trend. The goods of OW are consumed only after consuming the goods kept in RW. During the time interval (T', t_W'), the inventory in RW gradually decreases due to joint effect of constant demand and deterioration it vanishes at $t = t_W'$. At OW, the inventory is only depleted by the effect of deterioration. During (t_W', t_1') the inventory is depleted due to linearly increasing demand and deterioration.

Mathematically, the system can be represented by the following system of differential equations:

$$q_1'(t) = -(a_3 - b_3 t), \qquad\qquad 0 \le t \le T' \tag{20}$$
$$q_2'(t) + \theta_1(t)q_2(t) = -D_2, \qquad\qquad T' \le t \le t_W' \tag{21}$$
$$q_3'(t) + \theta_2(t)q_3(t) = 0, \qquad\qquad T' \le t \le t_W' \tag{22}$$
$$q_4'(t) + \theta_2(t)q_4(t) = -(a_4 + b_4 t), \qquad t_W' \le t \le t_1' \tag{23}$$

With the boundary conditions $q_1(0) = 0$, $q_2(t_W') = 0$, $q_3(T') = W$, $q_4(t_1') = 0$, one can arrive the following equations

$$q_1(t) = -\left[a_3 t - \frac{b_3}{2} t^2 \right], \qquad 0 \le t \le T' \tag{24}$$

$$q_2(t) = D_2 \left[(t_{W'} - t) + \frac{\theta_1}{6} \left(t_W'^3 - t^3 \right) - \frac{\theta_1}{2} \left(t_W' t^2 - t^3 \right) \right], \qquad T' \le t \le t_W' \tag{25}$$

$$q_3(t) = W e^{\frac{\theta_2}{2}(T'^2 - t^2)}, \qquad t_W' \le t \le t_1' \tag{26}$$

$$q_4(t) = \left[a_4 \left(t_1' - t \right) + \frac{\theta_2 a_4}{6} \left(t_1'^3 - t^3 \right) + \frac{b_4}{2} \left(t_1'^2 - t^2 \right) + \frac{\theta_2 a_4}{8} \left(t_1'^4 - t^4 \right) \right.$$
$$\left. - \frac{\theta_2 a_4}{2} \left(t_1' t^2 - t^3 \right) - \frac{\theta_2 b_4}{4} \left(t_1'^2 t^2 - t^4 \right) \right], \quad t_W' \le t \le t_1' \tag{27}$$

i **Ordering Cost** Since replenishment is done at the start of the cycle, the ordering cost per cycle is given by

$$A = C_0/t_1, \tag{28}$$

ii **Holding Cost for RW** The inventory holding cost in RW per cycle can be derived as

$$
\begin{aligned}
HC_{RW} &= \frac{C_{h2}}{t_1'} \left[\int_{T'}^{t_W'} q_2(t) \mathrm{d}t \right] \\
&= \frac{C_{h2}}{t_1'} \left[\left(\frac{t_W'^2}{2} - t_W' T' + \frac{T'^2}{2} \right) + \frac{\theta_1}{6} \left(\frac{3t_W'^2}{4} - t_W'^3 T' + \frac{T'^4}{4} \right) \right. \\
&\quad \left. - \frac{\theta_1}{6} \left(\frac{t_W'^4}{12} - t_W' T'^3 + \frac{T'^4}{4} \right) \right]
\end{aligned}
\tag{29}
$$

iii **Holding Cost for OW** The inventory holding cost in OW per cycle can be derived as

$$
\begin{aligned}
HC_{OW} &= \frac{C_{h1}}{t_1'} \left[\int_{T'}^{t_W'} q_3(t) \mathrm{d}t + \int_{t_W'}^{t_1'} q_4(t) \mathrm{d}t \right] \\
&= \frac{C_{h1}}{t_1'} \left[\left\{ W \left(t_W' - T' \right) + \frac{W \theta_2}{6} \left(T'^2 t_W' - \frac{t_W'^3}{3} - \frac{2T'^3}{3} \right) \right\} \right. \\
&\quad + a_4 \left(\frac{t_1'^3}{2} + t_1' t_W' + \frac{t_W'^2}{2} \right) + \frac{a_4 \theta_4}{6} \left(\frac{3t_1'^4}{4} + t_1'^3 t_W' + \frac{t_W'^4}{4} \right) \\
&\quad + \frac{b_4}{4} \left(\frac{2t_1'^3}{3} + t_1'^2 t_W' + \frac{t_W'^3}{3} \right) + \frac{b_4 \theta_2}{8} \left(\frac{4t_1'^5}{5} - t_1'^4 t_W' + \frac{t_W'^5}{5} \right) \\
&\quad \left. - \frac{a_4 \theta_2}{2} \left(\frac{t_1'^4}{12} + t_1' \frac{t_W'^3}{3} + \frac{t_W'^4}{4} \right) - \frac{b_4 \theta_2}{4} \left(\frac{2t_1'^5}{15} - t_1'^2 \frac{t_W'^3}{3} + \frac{t_W'^5}{5} \right) \right]
\end{aligned}
\tag{30}
$$

iv **Deterioration Cost** The deterioration cost per cycle can be derived as

$$
\begin{aligned}
DC &= \frac{C_D}{t_1'} \left[\int_{T'}^{t_W'} \theta_1 t . q_2(t) \mathrm{d}t + \int_{T'}^{t_W'} \theta_2 t . q_3(t) \mathrm{d}t + \int_{t_W'}^{t_1'} \theta_2 t . q_4(t) \mathrm{d}t \right] \\
&= \frac{C_D}{t_1'} \left[\left\{ D_2 \theta_1 \left(\frac{t_W'^3}{6} + t_W' \frac{T'^2}{2} + \frac{T'^3}{3} \right) + \frac{\theta_1}{6} \left(\frac{3t_W'^5}{10} + t_W'^3 \frac{T'^2}{2} + \frac{T'^5}{5} \right) \right. \right. \\
&\quad \left. - \frac{\theta_1}{2} \left(\frac{t_W'^5}{20} + t_W' \frac{T'^4}{4} + \frac{T'^5}{5} \right) \right\} \\
&\quad + \theta_2 W \left\{ \frac{1}{2} \left(t_W'^2 - T'^2 \right) + \frac{\theta_2}{2} \left(T'^2 \frac{t_W'^2}{2} - \frac{t_W'^4}{4} - \frac{T'^4}{4} \right) \right\}
\end{aligned}
$$

$$+ \theta_2 \left\{ \left(\frac{t_1'^3}{6} + t_1' \frac{t_W'^2}{2} + \frac{t_W'^3}{3} \right) + \frac{\theta_2 a_4}{6} \left(\frac{3t_1'^5}{10} + t_1'^3 \frac{t_W'^2}{2} + \frac{t_W'^5}{5} \right) \right.$$

$$+ \frac{b_4}{2} \left(\frac{t_1'^4}{4} + t_1'^2 \frac{t_W'^2}{2} + \frac{t_W'^4}{4} \right) + \frac{\theta_2 b_4}{8} \left\{ \left(\frac{2t_1'^6}{6} + t_1'^4 \frac{t_W'^2}{2} + \frac{t_W'^6}{6} \right) \right.$$

$$\left. - \frac{\theta_2 a_4}{2} \left(\frac{t_1'^5}{20} - t_1' \frac{t_W'^4}{4} + \frac{t_W'^5}{5} \right) - \frac{\theta_2 b_4}{4} \left(\frac{t_1'^6}{12} + t_1'^2 \frac{t_W'^4}{4} + \frac{t_W'^6}{6} \right) \right\} \right] \qquad (31)$$

v **Shortage Cost** The shortage cost per cycle can be derived as

$$SC = \frac{C_S}{t_1'} \left[\int_0^{T'} [-q_1(t)] \, dt \right] = \frac{C_S}{t_1'} \left[\frac{a_3}{2} T'^2 - \frac{b_3}{6} T'^3 \right] \qquad (32)$$

vi **Total Cost** Therefore, the annual total relevant cost for the retailer can be expressed as:

$$\mathrm{TC}(t_W', t_1', T') = ordering cost + stock holding cost in RW$$
$$+ stock holding cost in OW + deteriorating cost + shortage cost \qquad (33)$$

4 Appendix

To minimize total average cost per unit time (TC$_i$), i $= 1, 2$ for both of **Model 1.** and **Model 2**. The optimal values of t_W, t_1 and T for Model 1 and t_W', t_1' and T' for Model 2 can be obtained by solving the following equations simultaneously from the equations

$$\frac{\partial T C_i}{\partial t_W} = 0, \; \frac{\partial T C_i}{\partial t_1} = 0, \; \frac{\partial T C_i}{\partial T} = 0, \qquad (34)$$

$$\frac{\partial T C_i}{\partial t_W'} = 0, \; \frac{\partial T C_i}{\partial t_1'} = 0, \; \frac{\partial T C_i}{\partial T'} = 0, \qquad (35)$$

Provided, they satisfy the following conditions

$$\frac{\partial^2 T C_i}{\partial t_W^2} > 0, \; \frac{\partial^2 T C_i}{\partial t_1^2} > 0, \; \frac{\partial^2 T C_i}{\partial T^2} > 0, \qquad (36)$$

$$\frac{\partial^2 T C_i}{\partial t_W'^2} > 0, \; \frac{\partial^2 T C_i}{\partial t_1'^2} > 0, \; \frac{\partial^2 T C_i}{\partial T'^2} > 0, \qquad (37)$$

5 Solution Procedure

We use the classical optimization techniques for finding the minimum value of the total cost. The Eq. (34) consists of different equations for both cases of Model 1 and Eq. (35) consists of different equations for both cases of Model 2 are highly nonlinear in the continuous variable t_W, t_1, T for Model 1 and t'_W, t'_1, T' for Model 2. We have used the mathematical software **MATLAB 7.0.1.** to arrive at the solution of our system. We obtained the optimal values. With the use of these optimal values Eqs. (19) and (33) provides minimum total average cost per unit time of the system in consideration. Here, numerical illustration is to be given from 1st to 5th replenishment.

6 Numerical Illustrations and Analysis

To elucidate, by the preceding theory the following numerical data is given by:

Example 1 W = 500 units, C_0 = $100 per setup, C_{h1} = $0.3, C_{h2} = $0.6, C_D = $0.5 per unit, C_S = $3/ unit/unit time, a_1 = 200, b_1 = 5, I_P = $0.15 unit/year, I_e = $0.12 unit/year, c = $ 10/ unit, p= $15/ unit, a_2 = 220, b_2 = 10, D_1 = 500 units, a_3 = 220, b_3 = 10, a_4 = 200, b_4 = 5, D_2 = 500 unit, θ_1 = 0.002 , θ_2 = 0.005. Tables 1 and 2.

Table 1 Optimal solution for model 1

N	t_w	l_1	T	Total cost (TC)
1	6.32754	11.3479	14.7978	835.124
2	6.32754	11.3479	14.7978	835.113
3	6.32754	11.3479	14.7978	835.065
4	6.32754	11.3479	14.7978	835.043
5	6.32754	11.3479	14.7978	835.022

Table 2 Optimal solution for model 2

N	T'	t'_w	t'_1	Total cost (TC)
1	0.28193	1.36414	3.49076	169.856
2	0.28163	1.36618	3.46629	169.325
3	0.28145	1.36763	3.46474	169.186
4	0.28131	1.36423	3.46313	169.087
5	0.28124	1.36193	3.46190	169.056

7 Sensitivity Analysis

In Tables 3 and 4 some sensitivity analysis of the model is performed by changing the parameter values −50, −25, 25, and 50%, taking one at a time and keeping the remaining unchanged. The analysis is performed based on the adjusted results obtained in Tables 1 and 2. Some simple characteristics of parameter's impact on total cost (Figs. 4, 5).

7.1 Sensitivity Analysis for Model 1

7.2 Observations for Model 1

The following observations are made based on the above findings: The values of percentage variation in total costs are the highly sensitive to the following parameters: θ_1', 'a_1' and 'a_2', 'b_1', 'b_2'. The values of percentage variation in total costs are quite sensitive to the parameters 'D_1'. The values of percentage variation in total costs are not so sensitive to the following parameters: 'a_1', θ_2, and 'W'.

Table 3 Sensitivity analysis of optimal solution for model 1

Parameter	−50% Changed		−25% Changed		+25% Changed		+50% Changed	
	N	PCV	N	PCV	N	PCV	N	PCV
W	1	0.05	1	0.02	1	−0.03	1	−0.05
a_1	1	0.32	1	0.16	1	−0.17	1	−0.33
a_2	1	−12.95	1	−6.48	1	6.47	1	12.94
b_1	1	−22.17	1	−11.08	1	11.08	1	22.16
b_2	1	−21.07	1	−10.54	1	10.52	1	21.05
D_1	1	−3.42	1	−1.71	1	1.69	1	3.40
Θ_1	1	11.64	1	5.82	1	−5.82	1	−11.64
Θ_2	1	−0.75	1	−0.37	1	0.37	1	0.74

Table 4 Sensitivity analysis of optimal solution for model 2

Parameter	−50% Changed		−25% Changed		+25% Changed		+50% Changed	
	N	PCV	N	PCV	N	PCV	N	PCV
θ_1	1	−0.37	1	−0.19	1	0.19	1	0.38
θ_2	1	0.16	1	0.08	1	−0.08	1	−0.16
a_4	1	−1.28	1	−0.64	1	0.64	1	1.28
b_4	1	−20.31	1	−10.15	1	10.15	1	20.31
a_3	1	−1.35	1	−0.67	1	0.67	1	1.35
b_3	1	−1.36	1	−0.68	1	0.68	1	1.36
D_2	1	5.15	1	2.57	1	−1.93	1	−4.51
W	1	0.05	1	0.02	1	−0.02	1	−0.05

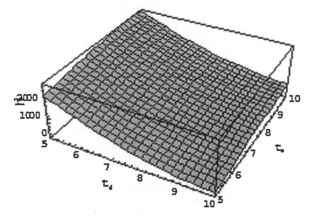

Fig. 4 Convexity of the total cost for model.1 when tW fixed

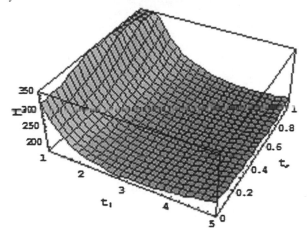

Fig. 5 Convexity of the total cost function when t'_W fixed

7.3 Sensitivity Analysis for Model 2

7.4 Observations for Model 2

The following observations are made based on the above findings: The values of percentage variation in total costs are the highly sensitive to the parameters 'b_4'. The values of percentage variation in total costs are quite sensitive to the following parameters: 'D_2', 'a_4', 'a_3', 'b_3'. The values of percentage variation in total costs are not so sensitive to the following parameters: θ_1, θ_2, and W.

8 Concluding Remarks

In this paper, we successfully provide a rigorous and efficient method to derive the optimal solution for the inventory models with deteriorating items and trapezoidal type demand rate under the permissible delay in payments. We study two alternative inventory models for determining the optimal replenishment schedule for two warehouse inventory problem under shortages, in which the inventory deteriorates at a variable rate over time. In this model deterioration rate at any item is assumed to time dependent.

The nature of demand of seasonal and fashionable products is increasing-steady-decreasing and becomes asymptotic. The demand pattern assumed here is found to occur not only for all types of seasonal products but also for fashion apparel, computer chips of advanced computers, spare parts, etc. The procedure presented here may be applied to many practical situations. Retailers in supermarket face this type of problem to deal with highly perishable seasonal products. Thus, to make a better combination of increasing- steady-decreasing demand pattern for perishable seasonal products.

In addition to the above mentioned facts, we have also tried to establish two possible shortage models. For each model the optimal policy is obtained. In general, Model 2 is less expensive to operate than Model 1 if the factors of trapezoidal type demand rate, shortages, deterioration, and permissible delay in payments are considered. A numerical assessment of the infinite planning horizon theoretical model has been done to illustrate the theory. The solution obtained has also been checked for sensitivity with the result that the model is found to be quite suitable and stable and discussed some interesting and important facts where the company's management need to take proper attention. The variations in the system statistics with a variation in system parameters have also been illustrated graphically.

Author Biographies

S.R. Singh Faculty of Mathematics in D.N. (P.G.) College Meerut (U.P.), has experience of 18 years in academics and research. His Ph.d. is from C.C.S. University, Meerut (U.P.). His areas of specialization are inventory control, supply chain management, cryptography and fuzzy set theory. He has attended various seminars/conferences and presented his research papers. Fifteen students have been awarded PhD under his supervision. He has published more than hundred research papers in reputed national and international journals. His research papers have been published in *Opsearch*, *International Journal of Operational Research*, *Fuzzy Sets and Systems* and *International Journal of Operations and Quantitative Management*. He is author/co-author of seven of books.

Monika Vishnoi has Ph.d. in Mathematics. She has completed her research in the field of Inventory Management. She has communicated several research papers to journals of international repute. Several of her papers have been published in journals of national and international repute, while a few are under review at journals of international repute.

References

1. Chung, K.J., Huang, T.S.: The optimal retailer's ordering policies for deteriorating items with limited storage capacity under trade credit financing. Int. J. Prod. Econ. **106**, 127–145 (2007)
2. Goyal, S.K., Giri, B.C.: Recent trends in modeling of deteriorating inventory. Eur. J. Oper. Res. 134(1), 1–16 (2001)
3. Giri, B.C., Jalan, A.K., Chaudhuri, K.S.: Economic order quantity model with Weibull deterioration distribution, shortage and ramp-type demand. Int. J. Syst. Sci. 34(4), 237–243 (2003)
4. Hill, R.M.: Optimal EOQ models for deteriorating items with time-varying demand. J. Oper. Res. Soci. **47**, 1228–1246 (1995)
5. Manna, S.K., Chaudhuri, K.S.: An EOQ model with ramp type demand rate, time dependent deterioration rate, unit production cost and shortages. Eur. J. Oper. Res. **171**, 557–566 (2006)
6. Ouyang, L.Y., Wu, K.S., Yang, C.T.: A study on an inventory model for non-instantaneous deteriorating items with permissible delay in payments. Comput. Ind. Eng. **51**, 637–651 (2006)
7. Rong, M., Mahapatra, N.K., Maiti, M.: A two warehouse inventory model for a deteriorating item with partially/fully backlogged shortage and fuzzy lead time. Eur. J. Oper. Res. **189**, 59–75 (2008)
8. Vishnoi, M., Shon, S.: An EPQ model with flexible production rate, inflation induced demand with partial backorder for two warehouses. Int. Trans. Appl. Sci. **2**(2), 211–224 (2010)
9. Singh, S.R., Vishnoi, M.: Optimal replenishment policy for deteriorating items with time dependent demand under the learning effect. Int. Trans. Math. Sci. Comput. **4**(2), 171–186 (2011)
10. Vishnoi, M., Shon, S.: Two levels of storage model for non-instantaneous deteriorating items with stock dependent demand, time varying partial backlogging under permissible delay in payments. Int. J. Oper. Res. Optim. **1**(1), 133–147 (2010)
11. Wu, K.S.: An EOQ model for items with Weibull distribution deterioration, ramp-type demand rate and partial backlogging. Prod. Plann. Control 12(8), 787–793 (2001)
12. Wu, K.S., Ouyang, L.Y., Yang, C.T.: An optimal replenishment policy for non-instantaneous deteriorating items with stock dependent demand and partial backlogging. Int. J. Prod. Econ. **101**, 369–384 (2006)

Development of an EOQ Model for Multi Source and Destinations, Deteriorating Products Under Fuzzy Environment

Kanika Gandhi and P. C. Jha

Abstract Business in the present highly competitive scenario emphasises the need to satisfy customers. Generally, uncertainty in demand is observed from customer side when products are deteriorating in nature. This uncertain demand cannot be predicted precisely, which causes fuzziness in related constraints and cost functions. Synchronizing inventory, procurement, and transportation of deteriorating natured products with fuzzy demand, and fuzzy holding cost at source and destination becomes essential in supply chain management (SCM). The current study demonstrates a fuzzy optimization model with an objective to minimize the cost of holding, procurement, and transportation of multi products from multi sources to multi destinations (demand point) with discount policies on ordered and weighted transportation quantity. A case study is illustrated to validate the model.

Keywords Supply chain management · Discount models · Transportation models · Deteriorating products · Fuzzy logics

1 Introduction

Deteriorating products are common in daily life. However, the academia has not reached a consensus on the definition of the deteriorating products. Deteriorating products can be classified into two categories. The first category refers to the products that become decayed, damaged, or expired through time, like meat, vegetables, fruit, medicine, etc; the other category refers to products that lose part or total value through

K. Gandhi (✉) · P. C. Jha
Department of Operational Research, Faculty of Mathematical Sciences, University of Delhi, Delhi 110007, India
e-mail: gandhi.kanika@gmail.com

P. C. Jha
e-mail: jhapc@yahoo.com

B. V. Babu et al. (eds.), *Proceedings of the Second International Conference on Soft Computing for Problem Solving (SocProS 2012), December 28–30, 2012*, Advances in Intelligent Systems and Computing 236, DOI: 10.1007/978-81-322-1602-5_142, © Springer India 2014

time because of new technology or the introduction of alternatives, like computer chips, mobile phones, fashion, and seasonal goods. Both the categories have the characteristic of short life cycle. For the first category, products have a short natural life cycle. After a specific period, the natural attributes of products will change and then lose useable value and economic value; for the second category, products have a short market life cycle. After a period of popularity in the market, the products lose the original economic value due to the changes in consumer preference, product upgrading, and other reasons. In the supply chain, decisions related to delivering ordered products in time become critical, as delivery or consumption delay leads to reduction in product's value. Because of the deteriorating nature of the products, demand of such products becomes highly volatile and cannot be forecasted precisely. Imprecision in demand forces total cost and holding costs to be imprecise and creates a fuzzy environment.

Many authors in the past have discussed concepts on deteriorating products and fuzzy environment due to fuzzy demand and cost. Initial discussion is with the assumption that the lifetime of a product is infinite while it is in storage. On the same lines, Silver and Meal [1] developed an approximate solution technique of a deterministic inventory model with time varying demand. Donaldson [2] first developed an exact solution procedure for products with a linearly increasing demand rate over a finite planning horizon. All these models are based on the assumption that there is no deterioration effect on inventory. But in reality, deterioration also depends on preserving facilities and environmental conditions of storage. So, due to deterioration effect, a certain fraction of the available quantity is either damaged or decayed and are not in a perfect condition to satisfy the future demand of customers for good items. Deterioration for such items is continuous and constant or time-dependent and is dependent on the on-hand inventory. A number of research papers have already been published on the above type of items by [3–5].

Due to deterioration, the demand of such products is uncertain, which develops a fuzzy environment. To convert fuzzy environment into crisp, Zimmermann [6] used the concept of fuzzy set in decision-making processes by considering the objective and constraints as fuzzy goals. He first applied fuzzy set theory with suitable choice of membership functions and derived a fuzzy linear programming problem. Lai and Hwang [7, 8] described the application of fuzzy sets to several operation research problems in two well-known books. Fuzzy set theory has also been used in few inventory models. Sommer [9] applied fuzzy dynamic programming to an inventory and production-scheduling problem. Kacprzyk and Staniewski [10] considered a fuzzy inventory problem in which, instead of minimizing the total average cost, they reduced it to a multi-stage fuzzy-decision-making problem and solved by a branch and bound algorithm. Lam and Wang [11] solved the fuzzy model of joint economic lot size problem with multiple price breaks. Roy and Maiti [12] solved the classical EOQ model in a fuzzy environment with fuzzy goal, fuzzy inventory costs, and fuzzy storage area by FNLP method using different types of membership functions for inventory parameters.

In the current study, a specific division of SCM is explained, where deteriorating natured multiple products are ordered to multi supplier by many buyers. The problem is faced by the suppliers not being able to forecast demand because of the deteriorating nature of the products. In the process, the buyer at demand point plays a major role, which provides a holistic approach by integrating all the holding, procurement, inspection, and transportation activities such that holding cost; ordered quantity to suppliers and its purchase cost; inspection cost on ordered quantity; transported weights and its freight cost. He would also take care of the holding cost at suppliers, because suppliers cannot keep goods in the warehouse for a long time. As the capacity of warehouse is limited and may incur more cost that may further affect the selling price. The paper presents a fuzzy optimization model, which integrates inventory, procurement, and transportation mechanism to minimize all the costs discussed above. The total cost of the model becomes fuzzy because of fuzzy holding cost and demand. The study also includes the aspect of discounts that benefits the buyer to avail discounts on bulk purchase and transported weights.

2 Problem Statement and Assumptions

The current study develops a fuzzy model which shows flow of deteriorating multi products from multi sources to multi demand points (destinations) with fuzzy demand and holding cost at source and destination. In this coordination, the cost is associated at every stage like purchase cost, distribution cost, inspection cost, and fuzzy holding cost. The objective of the current study is to minimize the total costs discussed above by integrating procurement and distribution phase with incorporating quantity discount policies at the time of purchase and freight discount at the time of transportation and maximum reduction in vagueness of the fuzzy environment. At the time of model development the following assumptions were taken:

- Demand is uncertain.
- Supply is instantaneous.
- Initial inventory of each product for the beginning of planning horizon is zero.
- Constant rate of deterioration, as a percentage of stored units.
- Constant inspection rate.

3 Proposed Model Formulation

3.1 Sets

Product set with cardinality P and indexed by i, whereas periods set with cardinality T and indexed by t, price discount break point set with cardinality L and indexed by l, and freight discount break point with cardinality K and indexed by k. Destination set with cardinality M indexed by m, and source set with cardinality J indexed by j.

3.2 Parameters

$\tilde{C}$ is fuzzy total cost, C_0 and C_0^* are the aspiration and tolerance level of fuzzy total cost respectively. $\tilde{h}_{ijt}$ and $\overline{h}_{ijt}$ are fuzzy and defuzzified holding cost per unit of product i in period t at source j. $\tilde{h}_{imt}$ and $\overline{h}_{imt}$ are fuzzy and defuzzified holding cost per unit of product i in period t at destination m. φ_{ijmt} is the unit purchase cost for product i from source j to destination m in period t. d_{ijmlt} *is the d*iscount factor is valid if more than a_{ijmlt} unit are purchased, $0 < d_{ijmlt} < 1$. β_{jmt} is the weight freight cost in tth period from source j to destination m. f_{jmkt} is the transportation freight discount factor from source j to destination m in period t at freight break k. $\tilde{D}_{imt}$ and $\overline{D}_{imt}$ are fuzzy and defuzzfied demand for product i in period t from mth destination. $\overline{D}_0^*$ and $\overline{D}_0$ are the aspiration and tolerance level of defuzzified demand, where $\overline{D}_0^* = \sum_{t=1}^{T} \sum_{m=1}^{M} \overline{D}_{imt}$. CR_{imt} is the consumption at destination m of product i in period t. a_{ijmlt} is the limit beyond which l^{th} price break becomes valid for destination m availed from source j in period t for product i. b_{jmkt} is the limit beyond which k^{th} freight break becomes from source j to destination m valid in period t. w_i is per unit weight of ith product. IN_{ij1} and IN_{im1} are initial inventory of the planning horizon at source j and destination m resp. for product i. λ_{ijmt} is per unit inspection cost of ith product in terms of quantity ordered. η is percentage of defective items of the stored units.

3.3 Decision Variables

X_{ijmt} is the amount of product i ordered in period t from source j for destination m. R_{ijmlt} is the binary variable, which is 1 *if the* ordered quantity falls in lth price break, otherwise zero. I_{ijt} and I_{imt} are inventory level at source j and destination$m$ for product i at the end of period t. Z_{jmkt} is the binary variable, which is 1 *if the* weighted quantity transported falls in k^{th} price break, otherwise zero. L_{jmt} *is the* total weighted quantity transported in period t from source j to destination m.

3.4 Fuzzy Optimization Model Formulation

In many real problems, input information is incomplete or unreliable to develop a crisp mathematical model, which can quantify uncertain parameters. This enables to employ fuzzy optimization methods and fuzzy parameters that provide more adequate solution of real problem for uncertain and vague environment. Therefore, we formulate fuzzy optimization model for vague aspiration levels on total cost, demand, and holding cost, the decision maker may decide his aspiration levels on the basis of past experience.

$$\text{Min } \tilde{C} = \sum_{t=1}^{T} \sum_{j=1}^{J} \sum_{i=1}^{P} \left[\tilde{h}_{ijt} I_{ijt} + \sum_{m=1}^{M} \left\{ \left(\sum_{l=1}^{L} R_{ijmlt} d_{ijmlt} \right) \varphi_{ijmt} X_{ijmt} + \lambda_{ijmt} X_{ijmt} \right\} \right]$$

$$+ \sum_{t=1}^{T} \sum_{m=1}^{M} \left[\tilde{h}_{imt} I_{imt} + \sum_{j=1}^{J} \left\{ \left(\sum_{k=1}^{K} Z_{jmkt} f_{jmkt} \right) \beta_{jmt} L_{jmt} \right\} \right] \qquad (1)$$

Subject to

$$I_{ijt} = I_{ij(t-1)} + \sum_{m=1}^{M} X_{ijmt} - \sum_{m=1}^{M} \tilde{D}_{imt} \quad i = 1, \ldots, P; j = 1, \ldots, J; t = 2, \ldots, T \qquad (2)$$

$$I_{ij1} = IN_{ij1} + \sum_{m=1}^{M} X_{ijm1} - \sum_{m=1}^{M} \tilde{D}_{im1} \quad i = 1, \ldots, P; j = 1, \ldots, J \qquad (3)$$

$$\sum_{t=1}^{T} I_{ijt} + \sum_{t=1}^{T} \sum_{m=1}^{M} X_{ijmt} \gtrsim \sum_{t=1}^{T} \sum_{m=1}^{M} \tilde{D}_{imt} \quad i = 1, \ldots, P; j = 1, \ldots, J \qquad (4)$$

$$I_{imt} = I_{im(t-1)} + \tilde{D}_{imt} - CR_{imt} - \eta I_{imt} \quad i = 1, \ldots, P; m = 1, \ldots, M; t = 2, \ldots, T \qquad (5)$$

$$I_{im1} = IN_{im1} + \tilde{D}_{im1} - CR_{im1} - \eta I_{im1} \quad i = 1, \ldots, P; m = 1, \ldots, M \qquad (6)$$

$$(1 - \eta) \sum_{t=1}^{T} I_{imt} + \sum_{t=1}^{T} \tilde{D}_{imt} \gtrsim \sum_{t=1}^{T} CR_{imt} \quad i = 1, \ldots, P; m = 1, \ldots, M \qquad (7)$$

$$X_{ijmt} \geq \sum_{l=1}^{L} a_{ijmlt} R_{ijmlt} \quad i = 1, \ldots, P; j = 1, \ldots, J; m = 1, \ldots, M; t = 1, \ldots, T \qquad (8)$$

$$\sum_{l=1}^{L} R_{ijmlt} = 1 \quad i = 1, \ldots, P; j = 1, \ldots, J; m = 1, \ldots, M; t = 1, \ldots, T \qquad (9)$$

$$L_{jmt} = \sum_{i=1}^{P} \left[w_i X_{ijmt} \sum_{l=1}^{L} R_{ijmlt} \right] \quad j = 1, \ldots, J; m = 1, \ldots, M; t = 1, \ldots, T \qquad (10)$$

$$L_{jmt} \geq \sum_{k=1}^{K} b_{jmkt} Z_{jmkt} \quad j = 1, \ldots, J; m = 1, \ldots.M; t = 1, \ldots, T \qquad (11)$$

$$\sum_{k=1}^{K} Z_{jmkt} = 1 \quad j = 1, \ldots, J; m = 1, \ldots, M; t = 1, \ldots, T \qquad (12)$$

$$X_{ijmt}, I_{imt}, I_{ijt}, L_{jmt} \geq 0 \quad R_{ijmlt} = 0 \text{ or } 1, \quad Z_{jmkt} = 0 \text{ or } 1$$

In the proposed model, Eq. (1) is the Fuzzy objective function to minimize the cost incurred in holding ending inventory at source, cost of purchasing the products, and cost of inspection on ordered quantity by destination m in period t reflected by the first term of the objective function; the combination of transportation cost from the source to the destination, and holding cost at destination is the second term. The cost is calculated for the duration of the planning horizon. The ordering cost is a fixed cost and is not affected by the ordering quantities and therefore is not the part of objective function. Constraints (2–7) are the balancing equations for sources and destinations where Eq. (2) finds total ending inventory at source j of ith product in t^{th} period is found by reducing the fuzzy demand of all the destinations from total of ending inventory of previous period and ordered quantity at t^{th} period of all the destinations. Equation (3) finds total ending inventory at source j of ith product in the first period by reducing the fuzzy demand of all the destinations from the total of initial inventory if the planning horizon and ordered quantity is at first period of all the destinations. Equation (4) shows that total fuzzy demand in all periods from all destinations is less than or equal to the total ending inventory and ordered quantity at source j in all periods, i.e., shortages are not allowed. Equation (5) calculates ending inventory at m^{th} destination for t^{th} period by reducing consumption rate and fraction of deteriorated ending inventory of the same destination from the combination of ending inventory of previous period and fuzzy demand of at m^{th} destination. Equation (5) calculates ending inventory for the first period at m^{th} destination by reducing consumption and fraction of deteriorated ending inventory of the same destination from the combination of initial inventory of planning horizon and fuzzy demand of mth destination. Equation (7) shows that the total consumption in all the periods at m^{th} destination is less than or equal to the total ending inventory and fuzzy demand at destination m in all the periods, i.e., shortages are not allowed. Equations (8–9) find out the order quantity of all products in period t which may exceed the quantity break threshold and avails discount on ordered quantity at exactly one quantity discount level. Equation (10) is the integrator for procurement Eqs. (2–9) and transportation Eqs. (11–12), which calculates per product weighted quantity to be transported from source j to destination m. Equations (11–12) find the weighted transport quantity of all products in period t which may exceed the freight break threshold, and avail discount on transportation quantity at exactly one freight discount break.

3.5 Price Breaks

As discussed above, variable R_{ijmlt} specifies the fact that when the order size at period t is larger than a_{ijmlt} it results in discounted prices for the ordered items for which the price breaks are defined as; Price breaks for ordering quantity are:

$$d_f = \begin{cases} d_{ijmlt} & a_{ijmlt} \leq X_{ijmt} \leq a_{ijm(l+1)t} \\ d_{ijmLt} & X_{ijmt} \geq a_{ijmLt} \end{cases}$$

$$i = 1, \ldots, P; \ j = 1, \ldots, J; \ t = 1, \ldots, T; l = 1, \ldots, L; \ m = 1, \ldots, M$$

Freight breaks for transporting quantity are:
Here b_{jmkt} is the minimum required quantity to be transported

$$d_f = \begin{cases} f_{jmkt} & b_{jmkt} \leq L_{jmt} \leq b_{jm(k+1)t} \\ f_{jmKt} & L_{jmt} \geq b_{jmKt} \end{cases}$$

$$j = 1, \ldots, J; m = 1, \ldots, M; t = 1, \ldots, T$$

4 Fuzzy Solution Algorithm

The following algorithm [7] specifies the sequential steps to solve the fuzzy mathematical programming problems.

Step 1. Compute the crisp equivalent of the fuzzy parameters using a defuzzification function. The same defuzzification function is to be used for each of the parameters. Let $\overline{D}_{imt}$ be the defuzzified value of $\tilde{D}_{imt}$ and $D^1_{imt}, D^2_{imt}, D^3_{imt}$ be triangular fuzzy numbers then, $\overline{D}_{imt} = (D^1_{imt} + 2D^2_{imt} + D^3_{imt})/4, \ i = 1, \ldots, P; t = 1, \ldots, T; m = 1, \ldots, M$

$\overline{D}_{imt}$ and C_0 are defuzzified aspiration levels of model's demand and cost. Similarly, $\overline{h}_{ijt}$ and $\overline{h}_{imt}$ are defuzzified aspiration levels of holding cost at source and destination.

Step 2. Employ extension principle to identify the fuzzy decision, which results in a crisp mathematical programming problem and on substituting the values for $\tilde{D}_{imt}$ as $\overline{D}_{imt}$; $\tilde{h}_{imt}$ as $\overline{h}_{imt}$; $\tilde{h}_{ijt}$ as $\overline{h}_{ijt}$, the problem becomes:

$$\text{Min } \overline{C} = \sum_{t=1}^{T} \sum_{j=1}^{J} \sum_{i=1}^{P} \left[\overline{h}_{ijt} I_{ijt} + \sum_{m=1}^{M} \left\{ \left(\sum_{l=1}^{L} R_{ijmlt} d_{ijmlt} \right) \varphi_{ijmt} X_{ijmt} + \lambda_{ijmt} X_{ijmt} \right\} \right]$$

$$+ \sum_{t=1}^{T} \sum_{m=1}^{M} \left[\overline{h}_{imt} I_{imt} + \sum_{j=1}^{J} \left\{ \left(\sum_{k=1}^{K} Z_{jmkt} f_{jmkt} \right) \beta_{jmt} L_{jmt} \right\} \right]$$

Subject to $X \in S = \{X \,| I_{ijt} = I_{ij(t-1)} + \sum_{m=1}^{M} X_{ijmt} - \sum_{m=1}^{M} \overline{D}_{imt} \quad \forall i, j, t = 2, \ldots, T$

$$I_{ij1} = IN_{ij1} + \sum_{m=1}^{M} X_{ijm1} - \sum_{m=1}^{M} \overline{D}_{im1} \quad \forall i, j; \; I_{imt}$$

$$= I_{im(t-1)} + \overline{D}_{imt} - CR_{imt} - \eta I_{imt} \quad \forall i, m, t$$

$I_{im1} = IN_{im1} + \overline{D}_{im1} - CR_{im1} - \eta I_{im1} \quad \forall i, m; \; X_{ijmt} \geq \sum_{l=1}^{L} a_{ijmlt} R_{ijmlt} \quad \forall i, j, m, t$

$$\sum_{l=1}^{L} R_{ijmlt} = 1 \quad \forall i, j, m, t; \; L_{jmt} = \sum_{i=1}^{P} \left[w_i X_{ijmt} \sum_{l=1}^{L} R_{ijmlt} \right] \quad \forall j, m, t$$

$$L_{jmt} \geq \sum_{k=1}^{K} b_{jmkt} Z_{jmkt} \quad \forall j, m, t \quad \sum_{k=1}^{K} Z_{jmkt} = 1 \quad \forall j, m, t\}$$

$$\sum_{t=1}^{T} I_{ijt} + \sum_{t=1}^{T} \sum_{m=1}^{M} X_{ijmt} \gtrsim \sum_{t=1}^{T} \sum_{m=1}^{M} \overline{D}_{imt} \quad \forall i, j;$$

$$(1 - \eta) \sum_{t=1}^{T} I_{imt} + \sum_{t=1}^{T} \overline{D}_{imt} \gtrsim \sum_{t=1}^{T} CR_{imt} \quad \forall i, m$$

$$\overline{C}(X) \lesssim C_0$$

$X_{ijmt}, I_{ijt}, I_{imt}, L_{jmt} \geq 0$ and integer, $R_{ijmlt}, Z_{jmkt} \in \{0, 1\}, \theta \in [0, 1]$

$i = 1, \ldots, P; j = 1, \ldots, J; m = 1, \ldots, M; t = 1, \ldots, T; l = 1, \ldots, L$

Step 3. Define appropriate membership functions for each fuzzy inequalities as well as constraints corresponding to the objective function. The membership function for the fuzzy is given as

$$\mu_C(X) = \begin{cases} 1 & ; \quad C(X) \leq C_0 \\ \frac{C_0^* - C(X)}{C_0^* - C_0} & ; \quad C_0 \leq C(X) < C_0^* \\ 0 & ; \quad C(X) > C_0^* \end{cases}$$

where C_0 is the restriction and C_0^* is the tolerance levels to the fuzzy total cost.

$$\mu_{I_{ijt}}(X) = \begin{cases} 1 & ; \quad I_{ijt}(X) \geq \overline{D_0^*} \\ \frac{I_{ijt}(X) - \overline{D_0}}{\overline{D_0^*} - \overline{D_0}} & ; \quad \overline{D_0} \leq I_{ijt}(X) < \overline{D_0^*} \ ; \\ 0 & ; \quad I_{ijt}(X) > \overline{D_0} \end{cases}$$

$$\mu_{I_{imt}}(X) = \begin{cases} 1 & ; \quad I_{imt}(X) \geq \overline{D_0^*} \\ \frac{I_{imt}(X) - \overline{D_0}}{\overline{D_0^*} - \overline{D_0}} & ; \quad \overline{D_0} \leq I_{imt}(X) < \overline{D_0^*} \\ 0 & ; \quad I_{imt}(X) > \overline{D_0} \end{cases}$$

where $\overline{D_0^*} = \sum_{m=1}^{M} \sum_{t=1}^{T} \overline{D}_{imt}$ is the aspiration and $\overline{D_0}$ is the tolerance level to inventory constraint.

Step 4. Employ extension principle to identify the fuzzy decision. While solving the problem its objective function is treated as constraint. Each constraint is considered to be an objective for the decision maker and the problem can be looked as crisp mathematical programming problem

Max θ subject to: $\mu_c(X) \leq \theta$; $\mu_{I_{ijt}}(X) \geq \theta$; $\mu_{I_{imt}}(X) \geq \theta$; $X \in S$ can be solved by the standard crisp mathematical programming algorithms.

5 Case Study

A dried fruit may deteriorate by oxidation, due to atmospheric oxygen and by water uptake. The oxidation may cause loss of color and undesirable odor changes. The case considers the influence of water on deterioration of dried fruits. Here, discussing the problems of a dried fruit company "Cocco" mainly visible in Gujarat (India). They purchase loose dried fruits from small wholesellers, pack, and brand them and further sell in their retail shops. The material is purchased from eight small wholesellers and shipped to their 18 retail shops. In the case, a tiny problem is explained with two wholesellers and three retail shops in Anand city (Gujarat), three products [Cashew Nuts (CN), Dry Dates (DD), and Dry Coconut (DC)] with packet size of 500 gms, 1 kg, and 1.5 kg, respectively, and 3 months period during rainy season (June, July, and August). In Anand city wholesale shops are located in Akriti Nagar (AN), and Chavdapura (CP) and retail shops are located at Swastik Vatika (SV), Adarsh Colony (AC), and Govardhan Nagar (GN). The company faces problems like uncertain (fuzzy) demand which leads to fuzzy inventory carrying cost; deterioration of ending inventory (at constant rate i.e. 6 %); inspection of each received packet at stores (₹3 for CN; ₹5 for DD and ₹7 for DC). Because of the company's basic problems, they desire to find out the optimum order quantity from different wholesellers so that they keep low inventory at shops. This decision is able to keep cost at optimum level. The cost to company at retail shops includes cost of purchasing, inspection cost, transportation cost, inventory carrying cost at retail shops, and inventory carrying cost of wholesellers as they don't keep big quantity. As far as the uncertain demand is concerned, the Co. has past idea of quantity demand, so it fixes three possible demands for each product. They also considered Aspired

Table 1 Holding cost at wholesale shop (source) (₹)

Period		June		July		August	
PDT	Source	Fuzzy	DF	Fuzzy	DF	Fuzzy	DF
CN	AN	30,25,16.4	24.1	27,30,26.6	28.4	30,24,30.8	27.2
	CP	30,25,24.4	26.1	27,25,27.8	26.2	26,27,21.2	25.3
DD	AN	15,14,9.4	13.1	19,18,18.2	18.3	16,15,22.8	17.2
	CP	20,17,18.8	18.2	16,18,12.8	16.2	19,14,13.4	15.1
DC	AN	18,16,14.4	16.1	21,22,19.8	21.2	20,21,18.4	20.1
	CP	22,23,16.8	21.1	20,21,15.2	19.3	20,21,10.8	18.2

where *PDT* Product Type; *DF* Defuzzified

Table 2 Holding cost at retail shop (destination) (₹)

Period		June		July		August	
PDT	Destination	Fuzzy	DF	Fuzzy	DF	Fuzzy	DF
CN	SV	26,25,21.6	24.4	25,26,27	26	35,34,39	35.5
	AC	30,28,29.6	28.9	25,26,19	24	22,20,24.4	21.6
	GN	28,27,28.4	27.6	23,25,15.8	22.2	26,25,23.2	24.8
DD	SV	27,28,23.8	26.7	23,25,17.4	22.6	33,31,35.8	32.7
	AC	27,25,30.6	26.9	27,25,30.6	26.9	25,26,21	24.5
	GN	26,27,23.6	25.9	21,22,18.2	20.8	25,26,22.2	24.8
DC	SV	24,26,18	23.5	26,24,29.6	25.9	16,15,17.2	15.8
	AC	24,24,30.8	25.7	20,19,15.6	18.4	14,13,14.8	13.7
	GN	15,16,11.4	14.6	17,16,17.4	16.6	24,23,24.4	23.6

Total Cost as ₹2692015 and Tolerance Cost as ₹3092015. Wholesellers provide quantity discounts on bulk purchase as well as transportation discounts are availed from the private carriage company. The data of the case are as follows (Tables 1, 2, 3, 4, 5, 6, 7, 8 and 9):

A LINGO code is generated to solve the proposed mathematical model by employing the case data. The total cost of the system is ₹2809260.605, which is comprising of ₹2359.5 as holding cost at all whole sellers, ₹2635993 as purchase cost, ₹85299 as inspection cost at retail stores, ₹81706.46 as transportation cost, and ₹3902.645 as holding cost at retail stores. We are discussing a small part of the results and the remaining results are shown in tubules in Tables 10, 11, 12, and 13. Ending inventory at Wholesale shop 'AN' is 0 pack of CN, 0 and 19 packs of DD and DC resp. Ending inventory at Retail store 'SV' are 2, 5, and 0 packs for product CN, DD, and DC resp. From wholesale shop AN, in the month of June, Co. ordered for retail store SV, and AC as 850 and 0 packs of CN with quantity discounts of 5 and 0% resp. 960 and 0 packs of DD with quantity discount of 6 and 0% for each. Transportation is done in weights (in kg). From wholesale point AN, 1692.5 and 501 kg is transported to retail store SV and AC each, 600 kg is transported to GN with transportation discount of 10, 0, and 0% respectively. As far as vagueness is concerned, Co. is able to minimize the uncertainty nearby 71%.

Table 3 Demand at retail shop (destination) (pack)

PDT, aspiration	Destintination	June Fuzzy	DF	July Fuzzy	DF	August Fuzzy	DF
CN, 2765	SV	240,270,300	270	290,340,290	315	290,340,310	320
	AC	280,290,340	300	340,390,440	390	210,240,190	220
	GN	240,270,340	280	300,310,440	340	300,310,400	330
DD, 3100	SV	210,240,150	210	300,310,480	350	430,460,450	450
	AC	380,390,440	400	380,390,160	330	330,350,330	340
	GN	380,390,240	350	460,490,480	480	200,170,220	190
DC, 2690	SV	200,170,180	180	340,380,340	360	380,390,400	390
	AC	380,390,360	380	240,270,260	260	230,220,250	230
	GN	380,390,280	360	280,290,300	290	240,230,260	240

Demand tolerance is always less than the aspired demand, which may vary.

Table 4 Consumption at retail shop (destination) for each product (in pack)

Destination	June CN	DD	DC	July CN	DD	DC	August CN	DD	DC
SV	268	205	180	300	349	360	320	450	388
AC	298	396	378	387	325	255	216	340	227
GN	276	342	358	333	480	288	327	190	240

Table 5 Purchase cost per pack ($\overline{\overline{\tau}}$)

	Period	June		July		August	
				Source			
PDT	Destination	AN	CP	AN	CP	AN	CP
CN	SV	241	261	284	262	272	253
	AC	254	221	233	263	214	201
	GN	352	323	212	241	232	243
DD	SV	131	182	183	162	172	151
	AC	153	124	131	164	113	104
	GN	253	221	112	141	132	141
DC	SV	161	212	212	193	201	182
	AC	183	154	164	191	143	134
	GN	282	251	142	173	161	172

Table 6 Transportation cost per weight ($\overline{\overline{\tau}}$)

Period	June		July		August	
			Source			
Destination	AN	CP	AN	CP	AN	CP
SV	3	6	5	7	4	6
AC	6	5	7	6	8	7
GN	3	4	4	4	5	5

Table 7 Quantity threshold and discount factor by source AN

CN		DD		DC	
Quantity threshold	Discount factor	Quantity threshold	Discount factor	Quantity threshold	Discount factor
0–100	1	0–200	1	0–205	1
100–200	0.97	200–320	0.98	205–310	0.92
200–320	0.96	320–480	0.96	310–400	0.87
320 and above	0.95	480 and above	0.94	400 and above	0.82

Table 8 Quantity threshold and discount factor by source CP

CN		DD		DC	
Quantity threshold	Discount factor	Quantity threshold	Discount factor	Quantity threshold	Discount factor
0–120	1	0–190	1	0–150	1
120–250	0.90	190–300	0.9	150–300	0.95
250–360	0.84	300–450	0.86	300–410	0.88
360 and above	0.79	450 and above	0.8	410 and above	0.8

Table 9 Weight threshold and discount factor

June		July		August	
Weight Threshold	Discount Factor	Weight Threshold	Discount Factor	Weight Threshold	Discount Factor
500–1000	1	600–900	1	600–1050	1
1000–1200	0.94	900–1400	0.91	1050–1500	0.9
1200 and above	0.9	1400 and above	0.87	1500 and above	0.85

Table 10 Ending inventory at wholesale shop (in packs)

Period	June		July		August	
PDT	AN	CP	AN	CP	AN	CP
CN	0	0	0	0	0	0
DD	0	1	0	0	0	0
DC	19	90	0	0	0	7

Table 11 Ending inventory at wholesale shop (in packs)

Period	June			July			August		
PDT	SV	AC	GN	SV	AC	GN	SV	AC	GN
CN	2	2	4	16	5	10	15	8	12
DD	5	4	8	5	8	7	5	7	8
DC	0	2	2	0	5	4	2	9	4

Table 12 Ordered quantity/discount % age from wholesale shop to retail store

Period		June		July		August	
				Source			
PDT	Destination	AN	CP	AN	CP	AN	CP
	SV	850/5	0	0	0	0	0
CN	AC	0	850/21	0	0	870/5	870/21
	GN	0	0	1045/5	1045/21	0	0
	SV	960/6	1/0	600/6	0	0	0
DD	AC	0	529/20	0	560/20	380/4	789/20
	GN	0	431/14	560/6	559/20	600/6	191/10
SV		205/8	333/12	400/18	0	400/18	410/20
DC	AC	334/13	631/20	0	410/20	460/18	184/5
	GN	400/18	46/0	491/18	410/20	0	273/5

Table 13 Transported weights/discount % age from wholesale shop to retail shop

Period		June			July			August	
					Destination				
Source	SV	AC	GN	SV	AC	GN	SV	AC	GN
AN	1692.5/10	501	600	600	600	1819/13	600	1,505/15	600
CP	500.5	1900.5/10	500	600	615	1696.5/13	615	1,500/15	600.5

6 Conclusion

Although there are many studies reported for procurement-distribution models, very few of them have incorporated fuzzy environment on deteriorating natured products. The objective of minimizing total incurred fuzzy cost during holding at sources and destinations, procurement, transportation with fuzzy demand and fuzzy holding cost in a supply chain network with both quantity and transportation discounts has not been addressed so far. So, the current study proposed a mathematical model for the literature gap identified and mentioned in this study. The proposed model was validated by applying to the real case study data, where the study is trying to reduce vagueness in fuzzy environment which converts the model in crisp form.

Acknowledgments Kanika Gandhi (Lecturer, Quantitative Techniques and Operations) is thankful to her organization "Bharatiya Vidya Bhavan's Usha and Lakshmi Mittal Institute of Management" to provide her opportunity for carrying research work.

References

1. Silver, E.A., Meal, H.C.: A heuristic for selecting lot-size quantities for the case of a deterministic time varying demand rate and discrete opportunities for replenishment. Prod. Invent. Manag. **14**, 64–74 (1973)

2. Donaldson, W.A.: Inventory replenishment policy for a linear trend in demand- an analytical solution. Oper. Res. Q. **28**, 663–670 (1977)
3. Sachan, R.S.: On (T, Si) policy inventory model for deteriorating items with time proportional demand. J. Oper. Res. Soc. **35**, 1013–1019 (1984)
4. Goswami, A., Chowdhury, K.S.: An EOQ model for deteriorating items with shortages and linear trend in demand. J. Oper. Res. Soc. **42**, 1105–1110 (1991)
5. Kang, S., Kim, I.: A study on the price and production level of the deteriorating inventory system. Int. J. Prod. Res. **21**(6), 899–908 (1983)
6. Zimmermann, H.J.: Description and optimization of fuzzy systems. Int. J. Gen. Syst. **2**, 209–215 (1976)
7. Lai, Y.J., Hwang, C.L.: Fuzzy mathematical programming, methods and applications. Springer, Heidelberg (1992)
8. Lai, Y.J., Hwang, C.L.: Fuzzy multiple objective decision making. Springer, Heidelberg (1994)
9. Sommer, G.: Fuzzy inventory scheduling. In: Lasker, G. (ed.) Applied Systems and Cybernetics, vol. VI. Academic Press, New York (1981)
10. Kacprzyk, J., Staniewski, P.: Long term inventory policy making through fuzzy decision making models. Fuzzy Sets Syst. **8**, 117–132 (1982)
11. Lam, S.M., Wang, D.C.: A fuzzy mathematical model for the joint economic lot-size problem with multiple price breaks. Eur. J. Oper. Res. **45**, 499–504 (1996)
12. Roy, T.K., Maiti, M.: A fuzzy inventory model with constraints. OPSEARCH **32**(4), 287–298 (1995)

A Goal Programming Model for Advertisement Selection on Online News Media

Prerna Manik, Anshu Gupta and P. C. Jha

Abstract Promotion plays an important role in determining success of a product/service. Out of the many mediums available, promotion through means of advertisements is most effective and is most commonly used. Due to increasing popularity of the Internet, advertisers yearn for placing their ads on web. Consequently, web advertising has become one of the major sources of income for many websites. Several websites provide free services to the users and generate revenue by placing ads on its webpages. Advertisement for any product/service is placed on the site considering various aspects such as webpage selection, customer demography, product category, page, slot, time, etc. Further, different advertisers bid different costs to place their ads on a particular rectangular slot of a webpage, that is, many ads compete with each other for their placement on a specific position. Hence, in order to maximize the revenue generated through the ads, optimal placement of ads becomes imperative. In this paper, we formulate an advertisement planning problem for web news media maximizing their revenue. Mathematical programming approach is used to solve the problem. A case study is presented in the paper to show the application of the problem.

Keywords Advertisement planning · Revenue maximization · Online news web

P. Manik (✉) · P. C. Jha
Department of Operational Research, University of Delhi, Delhi, India
e-mail: prernamanik@gmail.com

P. C. Jha
e-mail: jhapc@yahoo.com

A. Gupta
SBPPSE, Dr. B. R. Ambedkar University, Delhi, India
e-mail: anshu@aud.ac.in

B. V. Babu et al. (eds.), *Proceedings of the Second International Conference on Soft Computing for Problem Solving (SocProS 2012), December 28–30, 2012*, Advances in Intelligent Systems and Computing 236, DOI: 10.1007/978-81-322-1602-5_143, © Springer India 2014

1 Introduction

Advertising is an indispensable component of the marketing strategy for any firm. A well-designed advertisement campaign attracts a huge customer base, creates a brand name for the product and thereby enhances sales. Firms spend a large amount of capital to create effective exposure for its products by means of advertisements (ads). Today we see various media for advertising, starting from hoardings to television commercials, print ads to web media, and many more. Companies generally use a mix of various ad media to create maximum exposure for their products. Amongst all kinds of advertising media, the Internet has become the most famous and adopted media by the advertisers as well as consumers. And its popularity is increasing as information technology is reaching more and more people in the world and consumers stay connected to web for long hours. Other reasons for popularity of web advertising over other traditional media include traceability, cost effectiveness, reach, interactivity, etc. It is also capable of providing the dual features of both print and television media. The study in this paper focuses on a web ad scheduling problem where the objective is to maximize the revenue generated from placing ads on the multiple pages of a website. Maximization is achieved by selecting the ads on a slot from various competing ads in such a way that all the slots on every webpage under consideration are full at every time instant throughout the planning horizon. The proposed model surmounts one of the major limitations of the literature in the area.

Web ads commonly known as "banner ads" are devoted to promote, market, sell, or provide specific information about a product, service, or commercial event on web. Television and radio ads are expensive, short lived, and people tend to ignore them. However, in the case of web ads, customers have a choice as to whether or not they want to read or click on a web ad, while the ads may create an impression on the web users irrespective of their choice.

Though banner ads are the most popular ads on the web and constitute a major proportion of web advertising, there are also other ads that have been adopted by the advertisers such as are pop-up and pop-under ads, floating ads, unicast ads, etc. However, scope of this paper is limited to banner ads only. A banner ad is a small, typically rectangular, graphic image, which is linked to a target webpage. Many different types of banners with different sizes are being used in web advertisement. Rectangular-shaped banner ads are the most common type of banner ads. These banners usually appear on the side, top, or bottom of a screen as a distinct, clickable image [9]. For e.g., in Fig. 1, www.newswebsite.com displays a top banner ad of IBEF and two side banner ads of Artha Villas and CRAZEAL.

Due to constantly increasing popularity and power of web advertising, more and more websites and blogs are evolving that provide free services to their users. For e.g., websites such as Download.com, soft32, softpedia provide free software download facilities to the users. Also there are websites such as ApnaCircle and LinkedIn which help millions of professionals to connect and share their ideas for free. Then there are also websites such as hindustantimes.com and timesofindia.com, which provides their users a free service of e-paper, that is, a reader can read the newspaper

Fig. 1 Banner Ads on www.newswebsite.com (Date clicked: September 15, 2012)

online. Such websites generate major portion of their revenue by placing ads on their webpages. Hence, for such sites, optimal placement of ads on their webpages becomes imperative.

Many researchers have been working in the area of scheduling ads on web from past few years. One of the major focuses of the work has been on the effectiveness of web ads. Yager [14] described a general framework for the competitive selection of ads at web sites. A methodology was described in the paper for the use of intelligent agents to help in the determination of the appropriateness of displaying a given ad to a visitor at a site using very specific information about potential customers. Fuzzy system modeling was used for the construction of these intelligent agents. Dreze and Zufryden [3], Intern.com Corp. [5], Kohda and Endo [6], Marx [8] and Risden et al. [11] tackled the issue of increasing the effectiveness of web ads. Intern.com Corp. [5], McCandless [9], and Novak and Hoffman [10] described web advertising theories and terminologies. Researchers viz. Aggarwal et al. [2], Adler et al. [1], Kumar et al. [7] considered the issue of optimizing the ad space on the web. Aggarwal et al. [2] described a framework and provided an overview of general methods for optimizing the management of ads on web servers. They described a minimum cost flow model in order to optimize the assignment of ads to the predefined standard sizes of slots on webpages. Adler et al. [1] provided a heuristic called SUBSET-LSLF.

A major contribution in the area of ad scheduling has been done by Kumar et al. [7] and Gupta et al. [4]. Kumar et al. [7] addressed the problem of scheduling ads on a webpage in order to maximize revenue, for which they maximized the utilization of space available to place the ads. They used genetic algorithms to solve the problem. The major limitations of the model are that it considers only a particular side banner space on a specific page whose width is fixed and length can vary. The rectangular dimensions can thus be reduced to single dimension as the other dimension, i.e., width was assumed to be of unit size. All or some ads that can fit in this banner should therefore have the width of unit size. However, in practice banner ads that compete to be placed on a rectangular slot may be of varying rectangular dimensions. Second, in reality, the varying dimension of the slot for the banner can be a real value and need not necessarily be an integral multiple of the defined unit slot length. Therefore, the problem that maximizes the space utilization may not be the true representative of the revenue maximization problem. Gupta et al. [4] overcame the limitations of the model formulated by Kumar et al. [7]. They considered the set of ads competing to be placed on various rectangular slots (that may have varying rectangular dimensions) in a given planning horizon on various webpages of a news website in order to maximize the revenue, where the revenue is generated from the costs different advertisers pay to place their ads on the website. One of the limitations of the model formulated by Gupta et al. [4] is that it allows an ad to appear more than once on the same webpage. For instance, suppose that a webpage W_1 has three rectangular slots and suppose that an ad A_1 has appeared in slot 1 of this webpage at time period T_1. Then according to this model this ad A_1 can also appear in slot 2 and/or slot 3 of webpage W_1 at the same time period T_1.

In this paper, we formulate a web ad scheduling problem considering sets of ads competing to be placed on various rectangular slots (which may have different rectangular dimensions) in a given planning horizon on different webpages of a news website in order to maximize the revenue. The revenue is generated from the costs different advertisers pay to place their ads on the website. The proposed model restricts the selection of an ad on the same webpage more than once at any instant of time. We also discuss the solution methods for the proposed model, which is a 0-1 linear programming model. The model can be programmed and solved on LINGO [13] software. Depending on the available data the model may or may not be feasible. As the number of constraints increase the feasible area reduces and may tend to infeasibility. In this case, we use goal programming approach (GPA) [12] to obtain a compromised solution. The goal model of the problem can also be programmed and solved on LINGO [13].

The rest of the paper is organized as follows. In Sect. 2, we discuss the mathematical model formulation. A Case study has been discussed in Sect. 3. Section 4 concludes the paper.

2 Model Formulation

Notations

n : total number of webpages

m_j : number of rectangular slots on jth webpage

K : total number of ads

P : total number of time units in a day

Q : total number of days in a planning horizon

T : total number of time units over the planning horizon, where $T = P \times Q$

C_{ijk} : cost of kth ad competing for ith rectangular slot on jth webpage

S : set of K ads

S_{ij} : set of ads which compete for ith rectangular slot on jth webpage; $S_{ij} \subseteq S \forall i, j$

A_k : kth ad

w_k : minimum required time units for which kth ad appears in any rectangular slot

W_k : maximum time units for which kth ad appears in all the rectangular slots

D : total number of rectangular slots over a planning horizon (= Total number of rectangular slots on all the webpages × Length of planning horizon)

2.1 Web Ad Scheduling Problem

Web service providers endeavors to generate maximum revenue from the ads that are displayed on the webpages of their website. Therefore, optimal selection of the ads from the available sets of ads that compete to be placed on different rectangular slots of different webpages becomes critical.

We consider a set of K ads, $S = \{A_1, A_2, \ldots, A_K\}$ that compete to be placed on different rectangular slots of various webpages of a website in a planning horizon. The problem is formulated for a website consisting of n webpages, where jth webpage consists of m_j number of rectangular slots. A subset of ads S_{ij} competes to be placed on ith rectangular slot of jth webpage over a planning horizon. An advertiser k, where $A_k \in S_{ij}$, pays cost C_{ijk} to place his ad on ith rectangular slot of jth webpage with minimum frequency w_k and maximum frequency W_k.

Web ads are scheduled daily, fortnightly, weekly, monthly, or quarterly and so on depending on to the time units allocated to the ads. An ad which appears at any location stays there for some time and is then replaced by another ad. Consider for example, the minimum time for which an ad appears in any rectangular slot is one minute then, there will be $60 \times 24 = 1440$ time slots/units (P) in a day. And if the scheduling is to be done for say one week (i.e., $Q = 7$ days) then there will be a total of $1440 \times 7 = 10080$ time slots, i.e., the planning horizon would be $T = P \times Q = 1440 \times 7 = 10080$ time units.

Over a planning horizon, for each rectangular slot, web service provider selects ads which maximize their revenue and the unscheduled ads may compete for space in the next planning horizon with new ads. The set of ads assigned to all the slots for this time period is seen by the visitors who visit the site during that time interval and then the ads are updated according to their schedule. Now consider that we have in total $\sum_{j=1}^{n} m_j$ number of rectangular slots and a total of $T = P \times Q$ time units in the planning horizon, which can be considered as a scheduling problem of $D = T \times \sum_{j=1}^{n} m_j$ slots. Minimum frequency w_k represents the number of time units for which the ad A_k must appear when selected for some slot and maximum frequency W_k represents the number of time units for which the ad A_k must appear in all the rectangular slots over a planning horizon.

The problem to maximize the revenue generated by placing ads on the website over a planning horizon, which depends heavily on the costs different companies pay for placing their ads on ith rectangular slot of jth webpage is as follows:

$$\text{Maximize} \quad R = \sum_{i=1}^{m_j} \sum_{j=1}^{n} \sum_{k \in S_{ij}} \sum_{t=1}^{T} C_{ijk} x_{ijkt}$$

$$\text{Subject to} \quad \sum_{t=1}^{T} x_{ijkt} \geq w_k z_{ijk} \quad \forall\, i, j, k \in S_{ij}$$

$$\sum_{i=1}^{m_j} \sum_{j=1}^{n} \sum_{t=1}^{T} x_{ijkt} \leq W_k \quad \forall\, k \in S_{ij}$$

$$\sum_{i=1}^{m_j} \sum_{j=1}^{n} \sum_{k \in S_{ij}} \sum_{t=1}^{T} x_{ijkt} \leq D \tag{1}$$

$$\sum_{i=1}^{m_j} x_{ijkt} = 1 \quad \forall\, j, k \in S_{ij}, t$$

$$\sum_{k \in S_{ij}} x_{ijkt} \leq 1 \quad \forall\, i, j, t$$

$$\sum_{k \in S_{ij}} z_{ijk} \geq 1 \quad \forall\, i, j$$

where $x_{ijkt} = \begin{cases} 1, & \text{if } k\text{th ad is chosen to be placed on } i\text{th rectangular slot of } j\text{th} \\ & \text{webpage at } t\text{th time unit} \\ 0, & \text{otherwise} \end{cases}$

$z_{ijk} = \begin{cases} 1, & \text{if } k\text{th ad is placed on } i\text{th rectangular slot of } j\text{th webpage} \\ 0, & \text{otherwise} \end{cases}$

In the above problem $t = 1, \ldots, P \times Q = T$. Time slots are arranged in the ordinal manner i.e. the 1st P time units will correspond to 1st day, next P for 2nd day and so on.

Here, first constraint ensures that kth ad is assigned to at least w_k time slots. Second constraint guarantees that kth ad is assigned to not more than W_k number of slots over the planning horizon. Next constraint ensures the fullness of total number of rectangular slots over the planning horizon. Fourth constraint guarantees that if an ad is selected to be placed on any rectangular slot of a webpage at any given time period then that ad cannot appear on any other rectangular slot of that webpage at the same time unit. Next constraint ensures that at a particular time unit, on each rectangular slot on a webpage, not more than one ad can be placed. Last constraint ensures that number of times ad k appears on a particular rectangular slot over the planning horizon can be one or more than one.

Problem (1) can be solved using LINGO [13] software if a feasible solution to the problem exists. Otherwise for an infeasible solution, GPA [12] can be used to obtain a compromised solution.

2.2 Goal Programming Approach

In a simpler version of goal programming approach (GPA), management sets goals and relative importance (weights) for different objectives. Then an optimal solution is defined as one that minimizes both positive and negative deviations from set goals simultaneously or minimizes the amount by which each goal can be violated. First we solve the problem using rigid constraints only and then the goals of objectives are incorporated depending upon whether priorities or relative importance of different objectives are well defined or not. Problem (1) can be solved in two stages as follows:

$$
\textbf{Minimize} \quad g_0(\eta, \rho, x, z) = \sum_{i=1}^{m_j} \sum_{j=1}^{n} \sum_{k \in S_{ij}} \eta_{ijk}^1 + \sum_{k \in S_{ij}} \rho_k^2 + \rho^3
$$

$$
+ \sum_{j=1}^{n} \sum_{k \in S_{ij}} \sum_{t=1}^{T} \left(\eta_{jkt}^4 + \rho_{jkt}^4 \right) + \sum_{i=1}^{m_j} \sum_{j=1}^{n} \sum_{t=1}^{T} \rho_{ijt}^5
$$

$$
+ \sum_{i=1}^{m_j} \sum_{j=1}^{n} \eta_{ij}^6
$$

$$
\textbf{Subject to} \quad \sum_{t=1}^{T} x_{ijkt} + \eta_{ijk}^1 - \rho_{ijk}^1 = w_k z_{ijk} \quad \forall\, i, j, k \in S_{ij}
$$

$$
\sum_{i=1}^{m_j} \sum_{j=1}^{n} \sum_{t=1}^{T} x_{ijkt} + \eta_k^2 - \rho_k^2 = W_k \quad \forall\, k \in S_{ij}
$$

$$
\sum_{i=1}^{m_j} \sum_{j=1}^{n} \sum_{k \in S_{ij}} \sum_{t=1}^{T} x_{ijkt} + \eta^3 - \rho^3 = D \tag{2}
$$

$$
\sum_{i=1}^{m_j} x_{ijkt} + \eta_{jkt}^4 - \rho_{jkt}^4 = 1 \quad \forall\, j, k \in S_{ij}, t
$$

$$
\sum_{k \in S_{ij}} x_{ijkt} + \eta_{ijt}^5 - \rho_{ijt}^5 = 1 \quad \forall\, i, j, t
$$

$$
\sum_{k \in S_{ij}} z_{ijk} + \eta_{ij}^6 - \rho_{ij}^6 = 1 \quad \forall\, i, j
$$

$$
\eta, \rho \geq 0
$$

where, x_{ijkt} and z_{ijk} are as defined above and η and ρ are over-and under-achievement (positive- and negative-deviational) variables from the goals for the objective/constraint function and $g_0(\eta, \rho, x, z)$, is Goal objective function corresponding to rigid constraints.

The choice of deviational variable in the goal objective functions which has to be minimized depends upon the following rule. Let $f(X)$ and b be the function and its goal respectively and η and ρ be the over and under achievement variables then

if $f(X) \leq b$, ρ is minimized under the constraints $f(X) + \eta - \rho = b$,

if $f(X) \geq b$, η is minimized under the constraints $f(X) + \eta - \rho = b$,

if $f(X) = b$, $\eta + \rho$ is minimized under the constraints $f(X) + \eta - \rho = b$.

Let $(\eta^0, \rho^0, x^0, z^0)$ be the optimal solution for the problem (2) and $g_0(\eta^0, \rho^0, x^0, z^0)$ be its corresponding objective function value then finally GP problem can be formulated using optimal solution of the problem (2) through the problem (1) as follows:

$$\textbf{Minimize} \quad g(\eta, \rho, x, z) = \eta^7$$

$$\textbf{Subject to} \sum_{t=1}^{T} x_{ijkt} + \eta_{ijk}^1 - \rho_{ijk}^1 = w_k z_{ijk} \quad \forall\, i, j, k \in S_{ij}$$

$$\sum_{i=1}^{m_j} \sum_{j=1}^{n} \sum_{t=1}^{T} x_{ijkt} + \eta_k^2 - \rho_k^2 = W_k \quad \forall\, k \in S_{ij}$$

$$\sum_{i=1}^{m_j} \sum_{j=1}^{n} \sum_{k \in S_{ij}} \sum_{t=1}^{T} x_{ijkt} + \eta^3 - \rho^3 = D$$

$$\sum_{i=1}^{m_j} x_{ijkt} + \eta_{jkt}^4 - \rho_{jkt}^4 = 1 \quad \forall\, j, k \in S_{ij}, t \qquad (3)$$

$$\sum_{k \in S_{ij}} x_{ijkt} + \eta_{ijt}^5 - \rho_{ijt}^5 = 1 \quad \forall\, i, j, t$$

$$\sum_{k \in S_{ij}} z_{ijk} + \eta_{ij}^6 - \rho_{ij}^6 = 1 \quad \forall\, i, j$$

$$\sum_{i=1}^{m_j} \sum_{j=1}^{n} \sum_{k \in S_{ij}} \sum_{t=1}^{T} C_{ijk} x_{ijkt} + \eta^7 - \rho^7 = R^*$$

$$g_0(\eta, \rho, x, z) = g_0(\eta^0, \rho^0, x^0, z^0)$$

$$\eta, \rho \geq 0$$

where R^* is the aspiration level desired by the management on revenue and $g(\eta, \rho, x, z)$ is objective function of the problem (3). Problem (3) is solved using LINGO [13].

3 Case Study

In case of online news services, users spend long time on sites for reading news. In case of such websites, ads are updated periodically during this period, which is taken to be 1 h (length of one time slot) here.

We consider a news website which consists of five webpages. These pages have 3, 4, 2, 3, and 3 rectangular slots, respectively. A set of sixty ads, $S = \{A_1, A_2, \ldots, A_{60}\}$ compete to be placed on webpages of a news website in a planning horizon, which is taken as 1 week. Now, a week consists of 7 days and each day consists of 24 h. Since ads are updated every hour on the webpages, we refer to each hour as a time unit. Thus, in this case, we have $168(= 24 \times 7)$ time units to schedule ads. Ads need to be placed in $D = \sum_{j=1}^{5} m_j \times T(= 3 + 4 + 2 + 3 + 3) \times 168 = 2520$ slots. Sets of ads competing for ith rectangular slot on jth webpage and corresponding costs are as follows:

$S_{11} = \{A_1, A_3, A_6, A_8, A_{10}, A_{13}, A_{15}, A_{18}, A_{20}, A_{23}, A_{25}, A_{27}, A_{29}, A_{32}, A_{33}, A_{36}, A_{38}, A_{40},$
$\qquad A_{43}, A_{44}, A_{46}, A_{49}, A_{51}, A_{54}, A_{57}, A_{59}\}; C_{11k} = 2,500 \quad \forall k \in S_{11}$

$S_{21} = \{A_2, A_4, A_6, A_9, A_{11}, A_{13}, A_{14}, A_{16}, A_{19}, A_{22}, A_{24}, A_{26}, A_{28}, A_{31}, A_{33}, A_{35}, A_{37},$
$\qquad A_{39}, A_{42}, A_{44}, A_{47}, A_{49}, A_{50}, A_{52}, A_{54}, A_{56}, A_{57}, A_{60}\}; C_{21k} = 2,300 \quad \forall k \in S_{21}$

$S_{31} = \{A_1, A_2, A_5, A_8, A_{10}, A_{12}, A_{14}, A_{17}, A_{18}, A_{20}, A_{23}, A_{24}, A_{26}, A_{28}, A_{29}, A_{31}, A_{34}, A_{35},$
$\qquad A_{37}, A_{38}, A_{40}, A_{42}, A_{43}, A_{45}, A_{47}, A_{50}, A_{51}, A_{53}, A_{55}, A_{58}, A_{59}\}; C_{31k} = 2,000 \quad \forall k \in S_{31}$

$S_{12} = \{A_2, A_3, A_5, A_7, A_9, A_{11}, A_{13}, A_{15}, A_{17}, A_{19}, A_{21}, A_{22}, A_{24}, A_{25}, A_{27}, A_{30}, A_{32}, A_{35},$
$\qquad A_{36}, A_{38}, A_{39}, A_{41}, A_{43}, A_{44}, A_{46}, A_{47}, A_{49}, A_{52}, A_{54}, A_{56}, A_{60}\}; C_{12k} = 1850 \quad \forall k \in S_{12}$

$S_{22} = \{A_1, A_3, A_4, A_6, A_8, A_{11}, A_{14}, A_{16}, A_{18}, A_{19}, A_{21}, A_{23}, A_{26}, A_{27}, A_{28}, A_{29}, A_{33}, A_{34},$
$\qquad A_{37}, A_{39}, A_{41}, A_{44}, A_{45}, A_{47}, A_{48}, A_{50}, A_{51}, A_{54}, A_{55}, A_{57}, A_{59}\}; C_{22k} = 1800 \quad \forall k \in S_{22}$

$S_{32} = \{A_1, A_2, A_5, A_6, A_9, A_{12}, A_{15}, A_{17}, A_{18}, A_{20}, A_{22}, A_{24}, A_{25}, A_{26}, A_{28}, A_{30}, A_{34}, A_{35},$
$\qquad A_{38}, A_{40}, A_{42}, A_{46}, A_{48}, A_{51}, A_{55}, A_{58}\}; C_{32k} = 1700 \quad \forall k \in S_{32}$

$S_{42} = \{A_3, A_4, A_6, A_7, A_{10}, A_{12}, A_{14}, A_{16}, A_{17}, A_{19}, A_{22}, A_{25}, A_{27}, A_{29}, A_{30}, A_{32}, A_{35}, A_{36},$
$\qquad A_{37}, A_{40}, A_{42}, A_{45}, A_{46}, A_{47}, A_{48}, A_{50}, A_{52}, A_{54}, A_{56}, A_{58}, A_{60}\}; C_{42k} = 1600 \quad \forall k \in S_{42}$

$S_{13} = \{A_1, A_2, A_4, A_5, A_8, A_9, A_{11}, A_{13}, A_{15}, A_{17}, A_{18}, A_{20}, A_{21}, A_{23}, A_{24}, A_{28}, A_{31}, A_{33}, A_{35},$
$\qquad A_{37}, A_{39}, A_{41}, A_{43}, A_{45}, A_{48}, A_{51}, A_{52}, A_{53}, A_{54}, A_{55}, A_{57}, A_{59}\}; C_{13k} = 1500 \quad \forall k \in S_{13}$

$S_{23} = \{A_2, A_3, A_5, A_6, A_7, A_{10}, A_{12}, A_{14}, A_{16}, A_{18}, A_{19}, A_{21}, A_{23}, A_{26}, A_{28}, A_{29}, A_{31}, A_{34},$
$\qquad A_{36}, A_{38}, A_{40}, A_{42}, A_{44}, A_{46}, A_{49}, A_{53}, A_{54}, A_{58}, A_{60}\}; C_{23k} = 1400 \quad \forall k \in S_{23}$

$S_{14} = \{A_3, A_4, A_8, A_{10}, A_{11}, A_{14}, A_{17}, A_{22}, A_{24}, A_{25}, A_{26}, A_{29}, A_{32}, A_{34}, A_{38}, A_{41}, A_{44}, A_{46},$
$\qquad A_{49}, A_{51}, A_{53}, A_{56}, A_{59}\}; C_{14k} = 1300 \quad \forall k \in S_{14}$

$S_{24} = \{A_2, A_4, A_7, A_9, A_{12}, A_{14}, A_{16}, A_{18}, A_{21}, A_{23}, A_{26}, A_{28}, A_{29}, A_{31}, A_{33}, A_{35}, A_{37}, A_{39},$
$\qquad A_{42}, A_{45}, A_{47}, A_{49}, A_{52}, A_{54}, A_{56}, A_{59}, A_{60}\}; C_{24k} = 1200 \quad \forall k \in S_{24}$

$S_{34} = \{A_2, A_3, A_5, A_7, A_{10}, A_{13}, A_{15}, A_{17}, A_{19}, A_{22}, A_{25}, A_{26}, A_{29}, A_{32}, A_{33}, A_{36}, A_{38}, A_{40},$
$\qquad A_{43}, A_{44}, A_{46}, A_{48}, A_{51}, A_{53}, A_{55}, A_{58}, A_{60}\}; C_{34k} = 1100 \quad \forall k \in S_{34}$

$S_{15} = \{A_1, A_5, A_6, A_8, A_9, A_{11}, A_{12}, A_{16}, A_{18}, A_{20}, A_{22}, A_{23}, A_{25}, A_{27}, A_{30}, A_{34}, A_{35}, A_{39},$
$\qquad A_{41}, A_{43}, A_{45}, A_{46}, A_{47}, A_{48}, A_{51}, A_{54}, A_{55}, A_{57}, A_{59}\}; C_{15k} = 1000 \quad \forall k \in S_{15}$

$S_{25} = \{A_2, A_4, A_6, A_8, A_{10}, A_{13}, A_{15}, A_{17}, A_{19}, A_{21}, A_{24}, A_{26}, A_{28}, A_{31}, A_{33}, A_{35}, A_{36}, A_{38},$

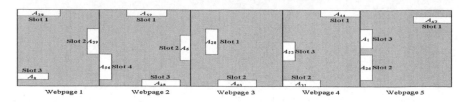

Fig. 2 Display of ads on the news website at time unit $t = 66$

$$A_{40}, A_{42}, A_{44}, A_{46}, A_{47}, A_{49}, A_{50}, A_{53}, A_{56}, A_{58}, A_{59}\}; C_{25k} = 850 \quad \forall k \in S_{25}$$
$$S_{35} = \{A_1, A_3, A_5, A_7, A_9, A_{11}, A_{12}, A_{14}, A_{18}, A_{20}, A_{21}, A_{23}, A_{25}, A_{27}, A_{30}, A_{32}, A_{34}, A_{37},$$
$$A_{39}, A_{41}, A_{43}, A_{45}, A_{48}, A_{50}, A_{52}, A_{54}, A_{57}, A_{59}\}; C_{35k} = 800 \quad \forall k \in S_{35}$$

Although different companies may pay different costs for placing their ads on a particular rectangular slot of a specific webpage but for the sake of simplicity, cost for all the ads competing to be placed on any rectangular slot is taken to be same here. Further, minimum and maximum frequencies of the ads are tabulated in Table 1.

When problem (1) is solved using the above data, we get an infeasible solution and hence it is imperative to use GPA to obtain a compromised solution. Aspiration desired on revenue is taken to be ₹ 40,00,000. The compromised solution obtained after applying GPA is given in Table 2.

The revenue generated by this placement of ads comes out to be ₹ 38,47,200. It can be seen from Table 2 that when users access webpage 3 of news website at 18th hour of day 3 i.e., at $t = 66$, ad A_{60} appears in second slot of that webpage. To give a clear picture of how the ads are actually displayed to the users in accordance with the schedule obtained in Table 2, a pictorial representation of the selected ads on the news website at time unit $t = 66$ is shown in Fig. 2.

Table 1 Frequency table

Ads (A_k)	Min. freq. (w_k)	Max. freq. (W_k)	Ads (A_k)	Min. freq. (w_k)	Max. freq. (W_k)	Ads (A_k)	Min. freq. (w_k)	Max. freq. (W_k)	Ads (A_k)	Min. freq. (w_k)	Max. freq. (W_k)	Ads (A_k)	Min. freq. (w_k)	Max. freq. (W_k)	Ads (A_k)	Min. freq. (w_k)	Max. freq. (W_k)
A_1	6	50	A_{11}	6	68	A_{21}	5	55	A_{31}	5	50	A_{41}	5	50	A_{51}	7	50
A_2	7	62	A_{12}	5	50	A_{22}	6	60	A_{32}	4	48	A_{42}	7	55	A_{52}	5	45
A_3	6	55	A_{13}	4	52	A_{23}	7	55	A_{33}	6	52	A_{43}	6	52	A_{53}	4	44
A_4	5	60	A_{14}	6	50	A_{24}	6	52	A_{34}	5	50	A_{44}	7	60	A_{54}	8	65
A_5	7	60	A_{15}	4	45	A_{25}	7	60	A_{35}	7	62	A_{45}	6	55	A_{55}	5	50
A_6	6	58	A_{16}	5	52	A_{26}	7	55	A_{36}	6	58	A_{46}	8	62	A_{56}	4	48
A_7	5	55	A_{17}	7	68	A_{27}	5	60	A_{37}	6	55	A_{47}	7	58	A_{57}	4	50
A_8	5	60	A_{18}	8	70	A_{28}	7	60	A_{38}	7	60	A_{48}	5	50	A_{58}	3	40
A_9	6	62	A_{19}	6	60	A_{29}	6	55	A_{39}	6	55	A_{49}	6	52	A_{59}	7	56
A_{10}	5	60	A_{20}	4	50	A_{30}	4	52	A_{40}	6	58	A_{50}	5	48	A_{60}	5	60

Table 2 Ads allocated to rectangular slot i of web page j at time unit t

Time slots	Rectangular slots														
	Slot 11	Slot 21	Slot 31	Slot 12	Slot 22	Slot 32	Slot 42	Slot 13	Slot 23	Slot 14	Slot 24	Slot 34	Slot 15	Slot 25	Slot 35
1	A_{59}	A_{26}	A_{12}	A_3	A_{26}	A_{34}	A_{32}	A_{51}	A_{18}	A_{11}	A_{54}	A_{32}	A_{11}	A_{40}	A_{39}
2	A_{33}	A_{39}	A_5	A_{52}	A_{59}	A_{51}	A_{30}	A_{48}	A_{14}	A_{44}	A_{52}	A_{26}	A_9	A_{38}	A_{37}
3	A_{59}	A_{35}	A_1	A_{47}	A_{55}	A_{35}	A_{29}	A_{45}	A_{12}	A_{41}	A_{49}	A_{22}	A_8	A_{36}	A_{34}
4	A_1	A_{49}	A_{59}	A_{44}	A_{51}	A_{17}	A_{27}	A_{43}	A_{10}	A_{38}	A_{14}	A_{15}	A_6	A_{35}	A_{32}
5	A_{59}	A_{16}	A_{55}	A_{41}	A_{48}	A_{12}	A_{25}	A_{41}	A_7	A_{53}	A_{45}	A_{13}	A_5	A_{33}	A_{30}
6	A_{40}	A_{33}	A_{51}	A_{19}	A_{51}	A_6	A_{22}	A_{39}	A_6	A_{49}	A_{39}	A_{10}	A_1	A_{28}	A_{27}
7	A_{29}	A_{24}	A_{47}	A_{35}	A_{41}	A_2	A_{19}	A_{37}	A_5	A_{46}	A_{37}	A_7	A_{16}	A_{26}	A_{11}
8	A_{33}	A_{16}	A_{43}	A_{30}	A_{37}	A_{22}	A_{17}	A_{35}	A_3	A_{49}	A_{35}	A_5	A_{57}	A_{24}	A_{23}
9	A_{18}	A_{26}	A_{40}	A_{25}	A_{33}	A_{17}	A_{16}	A_{33}	A_{42}	A_{38}	A_{33}	A_3	A_{55}	A_{21}	A_{27}
10	A_{43}	A_9	A_{37}	A_{22}	A_3	A_9	A_{14}	A_{31}	A_{38}	A_{29}	A_{31}	A_2	A_{54}	A_{19}	A_{20}
11	A_{25}	A_2	A_{34}	A_{19}	A_{47}	A_{26}	A_{12}	A_{28}	A_{34}	A_{24}	A_{29}	A_{60}	A_{51}	A_{17}	A_{18}
12	A_{57}	A_{50}	A_{29}	A_{13}	A_{39}	A_{58}	A_{10}	A_{39}	A_{29}	A_{17}	A_{28}	A_{58}	A_{48}	A_{15}	A_{14}
13	A_{59}	A_{49}	A_{37}	A_9	A_{29}	A_{48}	A_7	A_{35}	A_{26}	A_{11}	A_{26}	A_{55}	A_{47}	A_{13}	A_{12}
14	A_{40}	A_{47}	A_{31}	A_2	A_{19}	A_{40}	A_6	A_{39}	A_{40}	A_8	A_{56}	A_{53}	A_{46}	A_{10}	A_{11}
15	A_{23}	A_{44}	A_{24}	A_{27}	A_{59}	A_{34}	A_4	A_{31}	A_{34}	A_3	A_{52}	A_{51}	A_{45}	A_8	A_9
16	A_{51}	A_{42}	A_{18}	A_{60}	A_{54}	A_{25}	A_3	A_{21}	A_{28}	A_{59}	A_{52}	A_{55}	A_1	A_6	A_7
17	A_{10}	A_{39}	A_{26}	A_{54}	A_{48}	A_{20}	A_{60}	A_{15}	A_{19}	A_{53}	A_{47}	A_{48}	A_{59}	A_4	A_5
18	A_{44}	A_{37}	A_5	A_{49}	A_{44}	A_{15}	A_{58}	A_{33}	A_{14}	A_{49}	A_{23}	A_{44}	A_{57}	A_2	A_3
19	A_{40}	A_{35}	A_{29}	A_{46}	A_{37}	A_6	A_{56}	A_{31}	A_7	A_{44}	A_{21}	A_{43}	A_{55}	A_{59}	A_{27}
20	A_{54}	A_{60}	A_{55}	A_{43}	A_{29}	A_2	A_{54}	A_{28}	A_2	A_{41}	A_{18}	A_{40}	A_{54}	A_{58}	A_{23}
21	A_{27}	A_{56}	A_{37}	A_{39}	A_{23}	A_{55}	A_{52}	A_{55}	A_{60}	A_{22}	A_{16}	A_{38}	A_{51}	A_{56}	A_{20}
22	A_{57}	A_{52}	A_{34}	A_{36}	A_{21}	A_{46}	A_{50}	A_5	A_{53}	A_{34}	A_{14}	A_{36}	A_{48}	A_{53}	A_{14}
23	A_{18}	A_{49}	A_{29}	A_{32}	A_{21}	A_{38}	A_{48}	A_{51}	A_{44}	A_{32}	A_{12}	A_{33}	A_{47}	A_{50}	A_{11}

(continued)

Table 2 (continued)

Time slots	Rectangular slots														
	Slot 11	Slot 21	Slot 31	Slot 12	Slot 22	Slot 32	Slot 42	Slot 13	Slot 23	Slot 14	Slot 24	Slot 34	Slot 15	Slot 25	Slot 35
24	A_1	A_{22}	A_{40}	A_{27}	A_{18}	A_{30}	A_{47}	A_{45}	A_{40}	A_{29}	A_4	A_{32}	A_{46}	A_{49}	A_7
25	A_{57}	A_{39}	A_{36}	A_{49}	A_{14}	A_{24}	A_{46}	A_{41}	A_{34}	A_{26}	A_2	A_{29}	A_{45}	A_{47}	A_1
26	A_3	A_{33}	A_{51}	A_{21}	A_{14}	A_{18}	A_{45}	A_{37}	A_6	A_{25}	A_{60}	A_{26}	A_{43}	A_{46}	A_{59}
27	A_{43}	A_{28}	A_{50}	A_2	A_{11}	A_{12}	A_{58}	A_{33}	A_{28}	A_{24}	A_{59}	A_{25}	A_{41}	A_{15}	A_{54}
28	A_{57}	A_{24}	A_{40}	A_{54}	A_8	A_{51}	A_{56}	A_{24}	A_{21}	A_{38}	A_{56}	A_{22}	A_{39}	A_{13}	A_{50}
29	A_{36}	A_{16}	A_{31}	A_{49}	A_6	A_{58}	A_{54}	A_{21}	A_2	A_{17}	A_{54}	A_{19}	A_{35}	A_{10}	A_{41}
30	A_{49}	A_{13}	A_{18}	A_{46}	A_4	A_6	A_{52}	A_{18}	A_{60}	A_{14}	A_{52}	A_{17}	A_{34}	A_8	A_{37}
31	A_{46}	A_9	A_8	A_{43}	A_3	A_5	A_{50}	A_{15}	A_{58}	A_{11}	A_{49}	A_{15}	A_{30}	A_6	A_{32}
32	A_{54}	A_4	A_1	A_{39}	A_8	A_2	A_{48}	A_{11}	A_{49}	A_4	A_{47}	A_{10}	A_{27}	A_4	A_{21}
33	A_{46}	A_2	A_{55}	A_{38}	A_4	A_{12}	A_{47}	A_8	A_{44}	A_8	A_{45}	A_7	A_{25}	A_2	A_{21}
34	A_{44}	A_{60}	A_{45}	A_{35}	A_1	A_6	A_{46}	A_4	A_{42}	A_{10}	A_4	A_5	A_{23}	A_{59}	A_{18}
35	A_{43}	A_{57}	A_{42}	A_{32}	A_3	A_{26}	A_{45}	A_{51}	A_{36}	A_3	A_{39}	A_{33}	A_{22}	A_{58}	A_{12}
36	A_{40}	A_{56}	A_{38}	A_{30}	A_{54}	A_{55}	A_{42}	A_{55}	A_{31}	A_{59}	A_{37}	A_2	A_{20}	A_{56}	A_9
37	A_{38}	A_{54}	A_{34}	A_{27}	A_{51}	A_2	A_{40}	A_{52}	A_{28}	A_{56}	A_{35}	A_{60}	A_{18}	A_{53}	A_5
38	A_{44}	A_{52}	A_{17}	A_{25}	A_{59}	A_{40}	A_{37}	A_{43}	A_{23}	A_{53}	A_{33}	A_{58}	A_{16}	A_{50}	A_1
39	A_{36}	A_{50}	A_{26}	A_{24}	A_{55}	A_{35}	A_{36}	A_{37}	A_{18}	A_{51}	A_{31}	A_{55}	A_{12}	A_{49}	A_{59}
40	A_{32}	A_{49}	A_{23}	A_{22}	A_{50}	A_{30}	A_{35}	A_{31}	A_{14}	A_{49}	A_{29}	A_{53}	A_{11}	A_{47}	A_{57}
41	A_{27}	A_{47}	A_{18}	A_{21}	A_{47}	A_{25}	A_{32}	A_{23}	A_{10}	A_{46}	A_{28}	A_{51}	A_9	A_{46}	A_3
42	A_{27}	A_{44}	A_{14}	A_{19}	A_{44}	A_{35}	A_{30}	A_{8}	A_6	A_{44}	A_{26}	A_{58}	A_8	A_{44}	A_{52}
43	A_{25}	A_{42}	A_{10}	A_{17}	A_{41}	A_{28}	A_{29}	A_{13}	A_5	A_{41}	A_{56}	A_{44}	A_6	A_{42}	A_{50}
44	A_{23}	A_{39}	A_5	A_{15}	A_{39}	A_{20}	A_{27}	A_9	A_3	A_{38}	A_{52}	A_{43}	A_5	A_{40}	A_{48}
45	A_{40}	A_{37}	A_2	A_{13}	A_{37}	A_{15}	A_{25}	A_5	A_{54}	A_{34}	A_{47}	A_{40}	A_1	A_{38}	A_{45}
46	A_{36}	A_4	A_{58}	A_{11}	A_{34}	A_6	A_{22}	A_2	A_{49}	A_{32}	A_{42}	A_{38}	A_{59}	A_{36}	A_{43}

(continued)

Table 2 (continued)

Time slots	Rectangular slots														
	Slot 11	Slot 21	Slot 31	Slot 12	Slot 22	Slot 32	Slot 42	Slot 13	Slot 23	Slot 14	Slot 24	Slot 34	Slot 15	Slot 25	Slot 35
47	A_{32}	A_{33}	A_{53}	A_9	A_{33}	A_1	A_{19}	A_{54}	A_{19}	A_{29}	A_{37}	A_{36}	A_{57}	A_{35}	A_{41}
48	A_{27}	A_{31}	A_{50}	A_7	A_{29}	A_{58}	A_{17}	A_{48}	A_{40}	A_{10}	A_{33}	A_3	A_{55}	A_{33}	A_{39}
49	A_{20}	A_{28}	A_{45}	A_5	A_{28}	A_{51}	A_{16}	A_{41}	A_{36}	A_{25}	A_{29}	A_{32}	A_{54}	A_{31}	A_{37}
50	A_{15}	A_{16}	A_{51}	A_3	A_{27}	A_{55}	A_{14}	A_{35}	A_{31}	A_{24}	A_{23}	A_{29}	A_{51}	A_{28}	A_{34}
51	A_{10}	A_{13}	A_{38}	A_{24}	A_{26}	A_{42}	A_{12}	A_{45}	A_{28}	A_{53}	A_{18}	A_{26}	A_{48}	A_{26}	A_{32}
52	A_{51}	A_9	A_{36}	A_{21}	A_{33}	A_{35}	A_{10}	A_{39}	A_2	A_{49}	A_{14}	A_{25}	A_{16}	A_{24}	A_{30}
53	A_{49}	A_6	A_{31}	A_{17}	A_{16}	A_{28}	A_7	A_{33}	A_{21}	A_{44}	A_9	A_{22}	A_{12}	A_{21}	A_{27}
54	A_{54}	A_4	A_{28}	A_{21}	A_{23}	A_{12}	A_6	A_{24}	A_{38}	A_{38}	A_4	A_{19}	A_{11}	A_{19}	A_{25}
55	A_{43}	A_2	A_{50}	A_{11}	A_{19}	A_{22}	A_4	A_{20}	A_{31}	A_{32}	A_2	A_{17}	A_9	A_{17}	A_{50}
56	A_{33}	A_{60}	A_{45}	A_5	A_{16}	A_{18}	A_3	A_{15}	A_{26}	A_{25}	A_{52}	A_{15}	A_8	A_{15}	A_{45}
57	A_{13}	A_{57}	A_{42}	A_{17}	A_{26}	A_{15}	A_{60}	A_9	A_{16}	A_{22}	A_{49}	A_{13}	A_6	A_{13}	A_{41}
58	A_{13}	A_{56}	A_{38}	A_{56}	A_{29}	A_9	A_{58}	A_1	A_{10}	A_{14}	A_{47}	A_{10}	A_5	A_{10}	A_{37}
59	A_3	A_{54}	A_{36}	A_{47}	A_{27}	A_5	A_{56}	A_{59}	A_{54}	A_{10}	A_{45}	A_7	A_1	A_8	A_{32}
60	A_{20}	A_{52}	A_{31}	A_{43}	A_{21}	A_1	A_{54}	A_{54}	A_{49}	A_4	A_{42}	A_5	A_{59}	A_6	A_{27}
61	A_6	A_{50}	A_{28}	A_{38}	A_{18}	A_{58}	A_6	A_{52}	A_{46}	A_{29}	A_{21}	A_3	A_{57}	A_2	A_{21}
62	A_{44}	A_{49}	A_{24}	A_{11}	A_{16}	A_{51}	A_4	A_{48}	A_{40}	A_{56}	A_{26}	A_2	A_{55}	A_{59}	A_{18}
63	A_{40}	A_{47}	A_{18}	A_7	A_{14}	A_{48}	A_3	A_{43}	A_{36}	A_{51}	A_{49}	A_{60}	A_{54}	A_{33}	A_{12}
64	A_{25}	A_{44}	A_{17}	A_2	A_{11}	A_{46}	A_{60}	A_{39}	A_{31}	A_{49}	A_{45}	A_{58}	A_{51}	A_{31}	A_9
65	A_{10}	A_{42}	A_{12}	A_{56}	A_8	A_{42}	A_{58}	A_{33}	A_2	A_{46}	A_{35}	A_{55}	A_{48}	A_{28}	A_5
66	A_{29}	A_{39}	A_8	A_{52}	A_6	A_{48}	A_{56}	A_{28}	A_{60}	A_{44}	A_{31}	A_{53}	A_{47}	A_{26}	A_1
67	A_{46}	A_{37}	A_1	A_{47}	A_4	A_{38}	A_{54}	A_{31}	A_{53}	A_{41}	A_{28}	A_{51}	A_{46}	A_{24}	A_{41}
68	A_{32}	A_{35}	A_{59}	A_{44}	A_3	A_{35}	A_{52}	A_{21}	A_{44}	A_{38}	A_{26}	A_{53}	A_{45}	A_{21}	A_{39}
69	A_{15}	A_{57}	A_{55}	A_{41}	A_8	A_{30}	A_{50}	A_{17}	A_{38}	A_{34}	A_{49}	A_{46}	A_{43}	A_{19}	A_{37}

(continued)

Table 2 (continued)

Time slots	Rectangular slots														
	Slot 11	Slot 21	Slot 31	Slot 12	Slot 22	Slot 32	Slot 42	Slot 13	Slot 23	Slot 14	Slot 24	Slot 34	Slot 15	Slot 25	Slot 35
70	A_1	A_{54}	A_{37}	A_{38}	A_4	A_{40}	A_{48}	A_{11}	A_{29}	A_{32}	A_{45}	A_{44}	A_{41}	A_{17}	A_{34}
71	A_{57}	A_{50}	A_{47}	A_{36}	A_1	A_{22}	A_{47}	A_9	A_{23}	A_4	A_{39}	A_{43}	A_{39}	A_{15}	A_{32}
72	A_3	A_{22}	A_{43}	A_{35}	A_{57}	A_{12}	A_{46}	A_5	A_{18}	A_{26}	A_{33}	A_{40}	A_{35}	A_{13}	A_{30}
73	A_{43}	A_{19}	A_{40}	A_{32}	A_{54}	A_{34}	A_{45}	A_1	A_{12}	A_{59}	A_{29}	A_{38}	A_{34}	A_{10}	A_{27}
74	A_{38}	A_{31}	A_{37}	A_{30}	A_{51}	A_{28}	A_{42}	A_{59}	A_{58}	A_{53}	A_{28}	A_{36}	A_{30}	A_8	A_{25}
75	A_{33}	A_{37}	A_{31}	A_{27}	A_{59}	A_{26}	A_{40}	A_{55}	A_{53}	A_{49}	A_{21}	A_{32}	A_{27}	A_6	A_{50}
76	A_{27}	A_{24}	A_{28}	A_{25}	A_4	A_{55}	A_{37}	A_{53}	A_{46}	A_{44}	A_{14}	A_{29}	A_{25}	A_4	A_{45}
77	A_{23}	A_{26}	A_{24}	A_{24}	A_{50}	A_{48}	A_{36}	A_2	A_{42}	A_{38}	A_{12}	A_{26}	A_{23}	A_2	A_{41}
78	A_{18}	A_{14}	A_{20}	A_{22}	A_{47}	A_{34}	A_{35}	A_{45}	A_{38}	A_{32}	A_4	A_{25}	A_{22}	A_{59}	A_{37}
79	A_{13}	A_{11}	A_{17}	A_{21}	A_{44}	A_{38}	A_{32}	A_{43}	A_{34}	A_{25}	A_2	A_{22}	A_{20}	A_{58}	A_{32}
80	A_6	A_2	A_{12}	A_{19}	A_{41}	A_{34}	A_{30}	A_{41}	A_{29}	A_{22}	A_{59}	A_{19}	A_{18}	A_{56}	A_{27}
81	A_3	A_4	A_8	A_{17}	A_{39}	A_{25}	A_{29}	A_{39}	A_{26}	A_{14}	A_{26}	A_{17}	A_{16}	A_{53}	A_{21}
82	A_{29}	A_6	A_2	A_{15}	A_{37}	A_{24}	A_{27}	A_{37}	A_{60}	A_{10}	A_{52}	A_{13}	A_{12}	A_{50}	A_{18}
83	A_{54}	A_{60}	A_{59}	A_{13}	A_{34}	A_{20}	A_{25}	A_{35}	A_{53}	A_4	A_{49}	A_{10}	A_{11}	A_{49}	A_{12}
84	A_{44}	A_{57}	A_{55}	A_{11}	A_{33}	A_{17}	A_{22}	A_{39}	A_{44}	A_3	A_{47}	A_7	A_9	A_{47}	A_1
85	A_{38}	A_{56}	A_{51}	A_9	A_{29}	A_{12}	A_{19}	A_{37}	A_{23}	A_{56}	A_{45}	A_5	A_8	A_{46}	A_3
86	A_{32}	A_{54}	A_{47}	A_7	A_{28}	A_6	A_{17}	A_{35}	A_{31}	A_{51}	A_{42}	A_3	A_6	A_{44}	A_1
87	A_{23}	A_{52}	A_{43}	A_5	A_{27}	A_2	A_{16}	A_{33}	A_{28}	A_{49}	A_{60}	A_2	A_5	A_{42}	A_{59}
88	A_{15}	A_{39}	A_{40}	A_3	A_{26}	A_6	A_{12}	A_{31}	A_{21}	A_{46}	A_{56}	A_{60}	A_{30}	A_{40}	A_{57}
89	A_8	A_{35}	A_2	A_{22}	A_{29}	A_5	A_{10}	A_{28}	A_{18}	A_{44}	A_{52}	A_{58}	A_{27}	A_{38}	A_{54}
90	A_1	A_{56}	A_{59}	A_{19}	A_{27}	A_1	A_7	A_{24}	A_{14}	A_{41}	A_{16}	A_{55}	A_{25}	A_{36}	A_3
91	A_{59}	A_{52}	A_{55}	A_{15}	A_8	A_{51}	A_6	A_{23}	A_{10}	A_{38}	A_{39}	A_{53}	A_{23}	A_{35}	A_{50}
92	A_{51}	A_{49}	A_{24}	A_{11}	A_{55}	A_{48}	A_4	A_{21}	A_6	A_{34}	A_{35}	A_{51}	A_{22}	A_{33}	A_{48}
93	A_{43}	A_{24}	A_{47}	A_7	A_1	A_{46}	A_3	A_{30}	A_3	A_{32}	A_{31}	A_{48}	A_{20}	A_{31}	A_{45}

(continued)

Table 2 (continued)

Time slots	Rectangular slots														
	Slot 11	Slot 21	Slot 31	Slot 12	Slot 22	Slot 32	Slot 42	Slot 13	Slot 23	Slot 14	Slot 24	Slot 34	Slot 15	Slot 25	Slot 35
94	A_{36}	A_{39}	A_{43}	A_2	A_{57}	A_{42}	A_{60}	A_{51}	A_{42}	A_{51}	A_{28}	A_{46}	A_{18}	A_{26}	A_{43}
95	A_{49}	A_{33}	A_{40}	A_{60}	A_{54}	A_{55}	A_{14}	A_{28}	A_{36}	A_{26}	A_{21}	A_{44}	A_{16}	A_{24}	A_{41}
96	A_{27}	A_{26}	A_{37}	A_{24}	A_{50}	A_{48}	A_{12}	A_{39}	A_{29}	A_{59}	A_{18}	A_{43}	A_{12}	A_{21}	A_{39}
97	A_{25}	A_{24}	A_{34}	A_{21}	A_{47}	A_{40}	A_{10}	A_{31}	A_{21}	A_{25}	A_{16}	A_{40}	A_{11}	A_{19}	A_{37}
98	A_{23}	A_{19}	A_{29}	A_{17}	A_{44}	A_{35}	A_7	A_{20}	A_{16}	A_{24}	A_{14}	A_{38}	A_9	A_{17}	A_{34}
99	A_{43}	A_{14}	A_{26}	A_{13}	A_{39}	A_{30}	A_6	A_{11}	A_{10}	A_{22}	A_{12}	A_{36}	A_8	A_{47}	A_{32}
100	A_{20}	A_{11}	A_{51}	A_9	A_{34}	A_{18}	A_4	A_{57}	A_6	A_{17}	A_9	A_{33}	A_6	A_{46}	A_{30}
101	A_{32}	A_9	A_{45}	A_5	A_{29}	A_{25}	A_3	A_{54}	A_3	A_{14}	A_2	A_{32}	A_5	A_{44}	A_{27}
102	A_{27}	A_6	A_{40}	A_2	A_{27}	A_{26}	A_{60}	A_{52}	A_{31}	A_{11}	A_{37}	A_{29}	A_1	A_{42}	A_{25}
103	A_{20}	A_4	A_{36}	A_{56}	A_{23}	A_{22}	A_{58}	A_{45}	A_{28}	A_{10}	A_{33}	A_{26}	A_{59}	A_{40}	A_{50}
104	A_{15}	A_2	A_{29}	A_{52}	A_{19}	A_{20}	A_{56}	A_{41}	A_{23}	A_8	A_{60}	A_{25}	A_{57}	A_{38}	A_{45}
105	A_{10}	A_{60}	A_{24}	A_{47}	A_{16}	A_{18}	A_{54}	A_{37}	A_{21}	A_4	A_{59}	A_{22}	A_{55}	A_{36}	A_{18}
106	A_6	A_{57}	A_{17}	A_{44}	A_{11}	A_{17}	A_{52}	A_{33}	A_{58}	A_3	A_{56}	A_{19}	A_{54}	A_{35}	A_{11}
107	A_1	A_{56}	A_{10}	A_{41}	A_6	A_{15}	A_{50}	A_{20}	A_{54}	A_{59}	A_{54}	A_{17}	A_{51}	A_{33}	A_5
108	A_{29}	A_{54}	A_1	A_{38}	A_1	A_{12}	A_{48}	A_{23}	A_{53}	A_{56}	A_{52}	A_{15}	A_{48}	A_{31}	A_1
109	A_{57}	A_{52}	A_{59}	A_{36}	A_{59}	A_9	A_{47}	A_{20}	A_{49}	A_{53}	A_{49}	A_{13}	A_{47}	A_{28}	A_{57}
110	A_{54}	A_{50}	A_{53}	A_{35}	A_{55}	A_{15}	A_{46}	A_{17}	A_{46}	A_{51}	A_9	A_{10}	A_{46}	A_{26}	A_{52}
111	A_{51}	A_{49}	A_{47}	A_{32}	A_{51}	A_{12}	A_{45}	A_{13}	A_6	A_{49}	A_{45}	A_7	A_{45}	A_{24}	A_{50}
112	A_{49}	A_{47}	A_{40}	A_{30}	A_{59}	A_9	A_{42}	A_9	A_{26}	A_{46}	A_{42}	A_5	A_{43}	A_{21}	A_{48}
113	A_{49}	A_{44}	A_{36}	A_{27}	A_{55}	A_6	A_{40}	A_5	A_{40}	A_{44}	A_{29}	A_3	A_{41}	A_{19}	A_{45}
114	A_{44}	A_{42}	A_{29}	A_{25}	A_{50}	A_1	A_{37}	A_2	A_{38}	A_{41}	A_{49}	A_2	A_{39}	A_{17}	A_{43}
115	A_{40}	A_{39}	A_{23}	A_{24}	A_{47}	A_{35}	A_{36}	A_{57}	A_{36}	A_{38}	A_{45}	A_{58}	A_{35}	A_{15}	A_{41}
116	A_{36}	A_{37}	A_{26}	A_{22}	A_{44}	A_{30}	A_{35}	A_{53}	A_{34}	A_{34}	A_{37}	A_{55}	A_{34}	A_{13}	A_{39}
117	A_{32}	A_{35}	A_{14}	A_{21}	A_{41}	A_{58}	A_{32}	A_{41}	A_{10}	A_{32}	A_{33}	A_{53}	A_{30}	A_{10}	A_{37}

(continued)

Table 2 (continued)

Time slots	Rectangular slots														
	Slot 11	Slot 21	Slot 31	Slot 12	Slot 22	Slot 32	Slot 42	Slot 13	Slot 23	Slot 14	Slot 24	Slot 34	Slot 15	Slot 25	Slot 35
118	A_{25}	A_{54}	A_{10}	A_{19}	A_{39}	A_{55}	A_{29}	A_{39}	A_{40}	A_3	A_{29}	A_{51}	A_{27}	A_8	A_{34}
119	A_{20}	A_{50}	A_5	A_{17}	A_{37}	A_{25}	A_{27}	A_{37}	A_{36}	A_{46}	A_{26}	A_{53}	A_{25}	A_6	A_{32}
120	A_{15}	A_{24}	A_2	A_{15}	A_{34}	A_{48}	A_{25}	A_{35}	A_{29}	A_{51}	A_{23}	A_{46}	A_{23}	A_2	A_{30}
121	A_8	A_{42}	A_{58}	A_{11}	A_{33}	A_{46}	A_{22}	A_{33}	A_5	A_8	A_{21}	A_{40}	A_{22}	A_{59}	A_{27}
122	A_3	A_{37}	A_{53}	A_{11}	A_{29}	A_{42}	A_{19}	A_{31}	A_{19}	A_{41}	A_{18}	A_{43}	A_{20}	A_{58}	A_{25}
123	A_{36}	A_{31}	A_{18}	A_7	A_{28}	A_{53}	A_{17}	A_{28}	A_{18}	A_{34}	A_{16}	A_{40}	A_{18}	A_{56}	A_{52}
124	A_{27}	A_{26}	A_{14}	A_5	A_{27}	A_{24}	A_{16}	A_{24}	A_{16}	A_{29}	A_{14}	A_{38}	A_{46}	A_{53}	A_{48}
125	A_{20}	A_{22}	A_{10}	A_3	A_{48}	A_{18}	A_{14}	A_{23}	A_{14}	A_{24}	A_{12}	A_{33}	A_{45}	A_{50}	A_{43}
126	A_{13}	A_{39}	A_5	A_{25}	A_{44}	A_{38}	A_{12}	A_{21}	A_{12}	A_{17}	A_9	A_{32}	A_{43}	A_{49}	A_{39}
127	A_6	A_{16}	A_2	A_{22}	A_{37}	A_{34}	A_{10}	A_{20}	A_{10}	A_{11}	A_7	A_{29}	A_{41}	A_{47}	A_{34}
128	A_1	A_{31}	A_{58}	A_{19}	A_{29}	A_{28}	A_7	A_{18}	A_7	A_8	A_4	A_{26}	A_{39}	A_{46}	A_{30}
129	A_{54}	A_{24}	A_{53}	A_{15}	A_{23}	A_{26}	A_{22}	A_{17}	A_6	A_3	A_2	A_{25}	A_{35}	A_{44}	A_{23}
130	A_{46}	A_{19}	A_{50}	A_{39}	A_{18}	A_{51}	A_{19}	A_{15}	A_5	A_{59}	A_{60}	A_{22}	A_{34}	A_{42}	A_{20}
131	A_{40}	A_{14}	A_{45}	A_{36}	A_{11}	A_{46}	A_{17}	A_{13}	A_3	A_{53}	A_{59}	A_{19}	A_{30}	A_{40}	A_{14}
132	A_{33}	A_{11}	A_{42}	A_{32}	A_4	A_{30}	A_{16}	A_{11}	A_{58}	A_{51}	A_{56}	A_{17}	A_{27}	A_{38}	A_{11}
133	A_{25}	A_6	A_{38}	A_{27}	A_1	A_{38}	A_{14}	A_9	A_{53}	A_{49}	A_{54}	A_{15}	A_{25}	A_{36}	A_7
134	A_{20}	A_2	A_{36}	A_{22}	A_{57}	A_{34}	A_{12}	A_8	A_{46}	A_{46}	A_{54}	A_{13}	A_{23}	A_{35}	A_3
135	A_{13}	A_{60}	A_{31}	A_{19}	A_{54}	A_{25}	A_{10}	A_5	A_{42}	A_{44}	A_{52}	A_{10}	A_{22}	A_2	A_5
136	A_1	A_{56}	A_{28}	A_{15}	A_{48}	A_{24}	A_7	A_4	A_{38}	A_{41}	A_{49}	A_7	A_{20}	A_2	A_1
137	A_3	A_{52}	A_{23}	A_{11}	A_{44}	A_{20}	A_6	A_2	A_{14}	A_{38}	A_{47}	A_5	A_{18}	A_2	A_{57}
138	A_{54}	A_{49}	A_{20}	A_7	A_{41}	A_{17}	A_4	A_{17}	A_{12}	A_{34}	A_{45}	A_3	A_{16}	A_2	A_{52}
139	A_{43}	A_{44}	A_{17}	A_2	A_{37}	A_{12}	A_3	A_{13}	A_{10}	A_{32}	A_{42}	A_2	A_{12}	A_2	A_{50}
140	A_{49}	A_{39}	A_{12}	A_{60}	A_{33}	A_6	A_{48}	A_9	A_6	A_{29}	A_{39}	A_{60}	A_{11}	A_2	A_{48}
141	A_{46}	A_{35}	A_8	A_{54}	A_{28}	A_2	A_{58}	A_{53}	A_5	A_{26}	A_{37}	A_{58}	A_9	A_2	A_{45}
142	A_{44}	A_{47}	A_1	A_{49}	A_{26}	A_6	A_{56}	A_1	A_3	A_{59}	A_{35}	A_{55}	A_8	A_2	A_{43}
143	A_{43}	A_{39}	A_{59}	A_{46}	A_{21}	A_1	A_{54}	A_{59}	A_6	A_3	A_{33}	A_{53}	A_6	A_{44}	A_1

(continued)

Table 2 (continued)

Time slots	Rectangular slots														
	Slot 11	Slot 21	Slot 31	Slot 12	Slot 22	Slot 32	Slot 42	Slot 13	Slot 23	Slot 14	Slot 24	Slot 34	Slot 15	Slot 25	Slot 35
144	A_{40}	A_{31}	A_{55}	A_{43}	A_{18}	A_{55}	A_{52}	A_{55}	A_2	A_3	A_{31}	A_{51}	A_5	A_{42}	A_1
145	A_{38}	A_{26}	A_{36}	A_{39}	A_{14}	A_{51}	A_{50}	A_{53}	A_3	A_{26}	A_{29}	A_{48}	A_1	A_{40}	A_9
146	A_{36}	A_{22}	A_{47}	A_{38}	A_8	A_{48}	A_{60}	A_{51}	A_{54}	A_3	A_{28}	A_{46}	A_{59}	A_{38}	A_3
147	A_{33}	A_{16}	A_{43}	A_{36}	A_4	A_{20}	A_{47}	A_2	A_{53}	A_3	A_{26}	A_{44}	A_{57}	A_{36}	A_1
148	A_{32}	A_{13}	A_{38}	A_{35}	A_3	A_{42}	A_{46}	A_{43}	A_{49}	A_3	A_{23}	A_{43}	A_{55}	A_{35}	A_1
149	A_{29}	A_{11}	A_{26}	A_{32}	A_{57}	A_{58}	A_{45}	A_{41}	A_{46}	A_8	A_{21}	A_{40}	A_{54}	A_{33}	A_1
150	A_{57}	A_9	A_{45}	A_{30}	A_{50}	A_1	A_{42}	A_{48}	A_{44}	A_3	A_{18}	A_{38}	A_{51}	A_{31}	A_1
151	A_{51}	A_6	A_{40}	A_{25}	A_{44}	A_{58}	A_{37}	A_{45}	A_{26}	A_{59}	A_{16}	A_{36}	A_{48}	A_{26}	A_1
152	A_{46}	A_4	A_{34}	A_{24}	A_{39}	A_{20}	A_{36}	A_{43}	A_{49}	A_{53}	A_{14}	A_{33}	A_{47}	A_{24}	A_{25}
153	A_1	A_2	A_{28}	A_{22}	A_{33}	A_{34}	A_{35}	A_{41}	A_{44}	A_{49}	A_{12}	A_{32}	A_{46}	A_{21}	A_9
154	A_{32}	A_{60}	A_{23}	A_{21}	A_{28}	A_{24}	A_{32}	A_{39}	A_{40}	A_{44}	A_9	A_{29}	A_{45}	A_{19}	A_{18}
155	A_{43}	A_{57}	A_{17}	A_{19}	A_{34}	A_{55}	A_{30}	A_{37}	A_{36}	A_{38}	A_4	A_{26}	A_{43}	A_{17}	A_{12}
156	A_{36}	A_{56}	A_{10}	A_{17}	A_{27}	A_{42}	A_{29}	A_{35}	A_{31}	A_{32}	A_2	A_{25}	A_{41}	A_{15}	A_9
157	A_{29}	A_{54}	A_{12}	A_{15}	A_{21}	A_{46}	A_{27}	A_{33}	A_{29}	A_{25}	A_{60}	A_{22}	A_{39}	A_{13}	A_5
158	A_{20}	A_{52}	A_{10}	A_{13}	A_{14}	A_{34}	A_{25}	A_{31}	A_{28}	A_{22}	A_{59}	A_{19}	A_{35}	A_{10}	A_1
159	A_{54}	A_{50}	A_5	A_{11}	A_{44}	A_{28}	A_{22}	A_{28}	A_{26}	A_{14}	A_{56}	A_{15}	A_{34}	A_8	A_{57}
160	A_{51}	A_{49}	A_1	A_9	A_{41}	A_{30}	A_{19}	A_{24}	A_{23}	A_{10}	A_{54}	A_{60}	A_{59}	A_6	A_{52}
161	A_{59}	A_{47}	A_{58}	A_7	A_3	A_{24}	A_{17}	A_{23}	A_{21}	A_4	A_{52}	A_{55}	A_{57}	A_4	A_{48}
162	A_{57}	A_{44}	A_{53}	A_5	A_{33}	A_{18}	A_{16}	A_{21}	A_{19}	A_3	A_{49}	A_{53}	A_{55}	A_2	A_{43}
163	A_{10}	A_{42}	A_{50}	A_3	A_{29}	A_{15}	A_{32}	A_{20}	A_{18}	A_{56}	A_{47}	A_{51}	A_{54}	A_{59}	A_{39}
164	A_{36}	A_{35}	A_{45}	A_{49}	A_{28}	A_9	A_{30}	A_{18}	A_{16}	A_{51}	A_{45}	A_{48}	A_{51}	A_{58}	A_{34}
165	A_{13}	A_{33}	A_{42}	A_{46}	A_{27}	A_5	A_{29}	A_{17}	A_{14}	A_{26}	A_{42}	A_{46}	A_{48}	A_{56}	A_{30}
166	A_{23}	A_{31}	A_{38}	A_{43}	A_{26}	A_1	A_{27}	A_{15}	A_{12}	A_{41}	A_{39}	A_{44}	A_{47}	A_{53}	A_{25}
167	A_{15}	A_{28}	A_{34}	A_{38}	A_{48}	A_{55}	A_{25}	A_{13}	A_{10}	A_{34}	A_{37}	A_{43}	A_{46}	A_{50}	A_{32}
168	A_{13}	A_{26}	A_{29}	A_{35}	A_{44}	A_2	A_{22}	A_{11}	A_2	A_{29}	A_{35}	A_{40}	A_{45}	A_{49}	A_{23}

(continued)

4 Conclusion

In this paper, we have formulated a web ad scheduling problem to determine an optimal placement of ads that compete to be placed on rectangular slots in a given planning horizon on the various webpages of a news website in order to maximize the revenue generated from ads. The optimization model is a 0-1 linear programming model and restricts selection of an ad on the same webpage more than once at any instant of time. Problem is programmed on LINGO software to get the optimal solution. In case the problem results in an infeasible solution, goal programming method is used to solve the problem. A case study is presented in the paper to show the application of the problem.

References

1. Adler, M., Gibbons, P.B., Matias, Y.: Scheduling space-sharing for internet advertising. J. Sched. **5**(2), 103–119 (2002)
2. Aggarwal, C. C., Wolf, J. L., Yu, P.S.: A framework for the optimizing of WWW advertising. Trends in Distributed Systems for Electronic Commerce, pp. 1–10 (1998)
3. Dreze., X., Zufryden, F.: Testing web site design and promotional content. J. Advertising Res. **37**(2), 77–91 (1997)
4. Gupta, A., Manik, P., Aggarwal, S., Jha, P. C.: Optimal advertisement planning on online news media. In: Handa, S. S., Uma Shankar, Ashish Kumar Chakraborty (eds.) Proceedings of International Congress on Productivity, Quality, Reliability, Optimization and Modelling, vol. 2, pp. 963–978. Allied Publishers, New Delhi (2011)
5. Intern.com Corp.: Theories and methodology. http://webreference.com/dev/ banners /theories. html (1997)
6. Kohda, Y., Endo, S.: Ubiquitous advertising on the WWW: merging advertisement on the browser. Comput. Netw. ISDN Syst. **28**, 1493–1499 (1996)
7. Kumar, S., Jacob, V.S., Sriskandarajah, C.: Scheduling advertisements on a web page to maximize revenue. Eur. J. Oper. Res. **173**, 1067–1089 (2006)
8. Marx, W.: Study shows big lifts from animated ads. Business Marketing, vol. 1 (1996)
9. McCandless, M.: Web advertising. IEEE Intell. Syst. **13**(3), 8–9 (1998)
10. Novak, T.P., Hoffman, D.L.: New metrics for new media: Toward the development of web measurement standards. World Wide Web, **2**(1), 213–246 (Winter, 1997)
11. Risden, K., Czerwinski, M., Worley, S., Hamilton, L., Kubiniec, J., Hoffman, H., Mickel, N., Loftus, E.: Internet advertising: patterns of use and effectiveness. In: CIII' 98, Proceedings of the SIGCHI Conference on Human Factors in, Computing Systems, pp. 219–224 (1998)
12. Steuer, R.E.: Multiple Criteria Optimization: Theory Computation and Application. Wiley, New York (1986)
13. Thiriez, H.: OR software LINGO. Eur. J. Oper. Res. **12**, 655–656 (2000)
14. Yager, R.R.: Intelligent agents for World Wide Web advertising decisions. Int. J. Intell. Syst. **12**, 379–390 (1997)

An Integrated Approach and Framework for Document Clustering Using Graph Based Association Rule Mining

D. S. Rajput, R. S. Thakur and G. S. Thakur

Abstract Growth in number of documents increases day by day, and for managing this growth the document clustering techniques are used document clustering is a significant tool to allocating web search engines for data mining and knowledge discovery. In this paper, we have introduced a new framework graph-based frequent Term set for document clustering (GBFTDC). In this study, document clustering has been performed for extraction of useful information from document dataset based on frequent term set. We have generated association rules to perform pre-processing and then have applied clustering approach.

Keywords Document clustering · Text document · Association rule · Pre-processing

1 Introduction

Every day a huge number of document, reports, e-mails, and web pages are generated from different sources, such as Biomedical Software, Online media, Marketing, Academics, and Government Organizations, etc. This has resulted in demand for well-organized tools for turning data into important knowledge. To fulfill this necessity, researchers from several technological areas, like information retrieval [11], information extraction [17], Association Rule Mining [10, 35], pattern recognition

D. S. Rajput (✉) · R. S. Thakur · G. S. Thakur
Department of Computer Applications, M.A.N.I.T., Bhopal 462051, India
e-mail: dharm_raj85@yahoo.co.in

R. S. Thakur
e-mail: ramthakur2000@yahoo.com

G. S. Thakur
e-mail: ghanshyamthakur@gmail.com

B. V. Babu et al. (eds.), *Proceedings of the Second International Conference on Soft Computing for Problem Solving (SocProS 2012), December 28–30, 2012*, Advances in Intelligent Systems and Computing 236, DOI: 10.1007/978-81-322-1602-5_144, © Springer India 2014

[21], machine learning [16], data visualization [16], neural networks [21] etc., are using well-known approaches for knowledge discovery from different databases.

All these approaches have ushered us into resulted in an effective research area know as Data Mining [16, 17]. Data Mining is extraction of hidden predictive information from large databases; it is a powerful technology with great potential to help organizations focus on the most important information in their Data Warehouses [5, 11, 16, 21, 28]. It is popularly known as Knowledge Discovery in Databases (KDD), [11, 12, 21, 35] is the nontrivial extraction of implicit, previously unknown and potentially useful information from data in databases [11, 12]. Though, data mining and knowledge discovery in databases are frequently treated as synonyms. It is divided into two models [12, 18, 21, 24, 28, 31]; Predictive Model and Descriptive Model. In predictive model makes prediction about unknown data values by using the known values. Like classification [11], regression [16], time series analysis [24], prediction [16], etc. In descriptive model identifies the patterns or relationships in data and explores the properties of the data examined. like clustering [29], summarization [35], association rule [11], sequence discovery, etc.

The repeated existing property of internet, which allows textual documents to be shared over the cyber space, is remarkable. However, it also makes users face the information-overloading problem. When users create queries to WWW search engines, they usually confusingly receive a small number of relevant web pages combined with a large number of irrelevant web pages. To valuably manage the result of a search engine query, it is the motivation to the use of document clustering [2, 6].

Document clustering is an unsupervised technique [18, 22] for discovering valuable knowledge from data. It is the common data mining technique for finding hidden patterns in data. It is an important issue in the analysis and exploration of data. Document clustering in textual data is a progressively more significant research field, since the requirement of attaining knowledge from the massive amount of textual documents. It has become an increasingly important technique for enhancing search engine results, web crawling, unsupervised document organization, and information retrieval or filtering. It is widely applicable areas of science, technology, social science, biology, economics, medicine, and stock market. Clustering engage dividing a set of documents into a particular number of groups. A few familiar clustering methods are partitioning method [16, 20], hierarchical methods [3], density and grid-based clustering method [11].

Among the technique developed for data and text mining, association rule mining [23, 34, 37, 39] is one of the useful and successful techniques for discovering interesting rules. Rules generation for textual databases has been an important application area. Yet, its application on text databases still seems to be more promising, owing to the difference in characteristics of transaction databases with textual databases [1, 8, 21]. In this case, rules are deduced on the co-occurrences of terms in texts and therefore able to return semantic relations among the terms [41]. Rule generation using graph-based approach [10, 36] requires only one pass of database scan and it doesn't require costly candidate key generation method for creating new candidates as like Apriori [33, 38]. The graph theoretic algorithm takes less memory space for storing adjacency matrix. It also reduces I/O time and CPU overhead. Therefore, in this framework, we are using graph theoretic [10] algorithm to rule generation.

2 Literature Review

Document clustering [14] is a specialized data clustering problem, where the objects are in the form of documents. The objective of the clustering process is to group the documents which are similar in some sense like type of document, contents of document, etc., into a single group (cluster) [16]. The difficult part is to learn from a dataset, actually how many classes of such groups exist in the collection. Formally, a corpus of N unlabeled documents is given and a solution $C = \{C_j : j = 1, .., k\}$ is searched that partitions the document into k disjoint clusters document clustering aims to discover natural grouping among documents in such a way that documents with in a cluster are similar (high intra cluster similarity) to one another and are dissimilar to documents in other clusters (low inter cluster similarity). Exploring, analyzing, and correctly classifying the unknown nature of data in a document without supervision is the major requirement of document clustering method [30].

Traditionally, document clustering algorithm features like words, phrases, and sequences from the documents to perform clustering [14, 15, 20, 40]. Over the past few decades, several effective document clustering algorithms have been proposed to mitigate the hassle, including the K-means, Bisecting k-means [23], hierarchical agglomerative clustering (HAC) [32], and unweighted pair group method with arithmetic mean (UPGMA) [4]. K-means algorithm [18] for document clustering is easy to be implemented and works quickly in most situations, Table 5 shows the result of K-Mean Clustering for example of Table 2. Here the algorithm suffers from two major drawbacks, which make it incompatible for many applications. One is sensitivity to initialization and the other is convergence to local optima. To deal with the limitations that exist in traditional partition clustering methods especially K-means, recently new concepts and techniques have been entered into document clustering. In clustering of large document sets, it often concerned to some learning methods like optimization technique, which mostly intended to the lack of orthogonality (Table 1).

As pointed out by the literature survey [19–27], there are still challenges in improving the clustering quality, which are listed as follows: [6, 7]

1. Many document clustering algorithms work well on small document sets, but fail to deal with large document sets efficiently.
2. Some document clustering algorithms require users to specify the number of clusters as an input parameter. Though it seems complex to determine the number of clusters in advance. It doesn't provide good clustering accuracy.
3. Absence of Meaningful cluster labels is basic problem in document clustering.

Table 1 Literature review

Author	Proposed method	Overlapping	Cluster label	Semantic discovery	Cluster quality	Cluster efficiency
K. Lin. et al. in 2001 [27]	A word-based soft clustering algorithm for documents	Yes	No	No	–	Yes
F. Beil et al. in 2002 [13]	Frequent term-based text clustering	Yes	Yes	No	–	Yes
B. Fung et al. in 2003 [3]	Hierarchical document clustering using frequent Itemsets	No	Yes	No	Yes	Yes
D. R. Recupero in 2007 [9]	A new unsupervised method for document clustering by using WordNet lexical and conceptual relations	No	No	Yes	Yes	No
C. L. Chen et al. in 2009 [7]	An integration of fuzzy association rules and WordNet for document clustering	No	Yes	Yes	No	No
C. L. Chen et al. in 2010 [6]	Fuzzy-based multi-label document clustering (FMDC)	Yes	Yes	Yes	–	–

2.1 Existing Clustering Methods Using Frequent Itemsets

Most of the researchers worked in the area of finding association rules from textual data by using Apriori, FP growth, pincer, and many other algorithms and some of researchers have worked with clustering algorithms also they have solved many problems. A number of existing clustering methods using frequent itemsets are available for document clustering.

HFTC: frequent term-based clustering [26] has been a first algorithm in this regard. But it doesn't work with high dimensional document data and method isn't scalable.

FIHC: this method uses the notion of frequent itemsets, which comes from association rule mining, for document clustering. The drawbacks of FIHC are like it has a number of frequent itemsets which may be very large and redundant. This method used hard clustering approach, and wasn't comparable with previous methods.

Table 2 Document matrix

Document id	Word sequence
D_1	1,2,5,6,7
D_2	2,4,7
D_3	4,5
D_4	2,3,6,7
D_5	5,6
D_6	2,3,4,7
D_7	1,2,6,7

Table 3 Boolean representation of Table 2 after pre processing

Document id	Document vector
D_1	{1,1,0,0,1,1,1}
D_2	{0,1,0,1,0,0,1}
D_3	{0,0,0,1,1,0,0}
D_4	{0,1,1,0,0,1,1}
D_5	{0,0,0,0,1,1,0}
D_6	{0,1,1,1,0,0,1}
D_7	{1,1,0,0,0,1,1}

There are many other methods like FIHC, FTC, CMS, and CFWS, etc. Table 2 shows textual dataset $D = (D_1, D_2, D_3, D_4, D_5, D_6, D_7)$ with their term set.

Table 3 is Document representation using Boolean Matrix for Table 2.

Frequent term set extracted from document in Table 2 and their corresponding documents (Minimum Support = 2) using graph-based approach (Table 4).

Document Clustering produced by K-Mean with different number of clusters in Table 5.

Table 4 Frequent termset sets using graph based approach

Frequent termset	List of document
{1,2},{1,6},{1,7},{1,2,6}, {1,2,7},{1,6,7}, {1,2,6,7}	D_1, D_7
{2,3}, {3,7}, {2,3,7}	D_4, D_6
{2,4}, {4,7}, {2,4,7}	D_2, D_6
{2,6}, {2,7}, {2,6,7}	D_1, D_4, D_7
{5,6}	D_1, D_5

Table 5 Document clustering using K-mean

Number of clusters	Document in clusters
2	{ 3,5} {1, 4, 6, 7}
3	{3, 5} {1, 7} {4, 6}
4	{3} {5} {1, 7} {4, 6}

3 Proposed Method

This section proposed, a new framework for document clustering using frequent term set. Figure 1 shows the proposed framework which consist three modules, namely Document Pre-processing Module, Frequent Term extraction Module and Frequent Term based Cluster Extraction Module as discussed in Sects. 3.1–3.3, respectively. The proposed framework will received document dataset as input and document preprocessing module will perform pre-processing operation on document dataset and extract the select key termset, after that in second module apply graph-based rule mining technique to mine the document and find the frequent term set, finally the last module will use the frequent term set to design the similarity matrix and find the document clusters.

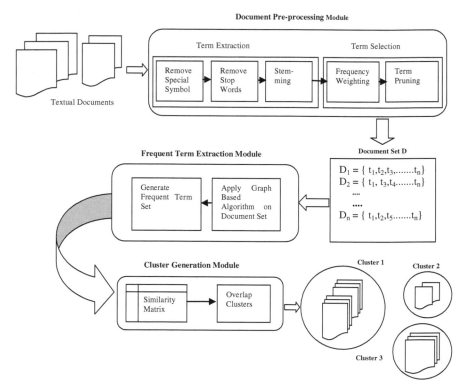

Fig. 1 Document clustering using graph based framework

3.1 Document Pre-Processing Module

Document Pre-processing Algorithms are specially designed for the properties of document data. Each document set has numerous stop words, special marks, punctuation marks, and spaces, first we remove all stop words and stemming, etc., and constructed term matrix for generating frequent term set. There are four steps, which are carried out when the document is provided to the pre-processing module of the system.

3.1.1 Stop Words and Special Symbol Elimination

Stop words are the words which don't have meaning with respect to the classification [10, 16, 40]. So these words are removing when the term matrix is created for the classification purpose. In short, the words are removed from the documents which are not necessary for the next stage. it removed all prepositions, conjunction and articles from dataset D.

Algorithm1: To remove stop words & special characters

Input: A Document database D and a well defined stop words list L.

$D=\{d_1,d_2,d_3,\ldots,d_n\}$,where $1<=n<=l$

t_{ij} is the jth term in ith document

Output: All valid text term in D
START
  1: for (all d$_i$ in D) do
  2: for (1 to j) do
        a. Extract t_{ij} from d_i
        b. If(t_{ij} in database d_i)
        c. Remove t_{ij} from d_i
  3: End for
  4: End for
END

3.1.2 Stemming

A stemming is a computational procedure which reduces all words with the same root (or, if prefixes are left untouched, the same stem) to a common form, usually by stripping each word of its derivational and inflectional suffixes [10, 25]. It is useful in many areas of computational linguistics and information-retrieval work. Researchers from the fields of Computational linguistics and information retrieval find it as, one of the necessary step to fulfill their research needs. It helps to decrease the size of the dictionary file.

Algorithm2: To remove the stemming word.

Input: A Document database *D* and List of stemming word list

$D= \{d_1, d_2, d_3,....,d_n\}$;where $1<=n<=i$

START

1. Determine where on ending list to begin searching for a match so that stem is at least two characters long.
2. Search ending list for a match to the last part of the word being stemmed.
3. If (ending is found) Then
4. {
5.   if context sensitive rule is satisfied
6. Remove ending
7. Recoding Procedure ();
8. Else
9. Go to step 3
10. }
11. Else
12. Recoding Procedure();

END

The Recoding procedure used by Stemming Algorithm is given below

Recoding Procedure ()

START

1. {
2. Undoable final consonant of stem, if applicable
3. Search list of transformations for match on remaining stem
4. If (match== found) Then
5. {
6. Recode stem according to rule
7. Output stem
8. }
9. Else (no rule applies)
10. Output stem
11. }

END

3.1.3 Feature Set Selection

In text classification, the important term selection is critical task for the classifier performance. With increasing number of documents, number of features also increases and to reduce the size of the dictionary, the threshold feature set reduction methods is used.

In this method, lower thresholds are decided according to the number of words in the dictionary. The term which below lower threshold are extracted from the document.

Table 6 BMM representation

Document id	Term 1	Term 2	......	Term n
D_1	0	1	...	1
D_2	1	0	...	1
...	...	...	...	...
D_3	0	0	...	1

Algorithm 3: Algorithm for feature selection

Input: A Document dataset D and y lower threshold value N is the counter
$D= \{d_1,d_2,d_3,....,d_n\}$;where $1<=n<=i$
t_{ij} is the jth term in i^{th} document
Output: Documents Dataset D after feature selection
START
  1: for (all d_i in D) do
  2: for (1 to j) do
          a. Count total occurrence of t_{ij} in document d_i
          b. Assign the total occurrence of t_{ij} in N
          c. If ($N<y$)
          d. Remove t_{ji} from the document d_i
  3: End for
  4: End for
END

3.1.4 Document Term Matrix and Normalization

The main principle of the pre-processing phase is to organize the document term matrix that will provide appropriate input to the next phase of the system. After selection process, we have limited terms in each document. Suppose we have n documents and maximum m steem words in a document. The binary matrix M is represented as [12, 14, 17] (Table 6).

$$M[d_i * w_j] = \begin{cases} 1 & \text{if } w_i \text{ in present in } d_i \\ 0 & \text{otherwise} \end{cases}$$

Where $i = 1, 2, 3 \ldots n$ and $j = 1, 2, 3 \ldots m$.

3.2 Frequent Term Extraction Module

After Sect. 3.1 we get structured vector, which will be treated as input for frequent term extract module. In this module, the graph-based algorithm [10, 36] is applied to generate the frequent termsets from term matrix. It reduces the CPU time because we are scanning only two times the preprocessed dataset and no need to generate candidate set like Apriori [16].

Algorithm 4: Graph Based Association Rule Mining Algorithm (GBARMA)

Input: The set of different Textual document D or BMM Table.

Output: Frequent Term Set.

1. Scan document table D and create directed graph G.
2. Create Adjacency Matrix A of G.
3. Update Value of each element $A_{ij}.list$ and $A_{ij}.count$ of matrix A.
4. Delete corresponding row and column of an element $A_{ij}.count=0$ only for diagonal elements.
5. Read each element A_{ij} of matrix A if $A_{ij}.count<$minimum Support then set $A_{ij}.Count=0$
6. Find 1- Frequent termset and 2- frequent termset from matrix.
7. Calculate other k-termsets from each column using logical AND operator.
8. END

Algorithm 4 generates frequent term sets based on predefined minimum support value $\ominus$ the minimum support of the frequent term set is usually in the range of 5–15 %. From a large textual document set. When the minimum support is too large, the total number of frequent terms would be very small, so that the resulting compact documents would not have enough information about the original dataset. In this case, a lot of documents will not be processed because they do not support any frequent word, and the final clustering result will not cover these documents.

3.3 Frequent Term-Based Cluster Generation Module

The objective of this module is to assign each document to multiple clusters and determine labels and assign them to document. If we are having the confidence to achieve these goals by some automatic process, we would completely bypass the expense of having humans assign labels, but the process known as document clustering is less than perfect. Document clustering assigns each of the documents in a collection to one or more smaller groups called clusters. Examinations show that the clusters should contain similar documents. The initial collection is a single cluster. After processing, the documents are distributed among a number of clusters, where ideally each document is very similar to the other documents in its cluster and much less similar to documents in other clusters.

We define a similarity matrix of all documents using frequent term set. We are providing frequent term set as an input is the matrix of similarities S. where $S(i, j)$ is the similarity matrix and $1 \geq i \geq n - 1$ and $1 \geq j \geq n$. The preferences should be placed on the half diagonal of the matrix.

Algorithm 5: Graph based Frequent Term Set based Document Clustering (GBFTDC)

Input: All Frequent Term Set of dataset D.
Output: Find clusters.
START
Step 1:- Construct Similarity Matrix S using all frequent term set they have in common.
Step 2:- Find minimum number in similarity matrix excluding zero.
Step 3:- Find Maximum value in S, and then finding all document pairs of unclustered Documents with the maximum value.
Step 4: If (Maximum value = Minimum value) **then**

> All documents in corresponding document pairs which do not attach to any cluster are used to construct a new cluster.
> **Else**
> All found document pairs with the maximum value, the following process is conducted. For all document pairs with maximum value, subsequent processes are conducted. First, for each document pair, if the documents are not belonging to any clusters, then they would be grouped together to form a document cluster. In other approach if one document of the pair belongs to an existing cluster, then the other unattached document will also be included in the same, finally similarities of found document pairs set to zero.

Step 5:- Go to step 3.
Step 6:- If there are any documents which do not attach to any cluster, then each of these documents is used to construct a new cluster.
END

3.4 An Illustrative Example

The working process of algorithms applies in document database and performs pre-processing face, finally we have Table 2. After that we find frequent termset using graph-based algorithm from Table 2, now apply frequent termset-based document clustering. Figures 2, 3 and 4 basically, shows the clustering procedure by similarity values between documents within clusters. It prefers to create new cluster other than adding documents to existing clusters in order to balance the amount of documents in a cluster. There is no overlap among document clusters produced by this method.

Step 1: Construct Similarity Matrix S using, frequent term set. (All frequent term set which are common in all document sets).

Step 2: finding minimum value of S when S ≥ 1.

Step 3: Document pair (1, 7) has largest similarity as 15, clustered pair (1, 7) set similarity as 0. Unclustering document set {2, 3, 4, 5, 6}.

Step 4: Document Pair (1, 4) (2, 6), (4, 6) (4, 7) are having largest similarity as 7, and sets Pair (1, 4) (2, 6), (4, 6) (4, 7) similarity is 0. Now add (2,4,6) with cluster (1, 7) and get (1,2,4,6,7) as new clustered pair. The unclustered document set is {3, 5}.

Now select the most frequent documents with maximum length for each cluster as its document topic (Table 7).

D. S. Rajput et al.

Fig. 2 Similarity matrix S using frequent term set

	1	2	3	4	5	6
1	.					
2	3	.				
3	1	1	.			
4	7	1	0	.		
5	3	0	1	1	.	
6	3	7	1	7	0	.
7	15	3	0	7	1	3

Fig. 3 Similarity matrix after step 3

	1	2	3	4	5	6
1	.					
2	3	.				
3	1	1	.			
4	7	1	0	.		
5	3	0	1	1	.	
6	3	7	1	7	0	.
7	0	3	0	7	1	3

Fig. 4 Similarity matrix after step 4

	1	2	3	4	5	6
1	.					
2	3	.				
3	1	1	.			
4	0	1	0	.		
5	3	0	1	1	.	
6	3	0	1	0	0	.
7	0	3	0	0	1	3

Table 7 Clustering results

Cluster id	Documents topics	List of documents
1	2, 6, 7	{ 1, 2, 4, 6, 7}
2	5	{3, 5}

Document topic reductions; if one frequent term belong in more than one cluster id (it is contained in the assigned topics of other clusters) then the frequent term should be eliminated from the document topics of the cluster. In this example, there are no document topic belonging to in more than one cluster so this step is not required here.

Table 8 Dataset description

Data set	Documents	Classes
20 newsgroup dataset	20000	20
Reuter's text categorization	8654	52

4 Experimental Evaluation

F-measure [16, 29] is employed to estimate performances of the proposed clustering methods. Further, clustering results are normalized into a fixed number of predefined clusters. The F-measure can be used to balance the contribution of false negatives by weighting recall through a parameter. F-measure compares the results to the pre-classified classes. Let precision and recall is defined as follows: The formula of F-measure is depicted as equations.

Precision: This is the percentage of retrieved documents that are in fact relevant to the query. It is calculated as:

$$P_i = TP/TP + FP \tag{1}$$

Recall: This is the percentage of documents that are relevant to the query and were, in fact, retrieved. It is calculated as

$$R_i = TP/TP + FN \tag{2}$$

We can calculate F-Measure by using to the formula

$$F = 2 * P_i * R_i / P_i + R_i \tag{3}$$

To test and compare proposed clustering algorithms, the pre-classified sets of documents are used. We have used 20 newsgroup dataset and Reuter's dataset, which are widely used in many publications. A summary description of these datasets is given in Table 8. The experiments were performed on an Intel core 2 Duo, 2.94 GHz system running Windows 7 professional with 2 GB of RAM.

We treat each cluster as, it were the result of a query and each class as if it were the relevant set of documents for a query. The proposed approach used these dataset as input and finally we get 20 clusters as result. For each cluster, we have computed the precision, Recall and F-measure with the help of Eqs. 1–3. The experimental results are shown in Table 9.

In this research study, we have choose some parameters to compare Performa of our proposed approach with other existing algorithms like K-mean, HFTC, FIHC. Therefore, the final comparison of these three algorithms are shown in Table 10.

Figures 5 and 6 shows overall F-Measure values for proposed GBFTDC and other existing algorithms with 20 newsgroup dataset and Reuter's dataset respectively. For both dataset we choose the Minimum threshold from the elements in {10, 15, 20, 25, and 30%}.

Table 9 Clustering performances obtained on 20 newsgroup dataset

Cluster	Precision	Recall	F-measure	Cluster	Precision	Recall	F-measure
C_1	0.8	0.5333	0.639976	C_{11}	0.17	0.65	0.26951
C_2	1	0.5	0.666667	C_{12}	0.9	0.7	0.7875
C_3	0.6666	0.1333	0.222172	C_{13}	0.63	1	0.7730
C_4	0.514	1	0.67899	C_{14}	0.5	0.4	0.4444
C_5	0.3333	0.0909	0.142843	C_{15}	0.231	0.745	0.35265
C_6	1	0.587	0.73976	C_{16}	1	0.0714	0.133284
C_7	0.51	0.29	0.36975	C_{17}	0.75	0.25	0.375
C_8	0.6	0.2784	0.38032	C_{18}	1	0.5	0.66667
C_9	0.84	0.45	0.58604	C_{19}	0.3333	0.0909	0.142843
C_{10}	0.744	0.12	0.21156	C_{20}	0.9285	1	0.962925

Table 10 Parameters list for our approach and the other three approaches

Parameter name	GBFTDC	K-means	HFTC	FIHC
Dataset	20 newsgroup, Reuter's	20 newsgroup, Reuter's	20 newsgroup, Reuter's	20 newsgroup, Reuter's
Stop word removal	Yes	Yes	Yes	Yes
Stemming	Yes	Yes	Yes	Yes
Length of smallest term(threshold)	5(Five)	5(Five)	5(Five)	5(Five)
Cluster count k	5, 10, 15, 20	Depend on value of k	–	–
Overlapping	No	Yes	Yes	No
Work with high dimensional data	Yes	No	Yes	Yes
Scalibity	Yes	No	No	Yes

Fig. 5 Overall F-measure comparison for 20 newsgroup dataset

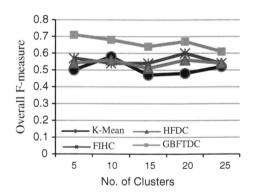

Fig. 6 Overall F-measure comparison for Reuter's dataset

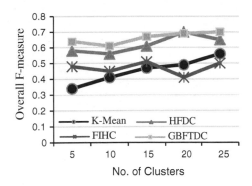

5 Conclusion

This paper presents a new framework design for document clustering using graph based frequent term set. First we review many papers and present a review method of document clustering using frequent termset. We take 20 news group data for document clustering and we do our pre-processing algorithm and find the frequent term set.

Advantage of this method is that it removes the overlapped clusters, and it reduces the CPU time because we are scan the original document set only twice to find the frequent term set and don't need to generate the candidate set. The method is applicable on large document databases. It uses half similarity matrix to find the similarity this save more memory.

Experiments have been conducted to evaluate the performance of proposed approach over the other methods like as HFTC and FIHC. The experimental results show that this method performs better because other methods follow apriori algorithm to find the frequent itemsets. As the Apriori algorithm have few disadvantages: like as generates candidate sets, and scans the dataset more number of time. HFTC isn't scalable and it doesn't work with high dimensional data. The FIHC method also works based on hard clustering approach.

A result shows the proposed method will perform well with high dimensional document data and no overlapping. With the above analysis, we may conclude that it has constructive quality in clustering documents using graph-based frequent term sets.

Acknowledgments This work is supported by research grant from MANIT, Bhopal, India under Grants in Aid Scheme 2010-11, No. Dean(R&C)/2010/63 dated 31/08/2010.

References

1. Kongthon, A.: A text mining framework for discovering technological intelligence to support science and technology management. Technical Report, Georgia Institute of Technology (2004)
2. Kalogeratos, A., Likas, A.: Document clustering using synthetic cluster prototypes. Data Knowl. Eng. **70**, 284–306 (2011)
3. Fung, B., Wang, K., Ester, M.: Hierarchical document clustering using frequent itemsets. In: Proceeding of SIAM International Conference on Data Mining (SDM'03), pp. 59–70 (2003)
4. Michenerand, C.D., Sokal, R.R.: A quantitative approach to a problem in classification. Evolution **11**, 130–162 (1957)
5. Chapman, P., Clinton, J., Kerber, R., Khabaza, T., Reinartz, T., Shearer, C., Wirth, R.: CRISP-DM 1.0 : Step-by-step data mining guide, NCR Systems Engineering Copenhagen (USA), DaimlerChrysler AG, SPSS Inc. (USA) and OHRA Verzekeringenen Bank Group B.V (Netherlands), (2000)
6. Chen, C.L., Frank, S.C.T., Liang, T.: An integration of wordnet and fuzzy association rule mining for multi-label document clustering. Data Knowl. Eng. **69**, 1208–1226 (2010)
7. Chen, C.L., Tseng, F.S.C., Liang, T.: An integration of fuzzy association rules and WordNet for document clustering. In: Proceeding of the 13th Pacific-Asia Conference on Knowledge Discovery and Data Mining (PAKDD-09), pp. 147–159 (2009)
8. Cutting, D.R., Karger, D.R., Pedersen, J.O., Tukey, J.W.: Scatter/Gather: A Cluster-based approach to browsing large document collections. In: Proceedings of the Fifteenth Annual International ACM SIGIR Conference, pp. 318–329, June 1992
9. Recupero, D.R.: A new unsupervised method for document clustering by using WordNet lexical and conceptual relations. Inf. Retrieval **10**(6), 563–579 (2007)
10. Rajput, D.S., Thakur, R.S., Thakur, G.S.: Rule generation from textual data by using graph based approach. In: International Journal of Computer Application (IJCA) 0975–8887, New york, ISBN: 978-93-80865-11-8, Vol. 31, No.9, pp. 36–43, Oct 2011
11. Dunham, M.H., Sridhar, S.: Data mining: introductory and advanced topics. Pearson Education, New Delhi, ISBN: 81-7758-785-4, 1st edn. (2006)
12. Fayyad, U., Piatetsky-Shapiro, G., Smyth, P.: From data mining to knowledge discovery in databases. AI Magazine, American Association for Artificial Intelligence (1996)
13. Beil, F., Ester, M., Xu, X.: Frequent term-based text clustering. In: Proceeding of International Conference on knowledge Discovery and Data Mining (KDD'02), pp. 436–442 (2002)
14. Fung, B.C.M., Wang, K., Ester, M.: Hierarchical document clustering using frequent itemsets. In: Proceedings of SIAM International Conference on Data Mining (2003)
15. Hammouda, K.M., Kamel, M.S.: Efficient phrase-based document indexing for web document clustering. IEEE Trans. Knowl. Data Eng. **16**, 1279–1296 (2004)
16. Han, I., Kamber, M.: Data Mining Concepts and Techniques, pp. 335–389. M. K. Publishers, Berlin (2000)
17. Haralampos, K., Christos, T., Babis, T.: An approach to text mining using information extraction. In: Proceeding Knowledge Management Theory Applications Workshop, (KMTA 2000), pp. 165–178. Lyon, Sept 2000
18. Hartigan, J.A., Wong, M.A.: A K-means clustering algorithm. Appl. Stat. **28**, 126–130(1979)
19. Hotho, A., Staab, S., Stumme, G.: Wordnet improves text document clustering. In: Proceeding of SIGIR International Conference on Semantic Web, Workshop, (2003)
20. Hung, C., Xiaotie, D.: Efficient phrase-based document similarity for clustering. IEEE Trans. Knowl. Data Eng. **20**, 1217–1229 (Sept 2008)
21. Introduction to Data Mining and Knowledge Discovery, 3rd edn. ISBN: 1-892095-02-5, Two Crows Corporation, 10500 Falls Road, Potomac, MD 20854, U.S.A., (1999)
22. Jain, A.K., Murty, M.N., Flynn, P.J.: Data clustering: a review. ACM Comput. Surv. **31**(3), 264–323 (1999)
23. MacQueen, J.B.: Some methods for classification and analysis of multivariate observations. In: Proceedings of 5th Berkeley Symposium on Mathematical Statistics and Probability, pp. 281–297 (1967)

24. Jensen, C.S.: Introduction to Temporal Database Research. http://www.cs.aau.dk/csj/Thesis/pdf/chapter1.pdf
25. Lovins, J.B.: Development of a stemming algorithm. Mech. Transl. Comput. Linguist. **11**(1, 2), 22–31, June 1968
26. Kiran, G.V.R., Ravi Shankar, Vikram Pudi: Frequent itemset based hierarchical document clustering using wikipedia as external knowledge. KES 2010, Part II, LNAI 6277, pp. 11–20. Springer, Berlin (2010)
27. Lin, K., Kondadadi, R.: A word-based soft clustering algorithm for documents. In: Proceedings of Computers and Their Applications, pp. 391–394. Seattle (2001)
28. Larose, D.T.: Discovering knowledge in data: an introduction to data mining, Wiley, Inc., 2005. International Journal of Distributed and Parallel systems (IJDPS) Vol. 1, No. 1, (2010)
29. Steinbach, M., Karypis, G., Kumar, V.: A comparison of document clustering techniques. KDD-2000 Workshop on Text Mining, pp. 109–110 (2000)
30. Rafi, Muhammad, Shahid Shaikh, M., Farooq, Amir: Document clustering based on topic maps. Int. J. Comput. Appl. **12**(1), 32–36 (2010)
31. Nasukawa, T., Nagano, T.: Text analysis and knowledge mining system. IBM Syst. J. **40**(4), 967–984 (2001)
32. Willett, P.: Recent trends in hierarchic document clustering: a critical review. Inf. Process. Manage. **24**(5), 577–597 (1988)
33. Lin, K., Kondadadi, R.: A word-based soft clustering algorithm for documents. In: Proceeding Computers and Their Applications, pp. 391–394 (2001)
34. Agrawal, R., Imielinski, T., Swami, A.: Mining association rules between sets of items in large databases. In: Proceedings of ACM SIGMOD International Conference on Management of Data, pp. 207–216 (1993)
35. Richards, A.L., Holmans, P., O'Donovan, M.C., Owen, M.J., Jones, L.: A comparison of four clustering methods for brain expression microarray data. BMC Bioinform. **9**, pp. 1–17 (2008)
36. Thakur, R.S., Jain, R.C., Pardasani, K.R.: Graph theoretic based algorithm for mining frequent patterns. In: IEEE World Congress on Computational Intelligence, pp. 629–633. Hong Kong (2008)
37. Thakur, R.S., Jain, R.C., Pardasani, K.R.: Fast algorithms for mining multi-level association rules in large databases. Asian J. Inf. Manage. USA **1**(1), 19–26 (2008)
38. Thakur, R.S., Jain, R.C., Pardasani, K.R.: MAXFP: a multi-strategy algorithm for mining maximum frequent pattern and their support counts. Trends Appl. Sci. Res. **1**(4), 402–415 (2006)
39. Vishnu Priya, R., Vadivel, A., Thakur, R.S.: Frequent pattern mining using modified CP-Tree for knowledge discovery. Advanced Data Mining and Applications, LNCS-2010, Vol. 6440, pp. 254–261. Springer, Berlin (2010)
40. Soon, M.C., John, D.H., Yanjun, L.: Text document clustering based on frequent word meaning sequences. Data Knowl. Eng. **64**, 381–404 (2008)
41. Valentina, C., Sylvie, D.: Text mining supported terminology construction. In: Proceedings of the 5th International Conference on Knowledge Management, pp. 588–595. Graz, Austria (2005)

Desktop Virtualization and Green Computing Solutions

Shalabh Agarwal and Asoke Nath

Abstract Greecomputing is now more than just being environmentally responsible. It is also the exercise of utilizing optimal IT resources in a more efficient way. It is realized by the computer professionals and also by the Scientists that one of the key enablers of Green computing is virtualisation. Virtual computing and management will enable toward environmentally sustainable ICT infrastructure. The desktop virtualisation enables to utilise the untapped processing power of today's high-power PCs and storage devices. The same or improved performance can be delivered with reduced operating expenses, a smaller carbon footprint and significantly curtailed greenhouse gas emissions. In this work the authors have made a complete study on Desktop virtualisation, Thin client architecture, and its role in Green computing.

Keywords Green computing · Virtualisation · ICT · Greenhouse gas · Thin client

1 Introduction

In the present scenario one cannot think of living without computers. At the same time it is important to find ways to use the computers in a way that can sustain the environment. Currently, personal computers (PC) consume far too much electricity and generate too much e-waste to be considered an eco-friendly solution by today's standards. With a typical PC taking approximately 110–240 W to run, and with well over 1 billion of them on the planet, it is easy to understand the carbon footprint and Green House Emissions generated by these PCs. E-waste is another issue that is the

S. Agarwal (✉) · A. Nath
St. Xavier's College [Autonomous], Kolkata WB 700016, India
e-mail: shalabh@sxccal.edu

A. Nath
e-mail: asokejoy1@gmail.com

B. V. Babu et al. (eds.), *Proceedings of the Second International Conference on Soft Computing for Problem Solving (SocProS 2012), December 28–30, 2012,* Advances in Intelligent Systems and Computing 236, DOI: 10.1007/978-81-322-1602-5_145, © Springer India 2014

fastest growing part of the waste stream and is harming the planet in more than one ways [1–3].

Today's PCs are so powerful that we no longer need one PC per person. We can utilise the excess capabilities in one PC and share it with many users. Desktop virtualisation thin client devices do just that and use just 1–5 W, last for a more than 10 years, and generate very little e-waste. Not only is this a simple solution to a complex problem, it is also very efficient. Desktop virtualisation saves 75 % on hardware, and since they draw very less power, we can reduce nearly 90 % energy footprint per user. These devices produce practically no heat, reducing the need for air-conditioning, which in turn saves power consumed by such cooling solutions.

Today, the biggest driving force for IT companies to adopt thin client desktop virtualisation is major savings. What has made server-based computing easier is that many desktop virtualisation solutions integrate seamlessly with existing virtualisation infrastructure and even use the same management tools, enabling IT managers to get started and feel comfortable with their new solution right away—realising benefits from day one [4]. Once deployed, desktop virtualisation solutions centralise management and enable easy software updates and security and patch rollouts. They also improve security and provide users with options like self-help and desktop mobility. Although the draw for IT to implement desktop virtualisation is often not green, more and more organisations have realised that with virtualisation, green benefits often correlate with cost savings. In turn, these cost savings result in a significant environmental impact that will lower an organisation's carbon footprint and e-waste volume while changing the future of green computing.

2 Green Computing

According to San Murugesan, the field of green computing is "the study and practice of designing, manufacturing, using, and disposing of computers, servers, and associated subsystems—such as monitors, printers, storage devices, and networking and communications systems—efficiently and effectively with minimal or no impact on the environment". It is about environmentally friendly use of computers and related technologies. Efforts to reduce the energy consumption associated with PC are often referred to as "green computing," which is the practice of using computing resources efficiently and in an environmentally sensitive manner. "Green IT" refers to all IT solutions that save energy at various levels of use. These include (i) hardware, (ii) software and (iii) services [1–3].

Green is used in everyday language to refer to environmentally sustainable activities. Green computing encompasses policies, procedures and personal computing practices associated with any use of information technology (IT). People employing sustainable or green computing practices strive to minimise green house gases and waste, while increasing the cost-effectiveness of IT, such as computers, local area networks and data centres. More directly it means using computers in ways that save the environment, save energy, and save money.

The greenhouse effect is a process by which thermal radiation from a planetary surface is absorbed by atmospheric greenhouse gases, and is re-radiated in all directions. Since part of this re-radiation is back towards the surface and the lower atmosphere, it results in an elevation of the average surface temperature above what it would be in the absence of the gases. Particularly since the arrival of industrialisation, the human race has intensified this effect, as combustion processes and other industrial processes release greenhouse gases which, with their increased concentration, influence the Earth's radiation balance and thus influence the greenhouse effect. Possible measures for reduction of greenhouse gas emissions include increasing the energy efficiency of machinery, e.g. via intelligent control programs, or replacing energy-intensive computer architectures with low-carbon, solutions. Here, green IT or server-based computing can make a significant contribution by using energy more efficiently than conventional IT equipment and thus saving greenhouse gases.

3 Thin Client Desktop Virtualisation

Wikipedia's definitions of a thin client is: A thin client (sometimes also called a lean or slim client) is a client computer or client software in client-server architecture networks which depends primarily on the central server for processing activities, and mainly focuses on conveying input and output between the user and the remote server. In contrast, a thick or fat client does as much processing as possible and passes only data for communications and storage to the server [4–8].

Consultants have been saying for a long time that thin clients are the future. Today, thin client technology finally has caught up with the vision. A few years ago, most companies had two or three models of thin clients, but today there are many models to choose from with varying CPU speeds, memory capacities, storage capacities and operating systems. Besides being more secure and easier to deploy, manage and maintain (than their PC counterparts) thin clients boast a longer life expectancy because they have no moving parts, small footprint on the desktop, lower power consumption and server-centralized data storage.

Desktop virtualisation is an implementation of thin client concept and is defined as a computing environment in which some or all components of the system, including operating system and applications, reside in a protected environment, isolated from the underlying hardware and software platforms. The virtualisation layer controls interactions between the virtual environment and the rest of the system. Essentially, servers host desktop environments specific to each user and stream applications and operating systems to the desktop (Fig.1).

Desktop virtualisation separates software from the basic hardware that provides it, putting the focus on what is being delivered, making the user unaware and unconcerned about how it is being delivered or from where it is coming. Virtualisation separates the fundamental operating system, applications and data from an end user's device and moves these components into the central server where they can be secured and centrally managed. This approach allows users to access their "virtual desktop" with a full personal computing experience across devices and locations. Desktop

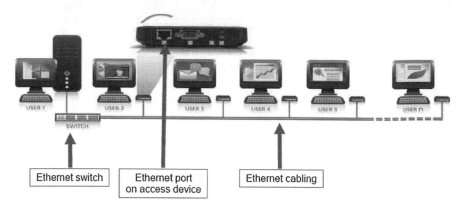

Fig. 1 Desktop virtualisation using N-computing. (*Source*: http://www.ncomputing.com)

virtualisation takes the efficiencies offered through a centralized processing environ-ment and merges it with the flexibility and ease of use found in a traditional PC. It is the concept of isolating a logical operating system instance from the client that is used to access it.

There are several different conceptual models of desktop virtualisation, which can broadly be divided into two categories based on whether or not the operating system instance is executed locally or remotely. It is important to note that not all forms of desktop virtualizstion involve the use of virtual machines which lead to more efficient use of computing resources, both in terms of energy consumption and cost-effectiveness.

Host-based forms of desktop virtualisation require that users view and interact with their desktops over a network by using a remote display protocol. Because processing takes place in a server, client devices can be thin clients, zero clients, smartphones and tablets.

Client-based types of desktop virtualisation require processing to occur on local hardware; the use of thin clients, zero clients, and mobile devices is not possible.

In this chapter, all the references to desktop virtualisation are related to host-based forms of desktop virtualisation where the client does not require any processor, memory or storage facilities. The solution can be implemented in a LAN as well as WAN environment. Whereas in the WAN scenario, the bandwidth and the speed becomes an important driving factor.

4 Virtualisation Benefits

4.1 Affordable to All

With virtualisation, the investment on PCs can be maximised by adding users for a small fraction of the cost. The client access devices are mostly thin or zero computing devices which are nearly three times cheaper than a PC. Hence 50–70 % increase in the number of computing seats can be implemented for the same budget. The savings enables to improve computer upgrade cycles, expand access or invest extra funds in other technology upgrades.

4.2 Compatible

The implementation of desktop virtualisation supports multiple operating system platforms like Microsoft Windows and Linux. So, the current applications can be used in the same environment without the need of any migration of technology.

4.3 Easy to Manage

Once deployed, desktop virtualisation solutions centralise management and enable easy software updates and security and patch rollouts. They also improve security and provide users with options like self-help and desktop mobility. Desktop virtualisation, creates a single "golden" image of the OS on a server and each user takes advantage of the same golden image. So, in effect, there is a need to manage only a single image, not hundreds or thousands. It is also a simple matter to create multiple such images for different groups, each setup with the applications that users in that group need.

4.4 Efficient to Operate

The efficiency of the virtualisation solution goes beyond getting more from the PC resources. The virtual desktop devices save space in the work area and save electricity by drawing substantially less power than a typical PC.

4.5 Simple to Deploy

Every virtualisation software and hardware product is designed to be easy to set up, secure and maintain by people with basic PC skills. The virtual desktops require no maintenance and do not contain sensitive components such as hard drives and fans. And with fewer PCs to manage, there will be fewer support issues.

5 Virtualisation Green Advantage

One of the best ways to reduce energy consumption is to consider the use of thin client technology. This is similar to server virtualisation in that one physical computer runs several workstations in the same way that several server instances can run on one physical server box using virtualisation tools [1].

This approach has been made possible by the increased power of modern computers. Processors are faster, they often have multicores, can run in 64 bit mode, and may be arranged in pairs or fours within a single machine. Memory has reduced in price considerably and, using 64 bit systems, is able to be addressed in more than 4 GB memory configurations thus dramatically increasing the performance of systems. Graphics and sound have also improved greatly with on board sound and graphics now more capable than the average user requires. These computer hardware improvements continue delivering a faster and richer experience to the user. The increased potential of modern PCs and the need to become more environmentally responsible has provided an option to run many workstations from one computer base unit.

The average person uses less than 5 % of the capacity of their PC. The rest is simply wasted. The Desktop Virtualisation solution is based on this simple fact that today's PCs are so powerful that the vast majority of applications only use a small fraction of the computer's capacity. Virtualisation tap this unused capacity so that it can be simultaneously shared by multiple users—maximizing the PC utilisation. Each user's monitor, keyboard, and mouse connect to the shared PC through a small and highly reliable virtualisation access device. These access devices use the concept of thin client or zero client. The zero client access device has no CPU, memory, or moving parts. Electricity consumption of a thin/zero client is less than 10 % that of a PC. A standard PC consumes 150–200 W compared to 5–15 W by thin/zero clients. Hence, virtualisation can reduce electricity consumption as well as cooling requirements, thus reducing both carbon emissions and cost.

Virtualisation reduces carbon emissions and have a significantly smaller footprint, with some solutions using less than one twentieth of the materials required for a traditional PC, which results in far less e-waste filling landfills. The elimination of the physical desktop PC lessens the landfill issues as zero-clients contain no processor, memory or other moving elements like hard disks [1–3].

Additionally, zero-client desktop virtualisation solutions can have a useful life more than twice the length of a traditional PC because they do not have an operating system, software or moving parts on the device which can fail or quickly become outdated or obsolete. As long the desktop device is capable of handling the software updates made on the server, the solution continues to work. Zero-clients can last for 8–10 years as opposed to 3–4 years of a conventional PC. Hence in case of Desktop virtualisation, e-waste reduction, as opposed to e-waste recycling, is definitely a grand step towards greener environment. PCs typically weigh about 10 kg and are disposed of in landfills after 3–5 years. Desktop virtualisation access devices weighs about 150 g, and easily last 5 years or more, for a 98 % reduction in electronic waste. So there's less to transport and less e-waste that will find its way to a landfill.

In addition to reduced energy use, emissions are also reduced dramatically with the concept of virtualisation. For example, 26 physical servers and a network would produce 210,269 pounds of carbon dioxide per year compared to a private cloud hosting solution that has the capacity of 35 servers and produces only 27,903 pounds of carbon dioxide per year. The Verdantix report, the result of a study, has found that if companies adopt cloud computing (which is an advanced implementation of virtualisation), they can reduce the energy consumption of their IT and save money on energy bills. The report, created by research firm Verdantix and sponsored by AT&T, estimates that cloud computing could enable companies to save $12.3 billion off their energy bills. That translates into carbon emission savings of 85.7 million metric tons per year by 2020 [4, 6, 7].

Virtualisation Green Benefits

Less energy consumption
Personal computers draw about 110 W of electricity, whereas virtual desktops only draw 1–5 W. The energy footprint per user declines by as much as 98 %.
Air-conditioning
Consume less energy and produces less heat as air-conditioning is not compulsory. Less air- conditioning compounds the savings on your electricity and releases less carbon footprint.
E-waste
The solutions are also the most eco-friendly on earth, and not just because of how little electricity they consume: they generate a negligible amount of e-waste, reducing contribution to the e-waste stream (Fig. 2).
Environmental impact
Because the products deliver green computing and cost less to buy and operate, they address the triple bottom line perfectly. That is, they save money while they help reduce climate change and hazardous e-waste.

As PC adoption grows globally, it is estimated that there will be more than two billion PCs in use by 2015. It took 30 years to reach one billion but will only take three more years to double that number. With this trend, something needs to change. If desktop virtualisation access systems are used at a ratio of 6 devices to each PC:

- Energy use would decline by over 143 billion kilowatt hours per year
- CO_2 emissions would decrease by 114 million metric tons. That's like planting 550 million trees!
- E-waste would be reduced by 7.9 million metric tons

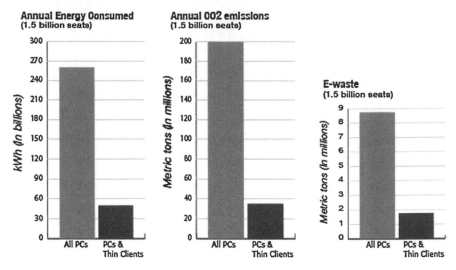

Fig. 2 Energy consumed, emissions and E-waste comparison of PCs and thin clients. (*Source*: http://www.ncomputing.com/company/green-computing)

6 The Global Warming Potential Effect

According to Wikipedia, Global-warming potential (GWP) is a relative measure of how much heat a greenhouse gas traps in the atmosphere. It compares the amount of heat trapped by a certain mass of the gas in question to the amount of heat trapped by a similar mass of carbon dioxide. A GWP is calculated over a specific time interval, commonly 20, 100 or 500 years. GWP is expressed as a factor of carbon dioxide (whose GWP is standardized to 1). For example, the 20 year GWP of methane is 72, which means that if the same mass of methane and carbon dioxide were introduced into the atmosphere, that methane will trap 72 times more heat than the carbon dioxide over the next 20 years [2, 3].

The GWP depends on the following factors:

- the absorption of infrared radiation by a given species
- the spectral location of its absorbing wavelengths
- the atmospheric lifetime of the species

Thus, a high GWP correlates with a large infrared absorption and a long atmospheric lifetime. The dependence of GWP on the wavelength of absorption is more complicated. Even if a gas absorbs radiation efficiently at a certain wavelength, this may not affect its GWP much if the atmosphere already absorbs most radiation at that wavelength. A gas has the most effect if it absorbs in a "window" of wavelengths where the atmosphere is fairly transparent.

Table 1 Global warming potential in kg CO_2eq

Phase	Desktop PC	Desktop virtualisation
Production phase	117.33	3.67
Manufacturing phase	21.04	0.66
Distribution phase	25.25	0.79
Operation phase	529.67	271.55
Total	693.29	276.67

According to a study "Thin Clients 2011—Ecological and economical aspects of virtual desktops", conducted by Fraunhofer Institute of Environmental, Safety and Energy Technology UMSICHT, compared to Desktop virtualisation, a Desktop PC user has nearly 2.5 times GWP. The following table shows the GWP at various phases.

As it is clear from Table 1, virtualisation has substantially less impact on global warming. This leads to a lots of savings in terms of CO_2 emissions. The following pie-chart shows the comparison (Fig. 3).

Fig. 3 Comparison of GWP of a desktop PC and virtualisation

7 Challenges of Virtualisation

However, desktop Virtualisation is not without drawbacks and challenges. Users that require multitasking, interaction with multimedia applications or support for local peripherals may not experience satisfactory performance. Traditional desktop virtualisation solutions are designed to support the "lowest common denominator" client device, often sacrificing user experience for improved access. In general thin client technology and desktop virtualisation is used in open access areas such as learning centres and libraries or general purpose classrooms. It is not used for activities requiring significant processing power such as graphics and audio editing tasks or data manipulation activities. For general purpose computing (word processing, e-mail, Web browsing) the user experience is unaffected. Some of the disadvantages relate to the reliability of the server, network bandwidth issues, less flexibility and being unsuitable for certain tasks such as multimedia.

As anyone who has undertaken a virtualisation project quickly discovers, many applications and hardware devices are not amenable to easy virtualisation; therefore, a detailed assessment of applications and hardware in use is a key prerequisite to successful virtualisation. Even if virtualisation is technically feasible, many vendors offer limited support for virtual instances of their software, and some even refuse to honor warranties and maintenance contracts if their apps are running in a virtual environment.

8 Conclusion

The specific recommendation for IT decision-makers is to consider thin clients as an alternative to the desktop PCs. This operating model offers ecological and economic advantages over conventional Client-Server models. In order to minimise the consumption of materials and energy, the IT infrastructure should be designed to suit the actual needs of the end-users and not just implement a heavily loaded solution in which the infrastructure is not utilised to the fullest capacity. Traditional terminal servers can be combined with desktop virtualisation solutions to provide each user equipment tailored to their specific requirements so as to optimise the capacity utilisation of the hardware.

Any organisation using for more than one PC should seriously consider the advantages of moving to desktop virtualisation. By taking advantage of today's low-cost yet ever-more-powerful computers, even the smallest organisation can appreciate immediate benefits without the high expense of mainframe computing or the complexity and performance limitations of server-based computing. Desktop virtualisation makes computing available to more people within the organisation for less cost. Best of all, it saves a lot of energy and minimises e-waste, thus contributing the sustainability of the environment.

References

1. Agarwal, S., Nath, A.: Green computing-a new horizon of energy efficiency and electronic waste minimization: a global perspective. In: Proceedings of IEEE CSNT-2011 held at SMVDU (Jammu), pp. 688–693, 3–6 June 2011
2. Agarwal,S., Nath, A.: Cloud computing is an application of green computing—a new horizon of energy efficiency and its beyond. In: Proceedings of International conference ICCA 2012 held at Pondechery, 27–31 Jan 2012
3. Agarwal, S., Nath, A., Chowdhury, D: Sustainable approaches and good practices in green software engiineering. Int. J. Res. Rev. Comp. Sci. 3(1), 2079–2557 (2012)
4. Knorr, E.: What desktop virtualization really means—depending on whom you talk to, desktop virtualization is either the hottest trend in IT or an expensive notion with limited appeal
5. Thin Clients 2011—Ecological and Economical Aspects of Virtual Desktops, a Study Conducted by Fraunhofer Institute of Environmental, Safety and Energy Technology UMSICHT
6. Chitnis, N., Bhaskaran, R., Biswas, T.: Going Green with Virtualisation. Setlabs Briefings, 9(1), 31–38 (2011)

7. Desktop Virtualisation in Higher Education. A Strategy Paper from Centre for Digital Education, 1–8
8. A. Orady, Going Green with Desktop Virtualisation
9. Freedman, R.: Investing in Virtualization has Green IT Payoffs

Noise Reduction from the Microarray Images to Identify the Intensity of the Expression

S. Valarmathi, Ayesha Sulthana, K. C. Latha, Ramya Rathan,
R. Sridhar and S. Balasubramanian

Abstract Microarray technique is used to study the role of genetics involved in the development of diseases in an early stage. Recently microarray has made an enormous contribution to explore the diverse molecular mechanisms involved in tumorigenesis. The end product of microarray is the digital image, whose quality is often degraded by noise caused due to inherent experimental variability. Therefore, noise reduction is a most contributing step involved in the microarray image processing to obtain high intensity gene expression results and to avoid biased results. Microarray data of breast cancer genes was obtained from National Institute of Animal Science and Rural Development Administration, Suwon, South Korea. Two algorithms were created for noise reduction and to calculate the intensity of gene expression of breast cancer susceptibility gene 1 (BRCA1) and breast cancer susceptibility gene 2 (BRCA2). The new algorithm successively decreased the noise and the expression value of microarray gene image was efficiently enhanced.

S. Valarmathi (✉) · R. Sridhar · S. Balasubramanian
DRDO-BU-CLS, Bharathiar University, Coimbatore, Tamil Nadu, India
e-mail: valar28aadarsh@gmail.com

R. Sridhar
e-mail: rmsridhar@rediffmail.com

S. Balasubramanian
e-mail: director_research@jssuni.edu.in

A. Sulthana · K. C. Latha
Department of Water and Health, JSS University, Mysore, Karnataka, India
e-mail: ayeshasulthanaa@gmail.com

K. C. Latha
e-mail: latha_tanvi23@yahoo.com

R. Rathan
Department of Anatomy, JSS University, Mysore, Karnataka, India
e-mail: ramirohith@ymail.com

B. V. Babu et al. (eds.), *Proceedings of the Second International Conference on Soft Computing for Problem Solving (SocProS 2012), December 28–30, 2012*, Advances in Intelligent Systems and Computing 236, DOI: 10.1007/978-81-322-1602-5_146, © Springer India 2014

Keywords Noise reduction · Image processing · Microarray · Algorithm · Gene expression · Breast cancer

1 Introduction

Thousands of individual DNA sequences are printed in parallel on a glass microscope slide of cDNA microarrays [19]. Microarray slide consists of a rectangular array of subgrids, each subgrid printed by one pin of the contact-printer. A subgrid consists of an array of spots, each spot containing a single cDNA probe. The hybridized arrays are imaged using a scanner and the output stored as 16-bit image files. Different dyes were used to visualize the image clearly. In the analysis of cDNA microarray images, the most important task involved is the gridding of the spots [9]. For this multiple methods related to array recognition; spot segmentation and measurement extraction have emerged over past several years [12]. Large numbers of commercial tools have been developed in microarray image processing. Microarrays measure expression levels of tens of thousands of genes simultaneously and suffer from the presence of inherent experimental noise. Detection of random noise causes inconsistencies in the image processing [15]. The noise generally results during the phases of sample preparation, hybridization, and scanning [16]. The sample preparation noise results from the process of RNA amplification, and the hybridization noise refers to the randomness in the process of RNA binding to the probes. The sources of scanning noise include leak of external light, variations in laser intensity, and presence of dirt [4]. Many types of distortions limit the quality of digital images during image acquisition, formation, storage, and transmission. Often, images are corrupted by impulse noise, thereby causing loss of image details. It is important to eliminate noise in images before using them for image processing techniques like edge detection, segmentation, and registration [13]. Removal of noise in the microarray technique involves three steps; gridding (addressing each spot), segmentation (separating spot pixels from background pixels), and quantification (putting spot intensity data into numerical form for comparison) [5]. Noise reduction framework enables efficient filtering of large and color images in real-time application with the preservation of their textural features [14]. An extensive quantitative measure of the efficiency of the noise estimation determines the quality of the JPEG file. Several authors have developed and discussed denoising methods in microarray data [3, 18].

Gene expression analysis were carried out using microarray for simultaneous interrogation of the expression of genes in a high-throughput fashion and offers unprecedented opportunities to obtain molecular signatures of the activity of diseased cells [10]. Stanford University was the first to describe and use microarray to study gene expression in various diseases including cancer [17]. Gene expression in cancer by microarray is fast gaining popularity in providing better prognostic and predictive information on the disease. Microarray-based gene expression profiling is used as a tool for Breast Cancer Management, in identifying genes whose expression has changed in response to pathogens or other organisms by comparing gene expression of infected to that of uninfected cells or tissues [1]. Heritable mutation influences the gene-expression profile of the breast cancer with BRCA1 and BRCA2 mutations [6]. Gene expression profiles evaluated by microarray-based quantification of RNA are used in studies of differential diagnosis and prognosis in cancer. Microarray data were evaluated using both unsupervised, and supervised multivariate statistical methods [8]. Differential gene expression was analyzed for gene oncology to identify important functional categories [2].

Our study is involved in developing an algorithm for the processing of microarray images with the following objectives:

I To reduce noise from a microarray image of breast cancer gene*.
II To identify the intensity of gene expression from a microarray image of a breast cancer gene.

2 Data and Methodology

For the present study the microarray data was obtained from National Institute of Animal Science and Rural Development Administration, Suwon, South Korea and was used to test our algorithm integrated with MATLAB. The main aim is to reduce the noise and to calculate the intensity of gene expression of BRCA1 and BRCA2 genes in the microarray image, rather to give interpretation on gene expression in microarray images of breast cancer genes. A fully automated method that removes the existing impulse of noise from the microarray plate with virtually no required user interaction or external information, greatly increasing efficiency of the image analysis was developed. For image processing of microarray data, two algorithms were developed in Visual Basic and integrated with MATLAB (Appendixes I and II):

(i) Algorithm for Noise reduction
(ii) Algorithm to calculate intensity of gene expression

3 Results

3.1 Removal of Noise in the Microarray Image

The proposed approach was used to test microarray image of the Breast Cancer Patient containing thousands of spots (Fig. 1). A section from the microarray image (Fig. 1) was cropped and it was named as original image (Fig. 2a). The noise from the microarray image was removed by using algorithm (i). Initially the original image (Fig. 2a) was converted into gray image. Using Otsu method (reduction of a gray level image to a binary image) the graythresh value [11] for the image was calculated. Using this graythresh value the image was then converted into Black and White image. From the image, the objects which have fewer than P pixels were removed and used to create a new structural binary image. This step was performed

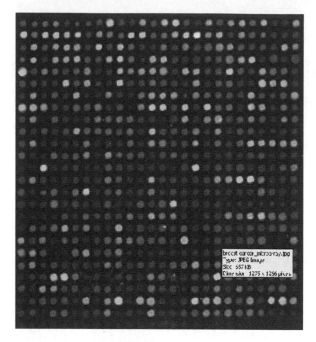

Fig. 1 Microarray image of the breast cancer patient

(a) (b) (c)

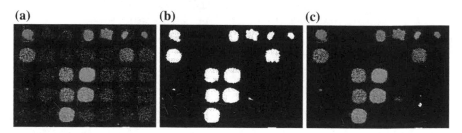

Fig. 2 Removal of noise from the microarray image **a** Original image **b** *Black* and *White* after Filtering **c** Image after noise removal

repeatedly until all the fewer objects from the image were removed. To this image, a flood-fill operation was performed to fill the background pixels and the output image was obtained (Fig. 2b).

The image obtained after filtering was compared with the original image (Fig. 2a) to produce the final image (Fig. 2c) using MATLAB environment. Thus the noise from the original image was removed and the obtained image can be used for calculating the intensity of gene expression and to find the level of risk of breast cancer using microarray. This method achieves an accuracy of more than 95 % and it outperforms the existing methods.

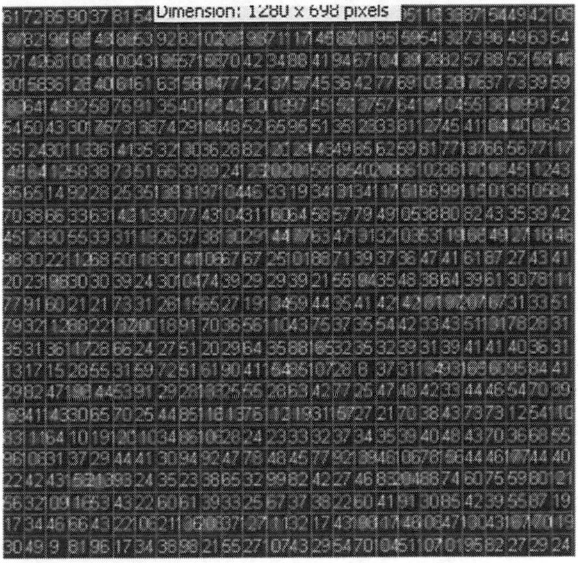

Fig. 3 *Red* intensity image of microarray plate

3.2 Gene Expression Analysis

The intensity was calculated from the 'noise removed microarray image' to find the gene expression value for each spot in the plate using algorithm (ii). The red intensity and green intensity in the microarray image was obtained based on the fluorescent tag used in the experimental analysis and are shown in the Figs. 3 and 4. The red spot (Cy5 fluorescent) indicates the expression value of the defective gene and the green spot (Cy3 fluorescent) indicates the expression value of the normal gene. The overall gene expression of a microarray plate was obtained by merging the image of the red and green intensity and is shown in the Fig. 5. In this image, the intensity of green spot shows that the expression of the gene is repressed. The red intensity spot shows that the gene expression is induced in the microarray plate, due to the presence of defective gene. The yellow (mixture of green and red) color spot in the image shows that gene expression is unchanged (normal and abnormal gene is equally expressed). This provides knowledge to the molecular biologist and particularly to oncologist in the diagnosis of defective genes. Therefore, the developed algorithm is a new attempt to predict the expression value of any microarray gene images.

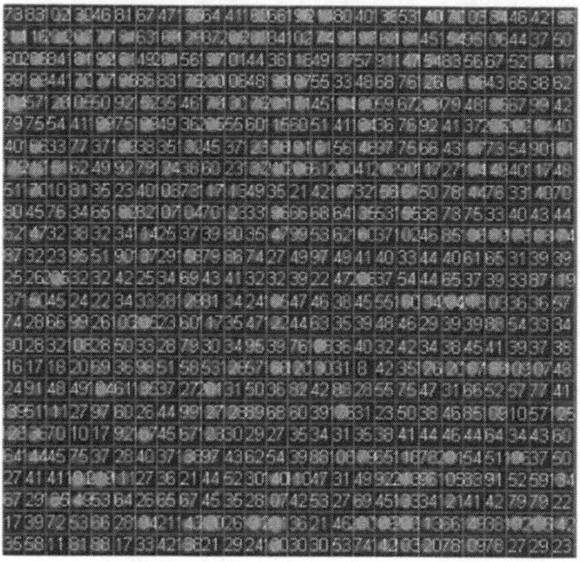

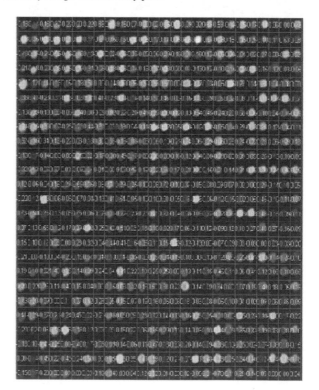

Fig. 4 *Green* intensity image of microarray plate

Fig. 5 Gene expression value in the spot of microarray plate

4 Conclusion

Microarray image quality is distorted by various source of noise like inherent technical variations and biological sample variations. Therefore the most critical aspect is to reduce the noise from microarray image to avoid the misinterpretation of results. In this study, the microarray image of a breast cancer patient was used to reduce the noise using the proposed algorithm developed in Visual Basic and integrated with MATLAB. Noise reduction of microarray image was achieved successfully without generating any negative effect on the data. The algorithm developed to identify the intensity of gene expression was implied on a noise reduced microarray image to identify the expression of the gene based on the intensity of the spot in the image. These algorithms can be used further to reduce noise in microarray images and to enhance the intensity of gene expression, this will lead to better analysis and interpretation of microarray images.

Appendix I

Algorithm for Noise Removal from the Microarray Plate

Step 1 : Read an RGB Microarray Image.
Step 2 : Convert the given image to gray.
Step 3 : Calculate the graythresh value of the given image using Otsu Method.
Step 4 : Convert the gray image to BW image using graythresh value.
Step 5 : Remove from a binary image all connected components (objects) that have fewer than P pixels, producing another binary image.
Step 6 : Create morphological structuring element.
Step 7 : Perform morphological closing on the binary image obtained from step 4 with the structuring element created in the step 6.
Step 8 : Perform a flood-fill operation on background pixels of the input binary image obtained from step 7. The output is saved as Picture 1.
Step 9 : The Picture 1 is compared with original image from which the final image is produced based on the white pixel shown in the Picture 1.

Algorithm to Convert RGB Image to Gray Image

```
For i = 0 To Picture1.ScaleWidth

    For j = 0 To Picture1.ScaleHeight
            tcol = GetPixel(Picture1.hdc, i, j)
    r = tcol Mod 256
    g = (tcol / 256) Mod 256
            b = tcol / 256 / 256
            colour = r * 0.3 + g * 0.59 + b * 0.11
            SetPixel Picture2.hdc, i, j, RGB(colour, colour, colour)
    Next

Next
```

Algorithm for Obtaining Graythresh Value

Otsu shows that minimizing the intraclass variance is the same as maximizing inter-class variance which is expressed in terms of class probabilities ωi and class means μi which in turn can be updated iteratively. This idea yields an effective algorithm.

Step 1 : Compute histogram and probabilities of each intensity level
Step 2 : Set up initial $\omega i(0)$ and $\mu i(0)$
Step 3 : Step through all possible thresholds maximum intensity
Step 4 : Update ωi and μi
Step 5 : Compute Desired threshold corresponds to the maximum

Algorithm to convert Gray Image to Black and White image
using Graythresh Value

```
For i = 0 To Picture1.ScaleWidth
    For j = 0 To Picture1.ScaleHeight
        tcol = GetPixel(Picture1.hdc, i, j)
        if tcol > graythresh_value then
        SetPixel Picture2.hdc, i, j, RGB(255, 255, 255)
        else
            SetPixel Picture2.hdc, i, j, RGB(0, 0, 0)
        endif
    Next
Next
```

Algorithm to Remove Small Objects

Step 1: Determine the connected components for BW image using 4-Connected Neighbourhood.
Step 2: Compute the area of each component which are expressed in terms of Pixel.
Step 3: Consider the only components whose pixels are greater than 10.

Create Morphological Elements using MATLAB

SE = strel('disk', R, N) creates a flat, disk-shaped structuring element, where R specifies the radius. R must be a nonnegative integer. N must be 0, 4, 6, or 8. When N is greater than 0, the disk-shaped structuring element is approximated by a sequence of N periodic-line structuring elements. When N equals 0, no approximation is used, and the structuring element members consist of all pixels whose centers are no greater than R away from the origin. If N is not specified, the default value is 4.

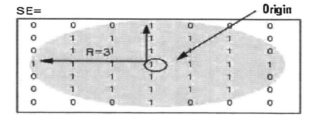

Algorithm for Black and White Image to RGB

Step 1:

```
For i = 0 To Picture1.ScaleWidth
  For j = 0 To Picture1.ScaleHeight
    tol = GetPixel(Picture2.hdc, i, j)
    If tol = 0 Then
            Grid2.TextMatrix(i, j) = 0
    Else
            Grid2.TextMatrix(i, j) = 1
    End If
  Next
Next
```

Step 2:

```
Picture3.BackColor = vbBlack
Picture3.height = Picture1.height
Picture3.width = Picture1.width
    For i = 0 To Picture1.ScaleWidth
            For j = 0 To Picture1.ScaleHeight
            If Grid2.TextMatrix(i, j) = 1 Then
                    SetPixel Picture3.hdc, i, j, val(Grid1.TextMatrix(i, j))
            End If
            Next

        Next
```

Appendix II

Algorithm for Calculating Expression Value for each spot in Microarray Plate

```
x = imread('MicroArraySlide.JPG');
        %x = imread('microarray_test6_agri.JPG');
        %x = imread('Composite-s416-.jpg');
        %x = imread('dna_microarray.JPG');
        %x = imread('t1.JPG');
        %x = imread('t9.jpg')
        %x = imread('t5.jpg')
        %x = imread('t10.jpg')
        %x = imread('microarray_test1_agri_rice.JPG');
        %x = imread('t4.JPG');
imageSize = size(x)
screenSize = get(0,'ScreenSize')
iptsetpref('ImshowBorder', 'tight')
```

```
    imshow(x)
title('original image')
```

Crop Specified Region

```
%y = imcrop(x, [622 2467 220 227]);
[y,RECT] = imcrop(x) %, [398 449 218 214]);
        %y = imcrop(x, [398 449 218 214]);
        f1 = figure('position', [40 46 285 280]);
imshow(y)
```

Display Red and Green Layers

```
f2 = figure('position', [265 163 647 327]);
subplot(121)
        redMap = gray(256);
        redMap(:, [2 3]) = 0;
        t = subimage(y(:,:,1), redMap)
axis off
title('red (layer 1)')
subplot(122)
        greenMap = gray(256);
        greenMap(:, [1 3]) = 0;
subimage(y(:, :, 2),greenMap)
axis off
title('green (layer 2)')
```

Convert RGB Image to Grayscale for Spot Finding

```
z = rgb2gray(y);
figure(f1)
imshow(z)
```

Create Horizontal Profile

```
xProfile = mean(z);

        f2 = figure('position', [39 346 284 73]);
plot(xProfile)

        title('horizontal profile')
axis tight
```

Estimate Spot Spacing by Autocorrelation

```
ac = xcov(xProfile);
f3 = figure('position', [-3 427 569 94]);
plot(ac)

        s1 = diff(ac([1 1:end]));
```

```
        s2 = diff(ac([1:end end]));
        maxima = find(s1>0 & s2<0);
        estPeriod = round(mean(diff(maxima)))
hold on

        plot(maxima,ac(maxima), 'r^')
hold off

        title('autocorrelation of profile')
axis tight
```

Remove Background Morphologically

```
seLine = strel('line', estPeriod, 0);

        %seLine = strel('disk', 2);
        xProfile2 = imtophat(xProfile,seLine);
        f4 = figure('position', [40 443 285 76]);
plot(xProfile2)
title('enhanced horizontal profile')
axis tight
```

Segment Peaks

```
level = graythresh(xProfile2/255)*255

        bw = im2bw(xProfile2/255, level/255);
        L = bwlabel(bw);
        f5 = figure('position', [40 540 285 70]);
plot(L)
axis tight
        title('labelled regions')
```

Locate Centers

```
stats = regionprops(L);

        centroids = [stats.Centroid];
        xCenters = centroids(1:2:end)
figure(f5)
hold on

        plot(xCenters, 1:max(L), 'ro')
hold off

        title('region centers')
```

Determine Divisions between Spots

```
gap = diff(xCenters)/2;
first = abs(xCenters(1) − gap(1));
xGrid = round([first xCenters(1:end) + gap([1:end end])])
```

Transpose and Repeat

```
yProfile = mean(z');
        ac = xcov(yProfile);
        p1 = diff(ac([1 1:end]));
        p2 = diff(ac([1:end end]));
        maxima = find(p1>0 & p2<0);
        estPeriod = round(mean(diff(maxima)))
        seLine = strel('line', estPeriod, 0);
%seLine = strel('disk', 2);
        yProfile2 = imtophat(yProfile, seLine);
        level = graythresh(yProfile2/255);
        bw = im2bw(yProfile2/255, level);
        L = bwlabel(bw);
        stats = regionprops(L);
        centroids = [stats.Centroid];
        yCenters = centroids(1:2:end)
        gap = diff(yCenters)/2;
        first = abs(yCenters(1)-gap(1));
```

List Defining Vertical Boundaries between Spot Regions

```
yGrid = round([first yCenters(1:end) + gap([1:end end])])
        f7 = figure('position', [52 94 954 425]);
        ax(1) = subplot(121);
subimage(y(:, :, 1), redMap)
        title('red intensity')
        ax(2) = subplot(122);
subimage(y(:, :, 2), greenMap)
        title('green intensity')
        f8 = figure('position', [316 34 482 497]);
ax(3) = get(imshow(y, 'notruesize'), 'parent');
        title('gene expression')
for i=1:3

    axes(ax(i))
    axis off
line(xGrid'*[1 1], yGrid([1 end]), 'color', 0.5*[1 1 1])
line(xGrid([1 end]), yGrid'*[1 1], 'color', 0.5*[1 1 1])
end

[X, Y] = meshgrid(xGrid(1:end-1), yGrid(1:end − 1));
[dX, dY]= meshgrid(diff(xGrid), diff(yGrid));
ROI = [X(:) Y(:) dX(:) dY(:)];
```

Segment Spots from Background by Thresholding

```
fSpots = figure('position', [265 163 647 327]);
subplot(121)
imshow(z)
title('gray image')
subplot(122)
        bw = im2bw(z, graythresh(z));
imshow(bw)
title('global threshold')
```

Apply Logarithmic Transformation then Threshold Intensities

```
figure(fSpots)
subplot(121)

        z2 = uint8(log(double(z) + 1)/log(255)*255);
imshow(z2)
title('log intensity')
subplot(122)
        bw = im2bw(z2, graythresh(z2));
imshow(bw)
title('global threshold')
```

Try local Thresholding Instead

```
figure(fSpots)
subplot(122)

  bw = false(size(z));
  for i=1:length(ROI)
        rows = round(ROI(i, 2)) + [0:(round(ROI(i, 4))-1)];
        cols = round(ROI(i, 1)) + [0:(round(ROI(i, 3))-1)];
        spot = z(rows, cols);
        bw(rows, cols) = im2bw(spot, graythresh(spot));
  end
imshow(bw)
title('local threshold')
```

Logically Combine Local and Global Thresholds

```
figure(fSpots)
subplot(121)

  bw = im2bw(z2, graythresh(z2));
  for i=1:length(ROI)
        rows = round(ROI(i, 2)) + [0:(round(ROI(i, 4))-1)];
        cols = round(ROI(i, 1)) + [0:(round(ROI(i, 3))-1)];
        spot = z(rows, cols);
        bw(rows, cols) = bw(rows, cols) | im2bw(spot, graythresh(spot));
  end
imshow(bw)

        title('combined threshold')
```

```
subplot(122)
imshow(z)

   title('linear intensity')
                    Fill Holes to Solidify Spots

figure(fSpots)
subplot(121)

          for i=1:length(ROI)
              rows = round(ROI(i, 2)) + [0:(round(ROI(i, 4))-1)];
              cols = round(ROI(i, 1)) + [0:(round(ROI(i, 3))-1)];
          end bw(rows, cols) = imfill(bw(rows, cols), 'holes');
seDisk = strel('disk', round(estPeriod));
L = zeros(size(bw));

          for i=1:length(ROI)
              rows = ROI(i, 2) + [0:(ROI(i, 4)-1)];
              cols = ROI(i, 1) + [0:(ROI(i, 3)-1)];
              rectMask = L(rows, cols);
              spotMask = bw(rows, cols);
              rectMask(spotMask) = i;
          end L(rows, cols) = rectMask;

spotData = [ROI zeros(length(ROI), 5)];

          for i=1:length(ROI)
              spot = imcrop(y, ROI(i, :));
              spot2 = imtophat(spot, seDisk);
              mask = imcrop(L, ROI(i, :))==i;
          for j=1:2
              layer = spot2(:, :, j);
              intensity(j) = double(median(layer(mask)));
          text(ROI(i, 1) + ROI(i, 3)/2, ROI(i, 2) + ROI(i, 4)/2, sprintf('%.0f',
          intensity(j)),...
              'color','y', 'HorizontalAlignment', 'center', 'parent', ax(j))
              rawLayer = spot(:, :, j);
          end rawIntensity(j) = double(median(layer(mask)));
          expression = log(intensity(1)/intensity(2));
          text(ROI(i, 1) + ROI(i, 3)/2, ROI(i, 2) + ROI(i, 4)/2, sprintf('%.2f',
          expression),...
              'color', 'w', 'HorizontalAlignment', 'center', 'parent', ax(3))
              drawnow
          end spotData(i, 5:9) = [intensity(:)' expression rawIntensity(:)'];
          xlswrite('microarray.xls', spotData)
```

References

1. Adomas, A., Heller, G., Olson, A., Osborne, J., Karlsson, M., Nahalkova, J., Vanzyl, L., Sederoff, R., Stenlid, J., Finlay, R., Asiegbu, F.O.: Comparative analysis of transcript abundance in *Pinus sylvestris* after challenge with a saprotrophic pathogenic or mutualistic fungus. Tree Physiol. **28**(6), 885–897 (2008)

2. Birnie, R., Bryce, D.S., Roome, C., Dussupt, V., Droop, A., Lang, H.S., Berry, A.P., Hyde, F.C., Lewis, L.J., Stower, J.M., Maitland, J.N., Collins, T.A.: Gene expression profiling of human prostate cancer stem cells reveals a pro-inflammatory phenotype and the importance of extracellular matrix interactions. Genome Biol. **9**(R83), 1–13 (2008)

3. Cai, X., Giannakis, G.B.: Identifying differentially expressed genes in microarray experiments with model-based variance estimation. IEEE Trans. Signal Process. **54**(6), 2418–2426 (2006)

4. Dror, R.: Noise models in gene array analysis. Report in fulfillment of the area exam requirement in the MIT Department of Electrical Engineering and Computer Science (2001)

5. Greenblum, S., Krucoff, M., Furst, J., Raicu, D.: Automated image analysis of noisy microarrays. Department of Biomedical Engineering and School of Computer Science, Telecommunications and Information systems, IL, USA (2006)

6. Hedenfalk, I., Duggan, D., Chen, Y., Radmacher, M., Bittner, M., Simon, R., Meltzer, P., Gusterson, B., Esteller, M., Kallioniemi, O.P., Wilfond, B., Borg, A., Trent, J., Raffeld, M., Yakhini, Z., Ben-Dor, A., Dougherty, E., Kononen, J., Bubendorf, L., Fehrle, W., Pittaluga, S., Gruvberger, S., Loman, N., Johannsson, O., Olsson, H., Sauter, G.: Gene-expression profiles in hereditary breast cancer. N. Engl. J. Med. **344**, 539–548 (2001)

7. Jain, A.N., Tokuyasu, T.A., Snijiders, A.M., Segraves, R., Albertson, D.G., Pinkel, D.: Fully automatic quantification of microarray image data. Genome Res. **12**, 325–332 (2002)

8. Jochumsen, K.M., Tan, Q., Dahlgaard, J., Kruse, A.T., Mogensen, O.: RNA quality and gene expression analysis of ovarian tumor tissue undergoing repeated thaw–freezing. Exp. Mol. Pathol. **82**(1), 95–102 (2007)

9. Larese, M.G., Gomez, J.C.: Automatic spot addressing in cDNA microarray images. JCS&T **8** (2008)

10. Macgregor, F.P., Squire, A.J.: Application of microarrays to the analysis of gene expression in cancer. Clin. Chem. **48**(8), 1170–1177 (2002)

11. Otsu, N.: A threshold selection method from gray-level histograms. IEEE Trans. Sys. Man. Cyber. **9**, 62–66 (1979)

12. Petrov, A., Shams, S.: Microarray image processing and quality control: genomic signal processing. J. VLSI Signal Process. **38**(3), 211–226 (2004)

13. PhaniDeepti, G., Maruti, V.B., Jayanthi, S.: Impulse noise removal from color images with Hopfield neural network and improved vector median filter. In: Proceedings of the 6th Indian Conference on Computer Vision, Graphics and Image Processing, pp. 17–24. IEEE Computer Society (2008)

14. Smolka, B., Lukac, R., Plataniotis, K.N.: Fast noise reduction in cDNA microarray images. In: Proceedings of the 23rd Biennial Symposium, pp 348–351. IEEE Xplore (2006)

15. StanislavSaic, B.M.: Using noise inconsistencies for blind image forensics. Image Vis. Comput. **27**(10), 1497–1503 (2009)

16. Tu, Y., Stolovitzky, G., Klein, U.: Quantitative noise analysis for gene expression microarray. Proc. Natl. Acad. Sci. **99**(22), 14031–14036 (2002)

17. Uma, S.R., Rajkumar, T.: DNA microarray and breast cancer-A review. Int. J. Hum. Genet. **7**(1), 49–56 (2007)

18. Vikalo, H., Hassibi, B., Hassibi, A.: A statistical model for microarrays, optimal estimation algorithms, and limits of performance. IEEE Trans. Signal Process. **54**(6), 2444–2455 (2006)

19. Yin, W., Chen, T., Zhou, X.S., Chakraborty, A.: Background correction for cDNA microarray images using the TV + L1. Advaced Access, publication February 22, 2005

A Comparative Study on Machine Learning Algorithms in Emotion State Recognition Using ECG

Abhishek Vaish and Pinki Kumari

Abstract Human-Computer-Interface (HCI) has become an emerging area of research among the scientific community. The uses of machine learning algorithms are dominating the subject of data mining, to achieve the optimized result in various areas. One such area is related with emotional state classification using bio-electrical signals. The aim of the paper is to investigate the efficacy, efficiency and computational loads of different algorithms scientific comparisons that are used in recognizing emotional state through cardiovascular physiological signals. In this paper, we have used Decision tables, Neural network, C4.5 and Naïve Bayes as a subject under study, the classification is done into two domains: *High Arousal and Low Arousal.*

Keywords PCA · Emotion classification · ECG and data mining algorithms

1 Introduction

Emotion is a psycho-physiological process triggered by conscious and unconscious perception of an object or situation and is often associated with mood, temperament, personality and disposition and motivation [1]. Emotions play an important role in human communication and can be expressed either verbally or by non-verbal cues such as tone of voice, facial expression, gesture and physiological behavior.

Interestingly, it has been observed through past researches that various fields are utilizing this as a key signal for system development in different context and some of the most applied areas are:

In the area of medical science many physiological disorders exists those are directly correlated with the one of the different class of emotions. According to the prior art of healthcare, numerous study has been conducted to recognize the early stage of stress to prevent the human's life before entering in danger zone. The outputs

A. Vaish (✉) · P. Kumari
Indian Institute of information Technology Allahabad, Allahabad, India
e-mail: abhishek@iiita.ac.in

B. V. Babu et al. (eds.), *Proceedings of the Second International Conference on Soft Computing for Problem Solving (SocProS 2012), December 28–30, 2012*, Advances in Intelligent Systems and Computing 236, DOI: 10.1007/978-81-322-1602-5_147, © Springer India 2014

of the studies are some kind of tools and algorithms which helps to detect the early stage of mental illness which is a manifestation of the fact that classification through machine learning is important.

In area of Multi-modal authentication system various bio-signals (ECG, EEG and SC etc.) are fused and interpreted for generation of unique identification factors. These factors have the capability like unique and robust and are strong enough to be cracked. This system could be used in securing highly sensitive areas like defense and banking section etc.

In the area of affective gaming the different levels of emotions are used to make the gaming software more affective and easy to use. For example the famous is NPC (Non-player character) in which each and character associated with the different emotions.

In view of above described areas of emotional state classification, the accuracy of a prediction is the only thing that really matters. Here, we are proposing an alternative method that will increase the efficacy and efficiency of the system using machine learning algorithms of data mining. The scope of the study has considered four machine learning algorithms among top 10 algorithms of data mining: Decision tables, neural network, C4.5 and Naïve Bayes and applied over the ECG data corpus with full features dataset and with reduced features datasets and examined the result's effect due to highly contained features. The ultimate contribution of the research work is related with analysis of the known classifiers and observed their performance. This would help the researchers to use the best model in the area of emotion classified through ECG data set.

2 Literature Review

A copious number of researches are present in the literature for recognizing human's emotions from the physiological signals. Recently, among researches, a great deal of attention has been received on the efficacy improvement. In this section, we would like to briefly review the dynamics of the ECG signal followed by the current state of the scientific contribution in the subject under study.

Electrocardiography is a tool which measures and records the electrical potentials of the heart. A complete ECG cycle can be represented in a waveform and is known as PQRST interval. A close analysis of the same reveals that two major orientations exist i.e. the positive orientation and the negative orientation, PRT is a positive orientation and QS is a negative orientation in a given PQRST interval and the same can be seen in Fig. 1. The description of ECG waves and intervals are vital to catch the state of emotions present in human body how he/she is feeling as negative feeling and stress feeling leads the dangers state of life Fig. 2.

Jang et al. [2] have compared few data mining algorithm such as SVM, CART, SOM and Naïve Bayes over few Bio-signals like ECG, EDA and SKT and got the accuracy like 93.0, 66.44, 74.93 and 37.67 %. Chang et al. [3] used two modal to classify the emotional state of human, one is facial expression and another is

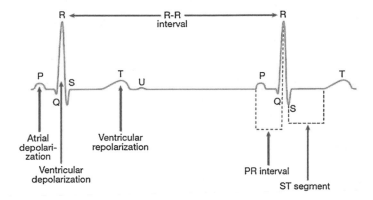

Fig. 1 ECG waveform generation from electrical activities of the heart

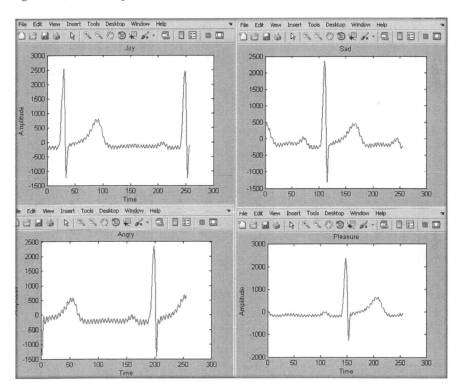

Fig. 2 Sample signals from AuBT data corpus

physiological signals of various subjects and then applied classification algorithm to recognize the different emotional state and got the accuracy somewhere around 88.33 % with the physiological signals. Gouizi et al. [4] have used support vector machine and obtained a result of 85 % recognition rate.

Li and Chen's [5] paper proposed to recognize emotion using physiological signals such as ECG, SKT, SC and respiration, selected to extract features for recognition and achieved accuracy such as 82 % with 17 features, 85.3 % with 22 features and same accuracy with 20 features. Siraj et al. [6], the presented paper says about classification of emotion of subjects in two classes i.e. active arousal and passive arousal using ECG pattern and achieved the accuracy around 82 %.

Li and Lu [7] described how emotion could be classified through EEG signals. They considered two types emotions: Happiness and sadness. Using common spatial patterns (CSP) and linear SVM classification has been done and achieved the satisfied accuracy. Murugappan [8] presents Electromyogram (EMG) signal based human emotion classification using K Nearest Neighbor (KNN) and Linear Discriminant Analysis (LDA). Five most dominating emotions such as: happy, disgust, fear, sad and neutral are considered and these emotions are induced through Audio-visual stimuli (video clips).

Sun [1], in this paper, we evaluate these tools' classification function in authentic emotion recognition. Meanwhile, we develop a hybrid classification algorithm and compare it with these data mining tools. Finally, we list the recognition results by various classifiers. Ma and Liu [9], wavelet transform was applied to accurately detect QRS complex for its advantages on time-frequency localization, in order to extract features from raw ECG signals. A method of feature selection based on Ant Colony System (ACS), using K-nearest neighbor for emotion classification, was introduced to obtain higher recognition rate and effective feature subset.

These research articles provide support for the conclusion that Bio-electrical signals carries generic information about the human behavior or different human emotions. There is however limited research which considers computational time effect by reducing the dimensionality of features of heart rhythms.

In the proposed article, we are trying to analyze the performance of the data mining algorithms with high feature dataset and reduced feature dataset.

3 Research Methodology

Person emotion recognition has lately evolved as an interesting research area. Physiological signals such as ECG, EEG, EMG and respiration rate are successfully utilized by quite a few researchers to attain the emotion classification [4]. In this research we have used simulative research design using quantitative data that has been collected using four bio-sensors. In this section, we would be highlighting the schematic diagram in Sect. 3.1 (Fig. 3) that has been followed to extract the results and also to make sure that the contamination in the research design could be avoided. Additionally, the description of data corpus in Sect. 3.2.

(a)   **(b)**

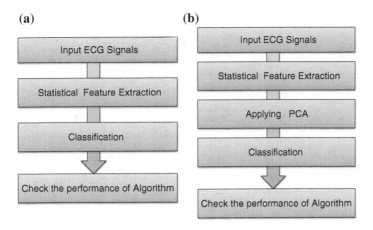

Fig. 3 Schematic diagram of proposed work. **a** Efficacy with high dimensionality. **b** Efficacy with reduced dimensionality

3.1 High Level Schematic Diagram

The whole of this high level diagram is categorized into three main steps. These are as follows:

- Extraction of statistical features—Help us to extract the maximum number of feature from the raw data set. Features are those data point that has potential information for output.
- Applying the PCA for Feature Reduction—Helps in extracting the best fit feature that without compromising the results improves the system performance.
- Classification algorithm to check the efficacy—Helps in the decision making system for accurate results.

3.2 Description of Subject

Each subject (participant) selects four favorite songs reminiscent of their certain emotional experiences corresponding to four emotion categories. Signals were collected with 25 subjects with four emotions within 25 days.

3.3 Mode of Collection Data

The physiological data were recorded through biosensor and the length of the recordings depends on the length of the songs, but was later cropped to a fixed length of the two minutes and ECG was sampled at 256 Hz.

3.4 Research Procedure

In the field of automatic emotion recognition probably the most often used features are based on statistically such as Mean, Median, Standard deviation, max min of all waves of ECG i.e. PQRST waveform which represents complete ECG cycle. For the feature extraction and feature selection of ECG signal in time domain we have used *Analysis of variance (ANOVA)* Method. After collecting the statistical features of cardiac signal then performed the different classification to investigate the efficacy of the classification algorithms of data mining such as Decision table, Neural network, Naïve Bayes and C4.5.

4 Experimental Results

In this section, the results are presented, the result are presented into different phases i.e. the feature selection and the classification of emotions, feature extraction with reduced dimensionalities and their classification, finally the computational load of each classification algorithm is presented.

Phase I
Figure 4 depicts the classification accuracy of different algorithms. The total numbers of feature extracted by inbuilt use of ANOVA were 82. It can be seen that multilayer perceptron is giving the best result among the four with approximately 60 % and Naïve Bayes is showing a response of 53 % which is considered to be the least. It can be interpreted with the result that the FAR is low and FRR is high.

Phase II
In order to optimize the result depicted in Fig. 4, we have used a dimensional reduction scheme and the most prevailing technique is the principle component analysis. It is pertinent to briefly discuss the equation used in PCA.

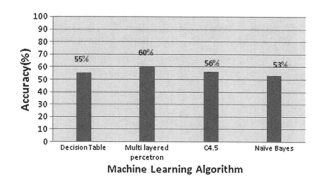

Fig. 4 Performance of machine learning algorithms with full features

Given the data, if each datum has N features represented for instance by x_{11} $x_{12} \ldots x_{1N}, x_{21}\ x_{22} \ldots x_{2N}$, the data set can be represented by a matrix Xn × m. The average observation is defined as:

$$\mu = \frac{1}{n} \sum_{i=1}^{n} x_i \qquad (1)$$

The deviation from the average is defined as:

$$\Phi_i = X_i - \mu/ \qquad (2)$$

Using Eqs. 1 and 2, we have extracted the following results as depicted in Table 1. The total numbers of extracted feature were 13 out of 82. The accuracy post extraction is depicted in Fig. 5. In this case C4.5 is giving us the most optimized result. So it can be inferred that the combination of PCA + C4.5 in the given data set is the maximum. However, if we look at the more modest form, it can be seen that performance of all classification algorithm has gone up to the average of 20 % using Eq. 3

$$\text{Impact of post PCA} = (\text{Post PCA} - \text{Pre PCA}) \qquad (3)$$

Table 2 depicted in this section give the specific of the scalar value of the classification algorithm. The tabulation has been arranged with row and column. The first column is populating with the machine learning algorithm and corresponding rows has the statistical measures like True-positive rate, precision, F measures and ROC area. The classes are labeled as High Arousal, Low Arousal and same can be correlated with Fig. 5.

Phase III
This section is interesting for the readers and researchers because the most fundamental question with development of decision making system lies with cost of decision and accuracy. The latter has been discussed in this phase. In Fig. 6 it can

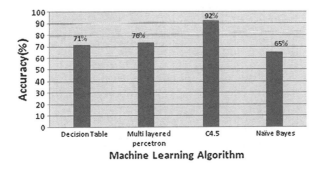

Fig. 5 Performance of machine learning algorithms with reduced features

Table 1 Reduced data corpus with 13 features

	Number of reduced features
1	ecgQ-mean
2	ecgS-range
3	ecgQS-max
4	ecgT-std
5	ecgSampl-std
6	ecgSampl-mean
7	ecgPQ-range
8	ecgHrvDistr-triind
9	ecgQS-std
10	ecgHrv-specRange1
11	ecgHrvDistr-min
12	ecgHrv-specRange1
13	ecgHrv-pNN50

Table 2 Efficacy of Machine Learning Algorithms for emotion classification

ML's algorithms	Class	TP Rate	Precision	F-Measure	ROC Area
C4.5	High Arousal	1	0.097	0.985	0.988
	Low Arousal	0.859	1	0.984	0.986
	Avg. weighted	**0.9295**	**0.985**	**0.984**	**0.987**
Neural network	High Arousal	0.722	0.885	0.898	0.812
	Low Arousal	0.807	0.912	0.886	0.823
	Avg. weighted	**0.7645**	**0.8985**	**0.892**	**0.8175**
Naïve Bayes	High Arousal	0.64	0.867	0.65	0.924
	Low Arousal	0.67	0.594	0.596	0.806
	Avg. Weighted	**0.655**	**0.7305**	**0.623**	**0.865**
Decision table	High Arousal	0.712	0.867	0.74	0.864
	Low Arousal	0.722	0.512	0.596	0.796
	Avg. weighted	**0.717**	**0.6895**	**0.668**	**0.83**

Fig. 6 Comparison of computational load with and without feature reduction

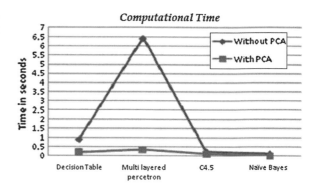

be seen that time taken to perform the task is categorizing in the domains i.e. with PCA and without PCA. The result is highly encouraging. Without PCA the time is reduced drastically. However, an important point worth observing that MLP is more computationally expensive among four in both the domains.

5 Conclusion and Future Works

The work aimed at showing the possibility of recognizing the two levels of emotional state: *High Arousal and Low Arousal*. We have presented an alternative method to investigate the performance of data mining algorithms to raise the efficacy of the emotional recognition system and also give solution the question how dimensionality reduction can save the burden of computing. The recognition rates increased after applying proposed methods are: 16 % with Decision table; 12 % with neural network; 36%, with C4.5 and 12 % with Naïve Bayes. All experiments have done with openly available tool (WEKA) for data mining and machine learning algorithms for data classification. There are few challenges we have faced which is the lack of data corpus quantum. In future, we would like to explore another data corpus available in this domain. Furthermore, we would like to work on investigation of those classification algorithms which have been used for ECG classification in two different levels of emotional states with other physiological signals such as EMG, EEG, SC and respiration. Another future work may be explored in area of affective computing like to automatically detect the stages of mental illness with good recognition rates through bio-signals. Bio-signals may also be fruitful in the area of multimodal authentication system or cognitive biometrics.

References

1. Sun, Y., Li, Z., Zhang, L., Qiu, S., Chen, Y.: Evaluating data mining tools for authentic emotion classification in Intelligent Computation Technology and Automation (ICICTA). 2010 International Conference, vol. 2, pp. 228–232 (2010)
2. Jang, E.-H., Park, B.-J., Kim, S.-H., Eum, Y., Sohn, J.-H.: Identification of the optimal emotion recognition algorithm using physiological signals. 2011 International Conference on Engineering and Industries (ICEI), 2011, pp. 1–6
3. Chang, C.-Y., Tsai, J.-S., Wang, C.-J., Chung, P.-C.: Emotion recognition with consideration of facial expression and physiological signals in computational intelligence in bioinformatics and computational biology. CIBCB '09 IEEE Symposium, 2009, pp. 278–283
4. Gouizi, K., Reguig, F.B., Maaoui, C.: Analysis physiological signals for emotion recognition in Systems, Signal processing and their Applications (WOSSPA). 2011 7th International Workshop , 2011, pp. 147–150.
5. Li, L., Chen, J.-H.: Emotion recognition using physiological signals from multiple subjects in intelligent information hiding and multimedia signal processing. IIH-MSP '06 International Conference, 2006, pp. 355–358
6. Siraj, F., Yusoff, N., Kee, L. C.: Emotion classification using neural network. International Conference on Computing and Informatics ICOCI '06, 2006, pp. 1–7

7. Li, M., Lu, B. -L.: Emotion classification based on gamma-band EEG in engineering in medicine and biology society. Annual International Conference of the IEEE, 2009, pp. 1223–1226
8. Murugappan, M.: Electromyogram signal based human emotion classification using KNN and LDA in System Engineering and Technology (ICSET). IEEE International Conference, 2011, pp. 106–110
9. Ma, C.-W., Liu, G.-Y.: Feature extraction, feature selection and classification from electro-cardiography to emotions in computational intelligence and natural computing. CINC '09 International Conference, vol. 1, pp. 190–193 (2009)
10. Wagner, J., Kim, J., Andre, E.: Signals, from physiological, to emotions: Implementing and comparing selected methods for feature extraction and classification. IEEE International Conference in multimedia and expo (ICME), 2005, pp. 940–943

Fault Diagnosis of Ball Bearings Using Support Vector Machine and Adaptive Neuro Fuzzy Classifier

Rohit Tiwari, Pavan Kumar Kankar and Vijay Kumar Gupta

Abstract Bearing faults are one of the major sources of malfunctioning in machinery. A reliable bearing health condition monitoring system is very useful in industries in early fault detection and to prevent machinery breakdown. This paper is focused on fault diagnosis of ball bearing using adaptive neuro fuzzy classifier (ANFC) and support vector machine (SVM). The vibration signals are captured and analyzed for different types of defects. The specific defects consider as inner race with spall, outer race with spall, and ball with spall. Statistical techniques are applied to calculate the features from the vibration data and comparative experimental study is carried using ANFC and SVM. The results show that these methods give satisfactory results and can be used for automated bearing fault diagnosis.

Keywords Fault diagnosis · Condition monitoring · Adaptive neuro fuzzy classifier · Support vector machine

1 Introduction

Rotating rolling element bearing is widely used in industrial machines. Bearing fault diagnosis is an even more challenging task in condition monitoring especially when the machine is operating in a noisy environment. For diagnosis of all types of fault either localized or distributed vibration analysis can be employed. A vibration

R. Tiwari (✉) · P. K. Kankar · V. K. Gupta
PDPM Indian Institute of Information Technology, Design and Manufacturing,
Jabalpur, India
e-mail: 1110508@iiitdmj.ac.in

P. K. Kankar
e-mail: kankar@iiitdmj.ac.in

V. K. Gupta
e-mail: vkgupta@iiitdmj.ac.in

B. V. Babu et al. (eds.), *Proceedings of the Second International Conference on Soft Computing for Problem Solving (SocProS 2012), December 28–30, 2012*, Advances in Intelligent Systems and Computing 236, DOI: 10.1007/978-81-322-1602-5_148, © Springer India 2014

signal-based fault diagnosis can be performed in time domain [6], frequency domain [8]. Artificial neural network [10], and support vector machines (SVM) [11] methods are also being used. Abasiona and Rafsanjani [1] have carried out multi-fault detection in such system with wavelet analysis and SVM.

This study is mainly focused on bearing fault classification using two methods, SVM and neuro-fuzzy, as both are capable in nonlinear classification. Four defects considered for the study includes inner race with spall, outer race with spall, ball with spall, and combined bearing component defect. Statistical methods are used to extract the features from the time domain signal. Selected features with their known class are used for training and testing of SVM and ANFC.

2 Support Vector Machines

SVM based on the statistical learning theory is a supervised machine learning method. It is used for classification of two or multi-class of data. The performance of SVM is not influenced by the dimension of feature space, which is why it is more efficient in large classification problem. Cristianini and Shawe-Tylor [3] have used SVM for pattern recognition and classification.

A simple example of two-class problem is shown in Fig. 1. In Fig. 1, square and triangles show two classes of sample points. These two classes are separated by hyper plane H. Planes passing through the sample points and nearest to H are given by H_1 and H_2 (shown by dashed line). SVM creates linear boundary in such a way that the margin between classes H_1 and H_2 will be maximum. This reduces the generalization error. Support vectors are the nearest data point used to define the margin.

3 Adaptive Neuro Fuzzy Classifier

The neuro fuzzy classifier is an adaptive network-based system in which fuzzy parameters are adapted with neural networks. In fuzzy classification feature space is

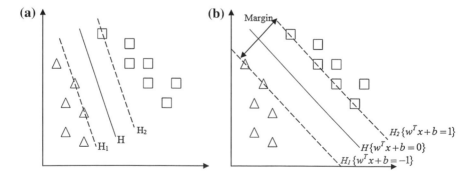

Fig. 1 Hyper plane classifying two classes: **a** small margin, **b** large margin

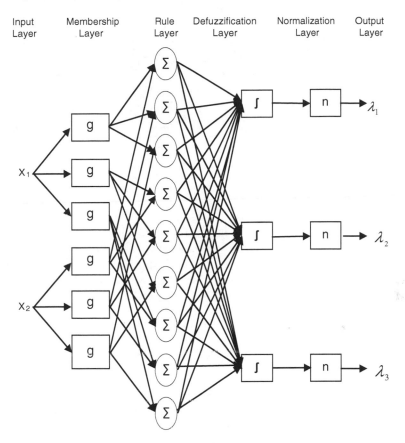

Fig. 2 Architecture of adaptive neuro fuzzy classifier

divided in to fuzzy classes. Sun and Jang [9] have applied fuzzy rules to control each fuzzy region. Cetisli [2] has developed an adaptive neuro fuzzy classifier using linguistic hedges and applied for different classification problems. Du and Er [4] have applied neuro fuzzy classifier for fault diagnosis in air-handling unit system.

In this study the proposed neuro fuzzy classifier is described by zero-order surgeon fuzzy model. Output level is a constant for a zero order surgeon model and weighted average operator [5] will gives the crisp output from the fuzzy rules. To initialize fuzzy rules k-mean algorithm is used. Fuzzy relation between variable are often created by fuzzy clustering [12]. In this architecture shown in Fig. 2, the first layer measures membership grade of each input to specified fuzzy region. Gaussian function is employed as membership function (MF).

The next layer is known as rule layer. This layer is responsible to calculate the firing strength of fuzzy rules using the membership value of the inputs. The third layer in architecture calculates the weighted outputs. Each class is decided according to the maximum firing strength of the rules.

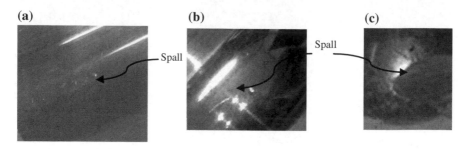

Fig. 3 Bearing components with faults induced in them. **a** Outer race with spall, **b** inner race with spall, **c** Ball with spall

After calculating the rule weights the next layer function is to normalize network output because the summation of the weight should not be more than 1.

After the normalization output layer is calculated the class labels. There are many methods for network parameter optimization. In this paper, scale conjugate gradient (SCG) method is used to adapt antecedent parameters [7].

4 Experimental Setup and Data Acquisition

Training and testing data sets are generated by the experimental test on a test rig. The rig is connected with the data acquisition system. At various speeds different faults are simulated on the bearing as shown in Fig. 3. Vibration data for healthy bearing operation is used as baseline data. These data can be used for comparison with faulty bearing signature data. Features are extracted from the vibration signal. These data are compiled to form a feature vector which is given as input to ANFC/SVM for training. Faults introduced in bearings are outer race with spall, inner race with spall, ball with spall, and combined faults in all bearing components. For acquiring training data from the data acquisition system cases considered are healthy bearing (HB), inner race defect (IRD), outer race defect (ORD), ball fault in bearing (BFB), and combined bearing component defect (CBD).

5 Results

Statistical features such as mean, standard deviation, skewness, kurtosis, min, max, and range are calculated from the vibration data. This will reduce the dimensionality of original vibration features. 9 attribute are selected including statistical features, speed, and loader.

After calculation of features, classification of faults is done using machine learning techniques, i.e., ANFC/SVM. Input features are compiled and given as input. The

Table 1 Confusion matrix

BFB		MFB		HB		ORD		IRD		Classified as
ANFC	SVM	ANFC	SVM	ANFC	SVM	ANFC	SVM	ANFC	SVM	
4	9	1	0	0	0	4	0	0	0	BFB
2	2	5	7	0	0	2	0	0	0	MFB
1	2	0	0	2	1	0	0	0	0	HB
0	3	1	2	0	0	8	4	0	0	ORD
0	0	0	0	0	0	0	0	10	10	IRD

Table 2 Evaluation of the success of the numeric prediction

Parameters	Values (ANFC)	Values (SVM)
Correctly classified instances	29 (72.50%)	31 (77.50%)
Incorrectly classified instances	11 (27.50%)	9 (22.50)

defects classified using ANFC/SVM are shown in Table 1. A total of 40 cases are considered for testing in which 9, 9, 3, 9, 10 cases of ball with spall, combined bearing component defects, healthy bearing, inner race with spall, and outer race with spall are taken.

Table 1 shows that ANFC has correctly classified 4, 5, 2, 8, and 10 cases, while SVM has predicted 9, 7, 1, 4, and 10 cases correctly for ball with spall combined bearing component defect, healthy bearing, inner race with spall, and outer race with spall respectively. Values of correctly and incorrectly classified instances with percentage accuracy are given in Table 2. Table 2 shows that SVM gives better classification efficiency of 77.5% than the ANFC.

6 Conclusions

In this study, bearing fault detection has been done by classifying them using two methods, ANFC and SVM. Statistical techniques are used to calculate the features from the time-domain vibration signals. It is observed that ANFC has less accuracy to predict the ball fault in bearing while SVM is not accurate to predict outer race defect in bearing. SVM has a better classification accuracy than ANFC. Results show that these methods can be apply to develop a reliable condition monitoring system and used to predict defect in early stage. This can avoid the machinery breakdown and reduce operating cost.

References

1. Abbasiona, S., Rafsanjani, A.: Rolling element bearing multi fault classification based on wavelet denoising and support vector machine. Mech. Syst. Signal Process. **2**, 2933–2945 (2007)
2. Cetisli, B.: Development of adaptive neuro-fuzzy classifier using linguistic hedges: Part-1. Expert Syst. Appl. **37**, 6093–6101 (2010)
3. Cristianini, N., Shawe-Tylor, N.J.: An introduction to support vector machines. Cambridge University press, Cambridge (2000)
4. Du, J., Er, M.J.: Fault diagnosis in air—handling unit system using dynamic fuzzy neural network. In: Proceeding of the 6th International FLINS Conference, pp. 483–488 (2004)
5. Jang, J.S.R., Sun, C.T., Mizutani, E.: Neuro-Fuzzy and Soft Computing. Upper Saddle River, prentice Hall (1997)
6. Kankar, P.K., Sharma, S.C., Harsha, S.P.: Fault diagnosis of ball bearings using machine learning methods. Expert Syst. Appl. **38**, 1876–1886 (2011)
7. Moller, M.F.: A Scaled conjugate gradient algorithm for fast supervised learning. Neural Netw. **6**(4), 525–533 (1993)
8. Robinson, J.C., Canada, R.G., Piety, K.R.: Peak vue analysis—new methodology for bearing fault detection. Sound Vib. **30**(11), 22–25 (1996)
9. Sun, C.T., Jang, J.S.: A neuro-fuzzy classifier and its applications. In: Proceedings of the IEEE International Conference on Fuzzy System, San Francisco, pp. 94–98 (1993)
10. Vyas, N.S., Satish kumar, D.: Artificial neural network design for fault identification in a rotor bearing system. Mech. Mach. Theory **36**, 157–175 (2001)
11. Widodo, A., Yang, B.S.: Review on support vector machine condition monitoring and fault diagnosis. Mech. Syst. Signal Process. **21**, 2560–2574 (2007)
12. Zio, E., Baraldi, P.: Evolutionary fuzzy clustering for the classification of transients in nuclear components. Prog. Nucl. Energy **46**(3–4), 282–296 (2005)

Designing a Closed-Loop Logistic Network in Supply Chain by Reducing its Unfriendly Consequences on Environment

Kiran Garg, Sanjam, Aman Jain and P. C. Jha

Abstract This paper examines the relationship between the operations of forward and reverse logistics and the environmental performance measures like CO_2 emission in the network due to transportation activities in closed-loop supply chain network design. A closed-loop structure in the green supply chain logistics and the location selection optimization was proposed in order to integrate the environmental issues into a traditional logistic system. So, we present an integrated and a generalized closed-loop network design, consisting four echelons in forward direction (i.e., suppliers, plants, and distribution centers, first customer zone) and four echelons in backward direction (i.e., collection centers, dismantlers, disposal centers, and second customer zone) for the logistics planning by formulating a cyclic logistics network problem. The model presented is bi objective and captures the trade-offs between various costs inherent in the network and of emission of greenhouse gas CO_2. Numerical experiments were presented, and the results showed that the proposed model and algorithm were able to support the logistic decisions in a closed loop supply chain efficiently and accurately.

Keywords Environmental performances · Cradle to cradle principle · Closed loop supply chain network · Trade off · Greenhouse gas

K. Garg · P. C. Jha (✉)
Department of Operational Research, University of Delhi, Delhi, India
e-mail: jhapc@yahoo.com

K. Garg
e-mail: mittalkiran12@gmail.com

Sanjam · A. Jain
Faculty of Management Science, University of Delhi, Delhi, India
e-mail: sanjam.s13@fms.edu

A. Jain
e-mail: aman.j13@fms.edu

B. V. Babu et al. (eds.), *Proceedings of the Second International Conference on Soft Computing for Problem Solving (SocProS 2012), December 28–30, 2012*, Advances in Intelligent Systems and Computing 236, DOI: 10.1007/978-81-322-1602-5_149, © Springer India 2014

1 Introduction

Supply chain consists of set of activities such as transformation and flow of goods, services, and information from the sources of materials to end-users. Due to the government legislation, environmental concern, social responsibility, and customer awareness, companies have been forced by customers not only to supply environmentally harmonious products but also to be responsible for the returned products. So, interest in supply chains lies in the recovery of products, which is achieved through processes such as repair, remanufacturing and recycling, which, combined with all the associated transportation and distribution operations, are collectively termed as *Reverse Chain activities.* In reverse logistics there is a link between the market that releases used products and the market for "new" products. When these two markets coincide, it is called *Closed Loop Network.* Thus the supply chain in which forward and reverse supply chain activities are integrated is said to be closed-loop, and research on such chains have given rise to the field of closed-loop supply chains (CLSCs) and Supply chain network design concerned with environmental issues, collectively named as *Green Supply Chain.*

At present, researcher's emphasis on green supply chain due to global warming and wants to minimize the waste at landfills. A closed-loop logistics management ensures the least waste of the materials by following the cradle to cradle principle and conservation law along the life cycles of the materials. In reverse logistics used products, either under warranty or at the end of use or at the end of lease are taken back, so that the products or its parts are appropriately disposed, recycled, reused, or remanufactured. Beside it they explicitly focus on significant sources of greenhouse gas emission, and one of those sources is transportation. CO_2 is very prominent in its hazardous consequences on human health. Transport is the second-largest sector of global CO_2 emission. CO_2 constraints in logistics markets will need to be realized in the near future as it was enforced by protocols, and a shift in freight transportation could be expected to reduce the CO_2 emissions within the reasonable cost and time constraints.

In this study, we model and analyze a CLSC for its operational and environmental performances, i.e., a multi-echelon forward–reverse logistics network model is described for the purpose of design with the reflection of the effects on environment of greenhouse gas emission. Objectives of the model is to maximize the total expected profit earned and minimizing CO_2 emission due to transporting material in forward and reverse logistics networks with the use of different type of vehicles for transport, each of which has its own emission rates and transportation costs. Using the proposed model and a numerical illustration result of computational experiments shed light on the interactions of various performance indicators, primarily measured by cost and then captures the environmental aspects.

2 Literature Review

This section presents a brief overview of the existing literature on closed loop supply chain (CLSC). Beamon [1] describes the challenges and opportunities facing the supply chain of the future and describes sustainability and effects on supply chain design, management and integration. Network chain members of a CLSC can be classified into two groups [2]: Forward logistics chain members and Reverse logistics chain members. But designing the forward and reverse logistics separately results in suboptimal designs with respect to objectives of supply chain; hence the design of forward and reverse logistics should be integrated [3–5]. This type of integration can be considered as either horizontal or vertical integration [6]. Manufacturers and demand nodes (i.e., customers) could be seen as 'junction' points where the forward and the reverse chains are combined to form the CLSC network. A closed-loop logistics model for remanufacture has been studied in [7], in which decisions relevant to shipment and remanufacturing of a set of products, as well as establishment of facilities to store the remanufactured products are taken into consideration [8]. Consider a reverse logistics network design problem which analyzes the impact of product return flows on logistics networks. A strategic and tactical model for the design and planning of supply chains with reverse flows was proposed by [9]. Authors considered the network design as a strategic decision, while tactical decisions are associated to production, storage and distribution planning. A general reverse logistics location allocation model was developed in [10] in a mixed integer linear programming form. The model behavior and the effect of different reverse logistics variables on the economy of the system were studied. Demand in this proposed model is deterministic. The problem of consolidating returned products in a CLSC has been studied in [11]. Kannan et al. [12] developed a multi-echelon, multiperiod, multi-product CLSC network model for product returns, in which decisions are made regarding material procurement, production, distribution, recycling, and disposal. For an excellent review of methodological and case study-based papers in reverse and closed-loop logistics network design, the reader is referred to [13].

Meixell and Gargeya [14] focused on the design of supply chains of production, purchasing, transportation, and profit and has neglected the environmental aspects. Given recent concerns on the harmful consequences of supply chain activities on the environment, and transportation in particular, it has become necessary to take into account environmental factors when planning and managing a supply chain. The list of environmental performance metrics of a supply chain includes emissions, energy use and recovery, spill and leak prevention, and discharges is discussed in [15]. A comprehensive survey of the field is provided by [16]. Sarkis [17] provides a strategic decision framework for green supply chain management, in which he investigates the use of an analytical network process for making decisions within the GrSC. Sheu et al. [18] present a multiobjective linear programming model for optimizing the operations of a green supply chain, composed of forward and reverse flows, including decisions pertaining to shipment and inventory [19]. Consider environmental issues within CLSCs and examine a supply chain design problem for refrigerators, offer a comprehensive mathematical model that minimizes costs associated with distribution, processing, and facility set-up, also takes into account the environmental costs of energy and waste.

The remainder of the paper is structured as follows. In Sect. 3 a CLSC model is proposed for single product, with the underlying assumptions. In Sect. 4, used methodology of goal programming is described. In Sect. 5, we present a numerical implementation in order to highlight the features of the proposed model. The paper ends with concluding remarks.

3 Model Description

The CLSC problem discussed in this paper is an integrated multiobjective multi-echelon problem in a forward/reverse logistic network, which requires more efforts to analyze than both forward and backward logistic simultaneously. Here we are considering the flow of a product in the network. The model considers modular product structure and every component of the product has an associated recycling rate, specifying the rate at which the component can be recycled. For instance, a rate of 100 % indicates that the used product can be fully recovered or transformed into a new one, whereas a rate of 50 % denotes that the product can only be partially recovered.

In the network suppliers are responsible for providing components to manufacturing plants. The new products are conveyed from plants to customers via distribution centers (d/c) to meet their demands. Returned products from customers are collected at collection centers where they are inspected. After testing in collection centers, the repairable and recyclable products are shipped to plants and dismantlers respectively, after completing the demand of secondary market of used products. At plant repairable used products are repaired and supplied back to distribution centers as new product. Dismantled components at dismantlers are drives back to suppliers if they are repairable else to disposal site to be disposed of.

The purpose of this paper is to evaluate a forward/reverse logistic system with respect to given objectives in order to determine the facility locations and flows between facilities using which type of transport. The transportation operations from one layer to another can be realized via a number of options. These options consist of different types of transport alternatives, e.g., different models of trucks. The proposed model considers the following assumptions and limitations:

1. Supplier and customer locations are known and fixed.
2. The demand of product is deterministic and no shortages are allowed.
3. The potential locations of manufacturing facilities, distribution centers, collection centers, and dismantlers are known.
4. The flow is only permitted to be transported between two consecutive stages. Moreover, there are no flows between facilities at the same stage.
5. The numbers of facilities that can be opened are restricted.

6. The other costs (i.e., operational costs and transportation costs) are known.
7. The estimated emission rate of CO_2 for all type of vehicle available is known.

Notations:
Sets:

S set of component's suppliers index by $s, s = 1, 2, \ldots, S$

P set of manufacturing plants index by $p, p = 1, 2, \ldots, P$

K set of distribution centers (d/c) index by $k, k == 1, 2, \ldots, K$

E set of first market customer zones index by $e, e = 1, 2, \ldots, E$

C set of collection centers (CC) index by $c, c == 1, 2, \ldots, C$

M set of dismantlers (d/m) position index by $m, m == 1, 2, \ldots, M$

H set of second market customer zones index by $h, h = 1, 2, \ldots, H$

F set of disposal sites (d/p) index by $f, f = 1, 2, \ldots, F$

A set of subassemblies index by $a, a = 1, 2, \ldots, A$

N set on nodes in the network ($N = S \cup P \cup K \cup E \cup C \cup M \cup H \cup F$)

Parameters:

SC_{sa} Unit purchasing cost of sub assembly a by supplier s

PC_p Unit production cost of product at manufacturing plant p

OC_k Unit operating cost of product at d/c k

IC_c Unit inspection cost of product at collection center c

RPC_p Unit repairing cost of used product at manufacturing plant p

DMC_m Unit dismantling cost of product at d/m position m

RCC_{sa} Unit recycling cost of sub assembly a at supplier s

D_e Demand of product at first customer e

D_h Demand of used product at second customer h

TPC^t Unit transportation cost per mile of product or component shipped from one node to another via type of truck t

D_{ij} Distance between any two nodes $i, j \in N$ of given CLSC network

CAP_{sa} Capacity of supplier s for sub assembly a

$PCAP_p$ Production capacity of plant p

$KCAP_k$ Capacity of distribution center k

$CCAP_c$ Capacity of collection center c

$MCAP_m$ Capacity of dismantler m

$FCAP_f$ Disposal capacity of disposal site f

$RPCAP_p$ Repairing capacity of plant p

$RCCAP_s$ Recycling capacity of supplier s

PF_a Unit profit made in the network from recycling component a

PF Unit profit made in the network from repairable product

PR_e Unit price of product at customer e

PR_h Unit price of product at customer e

ER^t Per mile emission rate of CO_2 gas from the type of transport $t \in T$

Rr Return ratio at the first customers

Rc_a Recycling ratio of component a

Rp   Repairing ratio

W   Weight of product in kg

W_a Weight of component $a \in A$ in kg

U_a Utilization rate of component $a \in A$

Decision variables:

x_{ija}^t Quantity of component a shipped from node i to node j, $i, j \in N$ in the network via transport of type $t \in T$

x_{ij}^t Quantity of product shipped from node i to node j, $i, j \in N$ in the network via transport of type $t \in T$

W_{ij}^t Weighted quantity transported from node i to node j, $i, j \in N$ in the network via transport of type $t \in T$

$$X_i = \begin{cases} 1, \text{if facility } i, (i \in P \cup K \cup C \cup M) \text{ is opened} \\ 0, \text{ otherwise} \end{cases}$$

$$L_{ij}^t = \begin{cases} 1, \text{ if a transportation link is established between any two locations} \\ \quad i \text{ and } j, i, j \in N \text{ via mode } t \\ 0, \text{ otherwise} \end{cases}$$

Model

Maximize, $Z_0 = \sum_k \sum_e \sum_t x_{ke}^t PR_e + \sum_c \sum_h \sum_t x_{ch}^t PR_h + \sum_m \sum_s \sum_a \sum_t PF_a x_{msa}^t$

$+ \sum_c \sum_p \sum_t PF x_{cp}^t - \left(\sum_s \sum_p \sum_a \sum_t x_{spa}^t SC_{sa} + \sum_p \sum_k \sum_t x_{pk}^t PC_p \right.$

$+ \sum_k \sum_e \sum_t x_{ke}^t OC_k + \sum_e \sum_c \sum_t x_{ec}^t IC_c + \sum_c \sum_p \sum_t x_{cp}^t RPC_p$

$+ \sum_c \sum_m \sum_t x_{cm}^t DMC_m + \sum_m \sum_s \sum_a \sum_t x_{msa}^t RCC_{sa}$

$+ \sum_m \sum_f \sum_a \sum_t x_{mfa}^t DPC_f + \sum_s \sum_p \sum_t TPC^t D_{sp} x_{sp}^t$

$+ \sum_p \sum_k \sum_t TPC^t D_{pk} x_{pk}^t + \sum_k \sum_e \sum_t TPC^t D_{ke} x_{ke}^t$

$+ \sum_e \sum_c \sum_t TPC^t D_{ec} x_{ec}^t + \sum_c \sum_p \sum_t TPC^t D_{cp} x_{cp}^t$

$+ \sum_c \sum_m \sum_t TPC^t D_{cm} x_{cm}^t + \sum_c \sum_h \sum_t TPC^t D_{ch} x_{ch}^t$

$\left. + \sum_m \sum_s \sum_a \sum_t TPC^t D_{ms} x_{msa}^t + \sum_m \sum_f \sum_a \sum_t TPC^t D_{mf} x_{mfa}^t \right)$

Minimize, $Z_1 = \sum_t ER^t \left(\sum_s \sum_p D_{sp} L_{sp}^t W_{sp}^t + \sum_p \sum_k D_{pk} L_{pk}^t W_{pk}^t + \sum_k \sum_e D_{ke} L_{ke}^t W_{ke}^t \right.$

$+ \sum_e \sum_c D_{ec} L_{ec}^t W_{ec}^t + \sum_c \sum_p D_{cp} L_{cp}^t W_{cp}^t$

$+ \sum_c \sum_m D_{cm} L_{cm}^t W_{cm}^t + \sum_c \sum_h D_{ch} L_{ch}^t W_{ch}^t + \sum_m \sum_s D_{ms} L_{ms}^t W_{ms}^t$

$\left. + \sum_m \sum_f D_{mf} L_{mf}^t W_{mf}^t \right)$

Subject To
(Flow balancing constraints)

$$\sum_s \sum_t x_{spa}^t + \sum_c \sum_t x_{cp}^t * U_a = \sum_k \sum_t x_{pk}^t * U_a \quad \forall p, a \tag{1}$$

$$\sum_p \sum_t x_{pk}^t = \sum_k \sum_e x_{ke}^t \quad \forall k \tag{2}$$

$$\sum_k \sum_e x_{ke}^t >= D_e \quad \forall e \tag{3}$$

$$\sum_c \sum_t x_{ec}^t = Rr * D_e \quad \forall e \tag{4}$$

$$\sum_c \sum_t x_{ch}^t <= D_h \quad \forall h \tag{5}$$

$$\sum_{p}\sum_{t} x_{cp}^t = Rp * \left(\sum_{e}\sum_{t} x_{ec}^t - \sum_{h}\sum_{t} x_{ch}^t \right) \quad \forall c \qquad (6)$$

$$\sum_{m}\sum_{t} x_{cm}^t = (1 - Rp) * \left(\sum_{e}\sum_{t} x_{ec}^t - \sum_{h}\sum_{t} x_{ch}^t \right) \quad \forall c \qquad (7)$$

$$\sum_{s}\sum_{t} x_{msa}^t = Rc_a * U_{a*} \sum_{c}\sum_{t} x_{cm}^t \quad \forall m, a \qquad (8)$$

$$\sum_{f}\sum_{t} x_{mfa}^t = (1 - Rc_a) * U_{a*} \sum_{c}\sum_{t} x_{cm}^t \quad \forall m, a \qquad (9)$$

Capacity constraints

$$\sum_{s}\sum_{t} x_{spa}^t <= CAP_{sa} \quad \forall s, a \qquad (10)$$

$$\sum_{k}\sum_{t} x_{pk}^t <= PCAP_p * X_p \quad \forall p \qquad (11)$$

$$\sum_{e}\sum_{t} x_{ke}^t <= KCAP_k * X_k \quad \forall e \qquad (12)$$

$$\sum_{e}\sum_{t} x_{ec}^t <= KCAP_k * X_k \quad \forall c \qquad (13)$$

$$\sum_{c}\sum_{t} x_{cp}^t <= RPCAP_p * X_p \quad \forall p \qquad (14)$$

$$\sum_{c}\sum_{t} x_{cm}^t <= MCAP_m * X_m \quad \forall m \qquad (15)$$

$$\sum_{m}\sum_{t} x_{msa}^t <= RCCAP_s \quad \forall s, a \qquad (16)$$

$$\sum_{m}\sum_{a}\sum_{t} x_{mfa}^t <= FCAP_s \quad \forall f \qquad (17)$$

$$w_{sp}^t = \sum_{a} x_{spa}^t * w_a \quad \forall s, p, t \qquad (18)$$

$$w_{pk}^t = x_{pk}^t * w \quad \forall p, k, t \qquad (19)$$

$$w_{ec}^t = x_{ec}^t * w \quad \forall e, c, t \qquad (20)$$

$$w_{ke}^t = x_{ke}^t * w \quad \forall k, e, t \qquad (21)$$

$$w_{cp}^t = x_{cp}^t * w \quad \forall c, p, t \qquad (22)$$

$$w^t_{ch} = x^t_{ch} * w \quad \forall c, h, t \tag{23}$$

$$w^t_{cm} = x^t_{cm} * w \quad \forall c, h, t \tag{24}$$

$$w^t_{ms} = \sum_a x^t_{msa} * w_a \quad \forall s, m, t \tag{25}$$

$$w^t_{mf} = \sum_a x^t_{mfa} * w_a \quad \forall m, f, t \tag{26}$$

Maximum number of activated locations constraints

$$\sum_p X_p <= P \tag{27}$$

$$\sum_k X_k <= K \tag{28}$$

$$\sum_c X_c <= C \tag{29}$$

$$\sum_m X_m <= M \tag{30}$$

Linking–shipping constraints

$$L^t_{sp} <= \sum_a x^t_{spa} \quad \forall s, p, t \tag{31}$$

$$L^t_{pk} <= x^t_{pk} \quad \forall p, k, t \tag{32}$$

$$L^t_{ke} <= x^t_{ke} \quad \forall k, e, t \tag{33}$$

$$L^t_{ec} <= x^t_{ec} \quad \forall e, c, t \tag{34}$$

$$L^t_{cp} <= x^t_{cp} \quad \forall c, p, t \tag{35}$$

$$L^t_{ch} <= x^t_{ch} \quad \forall c, h, t \tag{36}$$

$$L^t_{cm} <= x^t_{cm} \quad \forall c, m, t \tag{37}$$

$$L^t_{ms} <= \sum_a x^t_{msa} \quad \forall s, m, t \tag{38}$$

$$L^t_{mf} <= \sum_a x^t_{mfa} \quad \forall m, f, t \tag{39}$$

Shipping linking constraints

$$\sum_a x^t_{spa} <= MI * L^t_{sp} \quad \forall \, s, p, t \tag{40}$$

$$x^t_{pk} <= MI * L^t_{pk} \quad \forall \, p, k, t \tag{41}$$

$$x^t_{ke} <= MI * L^t_{ke} \quad \forall \, k, e, t \tag{42}$$

$$x^t_{ec} <= MI * L^t_{ec} \quad \forall \, e, c, t \tag{43}$$

$$x^t_{cp} <= MI * L^t_{cp} \quad \forall \, c, p, t \tag{44}$$

$$x^t_{cm} <= MI * L^t_{cm} \quad \forall \, c, m, t \tag{45}$$

$$x^t_{ch} <= MI * L^t_{ch} \quad \forall \, c, h, t \tag{46}$$

$$\sum_a x^t_{msa} <= MI * L^t_{ms} \quad \forall \, s, m, t \tag{47}$$

$$\sum_a x^t_{mfa} <= MI * L^t_{mf} \quad \forall \, m, f, t \tag{48}$$

$$x^t_{ija}, x^t_{ij} >= 0$$

$$X_i, L^t_{ij} \in \{0, 1\}$$

The first objective is to maximize the total profit including the total income and profit obtained by introducing recycled materials back into the (forward) supply chain (which is used as an incentive for the companies to choose and use recyclable products) minus the total cost which includes cost of purchasing components from suppliers, production cost incurred at plants, operating costs incurred at d/c, inspection cost for the returned products in collection centers, remanufacturing cost of recoverable products in plants, dismantling cost in dismantling the product, recycling cost at supplier and disposal costs for scrapped products. Second objective is to minimize the CO_2 emission by choosing various available type of transport.

Constraints are divided in five sets: first set is consisting of flow balancing constraints. Constraint (1) assures that the flow entering in the manufacturing plant is equal to the flow exit from it. Constraint (2) is for d/c. Constraint (3) insures that demands of all first customers are satisfied. Constraint (4) insures the flow entering in collection center through a customer will be equal to demand of the customer multiplied by return ratio. Constraint (5) insures that flow entering to each second customer from all collection centers does not exceed the second customer demand. Constraint (6) and (7) imposes that, the flow exiting from each collection center to all plants and dismantler is equal to the amount remaining at each collection center after satisfying second customer demand multiplied by the repairing ratio and (1-repairing ratio) respectively. Constraint (8) and (9) shows that, the flow exiting from each dismantler to supplier and disposal sites are equal to the flow entering from all CC multiplied by recycling ratio and (1-recycling ratio) respectively. Constraint (10–17) insures that flow either exiting or entering at any facility does not exceed the respective facility capacity. Constraints (27–30) limit the number of activated locations, where the sum of binary decision variables which indicate the number of activated locations, is less than the maximum limit of activated locations. Constraints (31–39) insure that there are no links between any locations without actual shipments during all periods. Constraints (40–48) ensure that there is no shipping between any non-linked locations.

4 Multiobjective Methodology: Goal Programming

The basic approach of **goal programming** is to establish a specific numeric goal for each of the objectives, formulate an objective function for each objective, and then seek a solution that minimizes both positive and negative deviations from set goals simultaneously or minimizes the amount by which each goal can be violated. There is a *hierarchy of priority levels* for the goals, so that the goals of primary importance receive first priority attention, those of secondary importance receive second-priority attention, and so forth.

Generalized model of goal programming is:

$$\min a = \{g_1 \left(\overline{\eta_1}, \overline{\rho_1} \right), \ldots, g_k \left(\overline{\eta_2}, \overline{\rho_2} \right)\}$$

$$s.t \quad f_i \left(\overline{x} \right) + \eta_i - \rho_i = b_i \quad \forall i = 1, 2, \ldots, m$$

$$\overline{x}, \overline{\eta}, \overline{\rho} \geq 0;$$

x_j is the jth decision variable, a is denoted as the achievement function; a row vector measure of the attainment of the objectives or constraints at each priority level, $g_k \left(\overline{\eta}, \overline{\rho} \right)$ is a function (normally linear) of the deviation variables associated with the objectives or constraints at priority level k, K is the total number of priority levels in the model, b_i is the right-hand side constant for goal (or constraint)i, $f_i(\overline{x})$ is the left-hand side of the linear goal or constraint i.

We seek to minimize the non-achievement of that goal or constraint by minimizing specific deviation variables. The deviation variables at each priority level are included in the function $g_k \left(\overline{\eta}, \overline{\rho} \right)$ and ordered in the achievement vector, according to their respective priority. Algorithm of sequential goal programming:

Step 1: Set $k = 1$ (k represents the priority level and K is the total of these).
Step 2: Establish the mathematical formulation as discussed above using positive and negative deviations for priority level k only.

Step 3: Solve this single-objective problem associated with priority level k and the optimal solution of g_k $(\bar{\eta}, \bar{\rho})$ is a*.
Step 4: Set $k = k + 1$. If $k > K$, go to Step 7.
Step 5: Establish the equivalent, single objective model for the next priority level (level k) with additional constraint g_k $(\bar{\eta}, \bar{\rho}) = a_s^*$.
Step 6: Go to Step 3.
Step 7: The solution vector x*, associated with the last single objective model solved, is the optimal vector for the original goal programming model.

5 Numerical Illustration

In this section, a numerical example is presented in order to demonstrate the applicability of the model. In considered CLSC, a product which is made up of six components say 1, 2, 3, 4, 5, and 6 with respective utilization rate of 1, 4, 1, 2, 1, and 3 and recycling rate of 1, 0.5, 7.5, 1, 0.3 and 0 flows between various facilities. In forward direction, there is a set of three suppliers that can provide components to two potential locations of manufacturing plants. Three potential location of d/cs are there in the network to cater the demand of 2000, 2700, 3250, 2550, and 2700 units from respective 5 zones of first customer market at a unit selling price of 11000, 10500, 10000, 10750, and 10500. In backward direction, potential locations of CC, dismantlers and disposal sites are 3, 2, and 1 respectively. Beside its demand of 500, 350, and 550 units of used product from respective three zones of second customers can be satisfied at unit selling price of 7500, 8000, and 7000. As for transportation, road-based transportation is used to carry out the shipping operations, for which there are three types of trucks available which are 0–3, 4–7, and 8–11 years old, respectively. We assume that the older the trucks, the cheaper their rental fees, but, at the same time, the greater their CO_2 emissions, due to decreasing engine efficiency. Unit transportation costs for the different types of trucks used are 1, 0.85 and 0.70 for truck types 1, 2, and 3, respectively. Emission rate of CO_2 found to be 1.3, 2.8, and 3.1 g/mi for truck types 1, 2 and 3 respectively. Profit raised in the network by repairing the product is 5500/unit and by recycling a unit of component 1, 2, 3, 4, and 5 are 250, 50, 90, 55, and 300 respectively.

Other parameters are set as follows: $Rr = 0.60$, $Rp = 0.25$, and $Rc_a = (1, 1, 1, 1, 1, 0)$. Set of unit purchasing costs of components (in order) from supplier 1, 2, and 3 are (460, 0, 190, 125, 0, 80), (480, 120, 200, 150, 650, 100) and (470, 95, 0, 0, 620, 90), respectively. Unit recycling costs of components (in order) at supplier 1, 2, and 3 are (20, 0, 60, 10, 0, 0), (25, 90, 55, 20, 390, 0) and (0, 65, 0, 0, 380, 0), respectively. Price 0 means that component service is not provided by respective supplier. 2500 and 3000 are unit production cost, and 1500 and 2200 are unit repairing costs of the product at plant 1 and 2, respectively. Unit operating costs at d/c 1, 2, and 3 are 500, 550, and 600 respectively. Unit Inspection costs at collection centers 1, 2, and 3 are 100, 100, and 120 respectively. Unit dismantling cost at d/m 1 and 2 are 125 and 110 respectively. Unit disposal cost of component 6 is 15.

Data on capacities at various facilities are as follows: Supplier 1 can supply at most of 8000, 0, 9000, 12000, 0, and 14000 units of component 1, 2, 3, 4, 5, and 6 respectively. Capacity of supplier 2 and 3 of components are (7500, 40000, 5000, 27000, 7700, 15000) and (0, 20000, 0, 0, 7500, 1400) respectively. Recycling capacities of supplier 1, 2 and 3 are (3000, 0, 2900, 4000, 0, 0), (2000, 15000, 2000, 6000, 2500, 0) and (0, 8000, 0, 0, 2500, 0) respectively. Production capacities of plants are 8000, 7500 and repairing capacities are 2000, 1800 respectively. Capacities of d/c's are 4800, 5000 and 5500, of CC's are 3500, 3000 and 2500; of dismantlers are 5000, 5000 and of disposal site is 250000.

Data on distance (in miles) between any two facilities is as follows:

$D_{ij} = \{D_{11}, D_{12}, D_{13}, \ldots, D_{21}, D_{22}, D_{23}, \ldots\}$

$D_{sp} = \{200, 190, 310, 350, 290, 280\},$

$D_{pk} = \{120, 100, 135, 170, 190, 200\},$

$D_{ke} = \{24, 17, 22, 21, 18, 29, 19, 21, 20, 31, 33, 25, 28, 15, 28\},$

$D_{ec} = \{6, 9, 8, 8.5, 7, 10, 11, 12, 13, 9, 8, 9.5, 11, 9, 8\},$

$D_{cp} = \{150, 120, 135, 110, 130, 100\}$

$D_{cm} = \{8.5, 9, 11, 12, 10, 11\},$

$D_{ch} = \{15, 21, 19, 24, 16, 18, 20, 22, 21\}$

$D_{ms} = \{100, 150, 120, 95, 154, 130\},$

$D_{mf} = \{80, 75\}$

The above data is employed to validate the proposed model. A LINGO code for generating the proposed mathematical models of the given data was developed and solved using LINGO11.0 [20]. Problem is solved individually with each objective subject to given set of constraints. Thus, Profit and amount of CO_2 emission would be 66625630 and 252121600 respectively. Which are set as the aspiration levels for profit and emission functions. Then multiobjective programming problem combining all the objectives and incorporating the individual aspirations is solved which results in infeasible solution hence goal programming technique has been used to obtain a compromise solution to the above problem. Giving weight age 0.5 and 0.5 to profit and CO_2 objective respectively, a compromised solution of allocation of facilities and transporting vehicle is obtained. Total profit thus generated in the network is Rs. 54, 240, 470 and amount of CO_2 emitted is 543, 833, 100. The flow between facilities using different type vehicles is given below.

$x_{spa}^t : x_{111}^3 = 3016, \quad x_{113}^3 = 4016, \quad x_{114}^3 = 12000, \quad x_{121}^2 = 4984, \quad x_{123}^2 = 4984,$

$x_{126}^1 = 14000, \quad x_{211}^2 = 3570, \quad x_{212}^2 = 12860, \quad x_{213}^2 = 2570, \quad x_{214}^2 = 1172,$

$x_{215}^2 = 4070, \quad x_{216}^2 = 5758, \quad x_{222}^1 = 19080, \quad x_{224}^1 = 9968, \quad x_{226}^1 = 952,$

$x_{312}^3 = 13484, \quad x_{315}^3 = 2516, \quad x_{316}^3 = 14000, \quad x_{322}^1 = 856, \quad x_{325}^1 = 4984$

$$x_{pk}^t : x_{11}^3 = 1086, \quad x_{13}^2 = 5500, \quad x_{21}^3 = 3714, \quad x_{22}^2 = 2900$$

$$x_{ke}^t : x_{13}^1 = 450, \quad x_{14}^1 = 1650, \quad x_{15}^1 = 2700, \quad x_{21}^1 = 2000, \quad x_{24}^1 = 900,$$

$$x_{32}^3 = 2700, \quad x_{33}^3 = 2800$$

$$x_{ec}^t : x_{13}^3 = 1200, \quad x_{21}^1 = 150, \quad x_{22}^3 = 1470, \quad x_{31}^2 = 1950, \quad x_{42}^3 = 1530,$$

$$x_{51}^3 = 1400, \quad x_{53}^3 = 220$$

$$x_{cp}^t : x_{12}^3 = 525, \quad x_{22}^1 = 750, \quad x_{32}^3 = 355$$

$$x_{cm}^t : x_{11}^3 = 1575, \quad x_{22}^1 = 2250, \quad x_{32}^1 = 1065$$

$$x_{ch}^t : x_{11}^1 = 500, \quad x_{12}^2 = 350, \quad x_{13}^2 = 550$$

$$x_{msa}^t : x_{121}^3 = 1575, \quad x_{122}^3 = 6300, \quad x_{123}^3 = 1575, \quad x_{124}^3 = 3150, \quad x_{125}^3 = 1575,$$

$$x_{211}^2 = 2890, \quad x_{214}^3 = 3780, \quad x_{221}^3 = 425, \quad x_{222}^2 = 8700,$$

$$x_{223}^3 - 425, \quad x_{224}^2 = 2850, \quad x_{225}^3 = 925, \quad x_{232}^1 = 4560, \quad x_{235}^1 = 2390$$

$$x_{mfa}^t : x_{116}^3 = 4725, \quad x_{216}^2 = 9945$$

6 Conclusions

One of the important planning activities in supply chain management (SCM) is to design the configuration of the supply chain network. Besides, due to the global warning recently attention has been given to reverse logistic in SCM. Modeling of a CLSC network design problem can be a challenging process because there is large number of components that need to be incorporated into model. Here in this paper, trade-offs between operational and environmental performance measures of shipping product were investigated. Due to global warming, this paper focused on CO_2 emissions, One of the main findings of this paper is that, costs of environmental impacts are still not as apparent as operational measures, as far as their relative importance in a emission rate function are concerned. Operational costs of handling products, both in forward and reverse networks, seem to be dominant ignoring emissions rate. Another interesting result is relevant to the promotion of reusable products, the use of which seems to lessen the operational costs of the chain, but places a burden on the environmental costs.

References

1. Beamon, B.M.: Sustainability and the future of supply chain management. Oper. Supply Chain Manag. **1**(1), 4–18 (2008)
2. Zhu, Q.H., Sarkis, J., Lai, K.H.: Green supply chain management implications for "closing the loop". Transp. Res. Part E **44**(1), 1–18 (2008)
3. Fleischmann, M., Beullens, P., Bloemhof-ruwaard, J.M., Wassenhove, L.: The impact of product recovery on logistics network design. Product. Oper. Manag. **10**, 156–173 (2001)
4. Lee, D., Dong, M.: A heuristic approach to logistics network design for end-of-lease computer products recovery. Transp. Res. Part E **44**, 455–474 (2008)
5. Verstrepen, S., Cruijssen, F., de Brito, M., Dullaert, W.: An exploratory analysis of reverse logistics in Flanders. Eur. J. Trans. Infrastruct. Res. **7**(4), 301–316 (2007)
6. Pishvaee, M.R., Farahani, R.Z., Dullaert, W.: A memetic algorithm for bi-objective integrated forward/reverse logistics network design. Comput. Oper. Res. **37**(6), 1100–1112 (2010)
7. Jayaraman, V., Guide Jr, V.D.R., Srivastava, R.: A closed-loop logistics model for remanufacturing. J. Oper. Res. Soc. **50**(5), 497–508 (1999)
8. Fleischmann, M., Beullens, P., Bloemhof-Ruwaard, J.M., Wassenhove, L.: The impact of product recovery on logistics network design. Prod. Oper. Manag. **10**(3), 156–173 (2001)
9. Salema, M.I., Barbosa-Póvoa, A.P., Novais, A.Q.: A strategic and tactical model for closed-loop supply chains, (pp. 361–386). EURO Winter Institute on Location and Logistics, Estoril (2007a)
10. El Saadany, Ahmed M. A., El-Kharbotly, Amin, K.: Reverse logistics modeling. In: 8th International Conference on Production Engineering and Design for Development, Alexandria, Egypt (2004)
11. Min, H., Ko, C.S., Ko, H.J.: The spatial and temporal consolidation of returned products in a closed-loop supply chain network. Comput. Ind. Eng. **51**(2), 309–320 (2006)
12. Kannan, G., Sasikumar, P., Devika, K.: A genetic algorithm approach for solving a closed loop supply chain model: a case of battery recycling. Appl. Math. Model. **34**(3), 655–670 (2010)
13. Aras, N., Boyaci, T., Verter, V.: Designing the reverse logistics network. In: Ferguson, M.E., Souza, G.C. (eds.), Closed-loop Supply Chains: New Developments to Improve the Sustainability of Business Practices, pp. 67–97. CRC Press, Taylor & Francis, Boca Raton (2010)
14. Meixell, M.J., Gargeya, V.B.: Global supply chain design: a literature review and critique. Transp. Res. Part E **41**(6), 531–550 (2005)
15. Hervani, A.A., Helms, M.M., Sarkis, J.: Performance measurement for green supply chain management. Benchmarking: An International Journal **12**(4), 330–353 (2005)
16. Srivastava, S.K.: Green supply-chain management: a state-of-the-art literature review. Int. J. Manag. Rev. **9**(1), 53–80 (2007)
17. Sarkis, J.: A strategic decision framework for green supply chain management. J. Cleaner Prod. **11**(4), 397–409 (2003)
18. Sheu, J.B., Chou, Y.H., Hu, C.: An integrated logistic operational model for green supply chain management. Transp. Res. Part E **41**(4), 287–313 (2005)
19. Krikke, H., Bloemhof-Ruwaard, J., Van Wassenhove, L.: Concurrent product and closed-loop supply chain design with an application to refrigerators. Int. J. Prod. Res. **41**(16), 3689–3719 (2003)
20. Thiriez, H.: OR software LINGO. Eur. J. Oper. Res. **12**, 655–656 (2000)

Optimal Component Selection Based on Cohesion and Coupling for Component-Based Software System

P. C. Jha, Vikram Bali, Sonam Narula and Mala Kalra

Abstract In modular-based software systems, each module has different alternatives with variation in their functional and nonfunctional properties, e.g., reliability, cost, delivery time, etc. The success of such systems largely depends upon the selection process of commercial-off-the shelf (COTS) components. In component-based software (CBS) development, it is desirable to choose the components that provide all necessary functionalities and at the same time optimize nonfunctional attributes of the system. In this paper, we have discussed the multiobjective optimization model for COTS selection in the development of a modular software system using CBSS approach. Fuzzy mathematical programming (FMP) is used for decision making to counter the effects of unreliable input information.

Keywords Commercial-off-the shelf (COTS) · Components-based software system (CBSS) · Fuzzy mathematical programming (FMP) · Intra-modular coupling density (ICD)

P. C. Jah (✉) · S. Narula
Department of Operational Research, University of Delhi, Delhi, India
e-mail: jhapc@yahoo.com

S. Narula
e-mail: sonam.narula88@gmail.com

V. Bali
Rayat Bahra Institute of Engineering and Bio-Technology, Mohali, Punjab, India
e-mail: vikramgcet@gmail.com

M. Kalra
National Institute of Technical Teachers Training and Research, Chandigarh, India
e-mail: malakalra2004@yahoo.co.in

B. V. Babu et al. (eds.), *Proceedings of the Second International Conference on Soft Computing for Problem Solving (SocProS 2012), December 28–30, 2012*, Advances in Intelligent Systems and Computing 236, DOI: 10.1007/978-81-322-1602-5_150, © Springer India 2014

1 Introduction

With the growing need of complex software systems, the use of commercial off-the-shelf (COTS) products has grown steadily. COTS products as a way of managing cost, developing time, and effort [1] requires less code that needs to be designed and implemented by the developers. Compared with traditional software development, COTS-based system development promises faster delivery with lower resource cost. However, the use of COTS products in software development can require a considerable integration effort [2].

In recent years, component-based approach to software development has become more and more popular [3]. Component-based software systems (CBSS) development focuses on the decomposition of a software system into functional and logical components with well-defined interfaces. It allows a software system to be developed using appropriate and suitable software components that are available in COTS components' market. The vendor of the COTS supplier provides information about cost and delivery time of the COTS products. The software development using CBSS approach has reduced significantly the software development time and also facilitates system with better maintainability. Optimization problems of optimum selection of COTS components are widely studied by many researchers in the literatures such as Belli and Jedrzejowicz [4], Berman and Ashrafi [5], Berman and Kumar [6], Cortellesa et al. [7], Gupta et al. [8, 9], Jha et al. [10], Kapur et al. [11], Kumar [12], Kwong et al. [13, 14], and Neubauer and Stummer [15]. The models proposed by the authors have been used to achieve the different attributes of quality along with the objective of minimizing the cost or keeping cost to a budgetary level.

CBSS development employs modular approach for the development of the software system. It also improves the flexibility and comprehensibility of the software [16]. In the development of modular-based conventional software systems, the criteria of minimizing the coupling and maximizing the cohesion of software modules were commonly used [16–19]. Coupling is about the measure of interactions among software modules while cohesion is about the measure of interactions among the software components which are within a software module. A good software system should possess software modules with high cohesion and low coupling. A highly cohesive module exhibits high reusability and loosely coupled systems enable easy maintenance of a software system [20].

In the previous studies of COTS selection discussed above it is assumed that COTS components within the set of alternative software components are often regarded to have the same functions in CBSS development. Since the COTS components are provided by multiple suppliers, functions performed by these components could be different from each other. Therefore, the functional contributions of the software components toward the functional requirements of a CBSS should be considered. Kwong et al. [21] presented a methodology for optimal selection of software components for CBSS development based on the criteria of simultaneously maximizing functional performance and intra-modular coupling density (ICD).

In this paper, we devise a multiobjective optimization model for COTS selection in the development of a modular software system using CBSS approach. The formulated model simultaneously maximizes the functional performance and also intra-modular coupling density which in turn maximizes cohesion and minimizes coupling in CBSS. The selection of COTS components is constrained using minimum threshold on the intra-modular coupling density and maximum allowable limit on budget and delivery time of acquiring all the COTS components for CBSS development. The proposed research can be considered as an extension of the optimization model proposed in [21]. The authors in their work have not considered budget and delivery time for the selection of COTS components. We cannot always have software system with highly cohesive and loosely coupled modules because of the limitations on budget and delivery time. The delivery time of the COTS component is the time of acquiring and integrating the components within and amongst the modules of the software system. The total delivery time includes integration and system testing. Similarly, we cannot spend as much as we want on component selection because in real life situation we have limitations on budget. Therefore, there is a need to maintain a trade-off between a highly cohesive software system and its cost. Moreover, in our work, we have also incorporated issue of compatibility amongst the components of the modules.

2 COTS Selection in CBSS

CBSS development starts with identification of modules in the software and each module must contain at least one COTS component. In order to select COTS components for modular software systems, the following criteria may be used.

2.1 Intra-modular Coupling Density

In this research, we have employed Abreu and Gaulao's [17] approach which yields the quantitative measures of cohesion and coupling. The authors in their work presented intra-modular coupling density (ICD) to measure the relationship between cohesion and coupling of modules in design of modular software system and is given as follows:

$$ICD = \frac{CI_{IN}}{CI_{IN} + CI_{OUT}} \tag{1}$$

where, CI_{IN} is the number of class interactions within modules, and CI_{OUT} is the number of interactions between classes of distinct modules.

Referring to Eq. (1), the ratio of cohesion to all interactions within the jth module can be expressed as ICD_j. However, it can be found if any module contains only one component, the values of ICD for that module becomes zero. To make up for the

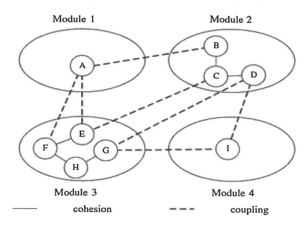

Fig. 1 Cohesion and coupling

deficiency 1 is added to the numerator of Eq. (1) to form another measure of ICD as follows:

$$ICD = \frac{(CI_{IN})_j + 1}{(CI_{IN})_j + (CI_{OUT})_j} \qquad (2)$$

where, ICD_j is the intra-modular coupling density for the jth module; $(CI_{IN})_j$ is the number of component interactions within the jth module; and $(CI_{OUT})_j$ is the number of component interactions between the jth module and other modules. Figure 1 (replicated from [21]) shows diagrammatic depiction of cohesion and coupling of software modules in the development of modular software system.

2.2 Functional Performance

Functionality of the COTS components can be defined as the ability of the component to perform according to the specific needs of the customer/organization requirements. The functional capabilities of the COTS components are different for different components as they are provided by different suppliers in the COTS market. We use functional ratings of the COTS components to the software module as coefficients in the objective function corresponding to maximizing the functional performance of the modular software system. These ratings are assumed to be given by the software development team.

2.3 Cost

Cost is the major factor in determining the selection of COTS components. Cost is one of the constraints in our proposed model. We have considered cost based on procurement and adaptation costs of COTS components.

2.4 Delivery Time

The delivery time of the COTS component is the time of acquiring and integrating the components within and amongst the modules of the software system. The total delivery time includes integration and system testing.

2.5 Compatibility

In the development of software system, sometimes the COTS product for one module is incompatible with the alternative COTS products for other modules due to problem such as implementation technology, interfaces, and licensing. Therefore, the issue of compatibility is also incorporated in the optimization model for COTS selection.

3 Formulation of Optimization Model for CBSS

Generally, a CBSS is developed based on a top-down approach. Based on this approach, functional/customer requirements are first identified. The number and nature of software modules are then determined. The next task is to integrate software components for modules. The selection of components should be in such a way so as to have maximum interactions of components within the software modules and minimum interactions of software components amongst the software modules.

Let S be a software architecture made of M modules, with a maximum number of N components available for each module. Figure 2 shows how CBSS can be developed using software components.

3.1 Notations

M the number of software modules

N the number of software components

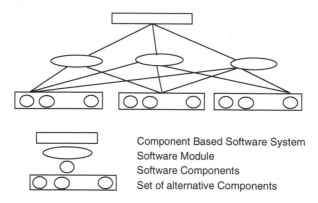

Fig. 2 A CBSS system

Sc_i the ith software component; $i = 1, 2, \ldots, N$

m_j the jth software module; $j = 1, 2, \ldots, M$

S_k a set of alternative software components for the kth functional requirement of a CBSS. Only one software component in S_k is selected to implement the kth requirement; $k = 1, 2, \ldots, L$

$i \in S_k$ denotes that Sc_i belongs to the kth set

$r_{ii'}$ the number of interactions between Sc_i and $\text{Sc}_{i'}$; $i; i' = 1, 2, \ldots, N$ as the coupling and cohesion are undirected relations, $r_{ii'} = r_{i'i}$

f_{ij} f_{ij} are real numbers ranging from 0 to 1 depicting the function rating of Sc_i to m_j; $i = 1, 2, \ldots, N$; $j = 1, 2, \ldots, M$

H a threshold value of ICD_j of each module that needs to be set by decision makers.

C_{ij} cost of ith component available for jth module

d_{ij} delivery time of ith component available for jth module

B maximum budget limit set by the decision makers

T maximum threshold given on delivery time of a component

Sc_{ij} the ith software component of jth software module, s.t. $\text{Sc}_{ij} = \text{Sc}_{ij'} = \text{Sc}_i$ for all $j, j' = 1, 2, \ldots, M$

x_{ij} binary variable $\begin{cases} 1, \text{ if } Sc_i \text{ is selected for } m_j \\ 0, \text{ otherwise} \end{cases}$

3.2 Assumptions

1. At least one component is supposed to get selected from each module.
2. A threshold value of ICD_j, budget and delivery time are set by the decision makers.
3. Redundancy not allowed i.e. exactly one software component in S_k may get selected to implement kth requirement.
4. The cost and delivery time of COTS components are given.
5. Interaction data for components is exactly same for all modules, irrespective of the selection happened.
6. Interaction associated is set by the software development team.

3.3 Optimization Model

The multiobjective optimization model for COTS selection using CBSS development can be formulated as follows:

$$\text{Max ICD} = \frac{\sum_{j=1}^{M} \sum_{i=1}^{N-1} \sum_{i'=i+1}^{N} r_{ii'} x_{ij} x_{i'j}}{\sum_{i=1}^{N-1} \sum_{i'=i+1}^{N} r_{ii'} \left(\sum_{j=1}^{M} x_{ij} \right) \left(\sum_{j'=1}^{M} x_{i'j'} \right)} \tag{3}$$

$$\text{Max F} = \sum_{j=1}^{M} \sum_{i=1}^{N} f_{ij} x_{ij} \tag{4}$$

Subject to $X \in S = \Big\{ x_{ij}$ is binary variable/

$$\frac{\sum_{i=1}^{N-1} \sum_{i'=i+1}^{N} r_{ii'} x_{ij} x_{i'j} + 1}{\sum_{i=1}^{N-1} \sum_{i'=i+1}^{N} r_{ii'} x_{ij} \sum_{j'=1}^{M} x_{i'j'}} \geq H; \ j = 1, 2, \ldots, M \ , \ j' = 1, 2, \ldots, M$$

$$\tag{5}$$

$$\sum_{j=1}^{M} \sum_{i=1}^{N} C_{ij} x_{ij} \leq B \tag{6}$$

$$\sum_{i \in S_k} \sum_{j=1}^{M} d_{ij} x_{ij} \leq T \; ; k = 1, 2, \ldots, L \tag{7}$$

$$\sum_{i \in S_k} \sum_{j=1}^{M} x_{ij} = 1 \; ; k = 1, 2, \ldots, L \tag{8}$$

$$\sum_{i=1}^{N} x_{ij} \geq 1 \; ; \; j = 1, 2, \ldots, M \tag{9}$$

$$x_{ij} \in \{0, 1\} \; ; \; i = 1, 2, \ldots, N \; ; \; j = 1, 2, \ldots, M \tag{10}$$

$$x_{rs} - x_{u_k h} \leq \bar{M} y_k \tag{11}$$

$$\sum_{k=1}^{z} y_k = z - 1 \tag{12}$$

$$y_k \in \{0, 1\} \; ; \; k = 1, 2, \ldots, z \Big\} \tag{13}$$

The objective function (3) is based on maximizing cohesion within software modules and minimizing coupling among software modules. Objective function (4) maximizes the functional performance of a software system to be developed. Constraint (5) shows the minimum threshold on ICD value. Constraint (6) is budget limitation. Constraint (7) is restriction on delivery time. Equation (8) denotes that only one software component can be selected from a set of alternative software components for a particular software module. Constraint (9) refers to the software module that contains at least one software component. Constraint (10) shows selection or rejection of a particular COTS product. Constraints (11)–(13) are used to deal with incompatibility amongst COTS components. The incompatibility constraint can be denoted by $x_{rs} \leq x_{u1t}$, that is if the module s chooses COTS component r, then the module t must choose the COTS component $u1$. This decision is called contingent decision constraint [22]. Suppose that there are two contingent decisions in the model, such as the COTS alternative for the module is only compatible with the COTS products $u1$ and $u2$ for the module t, i.e., either $x_{u1t} = 1$ if $x_{rs} = 1$ or $x_{u2t} = 1$ if $x_{rs} = 1$, these constraint can be represented as either $x_{rs} \leq x_{u1t}$ or $x_{rs} \leq x_{u2t}$. Since the presence of "either-or" constraint makes the optimization problem nonlinear, it can be linearized by binary variable y_k as follows:

$$y_k = \begin{cases} 0, & \text{if } k\text{th constraint is active} \\ 1, & \text{if } k\text{th constraint is inactive} \end{cases} .$$

Thus, only one out of z contingent decision constraints for any COTS products between two modules is guaranteed to be active if $\bar{M}$ is sufficiently large.

4 Fuzzy Approach for Solving Multiobjective Optimization Problem

Conventional optimization methods assume that all parameters and goals of an optimization model are precisely known. But for many practical problems there are incompleteness and unreliability of input information. This caused us to use fuzzy multiobjective optimization method with fuzzy parameters. Following steps are required to perform for solving fuzzy multiobjective optimization problem [9].

Step 1: Construct multiobjective optimization problem. Refer problem (P1)

Step 2: Solve multiobjective optimization problem by considering first objective function. This process is repeated for all the remaining objective functions. If all the solutions (i.e. $X^1 = X^2 = \cdots = X^k = x_{ij}, i = 1, \ldots, N; j = 1, \ldots, M$) are same, select one of them as an optimal compromise solution and stop. Otherwise, go to step 3.

Step 3: Evaluate the kth objective function at all solutions obtained and determine the best (worst) lower bound (L_k) and best (worst) upper bound (U_k) as the case may be.

Step 4: Define membership function of each objective of optimization model. The membership function for ICD is given as follows:

$$\mu_{\text{ICD}}(x) = \begin{cases} 1, & \text{if } \text{ICD}(x) \geq \text{ICD}_u, \\ \frac{\text{ICD}(x) - \text{ICD}_l}{\text{ICD}_u - \text{ICD}_l}, & \text{if } \text{ICD}_l < \text{ICD}(x) < \text{ICD}_u, \\ 0, & \text{if } \text{ICD}(x) \leq \text{ICD}_l \end{cases}$$

where ICD_l is the worst lower bound and ICD_u is the best upper bound of ICD objective function.

The membership function for functionality is given as follows.

$$\mu_F(x) = \begin{cases} 1, & \text{if } F(x) \geq F_u, \\ \frac{F(x) - F_l}{F_u - F_l}, & \text{if } F_l < F(x) < F_u, \\ 0, & \text{if } F(x) \leq F_l \end{cases}$$

where F_l is the worst lower bound and F_u is the best upper bound of functionality objective function.

Step 5: Develop fuzzy multiobjective optimization model.

Following Bellaman-Zadeh's maximization principle [23] and using the above defined fuzzy membership functions, the fuzzy multiobjective optimization model for COTS selection is formulated as follows:

$$\text{max} \quad \lambda$$
$$\text{subject to} \quad \lambda \le \mu_{\text{ICD}(x)}$$
$$\lambda \le \mu_{F(x)}$$
$$0 \le \lambda \le 1$$
$$X \in S$$

Solve the above model. Present the solution to the decision maker. If the decision maker accepts it, then stop. Otherwise, evaluate each objective function at the solution. Compare the upper (lower) bound of each objective function with new value of the objective function. If the new value is lower (higher) than the upper (lower) bound, consider it as a new upper (lower) bound. Otherwise, use the old values as it is. If there are no changes in current bounds of all the objective functions then stop otherwise go to step 4.

The solution process terminates when decision maker accepts the obtained solution and considers it as the preferred compromise solution which is in fact a compromise feasible solution that meets the decision maker's preference.

5 Case Study

A case study of CBSS development is presented to illustrate the proposed methodology of optimizing the selection of COTS components for CBSS development. A local software system supplier planned to develop a software system for small and medium size manufacturing enterprises. In this case, a software system is decomposed into three modules M_1, M_2 and M_3. A total of 20 software components (Sc_1–Sc_{20}) are available in market to make up ten sets of alternative software component (S_1–S_{10}) for each module. Total components available for selection are sixty and total twenty components are available for selection per module and may be represented as (Sc_1–Sc_{20}). Exactly one software components in each set of alternatives may get selected for a particular software module for fulfilling functional requirements. For example, Sc_1, Sc_2, Sc_3 and Sc_4 all belong to the set of alternative software components S_1. Hence, only one of the four components will be selected for fulfilling the S_1 functional requirement.

Individual functional requirements and their corresponding alternative software components, as well as the function ratings of software components corresponding to software modules, are shown in Table 1. The function ratings describe the degree of functional contributions of the software components toward the software modules. The function ratings range from 0 to 1 where 1 refers to a very high degree of contribution while 0 indicates zero degree of contribution.

Table 2 shows the degrees of interaction among software components. The range of the degrees is 1–10. The degree '1' means a very low degree of interaction while the degree '10' refers to a very high degree of interaction.

In Table 3, Cost (C_{ij} for all i, j) in 100\$ unit; and delivery time (d_{ij} for all i, j) in days associated with COTS components is given.

Table 1 Example description and functionality of COTS components

Functional requirements	S_k	Software components	Module1 (Front office)	Module2 (Back office)	Module3 (Finance)
Inventory control and management	S_1	Sc_1	0.68	0.51	0.00
		Sc_2	0.22	0.63	0.01
		Sc_3	0.15	0.79	0.00
		Sc_4	0.23	0.87	0.00
Payment collection and authorization	S_2	Sc_5	0.94	0.10	0.55
Sales	S_3	Sc_6	0.75	0.45	0.22
Automatic updates	S_4	Sc_7	0.08	0.94	0.01
		Sc_8	0.10	0.22	0.00
		Sc_9	0.00	1.00	0.20
		Sc_{10}	0.20	0.45	0.05
E-commerce	S_5	Sc_{11}	0.00	0.98	0.20
		Sc_{12}	0.00	0.31	0.10
Financial reporting	S_6	Sc_{13}	0.11	0.12	0.31
		Sc_{14}	0.05	0.07	0.71
Business rules and protocol	S_7	Sc_{15}	0.22	0.02	0.42
Shift wise reporting statistics	S_8	Sc_{16}	0.30	0.02	0.00
		Sc_{17}	0.80	0.10	0.00
Accounts	S_9	Sc_{18}	0.00	0.70	0.32
		Sc_{19}	0.00	0.06	0.78
Finance	S_{10}	Sc_{20}	0.00	0.00	0.18

6 Solution

Steps 1, 2 and *3* After forming multiobjective programming using the above data, the solution of each single objective problem is found and hence upper and lower bounds are obtained as follows:

	X^1	X^2
ICD	0.653	0.581
F	1.17	7.35

where $X^1 = (Sc_{201}, Sc_{142}, Sc_{152}, Sc_{192}, Sc_{33}, Sc_{53}, Sc_{63}, Sc_{103}, Sc_{113}, Sc_{173})$
$X^2 = (Sc_{51}, Sc_{61}, Sc_{161}, Sc_{32}, Sc_{92}, Sc_{112}, Sc_{143}, Sc_{153}, Sc_{193}, Sc_{203})$

Table 2 Interactions among COTS components

		S_1				S_2	S_3		S_4			S_5			S_6	S_7		S_8		S_9	S_{10}
		Sc_1	Sc_2	Sc_3	Sc_4	Sc_5	Sc_6	Sc_7	Sc_8	Sc_9	Sc_{10}	Sc_{11}	Sc_{12}	Sc_{13}	Sc_{14}	Sc_{15}	Sc_{16}	Sc_{17}	Sc_{18}	Sc_{19}	Sc_{20}
S_1	Sc_1	0	0	5	3	0	4	9	6	3	6	8	5	2	1	0	0	0	0	0	3
	Sc_2	0	0	3	1	0	4	5	3	7	8	5	3	4	2	0	0	1	0	0	2
	Sc_3	5	3	0	4	1	3	3	0	2	7	10	2	0	1	0	4	3	1	0	1
	Sc_4	3	1	4	0	2	2	7	8	6	9	2	3	0	0	0	0	0	2	1	1
S_2	Sc_5	0	0	1	2	0	10	1	2	3	2	1	2	0	0	0	0	0	2	1	2
S_3	Sc_6	4	4	3	2	10	0	1	2	2	2	2	0	0	0	0	8	7	1	1	2
S_4	Sc_7	9	5	3	7	1	1	0	0	3	7	3	2	2	1	0	0	0	1	2	1
	Sc_8	6	3	0	8	2	2	0	0	2	6	2	1	2	3	0	1	0	2	2	2
	Sc_9	3	7	2	6	3	2	3	2	0	4	10	7	2	1	0	0	0	2	1	1
	Sc_{10}	6	8	7	9	2	2	7	6	4	0	10	7	4	2	1	3	4	2	1	3
S_5	Sc_{11}	8	5	10	2	1	2	3	2	10	10	0	2	3	2	3	0	0	4	2	4
	Sc_{12}	5	3	2	3	2	0	2	1	7	7	2	0	4	1	2	0	0	3	1	5
S_6	Sc_{13}	2	4	0	0	0	0	2	2	2	4	3	4	0	0	8	1	0	10	7	10
	Sc_{14}	1	2	1	0	0	0	1	3	1	2	2	1	0	0	6	3	0	4	3	3
S_7	Sc_{15}	0	0	0	0	0	0	0	0	0	1	3	2	8	6	0	0	0	9	10	10
S_8	Sc_{16}	0	0	4	0	0	8	0	1	0	3	0	0	1	3	0	0	2	2	0	0
	Sc_{17}	0	1	3	0	0	7	0	0	0	4	0	0	0	0	0	2	0	0	0	0
S_9	Sc_{18}	0	0	1	2	2	1	1	2	2	2	4	3	10	4	9	2	0	0	1	10
	Sc_{19}	0	0	0	1	1	1	2	2	1	1	2	1	7	3	10	0	0	1	0	4
S_{10}	Sc_{20}	3	2	1	1	2	2	1	2	1	3	4	5	10	3	10	0	0	10	4	0

The value of D, H, and B are assumed as 9, 0.4, and 80, respectively.

Owing to the compatibility condition third module of second component is found to be compatible with the fifth component of first module.

Steps 4 and **5** Then fuzzy multiobjective is developed and solved using the LINGO software [24]. The following components are selected.

$X = (Sc_{51}, Sc_{61}, Sc_{161}, Sc_{32}, Sc_{92}, Sc_{112}, Sc_{143}, Sc_{153}, Sc_{193}, Sc_{203})$. The objective function values are: $\lambda = 0.919$, ICD $= 0.649$ and F $= 6.85$.

7 Conclusion

In this paper a fuzzy multiobjective optimization model is proposed for selection of components for CBSS development. The selection of components is based on the criteria of having maximum cohesion between components of modules and minimum coupling amongst the components of module. The model is also based on maximizing the function ratings of various COTS components. Compared with the previous studies on CBSS development, the model incorporates constraints on cost, delivery time and compatibility. To obtain the optimal solution a fuzzy algorithm

Table 3 Cost and delivery time data set of COTS components

Cost						Delivery time					
Module 1		Module2		Module3		Module 1		Module2		Module3	
C_{11}	10	C_{12}	9	C_{13}	8	d_{11}	3	d_{12}	5	d_{13}	6
C_{21}	9	C_{22}	8	C_{23}	9	d_{21}	4	d_{22}	6	d_{23}	4
C_{31}	8	C_{32}	7	C_{33}	6	d_{31}	6	d_{32}	7	d_{33}	9
C_{41}	8	C_{42}	10	C_{43}	7	d_{41}	6	d_{42}	4	d_{43}	7
C_{51}	7	C_{52}	7	C_{53}	8	d_{51}	7	d_{52}	7	d_{53}	6
C_{61}	9	C_{62}	8	C_{63}	9	d_{61}	5	d_{62}	6	d_{63}	4
C_{71}	6	C_{72}	9	C_{73}	6	d_{71}	8	d_{72}	4	d_{73}	8
C_{81}	7	C_{82}	6	C_{83}	7	d_{81}	7	d_{82}	8	d_{83}	7
C_{91}	8	C_{92}	10	C_{93}	10	d_{91}	6	d_{92}	3	d_{93}	3
C_{101}	10	C_{102}	8	C_{103}	8	d_{101}	3	d_{102}	6	d_{103}	6
C_{111}	9	C_{112}	8	C_{113}	7	d_{111}	4	d_{112}	6	d_{113}	7
C_{121}	9	C_{122}	9	C_{123}	9	d_{121}	4	d_{122}	5	d_{123}	5
C_{131}	10	C_{132}	7	C_{133}	8	d_{131}	4	d_{132}	7	d_{133}	6
C_{141}	8	C_{142}	6	C_{143}	9	d_{141}	6	d_{142}	8	d_{143}	4
C_{151}	7	C_{152}	8	C_{153}	6	d_{151}	7	d_{152}	6	d_{153}	8
C_{161}	6	C_{162}	9	C_{163}	7	d_{161}	9	d_{162}	4	d_{163}	7
C_{171}	7	C_{172}	7	C_{173}	10	d_{171}	7	d_{172}	7	d_{173}	3
C_{181}	8	C_{182}	8	C_{183}	7	d_{181}	6	d_{182}	6	d_{183}	7
C_{191}	9	C_{192}	9	C_{193}	8	d_{191}	5	d_{192}	4	d_{193}	6
C_{201}	9	C_{202}	10	C_{203}	6	d_{11}	3	d_{202}	3	d_{203}	8

was developed to get an optimal solution for selection of COTS components. A case study of manufacturing system is also discussed.

References

1. Gupta, P., Mehlawat, M., Mittal, G., Verma, S.: COTS Selection Using Fuzzy Interactive Approach, pp. 273–289. Springer (2010)
2. Ruiz, M., Ramos, I., Toro, M.: Using Dynamic Model and Simulation to Improve the COTS Software Process, pp. 568–581. Springer, Berlin (2004)
3. Szyperski, C., Pfister, C.: Component-oriented programming: WCOP'96 workshop report (Special issues in object-oriented programming). In: Workshop Reader of the 10th European Conference on Object-Oriented Programming, pp. 127–130 (1996)
4. Belli, F., Jadrzejowich, P.: An approach to reliability optimization of software with redundancy. IEEE Trans. Softw. Eng. **17**(3), 310–312 (1991)
5. Berman, O., Ashrafi, N.: Optimization models for reliability of modular software systems. IEEE Trans. Softw. Eng. **19**(11), 1119–1123 (1993)
6. Berman, O., Kumar, U.D.: Optimization models for recovery block schemes. Eur. J. Oper. Res. **115**, 368–379 (1999)
7. Cortellessa, V., Marinelli, F., Potena, P.: An optimization framework for "build-or-buy" decisions in software architecture. Comput. Oper. Res. (Elsevier Sci.) **35**(10), 3090–3106 (2008)

 8. Gupta, P., Verma, S., Mehlawat, M.K.: A membership function approach for cost-reliability trade-off of COTS selection in fuzzy environment. Int. J. Reliab. Qual. Safety Eng. **18**(6), 573–595 (2011)
 9. Gupta, P., Mehlawat, M.K., Verma, S.: COTS selection using fuzzy interactive approach. Opt. Lett. **6**(2), 273–289 (2012)
10. Jha, P.C., Kapur, P.K., Bali, S., Kumar, U.D.: Optimal component selection of COTS based software system under consensus recovery block scheme incorporating execution time. Int. J. Reliab. Qual. Safety Eng. **17**(3), 209–222 (2010)
11. Kapur, P.K., Bardhan, A.K., Jha, P.C.: Optimal reliability allocation problem for a modular software system (OPSEARCH). J. Oper. Res. Soc. India **40**(2), 133–148 (2003)
12. Kumar, U.D.: Reliability analysis of fault tolerant recovery blocks (OPSEARCH). J. Oper. Res. Soc. India **35**(4), 281–294 (1998)
13. Tang, J., Mu, L.F., Kwong, C.K., Luo, X.: An optimization model for software component under multiple applications development. Eur. J. Oper. Res. **212**, 301–311 (2011)
14. Wu, Z., Kwong, C.K., Tang, J., Chan, J.: Integrated models for software component selection with simultaneous consideration of implementation and verification. Comput. Oper. Res. (Elsevier) **39**, 3376–3393 (2012)
15. Neubauer, T., Stummer, C.: Interactive decision support for multiobjective COTS selection. In: Proceedings of the 40th Annual Hawaii International Conference on System Sciences (HICSS' 07). IEEE (2007)
16. Parsa, S., Bushehrian, O.: A framework to investigate and evaluate genetic clustering algorithms for automatic modularization of software systems. In: Lecture Notes in Computer Science, pp. 699–702 (2004)
17. Britoe Abreu, F., Goulao, M.: Coupling and cohesion as modularization drivers: are we being over-persuaded? In: Proceedings of the 5th European Conference on Software Maintenance and Reengineering (2001)
18. Carlo, G., Mehdi, J., Dino, M.: Fundamentals of Software Engineering. Prentice-Hall, Upper Saddle River (2001)
19. Ian, S.: Software Engineering. Addison-Wesley Longman, Reading (2001)
20. Seker, R., van der Merwe, A.J., Kotze, P., Tanik, M.M., Paul, R.: Assessment of coupling and cohesion for component-based software by using Shannon languages. J. Integr. Des. Process Sci. **8**(4), 33–43 (2004)
21. Kwong, C.K., Mu, L.F., Tang, J.F., Luo, X.G.: Optimization of software components selection for component-based software system development. Comput. Ind. Eng. **58**, 618–624 (2010)
22. Jung, H.W., Choi, B.: Optimization models for quality and cost of modular software systems. Eur. J. Oper. Res. **112**(3), 613–619 (1999)
23. Bellman, R.E., Zadeh, L.A.: Decision-making in a fuzzy environment. Manage. Sci. **17**(4), B141–B164 (1970)
24. Thiriez, H.: OR software LINGO. Eur. J. Oper. Res. **124**, 655–656 (2000)

Some Issues on Choices of Modalities for Multimodal Biometric Systems

Mohammad Imran, Ashok Rao, S. Noushath and G. Hemantha Kumar

Abstract Biometrics-based authentication has advantages over other mechanisms, but there are several variabilities and vulnerabilities that need to be addressed. No single modality or combinations of modalities can be applied universally that is best for all applications. This paper deliberates different combinations of physiological biometric modalities with different levels of fusion. In our experiments, we have selected Face, Palmprint, Finger Knuckle Print, Iris, and Handvein modalities. All the modalities are of image type and publicly available, comprising at least 100 users. Proper selection of modalities for fusion can yield desired level of performance. Through our experiments it is learnt that a multimodal system which is considered just by increasing number of modalities by fusion would not yield the desired level of performance. Many alternate options for increased performance are presented.

Keywords Multimodal · Feature level · Score level · Decision level · Fusion

1 Introduction

During the recent decades, identity theft has been dramatically increasing at an exponential rate. Cyber-crime is difficult to be prevented and traced. Hence, people are exploring reliable and authentic forms of identity security. Techniques that reliably

M. Imran (✉) · G. Hemantha Kumar
DoS in Computer Science, University of Mysore, Mysore, India
e-mail: emraangi@gmail.com

G. Hemantha Kumar
e-mail: ghk.2007@yahoo.com

A. Rao
Freelance Academician, 165, 11th main, S.Puram, Mysore, India
e-mail: ashokrao.mys@gmail.com

S. Noushath
Department of Information Technology, College of Applied Sciences Sohar, Sohar, Oman
e-mail: noushath.soh@cas.edu.om

B. V. Babu et al. (eds.), *Proceedings of the Second International Conference on Soft Computing for Problem Solving (SocProS 2012), December 28–30, 2012*, Advances in Intelligent Systems and Computing 236, DOI: 10.1007/978-81-322-1602-5_151, © Springer India 2014

identify people play a critical part in our everyday social and commercial activities. Within a scheme of access control to secure system, authorized users should be allowed with high levels of precision while unauthorized users should be denied [1]. Example of such an application includes physical access control to a secure facility as in, e-commerce which provides access to computer networking and welfare distribution. The primary task in an identity management system is in determining an individual's identity. Some of the traditional methods are password-based (knowledge) and ID card-based (token) [2]. In case of the password-based identification system most of us use obvious or randomly guessable passwords such as "password" or our pet names which are insecure and hackers can attack easily. The problem with ID card-based system is that it can be lost or stolen; therefore, they are unsuitable for identity verification in the modern world [3]. Only biometrics can solve all these problems by requiring an additional credential which is something associated with the person's own body traits. The advantages of biometrics over the traditional methods are the following: identification of the rightful owner, user convenience, elimination of repudiation claims, difficulty in being copied or forged, enhanced security, as well as the fact that it cannot be lost, forgotten, or transferred [4].

A number of biometric modalities are being used in various applications. Each biometric trait has its own pros and cons and, therefore the choice of a biometric trait for a particular application depends on a variety of issues besides its recognition rate. In general, several factors must be considered to determine the suitability of a physical or a behavioral trait to be used in a biometric application. When it comes to multimodal biometric systems, a number of issues arise such as how many modalities need to be fused to get 100 % GAR? Which are the modalities to be fused? Specifically, which combination of biometric traits gives the best accuracy? These subjective issues are the focus of investigation in this work.

2 Review of Literature

Poh et al. [5] in their study, carried out within the framework of the BioSecure DS2 (Access Control) evaluation campaign, organized by the University of Surrey that involved face, fingerprint and iris biometrics for person authentication, targeting the application of physical access control in a medium-size establishment with 500 persons. While multimodal biometrics is a well-investigated subject in the literature, there exists no benchmark for a fusion algorithm comparison. Working to achieve this objective, they designed two sets of experiments: quality-dependent and cost-sensitive evaluation. Loris Nanni et al. [6] presents a novel trained method for combining biometric matchers at the score level. This method is based on a combination of machine learning classifiers trained using the match scores from different biometric approaches as features. The parameters of a finite Gaussian mixture model are used for modeling the genuine and impostor score densities during the fusion.

Ajay Kumar et al. [7] investigated an information theoretic approach for formulating performance indices for the biometric authentication. Initially, they formulate the constrained capacity, as a performance index for biometric authentication system for the finite number of users. Like Shannon capacity, constrained capacity is formulated using signal-to-noise ratio, which is estimated from known statistics of users' biometric information in the database. Lorene Allano et al. [8] addressed the problem of measuring the dependency of multibiometric system scores using Kolmogorov-Smirnov and Mutual Information criteria and studying the validity of performance evaluation on chimeric persons on the NIST-BSSR1 database. Multibiometric systems can be evaluated on random chimeric persons. It shows that this is not valid for dependent scores and proposed protocol for building cluster-based chimeric persons maintaining the level of dependency between scores. Mingxing He et al. [9] proposed a new robust normalization scheme (Reduction of High-scores Effect normalization) which is derived from the Min–Max normalization scheme. They also show performance of sum rule-based score level fusion and support vector machines (SVM)-based score level fusion. Three biometric traits considered in this are fingerprint, face, and fingervein. Experiments on four different multimodal databases suggest that integrating the proposed scheme, in sum rule-based fusion and SVM-based fusion consistently leads to high accuracy.

Xin Geng et al. [10] proposed context-aware multi-biometric fusion, which can dynamically adapt the fusion rules to the real-time context. As a typical application, the context-aware fusion of gait and face for human identification in video are investigated. Two significant context factors that may affect the relationship between gait and face in the fusion are considered, i.e., view angle and subject-to-camera distance. Fusion methods adaptable to these two factors based on either prior knowledge or machine learning are proposed and tested.

3 Proposed Method

The performances of multimodal biometric systems generally give marginal to significant improvement over a unimodal system. This means some more additional information about that person is available that enable improved recognition. Experience shows that improvement in recognition rate ranges from Marginal to Significant. We use this to define the following [11]:

Complementary information: The information available in a multibiometric system that enables significant improvement in recognition rate over a unimodal system. Supplementary information: The information available in a multibiometric system that enables marginal improvement in recognition rate over a unimodal system [12].

While it is not a rule, it does imply relying on single (to elicit supplementary information) or heterogeneous (multiple) sensors (to elicit complementary information). For example in Images, Raw Image data can be seen as Supplementary information, whereas the texture data of the same can be seen as Complementary information of the same modality [13]. One of the challenges in multibiometric system seems to hinge on using just the required information for a given application. These can also be called as Weak (Supplementary) and Strong (Complementary) modalities [11].

Thus, many situations warrant a single sensor from which the supplementary information may be just adequate to get the desired FAR/FRR and it will be a low cost, easy to deploy system. The other extreme is the case of using multiple sensors to get enough Complementary information to get the desired FAR/FRR. These would be complex, expensive, extremely secure access situations like access to Nuclear Reactor site/Defense Laboratories, etc. These are difficult to deploy everywhere solutions [14].

Much of the current work in multimodal are based on fixed number of modalities since it is easy to adopt and yet flexible. There are some issues in multimodal which are still to be answered:

1. How many modalities are needed from the user to identify 100% correctly?
2. How many acquisition devices we need, to collect the user modalities?
3. Does same feature extractor works on all the modalities which we have taken from user, since some trait performs well to local features (ex: Fingerprint, Iris etc.) others perform well to global features (ex: Face, Palmprint etc.)
4. What mix and in what sequence the modalities that provide Complementary (strong biometric) or Supplementary (weak biometric) are needed to achieve a particular level of FRR/FAR
5. How to solve the curse of dimensionality problem especially at feature level fusion? Particularly so in multibiometric situation?

And some general issues like cost of deployment, enrollment time, throughput time, expected error rate, user habituation, etc., these issues really need to be solved in multimodal biometric systems. We will be answering some of these in this paper.

3.1 Feature Extraction

A number of feature extraction algorithms for biometrics have been used in research for identification like PCA, KICA, Gabor filter, SIFT features, etc. Every feature extraction algorithm has its own advantages and disadvantages depending on its usage on biometric modality. In our work, we have organized the vast range of biometric feature extraction algorithms into two different levels of features: (a) Appearance based and (b) Textures based, feature extraction algorithms.

In appearance-based feature extraction algorithms, we have used Principal Component Analysis (PCA), Linear Discriminant Analysis (LDA), Locality Preserving Projections (LPP), and Independent Component Analysis1 (ICA1) [15]. In texture-based feature extraction algorithms, we have used Local Binary Patterns Variance (LBPV), Local Phase Quantization (LPQ) and Gabor [16]. The above-mentioned feature extraction algorithms are well known to all and a lot of research has been done on extracting the features using these efficiently. Our main aim is to represent each modality in the optimal and informative form which is invariant to the deformations that are unavoidably present during extraction of these features.

3.2 Biometrics Fusion Strategies

Fusion of biometric systems, algorithms and/or traits is a well-known solution to improve authentication performance of biometric systems. Researchers have shown that multi-biometrics, i.e., fusion of multiple biometric evidences, enhances the recognition performance [12]. In biometric systems; fusion can be performed at different levels; Sensor Level, Feature Level, Score Level, and Decision Level Fusion [4].

Sensor Level Fusion: Entails the consolidation of evidence presented by multiple sources of raw data before they are subjected to feature extraction. Sensor level fusion can benefit multi-sample systems which capture multiple snapshots of the same biometric.

Feature Level Fusion: In feature-level fusion, the feature sets originating from multiple biometric algorithms are consolidated into a single feature set by the application of appropriate feature normalization, transformation, and reduction schemes. The primary benefit of feature-level fusion is the detection of correlated feature values generated by different biometric algorithms and, in the process, identifying a salient set of features that can improve recognition accuracy [14].

Score Level Fusion: The match scores output by multiple biometric matchers are combined to generate a new match score (a scalar).

Decision Level Fusion: Fusion is carried out at the abstract or decision level when only final decisions are available, this is the only available fusion strategy which fuses the output decision of matcher/classifier.

4 Experimental Results

The main objective of an evaluation is to provide consequences of level of fusion under different strategies of multimodal system. We have chosen Face, Palmprint (Pp), Handvein (Hv), and Finger Knuckle Print (FKP) modalities [17–20] and its corresponding best feature extraction algorithm which gives good performance. In all of our experiments, performance of levels of fusion is measured in terms of False Acceptance Rate (FAR) at values 0.01, 0.1 and 1 %, its related values of Genuine Acceptance Rate (GAR in %). First, we measure the performance of unimodal biometric system. Further, we evaluate the results for multimodal biometric system; Results obtained from all experiments are tabulated.

4.1 Performance Evaluation of Unimodal Systems

We have considered Face, Palmprint (Pp), Handvein (Hv), and Finger Knuckle Print (FKP) as physiological modalities. The performance analyses of the above-mentioned physiological modalities are obtained and tabulated.

Table 1 shows the performance of face modality with respect to appearance- and texture-based algorithms. Among distinct appearance-based feature extraction an algorithm, ICA1, performs relatively well on the face modality with highest values of GAR% for different values FAR%. The LDA also performs better for face modality compared to other feature extraction algorithms like LPP and PCA. The PCA gives lowest GAR% for different values of FAR%; it performs poorly among the appearance based algorithms. From Table 1, it can be observed that LPQ feature extraction algorithm gives the best performance on the face modality with highest GAR% for distinct values of FAR 0.1 and 1%. Typically, Gabor performs healthier for the face modality compared to appearance-based feature extraction method. The LBPV feature extraction algorithm performs worst compared to all other feature extraction algorithms, with lowest GAR% for different values of FAR%.

Performance of palmprint with respect to both appearance and texture based algorithms are shown in Table 2, ICA1 feature extraction algorithm gives the best performance to the palmprint modality with highest GAR% for distinct values of FAR 0.1 and 1%. Typically, the LDA also performs healthier for the palmprint modality in comparison to other appearance-based feature extraction methods. The LPP feature extraction algorithm performs worst among other appearance-based methods, with lowest GAR% for different values of FAR%. The performance of Palmprint modality with respect to texture-based algorithms. It can be seen from Table 2 that, among distinct texture-based feature extraction algorithms, LPQ performs relatively well to the Palmprint modality with highest values GAR% for different values of FAR%. The Gabor also performs best for Palmprint modality compared to other feature extraction algorithms.

The performance of appearance- and texture-based algorithms on the handvein modality is summarized in Table 3. It can be seen that PCA feature extraction method

Table 1 Performance of face modality on different appearance and texture-based feature extraction algorithms

Modality	Feature extraction algorithms	GAR%		
		0.01 %FAR	0.1 %FAR	1 %FAR
Face	PCA	26.0	40.5	56.0
	LDA	26.0	69.5	91.0
	LPP	28.5	46.0	73.0
	ICA1	67.5	81.0	91.5
	LBPV	7.5	9.0	28.5
	Gabor	48.5	65.0	80.5
	LPQ	32.0	80.5	92.0

Table 2 Performance of palmprint modality on different appearance- and texture-based feature extraction algorithms

Modality	Feature extraction algorithms	GAR%		
		0.01 %FAR	0.1 %FAR	1 %FAR
Palmprint	PCA	27.5	52.0	76.5
	LDA	–	58.5	69.5
	LPP	46.5	50.0	62.0
	ICA1	–	65.5	78.0
	LBPV	27.0	42.0	78.5
	Gabor	86.0	87.5	95.5
	LPQ	61.7	55.5	72.5

Table 3 Performance of handvein modality on different appearance- and texture-based feature extraction algorithms

Modality	Feature extraction algorithms	GAR%		
		0.01 %FAR	0.1 %FAR	1 %FAR
Handvein	PCA	21.5	34.0	54.0
	LDA	13.0	16.0	30.0
	LPP	3.0	6.0	13.5
	ICA1	32.5	46.0	60.5
	LBPV	5.0	17.0	37.0
	Gabor	33.5	54.0	74.5
	LPQ	26.5	38.0	55.0

performs slightly better than other methods with highest GAR% for different values of FAR%, but the LPP gives worst performance among all methods for handvein. It can be observed in Table 3 that Gabor feature extraction method performs better than other methods with highest GAR% for different values of FAR%, but the LBPV gives worst performance among all texture-based algorithms for handvein.

Table 4 shows the performance of various appearance and texture-based methods on the FKP. Initially at 0.01 and 0.1 %FAR LDA feature extraction algorithm outperforms the handvein modality with highest GAR%. The ICA1 also performs better for FKP modality among other feature extraction algorithms at 1 %FAR. LPP feature extraction algorithm underperforms with lowest GAR% for different values of FAR%. From the results listed in Table 4 that LPQ feature extraction algorithm outperforms on the handvein modality with highest GAR%. The Gabor also performs better for FKP modality among other feature extraction algorithms at 1 %FAR. LBPV feature extraction algorithm underperforms with lowest GAR% for different values of FAR%.

Table 4 Performance of FKP modality on different appearance and texture based feature extraction algorithms

Modality	Feature extraction algorithms	GAR%		
		0.01 %FAR	0.1 %FAR	1 %FAR
FKP	PCA	50.0	69.5	84.0
	LDA	73.5	78.0	85.0
	LPP	55.0	61.5	74.5
	ICA1	73.0	77.5	86.0
	LBPV	10.0	19.5	40.0
	Gabor	83.0	86.0	90.5
	LPQ	84.0	88.5	92.0

4.2 Performance Evaluation of Multimodal Systems

In this section, we have emphasized on comparative analysis of different levels of fusion in multimodal approach on fusion of two, three, and four modalities. The comparison is on sensor, feature, score, and decision level fusion with their relevant fusion rules. Each of the performance is measured in terms of False Acceptance Rate (FAR) at values 0.01, 0.1, 1 %, and its related values of Genuine Acceptance Rate (GAR in %) have been tabulated and discussed in details.

Table 5 shows the different levels of fusion and its strategies on fusion of Face and Palmprint modalities. When we compare each level of fusion, the sum rule in score level fusion outperforms with top GAR% for distinct values of FAR%. Further, the OR rule of decision level fusion perform relatively better; the sensor level fusion under performs with least GAR% for different values of FAR%. At feature level fusion, normalization rules such as, Min–max, Z-score, Median, and

Table 5 Comparative analysis on performance of different levels of fusion with different rules on fusion of Face (ICA1) and Palmprint (LPQ) in multimodal systems

Fusion	Rules	GAR%		
		0.01 %FAR	0.1 %FAR	1 %FAR
Sensor level	Wavelet based	35.0	48.5	70.5
Feature level	Min–Max	91.0	94.0	95.0
	Z-score	89.5	93.5	98.5
	Tanh	91.0	94.0	95.0
	Median	89.5	93.5	98.5
Score level	Min	85.0	89.5	97.0
	Max	93.8	94.5	99.0
	Sum	95.0	99.0	100
Decision level	OR	94.0	94.5	99.5
	AND	85.0	89.0	97.0

Table 6 Comparative analysis on performance of different levels of fusion with different rules on fusion of Face (ICA1), Palmprint (LPQ) and FKP (LPQ) in multimodal systems

Fusion	Rules	GAR%		
		0.01 %FAR	0.1 %FAR	1 %FAR
Sensor level	Wavelet based	16.0	39.5	65.5
Feature level	Min–Max	92.0	93.5	97.0
	Z-score	92.5	97.0	99.0
	Tanh	92.5	96.5	97.5
	Median	91.0	96.5	97.5
Score level	Min	95.0	89.5	93.5
	Max	99.5	94.5	99.5
	Sum	99.5	99.0	100
Decision level	OR	95.5	96.0	99.5
	AND	16.0	39.5	65.5

Tanh perform equally well. Hence, all levels of fusion which we have considered perform well compared to their unimodal case, except the sensor level fusion in multimodal approach on fusion Face and Palmprint.

A performance assessment of the results on different levels of fusion of three modalities namely Face, Palmprint, and Finger Knuckle Print is presented in Table 6. The result for each levels of fusion is as follows: at feature level fusion the normalization rules Min–Max, Z-score, Median, and Tanh achieve healthy GAR% for distinct values FAR%. The sum rule of score level fusion gives among the best GAR% for different values of FAR 0.01, 0.1 and 1% are 99.5, 99 and 100% respectively. The other rules of score level fusion perform equally better compared to their prior version of fusion of two modalities. At Decision level fusion OR and AND rules also perform well. Sensor level fusion is the only one fusion method which performs poorest amongst all the levels of fusion and also compared to its earlier fusion of two modalities.

Table 7 summarizes the overall performances of different levels of fusion and its fusion rules on fusion of four modalities namely Face, Palmprint, FKP, and Handvein. Each level of fusion can be compared in Table 7, wavelet base image fusion rule of sensor level is the only fusion scheme which is most abominable on fusion of the above-mentioned modalities. The other fusion schemes are as follows: at feature level there is no significant improvement in performance when compared to its previous fusion of three modalities with respect to its normalization schemes explicitly, Min–Max, Z-score, Median, and Tanh. The score level fusion sum rule achieves best GAR% of 99.5, 100, and 100 with distinct values of FAR% of 0.01, 0.1, and 1 respectively. The other rules of score level fusion, viz., Min and Max rules perform better for GAR% at values FAR% of 0.1 and 1. In decision level fusion both AND and OR rules execute less GAR% for different values of FAR% compared to its previous case fusion of three modalities.

Table 7 Comparative analysis on performance of different levels of fusion with different rules on fusion of Face (ICA1), Palmprint (LPQ), FKP (LPQ), and Handvein (Gabor) in multimodal systems

Fusion	Rules	GAR%		
		0.01 %FAR	0.1 %FAR	1 %FAR
Sensor level	Wavelet based	11.0	20.0	43.0
Feature level	Min–Max	92.0	96.5	97.0
	Z-score	92.5	97.0	99.0
	Tanh	92.5	96.5	97.5
	Median	92.5	96.5	97.5
Score level	Min	87.0	92.0	95.0
	Max	93.8	94.5	99.0
	Sum	99.5	100	100
Decision Level	OR	94.5	95.0	99.5
	AND	87.5	91.5	93.5

5 Conclusion

From the analysis of experimental results and observations one can conclude:

1. Choice of modality is application and context dependent.
2. Performance of the feature extraction algorithm is significantly subjective on the choice of biometric modality.
3. Although all the chosen modalities are of image type (Face, Palmprint, Finger Knuckle Print, and Handvein), the performance variation is quite significant. The worst being (handvein) and the best being (Finger Knuckle Print) in unimodal case.
4. While improved performance is available with increasing additional modalities, their choice and number are very critical.
5. One observes significant change from unimodal to biomodal case. However, further addition of modalities does not yield comparable improvement. There is tendency toward saturation with every additional modality.
6. Performance of multimodal system depends on the level of fusion.
7. In general, for two or more modalities under fusion the performance in decreasing order is as follows: score level, feature level, decision level, and sensor level.
8. Too much is too bad: It is necessary to recognize that, by simply increasing the number of modalities to be fused to get a desired level of performance is poor strategy. A proper selection of lesser number of modalities efficient feature extraction, levels of fusion and fusion rules all constitute to desire level of performance, even if the modalities are few.

Acknowledgments The research leading to these results has received Research Project Grant Funding from the Research Council of the Sultanate of Oman Research Grant Agreement No [ORG MoHE ICT 10 023].

References

1. Nandakumar, K., Jain, A.K., Ross, A.: Introduction to Biometrics. Springer, New York (2011)
2. Ross, A., Jain, A.K., Prabhakar, S.: An introduction to biometric recognition. IEEE Trans. Circuits Syst. Video Technol. **14**, 44–48 (2004)
3. Jain, A.K., Flynn, P.J., Ross, A.: Handbook of Biometrics. Springer, New York (2008)
4. Kumar, A., Zhang, D.: Combining fingerprint, palmprint and hand-shape for user authentication. ICPR **4**, 549–552 (2006)
5. Poh, N., Bourlai, T., Kittler, J., Allano, L., Alonso-Fernandez, F., Ambekar, O., Baker, J.P., Dorizzi, B., Fatukasi, O., Fiérrez-Aguilar, J., Ganster, H., Ortega-Garcia, J., Maurer, D.E., Ali Salah, A., Scheidat, T., Vielhauer, C.: Benchmarking quality-dependent and cost-sensitive score-level multimodal biometric fusion algorithms. IEEE Trans. Inf. Forensics Sec. **4**(4), 849–866 (2009)
6. Nanni, L.: Lumini, Alessandra, Brahnam, Sheryl: likelihood ratio based features for a trained biometric score fusion. Expert Syst. Appl. **38**(1), 58–63 (2011)
7. Bhatnagar, J., Kumar, A.: On estimating performance indices for biometric identification. Pattern Recognit. **42**(9), 1803–1815 (2009)
8. Allano, L., Dorizzi, B., Garcia-Salicetti, S.: A new protocol for multi-biometric systems' evaluation maintaining the dependencies between biometric scores. Pattern Recognit. **45**(1), 119–127 (2012)
9. He, M., Horng, S.-J., Fan, P., Run, R.-S., Chen, R.-J., Lai, J.-L.: Muhammad Khurram Khan, Kevin Octavius Sentosa: Performance evaluation of score level fusion in multimodal biometric systems. Pattern Recognit. **43**(5), 1789–1800 (2010)
10. Geng, X.: Smith-Miles, K., Wang, L., Li, M., Qiang, W.: Context-aware fusion: a case study on fusion of gait and face for human identification in video. Pattern Recognit. **43**(10), 3660–3673 (2010)
11. Imran, M.., Rao, A., Hemanthakumar, G.: Extreme subjectivity of multimodal biometrics solutions: role of strong and weak modalities/features information. In: Proceedings of IICAI 2011, pp. 1587–160
12. Kittler, J., Fumera, G., Roli, F., Muntoni, D.: An Experimental Comparison of Classifer Fusion Rules for Multimodal Personal Identity Verification System. Springer, Berlin (2002)
13. Yang, F., Ma, B.: A new mixed-mode biometrics information fusion based-on fingerprint, hand-geometry and palm-print. Fourth International Conference on Image and Graphics, 2007, ICIG 2007, 22–24 Aug 2007, pp. 689–693
14. Sim, T., Zhang, S., Janakiraman, R., Kumar, S.: Continuous verification using multimodal biometrics. IEEE Trans. Pattern Anal. Mach. Intell. **29**(4), 687–700 (April 2007)
15. Rao, A., Noushath, S.: Subspace methods for face recognition. Comput. Sci. Rev. **4**(1), 1–17 (2010)
16. Ojansivu, V., Heikkila, J.: Blur insensitive texture classification using local phase quantization. In: ICISP, pp. 236, 243 (2008)
17. Zhu, L.-q., Zhang, S.-y.: Multimodal biometric identification system based on finger geometry, knuckle print and palm print. Pattern Recogn. Lett. **31**(12), 1641–1649 (2010)
18. http://rvl1.ecn.purdue.edu/~aleix/aleix_face_DB.html
19. www.comp.polyu.edu.hk/~biometrics
20. Heenaye-Mamode Khan, M., Subramanian, R.K., Mamode Khan, N.A.: Low dimensional representation of dorsal hand vein features using principle component analysis (pca). World Academy of Science, Engineering and Technology, vol. 49, pp. 1001–1007 (2009)

An Adaptive Iterative PCA-SVM Based Technique for Dimensionality Reduction to Support Fast Mining of Leukemia Data

Vikrant Sabnis and Neelu Khare

Abstract Primary Goal of a Data mining technique is to detect and classify the data from a large data set without compromising the speed of the process. Data mining is the process of extracting patterns from a large dataset. Therefore the pattern discovery and mining are often time consuming. In any data pattern, a data is represented by several columns called the linear low dimensions. But the data identity does not equally depend upon each of these dimensions. Therefore scanning and processing the entire dataset for every query not only reduces the efficiency of the algorithm but at the same time minimizes the speed of processing. This can be solved significantly by identifying the intrinsic dimensionality of the data and applying the classification on the dataset corresponding to the intrinsic dataset only. Several algorithms have been proposed for identifying the intrinsic data dimensions and reducing the same. Once the dimension of the data is reduced, it affects the classification rate and classification rate may drop due to reduction in number of data points for decision. In this work we propose a unique technique for classifying the leukemia data by identifying and reducing the dimension of the training or knowledge dataset using Iterative process of Intrinsic dimensionality discovery and reduction using Principal Components Analysis (PCA) technique. Further the optimized data set is used to classify the given data using Support Vector Machines (SVM) classification. Results show that the proposed technique performs much better in terms of obtaining optimized data set and classification accuracy.

V. Sabnis (✉)
Maulana Azad National Institute of Technology, Bhopal, India
e-mail: vikrant_sabnis@rediffmail.com

N. Khare
VIT University Vellore, Vellore, Tamilnadu, India
e-mail: neelu.khare@vit.ac.in

B. V. Babu et al. (eds.), *Proceedings of the Second International Conference on Soft Computing for Problem Solving (SocProS 2012), December 28–30, 2012*, Advances in Intelligent Systems and Computing 236, DOI: 10.1007/978-81-322-1602-5_152, © Springer India 2014

Keywords Principle component analysis · Support vector machine · Dimensionality reduction · Eigen value · Local PCA

1 Introduction

Leukemia [8] is a type of cancer of the blood or bone marrow characterized by abnormal increase of white blood cells. Leukemia is a broad term covering a spectrum of diseases. It is generally detected from the blood cell structure and is the cause of significant increase in WBC (white blood Cells). It is generally categorized into two broader categories: Chronic and Acute. In this work we classify a given pattern into one of the acute Leukemia types, i.e., Acute lymphoblastic leukemia (ALL) and Acute myelogenous leukemia (AML). We use dataset of Leukemia provided by [7] for data training and classification.

The dataset has 7,200 dimensions for each pattern. Mining through entire dataset with such a huge number of dimension not only consumes time but at the same time reduces the efficiency of the classification to a great deal due to correlated and irrelevant data in the training and classification process.

Therefore, we adopt dimensionality reduction to mark the optimum dimensions and columns such that training and classification is fast and accurate.

One of the most common approach of representing the underneath pattern in any dataset is through certain kernels which are linear or nonlinear machines that can represent a dataset as a set of functional models and model coefficients [5]. The kernel techniques are nothing but mapping low-dimensional data with many columns into a high-dimensional feature space with fewer columns which are essentially model parameters for these functions. The higher the accuracy desired for patterns the higher would be dimensional plane of representation.

A unique way to efficiently deal with highly dense dataset is to reduce the number of dimensions obtained through a kernel through a mapping function such that the low dimensions retain the same information as their high-dimensional counterpart. The low dimensionality brings down computational complexity as well as spatial requirement. The concept is represented by Fig. 1.

High-dimensional data are difficult to visualize and therefore deriving an appropriate distance measure between two points in high dimensional datasets are difficult. For example, the distance between two points in x-y plane can be easily obtained through Euclidean distance but the distance between the same points over x, y and z plane are difficult to represent. This anomaly stands through for any type of dimensions representing any type of features. Therefore, a distance measure in the high dimensional data space stands a risk of returning a null set. The solution to the problem is known as multidimensional scaling [5] which is nothing but dimension reduction technique. The dimensionality reduction technique is further divided into two types (1) Linear mapping (2) Nonlinear mapping. A linear mapping is

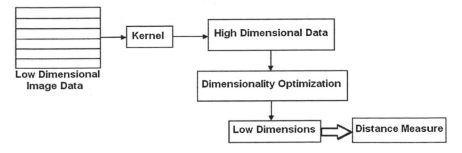

Fig. 1 Dimensionality mapping from high to low through Kernel-based technique

a method of transforming data objects from one dimension to another dimension through a linear quadratic equation or a direct mapping function or some look up tables. Real-time objects and the data are rarely bounded by linearity measures. Had it been so, there would not have any requirement for enhancing the dimension itself. Therefore, we assume that dimensional scaling or reduction is a nonlinear approach.

2 Methodology

2.1 Overall Technique

Identified intrinsic dataset and the reduced data do not always reciprocate the actual data behavior. Therefore, we use an iterative process to first dividing the entire dataset into two hypothetical parts. We use a part as training and another as testing. We calculate the intrinsic dimension of the training data and minimize the dataset using PCA to number of intrinsic dimension obtained. Now we apply the classification to the test dataset. The efficiency is calculated. If the efficiency is bellowing the desired threshold of 90%, we increase the dimension in the step of 1. This process is continued till the desired accuracy is obtained. Once the accuracy is obtained, we increase the hypothetical training data by 50% by migrating 50% of the test data and measure the efficiency once more. This process is continued till the optimum solution is reached. The iteration is stooped for optimum solution and the parameters are globally accepted as the trusted parameters for the classification process.

The training dataset is reduced with the number of dimension achieved through the above process and test data is classified against the reduced data.

2.2 MLE-Based Intrinsic Dimensionality Identification

The existing approaches to estimating the intrinsic dimension can be roughly divided into two groups: eigenvalue or projection methods, and geometric methods. Eigen value methods, from the early proposal of to a recent variant are based on a global or local PCA, with intrinsic dimension determined by the number of eigenvalues greater than a given threshold. Global PCA methods fail on nonlinear manifolds, and local methods depend heavily on the precise choice of local regions and thresholds. The eigenvalue methods may be a good tool for exploratory data analysis, where one might plot the eigenvalues and look for a clear-cut boundary, but not for providing reliable estimates of intrinsic dimension.

We first estimate intrinsic dimension of a dataset derived by applying the principle of maximum likelihood to the distances between close neighbors. We derive the estimator by a Poisson process approximation, assess its bias and variance theoretically and by simulations, and apply it to a number of simulated and real datasets. We also show it has the best overall performance compared with two other intrinsic dimension estimators.

Here, we derive the maximum likelihood estimator (MLE) of the dimension m from i.i.d. observations $X_1, ..., X_n$ in $\mathbb{X}^p$. The observations represent an embedding of a lower dimensional sample, i.e., $X_i = g(Y_i)$, where Y_i are sampled from an unknown smooth density f on $\mathbb{R}^m$, with unknown $m \leq p$, and g is a continuous and sufficiently smooth (but not necessarily globally isometric) mapping. This assumption ensures that close neighbors in $\mathbb{R}^m$ are mapped to close neighbors in the embedding.

The basic idea is to fix a point x, assume $f(x) \approx$ const in a small sphere $S_x(R)$ of radius R around x, and treat the observations as a homogeneous Poisson process in $S_x(R)$. Consider the inhomogeneous process $\{N(t, x), 0 \leq t \leq R\}$,

$$N(t, x) = \sum_{i=1}^{n} 1 \{X_i \in S_x(t)\} \tag{1}$$

which counts observations within distance t from x. Approximating this binomial (fixed n) process by a Poisson process and suppressing the dependence on x for now, we can write the rate $\lambda(t)$ of the process $N(t)$ as;

$$\lambda(t) = f(x)V(m)mt^{m-1} \tag{2}$$

This follows immediately from the Poisson process properties since $V(m)mt^{m-1}$ $= \dfrac{d}{dt}\left[V(m)t^m\right]$ is the surface area of the sphere $S_x(t)$. Letting $\theta = \log f(x)$, we can write the log-likelihood of the observed process $N(t)$ as

$$L(m, \theta) = \int_0^R \log \lambda(t) dN(t) - \int_0^R \lambda(t) dt \tag{3}$$

This is an exponential family for which MLEs exist with probability $\to 1$ as $n \to \infty$ and are unique. The MLEs must satisfy the likelihood equations

$$\frac{\partial L}{\partial \theta} = \int_0^R dN(t) - \int_0^R \lambda(t) dt = N(R) - e^\theta V(m) R^m = 0 \tag{4}$$

$$\frac{\partial L}{\partial m} = \left(\frac{1}{m} + \frac{V'(m)}{V(m)} \right) N(R) + \int_0^R \log t \, dN(t) - e^\theta V(m) R^m \left(\log R + \frac{V'(m)}{V(m)} \right) = 0 \tag{5}$$

Substituting (5) in (6) gives the MLE for m:

$$\hat{m}_R(x) = \left[\frac{1}{N(R, x)} \sum_{j=1}^{N(R,x)} \log \frac{R}{T_j(x)} \right]^{-1} \tag{6}$$

In practice, it may be more convenient to fix the number of neighbors k rather than the radius of the sphere R. Then the estimate in (7) becomes

$$\hat{m}_k(x) = \left[\frac{1}{k-1} \sum_{j=1}^{k-1} \log \frac{T_k(x)}{T_j(x)} \right]^{-1} \tag{7}$$

Note that we omit the last (zero) term in the sum in (6) and divide by $k - 1$ rather than k since that makes the estimator approximately unbiased, as we show below. Also note that the MLE of θ can be used to obtain an instant estimate of the entropy of f.

2.3 PCA-Based Dimension Reduction

Once the exact hidden dimension is obtained through maximum likelihood intrinsic dimensionality finding technique, the data needs to be reduced using PCA.

One of the most common forms of dimensionality reduction is PCA [6]. Given a set of data, PCA finds the linear lower dimensional representation of the data such that the variance of the reconstructed data is preserved. Intuitively, PCA finds a low-dimensional hyperplane such that, when we project our data onto the hyperplane, the variance of our data is changed as little as possible. A transformation that preserves variance seems appealing because it will maximally preserve our ability to distinguish between beliefs that are far apart in Euclidean norm. As we will see below, however, Euclidean norm is not the most appropriate way to measure distance between beliefs when our goal is to preserve the ability to choose good actions.

We first assume we have a data set of n beliefs $\{b_1, \ldots, b_n\} \in \mathbb{B}$, where each belief b_i is in $\mathbb{B}$, the high-dimensional belief space. We write these beliefs as column vectors in a matrix $B = [b_1 | \ldots | b_n]$, where $B \in \mathbb{R}^{|S| \times n}$. We use PCA to compute a low-dimensional representation of the beliefs by factoring B into the matrices U and $\tilde{B}$,

$$B = U \tilde{B}^T. \tag{8}$$

In Eq. (8), $U \in \mathbb{R}^{|S| \times l}$ corresponds to a matrix of bases that span the low-dimensional space of $l < |S|$ dimensions. $\tilde{B} \in \mathbb{R}^{n \times l}$ represents the data in the low-dimensional space. From a geometric perspective, U comprises a set of bases that span a hyperplane $\tilde{B}$ in the high-dimensional space of $\mathbb{B}$; $\tilde{B}$ are the co-ordinates of the data on that hyperplane. If no hyperplane of dimensionality l exists that contains the data exactly, PCA will find the surface of the given dimensionality that best preserves the variance of the data, after projecting the data onto that hyperplane and then reconstructing it. Minimizing the change in variance between the original data B and its reconstruction $U \tilde{B}^T$ is equivalent to minimizing the sum of squared error loss:

$$L(B, U, \tilde{B}) = \left\| B - U \tilde{B}^T \right\|_F^2 \tag{9}$$

2.4 Pre-Classification

Once the training or the knowledge data is reduced, the test dataset is classified. Now most important thing to notice here is that once a dataset is reduced to high dimension or fewer columns, these columns are no more linearly associated with the exact data values. Therefore, number of columns in the test data and the actual knowledge data and their dimensions will also be different. Therefore a normal data cannot be classified against a reduced dataset.

Therefore first the given test data needs to be considered as the part of the knowledge data and the entire set of knowledge data and one test data is subjected to dimensionality reduction as a single set. Once the reductionality is achieved, the test data is separated and is classified.

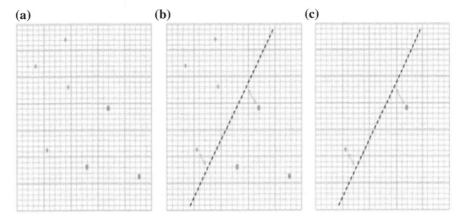

Fig. 2 Support vector machine as a maximum margin classifier

2.5 Support Vector Machine as Classification

SVMs is a supervised learning method which can be used for both classification and regression. In its simplest form, given two set of data points with same dimensions, SVM can form a decision model so that a new set of points can be classified as either of the two input points. If the examples are represented as points in space, a linear SVM model can be interpreted as a division of this space so that the examples belonging to separate categories are divided by a clear gap that is as wide as possible. New examples are then predicted to belong to a category based on which side of the gap they fall on (as explained in Fig. 2).

The (A) shows a decision problem for two classes, blue and green. (B) Shows the hyperplane which has the largest distance to the nearest training data points of each class. (C) Shows that only two data points are needed to define this hyperplane. Those will be taken as support vectors, and will be used to guide the decision process of new input data which needs to be classified.

A linear support vector machine is composed of a set of given support vectors **z** and a set of weights **w**. The computation for the output of a given SVM with N support vectors $z_1, z_2, \ldots, z_N$ and weights $w_1, w_2, \ldots, w_N$ is then given by:

$$F(x) = \sum_{i=1}^{N} w_i \langle z_i, x \rangle + b \qquad (10)$$

A decision function is then applied to transform this output in a binary decision. Usually, sign (.) is used, so that outputs greater than zero are taken as a class and outputs lesser than zero are taken as the other.

3 Results and Discussion

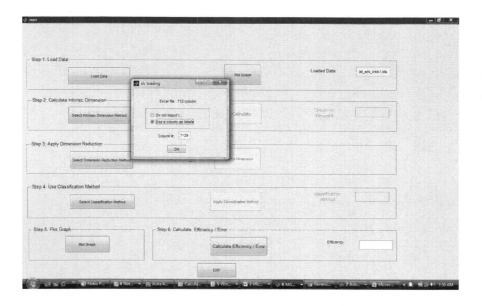

Loading of dataset with huge number of columns

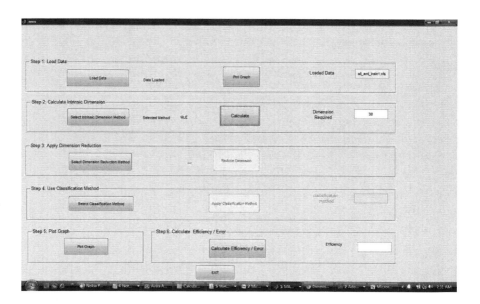

Intrinsic Dimension Obtained by MLE. 7,200 column of data is reduced to 30 Columns

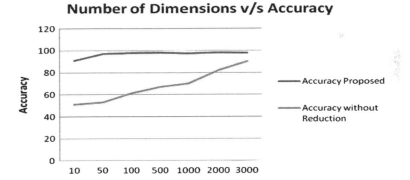

Classification Step

4 Graphs

Figure shows that higher number of dimensions results in better accuracy only in the case of linear mapping where numbers of dimensions are linearly dependent on each other. But once the data is reduced by a nonlinear mapping fewer number of data has as much accuracy as higher number of dataset.

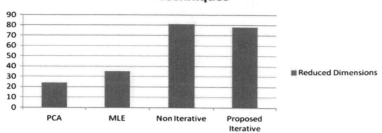

Shows that PCA and MLE finds out the optimum dimensions in a dataset. But the efficiency of the system for both is extremely low (69 %). Hence it can be concluded that though PCA and MLE together can resolve the intrinsic dimensionality of the data, they cannot identify the dimensionality which will result in better accuracy.

5 Conclusion

Dimensionality reduction in data mining is not a new task. Conventionally various techniques are being proposed toward this direction that reduces the data before processing. But in features data such as any medical patterns, dimensionality reduction may result in loss of accuracy in overall mining process due to the inherent relationship among the data set. Applying dimension reductionality must be combined with appropriate recognition rate in order to achieve good accuracy. The technique finds out the intrinsic dimensionality based on the MLE and by checking the efficiency of the system at each iteration. Therefore, the method not only succeeds to minimize the dimensionality for speeding up the processing but also preserves the accuracy of the recognition. A medical database was chosen to prove the theory due to sensitiveness of the data in such a database. Results show significant performance consistency over different iterations. Results also show that for optimum intrinsic dimensionality, the accuracy is better than nonreduced dataset. But the experiments also shows that due to nonlinear mapping of the dataset, linear kernels cannot be applied and polynomial kernels with appropriate optimization is better suited for the task. Another major observation was that the test vectors independently cannot be used to derive the data dependency and therefore needs to be processed with the knowledge data or the training vectors in order to acquire the dimensions of the test vectors also. Therefore a future work can be designed to obtain the actual mapping function for the knowledge data which can then be used to extract the needed dimensions from the test set without test set being simulated with the knowledge base for extracting the intrinsic dimensions of the test set.

References

1. http://en.wikipedia.org/wiki/Leukemia
2. http://en.wikipedia.org/wiki/Microarray_databases
3. http://www.ncbi.nlm.nih.gov/geo/
4. Jing, L., Shuzhong, L., Ming, L., Jianyun, N.: Application of dimensionality reduction analysis to fingerprint recognition. In: Proceedings of 2008 International Symposium on Computational Intelligence and Design, iscid, vol. 2, pp. 102–105 (2008)
5. Lespinats, S., Verleysen, M., Giron, A., Fertil, G.: DD-HDS: a method for visualization and exploration of high-dimensional data. IEEE Trans. Neural Netw. **18**(5), 1265–1279 (2007)
6. Segall, R. S., Pierce, R. M.: Data mining of Leukemia cells using self-organized maps. In: Proceedings of 2009 ALAR Conference on Applied Research in Information Technology, 13 February (2009)
7. Segall, R. S.: Data mining of microarray databases for the analysis of environmental factors on corn and maize. In: Proceedings of the 2005 Conference of Applied Research in Information Technology, Sponsored by Acxiom Laboratory for Applied Research (ALAR), University of Central Arkansas, 18 February (2005)
8. Segall, R.S.: Data mining of microarray databases for the analysis of environmental factors on plants using cluster analysis and predictive regression. In: Proceedings of the Thirty-sixth Annual Conference of the Southwest Decision Sciences Institute, vol. 36, no. 1, Dallas, TX, 3–5 March (2005)

Social Evolution: An Evolutionary Algorithm Inspired by Human Interactions

R. S. Pavithr and Gursaran

Abstract Inherent intelligent characteristics of humans, such as human interactions and information exchanges enable them to evolve more rapidly than any other species on the earth. Human interactions are generally selective and are free to explore randomly based on the individual bias. When the interactions are indecisive, individuals consult for second opinion to further evaluate the indecisive interaction before adopting the change to emerge and evolve. Inspired by such human properties, in this paper a novel social evolution (SE) algorithm is proposed and tested on four numerical test functions to ascertain the performance by comparing the results with the state-of-the-art soft computing techniques on standard performance metrics. The results indicate that, the performance of SE algorithm is better than or quite comparable to the state-of-the-art nature inspired algorithms.

Keywords Society and civilization · Social evolution · optimization

1 Introduction

Nature inspired computing is one of the main branches of natural computing techniques and an emerging computational paradigm for solving large-scale complex and dynamic real-world problems. Nature inspired computing builds on the principles of emergence, self-organization, and complex systems [1]. One of the main objectives of the nature inspired computing paradigm is to provide alternative stochastic, nature inspired search-based techniques to problems that have not been

R. S. Pavithr (✉) · Gursaran
Dayalbagh Educational Institute, Dayalbagh, Agra, India
e-mail: rspavithr@ieee.org

Gursaran
e-mail: gursaran.db@gmail.com

B. V. Babu et al. (eds.), *Proceedings of the Second International Conference on Soft Computing for Problem Solving (SocProS 2012), December 28–30, 2012*, Advances in Intelligent Systems and Computing 236, DOI: 10.1007/978-81-322-1602-5_153, © Springer India 2014

(satisfactorily) resolved by traditional deterministic algorithmic techniques, such as linear, nonlinear, and dynamic programming, etc [2].

Some of the interesting algorithms inspired by the natural phenomena are:

- Algorithms inspired by biology
- Algorithms Inspired by the behavior of groups of agents (Swarm)
- Algorithms inspired by human interactions and beliefs in the society.

For the past two decades, the research in the nature inspired evolutionary algorithms has been focused upon algorithms inspired by biological processes [3–7] and intelligent foraging behavior of social insects such as ants, birds, bees. For example, in the natural ant system, ants interact with each other indirectly by sharing the Pheromone trails. Swarm intelligence can be defined as "a property of a system of unintelligent agents of limited individual capabilities exhibiting collective intelligent behavior" [8]. Inspired by natural ant system, artificial ant colony optimization was designed and successfully implemented for many practical complex applications. Mimicking the behavior of flying birds, Kennedy and Ebergart [9] introduced another popular swarm-based evolutionary algorithm called particle swarm optimization. Artificial bee colony (ABC) algorithm is another stochastic search-based technique that belongs to class of swam intelligence-based algorithms, which is inspired by the intelligent behavior of the honeybees [10].

In 1994, Reynolds, designed a new algorithm called Cultural algorithm inspired by human interactions and beliefs and he argued that, "the cultural evolution enables the societies to evolve or adapt to their environments at rates that exceed that of biological evolution based on genetic inheritance only" [11]. Cultural algorithms are a class of computational models of cultural evolution that support dual inheritance, from belief space and individual interactions. This model of dual-inheritance is the key feature of Cultural Algorithms which is built on the principles of Renfrew's THINKS model [11] and which allows for a two-way system of learning and adaptation to take place. Reynolds and his team observed that, various knowledge sources like topographic knowledge, situational knowledge, and the fine-grained knowledge interact at the cultural level representing the Cultural Swarm, i.e., swarming behavior [12–14]. The early cultural evolution was modeled and studied based on single agent and multiagents to explore the impact of decision-making methods and resource sharing methods on population survival [15, 16].

Ray and Liew [17], inspired by nature's complex intra- and intersociety interactions, proposed an algorithm called society and civilization algorithm (SCA), in which, the artificial societies are build and the leader of the society is identified. The individuals collaborate with the leader and other individuals in the society to evolve. The leader will extend collaboration and communication with the other society leaders, in the civilization in order to improve. This may lead to migration of leaders and individuals to better performing societies. The SCA was implemented on engineering optimizations problems to demonstrate effectiveness of the algorithm.

Nature inspired evolutionary algorithms that emulate the behavior of living organisms and species integrated interactions among the agents for generating the next generation solution space. The outcome of these, interactions are either positive or negative based a random value compared with the algorithm specific control parameter. What if, the outcome of these interactions is indecisive? Indecisive interactions are also one of the outcomes of an interaction which may also need an attention to be accounted in the evolution process. Humans are intelligent species and are capable of withholding the information from the indecisive interaction in the memory and further evaluate, before adopting the change. The indecisive interactions drive the individual to seek a second opinion to build on the exchanged information from the previous interaction to emerge and evolve which is a common behavior with the people in the society.

In a society, in the process of evolution, individuals interact and exchange information. Human's interactions are generally selective and are free to explore randomly based on the individual bias. The individuals initially extend the interactions within the neighborhood because of affinity and trust worthiness. But, the interactions, does not get limited to neighborhood only. The individual is free to extend the interactions with the society in the process of evolution, particularly, when these interactions does not offer better prospects for growth or purely based on personal choice.

Motivated by such human characteristics, this paper introduces social evolution (SE) algorithm that employs the concept of second opinion and freedom of interactions to enable the individual's evolution leading to social evolution.

2 Social Evolution Algorithm

The pseudo code of SE algorithm is presented below:

1. Initialization:

   - Initialize Control parameters
     Maximum Cycle Number (MCN)
     Neighborhood Cooperation factor (NCR)
     Quality of Interaction (QI)
     Indecisive Factor (IFD)
     Second Opinion (Neighbor Best)–NB
     Second Opinion (Average Best)–AB
     Second Opinion (Society Best)–SB
   - Initialize the population

2. Evaluation Phase:

   - Evaluate the fitness of each individual
   - Calculate the probability of each individual and average solution
   - Store the best solution in the community

3. Cycle = 0
4. REPEAT when Cycle < Maximum cycle number

5. Interaction Phase:
   REPEAT (for all the individuals in the society)

   - Allow Individuals to interact based on their ability to interact (probabilistically).
   - Freedom of interaction:
     - Identify the neighbors based on von Neumann architecture
     - Store the best in the neighborhood
     - Choose between the random neighbor and a random individual based on the cooperation factor and fitness
   - Produce the new solution v_i for each individual using (1)

$$V_{ij} = X_{ij} + \emptyset_{ij}(X_{ij} - X_{kj}) \quad if \quad R_j < \text{QI},$$
$$\textit{Otherwise}, \text{evaluate IDF} \tag{1}$$

   [$\emptyset_{ij}$– is a Emotion Quotient—a random number in the range $[-1, 1]$. $k \in \{1, 2,... \text{SN}\}$ (SN: Number of individuals in a society) is randomly chosen index within the neighborhood/society of the individual. Although k is determined randomly, it has to be different from i. R_j is a randomly chosen real number in the range $[0, 1]$ and $j \in \{1, 2,... \text{D}\}$(D: Number of dimensions in a problem). [QI, Quality of interaction, is a control parameter].

   - Process of second opinion
     - Evaluate the Indecisive Factor (IDF)
       If Rand( ) > IDF, – $V_{ij} = X_{ij}$
       If Rand( ) < IDF – Look out for second opinion
     - Seek Second opinion and build on the previous interaction
       · If Rand( ) < NB

$$V_{ij} = V_{ij} + \emptyset_{ij}(V_{ij} - X_{ij}) \tag{2}$$

       $1 \in \{1, 2, ...\text{SN}\}$ is index of the neighborhood best
       · If Rand( ) < AB

$$V_{ij} = V_{ij} + \emptyset_{ij}(V_{ij} - A_j) \tag{3}$$

       A is the average individual and $j \in \{1, 2,... \text{D}\}$
       · If Rand( ) < SB

$$V_{ij} = V_{ij} + \emptyset_{ij}(V_{ij} - X_{mj}) \tag{4}$$

       $m \in \{1, 2, ...\text{SN}\}$ is index of the best in the society.
   - Evaluate the fitness of the individual before and after interaction and consider the best for next generation

6. UNTIL for all the individuals in the society are processed
7. Calculate the probability if each individual and average solution
8. Store the best solution in the society
9. Cycle = Cycle + 1

10. UNTIL (Maximum cycle number—The termination criteria is satisfied).

In the initialization phase, the basic control parameters such as number of maximum cycle number, quality of interaction, indecisive factor, second opinion ranges are initialized and the random population of individuals are generated. In the evaluation phase, each individual's fitness and its probability is calculated along with the average fitness of the individuals.

In the interaction phase, individuals interact with the neighbors probabilistically. In this algorithm, von Neumann neighborhood architecture is adopted for building the neighborhood. The individual first identifies the neighborhood individuals and randomly identifies a neighbor to interact. Before interaction, the individual evaluates the neighbor based on the cooperation factor and the ability or productivity of the neighbor. Based on the analysis or simply based on individual's bias, individual may interact with the identified neighbor or may consider a random individual in the society for an interaction. The individual interaction operator is inspired by the artificial bee colony optimization algorithm's employee/onlooker bee operator [10] as this operator exhibits the human interactions model. In the interaction operator, unlike the bee agents in ABC, the individual will not interact with any random solution in the society instead, they may interact more with the random neighbor in the von Neumann neighborhood architecture because of affinity and trust worthiness, but they are free to explore the society based on NCF. Also, once the individual is selected for the interaction, the individual solution interacts with the selected individual for all the dimensions of the problem unlike the operator used in ABC algorithm [10]. Once the interaction is performed, individual evaluate the quality of interaction (QI). If the quality of interaction is inferior, interaction's indecisive factor IDF is evaluated to decide on the interaction as negative or indecisive. All the indecisive interactions will undergo a second opinion process.

In the second opinion process, the individual can consult an expert either from the neighborhood or from the society or a non-existing individual with the average capabilities to further evaluate the indecisive interaction before adopting the change to emerge and evolve. After the interaction phase, evaluate the fitness of the updated solutions and compare with the respective original solution to consider the best for next generation. Before the above process is repeated until a termination condition (maximum cycle number), calculate the probabilities of the individuals, average solution the best in the society for the next generation.

The proposed SE algorithm is applied on four numerical benchmark problems and three to test the effectiveness and adoptability of the algorithm.

Table 1 Experimental parameters

Parameter	values
Population	100
Maximum cycle number (MCN)	10,000
Neighborhood cooperation factor (NCR)	0.75
Quality if interaction	0.8
Indecisive factor	0.5
Second opinion (Neighbor best)–NB	< 0.3
Second opinion (Average best)–AB	< 0.5
Second opinion (Best)–SB	< 1.0

3 Experiments

3.1 Unconstrained Benchmark Optimization Problems

Initially, the performance of the algorithm is tested on four standard numerical benchmark problems given in Table 2 using the experimental parameters presented in Table 1. Sphere is a convex, separable, unimodal function which has no local minimum except the global one. Schwefel is a multi model, non-separable function whose surface is composed of a great number of peaks and valleys. For this problem, many search algorithms get trapped in to the second best minimum far from the global minimum. Rastrigin is a multimode separable function, which was constructed from Sphere adding a modulator term. Its contour has a large number of local minima whose value increases with the distance to the global minimum. Dixon-Price is multimode, non-separable, and non-symmetric function.

Table 2 Benchmark functions

Function name	Interval	Function
Sphere	$[-100, 100]^n$	Min F $= \sum_{i=1}^{n} x_i^2$
Schwefel	$[-500, 500]^n$	Min F $= \sum_{i=1}^{n} \{-x_i \sin(\sqrt{|x_i|}\}$
Dixon–price	$[-10, 10]^n$	Min F $= (x_1 - 1)^2 + \sum_{i=2}^{n} i(2x_i^2 - x_{i-1})^2$
Rastrigin	$[-5.12, 5.12]^n$	Min F $= \sum_{i=1}^{n} \{(x_i^2 - 10\cos(2\pi x_i) + 10)\}$

Table 3 Results (Mean of the best values) of PSO, DE and ABC and SE algorithms on unconstrained numerical benchmark problems

Function name	PSO [18]	DE [18]	ABC [18]	SE
Sphere	0	0	0	0
Schwefel	−2654.030	−4177.990	−4189.830	−4189.830
Dixon–price	0.666	0.666	0	4.67E-01
Rastrigin	7.363	0	0	0

The social evolution algorithm executed 30 independent runs, to ascertain the performance of the algorithm on each of the listed problems with 10 dimensions. For easy comparisons, the experimental parameters population size and the maximum number of cycles are defined as per the parameters defined for DE, PSO, and ABC algorithms [18]. The mean of the best values for each of the problems is reported in Table 3 and compared against some of the evolutionary algorithms such as, differential evolution (DE), particle swarm optimization (PSO) and artificial bee colony optimization (ABC). The reported results suggest that, the algorithm finds the global optimum values for the three functions Sphere, Schwefel, and Rastrigin successfully. When compared with individual algorithms, the ABC algorithm performed better than SE on one function and SE performed better than PSO on three functions and better than DE on two functions.

4 Conclusions

This paper proposes a novel social evolution (SE) algorithm that mimics the human interactions, behavior, and their biases. This algorithm adopts two basic human characteristics. First, individual's ability and bias to evaluate the neighbor before establishing the interaction to evolve. Second, the ability that discriminate the quality of interaction and identifying indecisive interactions and further seeking a second opinion from an expert before adopting the change to emerge and evolve. The proposed algorithm mimics such characteristics of humans tested on four numerical test functions and compared the results with the state-of- the-art nature inspired algorithms. The results indicate that the proposed social evolution algorithm is better than or quite comparable to the existing state-of-the-art algorithms.

Acknowledgments Authors gratefully acknowledge the inspiration and guidance of the Most Revered Prof. P. S. Satsangi, the Chairman, Advisory Committee on Education, Dayalbagh, Agra, India.

References

1. Yoshida, Z.: Nonlinear Science: the Challenge of Complex Systems. Springer, Heidelberg (2010)
2. de Castro L.N.: Fundamentals of natural computing: an overview. Phys. Life Rev. **4**(1), 1–36 (2007)
3. Goldberg, D.E.: Genetic Algorithms in Search, Optimization and Machine Learning. Kluwer Academic Publishers, Boston, MA (1989)
4. Fogel, D.B.: Evolutionary computation: toward a new philosophy of machine intelligence (3rd edn). IEEE Press, Piscataway, NJ (2006)
5. Beyer, H.-G., Schwefel, H.-P.: Evolution strategies: a comprehensive introduction. J. Nat. Comput. **1**(1), 3–52 (2002)

6. Banzhaf, W., Nordin, P., Keller, R.E., Francone, F.D.: Genetic Programming: An Introduction: On the Automatic Evolution of Computer Programs and Its Applications. Morgan Kaufmann, Heidelberg (1998)
7. Korns, Michael: Abstract Expression Grammar Symbolic Regression, in Genetic Programming Theory and Practice VIII. Springer, New York (2010)
8. White, T., Pagurek, B.: Towards multi-swarm problem solving in networks, In: Proceedings of the 3rd International Conference on Multi-Agent Systems (ICMAS-98), pp. 333–40, (1998)
9. Kennedy, J., Eberhart, R.C.: Particle swarm optimization. In Proceedings of 1995 IEEE International Conference Neural Networks IV, pp. 1942–1948, (1995)
10. Karaboga, D., Basturk, B.: A powerful and efficient algorithm for numerical function optimization: artificial bee colony (ABC) algorithm. J. Global Optim. **39**, 459–471 (2007)
11. Reynolds, R.G.: An introduction to cultural algorithms. In: Proceedings of the Third Annual Conference on Evolutionary Programming, pp. 131–139. San Diego, California (1994)
12. Reynolds, R.G., Peng, B., Brewster, J.J.: Cultural swarms: knowledge-driven problem solving in social systems. IEEE Int. Conf. Syst. Man Cybern. **4**, 3589–3594 (2003)
13. Reynolds, R.G., Peng, B., Brewster, J.: Cultural swarms. Congr. Evol. Comput. **3**, 1965–1971 (2003A)
14. Reynolds, R.G., Jacoban, R., Brewster, J.: Cultural swarms: assessing the impact of culture on social interaction and problem solving. In: Proceedings of the 2003 IEEE Swarm Intelligence, Symposium, pp. 212–219 (2003b)
15. Reynolds, RG, Kobti, Z., Kohler, T.: The effect of culture on the resilience of social systems in the village multi-agent simulation. In: Proceedings of IEEE International Congress on Evolutionary Computation. Portland, OR, vol. 24, pp. 1743–1750, June 19 (2004)
16. Reynolds, R.G., Whallon, R., Mostafa, Z.A., Zadegan, B.M.: Agent-based modeling of early cultural evolution. IEEE Congress on Evolutionary Computation. pp. 1135–1142 (2006)
17. Ray, T., Liew, K.M.: Society and civilization: an optimization algorithm based on the simulation of social behavior. IEEE Trans. Evol. Comput. **7**(4), 386–396 (2003)
18. Akay B., Karaboga D.: Artificial bee colony algorithm for large-scale problems and engineering design optimization. J. Intel. Manuf. pp. 1–14 (2010). DOI: 10.1007/s10845-010-0393-4

A Survey on Filter Techniques for Feature Selection in Text Mining

Kusum Kumari Bharti and Pramod kumar Singh

Abstract A large portion of a document is usually covered by irrelevant features. Instead of identifying actual context of the document, such features increase dimensions in the representation model and computational complexity of underlying algorithm, and hence adversely affect the performance. It necessitates a requirement of relevant feature selection in the given feature space. In this context, feature selection plays a key role in removing irrelevant features from the original feature space. Feature selection methods are broadly categorized into three groups: filter, wrapper, and embedded. Filter methods are widely used in text mining because of their simplicity, computational complexity, and efficiency. In this article, we provide a brief survey of filter feature selection methods along with some of the recent developments in this area.

Keywords Text mining · Text categorization · Text clustering · Feature extraction · Feature selection · Filter methods

1 Introduction

It is difficult to extract relevant information in time from a heap of digital information, which is growing leap and bound with rapid development of Internet technology. It necessitates a need to organize available information in a well-structured format in order to facilitate quick and efficient retrieval of relevant information in time. Text mining is a key step to achieve this task. It discovers previously unknown information

K. K. Bharti (✉) · P. K. Singh
Computational Intelligence and Data Mining Research Lab, ABV-Indian Institute of Information Technology and Management Gwalior, Morena Link Road, Gwalior, Madhya Pradesh, India
e-mail: kkusum.bharti@gmail.com

P. K. Singh
e-mail: pksingh@iiitm.ac.in

B. V. Babu et al. (eds.), *Proceedings of the Second International Conference on Soft Computing for Problem Solving (SocProS 2012), December 28–30, 2012*, Advances in Intelligent Systems and Computing 236, DOI: 10.1007/978-81-322-1602-5_154, © Springer India 2014

by automatically extracting useful information from huge corpus. It is widely used in various areas, e.g., information technology, Internet, banks, business analytics, market analysis, pharmaceutical, healthcare.

Text classification and text clustering are two subfields of text mining. Text classification is a form of supervised learning where a new document is assigned to one of the predefined classes based on a defined model whereas text clustering is an example of unsupervised learning where the classes are not known apriori. Before organizing the documents based on their intrinsic characteristics, documents are represented into a common format, known as vector space model [22], where each dimension corresponds to a single term. It increases the number of dimensions of representation model unmanageably. It necessitates a requirement of relevant feature selection as a lot of features are irrelevant, redundant, and noisy which adversely affect the efficacy and efficiency of the underlying algorithm and sometimes even misguide them. Moreover, high-dimensional feature space also makes it difficult to apply computationally intensive algorithm on all datasets. In this context, dimension reduction plays a key role in reducing the number of features.

The primary aim of the dimension reduction methods is to generate low-dimensional feature space from high-dimensional space without scarifying the performance of the underlying algorithm. Feature selection and Feature transformation are two subcategories of dimension reduction methods. A brief overview of these methods is presented in Table 1.

Table 1 Comparative analysis of dimension reduction methods

Name	Key concept	Advantages	Disadvantages	Examples
Feature extraction (FE)	Summarize the dataset by creating linear combinations of the features	Preserves the original, relative distance between objects	Less effective in case of large number of irrelevant features	Principal component analysis [18]
		Covers latent structure	Sometime may be very difficult to interpret in the context of the domain	Latent semantic indexing [6]
Feature selection (FS)	Select a subset of relevant features based on defined criteria	Robust against irrelevant features	Feature selection criteria are hard to define	Information gain (IG) [21]
			May generate redundant features	Mutual information (MI) [5]
				Document frequency (DF) [12]

It can be observed from Table 1 that FE methods do not work well in case of high irrelevant features space. On the other hand, FS methods are more robust against irrelevant terms as selection of terms is not affected by irrelevant features. The FS methods efficiently reduce the high-dimensional irrelevant feature space into low-dimensional relevant subspace and hence increase performance of the underlying algorithm. Therefore, FS methods are more preferred for dimension reduction. These methods are broadly classified as filter, wrapper, and embedded. A brief description of these methods in terms of key concept, advantages, disadvantages, and examples is given in Table 2.

Table 2 Comparative analysis of feature selection methods

	Key concept	Advantages	Disadvantages	Examples
Wrapper	Use classifier for selecting features subset	Simple Consider interaction with classifiers Less prone to generate local optimal solution	Computationally expensive	Sequential forward Selection [20] Sequential backward elimination [20]
Filter	Use intrinsic properties of the data	Fast Scalable Computationally efficient than wrapper methods	No guarantee to generate discriminative set of features Ignore interaction with classifiers	MI [5] IG [21] DF [12] Term variance (TV) [12]
Embedded	Facilitate interaction between classifier and feature selection methods	Facilitate interaction with classifier Computationally less expensive than wrapper methods	Computationally expensive than filter methods Classifier dependent selection	Bayesian logistic Regression [2] Sparse logistic regression [24]

Filter methods evaluate quality of features based on intrinsic properties of the documents without iteratively testing them with classifier (refer, Table 2). This property makes it considerably less computational expensive to other methods. Therefore, they are prominently used for feature selection (dimension reduction) in text mining. In this paper, we present a brief summary of various traditional filter feature selection methods along with some recent developments in this area.

2 Filter Methods for Feature Selection

Based on the literature survey of filter FS methods for text mining, we propose taxonomy of such methods as shown in Fig. 1. These methods are categorized as ranking methods and space search methods based on the strategy they use to select features subset. The ranking and search methods are further subcategorized into two groups, supervised, and unsupervised methods, based on the fact that whether they use class label information or not for quantifying the relevancy of terms. A brief summary of supervised and unsupervised filter FS is mentioned in Table 3.

2.1 Ranking Methods

These methods rank every feature individually using defined scoring function and then sort features in decreasing order based on their relevance score. Scoring/ranking of these features is performed either on the basis of class label information (known as supervised ranking methods) or on the basis of intrinsic properties (known as unsupervised ranking methods) of documents. Finally, the top score features are selected while the rest are discarded.

Summary of ranking supervised filter feature selection methods is given in Table 4.

Summary of ranking unsupervised filter feature selection methods is given in Table 5.

2.2 Space Search Methods

These methods work on the concept of optimizing some defined objective function as used in wrapper and embedded methods. The objective functions are defined in terms of feature interaction and/or feature class interaction. In supervised filter feature

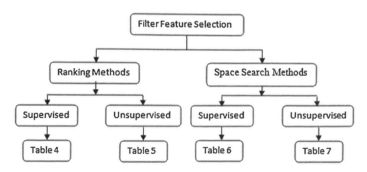

Fig. 1 Overview of proposed taxonomy

Table 3 Comparative analysis of supervised and unsupervised filter FS

Method	Key concept	Advantage	Disadvantages	Examples
Supervised FS	Use class label information for quantifying the relevancy of terms	Simple	May suffer from over fitting	IG [21]
			Require ground truth information	MI [5]
Unsupervised FS	Quantify the relevancy of terms based on the intrinsic properties of the document	Do not need ground truth information, i.e., class label	Less efficient than supervised FS	DF [12]
				TV [12]

selection methods the objective function works on the concept of feature to class interaction whereas in unsupervised filter feature selection methods the objective function works on feature to feature interaction.

Summary of space search supervised filter feature selection functions is given in Table 6.

Summary of space search unsupervised filter feature selection functions is given in Table 7.

Having covered here prominent traditional approaches for filter feature selection, we discuss some of the recent developments in this area in next section.

3 Recent Development in Filter Feature Selection

In this section, we discuss some of the recent developments in the area of filter feature selection methods. Based on the methodology, we categories the available literature as genetic algorithm (GA)-based feature selection methods, particle swarm optimization (PSO)-based feature selection methods, and Hybrid feature selection methods.

3.1 Genetic Algorithm-Based Feature Selection Methods

Chuang et al. [4] combine a filter FS method correlation-based feature selection (CFS) and a wrapper FS method taguchi-genetic algorithm (TGA) to form a new hybrid method for dimension reduction in gene analysis. They use k-nearest neighbor (KNN) with leave-one-out cross-validation (LOOCV) method as a classifier to judge the efficacy of their proposed method and observe that their proposed method

Table 4 Summary of ranking supervised filter feature selection methods

Name	Description	Main idea	Mathematical form	Remark
IG[21]	Information gain	Quantifies relevancy of terms with respect to the class	$\sum_{j=1}^{c} P(w\|c_j) \log \frac{P(w\|c_j)}{c_j} + \sum_{j=1}^{c} P(\bar{w}\|c_j) \log \frac{P(\bar{w}\|c_j)}{c_j}$	Considers presence and absence of terms in observed category
CC[17]	Correlation coefficient	Quantifies the positive dependency between term and class	$\sum_{j=1}^{c} \frac{\sqrt{n}(P(w\|c_j)P(\bar{w}\|\bar{c_j}) - P(w\|\bar{c_j})P(\bar{w}\|c_j))}{\sqrt{P(w)P(\bar{w})P(c_j)P(\bar{c_j})}}$	Get affected by size of the category
MI[5]	Mutual information	Measures dependency between target variable	$\sum_{j=1}^{k} P(c_j, w) \log \frac{P(c_j\|w)}{P(c_j)}$	Considers only positivity of term with respect to the class
MOR[1]	Multiclass odd ratio	Variant of MC-OR	$\sum_{j=1}^{c} \left\| \log \frac{P(w\|c_j)(1-P(w\|\bar{c_j}))}{P(w\|\bar{c_j})(1-P(w\|c_j))} \right\|$	Considers both positivity and negativity of term with respect to the class
CDM[1]	Class discriminating measure	Simplified form of MOR	$\sum_{j=1}^{c} \left\| \log \frac{P(w\|c_j)}{P(w\|\bar{c_j})} \right\|$	Considers both positivity and negativity of term with respect to the class

(continued)

Table 4 (continued)

Name	Description	Main idea	Mathematical form	Remark
GINI [23]	Gini index	Considers document frequency for quantifying relevancy of the terms	$\sum_{j=1}^{c} \left(\frac{A_i}{M_i}\right)^2 \left(\frac{A_i}{A_i+B_i}\right)^2$ Where A_i, number of documents with word w and belong to category B_i, number of documents with word w and do not belong to category C_i, number of document without word w and belong to category D_i, number of documents without word w and do not belong to category	Considers only document frequency of terms; ignores term frequency

Table 5 Summary of ranking unsupervised filter feature selection methods

Name	Description	Main idea	Formula	Remark
DF[12]	Document frequency	Frequent terms are more informative than non frequent terms.	df Total number of documents in which term appears	Easily biased by those common terms which have high document frequency but uniform distribution over different classes.
TV [12]	Term variance	Uses frequency of terms for quantifying the relevancy of terms	$\frac{1}{n}\sum_{j=1}^{n}\left(X_{ij}-\bar{X}_i\right)^2$	Measure deviation from mean
MM[8]	Mean median	Absolute difference between mean and median	$\left\|\bar{X}_i-median(X_i)\right\|$	Measure dispersion

Table 6 Summary of space search supervised filter feature functions

Name	Description	Formula	Remark
GCC [9]	Group correlation coefficient	$\frac{k\bar{r}_{cf}}{\sqrt{k+k(k-1)\bar{r}_{ff}}}$ r_{cf}, r_{ff} - the average feature to class/feature to feature correlation	Higher value is more preferable
MID[7]	Mutual information difference	$I\,(w,c)-\frac{1}{\|\Omega_s\|}\sum_t I\,(w_i,w_t)$ $I\,(w_i,w_t)$ the mutual information between feature $w_i \in \Omega_s$ and feature $w_t \in \Omega_s$	Need discretized feature inputs
MIQ [7]	Mutual information quotient	$\frac{I\,(w_i,c)}{\frac{1}{\|\Omega_s\|}\sum_t I\,(w_i,w_t)}$ $I\,(w_i,c)$ - the mutual information between feature $w_i \in \Omega_s$ and class label c $I\,(w_i,w_t)$ the mutual information between feature $w_i \in \Omega_s$ and feature $w_t \in \Omega_s$	Need discretized feature inputs

outperforms other competitive methods on 10 out of 11 classification profiles. Song and Park [25] propose, GAL, a genetic algorithm method based on LSI, for text clustering. As GA is computationally expensive for high dimensional space, the authors reduce the dimension of representational model using LSI. Additionally, they propose a variable string length GA which automatically evolves appropriated number of clusters for the given dataset. They show the superiority of their approach GAL

Table 7 Summary of space search unsupervised filter feature functions

Function	Formula	Remark
Maximum information compression index [16]	$var\,(w_i) + var\,(w_t) -$ $\sqrt{\dfrac{(var\,(w_i) + var\,(w_i))^2}{-4var\,(w_i)\,var\,(w_t)\,(1 - \rho(w_i,\,w_t)^2)}}$	Its zero value indicates that the features are linearly dependent.
Correlation coefficient [16]	$\dfrac{cov(w_i w_t)}{\sqrt{var(w_i)var(w_t)}}$	Its zero value indicates that the features are linearly related.
Absolute cosine [8]	$\left\lvert \dfrac{w_i, w_t}{w_i w_t} \right\rvert$	Its zero value indicates that the features are orthogonal.

over conventional GA applied in VSM model for Reuter-21,578 document clustering results.

Li et al. [11] use ReliefF in stage one to evaluate the quality of each individual feature and obtain sequential feature sets based on the marks evaluated by ReliefF. Further, they use cross validation technique to obtain candidate feature set from the sequential sets. In second stage, they use genetic algorithm to search a more compact feature set based on the candidate set. They use three different classifiers linear discriminant (LD), KNN, and naïve bayes (NB) to evaluate their proposed method. The results to the real gear fault diagnosis show that the proposed method obtains a higher performance with a small size feature set. Yang et al. [30] hybridize a filter method IG and a wrapper method GA for feature selection in microarray datasets. Initially, they rank each feature using IG to select discriminative feature subsets and then use GA to further reduce the dimensions of representation model. They use KNN with LOOCV to show the efficacy of their proposed model. Experimental results demonstrate that the proposed method simplify the number of gene expression level efficiently and obtain higher accuracy to other competitive feature selection methods.

Summary of the literature belonging to genetic algorithm-based feature selection methods is given in Table 8.

3.2 Particles Swarm Optimization-Based Feature Selection Methods

PSO has been widely studied for dimension reduction. Liu et al. [13] use PSO to select feature subset for classification task and train radial basis function (RBF) neural network simultaneously. They experiment with four datasets from UCI Repository of machine learning databases image segmentation, page-blocks, ionosphere, and wine to show that their method effectively selects features with high accuracy as well as small feature subset size. Tu et al. [26] show the use of PSO to implement feature selection. They use support vector machines (SVMs) with the one-versus-rest method

Table 8 Summary of the literature belonging to GA based feature selection

Authors	Key concept	Measures	Algorithms used	Remark
Chuang et al. [4]	CFS and TGA are combined to select discriminative features	CFS, TGA	KNN	Selects features highly correlated with class yet uncorrelated with each other
Song and Park [25]	Explore latent semantic structure along with decreasing the number of dimensions of representation model	LSI	GA	Creates appropriate number of clusters of documents without specifying them
Li et al. [11]	Use two step procedure for selecting discriminative set of features	ReliefF, GA	KNN, NB LDC	Improves the accuracy of classification with reduced feature set compared to original feature set
Yang et al. [30]	Use a hybrid approach for feature selection	IG GA	KNN	Effectively reduces the number of features and achieves a good classification accuracy

as a fitness function in PSO for the problem. The proposed method is applied to five classification problems vowel, wine, WDBC, ionosphere, and sonar from the UCI Repository of machine learning databases. They observe that their method simplifies features effectively, obtains higher classification accuracy, and requires minimum computational resources than the other competitive feature selection methods.

Liu et al. [14] modify PSO and name it modified multi-swarm PSO (MSPSO); it holds a number of subswarms scheduled by the multi-swarm scheduling module. The multi-swarm scheduling module monitors all the subswarms, and gathers the results from the subswarms. Further, they propose a two stage method improved feature selection (IFS) consisting of MSPSO, SVM, and F-score. In the stage one, both the SVM parameter optimization and the feature selection are dynamically executed by MSPSO. In the second stage, SVM performs classification using these optimal values and selects feature subsets via tenfold cross validation. The designed objective function also consists of two parts: one is classification accuracy rate and the other is the F-score. Both are summed to form a single objective function by linear weighting. Experimental results on 10 different datasets from the UCI machine learning and StatLog databases show the efficacy of the proposed method IFS.

Chuang et al. [3] propose an algorithm catfish binary particle swarm optimization (CatfishBPSO). In CatfishBPSO, a new particle initialized at extreme points in the search space, known as catfish particle, replaces the worst fit individual to

Table 9 Summary of the literature belonging to PSO based feature selection

Authors	Key concept	Measures	Algorithm used	Remark
Liu et al. [13]	Feature selection and neural network training are done simultaneously in each iteration	PSO	RBF Neural Network	Effectively selects features with high accuracy as well as small feature subset size
Tu et al. [26]	PSO is used to implement a feature selection, and SVM with the one-versus-rest method is used as evaluator for the PSO fitness	PSO	SVM	Obtains a good classification accuracy
Liu et al. [14]	Integrate multi-swarm PSO, SVM with F-score for selecting discriminative features	Multi-swarm PSO F-Score	SVM	A modified Multi-Swarm PSO to solve discrete problems
Chuang et al. [3]	Replace particles with the worst fitness by the catfish particle initialized at extreme points in the search space	CatfishBPSO	KNN	CatfishBPSO is effective in dimension reduction and/or classification accuracy
Unler et al. [29]	Use MI to weigh the bit selection probabilities in the discrete PSO	MI Discrete PSO	SVM	Consider feature-feature interaction for removing redundant features

avoid premature convergence if the gbest particle has not improved for a number of consecutive iterations. The experimental results on 10 different datasets from UCI machine learning database show that the proposed method efficiently reduces the dimension in the representation model and achieves highest accuracy in comparison to the other feature selection methods.

Unler et al. [29] propose a hybrid method maximum relevance minimum redundancy PSO (mr^2PSO) for feature selection. In this, they integrate filter method mutual information within wrapper method PSO for selecting the informative features set. Instead of just removing irrelevant features they also remove redundant features from original feature space. The experimental results show that the proposed model is effective and efficient in terms of classification accuracy and computational complexity to the competing methods.

Summary of literature belonging to particle swarm optimization based feature selection is given in Table 9.

3.3 Hybrid Approaches for Feature Selection Methods

Here, we discuss the literature, which hybrids filter-filter methods, filter-wrapper methods, and FS-FE methods to reduce the dimensions of representation model. Uguz[28] proposes two-stage filter-wrapper (IG-GA) and FS-FE method (IG-PCA) methods for dimension reduction to improve the performance of text categorization. First, each term in the document is ranked on the basis of its importance for classification using IG. In the second stage, a FS method (GA) and a FE method (PCA) is applied separately to reduce the dimension of representation model. To evaluate the effectiveness of both the two-stage methods, the author uses KNN and C4.5 decision tree algorithm on Reuters-21,578 and Classic3 datasets collection for text categorization. The experimental results show that the proposed methods are effective in terms of precision, recall and F-measure. A similar approach has been used by Uguz[27] for Doppler signal selection.

Meng et al.[15], also use a two-stage procedure to gradually reduce the high dimensional space into a low dimensional subspace. First, they apply feature contribution degree (FCD) to select relevant features, and then construct low semantic space from relevant feature space using feature extraction method latent semantic indexing (LSI). They show effectiveness of the proposed model on spam database categorization.

Hsu et al.[10] propose a two-stage hybrid feature selection method, which combines filter and wrapper methods to take advantage of both. In stage one, they apply computationally efficient filter methods IG and F-score to select candidate features subset. In the second stage, they refine the subspace using wrapper method inverse sequential floating search; here, first sequential backward search (SBS), and then sequential forward search (SFS) is applied to select discriminative features subset. They experiment with two bioinformatics problems protein disordered region prediction and gene selection in microarray cancer data and show that equal or better prediction accuracy can be achieved with a smaller feature set also.

Sometime identification of only relevant features do not help much in improving accuracy of the classifiers as there may be some redundant features in the given features space, which adversely affect performance of the underlying algorithm. Thus, it is necessary to remove redundant features also along with irrelevant features. One of the initial works based on this concept is due to Peng et al.[19]. They propose a method called maximum relevance and minimum redundancy (mRMR). Here, they use feature selection measure MI[5] to select relevant features and remove redundant features.

Summary of literature belonging to hybrid methods is given in Table 10.

Table 10 Summary of literature belonging to hybrid methods

Authors	Key concept	Measures	Algorithms used	Remark
Uguz [28]	Removes irrelevant features using two two-stage methods	IG PCA GA	KNN C4.5	Automatically selects number of dimensions in case of PCA
Uguz [27]	Hybridizes FS and FE methods for dimension reduction	IG PCA	SVM	Improves accuracy of underlying algorithm
Meng et al. [15]	Quantify the relevance by considering positivity of terms with respect the class.	FCD	SVM	Statistically derive conceptual indices to replace the individual terms.
		LSI		
Hsu et al. [10]	Utilize computational efficiency and accuracy of classifier for selecting discriminative set of features.	IG	SVM	Use two filter models for remove the most redundant or irrelevant features.
		F-Score SBS SFS		

4 Conclusion and Future Works

Identification of relevant feature subset in text mining is utmost important as irrelevant features not only increase the computational complexity but also adversely affect performance of the underlying algorithm. In this context, feature selection is considered as a prominent preprocessing step to remove irrelevant, redundant and noisy terms. Feature selection methods are broadly classified into three categories: filter, wrapper, and embedded. This survey states a brief summary of various filter feature selection methods as they are the most widely used methods in the text mining because of their simplicity, computational complexity and efficiency. In addition, we present a brief introduction to recent developments in the area, e.g., nature-inspired algorithms, hybrid methods, present in the literature.

Though feature selection is widely explored in the literature, maximum number of features, threshold value and minimum similarity in case of redundant term removal are still an open and challenging research problems.

References

1. Chen, J., Huang, H., Tian, S., Qu, Y.: Feature selection for text classification with Naïve Bayes. Expert Syst. Appl. **36**(3), 5432–5435 (2009)
2. Chen, X.: An improved branch and bound algorithm for feature selection. Pattern Recogn. Lett. **24**(12), 1925–1933 (2003)
3. Chuang, L.Y., Tsai, S.W., Yang, C.H.: Improved binary particle swarm optimization using catfish effect for feature selection. Expert Syst. Appl. **38**(10), 12699–12707 (2011)
4. Chuang, L.Y., Yang, C.H., Wu, K.C., Yang, C.H.: A hybrid feature selection method for DNA microarray data. Comput. Biol. Med. **41**(4), 228–237 (2011)
5. Church, K.W., Hanks, P.: Word association norm, mutual information and lexicography. J. Comput. Linguist. **27**(1), 22–29 (1990)
6. Deerwester, S.: Improving information retrieval with latent semantic indexing. In: Proceedings of the 51st Annual Meeting of the American Society for Information Science, Vol. 25, pp. 36–40 (1988)
7. Ding, C., Peng, H.: Minimum redundancy feature selection from microarray gene expression data. J. Bioinf. Comput. Biol. 185–205 (2005)
8. Ferreira, A.J., Figueired, M.A.T.: Efficient feature selection filters for high-dimensional data. Pattern Recogn. Lett. **33**(13), 1794–1804 (2012)
9. Hall, M.A.: Correlation-based feature selection for machine learning. Ph.D. Thesis. Department of Computer Science, University of Waikato (1999)
10. Hsu, H.H., Hsieh, C. W., Lu, M.D.: Hybrid feature selection by combining filters and wrappers. Expert Syst. Appl. **38**(7), 8144–8150 (2011)
11. Li, B., Zhang, P., Ren, G., Xing, Z.: A two stage feature selection method for gear fault diagnosis using reliefF and GA-wrapper. In: Proceedings International Conference on Measuring Technology and Mechatronics Automation, pp. 578–581 (2009)
12. Liu, L., Kang, J., Yu, J., Wang, Z.: A comparative study on unsupervised feature selection methods for text clustering. In: Proceedings of Natural Language Processing and Knowledge, Engineering, pp. 59–601 (2005)
13. Liu, Y., Qin, Z., Xu, Z., He, X.: Feature selection with particle swarms. In: Computational and Information Science, pp. 425–430. Springer, Heidelberg (2004)
14. Liu, Y., Wang, G., Chen, H., Dong, H., Zhu, X., Wang, S.: An improved particle swarm optimization for feature selection. J. Bionic Eng. **8**(2), 191–200 (2011)
15. Meng, J., Lin, H., Yu, Y.: A two-stage feature selection method for text categorization. Knowl.-Based Syst. **62**(7), 2793–2800 (2011)
16. Mitra, P., Murthy, C., Pal, S.: Unsupervised feature selection using feature similarity. IEEE Trans. Pattern Anal. Machine Intell. **24**(3), 301–312 (2002)
17. Ng, H. T., Goh, W. B., Low, K. L.: Feature selection, perception learning, and a usability case study for text categorization. In: Proceedings of the 20th ACM International Conference on Research and Development in, Information Retrieval, pp. 67–73 (1997)
18. Pearson, K.: On lines and planes of closest filt to systems of points in space. Phil. Mag. **1**(6), 559–572 (1901)
19. Peng, H., Long, F., Ding, C.: Feature selection based on mutual information: criteria of max-dependency, max-relevance, and min-redundancy. IEEE Trans. Pattern Anal. Mach. Intell. **27**(8), 1226–1238 (2005)
20. Pudil, P., Novoviciva, J., Kittler, J.: Floating search methods in feature selection. Pattern Recogn. Lett. **15**(11), 1119–1125 (1994)

21. Quinlan, J.R.: Induction of decision tree. Mach. learn. **1**(1), 81–106 (1986)
22. Salton, G., Wong, A., Yang, C. S.: A vector space model for automatic indexing. Commun. ACM**18**(11), 613–620 (1975)
23. Shang, W., Huang, H., Zhu, H., Lin, Y., Qu, Y., Wang, Z.: A novel feature selection algorithm for text clustering. Expert Syst. Appl. **33**(1), 1–5 (2007)
24. Shevade, S., Keerthi, S.: A simple and efficient algorithm for gene selection using sparse logistic regression. Bioinformatics **19**(17), 2246–2253 (2003)
25. Song, W., Park, S.C.: Genetic algorithm for text clustering based on latent semantic indexing. Comput. Math. Appl. **57**(11–12), 1901–1907 (2009)
26. Tu, C.J., Chuang, L.Y., Chang, J.Y., Yang, C.H.: Feature selection using PSO-SVM. In: Proceedings of Multiconferenc of Engineers, pp. 138–143 (2006)
27. Uguz, H.: A hybrid system based on information gain and principal component analysis for the classification of transcranial Doppler signals. Comput. Methods Programs Biomed. **107**(3), 598–609 (2012)
28. Uguz, H.: A two-stage feature selection method for text categorization by using information gain, principal component analysis and genetic algorithm. Knowl. Based. Syst. **24**(7), 1024–1032 (2011)
29. Unler, A., Murat, A., Chinnam, R.B.: mr^2PSO: A maximum relevance minimum redundancy feature selection method based on swarm intelligence for support vector machine classification. Inf. Sci. **181**(20), 4625–4641 (2011)
30. Yang, C.H., Chuang, L.Y., Yang, C.H.: IG-GA: a hybrid filter/wrapper method for feature selection of microarray data. J. Med. Biol. Eng. **30**(1), 23–28 (2009)

An Effective Hybrid Method Based on DE, GA, and K-means for Data Clustering

Jay Prakash and Pramod Kumar Singh

Abstract Clustering is an unsupervised classification method and plays essential role in applications in diverse fields. The evolutionary methods attracted attention and gained popularity among the data mining researchers for clustering due to their expedient implementation, parallel nature, ability to search global optima, and other advantages over conventional methods. However, conventional clustering methods, e.g., K-means, are computationally efficient and widely used local search methods. Therefore, many researchers paid attention to hybrid algorithms. However, most of the algorithms lag in proper balancing of exploration and exploitation of solutions in the search space. In this work, the authors propose a hybrid method DKGK. It uses DE to diversify candidate solutions in the search space. The obtained solutions are refined by K-means. Further, GA with heuristic crossover operator is applied for fast convergence of solutions and the obtained solutions are further refined by K-means. This is why proposed method is called DKGK. Performance of the proposed method is compared to that of Deferential Evolution (DE), genetic algorithm (GA), a hybrid of DE and K-means (DEKM), and a hybrid of GA and K-Means (GAKM) based on the sum of intra-cluster distances. The results obtained on three real and two synthetic datasets are very encouraging as the proposed method DKGK outperforms all the competing methods.

Keywords Evolutionary algorithm · Data clustering · Differential algorithm · Genetic algorithm · K-means

J. Prakash (✉) · P. K. Singh
Computational Intelligence and Data Mining Research Laboratory, ABV-Indian Institute of Information Technology and Management, Gwalior, India
e-mail: jayprakash.iiitm@gmail.com

P. K. Singh
e-mail: pksingh@iiitm.ac.in

B. V. Babu et al. (eds.), *Proceedings of the Second International Conference on Soft Computing for Problem Solving (SocProS 2012), December 28–30, 2012*, Advances in Intelligent Systems and Computing 236, DOI: 10.1007/978-81-322-1602-5_155, © Springer India 2014

1 Introduction

Clustering has been approached by many disciplines in the past few decades because of its wide applications. The clustering classifies or groups the objects of an unlabeled dataset on the basis of their similarity. Each group, known as cluster, consists of objects such that the objects belonging to the same cluster have more similarity than the objects belonging to the other clusters. Though several similarity measures, e.g., Manhattan distance, Euclidean distance, cosine distance, Mahalanabis distance, are available in the literature, the most widely used distance measure is the Euclidean distance [22].

Clustering can be performed in two different modes fuzzy or Hard [11]. In fuzzy clustering, each object may belong to each cluster with a certain fuzzy membership grade. In hard clustering, the clusters are disjoint and each object belongs to exactly one cluster. We can mathematically represent hard partitioning of a data set X, based on the description in [5] as follows. Consider X= $\{x_1, x_2, \ldots, x_N)$, where every data point x_i in X corresponds to a n-dimensional feature vector. Then, hard partitioning of X is a collection $C = \{C_1, C_2, \ldots, C_k\}$ of K number of nonempty and nonoverlapping groups of data such that

$$C_i \neq \phi \qquad I = 1, \ldots, K \quad and \qquad (1)$$

$$C_i \cap C_j = \phi \qquad i, j = 1, \ldots, K \quad \text{and} \quad i \neq j \qquad (2)$$

Though conventional clustering algorithms exhibit fast convergence and are computationally efficient, they have many problems, e.g., they are quite sensitive to the initialization of prototypes, they do not provide any guarantee to the global optimality rather they easily stuck into the local minima [10]. Specifically, for these reasons, powerful metaheuristics such as evolutionary algorithms offer to be more effective methods to overcome the deficiencies of the conventional clustering methods as they posses several desired key features, e.g., upgradation of the candidate solutions iteratively based on objective function (fitness function), decentralization, parallel nature, flexibility, robustness, no need of prior information about domain knowledge, self organizing behavior [21]. Consequently, many researchers found that hybridization of metaheuristics within and with conventional algorithms increases the efficiency and accuracy of the clustering. Kwedlo [13] proposes a method DEKM that combines DE and K-means to obtain clusters based on the sum of squared error (SSE) criterion when number of cluster is known. Here, DE is a global search algorithm with slow convergence and K-means is local search algorithm with fast convergence. Tvrd'ık and Kŕiv'y [20] empirically find that DE hybridized with K-means performs comparatively superior to the DE based on optimizing two basic criteria trace of within scatter matrix and variance ratio criterion. Tian et al. [19] use K-Harmonic mean in the place of K-means with DE when number of clusters is fixed. Here also the number of clusters is known *apriori*. As K-means is very sensitive toward the choice of initial centers, Laszlo and Mukharjee [14] use GA to obtain initial cluster centers for K-means.

Das et al. [6] propose a method to determine proper number of clusters during run time. They modify classical DE by tuning parameters in two different ways to improve its slow convergence property. First, the scale factor is changed in random manner in the range (0.5, 1); it helps to retain diversity as the search progresses. Second, the crossover rate (Cr) linearly decreases from maximum to minimum of crossover rate with iterations during the run. He et al. [9] propose a two-stage genetic clustering algorithm TGCA that Initially focuses on finding best number of clusters, and then gradually moves finding global optimal cluster centers. Chang et al. [2] develop a GA for automatic clustering based on dynamic niching with niche migration to overcome the downsides of the fitness sharing approach. However, exploration and exploitation capabilities of these algorithms are not up to the mark.

In this work, we present a novel hybrid algorithm DKGK, which takes advantage of characteristics of DE, GA, and K-means. Here, DE is used to provide diversity in the search space. The obtained candidate solutions are fine tuned using k-means. Further, heuristic crossover [17] in GA is applied for fast convergence of the solutions and the obtained solutions are again refined by K-means. We experiment on three real datasets from UCI machine learning databases and two synthetic datasets to judge the efficacy of the proposed method. The results are very encouraging; the proposed method outperforms all the competing methods in all the datasets.

Rest of the paper is organized as follows. Section 2 presents a brief introduction to the evolutionary algorithms DE and GA, a most widely used conventional clustering method K-means, and hybrid algorithms DEKM and GAKM. Section 3 presents the proposed hybrid method DKGK. Results and discussions are included in Sect. 4. Finally, Sect. 5 concludes with a hint on possible future research directions.

2 Algorithms Background

In this section, we present a brief introduction to the Differential Evolution (DE), Genetic Algorithm (GA), K-means and hybrid algorithms DEKM and GAKM.

2.1 Differential Evolution

Storn and Price [18] propose DE, an evolutionary algorithm for global optimization for continuous-valued problem. The basic DE strategy can be described as the notation DE/x/y/z, where x is the vector to be mutated (a random vector or the best vector), y is the number of difference vectors, and z is the crossover scheme [18]. Some of the more frequently used strategies comprise DE/rand/2/bin, DE/best/1/bin, and DE/best/2/bin. In each generation, for each individual X_1 in the current population, a mutation operator generates a mutant parameter vector T_1 by mutating target vector X_2 with difference vector of two individuals X_3 and X_4, making sure that all these four individuals are different as shown in equation (3).

$$T_{1j} = X_{2j} + \beta(X_{3j} - X_{4j}) \tag{3}$$

Here, j represents dimension of an individual and β is a scaling factor, which manages the magnification of difference vector. For smaller values of β, the algorithm converges slowly but it can be used to explore search in local area, whereas a larger value of β assists greater diversity in search space but it may cause the algorithm to escape good solutions. Therefore, the value of β requires a careful balancing. It has been shown empirically that solutions often converge prematurely for large values of β and population size [3] whereas $\beta = 0.5$ generally exhibits good performance [1, 18]. The generated trial vector by crossover operator through implementing a discrete recombination of the mutant vector T_1 and the parent vector X_1 is compared with the corresponding parent vector and the one with the higher fitness value is selected for the next generation. It is gaining attention of researchers, for solving optimization problems, as it is easy to implement, simple in design, and requires tuning of few parameters.

2.2 Genetic Algorithm

GA is a stochastic optimization method that mimics the process of natural evolution [7]. In GA, a solution to the problem is represented by a chromosome; a set of chromosomes is called a population. Typically, solutions of initial population are generated randomly and then three genetic operators crossover, mutation, and selection are applied on solutions of current generation to produce new candidate solutions. The candidate solutions are selected for the next generation based on their fitness value. The crossover, which usually operates on two parents and produces two offsprings with some crossover probability, is primarily responsible for diversification of solutions in the entire search space. The mutation alters the gene of a chromosome to find solutions near it with a very small probability. Selection operator selects solutions from current generation for next generation based on some selection strategy. This process of applying genetic operators to produce candidate solutions for next generation continues till termination criteria meets. GA provides optimal or near optimal solutions to a vast range of optimization problems as they are efficient, robust, self-adaptive, and parallel in nature. As GA is more suitable for discrete optimization problems, it is suitable for the problem.

2.3 K-means Algorithm

K-means [10] is a widely used local search clustering algorithm, which is computationally efficient and easy to implement. Since similarity among data points is measured using Euclidian distance, a data point having a smaller distance from a particular cluster center is associated with that cluster. Most popularly, K-means minimizes a measure SSE to determine the quality of solution [15]. The k-means is described below.

Step 1: Randomly initialize centroids $\{m_1, m_2, \ldots, m_k\}$ of the k clusters.
Step 2: Repeat steps (i) to (iii) until a stopping criteria is satisfied.

(i) Compute Euclidian distance of each data point to the centroid of each cluster and associate the data points with the clusters where they have minimum Euclidian distance.

$$ED(x_p, m_i) = \sqrt{\sum_{j=1}^{d} \left(x_{pj} - m_{ij}\right)^2} \tag{4}$$

Here, x_p denotes pth data point of the dataset consisting of n instances (data points); m_i denotes centroid of ith cluster; d indicates number of dimensions in the dataset.

(ii) Evaluate fitness value of the solution.
(iii) Recompute the centroids of clusters by taking average of all corresponding dimensions of data points belonging to that cluster.

$$m_i = \frac{1}{n_i} \sum_{\forall x_p \in c_i} x_p \tag{5}$$

Here, n_i is the number of data points in cluster c_i.

The termination criterion of K-means algorithm is based on user-specified parameter. However, in proposed algorithm, K-means algorithm is iterated single time after each of two algorithms DE and GA.

2.4 Hybrid Algorithms DEKM and GAKM

Since DE is a global search algorithm with slow convergence and K-means is local search algorithm with fast convergence, Kwedlo [13] proposes a method DEKM that combines DE and K-means to obtain clusters based on sum of SSE criterion. They use K-means twice before and after the DE; first, initial population for the DE is created using K-means and then, the best solution obtained by DE is refined by K-means. Although DEKM algorithm described in this paper is similar to the algorithm in [13], k-means is used only once to refine the solutions obtained by DE. Here, DE is used to perform clustering in global search space whereas K-means is used to refine the candidate solutions as local search heuristic. We run DE for user specified number of iterations and refine the obtained solutions by K-means in isolated single iteration.

Few researchers, e.g., [14], [12], hybridize GA and K-means; first, they run GA with random initial population to explore the search space and obtain good solutions, then they refine the fittest solution using K-means to exploit the search space locally to obtain the (near-) optimal solution. In this work, in GAKM, we use randomly generated initial population and heuristic crossover [17] in GA and refine the obtained solutions using K-means. Here, GA is run for user specified number of iterations and K-means is run in isolated single iteration.

3 DKGK Based Clustering

This section presents a description of the proposed algorithm DKGK. Here, we explain encoding scheme, initial population generation, fitness function, avoiding invalid solutions, and different steps of the DKGK.

3.1 Chromosome Representation

A chromosome is a representation of a solution in the evolutionary algorithms. In this work, we follow a chromosome representation as described in [13]. It is shown in Fig. 1. Every chromosome or candidate solution contains $k \times d$ dimensions (genes) where k represents the number of clusters and d indicates the number of dimensions in the data points.

Here, m_{ij} represents centroid of the jth cluster and i indicates ith dimension in the cluster's centroid.

3.2 Initialization of DE Population

Cluster centroids for initial population are randomly selected from the dataset. It means, k data points are selected to represent centroids of k clusters in a chromosome. For every algorithm, number of chromosomes for initial population is assigned 20 and remains fixed for every generation.

3.3 The Fitness Function

As sum of distances has a profound impact on the error rate, we use sum of intra-cluster distances as the fitness function [4]. Its lower value indicates a better quality

| m_{11} | m_{21} | .... | m_{d1} | m_{12} | m_{22} | .... | m_{d2} | ..... | m_{1k} | m_{2k} | ..... | m_{dk} |

Fig. 1 Chromosome representation

of solution and is represented as follows:

$$SICD = \sum_{i=1}^{k} \sum_{\forall x_p \in c_i} ||x_p - m_i|| \tag{6}$$

Here, x_p denotes pth data point of the dataset, m_i denotes centroid of the ith cluster, c_i denotes ith cluster and i denotes the number of clusters.

3.4 Avoiding Erroneous Particle

As data points are selected randomly from the dataset, there is a possibility that a data point, which is an outlier, is selected as centroid of a cluster. It is observed that a cluster represented by such a centroid generally does not consist of any data point or consists of too few data points to be regarded as a meaningful cluster. To overcome such a problem, respective chromosome is replaced a by new chromosome, which is created by taking average computation of n/k data points for each of the k clusters.

3.5 DKGK Algorithm

The DKGK algorithm is summarized in the following steps.
Inputs:
(a) Evolutionary algorithmic parameters
(b) $n \times d$ dataset, where n is the number of instances of data points and d is the dimension
(c) K-number of clusters

Output:
    A set of K clusters.

Algorithm:

 (i) Select initial chromosomes from the dataset randomly.
(ii) Run this step (DE) for 3/4*(total iterations) to obtain diverse solutions

   (a) Select three chromosomes X_{2j}, X_{3j}, X_{4j} randomly from population for chromosome T_{1j}, making sure that all four $T_{1j}, X_{2j}, X_{3j}, X_{4j}$ are different. Then, evaluate T_{1j} as follows:

$$T_{1j} = X_{2j} + \beta(X_{3j} - X_{4j})$$

Here, $\beta = R_1 (R_2 + R)$ is self-adaptive scaling factor used to evaluate mutant vector C_1, R_1 and R_2 are user defined constants, and R is random number between 0 and 1.

(b) Perform crossover with some crossover rate.

(c) Fitter solution in C_1 and X_1 is selected for further operations.

(iii) Refine the solutions obtained in step (ii) by K-means. The number of isolated iterations of K-means is subject to user-defined parameter (experimental study is based on single iteration).

(iv) Run this step (GA) for 1/4*(total iterations) for converging solutions

(a) Randomly select two different parents from the population and perform following heuristic crossover operation.

$$\text{Offspring1} = \text{Best\_Parent} + r(\text{Best\_Parent} - \text{Worst\_Parent}) \quad (7)$$

$$\text{Offspring2} = \text{Best\_Parent.} \quad (8)$$

Here, $r = R_3 (R_4 + R)$ is self-adaptive random number used to evaluate offspring1; R_3 and R_4 are user defined constant; R is random number between 0 and 1.

(b) Perform crossover with some crossover rate.

(c Candidate solutions move on to the next generation based on some selection strategy.

(v) Refine the solutions obtained in step (iv) by K-means algorithm. The number of isolated iterations for K-means is subject to user-defined parameter (experimental study is based on single iteration).

In DKGK algorithm, DE is used to provide diversity of solutions in global search space. The DE runs for ($3*total\_iteration/4$) iterations, where $total\_iteration$ is a user-defined parameter, which defines total number of iterations in DKGK. The solutions obtained by DE are refined by K-means in isolated single iteration. Further, using heuristic crossover [17] in GA, the solutions converge toward the quality solutions in last ($total\_iteration/4$) iterations. Thereafter, K-means is applied to the solutions for single isolated iteration to get final partitions in the dataset. As it is known empirically that $\beta = 0.5$ generally exhibits good performance, in our experiment, β varies from 0.25 to 0.75 randomly with mean value 0.50. Similarly, r also varies from 0.25 to 0.75 randomly with mean value 0.50. Note that inner-steps in steps (ii) and (iv) themselves execute 20 (size of population) times for one iteration of step (ii) and (iv) respectively.

4 Experimental Result and Discussions

This section presents experimental setup, datasets, and results and discussion. The experiments have been performed on a system with core $i5$ processor and 2 GB RAM in Windows 7 environment using programs written in C language.

4.1 Parameters Setup

As results of the nature-inspired algorithms are influenced by number of control parameters, these values should be chosen carefully. Here, we experiment with different sets of values and present the results for following set of parametric values as it produces the best results. These values are same for the GA, DE, DEKM, GAKM, and the proposed method DKGK for a fair comparison. It is shown in Table 1.

4.2 Datasets Descriptions

The datasets are in matrix of size $n \times d$ with real-valued elements and are partitioned into k nonoverlapping clusters. We consider three real datasets Iris, Vowel, Wisconsin Breast cancer (WBC) from the UCI machine learning repository [16] and two synthetic datasets 2d4c, 10d4c, which are generated from two generators developed by Handl and Knowles [8] and can be downloaded from http://personalpages.manchester.ac.uk/mbs/Julia.Handl/generators.html. A brief summary of these datasets is presented in Table 2. As WBC consists of 16 samples with some missing features, we remove them from the dataset and use only 683 samples out of originally 699 samples.

Table 1 Control parameters for GA, DE, DEKM, GAKM, and DKGK

Name of parameters	Value
Population size	20
Number of iterations	100
Number of independent runs	100
Scaling Factor of DE(β)	Self adaptive
Random number (r) in heuristic crossover of GA	Self adaptive
Crossover rate of DE	1
Crossover rate of GA	1
$R1, R2, R3, R4$	0.5

Table 2 Datasets descriptions

Name of dataset	Number of clusters	Number of dimensions	Number of instances
Iris	3	4	150
Vowel	6	3	871
WBC	2	9	683
2d4c	4	2	1572
10d4c	4	10	1289

Table 3 Best results obtained by different methods in 100 independent runs

Dataset	DE	GA	DEKM	GAKM	DKGK
Iris	97.65	98.97	97.07	97.06	96.99
Vowel	158521.75	159288.85	152281.03	150341.59	149227.54
WBC	3016.92	3011.45	2978.71	2978.50	2975.73
2d4c	3341.95	3348.88	3313.63	3319.67	3312.96
10d4c	18863.20	18752.88	17348.63	17340.57	17245.45

4.3 Experimental Results

The obtained results (sum of intra-cluster distances of the final clusters) by DE, GA, DEKM, GAKM, and the proposed method DKGK are shown in Table 3. The presented results are the best results obtained in 100 independent runs for each algorithm. The results have been generated by running all the algorithms with same initial population, on same set of parametric values as shown in Table 1, and on same fitness function for every test problem for a fair comparison. As it is a minimization problem, a minimum value of sum of intra-cluster distances represents a better solution. It is clear that DKGK consistently outperforms its competitive algorithms. A primary reason for better performance of proposed algorithm DKGK is separate choice of algorithms where DE takes care of exploration of solutions in the search space and the heuristic crossover of GA is responsible to converge to high quality solution. In addition, k-means does the refinement to obtain (near-) optimal solution.

5 Conclusion

As nature-inspired algorithms are more versatile and look for global optimal solutions in comparison to conventional algorithms for clustering, which usually look for local optimal solutions, they are increasingly attracting attention from researchers in diverse areas. In this paper, we propose a new hybrid algorithm DKGK, which maintains proper balance of exploration and exploitation of solutions in the search space taking advantages of DE, GA, and K-means. The empirical results on three real and two synthetic datasets suggest that performance of the proposed DKGK is comparatively superior to DE, GA, DEKM, and GAKM.

We wish to extend this humble beginning in various possible ways. This algorithm works when the number of clusters is known *apriori*; it may be adapted to decide proper number of clusters on its own instead of it being a user defined parameter. As a vast array of nature-inspired algorithms are available with different philosophical base and features, the DE and GA may be replaced by other algorithms, e.g., evolutionary and swarm, to provide better exploration and exploitation capability in search space. Further, the most frequently used K-means to hybridize nature-inspired algorithms may be replaced with a better conventional local search method.

References

1. Ali, M.M.: Törn, A.: Population set-based global optimization algorithms: Some modifications and numerical studies. Comput. Oper. Res. **31**(10), 1703–1725 (2004)
2. Chang, D., Zhang, X., Zheng, C., Zhang, D.: A robust dynamic niching genetic algorithm with niche migration for automatic clustering problem. Pattern Recogn. **43**, 1346–1360 (2010)
3. Chiou, J.-P., Wang, F.-S.: A hybrid method of di?erential evolution with application to optimal control problems of a bioprocess system, In: IEEE World Congress on Computational Intelligence,Proceedings of the IEEE International Conference on Evolutionary Computation, pp. 627–632. (1998)
4. Chuang, L.Y., Hsiao, C.J., Yang, C.H.: Chaotic particle swarm optimization for data clustering. Expert Syst. Appl. **38**, 14555–14563 (2011)
5. Cura, T.: A particle swarm optimization approach to clustering. Expert Syst. Appl. **39**, 1582–1588 (2012)
6. Das, S., Abraham, A., Konar, A.: Automatic clustering using an improved differential evolution algorithm. IEEE Trans. Syst. Man Cybern. Part A: Syst. Hum. **38**(1), 218–237 (2008)
7. Goldberg, D.E.: Genetic Algorithms-in Search, Optimization and Machine Learning. Addison-Wesley Publishing Company Inc., London (1989)
8. Handl, J., Knowles, J.: Improving the scalability of multiobjective clustering. In: Proceedings of the Congress on Evolutionary Computation, vol. 3, pp. 2372–2379 (2005)
9. He, H., Tan, Y.: A two-stage genetic algorithm for automatic clustering. Neurocomput. **81**, 49–59 (2012)
10. Jain, A.K., Dubes, R.C.: Algorithms for Clustering Data. Prentice-Hall, Engle-wood Cliffs, NJ (1988)
11. Jain, A.K., Murty, M.N., Flynn, P.J.: Data clustering: a review. ACM Comput. Surv. **31**(3), 264–323 (1999)
12. Kwedlo, W.,Iwanowicz, P.: Using genetic algorithm for selection of initial cluster centers for the K-Means method. In: Proceedings of 10th International Conference on Artificial Intelligence and Soft Computing. Part II, LNAI 6114, pp. 165–172, (2010)
13. Kwedlo, W.: A clustering method combining differential evolution with the K-means algorithm. Pattern Recogn. Lett. **32**, 1613–1621 (2011)
14. Laszlo, M.,Mukharjee, S.: A genetic algorithm that exchanges neighboring centers for k-means clustering. Pattern Recogn. Lett. **28**, 2359–2366 (2007)
15. MacQueen, J.: Some methods for classification and analysis of multivariate observations. In: Proceedings of the Fifth Berkeley Symposium on Mathematical Statistics and Probability, Vol. 1, 281–297, (1967)
16. Murphy, P., Aha, D.: UCI repository of machine learning data bases. (1995). URL http://www.sgi.com/tech/mlc/db

17. Peltokangas, R., Sorsa, A.: Real-coded genetic algorithms and nonlinear parameter identification. University of Oulu Control Engineering Laboratory Report, vol. 34, pp. 1–32 (2008)
18. Storn, R., Price, K.: Differential evolution—a simple and efficient heuristic for global optimization over continuous spaces. J. Global Optim. 11(4), 341–359 (1997)
19. Tian, Y., Liu, D., Qi, H.: K-harmonic means data clustering with differential evolution. In: Proceedings International Conference on Future BioMedical Information, Engineering, pp. 369–372, (2009)
20. Tvrdík, J., Křivý, I.: Differential evolution with competing strategies applied to partitional clustering. In: Proceedings Symposium on Swarm Intelligence and Differential Intelligence, LNCS 7269. pp. 136–144 (2012)
21. Velmurugan, T., Santhanam, T.: A survey of partition based clustering algorithms on data mining: an experimental approach. Int. Technol. J. 10, 478–484 (2011)
22. Xu, R., Wunsch II, D.: Survey of clustering algorithms. IEEE Trans. Neural Networks 16(3), 645–678 (2005)

Studies and Evaluation of EIT Image Reconstruction in EIDORS with Simulated Boundary Data

Tushar Kanti Bera and J. Nagaraju

Abstract Simulated boundary potential data for Electrical Impedance Tomography (EIT) are generated by a MATLAB based EIT data generator and the resistivity reconstruction is evaluated with Electrical Impedance Tomography and Diffuse Optical Tomography Reconstruction Software (EIDORS). Circular domains containing subdomains as inhomogeneity are defined in MATLAB- based EIT data generator and the boundary data are calculated by a constant current simulation with opposite current injection (OCI) method. The resistivity images reconstructed for different boundary data sets and images are analyzed with image parameters to evaluate the reconstruction.

Keywords EIT · Simulated boundary data · EIDORS · Image reconstruction · Resistivity images · Image parameters.

1 Introduction

Electrical Impedance Tomography (EIT) [1–6] is an image reconstruction technique in which a constant current is injected to the domain under test (DUT) and the electrical conductivity or resistivity distribution of the DUT is reconstructed from the boundary potentials using an image reconstruction algorithm [7–11]. EIT is a computed tomographic imaging modality which is being used in different fields of science and engineering due to its several advantages [12] over other conventional tomographic techniques [13]. Being a fast, portable, noninvasive, nonradiating, nonionizing, and inexpensive methodology, electrical impedance tomography has been extensively researched in medical diagnosis [14–18], biomedical

T. K. Bera (✉) · J. Nagaraju
Department of Instrumentation and Applied Physics, Indian Institute of Science,
Bangalore, Karnataka 560012, India
e-mail: tkbera77@gmail.com

B. V. Babu et al. (eds.), *Proceedings of the Second International Conference on Soft Computing for Problem Solving (SocProS 2012), December 28–30, 2012*, Advances in Intelligent Systems and Computing 236, DOI: 10.1007/978-81-322-1602-5_156, © Springer India 2014

engineering [19] and biotechnology [20]. Studying the image reconstruction process in EIT is essential to test, calibrate, and assess the system performance. Design and development of EIT instrumentation [21–26] as well as the practical phantoms [2, 27, 28, 30, 31] are very crucial in EIT as the improper design parameters of the instrumentation and phantoms adversely affect the boundary data. Hence, the development of the instrumentation and practical phantoms and the image reconstruction studies [2, 5, 6, 21, 26–28, 30–33] on practical phantoms and their boundary data obtained in real EIT systems becomes difficult in real case. Computer simulation [34, 35] in science and engineering is a study of a particular problem by developing its mathematical model in personal computer. Computer simulation is conducted to obtain the preliminary solution before studying the problem with practical experimentations. The simulation studies on EIT [36–38] are required to avoid the design and development works required in practical phantom studies or to reduce the errors in boundary data produced by phantoms and instrumentation. Studies on boundary data simulation and their EIT image reconstruction using standard reconstruction algorithm help the researchers to understand the physics and mathematics of EIT and its image reconstruction process.

In this paper, simulated data are generated with an EIT data generator and the resistivity reconstruction is studied with standard image reconstruction algorithm. In this direction, circular domains containing single or multiple circular subdomains are simulated as phantoms and the boundary data are generated in a MATLAB-based EIT data generator. The resistivity imaging in EIT is studied with simulated data using Electrical Impedance Tomography and Diffuse Optical Tomography Reconstruction Software (EIDORS) [39]. The resistivity reconstruction is conducted with Levenberg-Marquardt Regularization (LMR) technique [40]. A constant current injection is simulated with opposite current injection method in circular simulated domains and the simulated boundary data are calculated in PC. Simulated boundary data are generated for different domain geometries and the resistivity images are reconstructed. The image parameters are calculated from the reconstructed images and the images are studied to evaluate the reconstruction and boundary data quality.

2 Materials and Methods

2.1 Boundary Data Generation and Image Reconstruction

Circular simulated regions are simulated (Fig. 1a) as the DUT (Ω) in a MATLAB boundary data generator and a constant current injection is simulated to generate the boundary data. The circular DUTs (diameter D) are discretized with a finite element (FE) mesh (Fig. 1b) and the boundary nodes are separated to define the nodes at the electrode positions called the electrode nodes. The elements inside the inhomogeneity and the background are identified and the background and inhomogeneity elements are defined with two different resistivities. Inhomogeneity of required shape and size (diameter $d_i = 30$ mm, 40 mm, 50 mm etc.,) is made at different electrode positions inside the simulated domain by defining its center (P) with a polar co-ordinate (r, θ) as shown in Fig. 1a. Different circular DUTs are defined containing a single or multiple subdomains (diameter d) as the inhomogeneities near different electrode nodes. The elements within the inhomogeneity are assigned with a particular resistivity ($\rho_i = 50$ Ωm) while the rest of the elements in the surrounding region called background are assigned with a different resistivity ($\rho_b = 1.724$ Ωm). A constant current injection is simulated in circular domains defined as simulated phantoms in MATLAB and the simulated boundary data are calculated in PC using opposite current injection (OCI) method. Changing the value of r and θ, a number of phantom geometry can be obtained easily to generate the corresponding boundary data set.

1 mA 50 kHz constant current injection is simulated to the DUT boundary through the electrode nodes called current electrode nodes (CEN) and the voltages are calculated on the electrode nodes called voltage electrode nodes (VEN) using OCI protocol. In OCI method (Fig. 1c), the current is simulated through two opposite CEN and the electrode potentials are calculated on VEN, The potentials of all the FEM mesh nodes are calculated and the boundary node potentials are separated. The electrode node potentials are again separated from the boundary node potentials. The electrode node potentials are calculated for all the 16 current projections and the 256 electrode node potential data are saved as a .txt file in PC for the analysis and image reconstruction. As the boundary data for first eight current projections are sufficient to produce impedance image (due to reciprocity principle) and that is why the other eight current projections are not required [28]. For opposite current injection methods, the boundary potentials ($V_1, V_2, V_3, \ldots, V_{15}, V_{16}$) developed on the electrodes ($E_1, E_2, E_3, \ldots, E_{15}, E_{16}$ respectively) are collected for each current projection. Hence, in opposite current pattern (Fig. 1c), the boundary potentials on all the electrodes are measured for first eight current projections and the voltage data set containing 128 voltage data is obtained and it is then saved in a .txt file in PC for computation. Boundary potential data are generated for different phantom configurations and the resistivity images are reconstructed in EIDORS. The EIDORS discretizes the circular domains ($D = 150$ mm) with a FEM mesh containing 1,968 triangular elements and 1,049 nodes [28] to solve the forward problem and inverse problem. EIDORS reconstructs the resistivity from the simulated data sets using regularized Gauss-Newton method with LMR regularization [5]. The mean inhomogeneity resistivity (IR$_{Mean}$) [24], maximum inhomogeneity resistivity (IR$_{Max}$), contrast to noise ratio (CNR) [24], percentage of contrast recovery (PCR) [24], coefficient of contrast (COC) [24], and diametric resistivity profile (DRP) [24, 27] of the reconstructed images are studied to evaluate the reconstruction.

3 Results and Discussion

Result shows that the resistivity is successfully reconstructed from the boundary data generated by the the data generator with a circular domain (Fig. 2) using OC

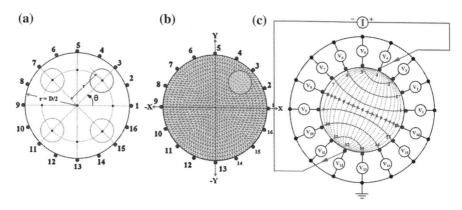

Fig. 1 a A two-dimensional circular domain (Ω) with inhomogeneity defined by the Cartesian and polar co-ordinates with a particular radius ($r = D/2$) in MATLAB, **b** Domain discretized by a FE mesh with triangular elements and the electrode nodes are located on boundary, **c** Boundary potential measurement for OCI in projection 4

It is observed that the reconstruction quality depends on the boundary data accuracy which depends on the geometric accuracy of the domain. Figure 2a shows the original resistivity distribution of the simulated domain with a single circular inhomogeneity near electrode number-3 ($D = 150$ mm, $\eta_{mr} = 4$, $d = 50$ mm, $r = 37.5$, mm, $\theta = 45°$, $\rho_i = 50 \,\Omega$m, $\rho_b = 1.724\,\Omega$m and $I = 1$ mA). Figure 2b, c show the resistivity image and DRP respectively.

Results show that the CNR of the reconstructed images for the inhomogeneity near electrodes 3 (Fig. 2a) is 5.1390 which indicates an efficient image reconstruction (CNR>3)[17, 22]. PCR and COC of the reconstructed image for the domain with inhomogeneity near electrode 3 (Fig. 2a) are 57.9489 % and 4.6199 respectively. IR$_{Mean}$, BR$_{Mean}$ and IR$_{Max}$ of the reconstructed image for the same domain are 35.7509 Ωm, 7.7385 Ωm and 56.7176 Ωm respectively. DRP (Fig. 2c) of the reconstructed images (Fig. 2b) for the domain with circular inhomogeneity near electrodes 3 is following the DRP of the original resistivity distribution. Results show that the reconstructed shape of the inhomogeneity is similar to that of the original one (Fig. 2a) and the reconstructed resistivity profile in Fig. 2c is almost similar to that of the original object in Fig. 2a.

Resistivity imaging of the simulated domain ($D = 150$ mm, $\eta_{mr} = 4$, $d = 50$ mm, $r = 37.5$, mm, $\theta = 45°$, $\rho_i = 50\,\Omega$m, $\rho_b = 1.724\,\Omega$m and $I = 1$ mA) with a circular inhomogeneity near electrode number 3 is studied (Fig. 3) in different iterations and all the images are evaluated in each steps. Results show that the image reconstruction starts with a poor quality image (Fig. 3a) and gradually it improves as the iteration goes on. It is observed that, in different iterations (Fig. 3a–p) the reconstructed image gradually becomes more localized and the resolution is improved. It is also observed that in each step the images are improved with a reduction in reconstruction errors appeared by the red color in the region of interface between the inhomogeneity and the background are minimized as the iterations goes on up to

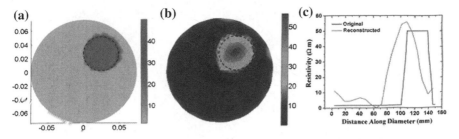

Fig. 2 **a** Domain with a circular inhomogeneity ($D = 150\,\text{mm}, r = 37.5\,\text{mm}, \theta = 45°, d = 50\,\text{mm}, \rho_i = 50\ \Omega\text{m}, \rho_b = 1.724\ \Omega\text{m}, \eta_{mr} = 4, I = 1\,\text{mA}$), **b** Resistivity image reconstructed from boundary data collected through opposite method of current injection, **c** DRP

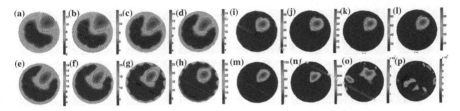

Fig. 3 Reconstructed images of a circular domain with a circular inhomogeneity ($D = 150\,\text{mm}, r = 37.5\,\text{mm}, \theta = 45°, d = 50\,\text{mm}, \rho_i = 50\ \Omega\text{m}, \rho_b = 1.724\ \Omega\text{m}, \eta_{mr} = 4, I = 1\,\text{mA}$) near electrode 3 for different numbers of iterations in inverse solver in EIDORS: **a** 1st iteration, **b** 2nd iteration, **c** 3rd iteration, **d** 4th iteration, **e** 5th iteration, **f** 6th iteration, **g** 7th iteration, **h** 8th iteration, **i** 9th iteration, **j** 10th iteration, **k** 11th iteration, **l** 12th iteration, **m** 13th iteration, **n** 14th iteration, **o** 15th iteration, **p** 16th iteration

12th iteration (Fig. 3a–l). Results show that, though the shape and resolution of all the reconstructed images in 9–12th iterations (Fig. 3i, j) are almost similar to that of the original object (shown by dotted circles in Fig. 3), but the resistivity is optimally reconstructed in 9th iteration (Fig. 3i).

Image parameter study (Fig. 4a–f) shows that the CNR (Fig. 4a), PCR (Fig. 4b) and COC (Fig. 4c) of the reconstructed images for the inhomogeneity near electrodes 3 (as shown in Fig. 3) are poor in 1st iteration and they get improved as the iteration goes on. The IR$_{\text{Max}}$ (Fig. 4d), BR$_{\text{Mean}}$ (Fig. 4e) and IR$_{\text{Mean}}$ (Fig. 4f) of the resistivity images are also improved as iteration goes but become over estimated after 9th iteration.

The DRPs of the reconstructed images (Fig. 5a–p) shows that the resistivity profile remains under estimated till the 8th iteration (Fig. 5a–h) and it becomes over estimated after 9th iteration (Fig. 5j–p). Therefore, it is observed that the reconstructed resistivity profile similar to that of the original is obtained only in the 9th iteration (Fig. 5i) and hence the 9th iteration is found as optimum iteration.

In 10–14th iterations (Fig. 5j–n) the resistivity is over estimated and the images are destroyed after 14th iteration (Fig. 5o, p). PCR becomes abnormal (>100%) after 11th iteration because of the over estimation of resistivity. In 15th iteratio

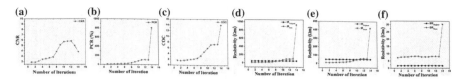

Fig. 4 Image parameters of the images obtained at different iterations as shown in Fig. 3 **a** CNR, **b** PCR, **c** COC, **d** IR$_{Max}$, **e** IR$_{Mean}$, **f** BR$_{Mean}$

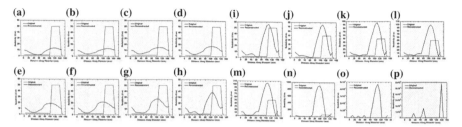

Fig. 5 DRP of the reconstructed images (shown in the Fig. 3) of the simulated domain ($D = 150$ mm, $\eta_{mr} = 4$, $d = 50$ mm, $r = 37.5$, mm, $\theta = 45°$, $\rho_i = 50$ Ωm, $\rho_b = 1.724$ Ωm and $I = 1$ mA) with a circular inhomogeneity near electrode number-3 in different iterations

the resistivity image is started to destroy (Fig. 5o) and in 16th iteration the image is completely lost (Fig. 5p). Hence the resistivity imaging studies with simulated boundary data show that boundary data generated by MatLAB-based boundary data generator is suitable for image reconstruction. Using simulated boundary data the resistivity imaging is successfully studied and evaluated using EIDORS. Image resistivity parameters and the image contrast parameters are calculated from the resistivity profiles of the reconstructed images and the image reconstruction process is studied and understood with image evaluation studies.

4 Conclusion

The EIT image reconstruction is studied and evaluated with simulated data generated by a MatLAB-based boundary data generator. Simulated boundary data are generated with circular domains for different configurations and the resistivity image reconstruction is studied in EIDORS. Results show that the resistivity images are successfully reconstructed from the simulated boundary data generated for all the domain configurations. Resistivity images reconstructed for simulated data are analyzed with their IR$_{Mean}$, IR$_{Max}$, CNR, PCR, COC and DRP. Results show that all the inhomogeneities are successfully reconstructed with their proper inhomogeneity and background profiles as well as with proper shapes and positions. Resistivity image reconstruction process is studied at different iterations and the results show that the reconstructed image quality varies with iteration steps and at a particular iteration the optimum reconstruction is obtained. It is observed that the resistivity images are easy to analyze and evaluate by image parameters to identify the suitable iteration step in the reconstruction process. Hence, it is concluded that the resistivity imaging is successfully studied and evaluated using EIDORS using simulated boundary data.

References

1. Webster, J.G.: Electrical Impedance Tomography. Adam Hilger Series of Biomedical Engineering, Adam Hilger, New York (1990)
2. Holder, D.S., Hanquan, Y., Rao, A.: Some practical biological phantoms for calibrating multifrequency electrical impedance tomography. Physiol. Meas. **17**, A167–77 (1996)
3. Denyer, C.W.L., Electronics for real-time and three-dimensional electrical impedance tomographs, PhD Thesis, Oxford Brookes University (1996)
4. Bera, T. K., Nagaraju, J.: Studies on thin film based flexible gold electrode arrays for resistivity imaging in Electrical Impedance Tomography. Measurement **47** 264–286 (2014)
5. Bera, T.K., Nagaraju, J.: Studying the resistivity imaging of chicken tissue phantoms with different current patterns in Electrical Impedance Tomography (EIT). Measurement **45**, 663–682 (2012)
6. Bera, T.K., Nagaraju, J.A.: Multifrequency Electrical Impedance Tomography (EIT) system for biomedical imaging. In: Proceedings of International Conference on Signal Processing and Communications (SPCOM 2012), pp. 1–5. India, IISc - Bangalore, Karnataka, India (2012)
7. Thomas, J.Y., John, G.W., Willis, J.T.: Comparng reconstruction algorthms for Electrcal Impedance Tomography. IEEE Trans. Biomed. Eng. BME-34(11), 843–852 (1987)
8. Bera, T.K., Biswas, S.K., Rajan, K., Nagaraju, J.: Improving conductivity image quality using Block Matrix-based Multiple Regularization (BMMR) technique in EIT: a simulation study. J. Electr. Bioimpedance **2**, 33–47 (2011)
9. Jing L., Liu S., Zhihong L., Meng S., An image reconstruction algorithm based on the extended Tikhonov regularization method for electrical capacitance tomography Original Research Article. Measurement, **42**(3), 368–376 (2009)
10. Bera, T.K., Biswas, S.K., Rajan, K., Nagaraju, J.: Improving image quality in Electrical Impedance Tomography (EIT) using Projection Error Propagation-Based Regularization (PEPR) Technique: a simulation study. J. Electr. Bioimpedance. **2**, 2–12 (2011)
11. Lionheart, W.R.B.: EIT reconstruction algorithms: pitfalls, challenges and recent developments, REVIEW ARTICLE. Physiol. Meas. **25**, 125–142 (2004)
12. Metherall, P.: Three dimensional electrical impedance tomography of the human thorax. PhD Thesis, University of Sheffield. January (1998)
13. Bushberg, J. T., Seibert, J. A., Leidholdt Jr. E. M., Boone, J. M.(eds.): The Essential Physics of Medical Imaging, 2nd edn. Lippincott Williams and Wilkins, Philadelphia (2001)
14. David, H. (ed.).: Clinical and Physiological Applications of Electrical Impedance Tomography, 1st edn. Taylor and Francis, UK (1993)
15. Li, Ying et al.: A novel combination method of Electrical Impedance Tomography inverse problem for brain imaging. IEEE Trans. Magn. **41**(5) (2005)
16. Bagshaw, A.P., et al.: Electrical impedance tomography of human brain function using reconstruction algorithms based on the finite element method. NeuroImage **20**, 752–764 (2003)
17. Murphy, D., Burton, P., Coombs, R., Tarassenko, L., Rolfe, P.: Impedance imaging in the newborn. Clin. Phys. Physiol. Meas. Suppl. A, **8**, 131–40 (1987)
18. Hope, T. A., Iles, S. E.: Technology review: the use of electrical impedance scanning in the detection of breast cancer. Breast. Cancer. Res. **6**, 69–74 (2004)
19. Brown, B. H.: Medical impedance tomography and process impedance tomography: a brief review. Meas. Sci. Technol. **12**, pp. 991–996 (2001)

20. Linderholm, Pontus et al.: Cell culture imaging using Microimpedance Tomography. IIEEE Trans. Biomed. Eng. **551** (2008)
21. Bera, T. K., Nagaraju, J. A.: Multifrequency constant current source for medical Electrical Impedance Tomography. In: Proceedings of the IEEE International Conference on Systems in Medicine and Biology 2010, pp. 278–283. Kharagpur, India (2010)
22. Bera, T.K., Nagaraju, J.: Surface electrode switching of a 16-Electrode wireless EIT system using RF-based digital data transmission scheme with 8 channel encoder/decoder ICs. Measurement **45**, 541–555 (2012)
23. Manuchehr, S.: Electrical impedance tomography system: an open access circuit design. Bio-Med. Eng. OnLine. **5**(28), 1–8 (2006)
24. Bera, T.K., Nagaraju, J.: Switching of a sixteen electrode array for wireless EIT system using a RF-based 8-Bit digital data transmission technique. Commun. Comput. Inform. Sci. **269**, 202–211 (2012)
25. Tong, I.O., Hun, W., Do, Y.K., Pil, J.Y., Eung J.W.: A fully parallel multi-frequency EIT system with flexible electrode configuration: KHU Mark2. Physiol. Meas. **32**(7), 835–49 (2007)
26. Bera, T.K., Nagaraju, J.: Switching of the surface electrodes array in a 16-electrode EIT system using 8-bit parallel digital data. In: Proceedings of IEEE World Congress on Information and Communication Technologies, pp. 1288–1293. Mumbai, India (2011)
27. Bera, T.K., Nagaraju, J.A.: Chicken tissue phantom for studying an Electrical Impedance Tomography (EIT) system suitable for clinical imaging, sensing and imaging. Int. J. **12**(3–4), 95–116 (2011)
28. Bera, T.K., Nagaraju, J.: Resistivity imaging of a reconfigurable phantom with circular inhomogeneities in 2D-Electrical Impedance Tomography. Measurement **44**(3), 518–526 (2011)
29. Bera, T.K., Nagaraju, J.: A reconfigurable practical phantom for studying the 2 D Electrical Impedance Tomography (EIT) using a FEM based forward solver. In: Proceedings of the 10th International Conference on Biomedical Applications of Electrical Impedance Tomography (EIT 2009), Manchester (2009)
30. Bera, T.K., Nagaraju, J. A.: Study of practical biological phantoms with simple instrumentation for Electrical Impedance Tomography (EIT). In: Proceedings of IEEE International Instrumentation and Measurement Technology Conference (I2MTC2009), pp. 511–516. Singapore (2009)
31. Bera, T.K., Nagaraju, J.A.: Simple instrumentation calibration technique for Electrical Impedance Tomography (EIT) Using A 16 Electrode Phantom.In: Proceedings of the 5th Annual IEEE Conference on Automation Science and Engineering, pp. 347–52. Bangalore (2009)
32. Bera, T. K., Nagaraju, J.: Studying The 2D Resistivity reconstruction of stainless steel electrode phantoms using different current patterns of Electrical Impedance Tomography (EIT). In: Proceedings of the International Conference on Biomedical Engineering, Biomedical Engineering, Narosa Publishing House, India, pp. 163–169 (2011)
33. Bera, T.K., Nagaraju J.: Studying the 2D-image reconstruction of non biological and biological inhomogeneities in Electrical Impedance Tomography (EIT) with EIDORS. In: Proceedings of the International Conference on Advanced Computing, Networking and Security: ADCONS 2011. India, NITK—Surathkal, India, 132–136 (2011)
34. Jochen, H., Leonard, M., Reindl.: A computer simulation platform for the estimation of measurement uncertainties in dimensional X-ray computed tomography, Measurement. **45**(8), 2166–2182 (2012)
35. Harvey, G., Jan, T., Wolfgang, C.: An Introduction to Computer Simulation Methods: Applications to Physical Systems (3rd Edn.), Addison-Wesley (January 19, 2006)
36. Tushar, K.B., Nagaraju, J.: A MATLAB Based Boundary Data Simulator for Studying The Resistivity Reconstruction Using Neighbouring Current Pattern, J. Med. Eng. vol. 2013, Article ID 193578, p. 15
37. Zlochiver, S., Radai, M.M., Abboud, S., Rosenfeld, M., Dong, X.Z., Liu, R.G., You, F.S., Xiang, H.Y., Shi, X.T.: Induced current electrical impedance tomography system: experimental results and numerical simulations. Physiol. Meas. **25**(1), 239–255 (2004)

38. Sadleir, R.J., Sajib, S.Z., Kim, H.J., Kwon, O.I., Woo, E.J.: Simulations and phantom evaluations of magnetic resonance electrical impedance tomography (MREIT) for breast cancer detection, J. Magn. Reson. **230**, 40–9 (2013)
39. Vauhkonen, M., Lionheart W.R.B., Heikkinen, L .M., Vauhkonen, P. J., Kaipio, J. P.: A MATLAB package for the EIDORS project to reconstruct two dimensional EIT images. Physiol. Meas. **22**, 107–111 (2001)
40. Bera, T.K., Biswas, S.K., Rajan,K., Nagaraju, J.: Image reconstruction in Electrical Impedance Tomography (EIT) with Projection Error Propagation-based Regularization (PEPR). In: A Practical Phantom Study, Lecture Notes in Computer Science: ADCONS 2011, vol. 7135/2012, pp. 95–105, Springer (2012)

Coverage of Indoor WLAN in Obstructed Environment Using Particle Swarm Optimization

LeenaArya and S. C.Sharma

Abstract Wireless communications is the fastest growing segment of the communications industry and over the recent years, it has rapidly emerged in the market providing users with network mobility, scalability, and connectivity. It is a flexible data communication system implemented as an extension to or as an alternative for, a wired LAN [1]. The placement of access points (AP) can be modeled as a nonlinear optimization problem. The work explores the measured data in terms of signal strength in the indoor WLAN 802.11 g at Malviya Bhavan, Boys Hostel Building, Indian Institute of Technology, Roorkee, Saharanpur Campus coverage using optimization technique. In the present study, an application of particle swarm optimization (PSO) is shown to determine the optimal placement of AP.

Keywords Wireless LAN · Access point · Path loss model · Obstructed environment · Particle swarm optimization.

1 Introduction

Wireless LANs enable users to access network resources and applications securely anytime and anywhere a wireless network is deployed [2]. Access points can nowadays be found in our daily environment, e.g., in many office buildings, public spaces, and in urban areas [1]. In the corporate enterprise, wireless LANs are usually employed as the final link between the existing wired network and a group of client computers, giving these users wireless access to the full resources and services of

L. Ary (✉)
Electronics and Communication Department, Lingaya's University, Faridabad, India
e-mail: leenaarya18@gmail.com

S. C. Sharma
Electronics and Computer Engineering Discipline Department, IIT Roorkee, UT, India
e-mail: scs60fpt@iitr.ernet.in

B. V. Babu et al. (eds.), *Proceedings of the Second International Conference on Soft Computing for Problem Solving (SocProS 2012), December 28–30, 2012*, Advances in Intelligent Systems and Computing 236, DOI: 10.1007/978-81-322-1602-5_157, © Springer India 2014

the corporate network across a building or campus [3]. WLAN networks have become very popular means for providing a wireless networking facility for home users, educational institutions, companies etc. due to their ease of installation and their high data rate provision, apart from providing, although limited, mobility to users [2].

If the APs are placed too far apart, they will generate a coverage gap; but if they are too close to each other, this will lead to excessive co-channel interferences and increases the cost unnecessarily [5]. In this paper, we use PSO to determine location in such a WLAN. In the indoor environment, the propagated electromagnetic signal can undergo the necessary three mechanisms of electromagnetic wave propagation—reflection, diffraction, and scattering.

The basic structure of a WLAN is called a Basic Service Set (BSS) which comes in two categories: Infrastructure BSSs and Independent BSSs. In infrastructure mode, the wireless network consists of at least one access point connected to the wired network infrastructure and a set of wireless end stations. In an independent BSSs (IBSS), stations communicate directly with each other and are usually composed of a small number of stations setup for a short period of time [3]. IBSSs are often referred to as ad hoc networks.

The rest of the paper is organized as follows: Notations are given in Sect. 2. Section 3 provides the mathematical model description and path loss model. Section 4 shows the working of PSO. Section 5 describes the method of testing, setup, and methodology. Analysis of results for obstructed environment on the first floor and measurements and analysis taken inside rooms with soft and hard partitions are presented in Sect. 6. Finally, Sect. 7 concludes the paper.

2 General Notations

Throughout this paper, the following notations are used:

a_j $j = 1 \dots N$ Access point (AP)
r_i $i = 1 \dots M$ Receiver/user
$d(a_j, r_i)$ Distance between AP and receiver
$g(a_j, r_i)$ Path loss from i_{th} user to access point j
g_{max} Maximum tolerable path loss
P_t Transmit power
P_r Received power
R_{th} Receive threshold
Ap Position of AP

It should be noted that a_j represents the unknown coordinates of APs. Their number N is not known either. The coordinates of users r_i are assumed to be known and these users can be distributed in design area according to the design specifications.

In the present analysis, the distance function assumed to be Euclidean, hence on the plane, the distance (d) between an AP a_j and a receiver r_i is given by [5]:

$$d(a_j, r_i) = \sqrt{(r_i^1 - a_j^1)^2 + (r_i^2 - a_j^2)^2}$$

where $a_j = a_j(a_j^1, a_j^2)$, and $r_i = r_i(r_i^1, r_i^2)$

3 Model Description

The aforementioned problem can be modeled as an optimization problem for which the objective function is to minimize the path loss. Mathematically it may be given as:

$$\min g(a_j, r_i) \leq g_{\max} \forall i = 1, \ldots, M \qquad (1)$$

Constraint (1) states that path loss is evaluated against the maximum tolerable path loss $g_{\max}$. This ensures that the quality of coverage at each receiver location is above the given threshold. This given value, $g_{\max}$ can be calculated by subtracting receiver threshold (R_{th}) from transmitter power (P_t).

$$g_{\max} = P_t - R_{th} \qquad (2)$$

The above inequality (1) can be expressed in the equality form as:

$$(\min_j g(a_j, r_i) - g_{\max})^+ = 0, \qquad (3)$$

3.1 Path Loss Model

The propagation of radio waves is characterized by several factors: (a) free space loss. (b) Attenuated by the objects on the propagation path such as windows, walls, table, chair, and floors of building. (c) The signal is scattered and can interfere with itself [7]. The basic propagation model is based on free space propagation. In general, the power received by an antenna that is separated from the transmitting antenna by the distance d in free space is given by [5, 6]:

$$P_r(a_j, r_i) = \frac{P_t G_t G_r \lambda^2}{(4\pi)^2 d(a_j, r_i)^2} \qquad (4)$$

where P_t is the transmitted power, G_t and G_r are the transmitter and receiver antenna gain, d is the distance between transmitter and receiver, and $\lambda = c/f$ is the wavelength of the carrier frequency, c is the speed of light (3×10^8 m/s), and f is the frequency of radio carrier in hertz. The path loss, which represents signal attenuation between the transmitted and the received power and is measured in dB (decibels), in free space environments, is given by [5, 6]:

$$g(a_j, r_i)[dB] = -10 \log \left[\frac{G_t G_r \lambda^2}{(4\pi)^2 d(a_j, r_i)^2} \right]$$

The above equation does not hold when points a_j and r_i are very close to each other. Therefore, large-scale propagation models use a close–in distance, d_0 which is known as the received power reference distance point. Therefore, path losses at reference distance assuming transmit and receive antenna with unity gain as described in [5, 6] can be calculated from:

$$g(a_j, r_i) = g(d_o)[dB] = 20 \log \frac{4\pi d_o f}{c} \tag{5}$$

Therefore, path loss function in free space at a distance greater than d_0 is given by

$$g(a_j, r_i)[dB] = g(d_o)[dB] + 10 \log \left(\frac{d(a_j, r_i)}{d_o} \right)^2 \tag{6}$$

The RF (radio frequency) path between transmitter and receiver is affected by the distance between the two terminals and the type and number of obstacles (walls, doors, windows, furniture, etc). Thus, including loss caused by partitions in path loss model, Eq. (6) can be written as [5, 6]:

$$g(a_j, r_i)[dB] = g(d_0)[dB] + \log \left(\frac{d(a_j, r_i)}{d_0} \right)^2 + \Sigma n_{SP} l_{SP} + \Sigma n_{HP} l_{HP} \tag{7}$$

where n_{SP} represents the number of soft partitions of a particular type and l_{SP} represents the loss in dB attributed to a particular soft-type partitions, n_{HP} represents the number of hard partitions related to a particular type and l_{HP} represents the loss in dB associated with a particular hard-type partitions. The soft partition consists of movable objects like furniture, users etc. While hard partitions comprises of fixed objects like walls, doors, window etc. A move around corner of the building or a wall can cause the received signal to drop suddenly.

4 Working of PSO

For a D-dimensional search space, the position of the ith particle is represented as $X_i = (x_{i1}, x_{i2}, \ldots, x_{id}, \ldots, x_{iD})$. Each particle maintains a memory of its previous best position $P_i = (p_{i1}, p_{i2}, \ldots, p_{id}, \ldots, p_{iD})$. The best one among all the particles in the population is represented as $P_g = (p_{g1}, p_{g2}, \ldots, p_{gd}, \ldots, p_{gD})$. The velocity of each particle is represented as $V_i = (v_{i1}, v_{i2}, \ldots, v_{id}, \ldots, v_{iD})$, is clamped to a maximum velocity $V_{max} = (v_{max,1}, v_{max,2}, \ldots, v_{max,d}, \ldots, v_{max,D})$ which is specified by the user. During each generation, each particle is accelerated toward the particle's previous best position and the global best position. The two basic equations which govern the working of PSO are that of velocity vector and position vector given by:

$$v_{id} = \omega^* v_{id} + c_1 r_1 (p_{id} - x_{id}) + c_2 r_2 (p_{gd} - x_{id}) \qquad (8)$$

$$x_{id} = x_{id} + v_{id} \qquad (9)$$

here c_1 and c_2 are acceleration constants.

5 Method of Testing

5.1 Setup for AP

We performed our experiment in first case without an obstacle and in the second case with obstacle in the design area which has the dimensions of 64 m x 60 m and has 400 users. It has 100 rooms and part of corridors. The data collection site building and google map of the site is given in Fig. 1.The layout of the floor is shown in Fig. 2. The entire wing of the first floor is covered by 10 access points installed at the locations indicated by red symbols in Fig. 2. Ten locations of measurement are chosen on the first floor of the Malviya Bhavan building as shown in Fig. 2 denoted as A, B, C, D, E, F, G, H, I, and J. The specification of the model of Access point is LINK (DWL-3200AP) and IEEE 802.11g standard are used to test the model. We conducted our experiments at the first and second floor of the Malviya Bhavan, IIT Roorkee, Saharanpur Campus. Our data collection system comprised of a laptop, running Windows 2007, an MSA 338 handheld spectrum analyzer, and an M304 omnidirectional dipole antenna.

5.2 Methodology

Once the priority area has been identified, the data has been collected nearby access points which require a connection to the wired LAN and a source of power. The signal strength has been measured using 3.3 GHs Spectrum Analyzer with omnidirectional dipole antennas at a number of points around the access point. The coverage and location of access points has been checked using optimization technique in those priority areas that are within range. While in other places the aim was to identify the points where the available bandwidth is likely to drop below the theoretical maximum, typically where the signal strength falls below -70 dBm [8].

The received signal strength calculated using the spectrum analyzer is shown in Fig. 3 and similarly signal strength have been calculated using spectrum analyzer for

Table 1 Optimal placement of APs for 100 users on first floor with $Pt = 21\,dBm$

Locations of access points	Received signal strength R_{th} (dBm)	Actual placement of access point coordinates	Access point coordinates by PSO
Location A	−38.4	(18.90, 10.93)	(22.16, 11.32)
Location B	−37.2	(24.72, 2.51)	(27.34, 4.54)
Location C	−27.1	(35.90, 1.59)	(38.25, 1.09)
Location D	−40.4	(42.19, 12.92)	(38.98, 10.25)
Location E	−32.4	(38.97, 26.24)	(40.54, 24.25)
Location F	−20.2	(41.88, 43.70)	(40.98, 44.72)
Location G	−41.6	(36.36, 52.43)	(32.22, 50.35)
Location H	−23.2	(26.10, 25.94)	(25.12, 24.23)
Location I	−42.0	(30.54, 12.00)	(32.45, 10.26)
Location J	−29.6	(33.60, 25.94)	(30.25, 24.31)

obstructed environment. The optimal placement of access points for 100 users on first floor with transmitter power of 21 dBm is shown in Table 1. The actual placement of access point coordinates is shown in Fig. 4 and the distribution of access point coordinates by using particle swarm optimization is shown in Fig. 5.

Fig. 1 Data collection site building

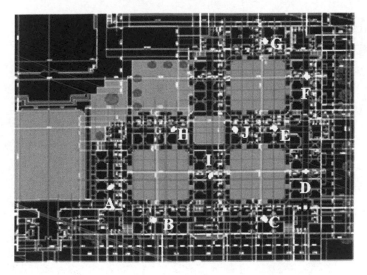

Fig. 2 Plan of the floor where the experiment was conducted. Readings were collected in the corridors

6 Analysis of Results for Obstructed Environment on First Floor

Results for obstructed environment on the first floor have been analyzed by using the optimization technique. The analysis of each is described in the following subsections:

For the present analysis, we consider the obstacles in indoor environment such as brick walls, wooden doors, windows, furniture, and moving human being. Reported available data for these obstructions is taken at 2.4 GHz as shown in Table 2.

Table 2 Signal attenuation through various obstructions at 2.4 GHz

Obstructions	Reported Signal Attenuation through various obstructions (dB)
Metal frame glass wall into building	6
Brick wall	6
Wooden block wall	4
Metal door in brick wall	12.4
Brick wall next to metal door	3
Window in a brick wall	2
Human	4

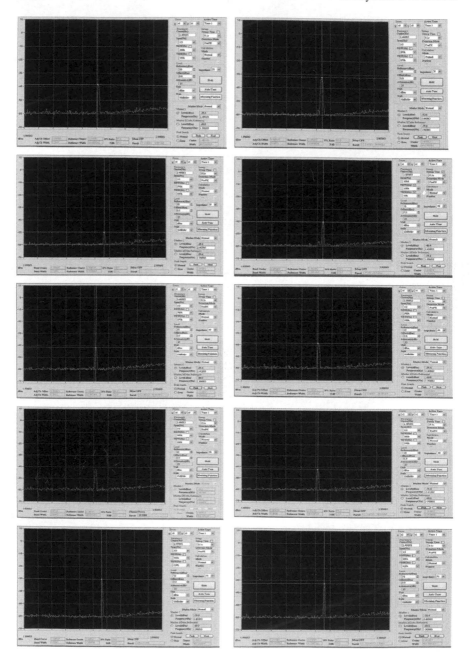

Fig. 3 Received signal strength calculated using spectrum analyser

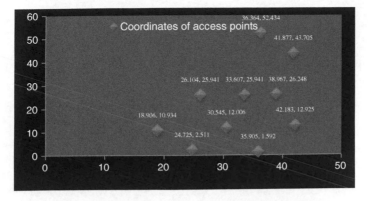

Fig. 4 Actual placement of access points

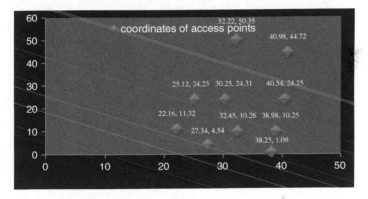

Fig. 5 Distribution of access points by using PSO

6.1 Measurements and Analysis Taken Inside Rooms with Soft and Hard Partitions on First Floor

Measurements taken inside rooms where the presence of a human being, desk, and table is possible and these are soft partitions, then

n_{SP} = number of soft partitions human being, desk, table = 1 each

l_{SP} = loss due to soft partitions human being, desk, table = 4 dB, 4 dB and 4 dB (as given in Table 2)

$$n_{SP}l_{SP} = (1 + 1 + 1) * (4 + 4 + 4) = 36\,\text{dB}$$

n_{HP} = number of hard partitions wall, glass window, almirah, and door=1 each

l_{HP} = loss due to hard partitions wall, glass window, almirah, and door =6 dB, 2 dB, 4 dB and 4 dB (as given in Table 2)

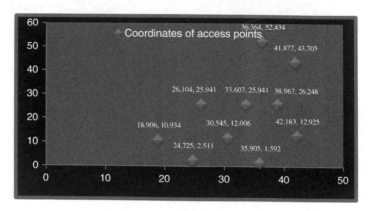

Fig. 6 Actual placement of access points

$$n_{HP}l_{HP} = (1 + 1 + 1 + 1) * (6 + 2 + 4 + 4) = 64\,\text{dB}$$

Now from the Eq. (6)

$$g(a_j, r_i)[dB] = g(d_o)[dB] + 10\log\left(\frac{d(a_j, r_i)}{d_o}\right)^2$$

$$n_{SP}l_{SP} + n_{HP}l_{HP} = 36 + 64 = 100\,\text{dB}$$

Now from the Eq. (7)

$$g(a_j, r_i)[dB] = g(d_o)[dB] + 10\log\left(\frac{d(a_j, r_i)}{d_o}\right)^2 + 100\,\text{dB}$$

The received signal strength calculated using the spectrum analyzer is shown in Fig. 6. Table 3 shows the measurements taken inside rooms taking wall, human, desk, table, almirah etc. as obstructions and closed door for different access points. The actual placement of access point coordinates is shown in Fig. 6 and the distribution of access point coordinates by using particle swarm optimization is shown in Fig. 7. Similarly, the results have been taken for the second floor as well as for the ground floor where there is no access point installed but due to floor attenuation factor (FAF) some access of the network is found on the staircase near ground floor.

7 Conclusion

In this paper, we have presented particle swarm optimization to predict the signal strength in indoor environments. It is observed that the size of the design area and the

Table 3 Measurements taken inside rooms with soft and hard partitions for different access points

S.No	Access point location	Positions	Distance between T-R separation in (m)	Signal Strength (dBm)	Actual placement of access point coordinates	Access point co-ordinates by PSO
1	Location A	Inside room A-200, human, desk, door open	4.41	−54.0	(18.90, 10.93)	(17.70, 11.94)
2	Location B	Inside room A-232, human, desk, door open	4	−54.0	(24.72, 2.51)	(23.42, 3.55)
3	Location C	Inside room A-249, on bed, door open	5.83	−50.0	(35.90, 1.59)	(36.79, 2.67)
4	Location D	Inside room A-254, table, almirah, human	4	−54.4	(42.19, 12.92)	(40.14, 15.39)
5	Location E	Inside room A-267, table, almirah door closed	8.7	−54.8	(38.97, 26.24)	(36.95, 26.78)
6	Location F	Inside room A-277, table, almirah, human, door closed	4	−54.4	(41.88, 43.70)	(40.66, 42.26)
7	Location G	Inside room A-280, table, bed, almirah, human, door closed	7.19	−54.8	(36.36, 52.43)	(33.77, 53.67)
8	Location H	Inside room A-219, table, almirah, human, door open	4.41	−51.2	(26.10, 25.94)	(27.67, 27.67)
9	Location I	Inside room A-226, desk, human, door open	4.41	−54.4	(30.54, 12.00)	(32.77, 13.11)
10	Location J	Inside room A-266, desk, human, door open	4.41	−54.8	(33.60, 25.94)	(34.56, 24.87)

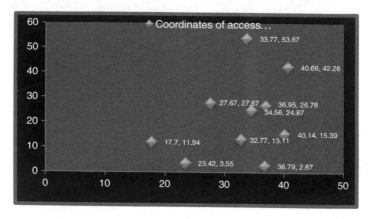

Fig. 7 Distribution of access points by using PSO

number of users and their locations have an effect on the location and the number of APs needed to cover users and path loss increases as a function of distance between the transmitter and user. This paper provides the results of comparison between the actual location of access points and the locations obtained by PSO in obstructed environment. The coverage area was varying between scenarios and there were different levels of signal strength for each receiver location depends on the obstacles between the receiver and the transmitter in the non-line of sight (NLOS) environment. It has been observed that the number of access points can be reduced so as to save the cost of installation of access points.

References

1. Traoré, Soungalo, Renfa, Li, Fanzi, Zeng: Evaluating and Improving Wireless Local Area Networks Performance. International Journal of Advancements in Computing Technology **3**(2), 156–164 (2011)
2. Wireless LAN Solutions At-A-Glance - Cisco www.cisco.com/web/AP/wireless/pdf/WirelessLAN.pdf, United States (2012)
3. IEEE 802.11b Wireless LANs http://www.3com.com/other/pdfs/infra/corpinfo/en_US/50307201.pdf
4. Particle Swarm Optimization (PSO) www23.homepage.villanova.edu/varadarajan.komanduri/PSO_meander-line.ppt
5. Kouhbor, S., Ugon, J., Kruger, A., Rubinov, A.: Optimal placement of access point in WLAN based on a new algorithm. In: proceedings of the International Conference on Mobile Business-IEEE/ICMB, pp. 592–598, Jul 2005
6. Shahnaz, K.: Physical security enhancement in WLAN systems. In: ISSNIP-IEEE, pp. 733–738 (2007)
7. Li, B., Dempster, A., Rizos, C., Barnes, J.: Hybrid method for localization using WLAN. In: Spatial Sciences Conference, www.gmat.unsw.edu.au/snap/publications/lib_etal2005c.pdf (2005)
8. GD/JANET/TECH.: Surveying wireless networks, www.ukema.ac.uk/documents/. . ./technical../ (2008)

Stereovision for 3D Information

Mary Ann George and Anna Merine George

Abstract Stereovision is a technique aimed at inferring depth from two or more cameras. It plays an important role in computer vision. Single image has no depth or 3D information. Stereovision takes two images of a scene from different viewpoints usually referred to as left and right images using two cameras. Stereovision is similar to the binocular (two-eyed) human vision capturing two different views of a scene and brain processing and matching the similarity in both the images and the differences allow the brain to build depth information. OpenCV Library is used to compute the output of stereovision process—disparity and depth map.

Keywords Calibration · Disparity · Rectification · Stereo-correspondence · Triangulation

1 Introduction

Vision begins with the detection of light from the world. Humans and most animals have two eyes with overlapping regions. Stereovision helps to perceive depth in the overlapping region. Perception, lighting and shadows make a 2D image look real. More than one 2D image reveals 3D information. Modern computer vision stereo algorithms are capable of producing real-time results which are accurate enough for simple applications.

Stereovision process involves four main steps namely Calibration, Rectification, Stereo-correspondence or Stereo Matching, and Triangulation. Camera calibration is important for relating camera measurements with measurements in the real,

M. A. George (✉) · A. M. George
M.I.T, Manipal 576104, India
e-mail: starmaryann1247@yahoo.com

A. M. George
e-mail: anna_thanku@yahoo.com

B. V. Babu et al. (eds.), *Proceedings of the Second International Conference on Soft Computing for Problem Solving (SocProS 2012), December 28–30, 2012*, Advances in Intelligent Systems and Computing 236, DOI: 10.1007/978-81-322-1602-5_158, © Springer India 2014

Fig. 1 Stereovision system

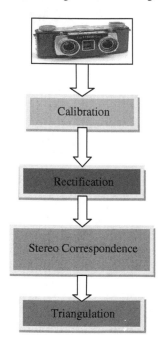

three-dimensional world. Rectification is a process of adjusting angle and distance between the cameras. Stereo-correspondence is a process of finding the same features in the left and right images. The output of this step is a disparity map. Disparity is the difference in x-coordinates on the image planes of the same feature viewed in the left and right cameras. Triangulation aims at calculating the depth or 3D position of points in the images from the disparity map and the geometry of the stereo setting. The output of this process is a depth map.

OpenCV stands for Open Source Computer Vision Library. It is an Intel-based library containing a collection of C and C++ functions to implement Image Processing and computer vision algorithms [1] (Fig. 1).

2 Overview of Stereovision System

2.1 Calibration

The camera calibration deals with both intrinsic and extrinsic parameters of the camera. Intrinsic parameters include both camera geometry and distortion model of the lens. For simplicity of analysis, we consider a pinhole camera model.

In a pinhole model, we do not consider the lens as a whole but only consider the camera as a point (optical center). Using this model, we calculate the focal length and form the camera intrinsic matrix.

There are two types of distortion namely tangential and radial distortion. Radial distortion arises as a result of the shape of lens, this distortion is zero at the (optical) center of the imager or image plane and increases as we move toward the periphery. Tangential distortions arise from the assembly process of the camera as a whole [2]. This distortion is due to manufacturing defects resulting from the lens not being exactly parallel to the imaging plane. Both these distortions are mathematically removed during the calibration process resulting in undistorted images. Extrinsic parameter deals with translation vector and rotational matrix. This is used to convert 3D object coordinates in meters (x, y, z) to camera coordinates in pixels (x, y).

For calibration we generally use a chess board as a calibration object. This pattern of alternating black and white squares ensures that there is no bias toward one side or the other in measurement. We take multiple views of this planar object.

2.2 Rectification

The basic geometry of a stereo imaging system is referred to as epipolar geometry. Here we consider two pin hole models one for each camera. Rectification process takes into account epiploar constraint. Given a feature in one image the corresponding feature in the other image will lie along the epipolar line. Epipolar constraint makes two-dimensional search for matching features across two imagers to a one-dimensional search along the epipolar lines leading to faster computation [2].

The Fig. 2 shows the epipolar geometry. The points P (actual viewed point), O_l and O_r (Center of projections) forms the epipolar plane and this plane intersects the left

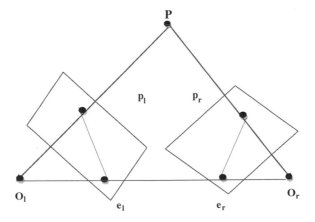

Fig. 2 Epipolar goemetry

and right image planes and form lines called the epipolar lines ($p_1 - e_1$ and $p_r - e_r$). p_1 and p_r are the projection of P on the left and right image plane respectively. e_1 and e_r are called the epipoles. The image of the center of projection $O_1(O_r)$ on the other image plane is an epipole$_r$ (e_1).

2.3 Stereo Matching

Stereo Matching or Stereo-correspondence aims at finding the homologous points in the two images of the same scene taken using two different cameras. Given a point A in the left image there will be a corresponding point B in the right image. The difference in the position between the corresponding points (A and B) in the two images is termed as the disparity. The output of this step is a disparity map. A pair of left and right image of the same scene taken using a stereo-camera is called a stereo-pair. There are two main classes of correspondence algorithm namely correlation-based and feature-based algorithm. In correlation-based method correspondence is achieved by matching the image intensities. This results in dense disparity map [2]. In feature-based method correspondence is achieved by matching image features like edges. This technique is faster and insensitive to illumination changes and produce sparse disparity map. Some challenges in stereo matching are color inconsistencies, untextured region, and occlusion problem.

2.4 Triangulation

It is the process of recovering the 3D position from the disparity map and geometry of the stereo setting. Triangulation is also called as reconstruction. In stereovision 3D location of any point is the intersection of two lines passing through their center of projection and projection of the point in each image.

In triangulation we are actually finding the depth from disparity. The depth and disparity are inversely related.

Figure 3 shows the depth calculation. P is the point in the 3D space with coordinates (X, Y, Z). This point is projected to left and right image plane as P_1 and P_r with coordinates (x_l, y_l) and (x_r, y_r) respectively. $x_1 - x_r$ is the disparity. x_1 and x_r are measured from the corresponding left and right optical axis. B is the base length and f is the focal length. Focal length is the distance from the optical center to the image plane. Considering the similar triangles in Fig. 2 and we derive an expression for depth Z as shown in Eq. 1

$$Z = \frac{Bf}{x_1 - x_r}$$

(1)

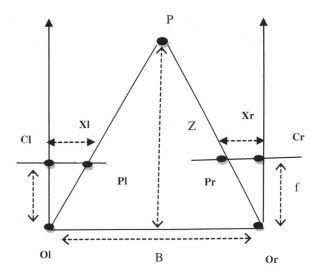

Fig. 3 Depth calculation

3 OpenCV

OpenCV means Intel ® Open Source Computer Vision Library. It contains C and C++ functions for high speed implementation of image processing and computer vision applications. It is Optimized and intended for real-time applications and Provides interface to Intel's Integrated Performance Primitives (IPP) with processor specific optimization (Intel processors).

Its key features include structural analysis, motion analysis, image processing, camera calibration, object recognition, etc. OpenCV module has CV, HighGUI, CXCORE, Cvaux, CvCam, and ML libraries [1].

Important OpenCV Functions used for stereovision are cvFindChessboard Corners() to find the chessboard corners, cvCalibrateCamera2() for calibration, cvStereoRectify() for rectification, cvFindStereoCorrespondenceBM() for stereo matching, and cvReprojectImageTo3D() for depth map.

4 Experimental Results

OpenCV 2.1 Software was used for stereovision. Figure 4 shows the input images (chess board patterns), such 10 pairs were used. Figure 5 shows images after corner detection. Figure 6 shows the rectified images. Figure 7 shows the disparity map and Fig. 8 shows the depth map.

Fig. 4 Input images

Fig. 5 Corner detected images

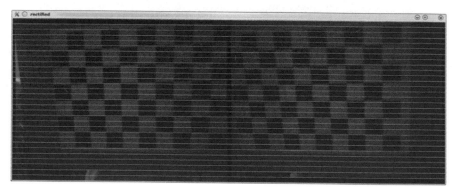

Fig. 6 Rectified images

Figure 9 shows the scale for the depth map. Red region shows the deeper region and Magenta region shows the less deep region.

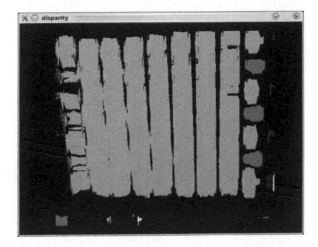

Fig. 7 Disparity map

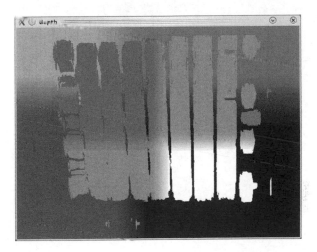

Fig. 8 Depth map

Fig. 9 Scale

5 Conclusion

Stereovision is of great importance in the field of machine vision, robotics, and image analysis. It is used to inspect infrastructure in human-inaccessible tunnels and pipes, and long sections of road and bridges. Stereovision can play an important role in the medical procedures like anthropometry and plastic surgery for capturing and reproducing 3D information about the human body. It can be used in mining where depth information is of greater importance. It is used by automobile manufacturers to automate driving and navigation of road vehicles. It can be used to extend capabilities of security monitoring systems, and improve human face-recognition and tracking algorithms. Research is also being done to create human-wearable Stereo Vision systems to assist the blind. In Energy harvesting and control applications stereovision can be used to replace occupancy sensors and photo sensors.

Acknowledgments This work has been carried out as a part of summer internship at NITK. The authors would like to acknowledge the support and guidance of Professor Ramesh. M. Kini.

References

1. Bradski, G., Kaehler, A.: O'Reilly learning OpenCV. 1st (edn), ISBN: 978-0-596-51613-0. O'Reilly Media Inc., NY, USA (2010).
2. Xu, G., Zhang, Z.: Epipolar geometry in stereo, motion and object recognition. Kluwer, Dordrecht (1996)
3. [OpenCV Wiki] Open Source Computer Vision Library Wiki. http://opencvlibrary.sourceforge. net/. Accessed 1 Oct 2008

About the Editors

Dr. B.V. Babu is the Director of IET at JK Lakshmipat University and prior to that he was with BITS Pilani. He has over 28 years of teaching, research, consultancy, and administrative experience. He did his Ph.D. from IIT Bombay. He held various administrative positions such as Dean of various divisions during over 15 years of his tenure at BITS Pilani. His research interests include Evolutionary Computation, Environmental Engineering, Biomass Gasification, Energy Integration, Artificial Neural Networks, Nanotechnology, and Modeling and Simulation. He is Editorial Board member of six International Journals, has published six books, and has around 220 research publications in International and National Journals and Conference Proceedings.

Professor Atulya K. Nagar holds the Foundation Chair, as Professor and Head of Mathematics and Computer Science at Liverpool Hope University. He is an internationally recognized scholar working at the cutting edge of theoretical computer science, natural computing, applied mathematical analysis, operations research, and systems engineering and his work is underpinned by strong complexity-theoretic foundations. He received a prestigious Commonwealth Fellowship for pursuing his Doctorate (D.Phil) in Applied Non-Linear Mathematics, which he earned from the University of York in 1996. He holds B.Sc. (Hons.), M.Sc., and M.Phil (with Distinction) from the MDS University of Ajmer, India.

Dr. Kusum Deep is a full-time Professor with the Department of Mathematics, Indian Institute of Technology Roorkee, India. She earned her Ph.D. from IIT Roorkee in 1988 and carried out Post Doctoral Research at Loughborough University, UK during 1993–1994. She has co-authored a book entitled "Optimization Techniques" and has 75 research publications in refereed International Journals and 60 research papers in International/National Conferences. She is on the editorial board of a number of International and National Journals. She is the Founder President of Soft Computing Research Society, India and the secretary of Forum of Interdisciplinary Mathematics. Her areas of specialization are numerical optimization and their applications to engineering, science, and industry.

Dr. Millie Pant is an Associate Professor in the Department of Applied Science and Engineering, Saharanpur Campus of IIT Roorkee. She did her Ph.D. from math-

B. V. Babu et al. (eds.), *Proceedings of the Second International Conference on Soft Computing for Problem Solving (SocProS 2012), December 28–30, 2012*, Advances in Intelligent Systems and Computing 236, DOI: 10.1007/978-81-322-1602-5, © Springer India 2014

ematics department of IIT Roorkee. She has supervised five Ph.D. theses and at present four Ph.D. students are working under her guidance. She has to her credit more than 100 research papers in various journals and conferences of national and international repute. Also she has completed sponsored projects of DST and MHRD. Dr. Millie Pant is a member of editorial board of 'International Journal of Mathematical Modeling, Simulation and Applications' and an executive editor of International Journal of Swarm Intelligence (IJSI).

Dr. Jagdish Chand Bansal is an Assistant Professor at South Asian University New Delhi. Dr. Bansal has obtained his Ph.D. in Mathematics from IIT Roorkee. He is the Editor-in-Chief of "International Journal of Swarm Intelligence (IJSI)" published by Inderscience. His Primary area of interest is Nature Inspired Optimization Techniques. He has published 42 research papers in various international journals/conferences.

Dr. Kanad Ray is an Associate Professor in Physics at Institute of Engineering and Technology of JK Lakshmipat University, Jaipur. He obtained his Doctoral Degree in Physics from Jadavpur University, Kolkata. In his academic career spanning over 19 years, he has published and presented research papers in several national and international journals and conferences in India and abroad. He has authored a book on Electromagnetic Field Theory. Dr. Ray's current research areas of interest include applied physics, communication, and cognitive neurodynamics.

Dr. Umesh Gupta is working as an Associate Professor in Mathematics at Institute of Engineering and Technology of JK Lakshmipat University, Jaipur. He has 13 years of teaching and research experience. He did his Ph.D. from University of Rajasthan. His areas of research are Fluid Mechanics, Optimization and Statistics. He has number of research publications in International and National Journals and Conference Proceedings to his credit. He has authored textbooks on Engineering Mathematics and Research Methods in Management. He has presented his papers in various conferences held in India and abroad.

Printed by Publishers' Graphics LLC